D1500006

INORGANIC BIOCHEMISTRY

INORGANIC BIOCHEMISTRY

VOLUME 1

Edited by

GUNTHER L. EICHHORN

National Institutes of Health, Gerontology Research Center,
Baltimore City Hospitals, Baltimore, U.S.A.

ELSEVIER SCIENTIFIC PUBLISHING COMPANY

AMSTERDAM — LONDON — NEW YORK

1973

ELSEVIER SCIENTIFIC PUBLISHING COMPANY
335 JAN VAN GALENSTRAAT
P.O. BOX 1270, AMSTERDAM, THE NETHERLANDS

AMERICAN ELSEVIER PUBLISHING COMPANY, INC.
52 VANDERBILT AVENUE
NEW YORK, NEW YORK 10017

LIBRARY OF CONGRESS CARD NUMBER: 72-135493

ISBN 0-444-41021-X

VOLUME 1: 114 ILLUSTRATIONS AND 81 TABLES

VOLUME 2: 200 ILLUSTRATIONS AND 58 TABLES

PRINTED IN THE NETHERLANDS

PREFACE

Until recently, the title "Inorganic Biochemistry" would have
appeared paradoxical to most, and it may even now appear so to many,
because biochemistry sounds "organic". The Wöhler synthesis of urea in
1828 may have demonstrated that organic chemistry need not be biochemistry, but for a century thereafter biochemical phenomena continued
to be associated more with organic chemistry than with the other branches
of chemistry. In recent years the compartmentalization between the
classical chemical disciplines, and between the scientific disciplines generally,
has broken down. In the process the usefulness of borderline fields between
the classical chemical departments has become evident.

The community of interest between inorganic chemistry and biochemistry has, of course, been apparent to concerned individuals for a long
time. Workers in the life sciences have encountered and solved problems of
inorganic chemistry in their research. Inorganic chemists have sometimes
realized the relevance of their work to biological processes and have
occasionally allowed this relevance to guide the subsequent direction of
their research. It was not until the late 1950s, however, that contacts
between inorganic chemists and biochemists were encouraged through
conferences on topics of mutual interest, and in the last few years such
conferences have become commonplace.

The widespread interest that inorganic biochemistry has recently
aroused makes the present work even more timely than it was when the
publishers first suggested such a book. What had been intended as a book by
one or two authors has evolved into a treatise involving 34 chapters by
45 authors. The vastness of the literature in this field makes it possible to
come closer to achieving comprehensive coverage in this way.

Although some symposia in this field have led to publications, and a
variety of reviews have previously appeared, the lack of a comprehensive
work in this field makes it necessary to define and to limit the territory that
such a work should comprise. Rather obviously, the inclusion of such
elements as sulfur, nitrogen, or phosphorus in biochemical substances does
not justify their discussion here, even though the chemistry of these
substances as determined by these elements has "inorganic" implication,
but such a course would identify inorganic biochemistry with all of biochemistry! The term "inorganic biochemistry" probably conveys the idea of
the involvement of metal ions in biological processes to anyone to whom it
conveys anything at all. We have therefore adopted an admittedly arbitrary
but hopefully justified and profitable definition of inorganic biochemistry
as the application of the principles of the coordination chemistry of metals

to biological problems. In order further to restrict the subject matter, pharmacological and nutritional uses of coordination compounds are not considered, except briefly in the introduction.

After defining the limits of the subject, it becomes necesary to organize it. Obviously numerous organizational schemes are possible. One that will occur to many is the treatment of the subject matter by metals; *i.e.* copper compounds would be considered together, as would iron compounds, zinc compounds, etc. Such a treatment would have unquestionable value. In this work, however, we have chosen to organize the subject on the basis of the structure of the ligands with the expectation that greater cohesiveness can be attained in this way. The subjects have been arranged to allow for a logical development based on structural similarities.

Many biological coordination compounds are macromolecules. It has been our aim to focus on the chemistry of the metal, but to understand this chemistry it is frequently essential to consider how the conformation of the macromolecule affects its metal component and therefore to pay considerable attention to the "organic" portions of the molecule. On the other hand, since coordination chemistry is basic to this book, some essentially "inorganic" chapters have been included. In this way the relationships between the biological coordination compounds and "model" inorganic complexes can be ascertained. In many chapters, biological substances and model compounds are considered together. The emphasis is on the biological substances.

Part I is designed to provide sufficient inorganic background for the biochemist to make it unnecessary for him to consult inorganic texts or literature to understand many (but of course not all) of the inorganic phenomena encountered in the later sections of this book or in his own research. Many clues to the elucidation of inorganic biochemical phenomena are provided. Chapter 1 is an introductory chapter on coordination chemistry and is not limited to the presently known "biological" metal ions, since structural and stereochemical features can be best understood by reference to the best examples, regardless of the metal ions used to furnish them. Moreover even platinum complexes can gain biological significance (see Introduction). In Chapter 2 the concepts relating structure and stability are developed, culminating in a comparison of the stabilities of biological complexes. Chapter 3 illustrates the usefulness of studies based on electronic configuration in the structural elucidation of some of the substances taken up later in the book.

Many biological coordination compounds are proteins and these are discussed in Parts II through VI. Part II begins with a systematic discussion, in Chapter 4, of the metal complexes of the protein monomers, the amino acids, and the simple oligopeptides. This chapter is followed by a review of some natural oligopeptides that have been specifically designed for iron (Chapter 5) and alkali metals (Chapter 6). Some of these substances are oligopeptides while others are not, but the coordination properties of the peptide and non-peptide compounds have many similarities. In Chapter 7, the interaction of metal ions with proteins is introduced to demonstrate the types of coordination sites to which metals can bind in the macromolecules.

In Part III are discussed some biological metalloproteins involved in the storage and transfer of iron (Chapters 8 and 9) and of copper (Chapter 10), as well as some oxygen-binding metalloproteins found in lower organisms (Chapters 11 and 12).

Part IV contains a consideration of the structure of metal enzymes and of the methods by which metal ions participate in enzymatic activity, particularly bond cleavage activity. In Chapter 13 are taken up the ways in which metal ions can induce chemical changes in ligands, without the benefit of the presence of proteins. Chapter 14 is an overview of metal enzymes not considered later in individual chapters. In Chapters 15 and 16 some metal enzymes are treated in depth partly because of the extent of information presently available on them, and other enzymes have been considered separately in Chapters 17 and 18 because of the relative ease of categorizing these groups of enzymes. Those enzymes that are primarily engaged in redox reactions or that involve porphyrins or other prosthetic groups are taken up later.

Enzymatic oxidation—reduction is the subject for Part V, which begins with a consideration, in Chapter 19, of oxidation—reduction in coordination compounds, with a theoretical treatment, and application to biological systems. Chapter 20 contains a classification of oxygenation reactions by metals in the presence or absence of enzyme. In Chapter 21 the different types of copper-containing oxidases are discussed. The ferredoxin-type proteins that engage in electron transfer are characterized in Chapter 22, followed by Chapter 23 on nitrogen fixation, in which the nitrogenase enzymes that require ferredoxins play an important role.

The metal ions in the proteins considered so far have been attached to the amino acid side chains of proteins. In Part VI we begin to take up proteins containing metal ions that are not attached directly to the protein, but to a "prosthetic group" or "coenzyme". The most versatile prosthetic group is the porphyrin; the discussion of the porphyrins in Chapter 24 is followed by Chapters 25—28 on the iron-porphyrin compounds, the hemo-proteins, and then by Chapter 29 on the magnesium porphyrin derivative, chlorophyll. Because of their structural resemblance to porphyrins, the corrins and the B_{12} coenzymes and enzymes follow in Chapter 30.

Part VII is devoted to the metal complexes of other prosthetic groups. It is not certain that the metal complexes of vitamin B_6, in Chapter 31, play a biological role, but they are excellent models of B_6-catalyzed enzymatic processes. The metal complexes of flavins and, briefly, metal flavoproteins are discussed in Chapter 32.

Part VIII, on the interaction of metal ions with nucleic acids, begins in

Chapter 33 with the metal complexes of nucleosides and nucleotides; these bear some resemblance to metal flavin complexes. Finally, in Chapter 34, are taken up complexes of polynucleotides and nucleic acids and their biological implications.

While this order of presentation has many advantages, other sequences can be proposed with different advantages. Thus one could argue for the incorporation of the cytochromes under oxidation—reduction, rather than the porphyrins. It appears preferable, however, to discuss porphyrins generally before the cytochromes; in fact, much of Part VI is concerned with oxidation—reduction, and therefore logically follows Part V. The placement of Part VIII somewhere ahead of Part IV could be justified because the nucleotide complexes discussed in Chapter 33 are so important in Chapter 18. However, Chapter 34 would then have been out of place. Obviously there is no completely satisfactory sequence, and for this reason a certain number of points have to be repeated in various places. Overlap has been eliminated whenever consistent with clarity. To achieve this and other objectives, there has been considerable interaction between authors, as well as between authors and editor.

The cooperation of the authors, some of whom have made valuable suggestions about parts of the book not written by them, is deeply appreciated. I have received helpful advice from many others; I hesitate to name them for fear of omission. I am indebted to my colleagues Nathan A. Berger, James J. Butzow, Patricia Clark, Jane Heim, Josef Pitha, Carmen Richardson, Joseph Rifkind, Yong A. Shin and Edward Tarien for their understanding during the preparation of this book, as well as for help with some of the editing. I am grateful to the National Institutes of Health for making this endeavor possible. I am greatly indebted to my secretary, Jacqueline Blake, for her dedicated help in all facets of editing this book. Finally, I must acknowledge the patience of my wife and children during the course of this project, which took so much more time than they or I had hoped.

GUNTHER L. EICHHORN

INTRODUCTION

The name "bioinorganic chemistry" has appeared as the title of a number of recent symposia as well as a new scientific journal. The name of the present work was conceived before the appearance of its linkage isomer (see Chapter 1, p. 46). It would have been possible to follow the popular trend, but it was decided instead to retain the name that had been originally intended for this book. One of the reasons for this decision is that the existence of both titles illustrates the different emphasis that can be placed upon the components of these names by the biochemist and the inorganic chemist. The former has looked upon inorganic chemistry to explain the chemical behavior of biological metal complexes. The latter has looked upon biochemistry to find relevant applications for his findings. The end result of the two approaches is, of course, the same, just as the "linkage isomer" names have the same meaning. However, the scientist's point of origin is of some importance in determining what he must learn to gain his objective and therefore what he will discover on the way.

A major difference between inorganic chemists and biochemists is that the former are accustomed to dealing with relatively small molecules that can sometimes be considered "models" of the more complex systems with which the latter are concerned. A fundamental question which inorganic biochemists face constantly is whether it is useful to study the model when the "natural" substance is at hand.

To answer this question it is required to make a realistic appraisal of the ways in which biological mechanisms are elucidated. The living cell is so complex that its workings can be understood only by isolating component parts, always with the inherent risk that the parts in isolation do not function as they do in the cell. The isolated components are therefore "models" of the components in the cell. For example, cytochrome c *in vitro* is really a model of cytochrome c in the cell. But it is most likely that isolated cytochrome c shares many characteristics with cellular cytochrome c, and eventually, after all the other molecules to which cytochrome c is bound in the cell have also been isolated and studied, it should be possible to understand the behavior of cytochrome c in the cell.

In the same way, an inorganic substance, such as a simple porphyrin complex that can engage in electron transfer, perhaps with approximately the same oxidation potential as isolated cytochrome c, can serve as a model for cytochrome c at a time when the mechanism of electron transfer in cytochrome c is not yet understood. Studies with the simple porphyrin complex can aid in understanding isolated cytochrome c, and studies with isolated cytochrome c can in turn aid in the understanding of cellular cytochrome c. From such a perspective, it seems that models at every level can be useful. It is necessary, of course, in the investigation of models always to remember the axiomatic limitation of the model — a model of a system is not the system itself. Inorganic chemists, as well as biochemists, can make a positive contribution when they remember this fact, and a negative one when they forget it.

Although the emphasis in this work is on the biological systems, many models are also considered, sometimes in separate chapters and sometimes alongside the more natural substances that they are designed to emulate. Correlations are frequently made, hopefully with the above limitation in mind. If inorganic biochemistry has any reason for existence, such correlations should lead to fruitful results.

What, then, is the importance of these results? What are the ultimate objectives of inorganic biochemistry?

These questions can be answered by considering this borderline field a part of biological science. There are two objectives in the study of biological science. The first, as in all discovery, is to study it because it is "there", to see what makes things tick. The second is to gain an understanding of normal and abnormal cellular processes, culminating in sufficient knowledge to lead to the conquest of disease. Inorganic biochemistry shares these objectives with the other biological sciences. The curiosity of the scientist and his desire for "relevance" at this point in history frequently causes him to be motivated by both of these objectives.

No attempt is made in this work to catalog the medical applications of inorganic biochemistry, partly because these are considered beyond the scope of this book. The pharmacological and nutritional aspects of coordination chemistry have been previously reviewed[1,2]. This introduction nevertheless provides an appropriate place to illustrate some of the useful consequences of the study of inorganic biochemistry.

The most obvious and widespread medical use of complexing agents is in the removal of undesired metal ions from the body. Such ligands as ethylenediamine tetraacetic acid, penicillamine, etc., have been employed in the treatment of diseases involving an overload of iron or copper (see Chapter 10) as well as in combatting the toxic effects of ingested metal ions[3-5].

Recently, it was discovered that the administration of penicillamine can lead to a pronounced decrease in taste acuity[6-8]. Such a decrease does not occur, however, in the treatment of Wilson's disease, in which an accumulation of copper(II) ions occurs in the body (Chapter 10). It was hypothesized that the decrease in taste acuity is due to the complexing of copper ions, and that the copper ions therefore are somehow involved in producing the sensation of taste. The failure to decrease taste acuity in Wilson's disease was attributed to the presence of such a high level of Cu^{2+} that the administered penicillamine cannot overcome the effect of the Cu^{2+}

If this hypothesis is correct, the administration of Cu(II) should restore taste acuity lost through the action of complexing agents. Experiments have shown that copper ions indeed restore taste; furthermore, Zn(II) and Ni(II) have effects similar to Cu(II) in restoring taste that is lost through treatment with penicillamine or through diseased states. Thus metal ions seem to be implicated in producing the sensation of taste[4].

It has recently been discovered that certain complexes of platinum exhibit potent anti-tumor activity[9]. Active complexes are cis-$[Pt^{II}(NH_3)_2 Cl_2]$, $[Pt^{II}en\ Cl_2]$, cis-$[Pt^{IV}(NH_3)_2 Cl_4]$ and $[Pt^{IV}en\ Cl_4]$*. On the other hand, $[Pt^{II}(NH_3)_4]Cl_2$ and $trans$-$[Pt^{IV}(NH_3)_2 Cl_4]$ are inactive. Tumor-inhibiting activity has been correlated with the inhibition of DNA replication[10] (see Chapter 34). Needless to say, these biological phenomena are intriguing to inorganic chemists.

It is to be expected that such discoveries as the use of metal ions in the recovery of taste and the use of metal complexes in tumor regression should help to stimulate activity in inorganic biochemistry.

REFERENCES

1 S. Chaberek and A.E. Martell, *Organic Sequestering Agents*, Wiley, New York, 1959, p. 416.
2 A, Schulman and F.P. Dwyer, in F.P. Dwyer and D.P. Mellor (Eds.), *Chelating Agents and Metal Chelates*, Academic Press, New York, 1964, p. 383.
3 M.J. Seven and L.A. Johnson (Eds.), *Metal Binding in Medicine*, Lippincott, Philadelphia, 1960.
4 L.A, Johnson and M.J. Seven (Eds.), *Federation Proc.*, 20 (3) (1961).
5 F. Gross (Ed.), *Iron Metabolism*, Springer Verlag, Berlin, 1964.
6 R.I. Henkin, H.R. Keiser, I.A. Jaffe, I. Sternlieb and I.H. Scheinberg, *Lancet*, (1967) 1268.
7 R.I. Henkin and D.F. Bradley, *Proc. Natn. Acad. Sci. U.S.*, 62 (1969) 30.
8 R.I. Henkin, P.P.G. Graziadei and D.F. Bradley, *Ann. Internal Med.*, 71 (1969) 791.
9 B. Rosenberg, L. van Camp, J.E. Trosko and V.H. Mansour, *Nature*, 222 (1969) 385.
10 H.C. Harder and B. Rosenberg, *Int. J. Cancer*, 6 (1970) 207. Earlier references can be found in this paper.

---

*en = ethylenediamine

LIST OF CONTRIBUTORS

Aisen, Philip, Department of Biophysics, Albert Einstein College of
    Medicine, New York (U.S.A.)

Angelici, Robert J., Department of Chemistry, Iowa State University, Ames,
    Iowa (U.S.A.)

Breslow, Esther, Department of Biochemistry, Cornell University Medical
    College, New York (U.S.A.)

Buckingham, David A., Research School of Chemistry, Australian National
    University, Canberra (Australia)

Burns, R.C., Central Research Department, Experimental Station, E.I. du
    Pont de Nemours & Co., Inc., Wilmington, Delaware (U.S.A.)

Caughey, Winslow S., Department of Chemistry, Arizona State University,
    Tempe, Arizona (U.S.A.)

Coleman, Joseph E., Department of Molecular Biophysics and Biochemistry,
    Yale University, New Haven, Connecticut (U.S.A.)

Eichhorn, Gunther L., Laboratory of Molecular Aging, Gerontology
    Research Center, National Institutes of Health, National Institute of
    Child Health and Human Development, Baltimore City Hospitals,
    Baltimore, Maryland (U.S.A.)

Freeman, Hans C., School of Chemistry, University of Sydney, Sydney,
    N.S.W. (Australia)

Gray, Harry B., Arthur Amos Noyes Laboratory of Chemical Physics,
    California Institute of Technology, Pasadena, California (U.S.A.)

Harbury, Henry A., Section of Biochemistry and Molecular Biology,
    Department of Biological Sciences, University of California, Santa
    Barbara, California (U.S.A.)

Hardy, Ralph W., Central Research Department, Experimental Station,
    E.I. du Pont de Nemours & Co., Inc., Wilmington, Delaware (U.S.A.)

Harrison, Pauline M., Department of Biochemistry, University of Sheffield,
    Sheffield (Gt. Britain)

Hemmerich, Peter, Fachbereich Biologie, Universität Konstanz, Konstanz
    (Germany)

Hill, H. Allen O., Inorganic Chemistry Laboratory, South Parks Road,
    Oxford (Gt. Britain)

Hix, Jr., James E., Department of Chemistry, East Carolina University,
    Greenville, North Carolina (U.S.A.)

Holm, Richard H., Department of Chemistry, Massachusetts Institute of
    Chemistry, Cambridge, Massachusetts (U.S.A.)

Hoy, T.G., Department of Biochemistry, University of Sheffield, Sheffield
    (Gt. Britain)

Jones, Mark M., Department of Chemistry, Vanderbilt University, Nashville, Tennessee (U.S.A.)

Katz, Joseph J., Chemistry Division, Argonne National Laboratory, Argonne, Illinois (U.S.A.)

Klotz, Irving M., Biochemistry Division, Department of Chemistry, Northwestern University, Evanston, Illinois (U.S.A.)

Lauterwein, J., Fachbereich Biologie, Universität Konstanz, Konstanz (Germany)

Lipscomb, W.N., Chemistry Department, Harvard University, Cambridge, Massachusetts (U.S.A.)

Lontie, René, Laboratorium voor Biochemie, Katholieke Universiteit te Leuven, Louvain (Belgium)

Ludwig, Martha L., Biophysics Research Division, Institute of Science and Technology, University of Michigan and Biological Chemistry Department, University of Michigan Medical School, Ann Arbor, Michigan (U.S.A.)

Malkin, Richard, Department of Cell Physiology, University of California, Berkeley, California (U.S.A.)

Marks, Richard H.L., Section of Biochemistry and Molecular Biology, Department of Biological Sciences, University of California, Santa Barbara, California (U.S.A.)

Martell, Arthur E., Department of Chemistry, Texas A & M University, College Station, Texas (U.S.A.)

Morell, Anatol G., Department of Medicine, Albert Einstein College of Medicine, New York (U.S.A.)

Neilands, J.B., Biochemistry Department, University of California, Berkeley, California (U.S.A.)

Okamura, Melvin Y., Biochemistry Division, Department of Chemistry, Northwestern University, Evanston, Illinois (U.S.A.)

Orme-Johnson, William H., Department of Biochemistry and Institute for Enzyme Research, University of Wisconsin, Madison, Wisconsin (U.S.A.)

O'Sullivan, William J., Department of Medicine, University of Sydney, Sydney, N.S.W. (Australia)

Parshall, G.W., Central Research Department, Experimental Station, E.I. du Pont de Nemours & Co., Inc., Wilmington, Delaware (U.S.A.)

Pressman, Berton C., Department of Pharmacology, University of Miami School of Medicine, Miami, Florida (U.S.A.)

Rifkind, Joseph M., Laboratory of Molecular Aging, Gerontology Research Center, National Institutes of Health, National Institute of Child Health and Human Development, Baltimore City Hospitals, Baltimore, Maryland (U.S.A.)

Saunders, B.C., University Chemical Laboratory, Lensfield Road, Cambridge (Gt. Britain)

Scheinberg, I. Herbert, Department of Medicine, Albert Einstein College of
    Medicine, New York (U.S.A.)
Schugar, Harvey J., Department of Chemistry, Rutgers University, New
    Brunswick, New Jersey (U.S.A.)
Scrutton, Michael C., Department of Biochemistry, Rutgers Medical School,
    Rutgers University, The State University of New Jersey, New
    Brunswick, New Jersey (U.S.A.)
Spiro, Thomas G., Department of Chemistry, Princeton University,
    Princeton, New Jersey (U.S.A.)
Sutin, Norman, Chemistry Department, Brookhaven National Laboratory,
    Upton, New York (U.S.A.)
Taqui Khan, M.M., Department of Chemistry, Texas A & M University,
    College Station, Texas (U.S.A.)
Wharton, David C., Section of Biochemistry and Molecular Biology,
    Division of Biological Sciences, Cornell University, Ithaca, New York
    (U.S.A.)
Witters, R., Laboratorium voor Biochemie, Katholieke Universiteit te
    Leuven (Belgium)

CONTENTS OF VOLUME 1

Part I. Coordination Chemistry

Chapter 1. Structure and Stereochemistry of Coordination
Compounds 3

by D.A. BUCKINGHAM

Chapter 2. Stability of Coordination Compounds 63

by R.J. ANGELICI

Chapter 3. Electronic Structures of Iron Complexes 102

by H.B. GRAY and H.J. SCHUGAR

Part II. Interaction of Metal Ions with Amino Acids, Peptides and Related Natural Chelators, and Proteins

Chapter 4. Metal Complexes of Amino Acids and Peptides 121

by H.C. FREEMAN

Chapter 5. Microbial Iron Transport Compounds (Siderochromes) . 167

by J.B. NEILANDS

Chapter 6. Alkali Metal Chelators — The Ionophores 203

by B.C. PRESSMAN

Chapter 14. Metal Enzymes 381

by M.C. SCRUTTON

Chapter 15. Carboxypeptidase A and Other Peptidases 438

by MARTHA L. LUDWIG and W.N. LIPSCOMB

Chapter 16. Carbonic Anhydrase 488

by J.E. COLEMAN

Chapter 17. Phosphate Transfer and its Activation by Metal Ions; Alkaline Phosphatase 549

by T.G. SPIRO

The following chapters are included in Volume 2.

PART 1

COORDINATION CHEMISTRY

Chapter 1

STRUCTURE AND STEREOCHEMISTRY OF COORDINATION COMPOUNDS

D. A. BUCKINGHAM

Research School of Chemistry, Australian National University, Canberra, Australia

1. INTRODUCTION*

A knowledge of structure and stereochemistry is important to all branches of chemistry, primarily because it allows ground state, and in some instances excited state, properties of chemical compounds to be more clearly understood. In inorganic coordination chemistry such information has been of immense use in the development of theories concerning the nature of the coordinate bond, that is the bond between a donor atom or donor atoms of the ligand and the metal atom. However, structural knowledge is in itself insufficient to define chemical bonding and at the present time this area of coordination chemistry is in an early stage of development. A more rewarding application of structural studies has been to the reactants, products and transition states of chemical reactions. It is in this area of chemical dynamics where structural and kinetic studies combine to give an appreciation of reaction mechanism that inorganic coordination chemistry has much to offer. This is particularly so when considering the topics discussed in this volume, where metal ions are found necessary for many biological processes but an understanding of the mechanism of their involvement is largely unknown. It is the purpose of this chapter to give a brief introduction to some of the more classical concepts developed by the inorganic coordination chemist in describing and discussing the stereochemistry of inorganic complexes so that these principles will be clear to the reader of the following, more specific, chapters.

In an introductory chapter it is impossible to describe in depth all aspects of the stereochemistry of inorganic complexes; such detail may be found in standard texts[1-7]. Therefore what follows gives examples of, rather than reviews, those concepts which are felt to be of most use to readers of this volume; coordination number and geometry, ligand type, factors influencing the stereochemistry of coordination complexes and the various types of coordination isomerism. Like other areas of inorganic chemistry, the stereochemical aspects of coordination complexes is currently under intensive investigation particularly in the areas of structure determination, conformational analysis and the thermodynamics and

---

* For abbreviations used in this Chapter, see page 62.

References pp. 61–62

4

kinetics of structural interconversion. No attempt will be made to rigorously discuss these areas and the reader is referred to recent texts and reviews for some of this information[2, 8-17].

Atoms or groups of atoms surrounding the central metal are called *ligands* and those atoms directly bonded to it are termed *donor atoms*. Donor atoms are usually less electronegative than the metal so that electrostatic attraction and to some extent charge redistribution towards the metal (covalency) are predominant features in their bonding. Ligands are commonly neutral or negatively charged molecules. Coordination complexes are characterized by retaining their identity, at least to some extent, in solution, although appreciable dissociation may occur. The spatial distribution of ligands about the central atom is termed the *configuration* of the complex, while some of the more detailed aspects of ligand stereochemistry (without reference to the metal atom) is usually discussed in terms of ligand *conformations*. The total number of donor atoms attached to the central metal atom is termed the *coordination number*. All donor atoms within bonding distance from the metal are included in this number, although some of these atoms may be further from the metal than others, and some of them contribute to the coordination of more than one metal, as in crystals and polymeric coordination compounds. The coordination number of a metal atom is defined once the stereochemistry is known, but the stereochemistry is not decided by the coordination number alone. Both properties are dependent on the nature of the metal-ligand bond, but it is outside the scope of this chapter to discuss this more fundamental property[18-22]. The structure of a metal coordination complex is clearly defined once the coordination number of the central metal atom, the stereochemistry, and the conformations adopted by the attached ligands are all known. At the present time such detailed information is confined to the solid state (diffraction techniques), and little is known about ligand structure in solution.

Perhaps the most singularly important concept which relates and controls the immense variety of coordination complexes is the Lewis acidity of the metal ion. This concept will be considered in Chapter 2, but suffice it to say here that complexes of the pretransition metals (Na^+, K^+, Ca^{2+}, Mg^{2+}, Ba^{2+}, Al^{3+}) are held together by electrostatic forces and their stereochemistry is decided almost exclusively by ligand size and the charge on the metal ion. The complex ion stabilities parallel the proton basicities of the ligands, and the effective role of the metal ion is similar to that of a proton. For the transition metals the stereochemistry is more complex and no satisfying empirical or theoretical model is at present available to describe in detail all aspects of their structure or even their stereochemistry. For many of these metals the ionic model is complicated by the non-spherical nature of the electron cloud (crystal field effects) and, as their name implies, by the very considerable deviations from ionic character which are brought about by the transition from ionic to covalent bonding. For such complexes charge neutralization as well as Lewis acidity is important, and ligand field and molecular orbital descriptions have been developed to describe their bonding[2,5].

2. COORDINATION NUMBER AND STEREOCHEMISTRY

The development of structural coordination chemistry dates back to about the end of the eighteenth century with the studies of two remarkable chemists, the Dane S. M. Jørgensen and the Swiss Alfred Werner. Before that time the stoichiometric combination of two or more inorganic compounds was known to form "complex compounds" but they were usually written as *double salts*, *e.g.* $ZnCl_2 . 2CsCl$, $2KCl.MgCl_2$, $Al_2(SO_4)_3 . K_2SO_4 . 24H_2O$, $Fe(CN)_2 . 4KCN$, $AlF_3 . 3KF$, $CoCl_2 . 2KCl$, although we know today that some of them should be formulated differently, *e.g.* K_2CoCl_4, K_2HgF_4, $K_2Fe(CN)_6$.[*]

Werner and Jørgensen prepared hundreds of coordination compounds, mostly ammines of Co(III), Pt(IV) and Pt(II), and studied their transformation, degrees of ionization, and the occurrence of isomers. As a direct result of the intense rivalry which developed between these two exceptional experimentalists the structural basis of modern day coordination chemistry was rapidly formulated. Werner in 1893 correctly interpreted his results in terms of primary and secondary valence. Essentially primary valence was considered as the normal electrovalence of the metal ion (*e.g.* 4 for Pt(IV); 2 for Pt(II) and 3 for Co(III)) and determined the total number of negative charges which must be carried by the anions present, whereas the secondary valence gave the spatial arrangement of bonds about the metal ion resulting in octahedral (Co(III), Pt(IV)) or square planar (Pt(II)) geometries. Secondary valence is now termed the coordination number.

The two most important principles developed by Werner were firstly *the assignment of a single coordination number for each metal ion which was satisfied by the attached ligands*, and secondly *that these coordination sites had a well defined stereochemical arrangement in space*. These two ideas were of the utmost importance in developing the stereochemical aspects of coordination chemistry. Thus the *praseo* and *violeo* forms of $CoCl_3 . 4NH_3$ were correctly formulated as the *trans* and *cis* isomers of

---

[*]Often it is not possible to distinguish between a double salt and a *coordination compound* by their behavior in aqueous solution. For example, X-ray data conclusively show that the four Cl⁻ ions are distributed tetrahedrally about the Co(II) ion in K_2CoCl_4, but this coordination complex dissociates rapidly in water to yield a pink solution of the $Co(H_2O)_6^{2+}$ ion so that its behaviour in solution resembles that of a double salt.

References pp. 61–62

$[Co(NH_3)_4Cl_2]Cl$. (The groups within the square brackets make up the octahedral coordination shell and the remaining chlorine atom is ionic and can be easily titrated, *e.g.* $[Co(NH_3)_4Cl_2]^+Cl^-$.) At the time it was found possible to classify all the known facts on metal complexes using this coordination theory and Werner's proposals have more recently been verified many times by X-ray crystallographic studies. However, they are now known to be not rigidly applicable in all cases.

Two other concepts introduced by Werner were those of *chelation* (or ring formation) and *optical isomerism*. Thus he recognized that the two amine functions of ethylenediamine could replace two of the NH_3 groups in $[Pt(NH_3)_4]Cl_2$ to form $[Pt(en)_2]Cl_2$ and in so doing form a 5-membered heterocyclic ring.

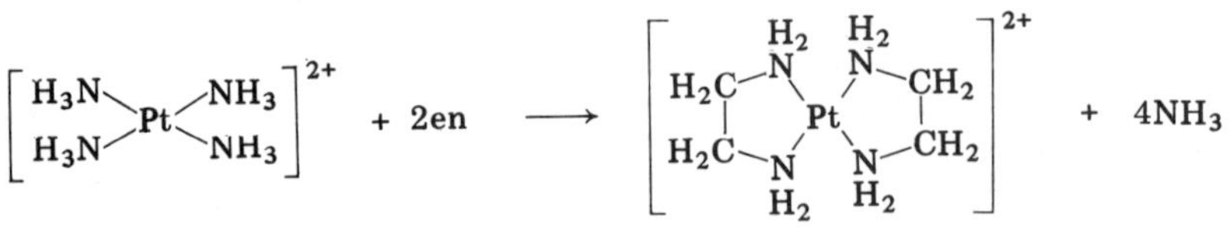

Similarly, when $[Co(en)_2Cl_2]Cl$ was treated with $Na_2C_2O_4$ the resulting compound, $[Co(en)_2(C_2O_4)]Cl$, was regarded as containing chelated oxalate by displacement of *cis* chlorine atoms.

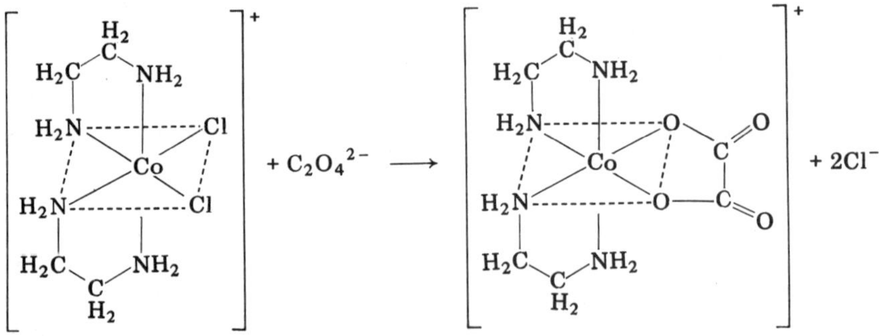

Werner also realized that tetrahedral and octahedral stereochemistries could also result in optical isomerism, as distinct from geometrical isomerism. Thus the *cis* isomer of $[Co(en)_2Cl_2]Cl$ has neither a plane nor a center of symmetry and should exist in non-superposable mirror image pairs, dextro and levo. Werner proved this point by resolving $[Co(en)_2NH_3Cl]^{2+}$ (1911) using *d*-bromocamphorsulfonate as the anion and a year later the $[Co(en)_2Cl_2]^+$, $[Co(en)_2(NO_2)_2]^+$ and $[Co(en)_3]^{3+}$ ions.

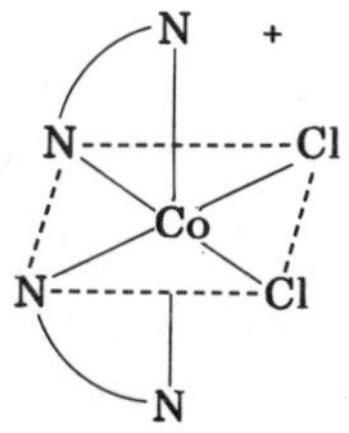

(1)

$(+)_{589}$-Λ-$[Co(en)_2Cl_2]^+$

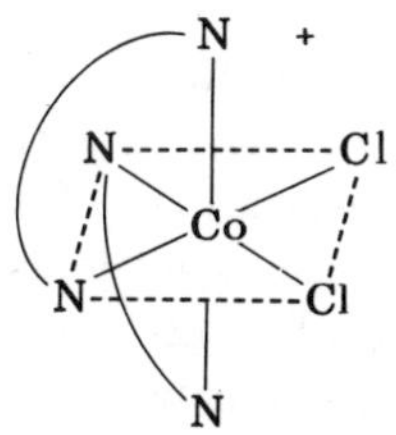

(2)

$(-)_{589}$-Δ-$[Co(en)_2Cl_2]^+$

To answer the criticism of Jørgensen that the optical activity in these complex ions was due in some way to asymmetry about the carbon atoms Werner prepared and resolved the purely inorganic complex ion (3) — a feat of

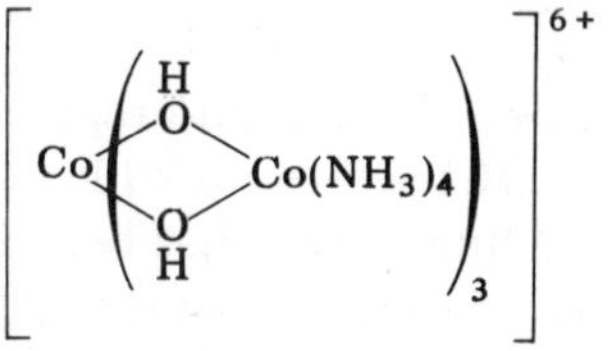

(3)

considerable magnitude at the time.

Werner's views on coordination number and stereochemistry are the basis of modern coordination chemistry[1], and few additional concepts are required to discuss even the most detailed aspects of the structure of metal complexes. Some slight modifications of Werner's proposals have been necessary in the light of the large number, and wide variety, of complexes now known; notably that some metal ions may exist with more than one coordination number and stereochemistry. Thus Cu(II) is found with coordination numbers of 4, 5 and 6, and actinide elements like U exhibit coordination numbers ranging from 5 to 20. Also, stereochemistry is not exclusively determined by the ligand and metal ion alone. Thus there are several Ni(II) complexes which exist as 4 coordinate tetrahedral, or 4 coordinate square plane structures depending on the temperature and/or solvent. Likewise several Ni(II), Zn(II) and Co(II) complexes are now known to exist in each of the geometries, tetrahedral, square planar and octahedral depending on the experimental conditions.

Since Werner's time a large amount of structural evidence has accumulated to define coordination numbers and the more detailed aspects of stereochemistry. Unequivocal direct evidence of structure in the solid state has largely resulted from X-ray diffraction studies, and this area of coordination chemistry has expanded tremendously in the last ten years; the inorganic literature at the present time contains 20 to 30 X-ray structure determinations each month and this number is growing rapidly. Also the degree of precision in these analyses is at a high level. There is not at present any direct means of gathering precise structural information in solution, but n.m.r. studies (^{1}H, ^{19}F, ^{31}P, ^{13}C nuclei in particular) and e.s.r. studies (paramagnetic complexes) are often definitive in their information, and indirect methods such as electronic spectra, dipole moments, magnetic moments and molecular weight measurements are used extensively for structural prediction. The following analysis briefly discusses coordination numbers and stereochemistries in a general way. Perhaps the most singularly distinguishing feature of coordination chemistry today is that the numerical superiority of four and six coordinate structures is being seriously eroded, and that "unusual" coordination numbers (five, seven, eight) and stereo-chemistries are becoming more and more common.

(A) Two coordination

Coordination number 2 is not very common, and is found regularly in only Cu(I), Ag(I), Au(I) and Hg(II) complexes. Of the two possible geometries, linear ($D_{\infty h}$) and angular (C_{2v}), only the linear form has been found, presumably because it provides minimum ligand–ligand repulsions. Typical examples are $[CuCl_2]^-$, $[AuCl_2]^-$ and $[Hg(CN)_2]$. Examples of more biochemical interest include $Ag(NH_3)_2^+$, Ag(gly), Ag(gly)0.5H$_2$O and Ag(glygly) (Chapter 4, Formulae III, IV, XI). In the latter structures Ag(I) binds readily to both the amino and carboxyl groups of glycine but only very weakly to the carbonyl oxygen of glycylglycine. It is very likely that Ag(I), Cu(I) and Au(I) will bind in a similar fashion to other amino acids and peptides.

Sometimes unusual ligands coordinate in unusual coordination geometries, *e.g.*

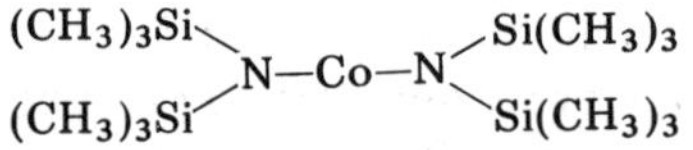

$$(CH_3)_3Si \diagdown \atop (CH_3)_3Si \diagup N-Co-N \diagup Si(CH_3)_3 \atop \diagdown Si(CH_3)_3$$

(4)

but these are rather uncommon. However, such cases serve to demonstrate that metal ions in even common oxidation states can have unusual coordinating properties.

(B) Three coordination

This is a very infrequent coordination number outside those molecules containing a relatively electronegative central atom, *e.g.* BF$_3$, NO$_3^-$ (trigonal plane); NH$_3$, ClO$_3^-$ (trigonal pyramid). Thus the compounds K$_2$[CuCl$_3$] and Cs$_2$[HgCl$_3$] contain infinite chains of MCl$_4$ tetrahedra while the halides MCl$_3$ (M = Cr, Fe, Mn, V) crystallize in lattices containing metal ions in octahedral sites. $Cs[CuCl_3]$ consists of infinite chains of $-Cl-CuCl_2-Cl-$ $CuCl_2-$ with four-fold coordination about the metal atom, while $AuCl_3$ is a dimer consisting of two planar $AuCl_4$ units sharing an edge. Indeed most simple metal compounds of composition MX_2 or MX_3 either dissociate readily in solution or give rise to $MX_2(H_2O)_4$ or $MX_3(H_2O)_3$ species, or exist as polyhedra of coordination number four or six. Two examples containing approximately planar equilateral triangle (D_{3h}) symmetry are tris(hexamethyldisilylamine)iron(III), (5), and the HgI_3^- anion in $(CH_3)_3S(HgI_3)$, (6). The planar T-geometry (C_{2v}) is found in the $(CH_3)_3Se^+$ cation, (7).

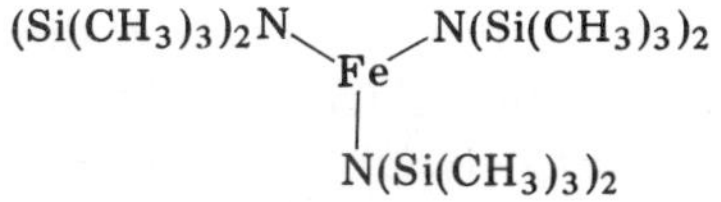

$$(Si(CH_3)_3)_2N \diagdown \underset{Fe}{} \diagup N(Si(CH_3)_3)_2$$
$$\underset{N(Si(CH_3)_3)_2}{|}$$

(5)

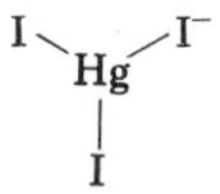

$$I \diagdown \underset{Hg}{} \diagup I^-$$
$$\underset{I}{|}$$

(6)

$$CH_3 - \overset{+}{Se} - CH_3$$
$$\underset{CH_3}{|}$$

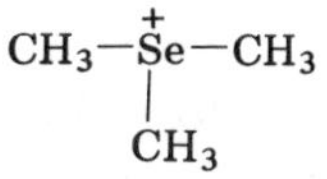

(7)

(C) Four coordination

(i) Tetrahedral

Many complexes contain this geometry. It is common among both the pre- and post-transition elements where its stability can be attributed partly to the use of covalent sp^3 metal hybrids and partly to the fact that from steric and electrostatic considerations the tetrahedral geometry is less hindered than any other involving 4-coordination. Thus in liquid ammonia Na(I) exists as $[Na(NH_3)_4]^+$ and X-ray scattering studies indicate a primary coordination number of four for K(I) in aqueous solution; B(III), Be(II), Zn(II), Cd(II), and Hg(II) in their MX_4 (X = F, Cl, CN) complexes are tetrahedral, as are many Be(II) and Hg(II) complexes with bidentate ligands.

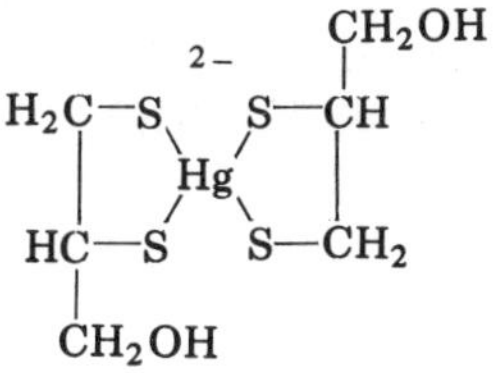

$$\overset{2-}{\underset{\underset{CH_2OH}{|}}{\underset{HC-S}{\overset{H_2C-S}{}}} \underset{Hg}{} \overset{\overset{CH_2OH}{|}}{\underset{S-CH_2}{\overset{S-CH}{}}}}$$

(8)

$$C_6H_5 \diagdown \underset{HC}{\overset{C=O}{}} \underset{Be}{\overset{O=C}{}} \diagup C_6H_5$$
$$HO_2C \diagup \underset{C=O}{} \underset{O=C}{} \diagdown CO_2H$$

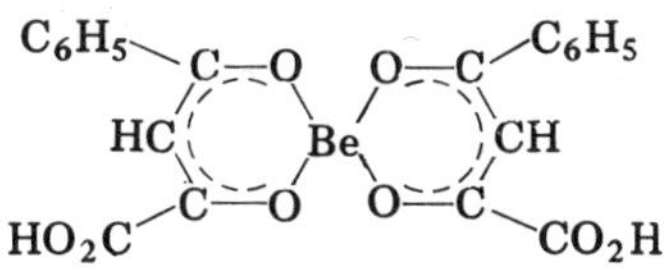

(9)

The bis(benzoylpyruvate)Be(II), (9), has been resolved into its optical forms and this establishes its tetrahedral geometry. In principle, only optical and not geometrical isomers are possible for tetrahedral complexes. However, the lability of the ligand groups, in contrast to the inertness of carbon bonds, has prevented the resolution into optical enantiomers of any tetrahedral complex containing four different monodentate groups.

Among the transition metal ions tetrahedral complexes are usually stable only under certain conditions. Notable exceptions are the ions $[FeCl_4]^{2-}$, $[CoX_4]^{2-}$ (X = Cl, Br, I, NCS) and some $[CoX_3(H_2O)]^-$ ions; these species exist in tetrahedral geometries in aqueous solution despite the fact that water might be expected to aquate them to form octahedral ions. The reasons for their stability are not known. However, the tetrahedral anions $[CuX_4]^{2-}$, $[NiX_4]^{2-}$, $[VX_4]^-$ and $[MnX_4]^{2-}$ (X = Cl, Br, I) are stabilized in solvents of low coordinating power and also in the crystalline state by large cations such as $[(C_6H_5)_3(CH_3)P]^+$, $[(C_6H_5)_4As]^+$ and $[(C_4H_9)_4N]^+$. None of them retains this geometry in coordinating solvents such as water or alcohols. Most tetrahedral complexes are anionic or neutral, $[MX_4]^{2-}$, $[MLX_3]^-$ or $[ML_2X_2]$ where M = Co, Ni, Fe and L is a neutral ligand (H_2O, pyridine, POR_3, $AsOR_3$, PR_3, AsR_3, NR_3) and X is an anion, usually halide. Very few tetrahedral cationic complexes are known, but $Co(H_2O)_4^{2+}$ occurs in small amounts in aqueous solution in equilibrium with $Co(H_2O)_6^{2+}$, and tetrahedral $Co(NH_3)_4^{2+}$ is found in $[Co(NH_3)_4](ReO_4)_2$. Bulky substituents can sometimes force bidentate ligands to form tetrahedral complexes with Co(II), and sometimes Ni(II), e.g. bis(N-isopropylsalicylaldiminato)cobalt(II), (10), and bis(dipivaloylmethane)nickel(II) (11), but usually polymerization

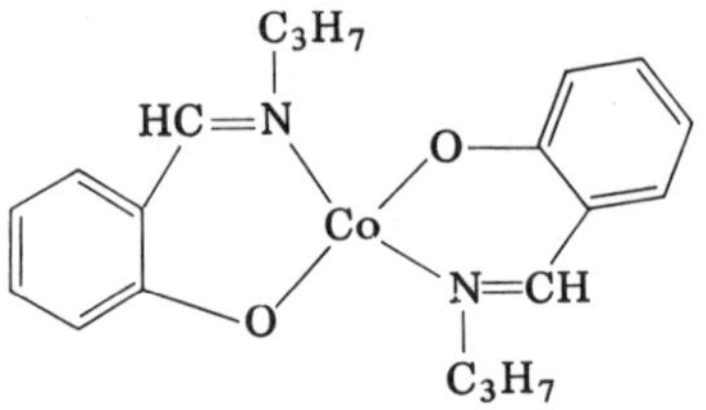

(10)

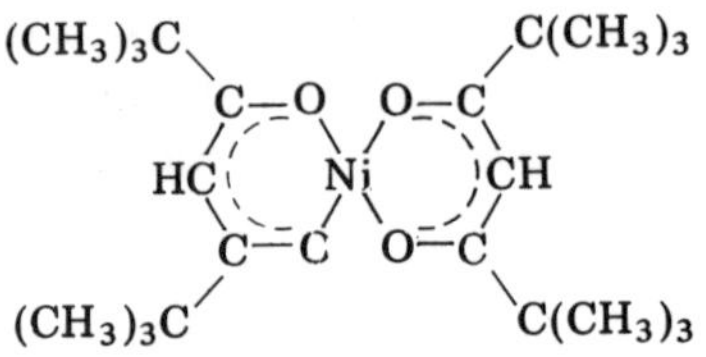

(11)

occurs in the non-sterically hindered cases with octahedral coordination about the metal, e.g. [Ni(acetylacetonato)$_2$]$_3$ and [Co(N-methylsalicylaldiminato)$_2$]$_2$. For those metal complexes where a square planar–tetrahedral equilibrium occurs, e.g. Ni(N-alkylaminotroponeimine)$_2$ (12), Ni(N-

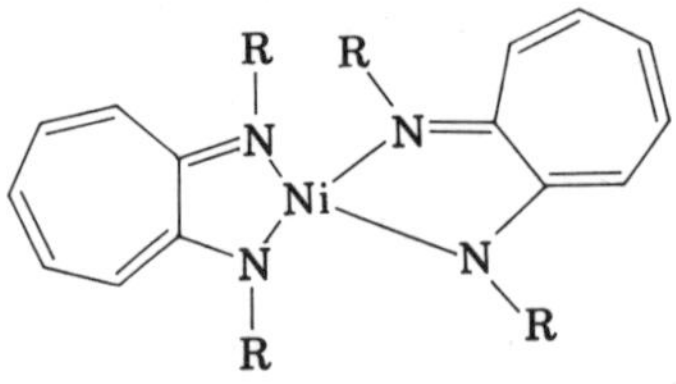

(12)

alkylsalicylaldiminato)$_2$, it appears that repulsion between the R groups in the square arrangement favors the tetrahedral structures, but this is not the universal reason, and electronic effects also appear to be important.

(ii) Square planar

This form of coordination is characteristic of certain metal oxidation states, but otherwise is not very common. It is the common structure found for Pt(II), Pd(II), Ag(II), Au(III), Rh(I), Ir(I), and it also occurs with Ni(II) and Cu(II); otherwise it is seldom, or never, observed. Geometrical isomerism, but not optical isomerism, is possible for square planar structures, However, a notable exception to this rule is provided by (13) which Mills and Quibell

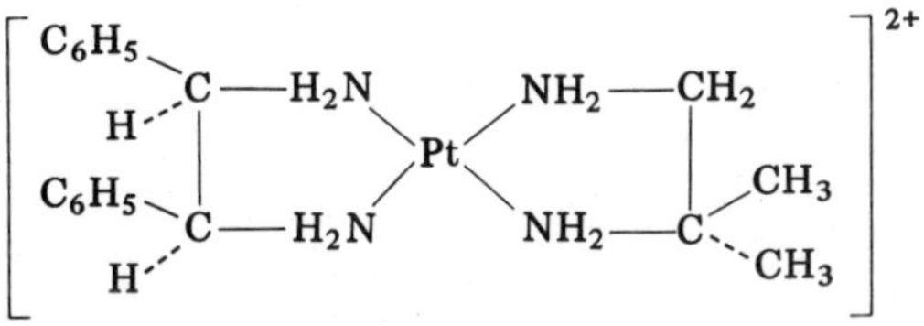

(13)

resolved into optical isomers in 1935 and thereby established that it was square planar and not tetrahedral. In this molecule the asymmetry does not reside in the metal atom or ligand itself, but in the disposition of the chelate rings. Square planar coordination in Pt(II) and Pd(II) complexes has been known from the time of Werner and occurs extensively, *e.g.* [Pt(glycinato)$_2$], (14), and [Pt(glycine-L-methionine)Cl] (Chapter 4, Structure 39). Also

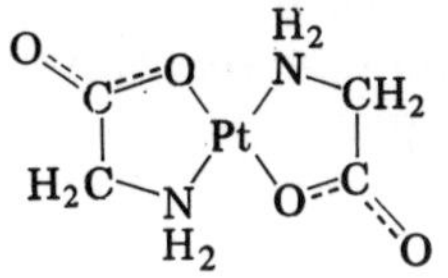

(14)

many Ni(II) and Cu(II) almost planar structures are known, usually of the neutral or anionic variety, *e.g.* Na$_2$[M(tetraglycinato)] (15) (M = Cu^{2+}, Ni^{2+}), and [Cu(1-aminocyclopentanecarboxylato)] (16). Square planar

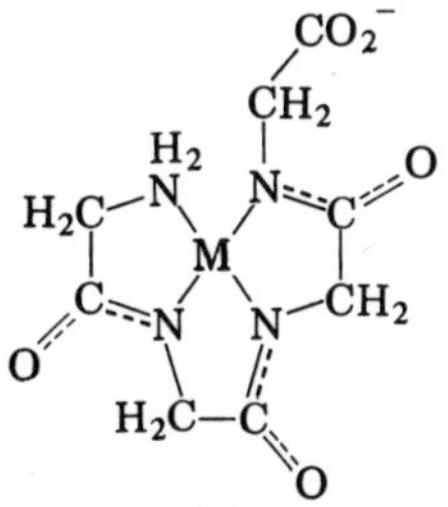

(15)

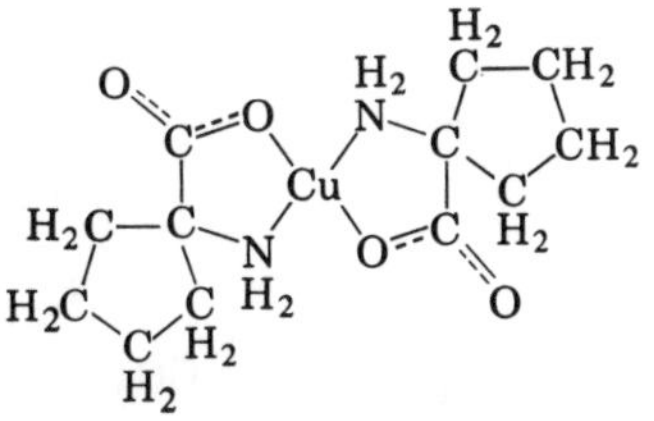

(16)

examples of biochemical interest are of course those associated with the phthalocyanine and porphyrin nucleus[24]. The metal phthalocyanine dyes (17) (M(II) = Be, Mn, Fe, Co, Ni, Cu, Pt) are all isostructural with the parent protonated base (space group $P2_1/a$), and chlorophyll (Mg(II)), vitamin B_{12} (Co(III)-octahedral) and hemoglobin (Fe(II)-5 coordinate) are important examples using the substituted porphyrin nucleus (18).

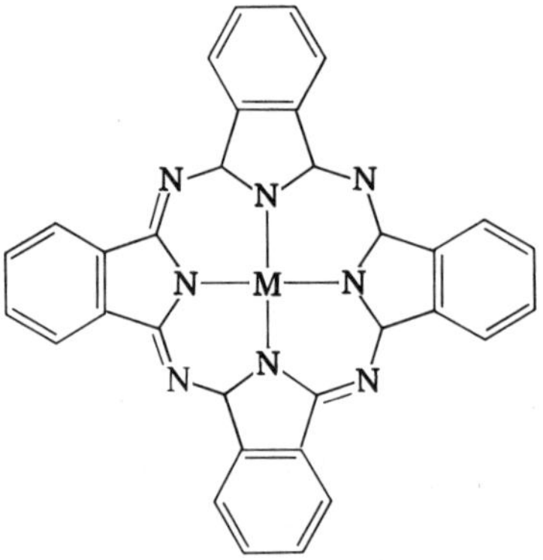

(17) 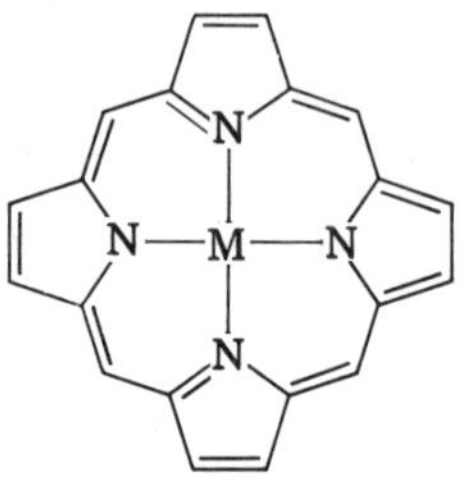   (18)

A very wide range of metal porphyrins have been studied by X-ray methods (Cu(II), Zn(II), Fe(III), VO(IV), Ni(II), Rh(III), Co(III), Au(III), Sn(IV)). Metal complexes of this type will be discussed in later chapters and it is sufficient to say here that the planar tetrapyrrole ring nucleus forces the metal to coordinate in a square planar or almost square planar arrangement.

(D) Five coordination

Though five-coordinate complexes have been historically of relatively minor importance they are much better known at the present time largely as a result of X-ray studies. Some result without apparent ligand constric-

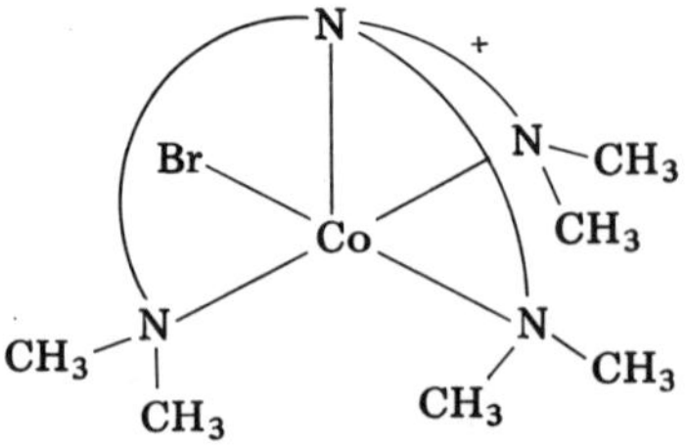

(19) 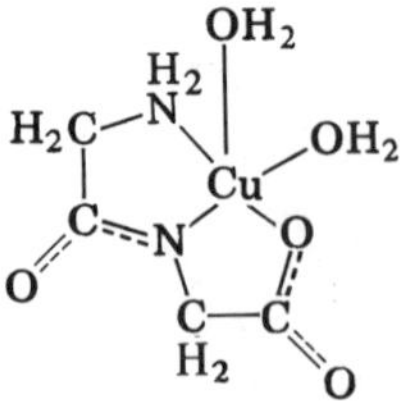   (20)

tions, *e.g.* $Fe(CO)_5$, $MnCl_5^{3-}$, $Ni_2Cl_8^{4-}$, but most five-coordinate complexes are forced by constraints imposed by the ligand, *e.g.* $[Co(Me_6 tren)Br]^+$, $(Ph_3P)_3RuCl_2$, $TiBr_3(NMe_3)_2$. Usually such species can be considered as being derived from one of higher or lower coordination number by exclusion (19) or addition (20) of a ligand. There are two fairly regular geometrical arrangements possible, the *trigonal bipyramid* (D_{3h}), (21) and *square pyramid* (C_{4v}), (22). Usually the trigonal bipyramid is found among low

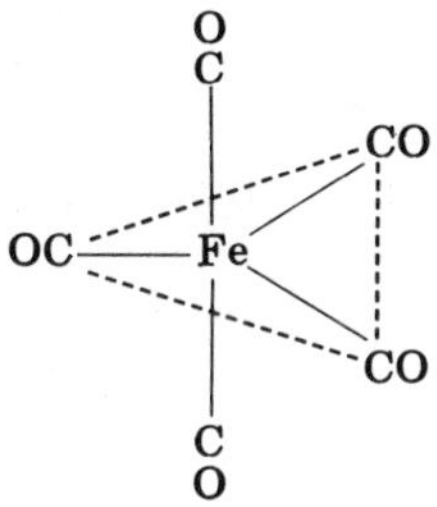

(21) $Fe(CO)_5$

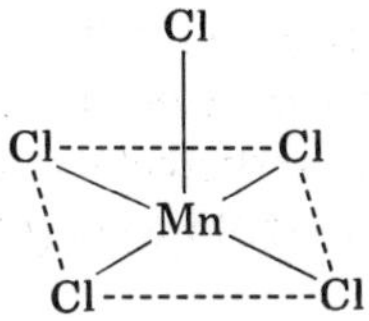

(22) $MnCl_5^{2-}$

valence state metals (Mn(-1), Fe(O), Co(I)) with good π-bonding ligands (CO, CNR, CN), while the square pyramidal arrangement is found with the more biologically significant oxidation states towards the right hand side of the transition series, *e.g.* Ni(II), Co(II), Zn(II), Cu(II). The latter complexes usually involve a much longer (weaker) bond to the apical position.

Finally, it is important to point out that the trigonal bipyramidal and square pyramidal geometries are not really very different and only slight deformations of bond angles are required to convert one into the other.

(E) Six coordination

This is by far the most common coordination number assumed by metal complexes. All metal ions, with the exception of the alkali metals Li^+, K^+ and possibly Na^+ (4 coordinate) and the very large lanthanide (9 coordinate) and uranium group ions (>10), coordinate six water molecules in aqueous solution. With few exceptions the octahedral geometry

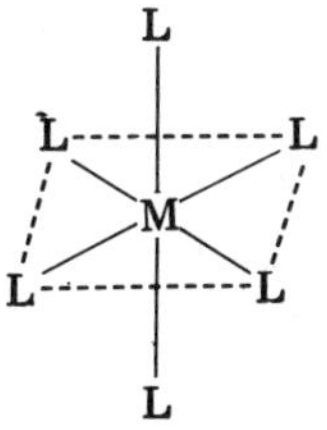

(23)

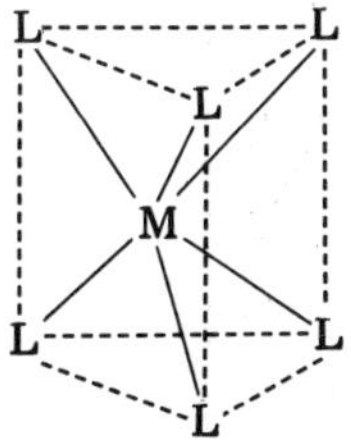

(24)

14

(23) is taken up, but the other possible regular geometry incorporating a *trigonal prism* (24) has been found in a number of cases with special ligands.

The octahedron is of high symmetry (O_h) and some metals form complexes with no or very little distortion from this. Thus Zn(II), Mn(II) (spin free), Cr(III) and Co(III) complexes all assume close to O_h symmetry. Other metal ions such as Cu(II), Ni(II) and Co(II) also assume octahedral geometries with a wide variety of ligands including water but with, in some instances, quite large distortions. Two forms of distortion are common, *trigonal* (25) and *tetragonal* (26). In a trigonally distorted molecule the octahedron is extended or compressed along one of its 3-fold axes ($O_h \rightarrow D_{3d}$); tetragonal distortion results from a similar extension or compression along a 4-fold axis ($O_h \rightarrow D_{4h}$).

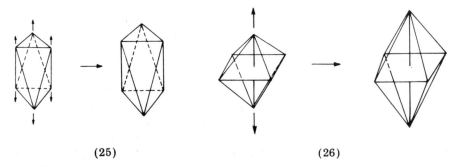

(25)   (26)

Obviously in the limit a tetragonally distorted molecule loses two *trans* ligands entirely and becomes four coordinate and square planar. Octahedral complexes may exhibit both geometrical and optical isomerism and this is taken up in Section 5 of this chapter.

(F) Seven coordination

Coordination numbers higher than six are rarely taken up by the first row transition metals, but some examples are known, *e.g.* [Fe(EDTA)(H$_2$O)]$^-$ (27) (pentagonal bipyramid) and [Mn(EDTA)(H$_2$O)]$^{2-}$ (28) (face-centered trigonal prism). Higher coordination numbers are usually reserved for the

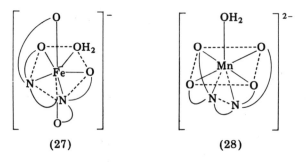

(27)   (28)

second and third row transition metals, and for lanthanide and actinide complexes.

Three geometrical arrangements are known, *pentagonal bipyramid* (D_{5h}), (27), *capped trigonal prism* (C_{2v}), (28), and *tetragonal base–trigonal prism* (C_S) (29).

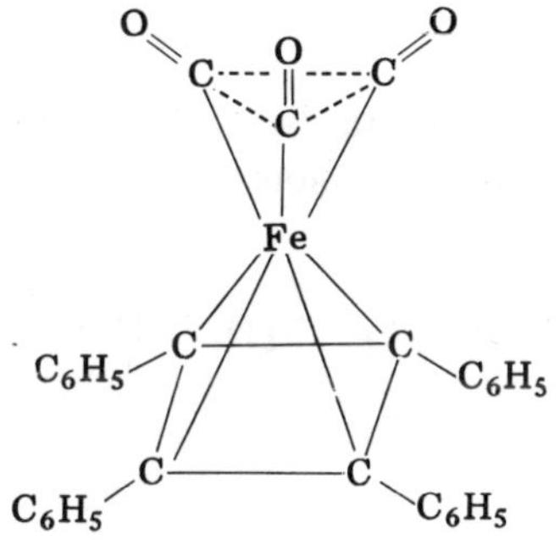

O=C---O---C=O
(with the structure around Fe)

C₆H₅-C / C-C₆H₅
C₆H₅-C----C-C₆H₅

(29)

(G) Eight coordination

A variety of geometries are possible but the most symmetrical arrangement, the *cube* (O_h), is not known for any discrete MX_8 molecules (it does occur in the CsCl lattice). This has been attributed to ligand repulsions resulting in more favorable distorted arrangements. Of the several observed geometries the *square antiprism* (D_{4h}) (30) and the *dodecahedron* (D_{2d}) (31) are the most common. Each are derived by appropriate distortions of

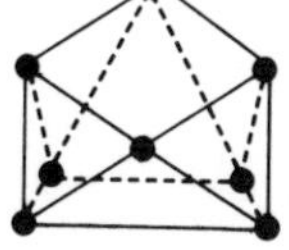

(30)

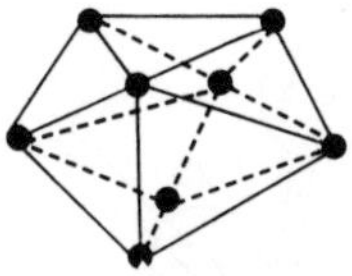

(31)

a cube. Interesting cases occur with $[Co(NO_3)_4]^{2-}$ and $[Zr(C_2O_4)_4]^{4-}$ in which the 4 and 5-membered nitrate and oxalate chelate rings respectively force dodecahedral geometry, while in $[Th(acac)_4]$ and $[Zr(acac)_4]$ with the larger 6-membered chelate rings antiprismatic structures obtain.

(H) Nine coordination

Several geometries are possible and have been observed in a few cases. The *face centered trigonal prism* (D_{3h}), (32), is the most abundant, and may

References pp. 61–62

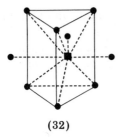

(32)

be considered as being derived from the trigonal prism by adding three ligand atoms outside the centers of the three vertical faces.

(I) Coordination numbers greater than nine

These occur only with the larger metal ions (*e.g.* Cs(I), La(III), U(III),

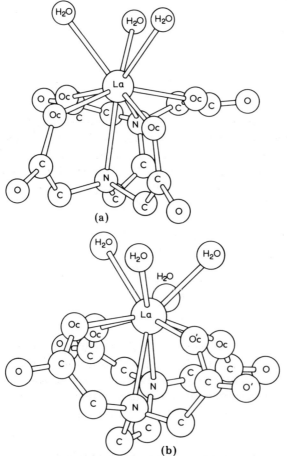

(a)

(b)

Fig. 1. (a) Structure of 9-coordinate $[La(OH_2)_3(EDTA)]^-$ anion[49]; (b) structure of 10-coordinate $[La(OH_2)_4(HEDTA)]$ complex[48].

Ce(III)) where the coordinate bond is much weaker than with the first row transition elements. In many instances the assignment of an exact coordination number is difficult and requires X-ray structural evidence. The resulting polyhedra are often irregular.

3. COORDINATION NUMBERS AND STEREOCHEMISTRIES OF THE COMMON TRANSITION METALS

The following Tables summarize the oxidation states, coordination numbers and geometries of complexes of the more commonly occurring transition metals. Those italicized are the more commonly encountered. The accompanying notes comment on their abundance and chemical behavior.

(A) *Titanium*

TABLE I

Oxidation state	$3d$ electron configuration	Coordination number	Stereochemistry	Example
Ti(−I)		6	Octahedral	$Tidipy_3^-$
Ti(O)		6	Octahedral	$Tidipy_3$
Ti(II)		6	Octahedral	$TiCl_2$
Ti(III)		6	Octahedral	TiF_6^{3-} $Ti(H_2O)_6^{3+}$, $Ti(urea)_6^{3+}$
Ti(IV)	d^0	4	*Tetrahedral*	$TiCl_4 (\pi\text{-}C_5H_5)_2TiCl_2$
		5	Distorted trigonal bipyramidal	$K_2Ti_2O_5$
			Square pyramidal	$TiO(acac)_2$
		6	*Octahedral*	TiO_2, TiF_6^{2-} $Ti(acac)_2Cl_2$ $[Ti(OC_2H_5)_4]_4$
		8	Dodecahedral	$TiCl_4 (diarsine)_2$

$$Ti^{2+} \xrightarrow{\ \sim 2v\ } Ti^{3+} \xrightarrow{\ -0.1v\ } TiO^{2+} \text{ (acidic solution)}$$

There is no aqueous chemistry of Ti^{2+}; Ti(IV) is the most stable and common oxidation state, and the extremely polarizing nature of the small Ti^{4+} ion results in considerable covalency, with Ti^{4+} having no real existence. Ti(IV) compounds readily hydrolyze in aqueous solution forming species with Ti—O bonds, many with octahedral coordination. Characteristically titanium prefers oxygen donors, forming polymeric octahedral structures.

References pp. 61–62

(B) Vanadium

TABLE II

Oxidation state	3d electron configuration	Coordination number	Stereochemistry	Example
V(−I)	d^5s^1 (?)	6	Octahedral	$V(CO)_6^-$, $[V(CN)_5NO]^{5-}$
V(O)	d^5 (?)	6	Octahedral	$[V(dipy)_3]$, $V(CO)_6$
V(I)	d^4	6	Octahedral	$[V(dipy)_3]^+$ $[V(CO)_4$ arene$]^+$
V(II)	d^3	6	Octahedral	$[V(H_2O)_6]^{2+}$, $[V(CN)_6]^{4-}$
V(III)	d^2	4	Tetrahedral	$[VCl_4]^-$
		5	Trigonal bipyramidal	$trans[VCl_3(SMe_2)_2]$
		6	Octahedral	$[V(NH_3)_6]^{3+}$, $[V(C_2O_4)_3]^{3-}$
V(IV)	d^1	4	Tetrahedral	VCl_4
		5	*Tetragonal pyramidal*	$VO(acac)_2$ $[VO(SCN)_4^2]^{2-}$
			Trigonal bipyramidal	$VOCl_2(N(CH_3)_3)_2$
		6	*Octahedral*	VO_2(rutile), $VO(acac)_2py$
		8	Dodecahedral	$[VCl_4(diarsine)_2]$
V(V)	d^0	4	Tetrahedral	$VOCl_3$, VO_4^{3-} (vanadates)
		5	Trigonal bipyramidal	VF_5, polymeric vanadates
		6	*Octahedral*	VO_6, octahedral in acidic vanadates

$$V^{2+} \xrightarrow{0.255v} V^{3+} \xrightarrow{-0.337v} VO^{2+} \xrightarrow{-1.00v} V(OH)_4^+ \text{ (acidic solution)}$$

V(IV) and V(V) are the most common oxidation states. Oxygen donors are preferred and polarizable ligands. The discrete V^{4+} and V^{5+} ions are unknown. The vanadyl ion $[VO(H_2O)_5]^{2+}$ and various V(V) polymeric species $[VO_3(OH)]^{2-}$, $[V_2O_6(OH)]^{3-}$, $[VO_2(OH)_2]^-$ are derived from the monomeric tetrahedral vanadate ion, VO_4^{3-} which exists only in very alkaline solutions (pH > 12); below pH ~7 V_2O_5 precipitates. The V(V) species are all probably 5-coordinate derived from the pervanadate ion, VO_2^+, using OH bridges. Complexes of V(IV) usually contain the vanadyl unit, VO^{2+}, with square or tetragonal pyramidal stereochemistries; e.g. $[VO(acac)_2]$ (33).

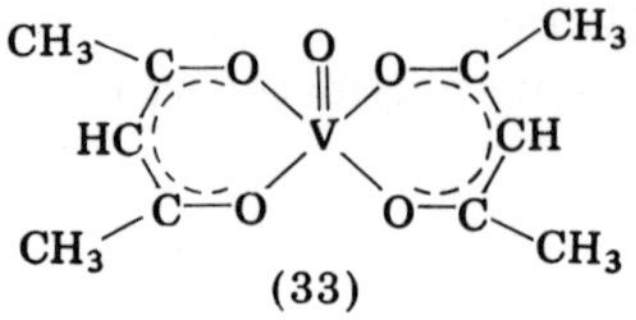

(33)

This provides the major feature of vanadium stereochemistry with chelating oxygen donors; $[\overset{\cdot}{V}O(C_2O_4)_2]^{2-}$, $[VO(C_2O_4)(H_2O)_2]$. Occasionally a weaker donor group may coordinate in the sixth position to form distorted octahedral structures; e.g. $[VO(acac)_2py]$, $[VO(acac)_2(imidazole)]$.

(C) Chromium

TABLE III

Oxidation state	3d electron configuration	Coordination number	Stereochemistry	Examples
Cr(O)	$d^5 s^1$ (?)	6	Octahedral	$Cr(CO)_6$, $Cr(dipy)_3$
Cr(I)	d^5	6	Octahedral	$[Cr(CN)_5NO]^{3-}$, $Cr(dipy)_3^+$
Cr(II)	d^4	6	*Octahedral* (distorted)	$CrCl_2$, CrS, $Cr(NCS)_6^{4-}$ $Cr(en)_3^{2+}$
Cr(III)	d^3	6	*Octahedral*	$Cr(H_2O)_6^{3+}$, $Cr(NH_3)_6^{3+}$ $Cr(acac)_3$, $Cr(CN)_6^{3-}$
Cr(IV)	d^2	4	Tetrahedral	Ba_2CrO_4
		6	Octahedral	K_2CrF_6
Cr(V)	d^1	4	Tetrahedral	CrO_4^{3-}
		6	Octahedral	$CrOCl_5^{2-}$
Cr(IV)	d^0	4	Tetrahedral	CrO_4^{2-}, CrO_2Cl_2, CrO_3

$$Cr^{2+} \xrightarrow{0.41v} Cr^{3+} \quad \text{(acidic solution)}$$

$$Cr(OH)_2 \xrightarrow{1.1v} Cr(OH)_3 \xrightarrow{0.13v} CrO_4^{2-} \quad \text{(basic solution)}$$

Only Cr(II) and Cr(III) assume significance in aqueous solution; Cr(IV) and (V) disproportionate readily into Cr(III) and Cr(VI) with the latter oxidation state being strongly oxidizing and existing only as oxo species, CrO_3, CrO_4^{2-}, CrO_2F_2, etc.

Cr(II) compounds are strong and rapid reducing agents and in aqueous solution can only be present by excluding oxygen. They are characteristically octahedral, e.g. $[Cr(NCS)_6]^{4-}$, $[Cr(CN)_6]^{2-}$. Cr(II) acetate, $[Cr(OCOCH_3)_2]_2$. H_2O, is however easily prepared and very stable (dimeric structure). Cr(III) complexes constitute the most stable and important oxidation state and an undistorted octahedral stereochemistry is preferred. Many hundreds of complexes are known, principally with O and N donors, and together with Co(III) they provide the most kinetically inert complexes of the first row transition metals.

(D) Manganese

TABLE IV

Oxidation state	3d electron configuration	Coordination number	Stereochemistry	Examples
Mn(−II)	?	4 (or 6)	Square	$[Mn(phthalocyanine)]^{2-}$
Mn(−I)	?	5	Trigonal bipyramidal	$Mn(CO)_5^-$
		4 (or 6)	Square	$[Mn(phthalocyanine)]^-$
Mn(O)	$d^6 s^1$	6	Octahedral	$Mn_2(CO)_{10}$
Mn(I)	d^6	6	Octahedral	$Mn(CN)_6^{5-}$, $Mn(CNR)_6^+$ $Mn(CO)_5Cl$
Mn(II)	d^5	4	Tetrahedral	$MnCl_4^{2-}$
		4	Square	$[Mn(H_2O)_4]SO_4 . H_2O$
		6	*Octahedral*	$Mn(H_2O)_6^{2+}$, $Mn(SCN)_6^{4-}$
			?	$[Mn(Me_5dien)X_2]$
		7	NbF_7^{2-}structure	$[Mn(H_2O)(EDTA)]^{2-}$
Mn(III)	d^4	6	Octahedral	$Mn(acac)_3$, $Mn(C_2O_4)_3^{3-}$
Mn(IV)	d^3	6	Octahedral	MnO_2, $MnCl_6^{2-}$
Mn(VII)	d^0	4	Tetrahedral	MnO_4^-, MnO_3F

$$Mn^{2+} \xrightarrow{-1.51v} Mn^{3+} \xrightarrow{-0.95v} MnO_2 \xrightarrow{-2.26v} MnO_4^{2-} \xrightarrow{-0.564v} MnO_4^-$$

(acidic solution)

$$Mn(OH)_2 \xrightarrow{-0.1v} Mn(OH)_3 \xrightarrow{0.2v} MnO_2 \quad \text{(basic solution)}$$

The aqueous chemistry is exclusively confined to that of Mn(II) and because of the large size of the Mn^{2+} ion compared to succeeding divalent metal ions (Fe^{2+} to Cu^{2+}) and its lack of any crystal field stabilization (spin free d^5), it forms only weak complexes which readily dissociate to $Mn(H_2O)_6^{2+}$ in water. However, those with chelating ligands such as ethylenediamine oxalate and EDTA may be isolated from aqueous solution.

The stereochemistry of Mn(II) compounds is usually octahedral; *e.g.* $Mn(NH_3)_6^{2+}$, $Mn(en)_3^{2+}$, $Mn(C_2O_4)_3^{2+}$. Some compounds have been shown to have irregular stereochemistries, *e.g.* 7-coordinate ($[Mn(OH_2)EDTA]^{2-}$) and 5-coordinate ($MnCl_5^{2-}$), while Mn(II) acetylacetonate $Mn(acac)_2$ achieves octahedral coordination by being trimeric. Some tetrahedral compounds are known *e.g.* $[Mn(PH_3PO)_2 Br_2]$ and $(R_4M)_2[MnX_4]$ (M = N, P, As and X = halogen) but these rapidly dissociate in water. Of the Mn(III) complexes, the octahedral acetate $Mn(C_2H_3O_2)_3 . 2H_2O$ is perhaps best known; all are easily reduced to Mn(II) in water and, in the absence of ligand effects, are expected to have a tetragonally distorted octahedral stereochemistry.

(E) Iron

TABLE V

Oxidation state	3d electron configuration	Coordination number	Stereochemistry	Examples
Fe(−II)	?	4	Tetrahedral	$Fe(CO)_4^{2-}$, $Fe(CO)_2(NO)_2$
Fe(0)	d^8 (?)	5	Trigonal bipyramidal	$Fe(CO)_5$, $Fe(PF_3)_5$
		6	Octahedral	$[Fe(CO)_6H]^+$
Fe(II)	d^6	4	Tetrahedral	$FeCl_4^{2-}$
		5	Trigonal bipyramidal	$Fe(Me_5 dien)X_2$
		6	Octahedral	$Fe(H_2O)_6^{2+}$, $Fe(CN)_6^{4-}$
Fe(III)	d^5	4	Tetrahedral	$FeCl_4^-$, Fe_3O_4
		6	Octahedral	Fe_2O_3, $Fe(acac)_3$, $Fe(C_2O_4)_3^{3-}$
		7	Pentagonal bipyramid	$[Fe(H_2O)EDTA]^-$
Fe(IV)	d^4	6	Octahedral	$[Fe(diarsine)_2Cl_2]^{2+}$

$$Fe^{2+} \xrightarrow{-0.771v} Fe^{3+} \qquad \text{(acidic solution)}$$
$$Fe(OH)_2 \xrightarrow{0.56v} Fe(OH)_3 \qquad \text{(basic solution)}$$

Both Fe(II) and Fe(III) complexes are stable in aqueous solution, and in the absence of ligand effects they adopt octahedral stereochemistries. Oxygen and sulphur donors are preferred over nitrogen, especially with Fe(III) where no simple amines (NH_3 are known. However, chelating N donors (en, phen, etc.) form very stable complexes with Fe(II) which may be readily oxidized to the Fe(III) state. Fe^{3+} polymerizes in aqueous solution of pH > 3 and $Fe(H_2O)_6^{3+}$ is only known in strong acid; $Fe(H_2O)_6^{2+}$, however, is far less acidic.

(F) Cobalt

TABLE VI

Oxidation state	3d electron configuration	Coordination number	Stereochemistry	Examples
Co(−I)	?	4	Tetrahedral	$Co(CO)_4^-$
Co(0)	?	4	Tetrahedral	$Co(CN)_4^{4-}$
Co(I)	d^8	4	Tetrahedral	$Co(CN)_3CO^{2-}$
		5	Tetragonal pyramidal	$(R_2CS_2)_2CoNO$
			Trigonal bipyramidal	$Co(NCR)_5^+$
		6	Octahedral	$Co(dipy)_3^+$
Co(II)	d^7	4	Tetrahedral	$CoCl_4^{2-}$
			Square	$[Co(Me_2edt)](ClO_4)_2$
		5	Trigonal bipyramidal	$[Co(Mesalen)_2]_2$
		6	Octahedral	$Co(H_2O)_6^{2+}$, $Co(CN)_6^{4-}$
Co(III)	d^6	6	Octahedral	$Co(en)_3^{3+}$, $Co(CN)_6^{3-}$ CoF_6^{3-}

$$Co^{2+} \xrightarrow{-1.82v} Co^{3+} \qquad \text{(acidic solution)}$$

$$Co(OH)_2 \xrightarrow{-0.14v} Co(OH)_3 \qquad \text{(basic solution)}$$

The oxidation states, Co(II) and Co(III), are most common, although synthetic Co(I) "cobaloxime" complexes have recently assumed some biochemical significance in relation to vitamin B_{12}[25]. $Co(H_2O)_6^{3+}$ is unstable in all but strong acids and Co(III) chemistry is confined to its complexes, which are with few exceptions octahedral (d^6, spin paired).

The stereochemistries of Co(II) complexes are extremely varied, and tetrahedral, square planar, and octahedral structures are important, but some trigonal bipyramidal and square pyramidal geometries occur.

Co^{2+} is the only d^7 ion of importance in the transition series, and a tetrahedral geometry is very common, sometimes in equilibrium with the octahedral form. This versatility is significant, has been intensively studied in recent years, and allows Co(II) to be studied with a wide variety of ligands and solvent systems. It is clear that the resulting stereochemistry is very dependent on steric requirements imposed on the donor atoms by the ligand groups. Tetrahedral geometry is preferred with monodentate anionic ligands (*e.g.* Cl^-, Br^-, I^-, SCN^-, N_3^-, OH^-) as in $[CoX_4]^{2-}$ or $[CoL_2X_2]$ (L = PR_3) and in some cases with bulky bidentate anions, (*N*-alkylsalicylalaninato, large β-diketonate anions). Planar Co(II) com-

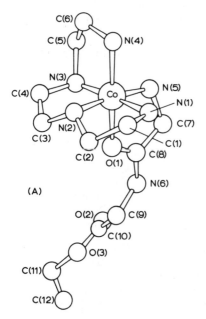

Fig. 2. Structure of β_2-[Co(trien)(glyglyOEt)]$^{3+}$ cation[28].

plexes occur with the dimethylglyoxime amino-oxalate, *o*-aminophenolate anions and in the interesting ethylenedithiolate complexes.

The inertness towards substitution of Co(III) complexes has resulted in this oxidation state being used for many studies, including those related to the mechanism of amino acid ester and peptide hydrolysis[27]. One active species in the latter case has been shown to involve coordination of the *N*-terminal amino acid residue via the amino nitrogen and carbonyl oxygen atoms (Fig. 2).

(G) Nickel

TABLE VII

Oxidation state	3d electron configuration	Coordination number	Stereochemistry	Examples
Ni(O)	d^{10} (?)	4	Tetrahedral	$Ni(CO)_4$, $Ni(CN)_4^{4-}$
Ni(II)	d^8	4	*Square*	$[NiBr_2(PEt_3)_2]$, $Ni(CN)_4^{2-}$
		4	*Tetrahedral*	
		5	Square pyramidal	$[Ni(CN)_5]^{3-}$
			Trigonal bipyramidal	$Ni(5Cl-Salen-N(C_2H_5)_2)_2$
				$[Ni(P(PrMe_2As)_3)CN]^+$
			Trigonal bipyramidal	$Ni(CN)_2(P(C_6H_5)(CH_3)_2)_3$
		6	*Octahedral*	$Ni(H_2O)_6^{2+}$, $Ni(NCS)_6^{4-}$
				$[Nidipy_3]^{2+}$
Ni(III)	d^7	5	Trigonal bipyramidal	$NiBr_3(P(C_6H_5)(CH_3)_2)_2$
		6	Octahedral	$[Ni(diars)_2Cl_2]^+$
Ni(IV)	d^6	6	Octahedral	$[Ni(diars)_2Cl_2]^{2+}$

Ni(II) complexes are by far the most important, and Ni^{2+} is the only simple ion found in aqueous solution. Octahedral, tetrahedral and square planar geometries occur, octahedral stereochemistry being preferred by uncharged ligands (H_2O, NH_3, en) and tetrahedral structures by monodentate anions or mixtures of anions and neutral monodentates (halides, As and P donors). The latter species sometimes exist in equilibrium with the square planar geometry and this aspect has been actively studied in recent years (L = alkyl, aryl phosphines, alkyl substituted salicylaldimato and aminotropoloneimine complexes).

$$NiL_2 \quad \rightleftharpoons \quad NiL_2$$

(tetrahedral)   (square planar)

It appears that the more open tetrahedral structure is preferred by more bulky ligands, alkyl-substituted salicylaldimato and aminotroponeimine

References pp. 61–62

complexes, while smaller substituents favor square geometries. With mixed ligand complexes of this type both tetrahedral and square planar geometries exist in solution, and sometimes together in the crystalline state.

Octahedral geometry may also occur in equilibrium with the square planar, especially when the added ligands are good electron donors ($A = H_2O$, pyridines).

$$NiL_2 + 2A \rightleftharpoons NiL_2A_2$$

yellow (square   blue (octahedral)
planar)

(L = salicylaldiminato type, alkyl- and aryl-substituted ethylenediamines, dialkylthioureas). Also, octahedral geometry may be achieved by polymerization such as in the trimer $[Ni(acac)_2]_3$, or dimerization to form 5-coordinate species (salicylaldimato complexes with R = H, OH). Thus the stereochemistry of Ni(II) complexes is extremely varied, and provides possibly the most unpredictable area for structural studies in transition metal chemistry.

(H) Copper

TABLE VIII

Oxidation state	3d electron configuration	Coordination number	Stereochemistry	Examples
Cu(I)	d^{10}	2	Linear	$Cu(NH_3)_2{}^+$, Cu_2O
		4	Tetrahedral	$Cu(CN)_4{}^{3-}$, CuI
Cu(II)	d^9	4	Tetrahedral (distorted)	$Cs[CuCl_4]$, $[Cu(N\text{-}Prsalen)_2]$
		5	Trigonal bipyramidal	$[Cu(dipy)_2I]^+$
			Square pyramidal	$[Cu(DMGH)_2]_2$
		4	Square planar	CuO, $[Cupy_4]^{2+}$
		6	Octahedral (distorted)	$Cu(EDTA)^{2-}$, $Cu(H_2O)_6{}^{2+}$

$$Cu^+ \xrightarrow{-0.153v} Cu^{2+} \qquad \text{(acidic solution)}$$

The most important stereochemistries are tetrahedral Cu(I), and considerably tetragonally distorted Cu(II). The relative stabilities of Cu(I) and Cu(II) complexes depend very strongly on the nature of the ligands present, and the equilibrium $2Cu(I) \rightleftarrows Cu(O) + Cu(II)$ can be displaced in either direction depending on the conditions. In aqueous solution Cu^+ is quite unstable but it can be stabilized by soft polarizable ligands (CN^-, I^-, R_2S) or by forming highly insoluble complexes, CuCN, CuCl.

The aqueous chemistry of copper is largely devoted to Cu(II) compounds. A wide variety of stereochemistries ranging from square planar to octahedral exist, and these may be related to the octahedral case with various degrees of elongation of two *trans* bonds (Jahn Teller distortion); a clear cut case of octahedral O_h symmetry has not been observed, and indeed is not expected to occur.

Occasionally, and of some generality in the peptide complexes of Cu(II), 5-coordinate square pyramidal geometries occur in the crystalline state (see Chapter 4). It appears that in the crystalline state Cu(II) can form a wide range of structures of coordination number 4, 5 and 6, the exact stereochemistry being decided by ligand conformations and lattice requirements; in solution it is very likely that a distorted octahedral stereochemistry prevails.

(I) Zinc

TABLE IX

Coordination number	Stereochemistry	Examples
4	*Tetrahedral*	$Zn(CN)_4^{2-}$, $Zn(NH_3)_4^{2+}$
5	Distorted trigonal bipyramidal or square pyramidal	$[Zn(acac)_2H_2O]$ $[Zn(terpy)Cl_2]$
6	*Octahedral*	$Zn(H_2O)_6^{2+}$, $Zn(NH_3)_6^{2+}$

Zn^{2+} is the only known aquated ion, and the only complexes known are those of Zn(II). Since this contains a full (d^{10}) configuration no crystal field effects occur. Thus the stereochemistry is decided entirely by electrostatic and covalent binding forces, and by ligand size. The tetrahedral geometry is most prevalent, followed by the octahedral, and in a few examples trigonal bipyramidal and square pyramidal structures occur.

(J) Molybdenum and tungsten

$$Mo^{3+} \xrightarrow{\sim 0.0v} MoO_2^+, \quad W^{3+} \xrightarrow{0.15v} WO_2 \quad \text{(acid solution)}$$

Mo and W are chemically very similar, but differ appreciably from Cr; the oxidation states (IV), (V) and (VI) are important whereas (II) is practically unknown. Both elements have a wide variety of stereochemistries with many examples of coordination numbers greater than six; *e.g.* $Mo(CN)_8^{4-}$, $Mo(CN)_8^{3-}$ (distorted triangular dodecahedra); polymeric oxides, sulfides. Discrete MoO_4^{2-} occurs in strongly alkaline solution, and on acidification forms a variety of isopolymolybdates or tungstates. These involve octahedral coordination by oxygen atoms and various types of

References pp. 61-62

26

TABLE X

Oxidation state	d electron configuration	Coordination number	Stereochemistry	Examples
Mo(O), W(O)	d^6	6	Octahedral	$Mo(CO)_6$, $Mo(diphos)_3$
Mo(I), W(I)	d^5	—	"Sandwich" bonding	$[(C_6H_6)_2Mo]^+$ $[\pi\text{-}C_5H_5Mo(CO)_3]_2$
Mo(II), W(II)	d^4	4	Tetrahedral (?)	$[Mo(OCOCH_3)_2]$
Mo(III), W(III)	d^3	6	Octahedral	$[Mo(SCN)_6]^{3-}$, WCl_6^{3-}
		8	Dodecahedral	$[Mo(CN)_7(H_2O)]^{4-}$
Mo(IV), W(IV)	d^2	6	Octahedral	$[Mo(SCN)_6]^{2-}$
		8	Dodecahedral	$[Mo(CN)_8]^{4-}$
Mo(V), W(V)	d^1	5	Trigonal bipyramidal	$MoCl_5$
		6	Octahedral	$WF_6^-[MoOF_5]^{2-}$
		8	Dodecahedral	$[Mo(CN)_8]^{3-}$ $[W(CN)_8]^{3-}$
Mo(VI), W(V)	d^0	4	Tetrahedral	MoO_4^{2-}, WO_4^{2-}
		6	Octahedral	WCl_6, MoF_6, MoO_6, WO_6 (in polyacids)

shared corners or edges making up the structure. Complexes are relatively rare in the (III) state, and those in the (IV), (V) and (VI) states are mostly anionic.

4. FACTORS AFFECTING COORDINATION NUMBER AND STEREOCHEMISTRY

(A) Metal ion effects

Metal ions can be divided into two rather distinct classes based on their bonding properties: (i) the alkali (Li^+, Na^+, K^+, Rb^+, Cs^+) and alkaline earth (Be^{2+}, Mg^{2+}, Ca^{2+}, Sr^{2+}, Ba^{2+}) metal ions; and (ii) the transition metals, especially those of the first transition series (Ti^{2+}, V^{2+}, VO^{2+}, Cr^{2+}, Cr^{3+}, Mn^{2+} Fe^{2+}, Fe^{3+}, Co^{2+}, Ni^{2+}, Cu^{2+}, Zn^{2+}).

(i) Alkali and alkaline earth metal ions

These metals are highly electropositive and with the exception of Be^{2+} which is unknown as the discrete Be^{2+} ion in the crystalline state or in solution★, the ions exist in their crystalline compounds as separate ions (ionic

★Mg^{2+}, and to some extent Li^+, because of their small size also have a very high polarizing influence and show some covalent character in their compounds.

lattices), and in solution as discrete aquated M^+ or M^{2+} ions. The compounds are held together almost exclusively by electrostatic forces operating between oppositely charged ions or ion–dipoles, and since the ions have the

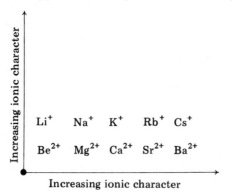

inert gas electronic configuration they are spherically symmetric in their bonding properties. Maximum coordination numbers (N) can be often calculated assuming a close packed arrangement of the ligands on the surface of the metal ion using the expression

$$N = \frac{2\pi}{(3)^{1/2}}\left(\frac{d}{r}\right)^2\left(\frac{1}{1 - r^2/8d^2}\right)$$

where d is the metal–ligand distance, and r is the van der Waals radius of the ligand. The attached ligands assume a regular stereochemistry dictated only by the intra-ligand repulsion energy; *i.e.* there is no steric requirement imposed on the ligands by the metal ions themselves and this is a major difference from the transition metals. This means that in aqueous solution the smaller cations, Li^+, Na^+, K^+ and Be^{2+} probably exist as tetrahedral $M(H_2O)_4$ ions; X-ray scattering results support this in the case of K^+. Since these ions are small, however, they have a very high surface charge density and secondary solvation is important. Thus the alkali metals Li^+, Na^+, and K^+ have large effective hydrated radii (Table XI) with that for Li^+_{aq} being greater than that for Cs^+_{aq}.

TABLE XI

	Li^+	Na^+	K^+	Rb^+	Cs^+
Crystal radii (Å)*	0.60	0.95	1.33	1.48	1.69
Hydrated radii (Å) (estimated)	3.40	2.76	2.32	2.28	2.28
	Be^{2+}	Mg^{2+}	Ca^{2+}	Sr^{2+}	Ba^{2+}
Crystal radii	0.34	0.65	0.94	1.1	1.3

The hydration number for Cs_{aq}^+ of about 10 probably results entirely from the primary hydration shell, with secondary hydration being unimportant. With the alkaline earth metals, however, the small highly charged ions prefer to adopt a higher coordination number in aqueous solution at the expense of direct cation–anion contact. Thus Mg^{2+}, while of smaller ionic radius than Na^+, exists as $Mg(H_2O)_6^{2+}$ in aqueous solution. However, the weakly electropositive Be^{2+} ion does occur as tetrahedral $Be(H_2O)_4^{2+}$, is very acidic compared to the other divalent and monovalent aquated ions, and the water molecules are strongly bound; *e.g.* $[Be(H_2O)_4]Cl_2$ loses no water over P_2O_5.

In the crystalline state, packing forces become very important. To obtain the lowest energy state a balance is maintained between the maximum coordination number calculated from the above equation and crystal packing forces resulting from the packing together of repetitive structural units. For the alkali metals in ionic complexes of the type M^+X^- this results in *cubic* 8-fold coordination when the ratio of the radii of the anion to cation (r^-/r^+) is less than 1.37 (CsCl structure), *octahedral* 6-fold coordinate when r^-/r^+ is less than 2.44 (NaCl structure) and *tetrahedral* 4-fold coordination when r^-/r^+ is less than 4.44 (ZnS structure). For ionic solids with different stoichiometries, *e.g.* M_2X, other lattice arrangements are taken up.

The aqueous solution chemistry of these metals is then almost exclusively concerned with M_{aq}^+ or M_{aq}^{2+} ions. Few simple complexes containing other ligands are known to exist in aqueous solution for the alkali metals (however see Chapter 6 for interesting exceptions), and the organo compounds of Na and K are essentially ionic and are vigorously hydrolyzed by water. Only the organo derivatives of Li^+ and Mg^{2+} (which are exceedingly useful as a source of R^+ in preparative organic chemistry) are covalent in character, being soluble in hydrocarbon and other non-polar solvents. Grignard reagents "RMgX" have been shown to consist of essentially tetrahedral

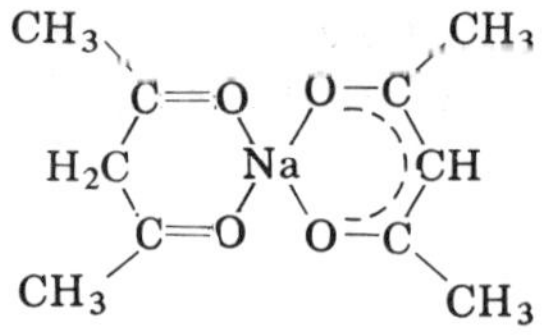

(34)

Mg in crystalline $C_6H_5MgBr \cdot 2Et_2O$, and a similar structure has been found for Na(acetylacetonate)$_2$ (34). Thus hydrogen bonding from coordinated water molecules and the size of the hydrated cation are important features in the biological function of Na_{aq}^+ and K_{aq}^+.

Of the alkaline earth metals only Be^{2+} and to some extent Mg^{2+} function as sufficiently strong electron acceptors to form rather weak covalent complexes. Owing to the small size of the Be^{2+} ion and its lack of an electronically induced stereochemistry preference the preferred coordination number is 4 with a tetrahedral dispositon of the ligands *e.g.* BeF_4^{2-} and

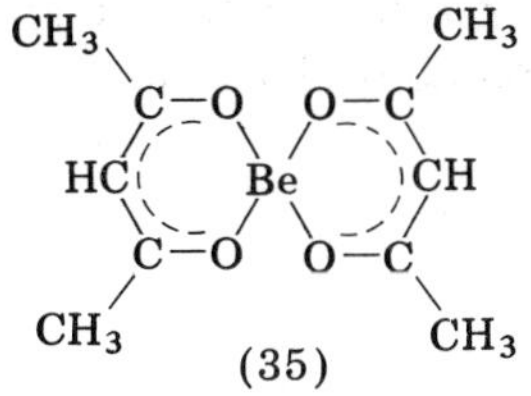

(35)

Be(acac)$_2$ (35). However, when the stereochemistry of the ligand is important, as it is in the porphyrin nucleus, other stereochemistries obtain; *e.g.* square planar, Mg(II) chlorophyll (Chapter 29, Fig. 1). Also octahedral $[Mg(NH_3)_6]Cl_2$ is known but it is readily hydrolyzed in water. Of the other alkaline earth metal ions only Ca^{2+} shows a tendency to form coordinated complexes with a decided preference towards carboxylate donor ligands. Thus the octahedral $[Ca(EDTA)]^{2-}$ ion assumes significance commercially (as a water "softener") and the complexing of Ca^{2+} and Mg^{2+} by the phosphate groups in ATP and ADP is of some importance in energy transfer in biological systems.

(ii) Transition metal ions

The coordination number of transition metals is basically decided by the size and the effective nuclear charge the ligand experiences, and by the properties of the ligand itself. These factors are together responsible for the large bond energies (50–500 kcal/mol) for M–L bonds, and provide a steadily increasing bond strength on traversing from the lighter metals (Ti^{2+}) to the heavier metals (Zn^{2+}). The ineffective screening of the nuclear charge by the d electrons leads to a higher effective charge at normal bond distances and a smaller ionic radius, both factors acting to lower coordination numbers along the series Ti–Zn and to increase the covalent character and strength of the coordinate bond.

However, the stereochemistries, and in many cases the coordination numbers also, often differ from those predicted by simple cation–anion size, charge, and polarizability relationships, simply because the disposition of electrons about the metal ion is no longer spherically symmetric. This results from the metal ion no longer having the inert gas configuration with a filling up of d electron shells ($3d$ for Sc–Zn, $4d$ for Y–Cd, $5d$ for Hf–Hg), and $4f$ and $5f$ shells for the lanthanide and actinide elements respectively. Further, and of utmost stereochemical significance since the partly filled valence shell extends out beyond the closed inert gas shells, the ligand atoms and consequently the ground state stereochemistries of the complexes, are very decidedly influenced by both the number and disposition of the d electrons present.

References pp. 61–62

It is not proposed to discuss here the theoretical aspects of bonding in transition metal chemistry[18-22], but the result is of considerable importance in determining stereochemistry. Briefly, in the electrostatic field of the ligands the five degenerate d-orbitals of the free gaseous metal ion split into different energy sets (Fig. 3). By allowing the metal d electrons to

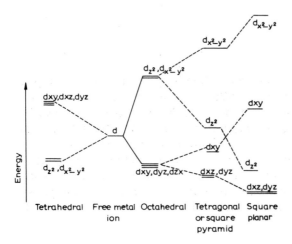

Fig. 3. Crystal field splittings of d-levels of a metal ion in regular geometries.

occupy these crystal field orbitals, lowest lying first with appropriate application of the Pauli Principle a nett gain in overall energy usually obtains. This is termed the crystal field stabilization energy (CFSE). Some representative CFSE values for various configurations are given in Table XII.[*]
By considering as well the energies of appropriate ligand orbitals, and by allowing for metal and ligand orbitals of similar energy and appropriate symmetry to combine, allowance can be made for electron redistribution between the ligand and the metal atom, $i.e.$ for covalent bonding. The CFSE values are only of the order of 10–50 kcal/mol which is small compared to a total bond energy of 300–3000 kcal/mol, but in many cases the resulting stereochemistry is decided by such effects. A summary of the more important stereochemical consequences of the crystal field theory is given in Table XIII. It is important to realize that the coordination number is not predicted by the crystal field theory, but only the resulting stereochemistry once the coordination number is known. The coordination number is determined largely by the size and charge on the metal ion and ligand.

---

[*]The *weak field* case corresponds to the rigid application of the Pauli Principle; the *strong field* case results when the energy separation between the oribital sets (Δ) is larger than the spin pairing energy. This results in preferential complete occupancy of the lower lying orbital set first.

TABLE XII

CRYSTAL FIELD STABILIZATION ENERGIES FOR d^n COMPLEXES
(IN Dq UNITS[a])

d^n system	Examples	Octahedral		Tetrahedral		Square planar	
		Electrons unpaired	Electrons paired	Electrons unpaired	Electrons paired	Electrons unpaired	Electrons paired
d^0	Ca^{2+}, Sc^{3+}	0	0	0	0	0	0
d^1	Ti^{3+}, U^{4+}	4	4	2.67	2.67	5.14	5.14
d^2	Ti^{2+}, V^{3+}	8	8	5.34	5.34	10.28	10.28
d^3	V^{2+}, Cr^{3+}	12	12	3.56	8.01	14.56	14.56
d^4	Cr^{2+}, Mn^{3+}	6	16	1.78	10.68	12.28	19.70
d^5	Mn^{2+}, Fe^{3+} Ru^{2+}, Os^{2+}	0	20	0	8.90	0	24.84
d^6	Fe^{2+}, Co^{3+} Rh^{3+}, Ir^{3+}	4	24	2.67	7.12	5.14	29.12
d^7	Co^{2+}, Ni^{3+}, Rh^{2+}	8	18	5.34	5.34	10.24	26.84
d^8	Ni^{2+}, Pd^{2+}, Pt^{2+}, Au^{3+}	12	12	3.56	3.56	14.56	24.56
d^9	Cu^{2+}, Ag^{2+}	6	6	1.78	1.78	12.28	12.28
d^{10}	Cu^+, Zn^{2+}, Hg^{2+}, Ag^+	0	0	0	0	0	0

[a] Dq (tetrahedral) $\simeq \frac{4}{9}$ Dq (octahedral)

It is useful to combine the stereochemical consequences of the crystal field perturbations with other properties of the transition metal ions.

(a) Oxidation state stability. For the first half of the first transition series, Sc to Mn, the maximum valences *Sc(III)*, *Mn(IV)*, *V(V)*, Cr(VI) and Mn(VII) all exist, with those italicized being the most stable and abundant. However, the d electron screening of the nucleus is inefficient, and lower valence states become progressively more abundant. Thus Cr(VI) and Mn(VII) exist only in their oxides or oxo-anions; Cr(III) is the only significant oxidation state in aqueous solution; Mn(III) and Mn(II) are about equally important; Fe(II) is slightly favored over Fe(III); Co(II) more so over Co(III) (which is abundant only in its complexes where it is spin paired); Ni(II) is the only important oxidation state of nickel; Cu(II) and Cu(I) occur with the cupric state of major significance, while with zinc only Zn(II) with the d^{10} configuration occurs.

(b) Metal ion preferences. Generally complexes of the earlier transition metals are ionic in nature (Sc^{2+}, Ti^{2+}, V^{2+}), while those of metals later in

the series (Cu^{2+}, Ni^{2+}) are more covalent especially with polarizable ligands (O, S). This results from the increasing effective nuclear charge on traversing the transition series. A more covalent bond does not necessarily mean

TABLE XIII

STEREOCHEMICAL PREDICTIONS OF CRYSTAL FIELD THEORY[30]

Number of non-bonding d electrons	Unpaired electrons	Four coordinate	Six coordinate
		No electron pairing (weak field—high spin)	
0, 5, 10	0 or 5	tetrahedral	octahedral
4 or 9	1	square planar	tetragonal
3 or 8	2	distorted tetrahedral	octahedral
2 or 7	3	tetrahedral	almost regular octahedral
1 or 6	4	almost regular tetrahedral	almost regular octahedral
		Electron paired (strong field—low spin)	
1 or 2	—	—	—
3	1	almost regular tetrahedral	—
4	0	tetrahedral	almost regular octahedral
5	1	distorted tetrahedral	almost regular octahedral
6	0	distorted tetrahedral	octahedral
7	1	square planar	tetragonal
8	0	square planar	tetragonal

increased stability however. Also the degree of covalency increases markedly for the second and third row transition metal complexes for a similar reason, but here usually also results in increased kinetic stability.

The stereochemical consequences of the crystal field theory are several, and may be summarized as follows. (i) Higher coordination numbers are preferred by the lighter transition metals, and lower coordination numbers by the heavier metals. Thus d^0–d^6 ions favor octahedral coordination, while d^7 (Co(II), Rh(II)), d^8 (Ni(II), Pd(II), Pt(II)) and d^9 (Cu(II), Au(III)) complexes prefer square planar structures. This is particularly evident for the second and third row transition elements where electron pairing occurs readily (low-spin complexes). (ii) Higher coordination numbers are usually associated with cationic complexes ($Co(NH_3)_6^{2+}$) and lower coordination numbers in anionic complexes ($CoCl_4^{2-}$). Thus no complex fluorides or oxides of divalent ions result in isolated octahedral anions in the solid state. (iii) Higher coordination numbers are sometimes associated with higher oxidation states; e.g. $CuCl_2^-$, $CuCl_4^{2-}$, CuF_6^{3-}; $FeCl_4^{2-}$, $FeCl_6^{3-}$. (iv) Coordination number sometimes falls with increasing polarizing power of the metal atom; e.g. $Ag(CN)_2^-$ (ionization potential 9.22 eV) and $Cu(CN)_4^{3-}$

(ionization potential 7.72 eV). (v) Crystal field stabilization energies (Table XII) are a good guide to the stereochemical preference of a metal ion for a particular coordination number. Thus square planar complexes are found most readily with d^8 (Ni, Pd, Pt) and d^9 (Cu^{2+}) ions, and to a lesser extent with d^7 (Au(III)) ions. (vi) Square pyramidal structures (coordination number 5) have crystal field stabilization energies intermediate between regular octahedral and square planar structures. Thus for d^7 systems (Co(II), Ni(III)) this structure is favored, e.g. $[Co(CN)_5]^{3-}$, [Co(triarsine)-I_2], $[Ni(Et_3P)_2Br_3]$. (vii) Systems with no CFSE (d^0, d^5, d^{10}) form tetrahedral complexes readily; Fe(III), Zn(II), Al(III), Cd(II), Mn(VII). (viii) The Co(II) ion (d^7) favors octahedral over tetrahedral coordination to a smaller extent than any other d^n configuration. (ix) Theory predicts that distortions from perfect octahedral geometry might be expected for: (a) high-spin Cr(II) and Mn(III) complexes (d^4); (b) low-spin Ni(III) and Co(II) complexes (d^7); (c) Cu(II) (d^9), and that these take the form of elongation of the octahedron along one axis — tetragonal distortion, e.g. $CrCl_2$ ($4Cl^-$ at 2.39 Å, $2Cl^-$ at 2.90 Å); (x) For four coordinate chelate structures crystal field preferences seem to hold. Thus for complexes of the type (36) the following sequence occurs[31]: Cr(II) (d^4) planar; Cu(II)

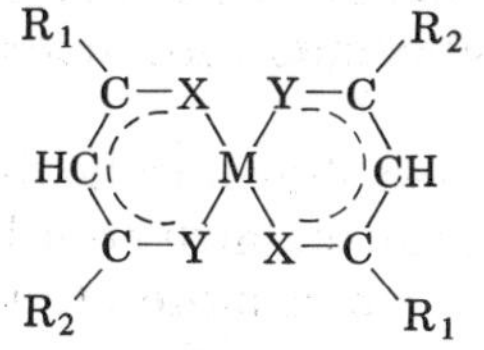

$X, Y = O, S, NR$

$R = $ alkyl, aryl

(36)

(d^9) planar in all but sterically hindered systems; Ni(II) (d^8) planar mostly but planar/tetrahedral equilibria occur for some systems (O,NCH_3; S,NCH_3); Co(II) (d^7) mostly tetrahedral but planar/tetrahedral equilibria and purely planar situations (N,S) do occur; Fe(II) (d^6) tetrahedral; Zn(II) (d^{10}) tetrahedral.

(B) Ligand properties

The ligand, being the copartner in the formation of the coordinate bond, assumes considerable importance in determining stereochemistry. For monodentate ligands coordination number of a metal ion is determined largely by ligand size and for polydentate ligands by ligand size and the number of potential donor atoms, whereas for a particular coordination number the detailed stereochemistry is controlled in most cases by metal ion preferences (crystal field effects) and to varying extents by ligand stereochemistry, chelating properties and type of donor atoms involved.

(i) Donor atoms

The combinations "hard" base–"hard" acid, "soft" base–"soft" acid are useful in predicting the characteristics of complexes. Here "acid" refers to the metal ion in its formal oxidation state (Lewis acid) and "base" to the ligand donor atom. A "hard" acid or base is one which is not easily polarized, while a "soft" acid or base is more easily distorted. The consequence of hardness and softness of metal ions and ligands is taken up in Chapter 2, pages 69–72 (Tables II and III). As shown there, for the hard metals, *class A* type, the ligand preference usually follows the order:

$$F > Cl > Br > I$$

$$O \gg S \; (>Se)$$

$$N \gg P \; (>As)$$

whereas for the soft *class B* metals the usual stability order is:

$$S \sim C > I > Br > Cl > N > O > F$$

As a general rule the coordination number falls as the polarizability of the ligand increases, *e.g.* FeF_6^{3-}, $FeBr_4^-$; a similar trend occurs with increased polarizability of the metal, *e.g.* $MnCl_6^{4-}$, $ZnCl_4^{2-}$.

The size of a donor atom rarely, if ever, has been shown to be directly responsible for changes in coordination number. Thus the high coordination adopted by F^-, and H^- ions, in IF_5, TaF_8^{3-}, ReH_9^{2-}, *etc.* is more probably due to poor polarizability of the ligand accompanied by a high oxidation state of the acceptor [I(V), Ta(V), Re(VII)], rather than by small donor atom size. It is clear, however, that large donor atoms or donor groups will not permit abnormally high coordination numbers.

Stereochemical preference of the donor atom is important. This is clearly demonstrated by several N, S and O chelates where S distinctly prefers pyramidal coordination, whereas N and O are not so demanding [H_2S bond angle 92°; H_2O (105°); NH_3 (107°)]. Thus all three geometrical structures (37–39) have been observed where Y = N, but only (37) and (38)

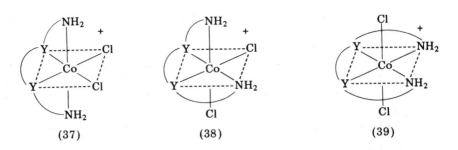

(37) (38) (39)

when Y = S. Similarly both bent (40) and linear (41) chelate junctions are possible with the sexadentate

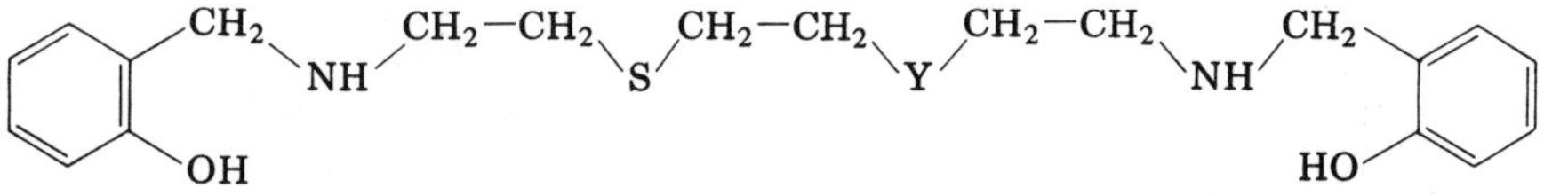

where Y = O or NH but only the bent stereochemistry (40) when Y = S:

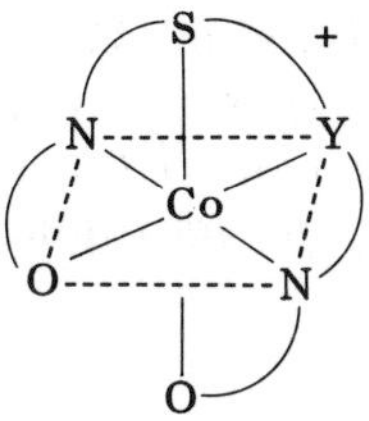

(40) 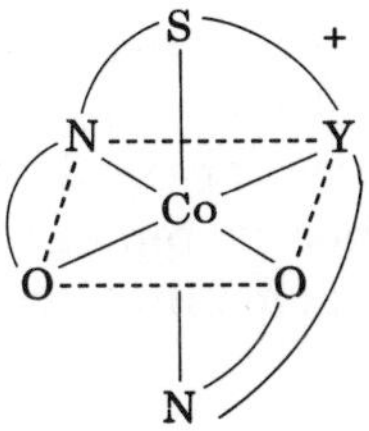   (41)

(ii) Size of the ligand

Ligand size can be very important in deciding the coordination number, and in some cases the stereochemistry. Other factors being approximately equal, bulky ligands prefer lower coordination numbers. Examples include:

[Co(acac)$_2$]$_4$ and   Co(dipivaloylmethane)$_2$
(octahedral coordination   (tetrahedral coordination of
of acetylacetone)   dipivaloylmethane)

[Co(N(CH$_2$CH$_2$NH$_2$)$_3$)Br$_2$] and [Co(N(CH$_2$CH$_2$N(CH$_3$)$_2$)$_3$)Br]Br
(octahedral)   (trigonal bipyramidal)

Changes in stereochemistry in the Schiff base complex (42) occur depending on whether R = H (*square planar*) or R = butyl (*tetrahedral*), presumably

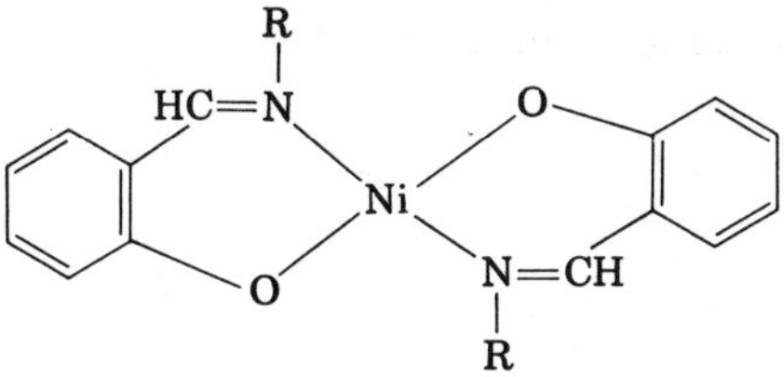

(42)

because the tetrahedral stereochemistry is more open. Also it appears that increasing the ring size in the quadridentate $NH_2(CH_2)_x NH(CH_2)_y NH(CH_2)_x$-$NH_2$ from y (or x) = 2 to 3 in [Co(amine)Cl$_2$]$^+$ complexes stabilizes structure (44) over structure (43), probably because the 6-membered ring allows for less valence angle strain about the pyramidal secondary nitrogen centers.

References pp. 61–62

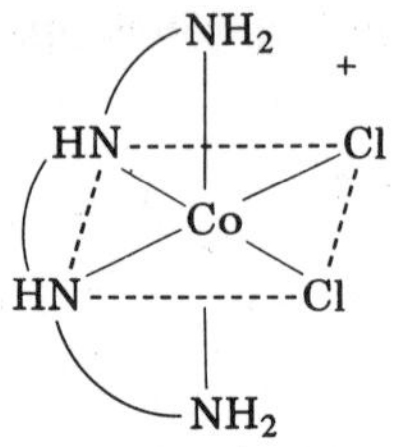

(43)

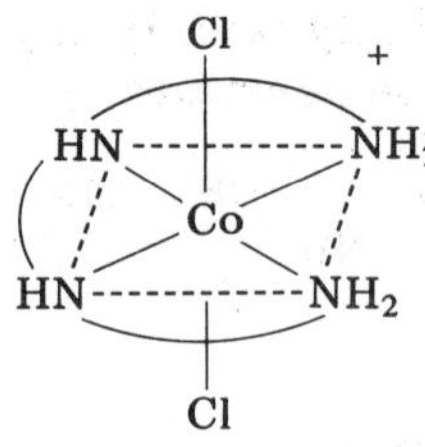

(44)

Also the unusual trigonal prism geometry (45) is found in [Re(cis-1,2-diphenylethene-1,2-dithiolate)$_3$] presumably because steric repulsions between the large phenyl rings destabilize the more usual octahedral stereochemistry. Many other examples of ligand size affecting the coordination number and/or stereochemistry have been demonstrated.

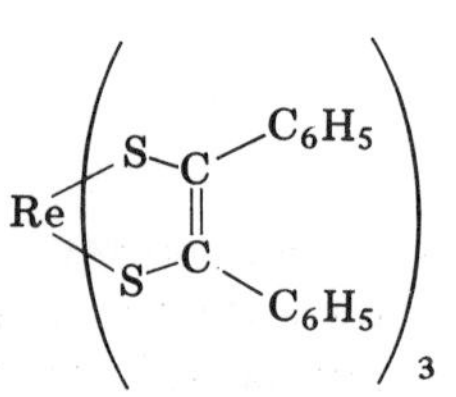

(45)

(iii) Stereochemistry of the ligand

Polydenate ligands sometimes generate a geometry which is relatively insensitive to the metal ion. In fact almost any suitable geometry can be established by suitable design of the ligand[32,33]. A simple case of high coordination generated by the ligand requirements is eight-fold coordination found in $Co(NO_3)_4^{2-}$ where the ligand "bite" is only 53°, (46). A more recent example is given by the ligand 1,8-naphthyridine (47) which assumes 8-coordination with divalent Mn, Fe, Ni, Cu, Zn, Pd, and Cd metal ions through a constrained ligand–metal–ligand bond angle.

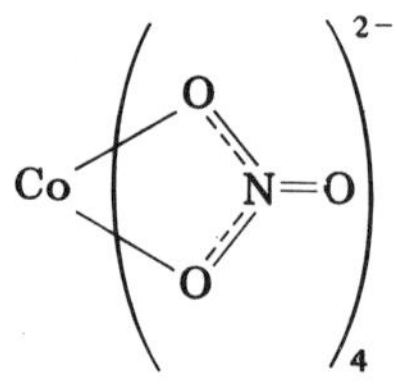

(46)

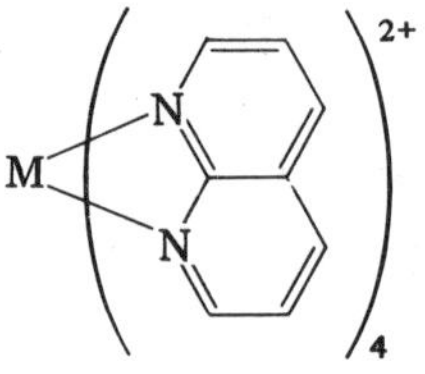

(47)

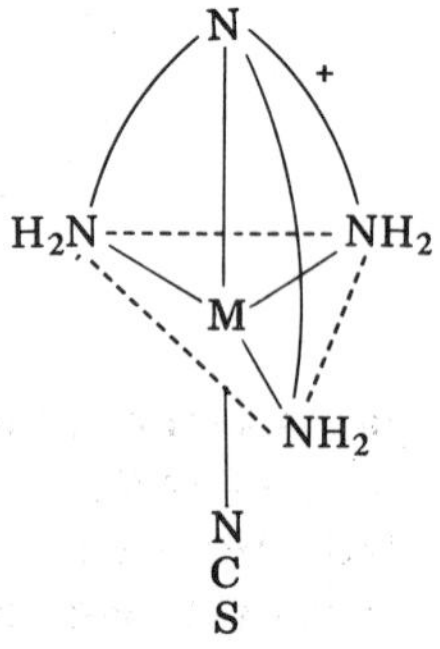

(48)

The ligand $N(CH_2CH_2NH_2)_3$ (tren) forces trigonal bipyramidal geometry on ions preferring square planar or tetrahedral geometry as a result of the trigonal nature of the tertiary N atom. Thus Cu(II) and Zn(II) in $[M(tren)X]^+$ (X = Cl, Br, NCS) have this structure (48). There are many related examples of this type including, in addition to N, tertiary P and As quadridentates. Also the uncommon trigonal prismatic geometry has been observed with suitably designed ligands; for example $[Zn(py_3tach)](ClO_4)_2$ has been shown to have this structure (49):

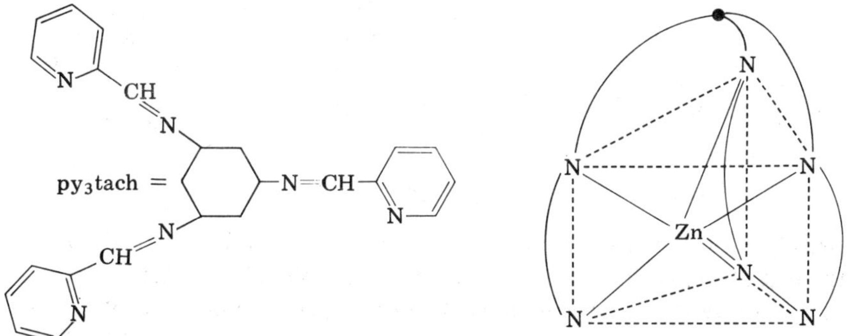

(49)

Unsaturation in the ligand can result in forced square planar geometries as in [Co(salen)] (50), and the many similar cyclic quadridentates of the type involving coordination of imine N (134–136), (51) or saturated N (52).

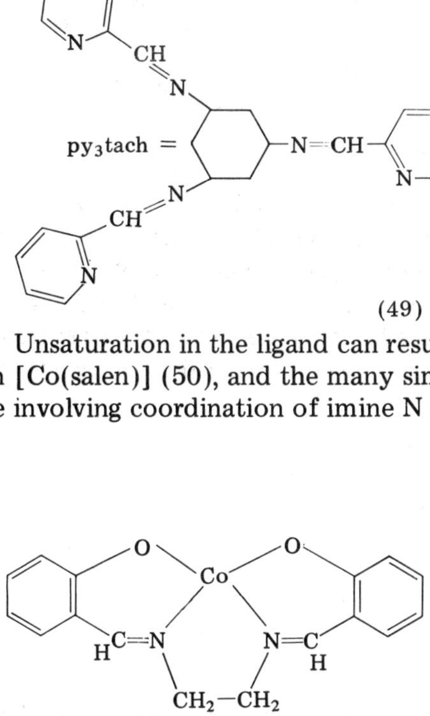

(50)

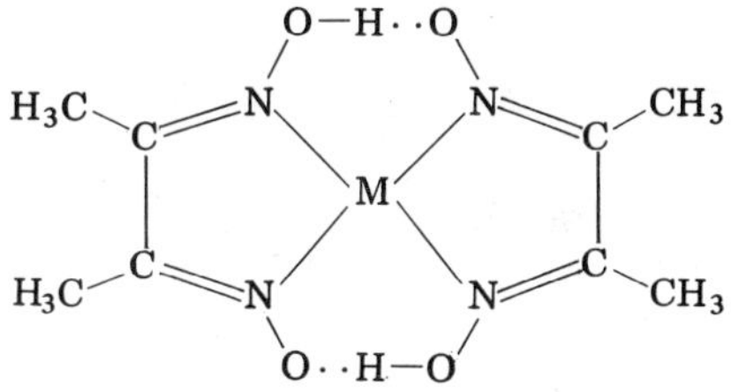

(51)

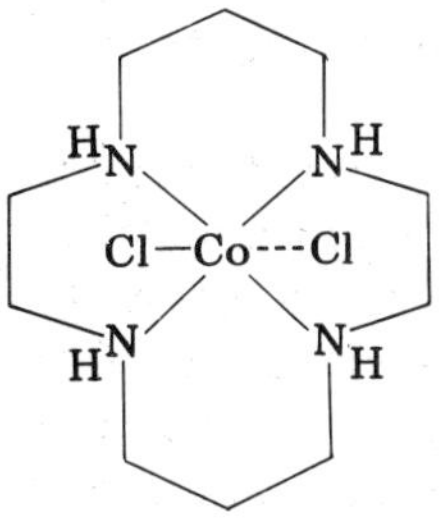

(52)

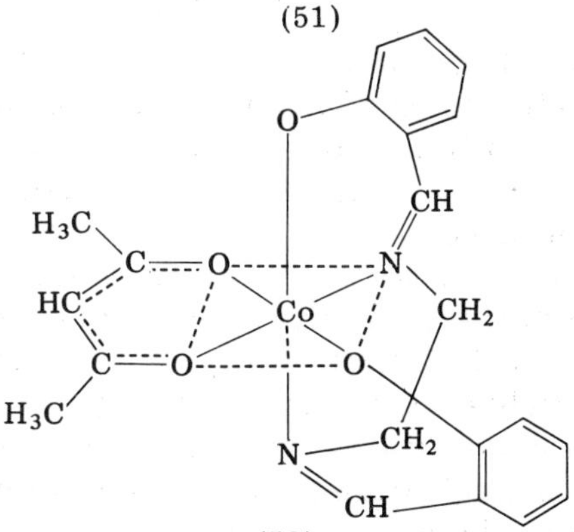

(53)

However, occasionally such ligands can be made to deform to meet coordination requirements as has been shown in a recent X-ray structure of [Co(salen)(acac)].H_2O (53)[34]. Yet all known structures incorporating 2,2′,2″-terpyridyl are linear because sp^2 hybridization at carbon forces the donor N atoms to remain essentially coplanar, e.g. [Zn(terpy)Cl$_2$] (trigonal bipyramidal). A similar situation occurs in other unsaturated groupings of the type (54–56), and amino acids form almost flat 5-membered chelate

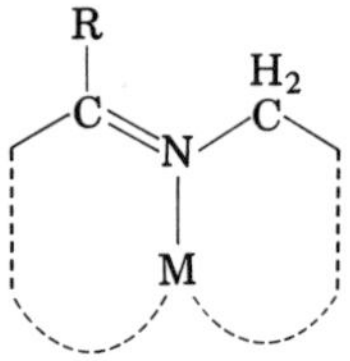

(54) 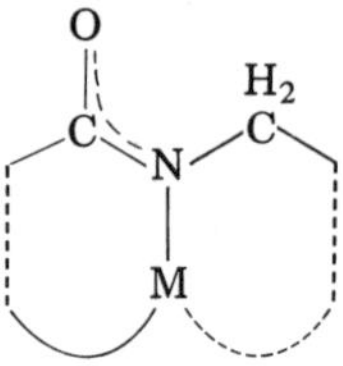   (55) 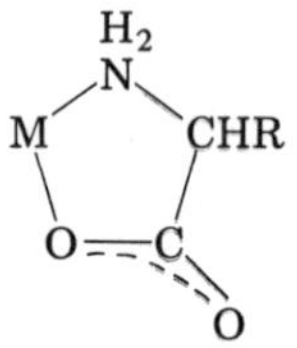   (56)

rings (56), the central carbon being no more than 6° out of the plane in the chelated glycinate anion. Another example of this class occurs with gly-cylglycinate tridentates such as [Co(glygly)$_2$]$^-$, (Chapter 4, Structure XX). The directed valence of carbonyl oxygen makes more than bidentate chelation by small amino bound peptides to the same metal atom virtually impossible when amide oxygen coordination is involved (57) since the

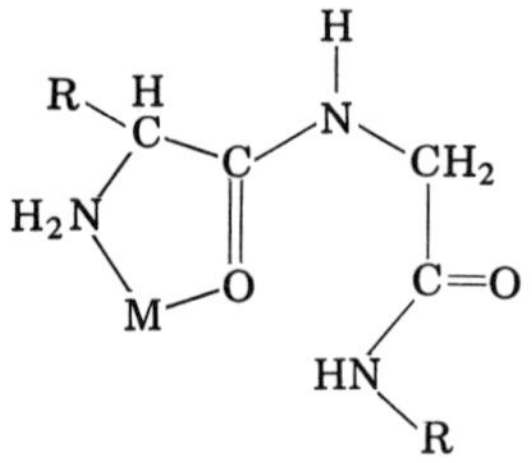

(57)

amide group must remain essentially planar; only when deprotonated amide nitrogen coordination is involved can more than one chelate ring be formed. A similar situation should occur in longer peptides so that carbonyl oxygen coordination is unlikely when more than a single chelate ring to the same metal centre is involved.

It is apparent from these examples that divalent transition metal ions will readily assume a variety of coordination numbers or geometries with suitable ligands. Perhaps the most striking feature of recent crystallographic studies is that regular four- and six-coordinate geometries are rarely achieved, and distortions or unusual stereochemistries are very common indeed. This obviously has significant implications with respect to metal binding with proteins and active sites in enzyme systems.

(iv) Chelate rings

As well as imparting added stability to the complex (Chapter 2), the formation of chelate rings acts to stabilize high coordination numbers. This might be anticipated from the intramolecular nature of ring closure with $K_1 > K_2$, and in some cases from a restriction imposed by the chelate span on the donor atom "bite" on the metal.

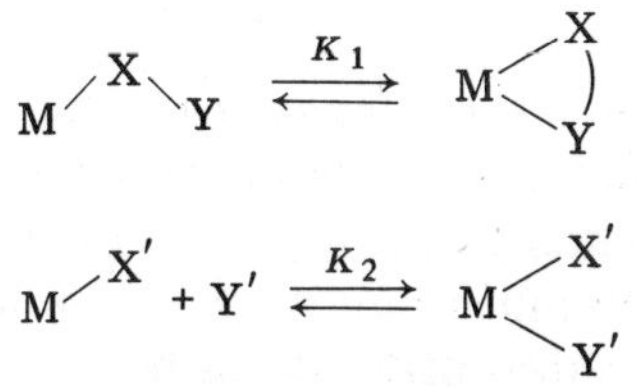

Examples of high coordination numbers are $Zn(en)_3^{2+}$ (octahedral) and $[Zn(terpy)Cl_2]$ (trigonal bipyramidal), compared to the usual 4-coordinate tetrahedral stereochemistry, *e.g.* $Zn(NH_3)_4^{2+}$. Examples of high "unusual" coordination numbers are 8-coordinate $Co(NO_3)_4^{2-}$ (46) and $Zr(C_2O_4)_4^{4-}$ (58) (dodecahedral) and 7-coordinate Mn(II) and Fe(III) structures incorporating the sexadentate ligand ethylenediaminetetraacetic acid, $[M(EDTA)(H_2O)]^{-/2-}$ (27) and (28). Likewise the La(III) complex with

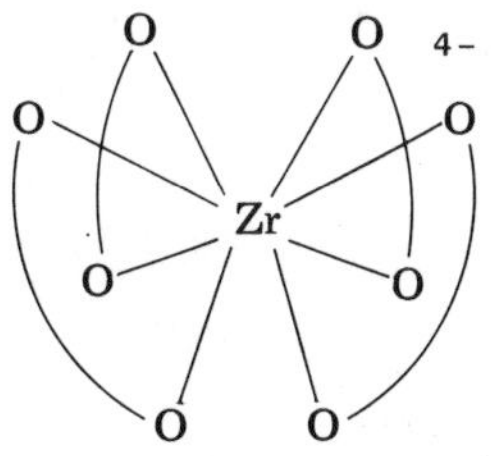

(58) 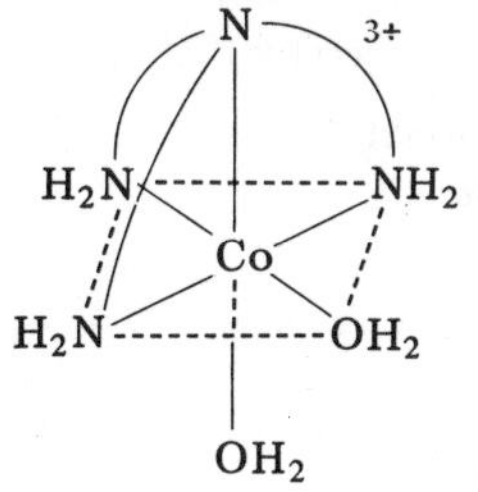   (59) 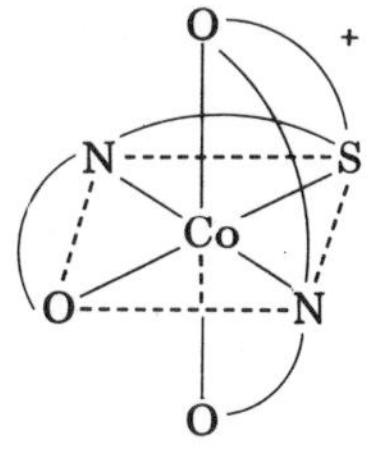   (60)

EDTA is 10-coordinate and has the composition $[La(HEDTA)(H_2O)_4]$ (Fig. 1(b), page 16).

Another consequence of chelating ligands is to force coordination of otherwise poor donor atoms. For example, tertiary nitrogen has repeatedly failed to coordinate to Co(III) when monodentate ligands are involved, but it is found in the $[Co(tren)(H_2O)_2]^{3+}$ ion, (59). Similarly the ether oxygen grouping in the sexadentate

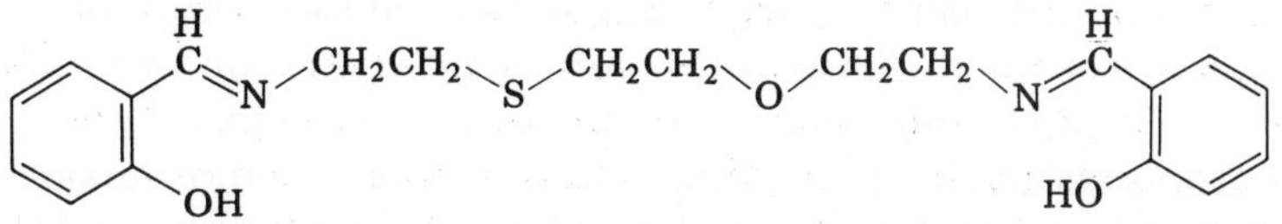

is forced to coordinate by the disposition of donor atoms of the adjacent

References pp. 61–62

chelate rings (60). Many other examples of this type occur in simple co-ordination complexes, and it is therefore very likely that the same tendency will occur in binding at protein sites where the fixed geometry of potential donor atoms may be sufficient to force coordination.

5. STEREOISOMERISM

The origin of stereoisomerism in inorganic complexes lies in the different spatial arrangements of ligands or ligand groups about the central metal atom. All complexes having the same chemical composition and which can be converted into each other by permutations of the ligands, are called *configurational* isomers. Within this definition it has been found useful to single out for special mention those isomers which differ only in torsion angles about bonded pairs of atoms in the same ligand, and call them *conformational* isomers.

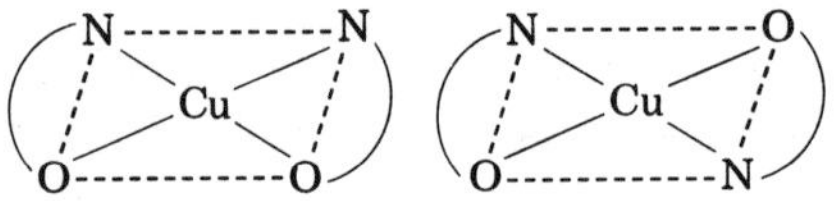

(*cis*)   (*trans*)

configurational
isomers

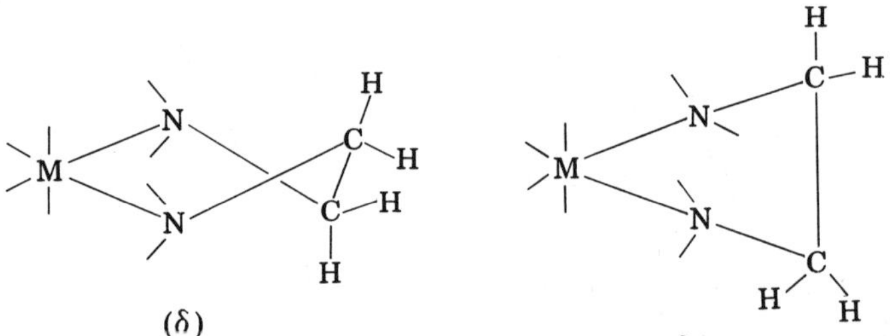

(δ)

(λ)

conformational
isomers

Thus conformational isomers arise naturally from configurational effects and it is occasionally difficult to make a clear distinction between the two classes. For the purpose of this survey, however, conformational isomers will be kept separate since they do not require rearrangement of the donor atoms about the metal atom and thus their interconversion is usually rapid and probably largely independent of the properties of the metal.

(A) Configurational isomers

Two classes of configurational isomers are possible, *diastereoisomers* and *optical isomers*. Diastereoisomers have different physical and chemical properties and can in theory be separated by fractional crystallization or by chromatography, or detected in admixture by a physical technique (*e.g.* u.v. or visible spectra, ^{1}H, ^{19}F, ^{13}C, e.s.c.a. or electron emission spectroscopy). They include different geometrical and structural forms of the same molecule and several isomers may be possible. Optical isomers differ only in that they are non-superposable mirror image forms of the same molecule and this limits their number to two. These are usually called *enantiomers* or *enantiomeric forms* and sometimes *catoptromers*.

(i) Diastereoisomers

(a) Geometrical isomers. Geometrical isomerism results from a different spatial distribution of the donor atoms or chelate rings about the metal center. It is found widely in square planar or octahedral complexes, but is not necessarily restricted to these geometries. Classical examples occur in Pt(II), Pd(II) square planar (61 and 62), and Cr(III), Co(III) octahedral (63, 64) complexes. These metals, as well as Ru(II), Rh(III), Cu(II),

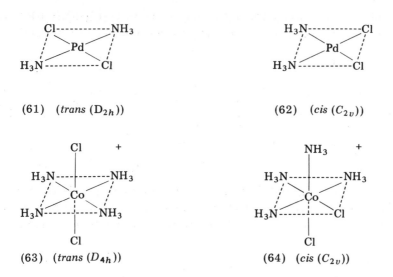

    (61) (*trans* (D_{2h}))   (62) (*cis* (C_{2v}))

    (63) (*trans* (D_{4h}))   (64) (*cis* (C_{2v}))

Ni(II) have many geometrical isomers of various degrees of complexity, and they are usually easily distinguished spectrally (*e.g.* above) or able to be separated chemically. Other examples include the *cis* (65) and *trans* (66) forms of [Cu(alanine)$_2$] and [Ni(thiourea)$_2$Cl$_2$] (67, 68), and [Pt(NH$_3$)-pyClBr] which has been separated into its three possible forms (69–71).

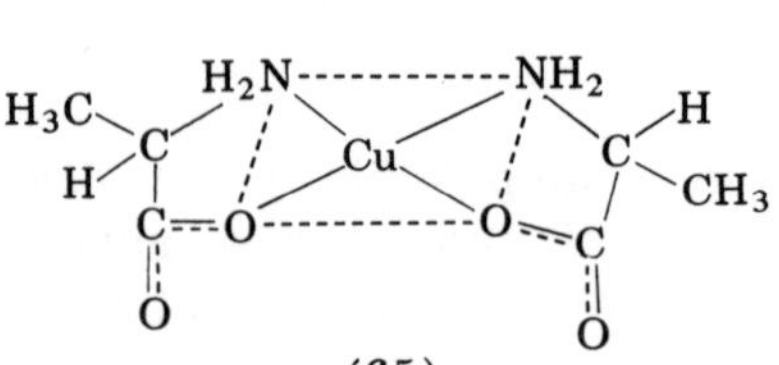

(65)

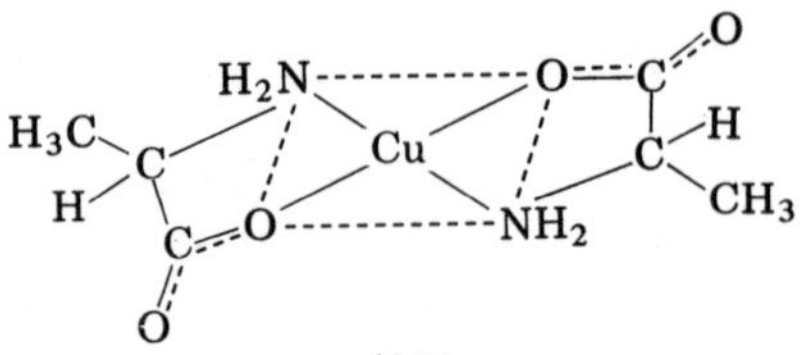

(66)

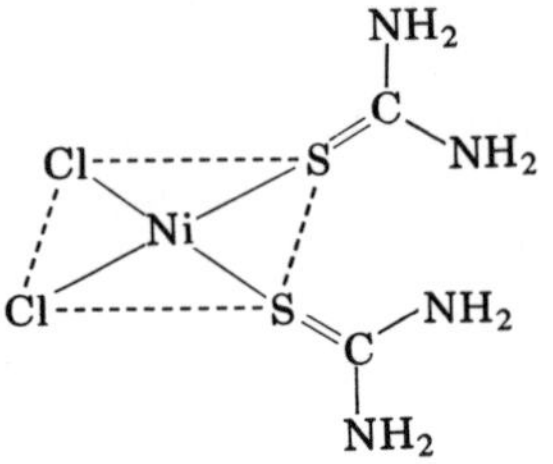

(67)

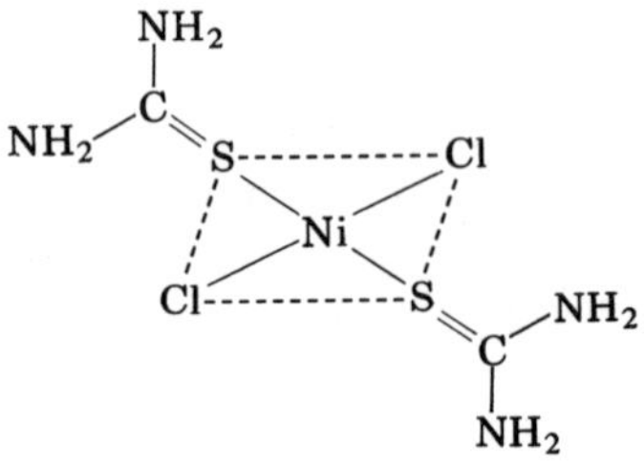

(68)

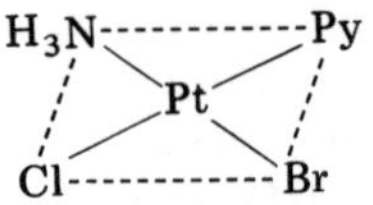

(69)

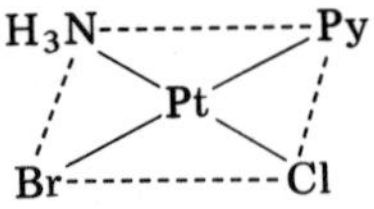

(70)

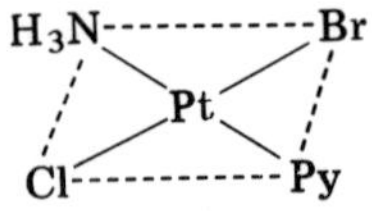

(71)

An example involving bridged metal atoms occurs in $[PtCl(SC_2H_5)(PPr_3)]_2$, of which the two symmetrical forms (72, 73) have been characterized. An

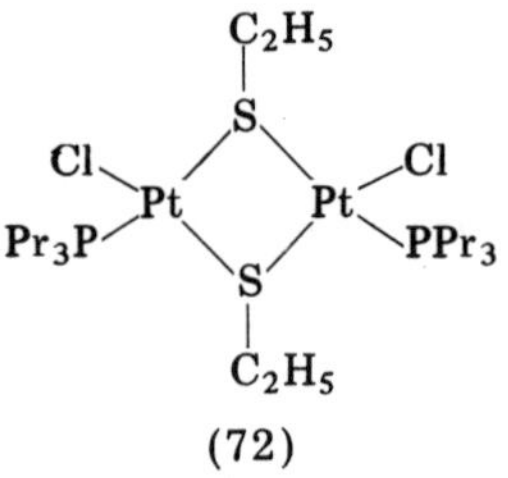

(72)

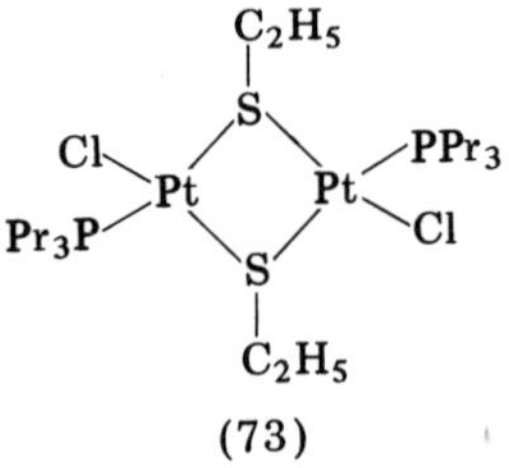

(73)

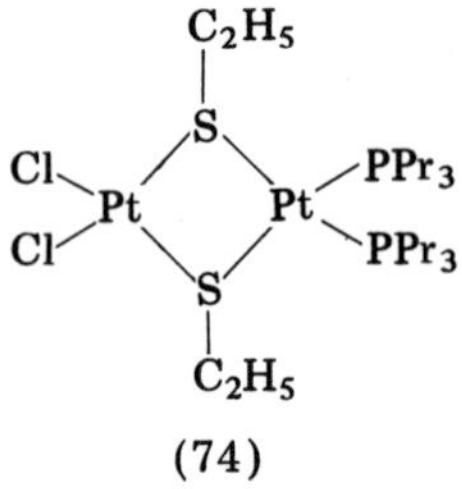

(74)

octahedral example containing chelate rings is the facial *cis* (75) and meri-dional *trans* (76) arrangement of $[Co(glycinato)_3]$. Both forms have been isolated in the solid state, as have the corresponding alaninato isomers.

(75)

(76)

Another class of geometrical isomer results from the different arrangements of chelate rings about a metal atom in multidentate ligands. Many examples of this type are known and recent cases include the [Co(triethylenetetramine)(glycinato)]$^{2+}$ ion which has been obtained in three arrangements (77–79) and by the [Co(tetraethylenepentamine)Cl]$^{2+}$ ion,

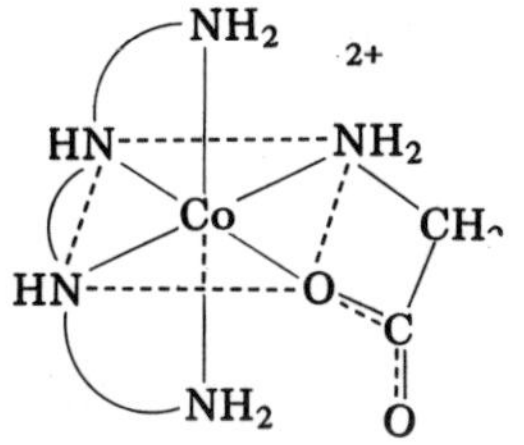

(77) (α-isomer) 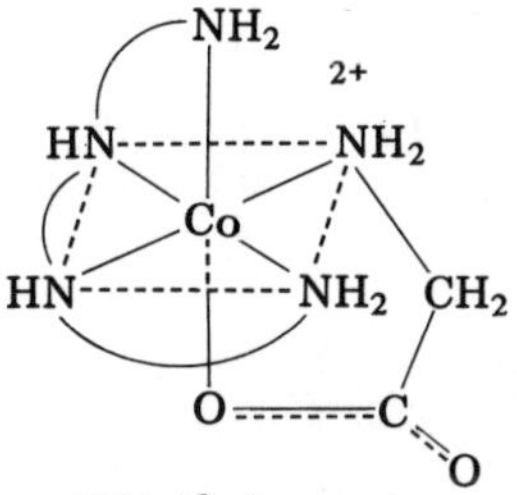 (78) (β$_2$-isomer)

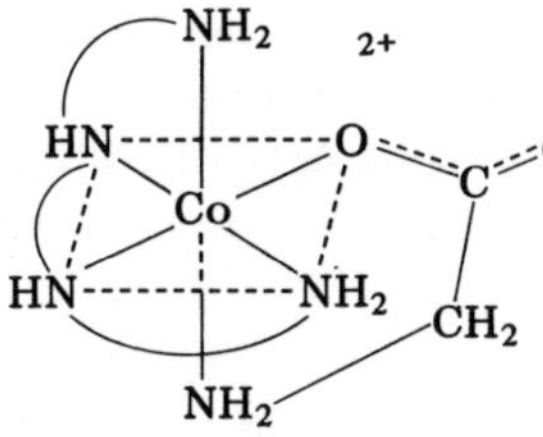

(79) (β$_1$-isomer) 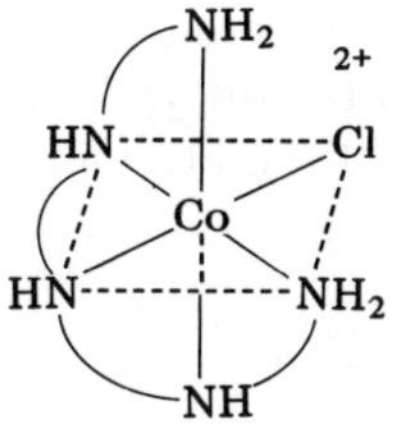 (80) (αα-isomer)

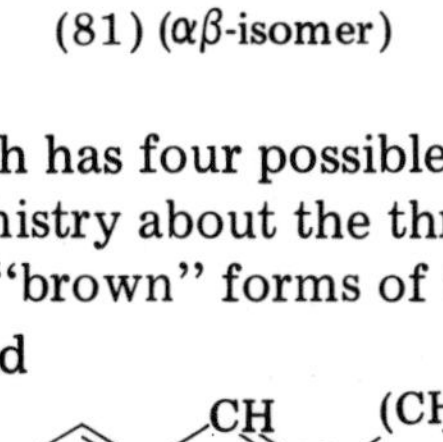

(81) (αβ-isomer) 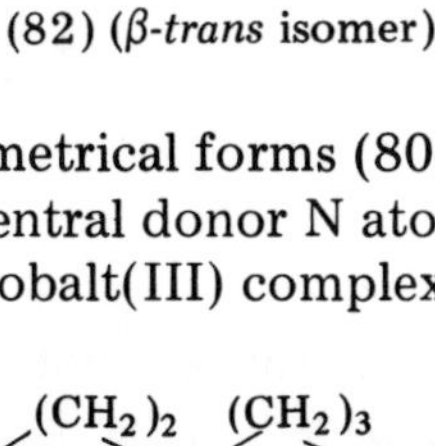 (82) (β-*trans* isomer) 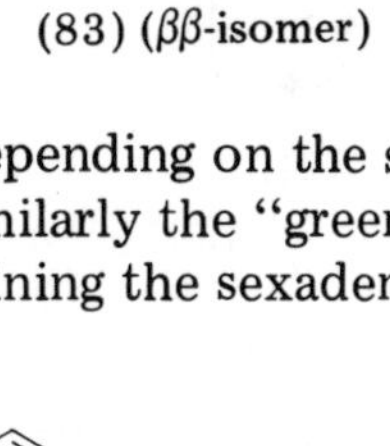 (83) (ββ-isomer)

which has four possible geometrical forms (80–83) depending on the stereochemistry about the three central donor N atoms. Similarly the "green" and "brown" forms of the cobalt(III) complex containing the sexadentate ligand

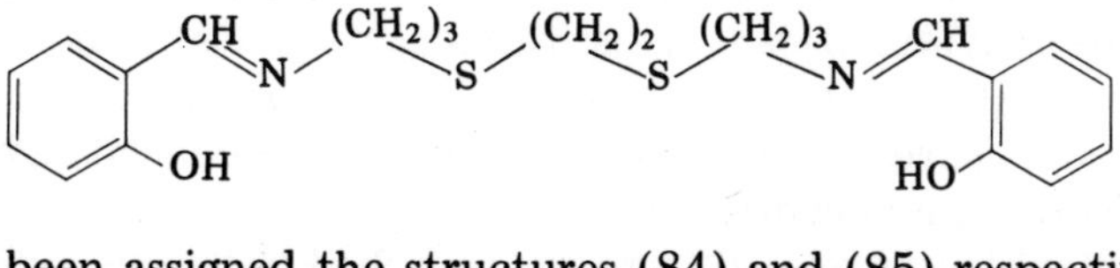

have been assigned the structures (84) and (85) respectively; and by adding further flexibility to this ligand, by saturation or by larger chelate rings, the other two possible geometries (86) and (87) are also feasible. Such isomers are usually easily separated by fractional crystallization or ion-exchange chromatography, but it is often a more difficult task to determine

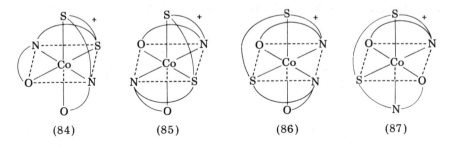

(84)   (85)   (86)   (87)

which is which; usually X-ray crystallographic analysis provides the only unequivocal assignment.

A slightly different type of isomer results from substitution in the ligand. Thus $[Pt(L-pn)_2]^{2+}$ exists in both *cis* (88) and *trans* (89) forms, and both "facial" (90) and "meridional" (91) forms of Λ-$[Co(L-pn)_3]^{3+}$ have

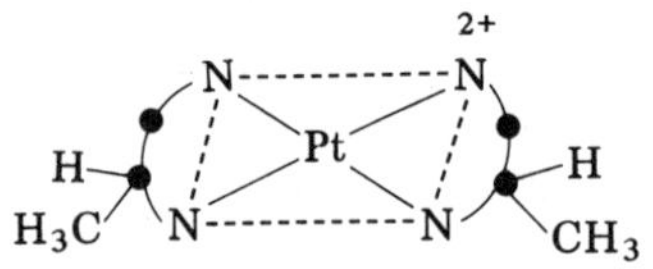

(88) 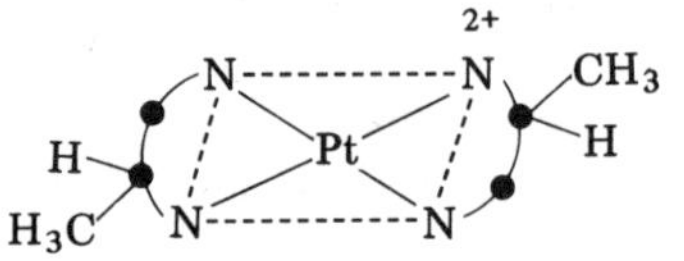   (89)

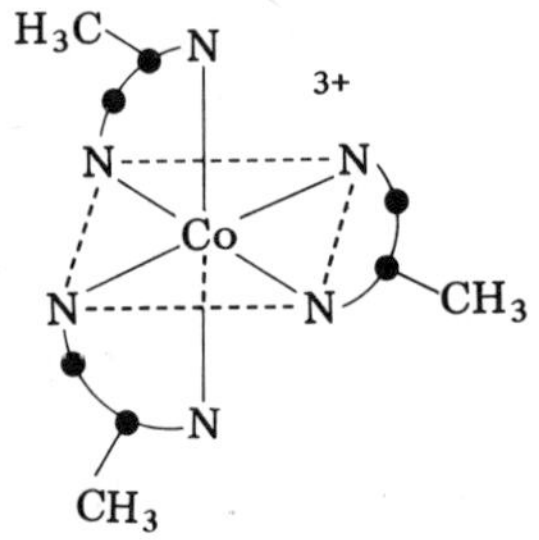

(90) 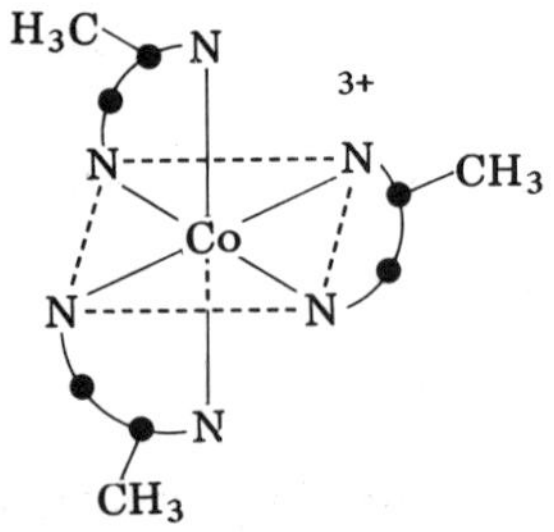   (91)

been identified after chromatographic separation. Other examples of this type include various unsymmetrically substituted acetylacetone chelates, *e.g.* $[Co(tfa)_3]$ (tfa = trifluoroacetylacetone), and such complexes containing unsymmetrical chelates have proved useful in studies on the mechanism of inversion at asymmetric metal centers.

A class of isomer closely related to those described in the preceding paragraph is where the orientation of a proton can result in configurational

effects. Thus the ion [Co(dien)(en)Cl]$^{2+}$ can occur in two geometrical forms
(92) and (93) both of which have been isolated, and β_2-[Co(trien)(gly)]$^{2+}$

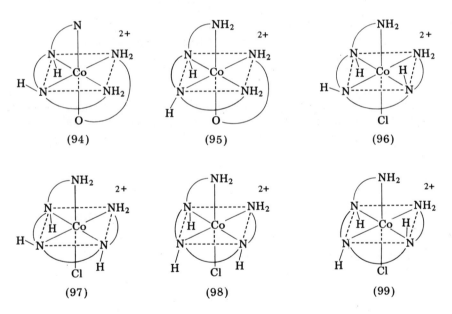

has been separated on Dowex cation-exchange resin into the two forms (94)
and (95). A more complicated example occurs with an isomer of [Co(tetraethylenepentamine)Cl]$^{2+}$ which could possibly occur in four arrangements
(96–99) depending on the configuration about the two secondary N atoms
joining chelate rings in the same plane.

The latter examples incur asymmetry on the N atom as well, and this aspect
is discussed in more detail in Section 5A(ii) b below. Other examples occur
with saturated P and As donors bridging chelate rings in the same coordination plane. In the case of the above examples incorporating N,
equilibration is easily achieved in alkaline solution where NH proton
exchange is rapid, but the isomers are stable in acidic solutions and in many

cases are easily separated and distinguished (visible or p.m.r. spectroscopy).

(b) *Structural isomers.* This class includes *linkage isomers* containing alternative modes of coordination of the same ligand, and *ligand isomers* which are themselves structurally different.

Many cases of linkage isomerism exist for monodentate ligands containing two different donor atoms (amphiphiles). Thus M-ONO(nitrito), M-NOO(nitro); M-SCN(thiocyanato), M-NCS(isothiocyanato); M-OCN(cyanato), M-NCO(isocyanato); M-CN(cyano), M-NC(isocyano) are classical examples in Co(III) chemistry. Similarly the $[Co(NH_3)_5$-(glycinato)$]^{2+}$ ion exists in both the N (100) and O (101) bonded forms. Also acetamide in $[Co(NH_3)_5$ (acetamide)$]^{3+}$ has been found in both the O (102) and N (103) bonded varieties, and the different coordination in this case requires proton rearrangement.

$$(NH_3)_5Co-NH_2CH_2CO_2{}^{2+}$$

(100)

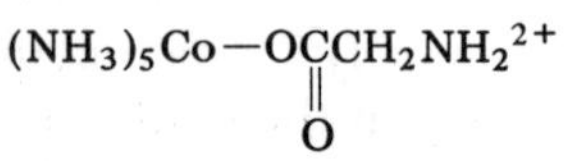

$$(NH_3)_5Co-OCCH_2NH_2{}^{2+}$$
$$\overset{\|}{O}$$

(101)

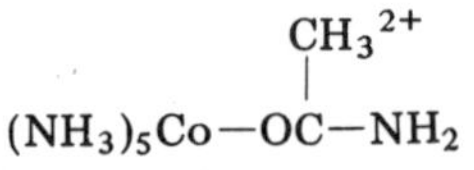

$$\overset{CH_3{}^{2+}}{\underset{}{|}}$$
$$(NH_3)_5Co-OC-NH_2$$

(102)

$$\overset{CH_3{}^{2+}}{\underset{}{|}}$$
$$(NH_3)_5Co-NH=C-OH$$

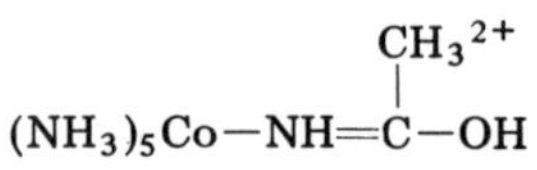

(103)

Bidentate examples include the N and O isomers of chelated glycine amide found in $[Co(en)_2(glycinamide)]^{3+}$ (104, 105), the latter having a $pK_a \sim 0.4$ for the enol proton, and the peculiar complex (106) which is believed to contain two different sized chelate rings incorporating the same ligand.

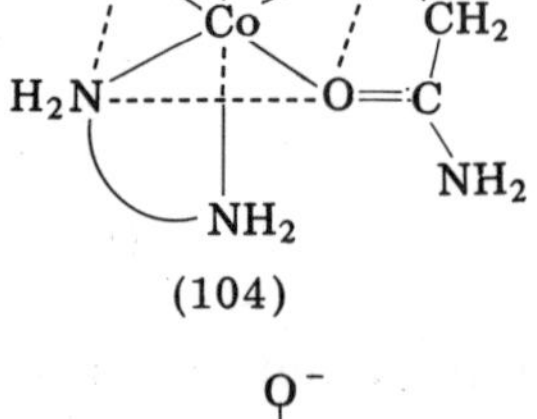

(104)

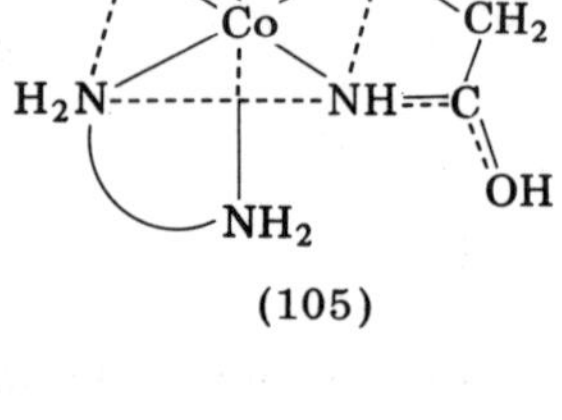

(105)

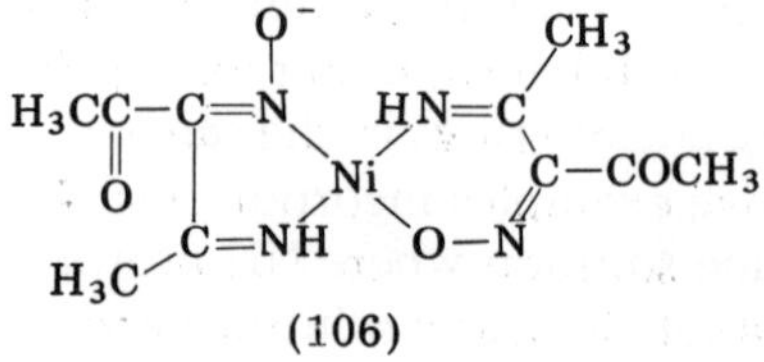

(106)

Ligand isomers are also fairly numerous, as they simply result from entirely different ligands having the same chemical composition. Classic examples occur with 1,2-diaminopropane (pn) and trimethylenediamine (tn) as in $[Pt(tn)_2]^{2+}$ (107) and $[Pt(pn)_2]^{2+}$ (108) respectively, and with $M-NCCH_3$ (methylcyanide) and $M-CNCH_3$ (methylisocyanide) complexes.

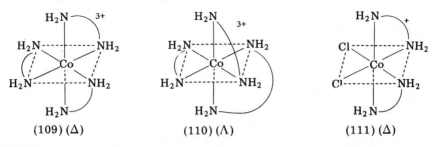

(ii) Optical isomers

Optical isomers arise when the molecule and its mirror image are non-superposable. Such a molecule is said to be dissymmetric and dissymmetry arises whenever a molecule has neither a center of symmetry, a plane of symmetry, nor an improper axis of symmetry.★ Thus asymmetric structures always occur in optically active forms, but so too do molecules with some elements of symmetry — notably two, three or four-fold axes of symmetry.

It is convenient to classify optical effects in coordination complexes as resulting from three causes: (a) the geometrical arrangement of ligand groups about the metal (*configurational* effects), (b) asymmetry present or generated in the ligand or its donor atoms by coordination, (*vicinal* effects), and (c) *conformational* properties of the coordinated ligand. Since optical activity arising from conformational effects results from different torsion angles about bonded pairs of atoms it will be considered in Section 5B.

(a) Configurational enantiomers. These were first demonstrated by Werner in a defence of his theory of coordination. Thus he demonstrated the octahedral geometry of $[Co(en)_3]^{3+}$ and *cis*-$[Co(en)_2Cl_2]^+$ by resolution into their enantiomeric forms (109), (110) and (111), (112) respectively, but the ions $[Co(en)(NH_3)_4]^{3+}$ (113) and *trans*-$[Co(en)_2Cl_2]^+$ (114) having planes of symmetry were incapable of resolution.

★Since a center of symmetry (*i*) is equivalent to two-fold improper axis (S_2), and a plane of symmetry (σ) can be considered as an S_1 symmetry element, a more succinct definition of dissymmetry is "when a molecule fails to have an improper axis S_n, with $n \geqslant 1$".

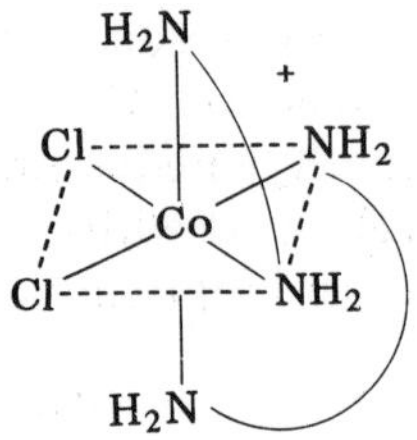

(112) (Λ) 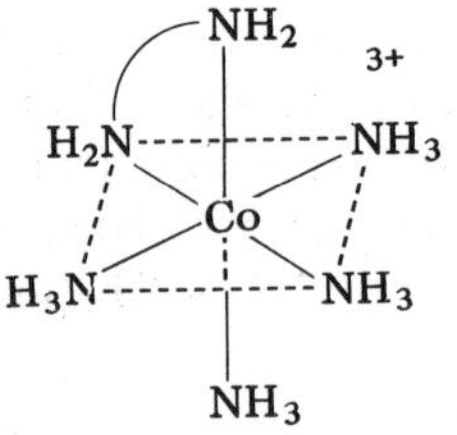   (113) 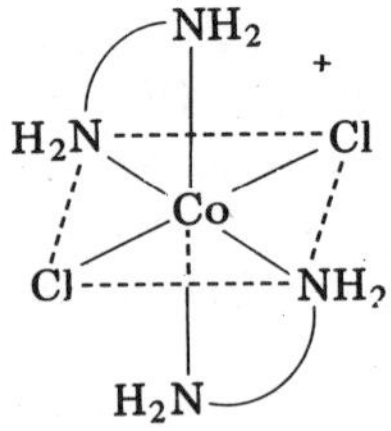   (114)

Since Werner's time many Co(III) and Cr(III) complexes in particular have been resolved and their absolute configuration has been demonstrated by X-ray crystal structure analysis[35,51]. Such optically active complexes have led to substantial advances in our understanding of two major areas of inorganic coordination chemistry: (a) the steric course and mechanism of substitution and racemization processes; and (b) through ORD (optical rotary dispersion) and CD (circular dichroism) studies the assignment of electronic absorption bands and the refining of our understanding of the electronic structures of complexes. A discussion of these topics is outside the scope of this chapter and the reader is directed to other sources[2,36,37].

Conventional resolution procedures[10,38] may be applied to kinetically inert complexes which are stable, or undergo only slow racemization in aqueous solution. This applies to Co(III), Cr(III), Rh(III), Ir(III), Os(II) and Pt(IV) complexes in general, and in some cases to Fe(II), Ni(II) and Co(II) octahedral complexes. For cationic and anionic complexes this usually means formation of diastereoisomeric salts with an optically active organic or inorganic anion or cation, but neutral complexes are more difficult to resolve and most common techniques use chromatographic or zone melting methods. For kinetically labile octahedral or tetrahedral complexes of Fe(II), Fe(III), Ni(II), Co(II), Mn(II), Cu(II), Zn(II), *etc.* the enantiomeric forms can only in exceptionally stable cases be isolated, and usually studies must be carried out in an asymmetric environment such as an optically active alcohol (2-butanol) or in aqueous solution containing an added asymmetric salt (ammonium-*d*-π-bromocamphor sulfonate, sodium antimonyl-*d*-tartrate, cinchonine, strychnine, *etc.*). Such "configurational activity" effects arise from differences in the activities of the enantiomeric forms in the asymmetric environment.

No tetrahedral complexes of the type M(ABCD) simulating classical carbon chemistry have been resolved into optical isomers, since they are difficult to prepare and are probably too labile for study by conventional methods. A few tetrahedral complexes of the type M(A–B)$_2$ have been resolved, *e.g.* [B(salicylaldehydo)$_2$]$^+$ (115) and [Be(benzoylpyruvato)$_2$]

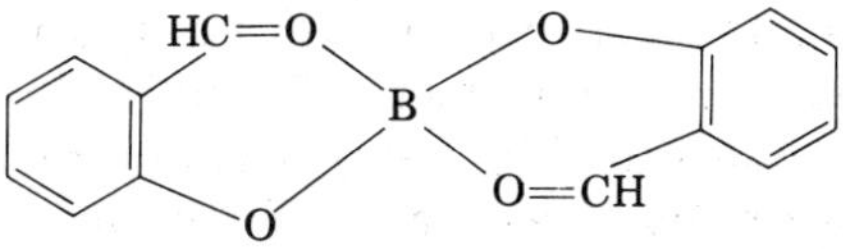

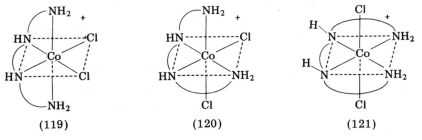

(115)

(9), but usually such complexes are too labile towards dissociation to study by conventional methods.

Square planar coordination usually results in the absence of optical isomers, but an unusual example which demonstrates square planar rather than tetrahedral coordination is provided by (13). If tetrahedral coordination were involved the complex would have a plane of symmetry.

By far the most extensive studies have been carried out on octahedral complexes of Co(III) or Cr(III) and Rh(III) containing chelated ligands. Thus, not only have dissymmetric complexes such as $[Co(en)_3]^{3+}$, $[Cr(en)_3]^{3+}$ $[Co(oxalato)_3]^{3-}$ and $[Co(glycinato)_3]$ been resolved but so have many less symmetrical molecules. For example, of the three possible geometrical forms of $[Co(dien)_2]^{3+}$ (116–118), (116) having C_{2h} symmetry is not dissymmetric, but (117) and (118) with C_2 symmetry have been resolved into

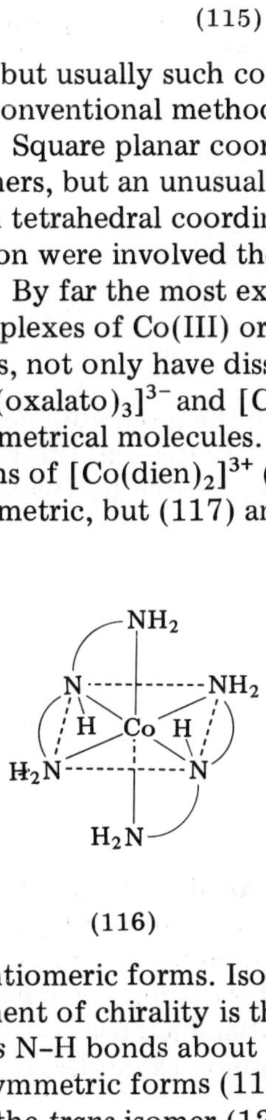

(116) 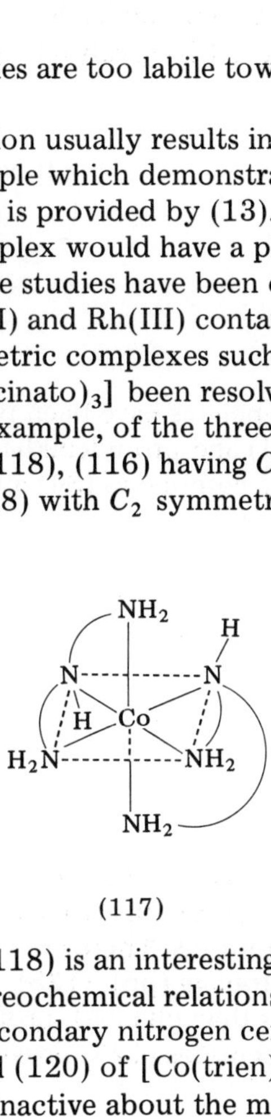 (117) 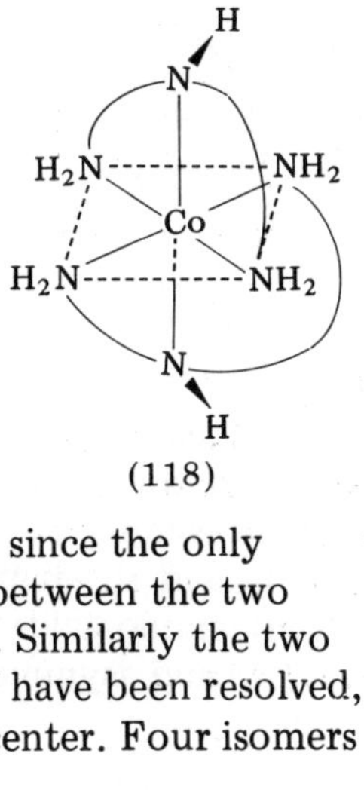 (118)

enantiomeric forms. Isomer (118) is an interesting case since the only element of chirality is the stereochemical relationship between the two *trans* N–H bonds about the secondary nitrogen centres. Similarly the two dissymmetric forms (119) and (120) of $[Co(trien)Cl_2]^+$ have been resolved, but the *trans* isomer (121) is inactive about the metal center. Four isomers

(119) (120) (121)

References pp. 61–62

50

of [Co(tetraen)Cl]⁺ are structurally possible (80–83), but (83) has a two-fold axis of symmetry and is hence inactive. The isomers (80) and (81) have been resolved and their configurations confirmed by X-ray. Symmetrical planar tridentates such as terpyridyl form inactive octahedral complexes (122), but unsymmetrical tridentates such as glycylglycine form optically active complexes, such as the [Co(glygly)₂]⁻ anion (123).

(122) (123)

Several sexadentate chelates have been resolved, the most notable being those of EDTA (124) and propylenediaminetetracetic acid (PDTA), and the Co(III) Schiff base complexes (84) and (85) prepared by Dwyer and Lions[39].

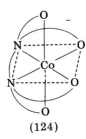

(124)

(b) *Vicinal effects.* This type of dissymmetry is provided by ligands which are themselves asymmetric, or by ligands which when coordinated provide asymmetric donor atoms. There are many examples of asymmetric ligands, and 1,2-diaminopropane (pn), (125), *trans*-1,2-diaminocyclohexane (chxn), amino acids (other than glycine), propylenediaminetetracetic acid (PDTA) and asymmetric peptides (126) provide common examples. It is

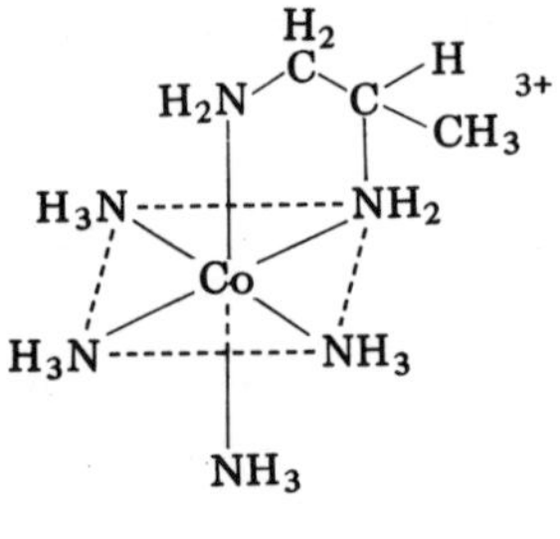

(125) 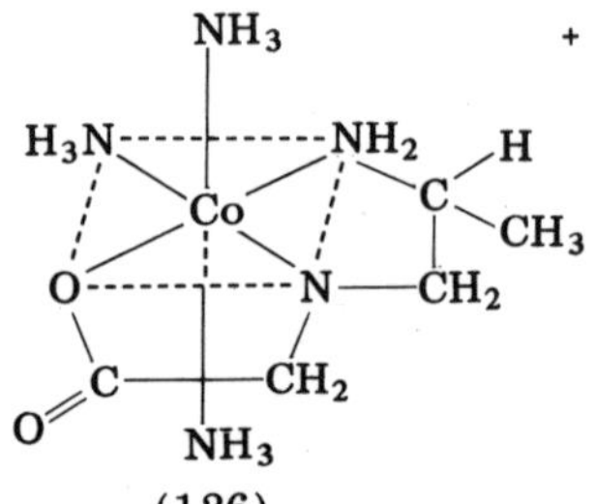 (126)

obvious that since the ligand is asymmetric so too is the complex, and this results in asymmetric contributions to the d-electronic transitions of the metal. Two such examples are (127) and (128) containing monodentate

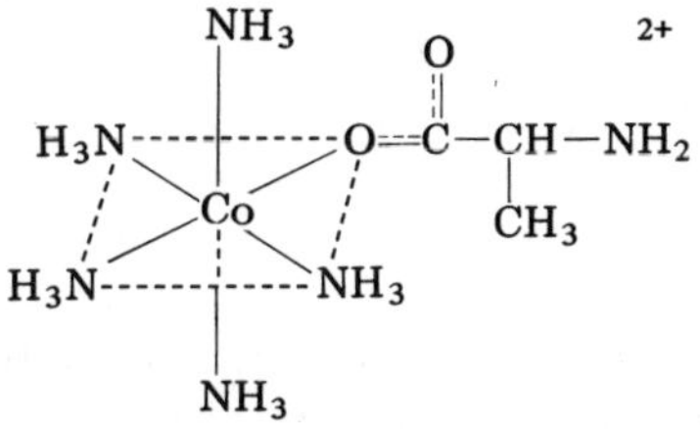

(127) 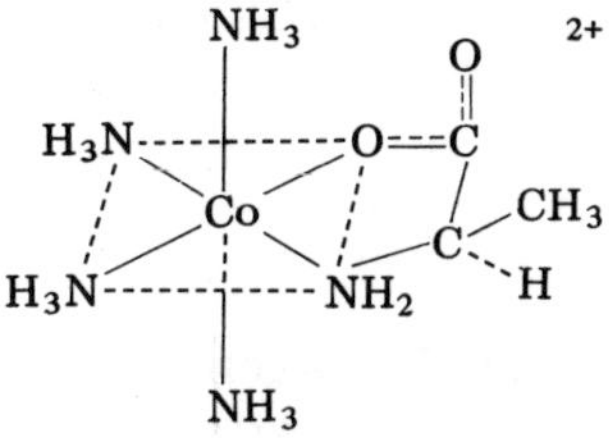   (128)

and bidentate alanine respectively. The contribution of the asymmetric C center to the circular dichroism associated with the metal ion is similar in the two complexes. In some cases, however, the presence of two asymmetric chelates may give rise to internally compensated, and hence, inactive isomers. Thus the two complexes(129) and (130 are inactive, the *trans*

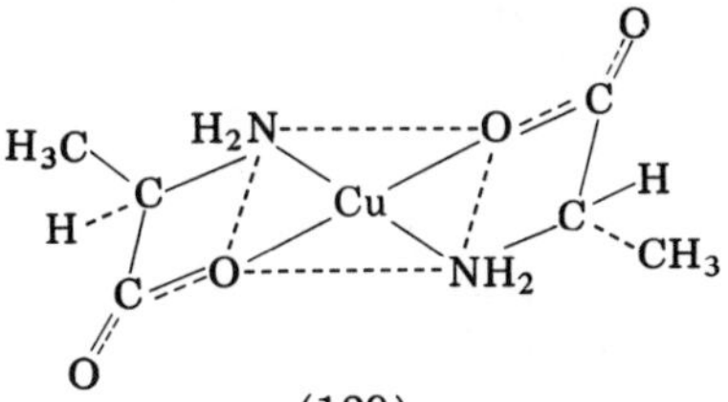

(129) 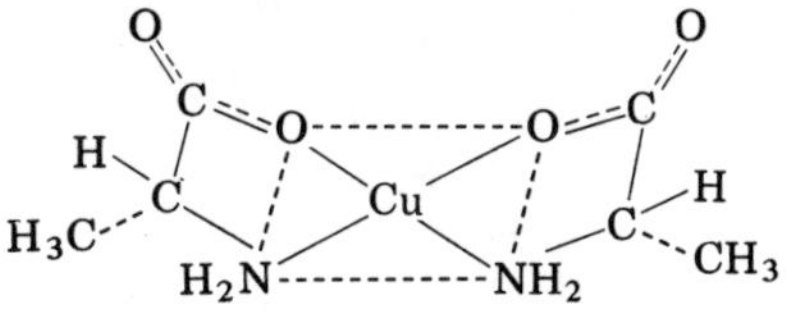   (130)

isomer (129) having a center of symmetry and the *cis* (130) a plane of symmetry. Mutarotation of asymmetric centers adjacent to coordinated carbonyl or carboxyl functions in such complexes is greatly facilitated by coordination, and may be ascribed to the increased acidity of the proton. Thus the chelated alanine in (131) mutarotates rapidly at pH 12, while L-alanylglycine in (132) mutarotates at pH 10.

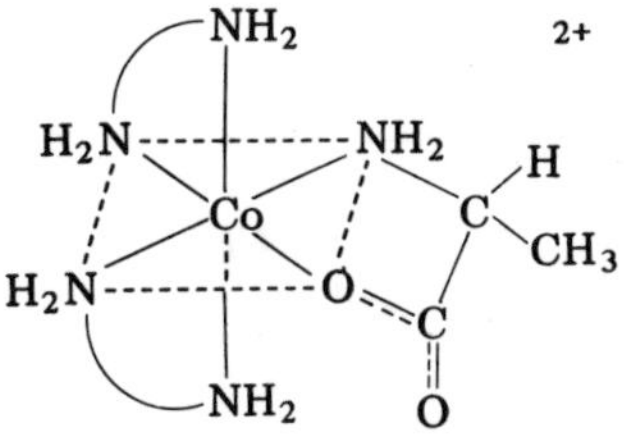

(131) 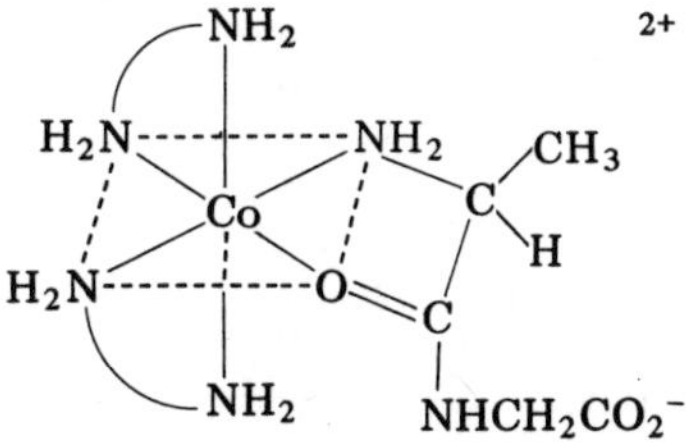   (132)

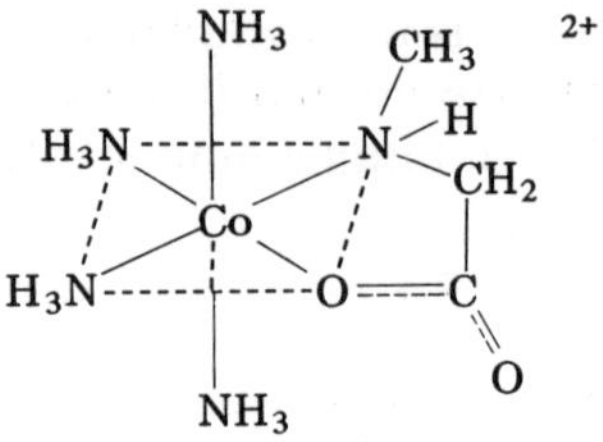

(133)

Vicinal effects generated at donor atoms by coordination have been the center of considerable recent activity, although the effect has long been known. For example, chelated sarcosine in the $[Co(NH_3)_4(sar)]^{2+}$ ion (133) is asymmetric by virtue of the 4-covalent nitrogen and this ion has been resolved into its optical enantiomers. Likewise the (±)*trans*-[14]diene complex of Ni(II) (134), containing a quadridentate Schiff base is essentially planar but has been resolved into optical forms. The *meso* isomer (135) containing secondary NH protons on opposite sides of the coordination plane

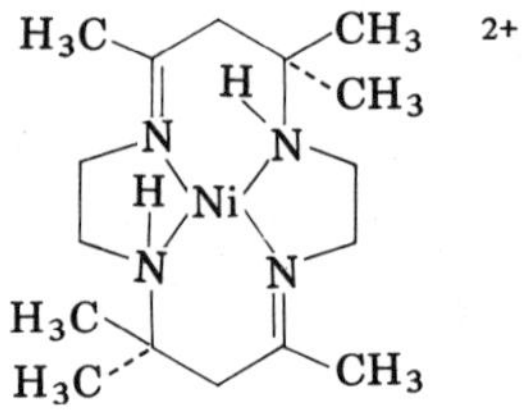

(134) (*trans*-[14]diene)

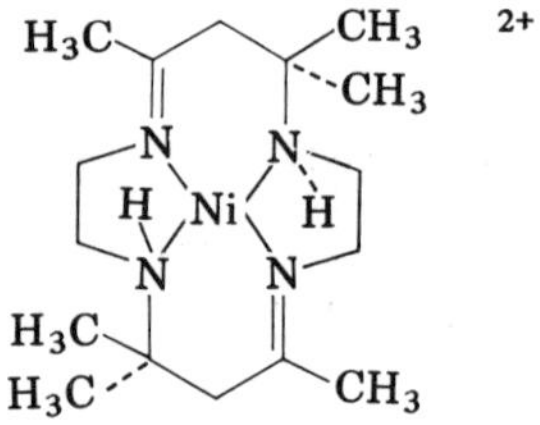

(135) (*meso-trans*-[14]diene)

is internally compensated and hence inactive. Crystal structures are available on both complexes[40]. Similarly the *cis*-[14]diene (136) occurs in active and *meso* forms by virtue of the N–H protons. Many other examples have since been resolved into enantiomeric forms including *trans*-[Co(trien)Cl$_2$]$^+$ (138), *trans*-[Co(Meen)$_2$Cl$_2$]$^+$ (139), [Pt(Meen)$_2$]$^{2+}$ (137) and [Ni(cyclam)]$^{2+}$ (140).

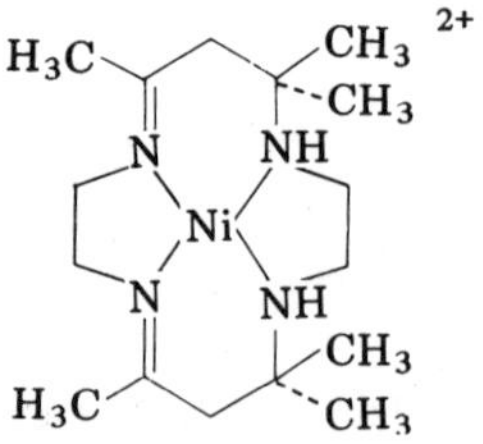

(136)

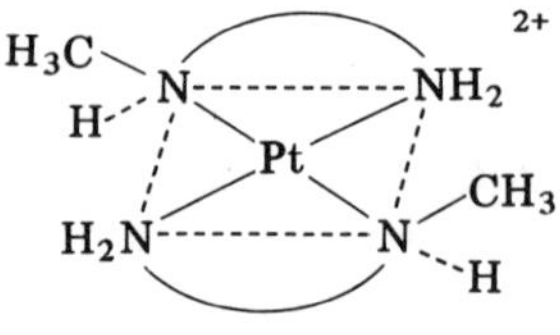

(137)

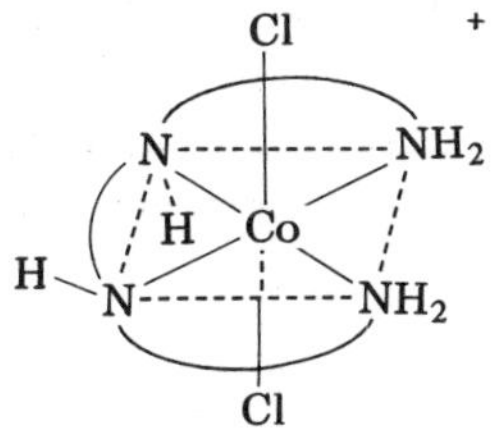

(138) 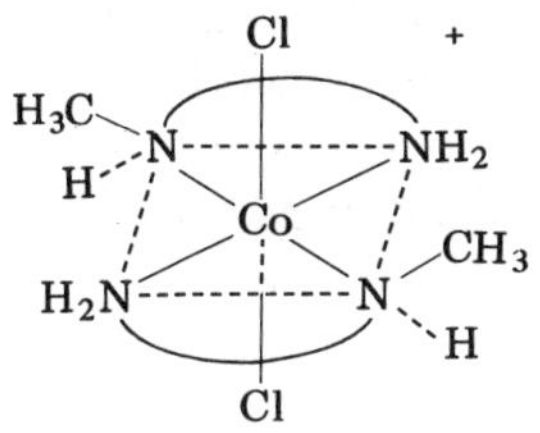   (139) 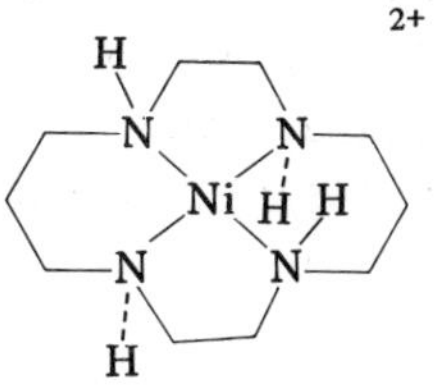   (140)

The inertness of the N centers towards racemization contrasts sharply with the rapid rates of inversion in organic amines or substituted ammonium ions under the same conditions, and the complex isomers are optically stable in weakly acidic solutions. It appears that racemization occurs only when the N–H proton is removed, and moreover that the rate of racemization is 10^2 to 10^6 times slower than H-exchange in the Pt(II), Pt(IV) and Co(III) complexes[26]. This has been interpreted in terms of a large retention factor for the amine conjugate base. Both H-exchange and nitrogen inversion obey the rate law $k_{obs} = k[OH^-]$ in complexes of this type.

Coordination by S, P and As also generates asymmetry and, as might be expected, retention of configuration is usually more pronounced here than in the case of asymmetric nitrogen. Thus the Pt(IV) salt (141) was

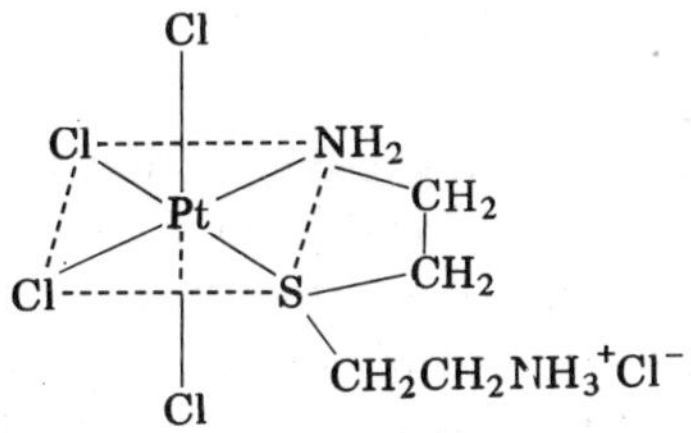

(141)

resolved as early as 1930[41], and recently the arsenic base $As(CH_3)(C_2H_5)(C_6H_5)$ was resolved via the Pt(II) complex (142)[42] and a similar Pt(II) complex has been used to resolve $P(CH_3)(C_6H_5)(But)$[43].

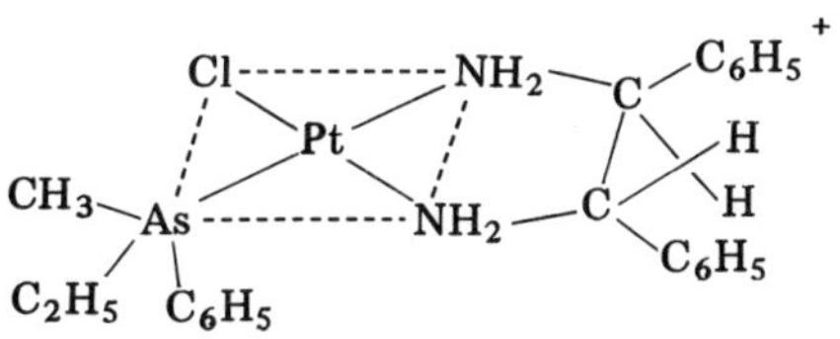

(142)

Obviously this effect leads to a more complex picture of isomerism in inert metal complexes. For example while the α-[Co(trien)gly]$^{2+}$ ion has

54

stereospecifically directed N atoms in the Λ and Δ-enantiomers (143), the β_1-[Co(trien)gly]$^{2+}$ ion exists in two internal diastereoisomeric forms (of similar stability) for each of the Λ and Δ configurations about the metal (144), (145). A similar situation is found with the third geometrical isomer,

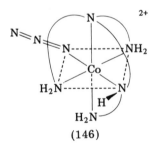

(143) (a-Λ-RR)

(144) (β_1-Δ-RR)

(145) (β_1-Δ-RS)

β_2-[Co(trien)(gly)]$^{2+}$ (9:1 stability relationship in this case), so that ten possible configurational isomers have been realized for this complex. The internal diastereoisomeric pairs can be readily interconverted in basic solution where H-exchange is rapid. Likewise the ion (146) has a plane of symmetry if the asymmetric N atom and chelate ring conformations are neglected, but this ion has also been resolved into its mirror image forms demonstrating the importance of isomerism of this type.

(146)

Such considerations often lead to stereospecific effects. This is demonstrated by (143); N and P donors joining 5-membered chelate rings in different coordination planes adopt the configuration forced on them by

the ring orientations. A less obvious, but equally specific effect is demonstrated by the Δ-[Co(trien)(sarcosinato)]$^{2+}$ ion (147) where the amino acid
moiety adopts the S-configuration exclusively. The alternative R-config-
uration (148) is not found in this ion (it occurs in the Λ-complex) as non-
bonded interactions between the methyl group and the apical chelate ring
destabilize it by a calculated 3.3 kcal mol^{-1} [12].

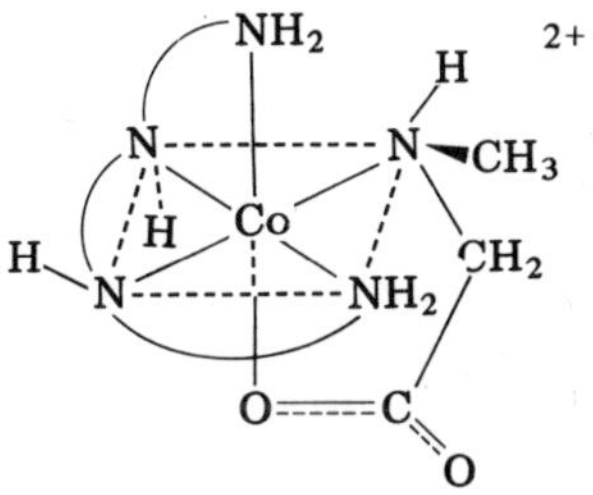

(147) (Δ-β_2-RRS)

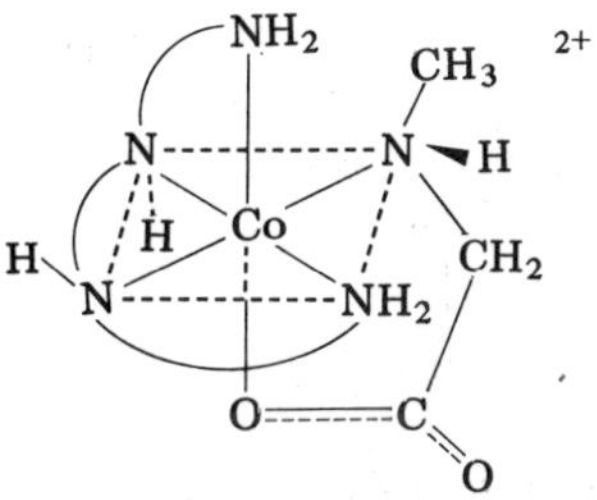

(148) (Δ-β_2-RRR)

(B) Conformational isomerism

(i) General

Conformational isomers arise from non-superposable spatial arrangements of atoms or groups of atoms in a molecule with the same chemical
composition and configuration, but with different torsional angles about
bonded pairs of atoms. Examples of the substituted diphenyl type have
been known for a long time with a classic example being that incorporating
restricted rotation in 2,2'-diamino-6,6'-dimethyldiphenyl (149). The Cu(II)
complex of the Schiff base adduct with salicylaldehyde (150) was prepared

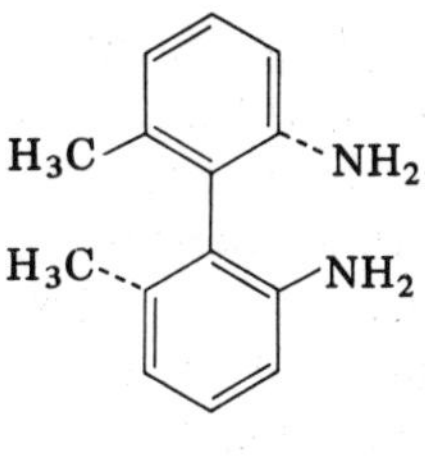

(149)

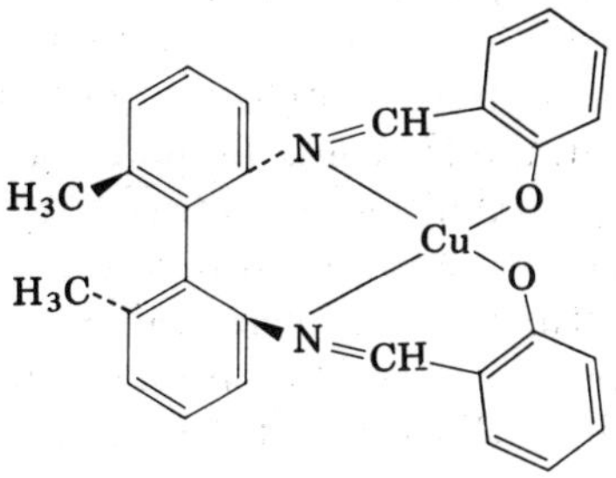

(150)

in optically active forms using the optical forms of the diamine. The Cu
atom adopts tetrahedral coordination due to the non-planar nature of the
benzene rings. Usually, however, the activation energy for the interconversion of conformational isomers is small, since only a twisting motion
is involved, and such processes are therefore rapid. Thus both δ (151) and

λ (152) mirror image conformations of chelated ethylenediamine occur in the crystalline state but in solution interconversion is rapid in Co(III), Pt(II), Rh(III) and Pt(IV) complexes, and the two conformations have not yet been separately identified.

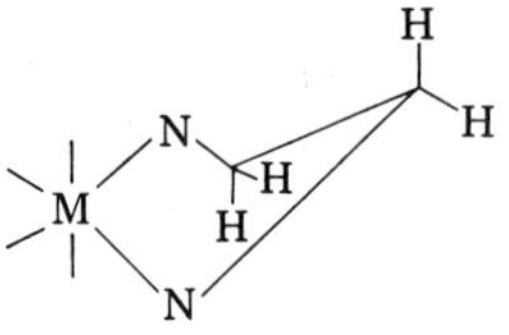

(151) 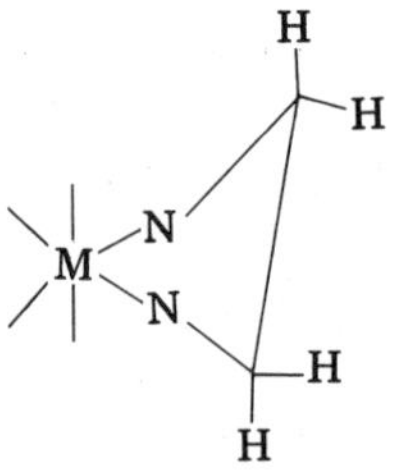   (152)

(ii) Conformational isomerism in chelates
(a) Five-membered rings. By far the largest number of chelate ring systems are either five or six-membered. Many of these such as oxalate, amino acids, peptide chelates and corrin type quadridentates, are planar or almost planar and lack much conformational character. In square planar and octahedral geometries, ligands such as ethylenediamine containing four tetrahedral skeletal atoms give rise to chelate rings with pronounced conformational properties, and this will occur in most chelate ring systems containing tetrahedral carbon atoms. Such ring systems differ from analogous five-membered carbocyclic systems in having bond angles at the metal of close to 90°. A consequence of this is that the remainder of the molecule must skew to prevent unreasonable distortion of tetrahedral centers. Several structures arise depending on the precise metal–ligand angles, bond lengths, and the size and type of donor atom involved. For ethylenediamine the structure usually found is depicted by (151, 152) with the two carbon atoms equidistant from, but on opposite sides of the M–N–N plane. In this half-chair arrangement all the H atoms are more or less staggered, and those on C are almost strictly "axial" or "equatorial" in the cyclohexane sense; the H atoms on N are less conformationally distinct. This arrangement is found in all M–en structures so far examined crystallographically and the other possible conformation, containing both C atoms on the same side of the M–N–N plane ("eclipsed" form) (153), has not yet been identified[52].

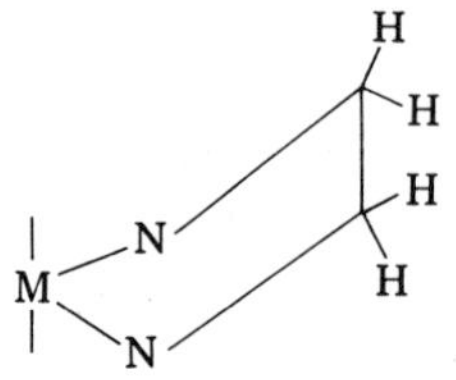

(153)

This structure might be expected to be stabilized with shorter M–N bond lengths (~ 1.75 Å) provided the N–M–N angle remains unaltered. For larger donor attoms such as P or As, larger M–donor bond lengths are expected and the half-chair conformation will be preferred. For S as the donor, there should be a pronounced preference for a smaller than tetrahedral M–S–C angle ($\sim 92°$), and a less planar conformation as measured by an increase in the dihedral angle about the C–C bond. A similar but smaller effect is to be expected for O as the donor. Puckering of chelate rings is also found in structures containing multidentate 5-membered chelate rings such as diethylenetriamine (dien), (116–118), triethylenetetramine (trien), and tetraethylenepentamine (tetraen) (80–83). In these structures adjacent chelate rings in the same coordination plane have their conformations determined by the configuration about the common asymmetric N center, but apical chelate rings have the possibility of existing in δ or λ forms. A recent crystal structure of β_2-[Co(trien)-(glycinato)]I$_2$ has confirmed the presence of the $\delta\delta\lambda$ (154) and $\lambda\delta\lambda$ (155) conformations in the triethylenetetramine

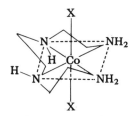

(154)   (155)

chelate. Usually however, one conformation is sufficiently stable over the other to occur exclusively ($> 99\%$) under equilibrium conditions. The apparent absence of the *meso* [Co(trien)X$_2$]$^+$ structure (157) can similarly be accounted for by the central ring being forced into an eclipsed "envelope" conformation[52]. Similar considerations apply in other ring systems involving N, O or S donor atoms.

(156) (preferred conformation)   (157) (unstable envelope conformation)

For tetrahedral structures, five-membered ethylenediamine-type chelate rings must be strained because the N–M–N angle is increased from 90° towards 109°. Although the chelate ring will prefer to remain puckered it might be expected to be extended towards the planar state. In trigonal bipyramidal structures chelation from the trigonal plane to the apical positions will be preferred over chelation within the trigonal plane for both five and six-membered chelate rings.

Methyl substitution on carbon in these 5-membered rings may occur in either the "axial" or "equatorial" positions. For R-1,2-diaminopropane this results in the δ (159) and λ (158) conformations respectively and crystal structures and theoretical considerations have established that the λ

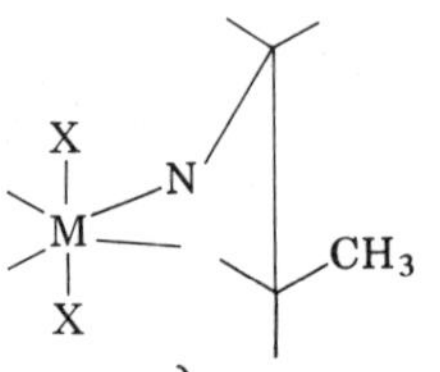

λ

(158)

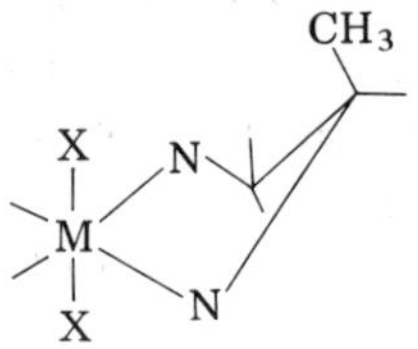

δ

(159)

conformation is stabilized over the δ conformation by about 2 kcal mol^{-1}. This is probably due to the substantial interactions between the axial methyl group and the axial X groups on the metal. If S-1,2-diaminopropane is considered then the δ conformation is similarly stabilized over λ. Thus in both the Λ and Δ forms of [Co(R(−)pn)$_3$]$^{3+}$ the chelate rings adopt λ conformations. Larger substituents will clearly accentuate this effect as will symmetrical disubstitutions such as in R,R-2,3-diaminobutane (160) and here the ligand appears to be stereospecific for the Δ-[Co((−)2,3-butanediamine)$_3$]$^{3+}$ ion[45]. However if the substituents are on the same C atom there will always be one axial and one equatorial substituent or some compromise between them. The same is true for meso-2,3-diaminobutane (161) but little is known about these systems.

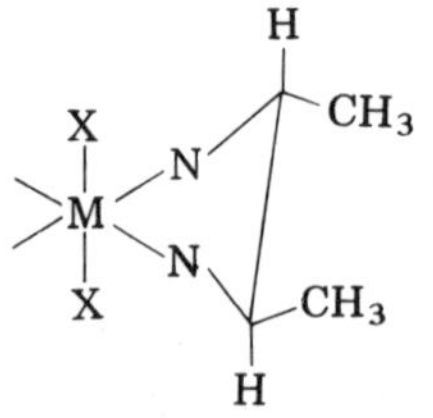

(160) (RR(−)bn)

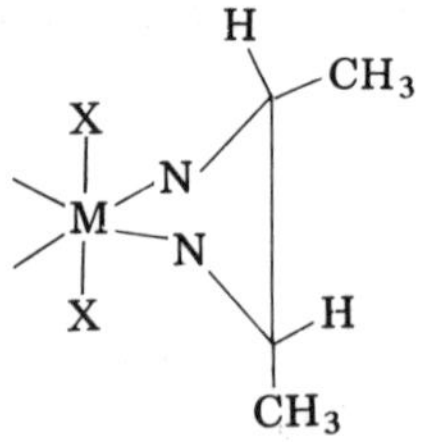

(161) (RS-meso bn)

Substitution on the donor atoms produces similar results. In N-methyl-1,2-diaminoethane (Meen) the difference between the axial (163) and

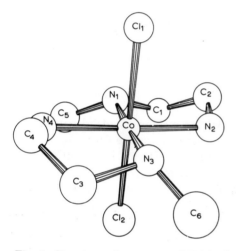

(162)   (163)

equatorial (162) methyl groups is less pronounced ($\sim 20°$), but the substituent is now closer to the axial X group attached to the metal, and non-bonded interactions should now be larger. As expected, the crystal structure of *trans*-RR-[Co(Meen)$_2$Cl$_2$]ClO$_4$ (Fig. 4) demonstrates equatorial

Fig. 4. Structure of *trans,trans*-RR-[Co(Meen)$_2$Cl$_2$]$^+$ cation[50].

methyl groups and δ ring conformations. [Force field calculations predict only a small energy difference between this and the axial conformation (~ 1 kcal mol^{-1})]. Symmetrical disubstitution on both N atoms appear to further stabilize the δ ring and the p.m.r. spectrum of the *trans*-[Co(Me$_2$en)$_2$-Cl$_2$]$^+$ ion exhibits a well resolved AA'BB' pattern expected for a frozen conformation in solution. It is clear then that donor atom or ligand atom substitution in 5-membered chelates results in a preferred, and in many instances exclusive, ring stereochemistry.

(b) Six-membered rings. Most six-membered chelates show pronounced conformational properties. Examples include 1,3-diaminopropane (trimethylenediamine), malonate ion, β-alanine, and 1,3-diarsenopropane or 1,3-diphosphinopropane with substituted As or P atoms. The chelate systems are analogous to cyclohexane and chair (164), boat (165, 166) and

60

skew boat (167) conformations are possible. The structural analysis of
Λ-[Co(trimethylenediamine)$_3$] Br.H$_2$O^{46} indicates that in contrast to Co–en
rings the N–Co–N angle is greater than 90° (92–96°) and this presumably
influences the Co–N–C angle which expands to 117.4°. The C–C–C angle

(164)

(165)

(166)

(167)

(112°) is closer to the tetrahedral value. All three rings adopt "distorted
chair" conformations (164) and similar conformations are found in
[Co(trimethylenediamine)$_2$CO$_3$]ClO$_4$. However, flattened boat conformations are found for the malonate ion in [Co(en)(mal)$_2$]$^-$. It would appear
that the extreme boat forms of the trimethylenediamine type chelates are
unlikely as large non-bonded interactions are involved between one H atom
of the central methylene group and the axial substituent on the metal (165)
and between H atoms on N and C, and C and C (166). However, not enough

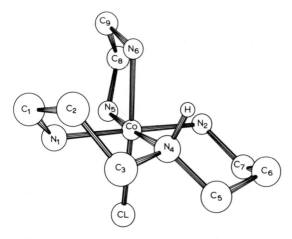

Fig. 5. Structure of [Co(en)(ditn)Cl]$^{2+}$ cation^{47}.

X-ray structures have been determined to allow a detailed analysis in these systems yet, and it is likely that intermediate conformations will occur due to the increased ring size. An indication of this is found in the structure of one isomer of $[Co(en)(ditrimethylenediamine)Cl]^{2+}$ where one of the 6-membered rings has a chair conformation, and the other is distorted boat (Fig. 5)[47]. Packing forces and cation–anion interactions are probably important factors in such ring systems and conformations in the crystalline state are in many cases unlikely to be preserved in solution.

REFERENCES

1 J. C. Bailar, *The Chemistry of Coordination Compounds*, A.C.S. Monograph No. 131, Reinhold, New York, 1956.
2 F. Basolo and R. G. Pearson, *Mechanism of Inorganic Reactions*, 2nd edn., Wiley, New York, 1967.
3 S. Charberek and A. E. Martell, *Sequestering Agents*, Wiley–Interscience, New York, 1959.
4 F. P. Dwyer and D. P. Mellor (Eds.) *Chelating Agents and Metal Chelates*, Academic Press, New York, 1964.
5 F. A. Cotton and G. Wilkinson, *Advanced Inorganic Chemistry*, 2nd edn., Interscience, New York, 1966.
6 J. Lewis and R. G. Wilkins (Eds.) *Modern Coordination Chemistry, Principles and Methods*, Interscience, New York, 1960.
7 A. E. Martell and M. Calvin, *Chemistry of Metal Chelates*. Prentice-Hall, New York, 1952.
8 Y. Saito, *Pure Appl. Chem.*, **17** (1968) 21.
9 H. C. Freeman, *Adv. Protein Chem.*, **22** (1967) 337.
10 A. M. Sargeson, Chapter 5 in ref. 4.
11 A. M. Sargeson, in R. L. Carlin, *Transition Metal Chemistry*, Vol. 3, Marcel Dekker, New York, 1966, p. 303.
12 D. A. Buckingham and A. M. Sargeson, in Eliel and Allinger, *Topics in Stereochemistry*, Vol. 6, 1971, p.219.
13 J. O. Edwards, *Inorganic Reaction Mechanisms*, Benjamin, New York, 1964.
14 F. Basolo, *Survey Progr. Chem.* 2 (1964) 1.
15 J. P. Hunt, *Metal Ions in Aqueous Solution*, Benjamin, New York, 1963.
16 M. Eigen and R. G. Wilkins, in *Mechanisms of Inorganic Reactions*, Adv. in Chem. Series, No. 49, Am. Chem. Soc., Washington, D.C., p. 55.
17 E. Cartmell and G. W. A. Fowles, *Valency and Molecular Structure*, 2nd edn., Butterworths, London, 1961.
18 D. P. Craig and R. S. Nyholm, Chapter 2 in ref. 4.
19 H. B. Gray, *J. Chem. Educ.*, 41 (1964) 2.
20 T. M. Dunn, D. S. McClure and R. G. Pearson, *Some Aspects of Crystal Field Theory*, Harper and Row, New York, 1965.
21 H. B. Gray, *Electrons and Chemical Bonding*, Benjamin, New York, 1964.
22 A. D. Liehr, *J. Chem. Educ.*, 39 (1962) 135.
23 D. C. Bradley, M. B. Hursthouse and P. F. Rodesiler, *Chem. Commun.*, (1969) 14.
24 E. B. Fleischer, *Accounts Chem. Res.*, 3 (1970) 105.
25 G. N. Schrauzer, *Accounts Chem. Res.*, 1 (1968) 97.

62

26 D. A. Buckingham, L. G. Marzilli and A. M. Sargeson, *J. Am. Chem. Soc.*, 91 (1969)
    5227.
27a D. A. Buckingham, D. M. Foster and A. M. Sargeson, *J. Am. Chem. Soc.*, 90 (1968)
    6032.
27b *Ibid. J. Am. Chem. Soc.*, 92 (1970) 5571.
28 D. A. Buckingham, P. A. Marzilli, I. E. Maxwell, A. M. Sargeson, M. Fehlmann and
    H. C. Freeman, *Chem. Commun.*, (1968) 488.
29 A. F. Wells, in *Structural Inorganic Chemistry*, 3rd edn., Clarendon Press, Oxford,
    1962, p. 71.
30 R. S. Nyholm, *Proc. Chem. Soc.*, (1961) 273.
31 D. H. Gerlach and R. H. Holm, *Inorg. Chem.*, 9 (1970) 588.
32 F. Lions, *Rec. Chem. Progr.*, 22 (1961) 69.
33 F. Lions, *Rev. Pure Appl. Chem. (Aust).*, 19 (1969) 177.
34 M. Calliganis, G. Nardin and L. Randaccio, *Chem. Commun.*, (1969) 1248.
35 Y. Saito, *Pure Appl. Chem.*, 17 (1968) 21.
36 R. D. Gillard, *Progr. Inorg. Chem.*, 7 (1966) 215.
37 R. D. Gillard and P. R. Mitchell, in *Structure and Bonding*, 7 (1970) 46.
38 R. G. Wilkins and M. J. G. Williams, Chapter 3 in ref. 6.
39 F. P. Dwyer and F. Lions, *J. Am. Chem. Soc.*, 69 (1947) 2917; *J. Am. Chem. Soc.*,
    72 (1950) 1546.
40 M. F. Bailey and I. E. Maxwell, *Chem. Commun.*, (1966) 908.
41 F. G. Mann, *J. Chem. Soc.*, (1930) 1745.
42 B. Bosnich and S. B. Wild, *J. Am. Chem. Soc.*, 92 (1970) 459.
43 J. H. Chan, *Chem. Commun.*, (1968) 895.
44 I. E. Maxwell, *Thesis*, Australian National University, 1969.
45 F. Woldbye, *Studier over Optisk Activet*, Polyteknisk Forlog, København, 1969,
    p. 176 *et seq.*
46 T. Nomura, F. Marumo and Y. Saito, *Bull. Chem. Soc. Japan*, 42 (1969) 1016.
47 P. R. Ireland, D. A. House and W. T. Robinson, *Inorg. Chim. Acta*, 4 (1970) 137.
48 M. D. Lind, B. Lee and J. L. Hoard, *J. Am. Chem. Soc.*, 87 (1965) 1611.
49 J. L. Hoard, B. Lee and M. D. Lind, *J. Am. Chem. Soc.*, 87 (1965) 1612.
50 D. A. Buckingham, G. Chandler, L. G. Marzilli and A. M. Sargeson, *Chem. Commun.*,
    (1969) 539.
51 O. Kennard and D. G. Watson (Eds.), *Molecular Structures and Dimensions*, International Union of Crystallography, Vol. 2, 1970 and Vol. 3, 1971.
52 Note added in proof. Since the time of writing, the *meso* form of [Ni trien](ClO$_4$)$_2$
    has been found where the central chelate ring adopts the eclipsed configuration
    (A. McPherson, M. G. Rossmann, D. W. Margerum and M. R. James, *J. Coord. Chem.*,
    1 (1971) 39.

The following abbreviations are used in this Chapter:

en = ethylenediamine	gly = glycine
pn = 1,2-diaminopropane	glygly = glycylglycine
dien = diethylenetriamine	py = pyridine
trien = triethylenetetramine	dipy = dipyridyl
tetraen = tetraethylenepentamine	terpy = terpyridyl
acac = acetylacetone	diars = diarsine
salen = salicylalethylenediamine	EDTA = ethylenediamine-
me = methyl	tetraacetic acid
et = ethyl	meen = N-methyl-ethylenediamine

Chapter 2

STABILITY OF COORDINATION COMPOUNDS

ROBERT J. ANGELICI

Department of Chemistry, Iowa State University, Ames, Iowa 50010, U.S.A.

I. INTRODUCTION

Although the word *stability* has been used to describe a variety of chemical properties of compounds, its use in this chapter refers to the equilibrium constant for the reaction of a metal ion (M^{+m}) and a ligand (L^{-n}) to form the metal–ion complex, $ML^{+(m-n)}$:

$$M^{+m} + L^{-n} \rightleftharpoons ML^{+(m-n)} \tag{1}$$

This equilibrium constant is called a stability constant (or sometimes a formation constant). In general M^{+m} usually will have a $+1$, $+2$ or $+3$ charge. Since Chapter 6 deals specifically with metal–ion complexes of the alkali metal ions, our concern here will be primarily with transition metal ions (*e.g.* Fe^{2+}, Fe^{3+}, Ni^{2+}, Cu^{2+}, Zn^{2+}, *etc.*) and to a lesser extent with the lanthanide metal ions (*e.g.*, Ce^{3+}, Sm^{3+}, Gd^{3+}, Ho^{3+}, Tm^{3+}, Lu^{3+}, *etc.*). The ligands, L^{-n}, will normally have charges of 0, -1, or -2. Only space limits the types of ligands which will be considered; ligands such as NH_3, Cl^-, Br^-, I^-, PO_4^{3-}, $P_2O_7^{4-}$, amino acidates, small peptides, nucleotides, and porphyrins are within the scope of this chapter. Although there are a few known complexes, *e.g.*, those of the tartrate anion[1], which contain two metal ions bridged by one or more ligands, these binuclear complexes will not be discussed.

The purpose of this chapter is to point out some of the major factors affecting stability constants and to provide a sufficient breadth of general references to give the reader ready access to the extensive literature in this area. No attempt is made to provide comprehensive coverage of the topic. Unless indicated otherwise, stability constants used in this chapter refer to measurements made at 25°C.

Since the vast majority of studies of reaction (1) have been made in aqueous solutions, it is important to point out that M^{+m} exists in solution as the aquo complex. For the first row transition element ions, $Mn(OH_2)_6^{2+}$, $Fe(OH_2)_6^{2+}$, $Co(OH_2)_6^{2+}$, $Ni(OH_2)_6^{2+}$, and $Zn(OH_2)_6^{2+}$ are the species present in solution. For the lanthanides 9 or 10 coordinate ions[2-6] such as $La(OH_2)_9^{3+}$, $Pr(OH_2)_9^{3+}$, $Er(OH_2)_9^{3+}$, or $Y(OH_2)_9^{3+}$ appear to be the aquated form of the metal ions. Thus reactions of type (1) are more fully written as substitution reactions:

$$M(OH_2)_x{}^{+m} + L^{-n} \rightleftharpoons M(OH_2)_{x-1}(L)^{+(m-n)} + H_2O \tag{2}$$

If L were a bidentate or tridentate ligand then two or three molecules of H_2O would be displaced from the coordination sphere of the complex.

The thermodynamic equilibrium constant, $K°$, for reaction (1)

$$K° = \frac{\{ML^{+(m-n)}\}}{\{M^{+m}\}\{L^{-n}\}} = \frac{[ML^{+(m-n)}]}{[M^{+m}][L^{-n}]} \cdot \frac{\gamma_{ML}{}^{+(m-n)}}{\gamma_M{}^{+m}\gamma_L{}^{-n}} \tag{3}$$

may be expressed in terms of the activities of the species, $\{ML^{+(m-n)}\}$, $\{M^{+m}\}$, and $\{L^{-n}\}$. Note that the activity of water is considered constant and is included as part of $K°$ because in dilute solutions the activity of water remains essentially constant regardless of the position of equilibrium (2). Since the activity of a species is defined as the product of its concentration and activity coefficient, e.g., $\{M^{+m}\} = [M^{+m}]\gamma_M{}^{+m}$, equation (3) may be expressed in terms of concentrations and activity coefficients as shown on the right side of equation (3). Since experimentally one usually measures concentrations of the species involved, the thermodynamic stability constants of these reactions can be determined only either by knowing the activity coefficients or by measuring the equilibrium constants for the reaction at several ionic strengths and then extrapolating to zero ionic strength where the activity coefficients are defined as unity. These techniques are either unreliable or very time-consuming or both; hence in practice relatively few investigators have attempted to evaluate thermodynamic stability constants but instead have simply used concentrations in the equilibrium expression.

Thus the vast majority of stability constants reported in the literature are stoichiometric stability constants (sometimes called concentration stability constants), K, in which the ratio of the

$$K = \frac{[ML^{+(m-n)}]}{[M^{+m}][L^{-n}]} \tag{4}$$

activity coefficients (in equation (3)) is included in the constant, K. The only important condition for determinations of K is that the activity coefficients of the species be constant regardless of the equilibrium position of the reaction being studied. Hence a high concentration of an unreactive ionic salt is added to the solution to maintain a constant ionic strength. This ensures that the activity coefficients of the reacting species in equation (1) will remain constant. It is, of course, desirable that the salt does not itself react with the species involved in the reaction. Salts which are commonly chosen are $LiClO_4$, $NaClO_4$ or KNO_3. Salts of Cl^- are inferior choices because of the known tendency of this anion to coordinate to metal ions in rather concentrated aqueous solutions. In this connection, it should be mentioned that the basic component of virtually all buffers likewise forms metal-ion complexes of varying degrees of stability.

Although a constant ionic strength does allow the determination of a reliable stoichiometric stability constant, this constant is only valid at the ionic strength of the measurement. An example of a stability constant which should depend quite strongly on the ionic strength would be that obtained[7] for the chelation of Cu^{2+} by the -2 charged anion, phthalate:

$$Cu^{2+} + C_6H_4(CO_2)_2{}^{2-} \overset{K}{\rightleftharpoons} [C_6H_4(CO_2)_2]Cu \tag{5}$$

As expected the stoichiometric stability constant decreases with increasing concentration of the supporting electrolyte, Na_2SO_4 (Table I). A Debye–Hückel limiting law plot of $\log K$ against $(\mu)^{\frac{1}{2}}$ gives a straight line. Other

TABLE I

IONIC STRENGTH(μ) DEPENDENCE OF K FOR REACTION (5)

μ	$10^{-3}K$
0.03	1.84
0.06	1.41
0.12	1.04
0.25	0.80

electrolytes such as $NaClO_4$, $NaNO_3$, or $NaCl$ gave values of K which were almost identical with those observed using Na_2SO_4. Because few other ligands are more negatively charged than -2, most stability constants[8] exhibit a smaller dependence on the ionic strength than noted for reaction (5). In the present chapter, ionic strength effects on stability constants are generally small as compared to other effects of interest.

Since metal ions usually have four or more coordination positions, they frequently coordinate to more than one ligand. Successive coordination of ligands to a metal ion occurs in steps, and stepwise stability constants may be written and determined experimentally for each of the equilibria. For a metal ion which has four coordination sites, such as Cu^{2+}, the following equilibria are possible:

$$M^{2+} + L^- \rightleftharpoons ML^+; \; K_1 = \frac{[ML^+]}{[M^{2+}][L^-]}$$

$$ML^+ + L^- \rightleftharpoons ML_2; \; K_2 = \frac{[ML_2]}{[ML^+][L^-]}$$

$$ML_2 + L^- \rightleftharpoons ML_3{}^-; \; K_3 = \frac{[ML_3{}^-]}{[ML_2][L^-]}$$

$$ML_3{}^- + L^- \rightleftharpoons ML_4{}^{2-}; \; K_4 = \frac{[ML_4{}^{2-}]}{[ML_3{}^-][L^-]}$$

66

Overall stability constants, β, have also been used to express these successive equilibria; they are simply the products of the stepwise stability constants, i.e., $\beta_1 = K_1, \beta_2 = K_1 K_2, \beta_3 = K_1 K_2 K_3$ and $\beta_4 = K_1 K_2 K_3 K_4$.

For statistical[9], steric, and electrostatic reasons, the successive addition of negative ligands to M^{2+} is expected to become less and less favorable. Indeed this is generally true, as for example, for the equilibria

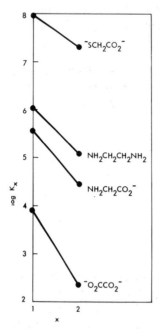

Fig. 1. Log K_1 and log K_2 values for the reactions of Zn^{2+} with some bidentate ligands at 25°C[22,23].

involving Cu^{2+} and the monodentate ligand acetate: log $K_1(1.67) >$ log $K_2(0.98) >$ log $K_3(0.42) >$ log $K_4(-0.19)$. Figure 1 illustrates additional examples of this trend; thus regardless of the donor atom $K_1 > K_2$ for a variety of bidentate ligands in their reactions with Zn^{2+}.

Various aspects of stabilities have been reviewed previously. General experimental techniques have been surveyed by several authors[8-12]. More specific discussions of stability constant determinations by calorimetry[13], pH methods[14,15], polarography[16], ion exchange[17-20], solvent extraction[19], and solubility[21] have been published. A very useful compilation of stability constants taken from the literature prior to 1968 has been assembled by Sillén and Martell[22]; a less comprehensive volume by Yatsimirskii and Vasilev[23] is also available. Most of the published stability data involve metal ions which react very rapidly[24] (within seconds) with ligands.

Considerably less is known about equilibria involving inert metal ions, which are experimentally less conveniently studied. Trends in stability constants have been analyzed and reviewed by many authors[5,8,9,23,25-29]. Sigel and McCormick[29a] have recently published a brief but pertinent account of equilibrium interactions of metal ions with ligands of biological significance. Enthalpies and entropies of complex formation have been summarized and interpreted in several recent reviews[8,25,30,31].

II. FACTORS AFFECTING THE STABILITY OF METAL-ION COMPLEXES

(A) Nature of the metal ion

(1) Charge and size of the metal ion

Possibly the simplest approach to an understanding of the interaction between a metal ion and its ligands is to consider the metal–ligand bond as being entirely ionic or electrostatic. Thus, for example, the coordination of a fluoride ion, F^-, by a metal ion would depend upon the charges on the ions involved and the distance(r) separating their nuclei. It is assumed that the ions are hard spheres

$$\underbrace{M^{+m}\ F^-}_{r}$$

with the charges located at their nuclei. The calculated electrostatic energy of such a bond is $E = e^2 Z_{M^{+m}} Z_{F^-}/r$, where e is the charge on the electron, and $Z_{M^{+m}}$ and Z_{F^-} are the integral charges on the ions.

From the expression it would be expected that a high charge on M^{+m} and a small internuclear distance (*i.e.*, a small radius for M^{+m}) would favor the formation of a stable complex. To be sure this simple model ignores entropy and solvation effects and any covalent bonding. For complexes of very electropositive metal ions such as Li^+, Mg^{2+}, La^{3+}, *etc.*, and very electronegative ligand donor atoms such as F^- and O^{2-}, covalent bonding is small and the bond energy is dominated by the electrostatic attraction of the metal ion and ligand. Metal ions and ligands of this type have been called *hard* (or class (a)) acids and bases, respectively. This system of classification is discussed in more detail in Section II(A)(2).

If only hard metal ions and ligands are considered, it is generally true that for metal ions with nearly the same radii, those with the highest positive charge form the most stable complexes. Figure 2 shows how log K for $M^{+m} + L^{-n} \rightleftharpoons ML^{+(m-n)}$ increases with the charge on the metal ion in the order: $Li^+(0.68$ Å$) < Mg^{2+}(0.65$ Å$) < Fe^{3+}(0.53$ Å$)$; ionic radii are given in Ångstroms in parentheses. It should also be noted that there is no corresponding increase in K with an increasing charge on the ligand. The

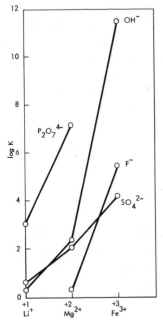

Fig. 2. Increasing values of log K with increasing charge on the metal ion[22].

ligand behavior is complicated by the possibility that SO_4^{2-} may be either monodentate or bidentate, whereas $P_2O_7^{4-}$ is very probably bidentate.

Values of K also increase as the size of the metal ion *decreases*. Thus, for example, in the complexation of the alkaline earth ions[22] by the pyrophosphate ion, $P_2O_7^{4-}$, the K values decrease on descending the group:

	Mg^{2+}	Ca^{2+}	Sr^{2+}	Ba^{2+}
radius (Å)	0.65	0.94	1.10	1.29
log K	7.2	6.8	5.4	4.6

This trend is also illustrated (Fig. 3) by the +3 lanthanides whose radii decrease regularly from 1.06 Å for the lightest element, La^{3+}, to 0.85 Å for the heaviest element, Lu^{3+}. The upper data in Fig. 3 refer to the stability constants for the glycolate, $HOCH_2CO_2^-$, anion which coordinates bidentately through a carboxylate oxygen and the alcoholic oxygen. The lower data refer to the reaction with acetate ion. The simple electrostatic model would predict a regular increase in log K from La^{3+} to Lu^{3+}. More specifically, a plot of log K *versus* $1/r$ should give a straight line. Such plots do not give straight lines, and even qualitatively, there is not a regular increase in log K values. Indeed the lanthanides behave as if they consisted of two different groups with the division occurring at Gd^{3+}. Various reasons such as crystal field stabilization energy or steric factors have been used[5] to account for this division, but no completely satisfactory explanation has been found. The type of plot observed for glycolate is commonly observed for equilibria involving the lanthanide ions and numerous other ligands[5,32]. The behavior of acetate (Fig. 3) in the last half of the lanthanide series is unusual.

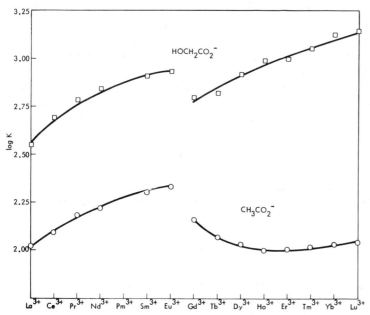

Fig. 3. Variations in log K for the reactions of glycolate ($\square$) and acetate ($\bigcirc$) ions with the lanthanide (+3) ions (20°C, 0.1M $NaClO_4$)[5].

As demonstrated by the lanthanide series, relatively small changes in ionic radius do not always give increased values of K, even when the discussion is limited to *hard* metal ions and donor atoms. Yet, large decreases in radius appear to usually give more stable complexes. The effect of increased positive charge on the metal ion is clearer: for hard metal ions and ligand donor atoms, complex stability increases with increasing positive charge on the metal ion.

(2) Hard and soft nature of metal ions and ligand donor atoms
Many attempts have been made to correlate stability constants with properties of the metal ions and ligands. As noted in Section II(A)(1), limited correlations with the charge and size of the metal ion have been observed. However, if one chooses a broad range of metal ions and ligands,

it is clear that there are very few correlations. There are many examples where trends are even reversed. For example, the stability constants, K, for halide complexation with Fe^{3+} decrease in the order: Fe^{3+}: $F^- >$ $Cl^- > Br^- > I^-$. In contrast, Hg^{2+} forms more stable complexes with the heavier halides and K decreases as follows:

$$Hg^{2+}: I^- > Br^- > Cl^- > F^-$$

From this and many other examples, it was concluded by Schwarzenbach[33] and then Ahrland, Chatt, and Davies[34] that some metal ions behave like Fe^{3+} and others like Hg^{2+} toward halide ions. The former were called class (a) and the latter class (b) metal ions or acids. By comparing stability constants for reactions of other ligands with different metal ions, it was possible to even further characterize the two classes of metal ions. Assembled in Table II are trends in stability constants which are commonly observed for class (a) and class (b) metal ions with different ligand donor atoms. Thermodynamic parameters[31] for these reactions have also been of value in classifying metal ions.

TABLE II

DEFINITIONS OF HARD (CLASS (a)) AND SOFT (CLASS (b)) METAL IONS (ACCORDING TO TRENDS OF COMPLEX STABILITIES WITH DIFFERENT LIGAND DONOR ATOMS)

Hard or class (a) metal ion trends	Soft or class (b) metal ion trends
F > Cl > Br > I	F < Cl < Br < I
O ≫ S > Se > Te	O ≪ S ~ Se ~ Te
N ≫ P > As > Sb	N ≪ P > As > Sb

Metal ions which are experimentally found to belong to these classes are designated[35] in Table III. It should be pointed out that there are some

TABLE III

CLASSIFICATION OF METAL IONS

Hard or class (a)	Borderline	Soft or class (b)
H^+, Li^+, Na^+, K^+	Fe^{2+}, Co^{2+}, Ni^{2+}	Cu^+, Ag^+, Au^+, Tl^+
Be^{2+}, Mg^{2+}, Ca^{2+}, Sr^{2+}	Cu^{2+}, Zn^{2+}, Pb^{2+}	Hg_2^{2+}, Hg^{2+}, Pd^{2+}, Pt^{2+}
Mn^{2+}, Al^{3+}, Sc^{3+}, Ga^{3+}	Sn^{2+}, Sb^{3+}, Bi^{3+}	Pt^{4+}, Tl^{3+}
In^{3+}, La^{3+}, Gd^{3+}, Lu^{3+}		
Cr^{3+}, Co^{3+}, Fe^{3+}, Si^{4+}		
Ti^{4+}, Sn^{4+}, WO^{4+}, VO^{2+}		

metal ions which obey some trends of class (*a*) and some of class (*b*) metal ions; these are placed in the borderline classification. It might be noted that many of the borderline metal ions are the +2 transition metal ions.

On examining the class (*a*) metal ions, they have many similar properties. They have small radii; they have high positive charges; and they usually have no unshared electron pairs in their valence shell (and many have noble gas electron configurations). The small sizes and high positive charges of these metal ions result in low polarizabilities; for this reason they have been called[35] "hard". These metal ions form their most stable complexes with very electronegative atoms such as O, N, and F (Table II). The bonding in these complexes is largely ionic.

Class (*b*) metal ions are much larger, have lower positive charges and have unshared valence *p* or *d* electrons. These properties make class (*b*) metal ions much more polarizable; hence they have been called "soft". These metal ions form their most stable complexes with ligand donor atoms (P, S, and I) which have low electronegativities and are highly polarizable. Covalent bonding contributes substantially to bonding in these complexes.

As noted above hard metal ions complex most strongly with donor atoms (N, O, and F) which possess high electronegativities, low polarizabilities, small radii, and are difficult to oxidize. Because of these properties, they have been designated as hard Lewis bases or donor atoms. On the other hand, those ligand atoms (P, S, and I) which form the most stable complexes with soft metal ions are large, have low electronegativities, high polarizabilities, and are relatively easily oxidized; they are called soft ligands. Some typical hard, soft, and borderline ligands are listed[35] in Table IV.

TABLE IV

CLASSIFICATION OF LIGANDS

Hard or class (*a*)	Borderline	Soft or Class (*b*)
H_2O, OH^-, F^-, Cl^-	$C_6H_5NH_2$, C_5H_5N	R_2S, RS^-, I^-
$CH_3CO_2^-$, PO_4^{3-}, SO_4^{2-}	N_3^-, Br^-, NO_2^-	SCN^-, $S_2O_3^{2-}$
CO_3^{2-}, ClO_4^-, NO_3^-	SO_3^{2-}, N_2	R_3P, $(RO)_3P$, R_3As
ROH, R_2O, NH_3		CN^-, RNC, CO,
RNH_2, N_2H_4		H^-, R^-, C_2H_4

In terms of our new nomenclature, the trends in Table II may be summarized as follows[40]: *Hard metal ions prefer to coordinate hard ligands, and soft metal ions prefer to coordinate soft ligands.* Although many attempts[35-38] have been made to justify this rule theoretically, it must be regarded as empirical. A semi-empirical treatment[39] using the ionic radii, charges, ionization potentials, and electronegativities of metal ions has yielded a quantitative parameter for the softness of a metal ion. This together with other metal and ligand parameters correlates a fairly large number of metal complex stability constants. A more theoretical approach[40] has also had considerable success in correlating hard and soft properties.

Qualitatively this rule is helpful in accounting for interactions observed in biological systems. Pearson[41] has pointed out that biological complexes are largely composed of hard metal ions and ligands. Oxygen and nitrogen are the predominant ligand donor atoms and hard alkali and alkaline earth metal-ions are present in great abundance. To be sure, there are relatively soft metal ions (*e.g.* Cu^{2+}) and ligands (*e.g.*, sulfur donors) present but they are relatively immobile and available only in low concentrations. Pearson also notes that poisons to living systems are frequently soft, as for example, such ligands as CO, CN^-, and H_2S and such metal ions as Hg^{2+}. These soft poisonous species when present in relatively high concentrations combine with soft metal ions and ligands in the biological systems and prevent them from carrying out their functions. Although these generalizations are undoubtedly not applicable to all aspects of living units, they do provide another starting point for examining these exceedingly complicated systems.

The hard-soft principle may be applied to the coordination of transition metal ions to nucleotide-di and -triphosphates[42]. Hard metal ions such as the alkaline earths and Mn^{2+} coordinate mainly through the hard phosphate oxygen atoms whereas the softer Cu^{2+} coordinates to both the soft nitrogen atom of the aromatic purine or pyrimidine base and to a phosphate group. Similarly the ligand $CH_3CH_2SCH_2CO_2^-$ coordinates to the relatively hard metal ions Mn^{2+} and Zn^{2+} through an oxygen atom only whereas Cu^{2+} is chelated by coordination through both an oxygen as well as the soft sulfur[43].

Our considerations of metal ions have assumed that water occupies coordination sites not occupied by other ligands. As has been pointed out[35], the nature of the ligands bound to a metal ion also influence its hard or soft character. It appears that hard ligands make the metal ion harder and soft ligands make it softer toward additional ligands. Thus, for example, metal ions bearing sulfur ligands or relatively soft porphyrin rings will have a greater tendency to form complexes with additional soft ligands than would the hydrated metal ion $M(OH_2)_6^{2+}$ which is bound by hard water ligands. The influence of ligands may explain why the iron in heme coordinates so strongly to such soft ligands as CO and CN^- even in the presence of large amounts of the hard ligand water. It may also account for Fe^{2+} (and apparently even Fe^{3+}) having such a high-affinity for sulfur in non-heme iron–sulfur proteins[44,45].

(3) Irving–Williams order of stability for first row transition metal–ion complexes

As noted in the previous section, the first row transition metal ions have borderline hard-soft properties. For this reason, it is frequently difficult to predict the stabilities of their complexes. However, it has been known[46-48] for many years that the stability constants of the +2 ions of these elements with a given ligand follow the general trend:

$$Mn^{2+} < Fe^{2+} < Co^{2+} < Ni^{2+} < Cu^{2+} > Zn^{2+}$$

This trend is frequently called the Irving–Williams order of stability and is illustrated for several ligands in Fig. 4. The persistence of this trend in not

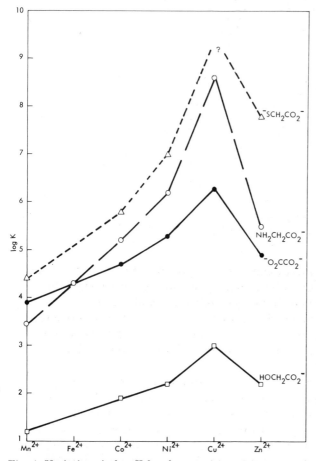

Fig. 4. Variations in log K for the reactions of first row transition metal ions (M^{2+}) with several bidentate ligands at 25°C[22].

References pp. 98–101

only K_1 but K_2 and for the most part also in K_3 is shown in Fig. 5. The abnormally low value of K_3 for Cu^{2+} is attributed to the low tendency of this ion to coordinate to more than four donor atoms. Although the Irving–Williams order also appears to hold for K_2, K_3, etc. in other systems, there is a general shortage of reliable data to confirm this trend.

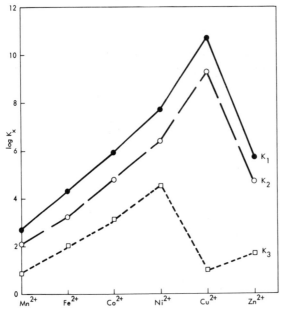

Fig. 5. Variations in log K_1, log K_2, and log K_3 for the reactions of first row transition metal ions (M^{2+}) with ethylenediamine ($NH_2CH_2CH_2NH_2$) at $30°C^{22}$.

For ligands which contain amino donor groups, the Irving–Williams order of stabilities is largely determined by the enthalpies (ΔH) of the reactions. Thus the ΔH values for ethylenediamine and glycine are negative and follow the expected order (Fig. 6). In contrast, the reactions of carboxylate donor ligands are slightly endothermic and the entropy (ΔS) contributes substantially to the observed overall Irving–Williams order for the stability constants.

Since the Irving–Williams order is determined at least in many instances by the ΔH of the interaction between the metal ion and the ligand, attempts to account for the observed stabilities have focused on the metal–ligand bond energies. Although variations in the ionic radii of the metal ions have been used to account for the observed trend using an electrostatic model [see Section II(A)(1)], these changes in radius are really too small to account adequately for the large variations in K values.

Another approach[51] has involved correlating the stability constants with the ionization potentials of the metal ions.

More recently the successes of crystal field and ligand field theories in accounting for the thermodynamics and kinetics of reactions[24] of transition metal complexes have led to their application to the Irving–Williams order of complex stabilities[52]. Calculated crystal field stabilizations

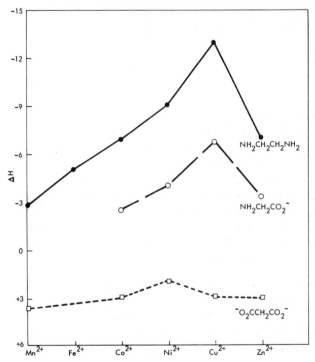

Fig. 6. Variations in ΔH for the reactions, M + L $\rightleftharpoons$ ML, where L = ethylenediamine (●), glycinate (○) or malonate (□) at 25°C[49,50].

together with electrostatic bonding energies qualitatively account for the Irving–Williams order for high spin complexes of these metal ions. Since covalent bonding, solvation energies and entropies are not considered in these calculations, it is not surprising that quantitative agreement is not possible.

Although the Irving–Williams order is followed by the vast majority of ligands, there are some exceptions: (1) when a ligand has donor atoms which may all coordinate in an octahedral complex (e.g., Ni^{2+}), but may not all coordinate in the square planar Cu^{2+} complex. Thus, for example, with hexadentate EDTA (ethylenediaminetetraacetate, $(^-O_2CCH_2)_2$-

$NCH_2CH_2N(CH_2CO_2^-)_2)$ K_1 for Cu^{2+} is only slightly larger than for Ni^{2+}. Copper is not able to coordinate strongly with all six donor atoms; (2) when ligands which have a sufficiently high ligand field strength to change the metal ion from high to a low spin state are used. This primarily occurs for complexes of Fe^{2+} with such ligands[53] as o-phenanthroline, 2,2'-dipyridyl and cyanide. Unusually stable complexes are formed. Similar changes in stability order occur for two ligands of more biochemical interest, imidazole and histamine. The tendency to form Fe^{2+} complexes of extraordinary stability appears to occur most commonly in ligands which contain aromatic heterocyclic nitrogen donor atoms.

(B) Nature of the ligand

(1) Ligand basicity

As was noted previously [Section II(A)(1)], the electrostatic model of bonding metal–ion complexes has limited applicability in accounting for trends in stability constants with different metal ions and almost no applicability to stability trends with changes in the ligand. Because of the importance of solvation, entropy and covalent bonding, this is not surprising. Hence correlations of other ligand properties with stability constants have been examined. One of the most successful correlations has been with the basicity (pK_a) of the ligand[54]. The reasoning is based on the fact that H^+ and metal ions, M^{+m}, both act as Lewis acids towards Lewis base ligands. The pK_a (i.e., $- \log K_a$) for the aqueous reaction,

$HL \; \overset{K_a}{\rightleftharpoons} \; H^+ + L^-$, is taken as the measure of the basicity of the ligand, L^-, toward the proton; the higher its pK_a the stronger is the base L^- toward H^+ and presumably also toward M^{+m}. Indeed there is frequently a linear free energy correlation between the pK_a and the logarithm of the stability constant, K, for complex formation.

In Fig. 7 are shown several such correlations[50] where the Lewis acids are $Cu(dipy)^{2+}$, Cu^{2+}, $Zn(dipy)^{2+}$, and Zn^{2+}. The ligands are monodentate carboxylate anions, RCO_2^-. As indicated in the Figure, some of these ligands are derivatives of the aromatic benzoate anion, while others (II, V, and VII) are not. As indicated for Cu^{2+}, the stability constants for both the aromatic and non-aromatic carboxylates correlate well with their pK_a values. On the other hand, only the aromatic benzoates have been included in the Zn^{2+} and $Zn(dipy)^{2+}$ correlations and only the non-aromatic carboxylates with $Cu(dipy)^{2+}$. This has been done because the aromatic and non-aromatic carboxylates do, in these cases, have slightly different slopes and both groups of carboxylates do not fit the same line. As is generally true[56] for such pK_a against $\log K$ correlations, the closer the structures of the ligands in the series the better the correlations. For ligands having the same donor atom but quite different structures, as in

NH₃, aniline, pyridine and imidazole, correlations are so poor as to be of no value in predicting unknown stability constants from pK_a values.

For ligands which have different donor atoms such as RNH_2, RCO_2^-, and RS^- there is no correlation with the pK_a of the protonated ligand. In view of the earlier discussion [Section II(A)(2)] concerned with matching the hard and soft properties of metal ions and ligand donor atoms, one cannot expect any correlation of $\log K$ with pK_a for different donor atoms.

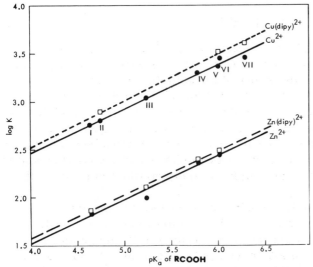

Fig. 7. Dependence of $\log K$ for the reaction, $M^{2+} + RCO_2^- \rightleftharpoons RCO_2M^+$, on the pK_a of RCO_2H (where R = p-NO₂C₆H₄⁻, I; H⁻, II; m-ClC₆H₄⁻, III; C₆H₅⁻, IV; CH₃⁻, V; p-CH₃C₆H₄⁻, VI; CH₃CH₂⁻, VII) in 50% dioxane–water at 25°C[55].

To take into account the variations in hard and soft character of metal ions and ligands, reference acids other than H⁺ have been used. Thus stability constants ($\log K$) for complex formation (ML) of a reference metal ion have been used as a measure of the donor properties of the ligand, L^-. These $\log K$ values have been correlated[57,58] with stability constants for the formation of $M'L$ using another metal ion (M') over a series of ligands, L^-, with varying donor properties.

Plots such as those in Fig. 7 fit the equation:

$$\log K = a\ pK_a + b \tag{6}$$

The slope, a, of the lines for all of the metal ions in the Figure is approximately 0.5. Although the positive slope indicates that the more basic ligands form the more stable complexes, the availability of H⁺ in aqueous solutions means that the more basic ligands will also bind more strongly to H^+.

Thus the two equilibria,

$$H^+ + L^- \rightleftharpoons HL, \; 1/K_a \tag{7}$$

$$M^{+m} + L^- \rightleftharpoons ML^{+(m-1)}, \; K \tag{8}$$

are both important in solution. Subtracting equation (7) from (8) gives the predominate equilibrium at low pH where the L^- concentration is very small.

$$HL + M^{+m} \rightleftharpoons ML^{+(m-1)} + H^+, \; K_a K \tag{9}$$

Thus the equilibrium constant for reaction (9) is $K_a K$. Taking K_a and K data from Fig. 7 for the reactions of Cu^{2+} with the weak base HCO_2^- and the strong base $CH_3CO_2^-$, the values of the equilibrium constants for reaction (9) are $(10^{-4.75})(10^{2.8}) = 10^{-1.95}$ for HCO_2^- and $(10^{-6.01})(10^{3.36}) = 10^{-2.65}$ for $CH_3CO_2^-$. Hence the equilibrium constants for complex formation (equation (9)) are actually lower for the more basic ligand, *i.e.*, more formate would be complexed to Cu^{2+} than acetate. It must be remembered that in aqueous solution, the amount of complex actually *present* in solution will not only depend upon the complex stability constant but also on the pK_a of the ligand.

With very basic ligands, it is sometimes necessary to make the solutions relatively basic, by adding OH^-, in order to provide enough unprotonated ligand for complexation to occur with the metal ion. Sometimes the OH^- concentration required exceeds the solubility product of the metal hydroxide, $M(OH)_2$, and it precipitates from solution which eliminates any possibility of forming a complex. This occurs in the reactions of Ni^{2+} and Zn^{2+} with 1,1,7,7-tetraethyldiethylenetriamine; this ligand is a very strong base but not a very strong ligand. At the pH where there is sufficient unprotonated triamine available for coordination to the metal ion, the concentration of OH^- is high enough to precipitate the metal hydroxide, $M(OH)_2$. This precipitation prevents these metal ions from forming complexes of this ligand in aqueous solution. On the other hand, Cu^{2+} forms a sufficiently stable complex of the ligand to be studied[59]. Solubilities of metal hydroxides are given in ref. 60.

For a series of ligands obeying equation (6), if a is less than 1, the amount of metal–ion–ligand complex actually present in solution will be less for the more basic ligands, as noted for those in Fig. 7. If $a = 1$, the amount of complex will not change with an increase in ligand pK_a; if a is greater than 1, an increase in ligand pK_a will also increase the amount of metal ion complex in solution.

Although it would be of interest to know what properties of the metal ions and ligands determine the values of a, insufficient data are available at present. Several authors[61,62] have attempted to account for those values which are known, but there seem to be no general conclusions. Values of a appear to range[62] from 0.5 (as in Fig. 7) to 1.5; many are near 1.0. As found for Cu^{2+} and Zn^{2+} shown in Fig. 7, a values are very similar for the first row transition metal ions but larger than those for Mg^{2+} or Ca^{2+}.

Such pK_a against $\log K$ plots have been of particular value in establishing the coordination of additional groups in the ligand. For example, the acetate derivative ligands, $HOCH_2CO_2^-$ and $CH_3CH_2SCH_2CO_2^-$, form much more stable complexes with Cu^{2+} than is expected from their pK_a values[43]. This suggests that the hydroxyl oxygen and sulfur atoms of the two ligands are coordinating to the metal ion to give the complexes unusually high stability. Similarly it was shown that the sulfur atom of $CH_3CH_2SCH_2CO_2^-$ does not coordinate to Mn^{2+} or Zn^{2+}.

Stability constants for the reaction of various metal ions with iminodiacetic acid derivatives, $RN(CH_2CO_2^-)_2$, have been correlated[32,63] with the pK_a for dissociation of H^+ from the nitrogen atom of the ligand. Non-coordinating R groups such as CH_3-, $(CH_3)_3C-$, and C_6H_5- were used to establish the pK_a against $\log K$ relationship. Stability constants of ligands with R groups which contained potential donor atoms were compared with the pK_a against $\log K$ plot. Such R groups as $HOCH_2CH_2-$, $CH_3OCH_2CH_2-$, $HSCH_2CH_2-$, $^-SCH_2CH_2-$, $NH_2CH_2CH_2-$, $^-O_2CCH_2-$, $NH_2(O)CCH_2-$, $N{\equiv}CCH_2-$, and $(CH_3)_3\overset{+}{N}CH_2CH_2-$ were used. Depending on the metal ion and the donor atom in the R group, some stability constants were much greater than expected from the pK_a of the ligand. It was assumed that this was caused by some interaction between donor atoms in the R group and the metal ions. It was found that R groups with soft donor atoms (*e.g.*, sulfur) generally coordinate to soft metal ions (*e.g.*, Cu^{2+}) and hard donor atoms (*e.g.*, O) to hard metal ions (*e.g.*, Mg^{2+}) [see Section II(A)(2)]. Similar stability constant studies[64] of $C_2H_5O_2CCH_2N(CH_2CO_2-)_2$ with various metal ions suggested a lack of interaction between the metal ion and the ester group. This information was used in the interpretation of the mechanism of ester hydrolysis as catalyzed by metal ions.

(2) Chelate effect

Chelate formation. Experimentally it is observed that complexes of chelating ligands are more stable than those of comparable monodentate ligands. For example, the stability constant for the complexation of Ni^{2+} by ethylenediamine(en) is larger than for complexation by two NH_3 molecules.

$$Ni^{2+} + en \rightleftharpoons Ni(en)^{2+}, \log K = 7.5 \tag{10}$$

$$Ni^{2+} + 2NH_3 \rightleftharpoons Ni(NH_3)_2^{2+}, \log K = 5.0 \tag{11}$$

Combining these equations gives

$$Ni(NH_3)_2^{2+} + en \rightleftharpoons Ni(en)^{2+} + 2NH_3, \log K = 2.5 \qquad (12)$$

$$\Delta H° = -1.9 \text{ kcal/mole}, \Delta S° = +6.2 \text{ cal/mole deg.}$$

Ethylenediamine and NH_3 are compared because their pK_a values (9.6 for $H_2NCH_2CH_2NH_3^+$ and 10.0 for NH_4^+) are very similar and differences in their basicities [see Section II(B)(1)] will not account for their different complex stabilities. This enhanced stability of chelates as compared to that of analogous monodentate ligands is known as the chelate effect[65].

For the reaction involving the displacement of two NH_3 ligands by en (equation (12)), it is clear that both the negative $\Delta H°$ and positive $\Delta S°$ contribute to the enhanced stability of the chelate complex. The formation of three moles of products as compared to two of reactants would give a positive translational entropy, which would contribute to the overall positive $\Delta S°$. Hence the chelate effect is, at least in part, caused by a favorable entropy of reaction. It should be noted, however, that the magnitude of $\Delta S°$ depends upon the standard state used in its evaluation. If unit mole fractions are used in place of the usual one molal standard states, the calculated entropies for reactions such as (12) are lower and, in fact, are quite close to zero. Thus the use of this standard state tends to minimize the translational entropy effect arising from the formation of 3 moles of products from 2 moles of reactants. Moreover, the dependence of $\Delta S°$ on the choice of standard states makes any quantitative interpretation of $\Delta S°$ values difficult.

It should also be noted in equation (12) that $\Delta H°$ is favorable (negative) for chelation. This is generally[65] attributed to repulsions between the two donor atoms when they approach each other in forming the complex. That is, for en the two NH_2 groups in the ligand presumably experience mutual repulsion before complex formation; in forming the complex little additional repulsion is generated. On the other hand with NH_3, there is no repulsion between the two NH_3 molecules in solution but on complexation they do repel each other. This largely electrostatic repulsion of the NH_3 groups in the complex makes the $\Delta H°$ less favorable. Regardless of how one accounts for the change in $\Delta H°$ on chelation, it, as well as $\Delta S°$, contributes substantially to the chelate effect[66].

Chelate ring size. In general, the stability of a chelate complex decreases[66] as the size of the chelate ring increases from 5 to 6 to 7 members. Hence in the series of metal-ion complexes of dicarboxylic acids, the most stable is

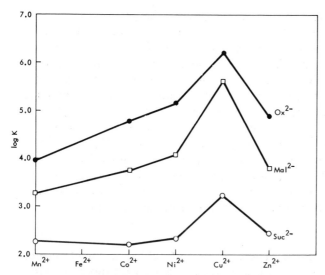

Fig. 8. Dependence of log K (at 25°C) on the size of the chelate ring. The bidentate ligands are oxalate (Ox^{2-}), malonate (Mal^{2-}) and succinate (Suc^{2-})[8].

that of the oxalate ion. The decreasing stability of 5, 6, and 7 membered chelate complexes are illustrated for oxalate, malonate, and succinate ions with the first row transition metal ions in Fig. 8. Stability data are given for other five and six-membered chelate complexes in Table V.

TABLE V

STABILITY DATA FOR FIVE AND SIX-MEMBERED CHELATE COMPLEXES[a] AT 25°C[8,50]

$Ni^{2+} + L^{-n} \rightleftharpoons NiL^{+(2-n)}$

	log K	ΔH(kcal/mole)	ΔS(cal/mole deg.)
$Ni^{2+} + Ox^{2-}$	5.2	0.15	24.2
$Ni^{2+} + Mal^{2-}$	4.1	1.88	25.0
$Ni^{2+} + Gly^-$	6.2	−4.14	14.4
$Ni^{2+} + \beta$-Ala$^-$	4.7	−3.81	10.2
$Ni^{2+} + en$	7.7	−9.05	4.0
$Ni^{2+} + pn$	6.3	——	——

[a] Abbreviations: oxalate (Ox^{2-}); malonate (Mal^{2-}); glycinate (Gly^-); β-alaninate (β-Ala$^-$); ethylenediamine (en); 1,3-propylenediamine (pn).

There are several trends that should be noted in this Table. First, the log K for five-membered rings is 1.1–1.5 log units greater than for the analogous six-membered ring complexes. Second, the difference in log K

values is caused by changes in both ΔH and ΔS. Third, the enthalpies become more exothermic with the ligand as follows: dicarboxylate $<$ amino acidate $<$ diamine. The ΔH associated with Ni^{2+}–carboxylate coordination is near zero or slightly endothermic, whereas amino group coordination gives negative ΔH values. Thus the ΔH of reaction greatly stabilizes complexes of ligands bearing amino donor groups as compared to those with carboxylate groups. Fourth, the stability of carboxylate complexes is largely determined by the large positive entropies. These decrease markedly from the dicarboxylates to the diamine ligands. The high positive ΔS for the dicarboxylate reactions is presumed to result in part from the release of solvent water molecules from the highly charged Ni^{2+} and Ox^{2-} (or Mal^{2-}) ions as they form the neutral complex $NiOx$. Lesser charge neutralization occurs with the mono-anionic amino acidates.

Number of chelate rings. In view of the well-established chelate effect for bidentate ligands it is not surprising that a ligand with more chelate rings will give even more stable complexes. This assumes that the geometry of the ligand and the metal ion allows the coordination of all the donor atoms. Figure 9 shows how log K for a series of amino-carboxylate ligands increases with the number of chelate rings and donor atoms: $Gly^- <$ $IMDA^{2-} < NTA^{3-}$. Likewise in the amine–ligand series, the stability constants increase as: en $<$ dien $<$ trien. Table VI summarizes ΔH and ΔS data for their complexes of Ni^{2+}.

TABLE VI

STABILITY DATA FOR MULTIDENTATE LIGANDS[a] WITH Ni^{2+} AT 25°C[8].

$Ni^{2+} + L^{-n} \rightleftharpoons NiL^{+(2-n)}$

	log K	ΔH(kcal/mole)	ΔS(cal/mole deg.)
$Ni^{2+} + Gly^-$	6.2	-4.14	14.4
$Ni^{2+} + IMDA^{2-}$	8.0	-5.05	20.0
$Ni^{2+} + NTA^{3-}$	11.3	-2.53	44.0
$Ni^{2+} + en$	7.7	-9.05	4.0
$Ni^{2+} + dien$	10.5	-11.85	8.5
$Ni^{2+} + trien$	13.7	-14.00	16.0

[a] See ligand abbreviations in Fig. 9.

As the number of donors in the ligand increases the ΔS of complex formation increases. This is expected because of the increasing number of H_2O ligands released from the Ni^{2+}. In addition to the entropy stabilization of the complexes, the ΔH values become more negative with increasing numbers of amine donor groups. In contrast, ΔH for the amino–carboxylate complexes varies only slightly and irregularly which perhaps results from the previously noted tendency of carboxylate ligands to give near zero or

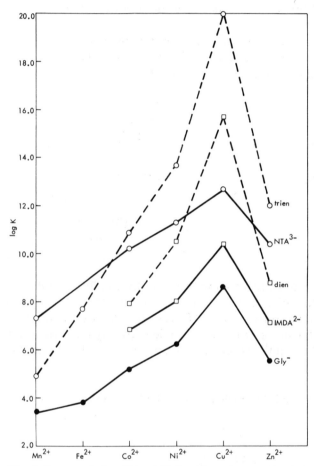

Fig. 9. Increase in log K (at 25°C) with an increase in the number of chelate rings. The multidentate ligands are $NH_2CH_2CO_2^-$, Gly$^-$; $NH(CH_2CO_2^-)_2$, IMDA^{2-}; $N(CH_2CO_2^-)_3$, NTA^{3-}; $NH_2CH_2CH_2NHCH_2CH_2NHCH_2CH_2NH_2$, trien[8].

positive enthalpies of coordination. These two series of ligands again demonstrate the observation that amino nitrogen donors make much larger contributions to the ΔH of complexation than do carboxylate donor groups; in contrast carboxylate groups are primarily responsible for the high positive ΔS values of amino-carboxylate ligand coordination. As will be noted later, these trends are particularly important for metal ion complexation by amino acids.

References pp. 98–101

(3) Macrocyclic effect

While the chelate effect has been known for decades, the stability of a complex containing a macrocyclic ligand of the type found in porphyrins has been essentially unknown. A major problem in determining these stability constants is the very slow rate at which metal ions and macrocyclic ligands react.

Recently Cabbiness and Margerum[67] measured stability constants for Cu^{2+} complex formation with the non-cyclic tetraamine ligand (2,3,2-tet; I) and a similar macrocyclic tetraamine ligand (tet a; II).

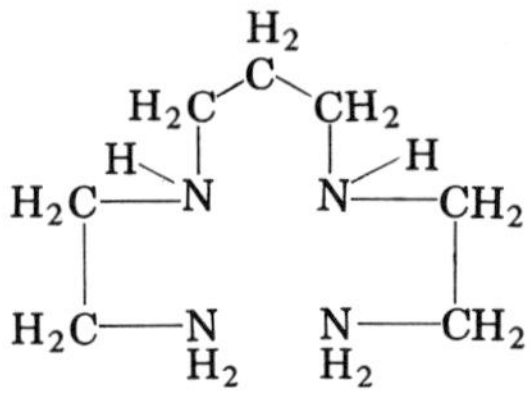

(I)

2,3,2-tet

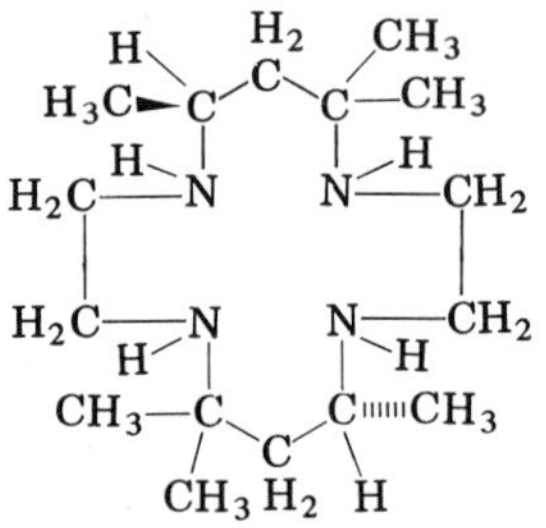

(II)

tet a

Because of the slow rate of reaction with tet a, the stability constant measurements were conducted at 50°, 75°, 100°, and 130°C. For comparison with other data, the stability constant at 25°C was obtained by extrapolation. This constant with those for 2,3,2-tet and trien (triethylenetetraamine) are given in Table VII. The stability constant of $Cu(tet\ a)^{2+}$ is 4.1 log units

TABLE VII

LOG K VALUES FOR Cu^{2+} COMPLEXES OF TETRAAMINE LIGANDS AT 25°C[67]

	log K
Cu^{2+} + tet a	28
Cu^{2+} + 2,3,2-tet	23.9
Cu^{2+} + trien	20.1

larger than that for non-cyclic $Cu(2,3,2\text{-tet})^{2+}$. The addition of a $-CH_2CH_2CH_2-$ bridge across the terminal $-NH_2$ groups of 2,3,2-tet to form tet a has enhanced the stability of the complex by more than would be expected from the chelate effect alone. Although little is really known about the origin of this enhanced stability, it is called the macrocyclic effect. In addition to favorable bonding in the metal complex, the configuration and solvation of the free macrocycle as compared to the non-cyclic ligand almost certainly contribute to the high stability of $Cu(tet\ a)^{2+}$.

Also of interest in Table VII is the difference in stability constants for $Cu(2,3,2\text{-tet})^{2+}$ and $Cu(\text{trien})^{2+}$. It has been pointed out[68] that the geometry of the ligand does not allow strain-free square planar coordination of trien to copper(II). Unlike trien which has a $-CH_2-CH_2-$ group connecting the two secondary nitrogen atoms in the ligand, the ligand 2,3,2-tet has a $-CH_2CH_2CH_2-$ group which apparently reduces the strain in coordinating to the Cu^{2+}. This is supported by the substantial increase in log K for 2,3,2-tet as compared to trien.

The macrocyclic effect has especial importance for the metalloporphyrins. Although very few stability constant data are available for this type of complex, a log K = 29 has been estimated[69] for the interaction of Zn^{2+} with the dianionic form of mesoporphyrin dimethyl ester. Only in complexes of macrocyclic ligands have such high values been observed. Stabilities of metalloporphyrins are discussed in more detail in Section III(D).

(4) Mixed (or ternary) complexes

As noted in the Introduction, stepwise stability constants usually decrease as the number of ligands on the complex increases (for an exception, see Fig. 7). Thus K_{MA}^{M} is larger than $K_{MA_2}^{MA}$ for the following complexation equilibria:

$$M + A \rightleftharpoons MA, \quad K_{MA}^{M} \tag{13}$$

$$MA + A \rightleftharpoons MA_2, \quad K_{MA_2}^{MA} \tag{14}$$

(The equilibrium constant will be specified by indicating the reactant complex as the superscript and the product complex as the subscript.) This trend is expected on the basis of statistical, steric, and electrostatic considerations.

If instead of using just one ligand A, a second ligand B is introduced, the possible equilibria increase substantially. In addition to those involving the simple formation of binary complexes such as MA, MA_2, MB, and MB_2, there is also the related displacement reaction:

$$MA + B \rightleftharpoons MB + A \tag{15}$$

This type of reaction would be important when one or both of the ligands, A and B, occupies all or almost all of the metal-ion coordination sites — as for example where $A = NH_2CH_2CO_2^-$ and $B = EDTA^{4-}$. In this case the hexadentate nature of $EDTA^{4-}$ would exclude the simultaneous coordination of glycinate. Hence there would be no formation of mixed or ternary complexes with two or more different ligands. Equilibrium constants for reactions of the type in equation (15) are simple ratios of the stability constants for the formation of MB and MA. The factors involved in these reactions have already been discussed.

We now turn to the problem of mixed complexes. (The terms mixed or ternary will be used interchangeably in describing these complexes.) When A and B have a sufficiently small number of donor atoms so that both ligands may be coordinated to a metal-ion simultaneously, mixed complex formation is possible. For the case in which one mole of M reacts with one mole each of A and B, a possible equilibrium is:

$$MA + B \rightleftharpoons MAB, \quad K_{MAB}^{MA} \tag{16}$$

A knowledge of K_{MAB}^{MA}, K_{MA}^{M} and K_{MB}^{M} allows the calculation of the other mixed stability constant, K_{MAB}^{MB}:

$$MB + A \rightleftharpoons MAB, \quad K_{MAB}^{MB} = K_{MAB}^{MA} K_{MA}^{M} / K_{MB}^{M} \tag{17}$$

Equilibrium constants such as those for reactions (16) and (17) have not been studied extensively, but the importance of ternary complexes to biochemical systems provides a strong impetus for increasing interest in this area. A recent comprehensive review[70] of mixed complexes summarizes most of the data which are now available.

It is of particular interest to determine how, in equation (16), the coordinated ligand A affects the coordination tendency of ligand B; *i.e.*

TABLE VIII

STABILITY CONSTANTS FOR THE FORMATION OF MIXED COMPLEXES[a]
(at 37°C, and $\mu = 0.15$ KNO$_3$ [71-73])

	MA	B	Co^{2+} log K	Ni^{2+} log K	Cu^{2+} log K	Zn^{2+} log K
I	M^{2+}	+ en	5.30	6.98	10.18	5.53
	Men^{2+}	+ en	4.28	5.81	8.77	4.75
	Mhm^{2+}	+ en	4.42	5.78	8.15	5.34
	MSer^{+}	+ en	4.84	6.57	9.31	5.39
II	M^{2+}	+ hm	4.89	6.60	9.28	5.03
	Men^{2+}	+ hm	4.01	5.40	7.86	4.84
	Mhm^{2+}	+ hm	3.54	4.84	6.30	4.78
	MSer^{+}	+ hm	4.41	5.77	8.71	5.20
III	M^{2+}	+ Ser^{-}	4.20	5.21	7.57	4.47
	Men^{2+}	+ Ser^{-}	3.74	4.80	6.70	4.33
	Mhm^{2+}	+ Ser^{-}	3.72	4.38	6.94	4.64
	MSer^{+}	+ Ser^{-}	3.36	4.38	6.45	3.84

[a] en = ethylenediamine; hm = histamine; Ser^{-} = serine anion.

how does K_{MAB}^{MA} vary with the nature of A? In Table VIII are some recent data[71-73] for mixed complexes of ethylenediamine, histamine, and the serine anion. There are several trends which should be noted. First, like the first stability constants, K_{MA}^M, all of the mixed stability constants, K_{MAB}^{MA}, follow the Irving–Williams order [see Section II(A)(3)] of stability as a function of the metal ion, $i.e.$, $Co^{2+} < Ni^{2+} < Cu^{2+} > Zn^{2+}$. In any row of the Table this order is found. This trend appears to hold for almost all stability constants, whether for binary or ternary (mixed) complexes.

Second, with two exceptions ($i.e.$, $Ni(en)^{2+}$ and $Cu(en)^{2+}$ + en) the stability constants for the binary complexes, $K_{MB_2}^{MB}$, are smaller than the stability constants for any of the mixed complexes, K_{MAB}^{MA}. For example, in the en reactions (II in Table VIII) the $Co(en)^{2+}$ + en stability constant, $K_{MB_2}^B$, is lower than constants for the reactions of $Co(hm)^{2+}$ or $Co(Ser)^+$ with en, K_{MAB}^{MA}. Hence it appears that the addition of another B ligand to a complex MB which already bears a B ligand is more difficult than adding B to a complex MA which contains a different ligand, A. This group of data (Table VIII) thus illustrates the unusual stability of ternary complexes as compared to binary complexes. Although it is not clear why this should be so, it is an observation which continues to reappear and will be discussed in more detail below.

Third, the stability constants, K_{MB}^M, for the reaction of the metal ions, M^{2+}, with different ligands B decrease with B in the order: $en > hm > Ser^-$. Moreover, this same order is generally followed for the addition of B to a given complex MA. Hence the presence of A has not altered the order of complexing ability of the B ligands. Because of the many factors which determine the complexing ability of ligands it is not possible to explain the observed trend. (The few deviations from this trend occur for the formation of binary complexes where the stability constants, $K_{MB_2}^{MB}$, are unusually low.)

Fourth, for the reactions where B = en or hm, the stability constants, K_{MAB}^{MA}, decrease with MA in the order: $M^{2+} > MSer^+ > Men^{2+} > Mhm^{2+}$. Exceptions to this order again occur for the low binary complex stability constants, $K_{MB_2}^{MB}$. Again it is not clear why the above order is followed. A simple electrostatic argument would suggest that $MSer^+$ would form less stable mixed complexes than any of the +2 charged species; even this is not observed.

In another study[75], stability constants for the complexation of glycinate anion by a series of Ni^{2+} complexes have been determined (Table IX). Again the constants do not appear to correlate well with the charge on the complex. Jackobs and Margerum[75] have used seven parameters to account empirically for variations of K in these as well as some related equilibria. Among their parameters are the number of bound nitrogen atoms and the number of bound carboxylate groups in the initial Ni^{2+} complex, and the charges on the complex and on Gly^-. All three of these factors contribute importantly to the magnitudes of the constants.

TABLE IX

GLYCINE ANION COORDINATION BY NICKEL COMPLEXES[a]
(at 25°C, $\mu = 0.5$ (NaCl)[75])

	$\log K^{MA}_{MAB}$
$Ni^{2+} + Gly^-$	5.77
$Nidien^{2+} + Gly^-$	5.13
$Ni(NTA)^- + Gly^-$	4.89
$NiGly^+ + Gly^-$	4.80

[a] Abbreviations: $Gly^- = NH_2CH_2CO_2{}^-$; dien $= HN(CH_2CH_2NH_2)_2$; $NTA = N(CH_2CO_2{}^-)_3$

Extensions of this approach might be helpful in elucidating factors which determine stability constants of mixed complexes.

Although we have spoken of the formation of mixed complexes as reactions which were independent of other equilibria in solution, mixed complexes are, of course, in equilibrium with the corresponding binary complexes. This equilibrium may be written as:

$$MA_2 + MB_2 \rightleftharpoons 2MAB, \quad K = [MAB]^2/[MA_2][MB_2] \tag{18}$$

For the mixed complexes of ethylenediamine, histamine, and the serine anion in Table VIII, values of the dimensionless mixing constant K may be calculated. These are recorded in Table X. Probably the most interesting

TABLE X

MIXING CONSTANTS, K, FOR REACTION (18)[a]

	Co^{2+} $\log K$	Ni^{2+} $\log K$	Cu^{2+} $\log K$	Zn^{2+} $\log K$
$M(en)(hm)^{2+}$	0.61	0.53	1.53	0.65
$M(en)(Ser)^+$	0.94	1.18	0.77	1.13
$M(hm)(Ser)^+$	1.23	0.95	2.94	1.22

[a] Same conditions, abbreviations and references as in Table VIII.

aspect of these values is the fact that they are all positive, indicating that all equilibria favor the formation of the mixed complexes. The unusual stability of the mixed complexes was noted earlier and appears to be a quite general phenomenon, although there are exceptions[70].

Undoubtedly one reason why equilibrium (18) lies to the right is the 2:1 statistical advantage of MAB formation over that of MA_2 or MB_2. In equation (18), this leads to the conclusion that $\log K$ will be 0.60 (= log 4) if only statistics were responsible for the mixing constant. Thus the log K values in Table X which lie near 0.6 may be accounted for in terms of statistics only. Mixing constants have similarly been calculated for more complicated mixed complexes[76]. In general, however, these constants differ from the statistical values.

The paucity of stability constant data for mixed complexes and the dependence of these data on ionic strength and other solution conditions[70] leaves this area of metal complex stabilities as one which is at present poorly understood.

(5) Optically active ligands and stereoselectivity

Related to the discussion of mixed complexes is the consideration of mixed complexes with two ligands which differ by the configuration around an asymmetric carbon atom. For example, one might compare equilibrium constants for the following reactions involving optically active amino acidates (*i.e.*, anions of amino acids) of differing chiralities:

$$Cu(L\text{-}A)^+ + L\text{-}A^- \rightleftharpoons Cu(L\text{-}A)_2 \qquad (19)$$

$$Cu(L\text{-}A)^+ + D\text{-}A^- \rightleftharpoons Cu(L\text{-}A)(D\text{-}A) \qquad (20)$$

It is found experimentally[77] that stability constants for equations (19) and (20) are identical when the amino acidates are alanine, phenylalanine, valine or proline. For the anion ligand of asparagine, the equilibrium constant[78] for reaction (19) (log K = 6.45) is reported to be larger than for reaction (20) (log K = 5.85). However, a more recent study[78a] of asparagine with Cu(II), Ni(II) and Co(II) indicates that stability constants for reactions (19) and (20) are identical. Identical stability constants also were found for the reactions of optical isomers of glutamic acid, aspartic acid, and glutamine with Cu(II), Ni(II), and Co(II).

In constrast, the equilibrium constant for the reaction[78b],

$$Co(D\text{-}Hist)_2 + Co(L\text{-}Hist)_2 \rightleftharpoons 2Co(D\text{-}Hist)(L\text{-}Hist) \qquad (21)$$

involving Co(II) and D,L-histidine is log K = 1.1, as determined by proton magnetic resonance. This stereoselectivity favoring the mixed Co(D-Hist)-(L-Hist) isomer was confirmed later[78c] by potentiometric investigations. A similar stereoselectivity favoring Ni(L-Hist)(D-Hist) was also observed. Crystals of Co(D-Hist)(L-Hist) have been isolated from solutions of Co(II) and D,L-histidine. The structures of both Co(D-Hist)(L-Hist)[79] and Co(L-Hist)$_2$[80] have been established by X-ray studies. The histidine ligands are tridentate in both structures, but the differences in configurations in the optically active ligands cause substantial changes in the relative positions of the histidine molecules in the coordination sphere. It is interesting that crystals of only Ni(L-Hist)$_2$ and Ni(D-Hist)$_2$, and not Ni(D-Hist)(L-Hist) may be isolated[81] from solutions of Ni^{2+} and D,L-histidine (see page 146).

Stability constants for the coordination of optically active amino acidates to (L-valine-*N*-monoacetato)copper(II), Cu(L-ValMA), have also been determined[82]. These values are given in Table XI for the reaction:

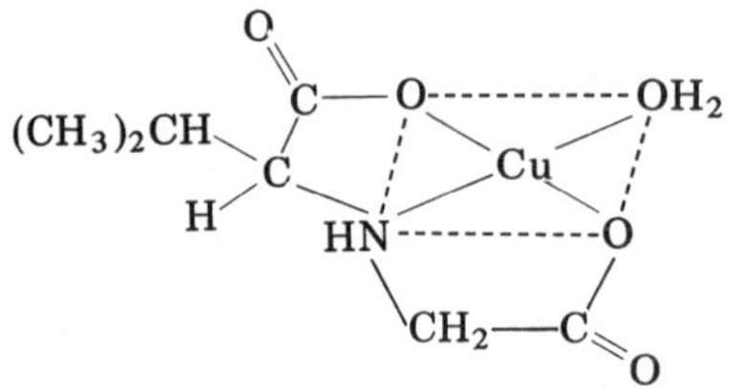

$$\text{Cu(L-ValMA)} + \text{L- or D-A}^- \rightleftharpoons \text{Cu(L-ValMA)(A)}^- \qquad (22)$$

TABLE XI

STABILITY CONSTANTS FOR THE REACTION OF Cu(L-ValMA) WITH OPTICALLY ACTIVE AMINO ACIDATES ACCORDING TO EQUATION (22) AT 25°C[82]

	log K	
A$^-$	L-A$^-$	D-A$^-$
Leu$^-$	5.74	4.93
PhAla$^-$	5.52	4.92
Ala$^-$	5.29	4.74
Ser$^-$	5.55	5.03

The favored coordination of L-amino acidates as compared to the D-enantiomers must be explained in terms of the geometry of the Cu(L-ValMA)(A)$^-$ complexes. Little is known of their structure but the magnitudes of the stability constants suggest that the amino acidate is a bidentate ligand. Although it is possible to propose structures in which L-A$^-$ coordination would be sterically preferred over D-A$^-$, there is no independent evidence to support such structures. It has also been shown[82] that Cu(L-ValMA) stereoselectively coordinates to optically active amino acid esters (presumably through the N-atom only). As a result, Cu(L-ValMA) stereoselectively catalyzes the hydrolysis of D- and L-amino acid esters.

The vast majority of work[83,84] on stereoselectivity in metal-ion complexes has come from studies of Co(III). But because of the inertness of Co(III) complexes, few equilibrium studies have been carried out; thus it has not been possible to determine whether many examples of stereoselectivity in Co(III) chemistry are due to kinetic or thermodynamic effects. Obviously much remains to be done on stereoselectivity of metal-ion complexes, particularly those of labile metal ions.

III. SOME BIOCHEMICAL LIGANDS

(A) Amino acids

Although stability constants for the complexation of glycine by metal ions have been discussed earlier in this chapter, it is of interest to examine stability constants of other amino acids. A review by Gurd and Wilcox[60] summarized the early data in this area.

Although the normal mode of amino acid coordination is through the amino and carboxylate groups, Childs and Perrin[85] have reported a stability constant for coordination of the glycine zwitterion through the carboxylate group only to Cu^{2+}.

$$Cu^{2+} + {}^-O_2CCH_2NH_3{}^+ \rightleftharpoons Cu(O_2CCH_2NH_3)^{2+}, \log K = 1.53$$

The $\log K$ value of 1.53 is smaller than 1.84 for the Cu^{2+} reaction with $CH_3CO_2{}^-$; presumably the lower basicity [Section II(B)(1)] of the glycine zwitterion accounts for its lower stability constant. To be sure, these are very small constants when compared to others which we have discussed. In fact, other researchers[86,87] have found no evidence for complex formation between metal ions and amino acid zwitterion.

The amino acidate anions form stable complexes of a wide variety of metal ions, particularly those of the transition elements. In most cases the amino acidates act as bidentate ligands, coordinating through the amino and carboxylate groups. Among the amino acids which appear to behave bidentately toward such metal ions as Ni^{2+}, Cu^{2+}, and Zn^{2+} are glycine, alanine, valine, proline, leucine, sarcosine, tyrosine[88], serine[87], threonine[87], lysine[89,90], arginine[22,89,91], tryptophan, asparagine, and methionine. The similarity of their stability constants (Table XII) in the Zn^{2+}, Fe^{2+} and Fe^{3+} complexes suggests that they all coordinate only through the amino and carboxylate groups. The unusually low stability constants for methionine and arginine (whose guanidino group is protonated) probably reflect the lower basicity of the $-NH_2$ group in these amino acids.

It should be noted that methionine may coordinate to soft metal ions[91a] such as Pd(II) and Hg(II) through the sulfur atom. Perrin[92,93] has correlated the Fe^{2+} and Fe^{3+} $\log K$ values of the amino acids in Table XII with their $-NH_2$ group pK_a values [see Section II (B)(1)]. These data fit the equation (6) where $a = 0.4$ for Fe^{2+} and $a = 1.8$ for Fe^{3+}. This large difference in slope a indicates that $\log K$ is a much more sensitive function of the pK_a for Fe^{3+} than for Fe^{2+}. On the basis of deviations from $\log K$ against pK_a plots, Perrin suggests that the aspartic acid dianion exhibits substantial tridentate character in the Fe^{2+} and Fe^{3+} complexes. From the data in Table XII, this appears to be true for Zn^{2+} also. Other data[94,95]

TABLE XII

STABILITY CONSTANTS FOR THE REACTIONS OF AMINO ACIDATES WITH Zn^{2+}, Fe^{2+} AND Fe^{3+}

$$M^{+m} + A^- \rightleftharpoons MA^{+(m-1)}$$

		log K		
Amino acidate	pK_a of -NH$_2$	$Zn^{2+,a}$	$Fe^{2+,b}$	$Fe^{3+,c}$
Arginine	9.21	4.2	3.2	8.7
Methionine	9.13	4.4	3.2	9.1
Threonine	8.86	—	3.3	8.6
Asparagine	8.79	—	3.4	8.6
Tryptophan	9.43	—	3.4	9.0
Serine	9.12	—	3.4	9.2
Valine	9.59	5.0	3.4	9.6
Leucine	9.62	4.9	3.4	9.9
Alanine	9.79	5.1	3.5	10.4
Sarcosine	10.02	—	3.5	9.7
Glycine	9.76	5.0	3.8	10.0
Proline	10.52	—	4.1	10.0
Glutamic acid	9.54	5.4	3.5	12.1
Aspartic acid	9.56	5.8	4.3	11.4
Histidine	9.20	6.6	5.9	4.7
Cysteine	10.78	9.9	6.2	—

[a] Ref. 22.
[b] Ref. 92.
[c] Ref. 93.

suggest that aspartic acid is tridentate toward Cu^{2+} and Ni^{2+}; X-ray structural studies[96] of aspartic acid complexes of Co^{2+}, Ni^{2+}, and Zn^{2+} clearly indicate tridentate bonding in the solids. In contrast X-ray studies[96] show that glutamic acid coordinates to one metal ion via one -NH$_2$ and one -CO$_2^-$ group, whereas the second -CO$_2^-$ binds to a second metal ion (p. 129). In solution evidence[94,95] suggests that the second -CO$_2^-$ dangles free in solution. Except for its complexes of Fe^{3+} (Table XII), stability constants suggest that glutamic acid behaves primarily as a simple bidentate amino acid.

Stability constants for the last two amino acids, histidine and cysteine, in Table XII indicate that they do not coordinate in the same manner as the other amino acids. The similarity of the stability constants for cysteine and the anion of mercaptoethylamine, $NH_2CH_2CH_2S^-$ (e.g., log K = 9.9 for Zn^{2+}) strongly suggests that the first row transition metal ions coordinate to cysteine through the -NH$_2$ and -S$^-$ groups, and the -CO$_2^-$ is uncoordinated.

More definitive structural work indicates that this is also the mode of bonding in Co^{3+} complexes[97]. Spectral studies[98] on Fe^{3+} complexes of cysteine indicate that the carboxylate group is probably also involved in coordination under certain conditions. Like cysteine, the related ligand penicillamine, $^-SC(CH_3)_2CH(NH_2)CO_2{}^-$, coordinates to Ni^{2+} and Zn^{2+} through the $-NH_2$ and $-S^-$ groups. Its log K values are very similar to those of cysteine. Penicillamine and cysteine are among the few ligands[99] in which the second stability constant, K_2, is larger than the first, K_1.

The coordination of histidine to transition metal ions is a somewhat more complicated system, which is considered in Chapter 4. Complexes of Cu^{2+}, some of which contain the neutral protonated histidine ligand and others containing the anionic histidine ligand, are discussed in that chapter.

Enthalpies and entropies for amino acid complex formation have been determined for a variety of amino acids. Some of these data are given in Table XIII. They show that ΔG, ΔH, and ΔS are very similar for all the

TABLE XIII

ΔG, ΔH AND ΔS VALUES FOR THE REACTION $M^{2+} + A^- \rightleftharpoons MA^+$ AT 25°C (A$^-$ IS AN AMINO ACIDATE)

Amino acidate	Cu^{2+}			Ni^{2+}		
	ΔG (kcal/ mole)	ΔH (kcal/ mole)	ΔS (cal/mole deg)	ΔG (kcal/ mole)	ΔH (kcal/ mole)	ΔS (cal/mole deg)
Glycine[a]	−11.71	−6.76	16.6	−8.43	−4.14	14.4
Serine[b]	−10.71	−5.51	17.5	−7.43	−3.76	12.3
Threonine[b]	−10.83	−5.56	17.7	−7.45	−3.81	12.2
Tyrosine[c]	−10.80	−5.42	18.1			

[a] Ref. 50.
[b] Ref. 87.
[c] Ref. 88.

listed amino acids. As was noted in a previous section [Section II (B)(2)], the enthalpies of these reactions are negative and largely determined by the amino nitrogen coordination. Thus the observed enthalpies in Table XIII are roughly 5.8 kcal/mole for Cu^{2+} and 3.9 for Ni^{2+}. If these values are indeed dominated by $-NH_2$ coordination, they should be similar to $\frac{1}{2}\Delta H$ for the formation[8] of $Cu(en)^{2+}$, 6.5 kcal/mole, and $Ni(en)^{2+}$, 4.5 kcal/mole, as in fact they are. Likewise, it has been noted [Section II (B)(2)] that the positive ΔS of these reactions is dominated by coordination of the carboxylate group in the amino acidate. Indeed the entropies for the formation of CuA^+, 17.5 e.u., and NiA^+, 12.9 e.u., are similar to

$\frac{1}{2}\Delta S$ for the formation[8] of the 1:1 oxalate complexes: Cu(Ox), 14.1 e.u., and Ni(Ox), 12.1 e.u.

Similar correlations of thermodynamic data for the formation of the bis(amino acidate) complexes[100], $M^{2+} + 2A^- \rightleftharpoons MA_2$, have also been made. In addition, it was found that the ΔH (21.3 kcal/mole) for the formation of Cu(Hist)$_2$, in which both amino and imidazole nitrogen atoms are presumably coordinated, could be closely approximated by summing ΔH for Cu(Gly)$_2$, 12.8 kcal/mole, and $\frac{1}{2}\Delta H$ for Cu(2,2'-dipyridyl)$_2$$^{2+}$, 8.3 kcal/mole. This approximation was also found to hold for the analogous Ni^{2+} and Zn^{2+} complexes. Finally, it was noted that for different metal ions the enthalpies of reaction with amino acidates generally follow the Irving–Williams order [see Section II (A)(3)].

(B) Peptides

Simple di-, tri-, and tetra-peptides form complexes with transition metal ions. One of the most thoroughly studied systems is that of Cu^{2+} with glycylglycine. Extensive investigations have shown that the mono-anion, glygly$^-$, forms a 1:1 complex with Cu^{2+} below approximately pH 4. A proton dissociates from this complex with pK_a = 4.38. These two equilibria are summarized below:

$$Cu^{2+} + glygly^- \rightleftharpoons Cu(glygly)^+, \log K = 5.42$$

$$Cu(glygly)^+ \rightleftharpoons Cu(glygly-H) + H^+, \log K_a = -4.38$$

These Cu(II) complexes, along with their stabilities and structures, are discussed fully in Chapter 4, page 121 *et seq.* and Tables I–IV. Stability constant studies have been carried out on analogous glycine peptide complexes of Co(II)[101], Ni(II)[102–104], and Pd(II)[90]. The ease of ionization of the peptide proton increases in the order: Co(II) < Ni(II) < Cu(II) < Pd(II). Studies[105] of Zn^{2+} with a synthetic model peptide, *N,N'*-diglycylethylenediamine, indicate that it is ineffective in promoting peptide proton ionization.

Attempts have been made to determine metal-ion binding sites in large proteins by comparing stability constants for the interaction of the protein and various metal ions with stability constants of model ligands with the same metal ions. Perhaps the best characterized example is that of the metalloenzyme carboxypeptidase A, CPA, which in nature contains a Zn^{2+} ion and approximately 300 amino acid residues. Stability constants for the coordination of Zn^{2+} and several other metal ions by apoCPA (the enzyme with the metal ion removed) have been determined and compared with those of model ligands. Together with spectral and other chemical data, the stability constants[106] suggested that the Zn^{2+} was coordinated to one –NH$_2$ and one –S$^-$ group. Because of the large number of possible coordination sites and the possibility that different metal ions may chose different sites[107], it was recognized that the stability constant results could be in error. With the recent refinement of the X-ray data[108] taken on CPA, it is now clear that this is true. In fact, the Zn^{2+} is coordinated to two imidazole groups of histidine residues and a carboxylate group of a glutamic acid residue. Based on equilibrium studies of metal ions with glycine peptides as noted above, it might be anticipated that the metal ion would bind to the terminal $-NH_2$ of the protein. This is also not observed.

The difficulties associated with comparing stability constants of proteins and relatively simple model ligands have also been illustrated by studies of sperm whale myoglobin[109,110] and bovine serum albumin[111,112]. The lack of stability constant data for the coordination of metal ions by small peptides bearing coordinating side chains (*e.g.*, histidine, lysine, or glutamic acid residues) has greatly retarded our understanding of the much more complicated interactions of metal ions with proteins.

(C) Nucleotides and nucleic acids

Nucleotides consist of three structural units — the purine or pyrimidine base, the ribose unit, and a mono-, di- or tri-phosphate group. The structure of the nucleotide adenosine triphosphate (ATP) is shown below with each of these units labeled.

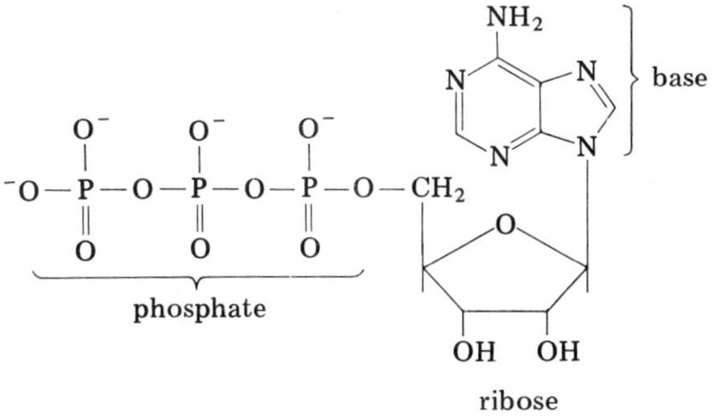

While it is known that nucleotides form rather stable complexes with metal ions, the site of coordination is difficult to establish unambigously, although a number of strides in that direction have recently been made. The poor ligand properties of sugars toward transition metal ions as well as alkali and alkaline earth metal ions suggests that the ribose portion of nucleotides will not generally be a strong site of coordination. The known, but weak, coordinating ability of nucleosides[42] (a pyrimidine or purine base bound to a ribose residue) to metal ions leaves the phosphate group as the strongest site for metal binding in nucleotides. The coordination sites are discussed in detail in Chapter 33.

The nucleotide which has been studied most extensively[113] is adenosine triphosphate(ATP); it contains four ionizable hydrogens on the triphosphate chain of the molecule.

The -4 anion, ATP^{4-}, and the monoprotonated form, $ATPH^{3-}$, both form metal-ion complexes in solution. The stability constants for both the ATP^{4-} and the $ATPH^{3-}$ ligands increase with the metal ion in the order[114]: Ba(II) < Sr(II) < Mg(II) < Co(II) < Mn(II) < Zn(II) < Ni(II) < Cu(II). Except for the reversal of Co(II) and Mn(II), this is the usual Irving-Williams order. A variety of experimental techniques[42] have inferred the coordination of these metal ions (except Cu^{2+}) to the triphosphate portion of ATP. This is supported by stability constants of analogous nucleotides with different purine or pyrimidine bases; these constants are largely independent of the base in the molecule. Moreover, the stability constants increase as the number of phosphates in the nucleotide increase. Thus for adenosine monophosphate (AMP^{2-}), adenosine diphosphate (ADP^{3-}), and ATP^{4-}, the log K values for the reaction with Cu(II) increase as follows: AMP^{2-} (3.2) < ADP^{3-} (5.9) < ATP^{4-} (6.1). The values for ADP^{3-} and ATP^{4-} are similar to those for monoprotonated polyphosphates[115]: $HP_2O_7^{3-}$ (5.4) < $HP_3O_{10}^{4-}$ (5.7).

Although the trend in stability constants follows a familiar order, the differences in log K values are very small; for example with ATP^{4-}, they are Mg(II), 4.2; Co(II), 4.7; Mn(II), 4.8; Zn(II), 4.8; Ni(II), 5.0; Cu(II), 6.1. The similarity of the constants has been interpreted[113,114] to mean that the phosphate groups are not coordinated directly to the metal ion, but a hydrated ion pair is formed whose stability largely depends on only the charge of the metal ion. The somewhat higher stability of the Cu(II) complex and other spectral data suggest that Cu(II) binds somewhat to the adenine base as well as to the phosphate chain. All of the other metal ions appear to bind only to the phosphate chain. A model ligand, 2-pyridylmethyl phosphate, containing pyridine and phosphate donor groups, likewise coordinates to metal ions only through the phosphate group, except for Cu(II) which binds to the pyridine nitrogen atom as well[116].

The interaction of metal ions with polynucleotides or nucleic acids, such as RNA and DNA, is much the same as observed with the nucleotides[42,117], as will be discussed in Chapter 34.

(D) Porphyrins

Porphyrins[118] are tetrapyrrole macrocycles which lose two protons on coordination to metal ions. The tetradentate dianionic ligand, P^{2-}, forms metal-ion complexes[119] which are basically planar[120] but because of some flexibility may be somewhat distorted.

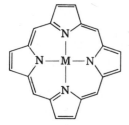

metalloporphyrin

Only one stability constant for the formation of a metalloporphyrin has been reported[69]: this was for the reaction of Zn(II) with mesoporphyrin dimethylester, P^{2-}, to form ZnP. The log K value is approximately 29, a figure which is only achieved by macrocyclic ligands [see Section II (B)(3)]. The paucity of equilibrium data for these reactions is undoubtedly due to their extreme slowness, months often being required to reach equilibrium at room temperature.

In the absence of stability constants for metalloporphyrin formation, a variety of techniques such as metal ion displacement reactions, reactions involving displacement of the metal ion by acid, and spectroscopic studies have been used[121] to establish a qualitative order of stability: $Ag(I)_2 <$ $K_2 < Ba(II) < Na_2 < Li_2 < Sn(II) < Cd(II) < Mg(II) < Zn(II) < Cu(II) <$ $Ag(II) < Co(II) < Ni(II) < Pd(II) < Pt(II)$. It has been suggested[121] that the unusually low stability of the Ba(II) complex is due to its large size which prevents it from fitting into the planar ring. The reason why the stabilities of the first row transition elements do not follow the Irving-Williams order [see Section II(A)(3)] is not clear.

The square planar metalloporphyrins, as well as the related phthalocyanine[122] complexes, have a tendency to add one or two monodentate ligands to give five or six coordinate complexes. In equilibrium studies of Zn(II), Cd(II) and Hg(II) complexes[123,124] of $\alpha,\beta,\gamma,\delta$-tetraphenylporphyrin with substituted pyridines, one pyridine ligand is added to the metal ion. The analogous Cu(II) and Ni(II) complexes[125], however, have only a small tendency to add pyridine. Log K values are similar for Zn(II) and Cd(II) but are lower for Hg(II); with pyridine they are 3.78 (Zn), 3.51 (Cd) and 1.21 (Hg). With all three metal ions, the log K values increase with increasing basicity of the pyridine, and plots of log K against pK_a of the pyridine are linear [see Section II(B)(1)]. Unfortunately a similar correlation is not observed[126] in the formation of bis-pyridine complexes of several different Fe(II) porphyrins.

The introduction of electron-releasing or -withdrawing groups into the porphyrin ring also alters the tendency of metalloporphyrins to add ligands. In general[118], electron-releasing groups diminish coordination of amines whereas withdrawing groups enhance coordination. This has been particularly clearly demonstrated[127] for a series of substituted deutero-porphyrin IX dimethyl ester complexes of Ni(II). These metalloporphyrins react with two moles of piperidine according to the equation, NiP + 2L $\rightleftharpoons$ NiPL$_2$. Log K values ranged from –2.0 for the 2,4-diethyl substituted metalloporphyrin to 0.15 for the 2,4-diformyl complex. The basicity (pK_a) of the diethyl porphyrin (5.8) is much greater than that of the diformyl derivative (2.8). Thus the more basic the porphyrin the less stable the NiPL$_2$ complex. In fact there is an inverse linear correlation between log K and the pK_a of the porphyrin.

REFERENCES

1 R. E. Tapscott, R. L. Belford and I. C. Paul, *Coord. Chem. Rev.*, 4 (1969) 323.
2 D. R. Fitzwater and R. E. Rundle, *Z. Krist.*, 112 (1959) 362.
3 E. B. Hunt, R. E. Rundle and A. J. Stosick, *Acta Cryst.*, 7 (1954) 106.
4 J. E. Powell and D. L. G. Rowlands, *Inorg. Chem.*, 5 (1966) 819.
5 T. Moeller, D. F. Martin, L. C. Thompson, R. Ferrus, G. R. Feistel and W. J. Randall, *Chem. Rev.*, 65 (1965) 1.
6 R. J. Hinchey and J. W. Cobble, *Inorg. Chem.*, 9 (1970) 917.
7 D. P. Graddon, *J. Inorg. Nucl. Chem.*, 5 (1958) 219.
8 G. H. Nancollas, *Interactions in Electrolyte Solutions*, Elsevier Publishing Co., Amsterdam, 1966.
9 M. M. Jones, *Elementary Coordination Chemistry*, Prentice-Hall, Englewood Cliffs, N.J., 1964, p. 333.
10 S. Fronaeus, in *Technique of Inorganic Chemistry*, Vol. I, Interscience Publishers, New York, 1963, p. 1.
11 F. J. C. Rossotti and H. Rossotti, *The Determination of Stability Constants*, McGraw-Hill, New York, 1961.
12 P. J. Lingane and Z. Z. Hugus, Jr., *Inorg. Chem.*, 9 (1970) 757.
12a M. T. Beck, *Chemistry of Complex Equilibria*, Van Nostrand Reinold, London, 1970.
13 D. J. Eatough, *Anal. Chem.*, 42 (1970) 635.
14 D. E. Goldberg, *J. Chem. Educ.*, 39 (1962) 328.
15 R. J. Angelici, *Synthesis and Technique in Inorganic Chemistry*, W. B. Saunders Co., Philadelphia, 1969, p. 105.
16 D. R. Crow and J. V. Westwood, *Quart. Rev.*, 19 (1965) 57.
17 J. Schubert, *Methods of Biochemical Analysis*, 3 (1956) 247.
18 Y. Marcus, *Ion Exchange*, 1 (1966) 101.
19 Y. Marcus and A. S. Kertes, *Ion Exchange and Solvent Extraction*, John Wiley and Sons, Inc., New York, 1969.
20 F. Helfferich, *Ion Exchange*, McGraw-Hill, New York, 1962, p. 221.
21 L. Johansson, *Coord. Chem. Rev.*, 3 (1968) 293.
22 L. G. Sillén and A. E. Martell, *Stability Constants of Metal-Ion Complexes*, Special Publication Nos. 17 and 25, The Chemical Society, London, 1964 and 1971.
23 K. B. Yatsimirskii and V. P. Vasilev, *Instability Constants of Complex Compounds*, Consultants Bureau, New York, 1960.
24 F. Basolo and R. G. Pearson, *Mechanisms of Inorganic Reactions*, 2nd edn., John Wiley and Sons, New York, 1967.

25 F. J. C. Rossotti in J. Lewis and R. G. Wilkins, *Modern Coordination Chemistry*, Interscience Publishers, New York, 1960, p. 1.

26 G. Schwarzenbach, *Adv. Inorg. Chem. Radiochem.*, 3 (1961) 257.

27 S. Chaberek and A. E. Martell, *Sequestering Agents*, Wiley–Interscience, New York, 1959.

28 D. P. Mellor in F. P. Dwyer and D. P. Mellor, *Chelating Agents and Metal Chelates*, Academic Press, New York, 1964, p. 1.

29 C. S. G. Phillips and R. J. P. Williams, *Inorganic Chemistry*, Vol. II, Oxford University Press, New York, 1966, pp. 267–83, 517–49.

29a H. Sigel and D. B. McCormick, *Accounts Chem. Res.*, 3 (1970) 201.

30 G. Beech, *Quart. Rev.*, 23 (1969) 410.

31 S. Ahrland, *Structure and Bonding*, 5 (1968) 118.

32 L. C. Thompson, B. L. Shafer, J. A. Edgar and K. D. Mannila, *Advances in Chemistry Series*, No. 71, American Chemical Society, Washington, D.C., 1967, p. 169.

33 G. Schwarzenbach, *Experentia Suppl.*, 5 (1956) 162.

34 S. Ahrland, J. Chatt and N. R. Davies, *Quart. Rev.*, 12 (1958) 265.

35 R. G. Pearson, *J. Chem. Educ.*, 45 (1968) 581, 643.

36 R. J. P. Williams and J. D. Hale, *Structure and Bonding*, 1 (1966) 249.

37 S. Ahrland, *Structure and Bonding*, 1 (1966) 207.

38 R. S. Evans and J. E. Huheey, *J. Inorg. Nucl. Chem.*, 32 (1970) 777.

39 M. Misono, E. Ochiai, Y. Saito and Y. Yoneda, *J. Inorg. Nucl. Chem.*, 29 (1967) 2685.

40 G. Klopman, *J. Am. Chem. Soc.*, 90 (1968) 223.

41 R. G. Pearson, *Science*, 151 (1966) 172.

42 U. Weser, *Structure and Bonding*, 5 (1968) 41.

43 H. Sigel, R. Griesser, B. Prijs, D. B. McCormick and M. G. Joiner, *Arch. Biochem. Biophys.*, 130 (1969) 514. H. Sigel, D. B. McCormick, R. Griesser, B. Prijs and L. D. Wright, *Biochemistry*, 8 (1969) 2687.

44 T. Kimura, *Structure and Bonding*, 5 (1968) 1.

45 T. G. Spiro and P. Saltman, *Structure and Bonding*, 6 (1969) 116.

46 D. P. Mellor and L. Maley, *Nature*, 159 (1947) 370; 161 (1948) 436.

47 H. Irving and R. J. P. Williams, *Nature*, 162 (1948) 746; *J. Chem. Soc.*, (1953) 3192.

48 M. Calvin and N. C. Melchior, *J. Am. Chem. Soc.*, 70 (1948) 3270.

49 Ref. 8, p. 183.

50 S. Boyd, J. R. Brannan, H. S. Dunsmore and G. H. Nancollas, *J. Chem. Eng. Data*, 12 (1967) 601.

51 H. Irving and R. J. P. Williams, *J. Chem. Soc.*, (1953) 3192.

52 Ref. 24, p. 77.

53 Ref. 29, p. 269.

54 A. E. Martell and M. Calvin, *Chemistry of the Metal Chelate Compounds*, Prentice-Hall, Englewood Cliffs, N.J., 1952, p. 151.

55 R. Griesser, B. Prijs and H. Sigel, *Inorg. Nucl. Chem. Lett.*, 4 (1968) 443; R. Griesser, B. Prijs and H. Sigel, *Inorg. Nucl. Chem. Lett.*, 5 (1969) 951.

56 Ref. 9, p. 339.

57 H. Irving and H. Rossotti, *Acta Chem. Scand.*, 10 (1956) 72.

58 H. Irving and J. J. R. F. da Silva, *J. Chem. Soc.*, (1963) 945.

59 D. W. Margerum, B. L. Powell and J. A. Luthy, *Inorg. Chem.*, 7 (1968) 800.

60 F. R. N. Gurd and P. E. Wilsox, *Adv. Protein Chem.*, 11 (1956) 311.

61 Ref. 25, p. 56.

62 J. G. Jones, J. B. Poole, J. C. Tomkinson and R. J. P. Williams, *J. Chem. Soc.*, (1958) 2001.

100

63 G. Schwarzenbach, G. Anderegg, W. Schneider and H. Senn, *Helv. Chim. Acta*, 38 (1955) 1147.
64 R. J. Angelici and B. E. Leach, *J. Am. Chem. Soc.*, 90 (1968) 2499.
65 A. E. Martell, in *Advances in Chemistry Series*, No. 62, American Chemical Society, Washington, D.C., 1967, p. 272.
66 Ref. 25, p. 57.
67 D. K. Cabbiness and D. W. Margerum, *J. Am. Chem. Soc.*, 91 (1969) 6540.
68 L. Sacconi, P. Paoletti and M. Ciampolini, *J. Chem. Soc.*, (1961) 5115; P. Paoletti, M. Ciampolini and L. Sacconi, *J. Chem. Soc.*, (1963) 3589.
69 B. Dempsey, M. B. Lowe and J. N. Phillips, in J. E. Falk, R. Lemberg and R. K. Morton, *Haematin Enzymes*, Pergamon Press, London, 1961, p. 29.
70 Y. Marcus and I. Eliezer, *Coord. Chem. Rev.*, 4 (1969) 273.
71 D. D. Perrin, I. G. Sayce and V. S. Sharma, *J. Chem. Soc. (A)*, (1967) 1755.
72 D. D. Perrin and V. S. Sharma, *J. Chem. Soc. (A)*, (1968) 446.
73 D. D. Perrin and V. S. Sharma, *J. Chem. Soc. (A)*, (1969) 2060.
74 R. Griesser and H. Sigel, *Inorg. Chem.*, 9 (1970) 1238.
75 N. E. Jackobs and D. W. Margerum, *Inorg. Chem.*, 6 (1967) 2038.
76 V. S. Sharma and J. Schubert, *J. Chem. Educ.*, 46 (1969) 506.
77 R. D. Gillard, H. M. Irving, R. M. Parkins, N. C. Payne and L. D. Pettit, *J. Chem. Soc. (A)*, (1966) 1159; R. D. Gillard, H. M. Irving and L. D. Pettit, *J. Chem. Soc. (A)*, (1968) 673. V. Simeon and O. A. Weber, *Croat. Chem. Acta*, 38 (1966) 161.
78 W. E. Bennett, *J. Am. Chem. Soc.*, 81 (1959) 246.
78a J. H. Ritsma, G. A. Wiegers and F. Jellinek, *Rec. Trav. Chim.*, 84 (1965) 1577.
78b C. C. MacDonald and W. D. Phillips, *J. Am. Chem. Soc.*, 85 (1963) 3736.
78c J. H. Ritsma, J. C. Van de Grampel and F. Jellinek, *Rec. Trav. Chim.*, 88 (1969) 411.
79 R. Candlin and M. M. Harding, *J. Chem. Soc. (A)*, (1970) 384.
80 M. M. Harding and H. A. Long, *J. Chem. Soc. (A)*, (1968) 2554.
81 K. A. Fraser and M. M. Harding, *J. Chem. Soc. (A)*, (1967) 415.
82 B. E. Leach and R. J. Angelici, *J. Am. Chem. Soc.*, 91 (1969) 6296.
83 J. H. Dunlop and R. D. Gillard, *Adv. Inorg. Chem. Radiochem.*, 9 (1966) 185.
84 A. M. Sargeson, in F. P. Dwyer and D. P. Mellor, *Chelating Agents and Metal Chelates*, Academic Press, New York, 1964, p. 183.
85 C. W. Childs and D. D. Perrin, *J. Chem. Soc. (A)*, (1969) 1039.
86 A. F. Pearlmutter and J. Stuehr, *J. Am. Chem. Soc.*, 90 (1968) 858.
87 J. E. Letter, Jr. and J. E. Bauman, Jr., *J. Am. Chem. Soc.*, 92 (1970) 437.
88 J. E. Letter, Jr. and J. E. Bauman, Jr., *J. Am. Chem. Soc.*, 92 (1970) 443.
89 J. P. Greenstein and M. Winitz, *Chemistry of the Amino Acids*, Vol. 1, John Wiley and Sons, New York, 1961, p. 618.
90 E. W. Wilson, Jr. and R. B. Martin, *Inorg. Chem.*, 9 (1970) 528.
91 E. R. Clarke and A. E. Martell, *J. Inorg. Nucl. Chem.*, 32 (1970) 911.
91a N. C. Stephenson, J. F. McConnell and R. Warren, *Inorg. Nucl. Chem. Lett.*, 3 (1967) 553; D. F. S. Natusch and L. J. Porter, *Chem. Comm.*, (1970) 596.
92 D. D. Perrin, *J. Chem. Soc.*, (1959) 290.
93 D. D. Perrin, *J. Chem. Soc.*, (1958) 3125.
94 K. M. Wellman, T. G. Mecca, W. Mungall and C. R. Hare, *J. Am. Chem. Soc.*, 90 (1968) 805.
95 F. F. L. Ho, L. E. Erickson, S. R. Watkins and C. N. Reilley, *Inorg. Chem.*, 9 (1970) 1139.
96 H. C. Freeman, *Adv. Protein Chem.*, 22 (1967) 258.
97 V. M. Kothari and D. H. Busch, *Inorg. Chem.*, 8 (1969) 2276.
98 A. Tomita, H. Hirai and S. Makishima, *Inorg. Chem.*, 7 (1968) 760.

99 D. D. Perrin and I. G. Sayce, *J. Chem. Soc. (A)*, (1968) 53.
100 W. F. Stack and H. A. Skinner, *Trans. Faraday Soc.*, 63 (1967) 1136.
101 M. S. Michailidis and R. B. Martin, *J. Am. Chem. Soc.*, 91 (1969) 4683.
102 B. Sarkar and Y. Wigfield, *J. Biol. Chem.*, 242 (1967) 5572.
103 R. B. Martin, M. Chamberlin and J. T. Edsall, *J. Am. Chem. Soc.*, 82 (1960) 495.
104 M. K. Kim and A. E. Martell, *J. Am. Chem. Soc.*, 89 (1967) 5138.
105 K. S. Bai and A. E. Martell, *J. Am. Chem. Soc.*, 91 (1969) 4412.
106 B. L. Vallee, R. J. P. Williams and J. E. Coleman, *Nature*, 190 (1961) 633.
107 A. E. Dennard and R. J. P. Williams, *Transition Metal Chem.*, (1966) 115.
108 W. N. Lipscomb, *Accounts Chem. Res.*, 3 (1970) 81.
109 F. R. N. Gurd and G. F. Bryce, in J. Peisach, P. Aisen and W. E. Blumberg, *The Biochemistry of Copper*, Academic Press, New York, 1966, p. 115.
110 F. R. N. Gurd, K. Falk, B. G. Malmström and T. Vänngard, *J. Biol. Chem.*, 242 (1967) 5724.
111 R. A. Bradshaw, W. T. Shearer and F. R. N. Gurd, *J. Biol. Chem.*, 242 (1967) 5451; 243 (1968) 3817.
112 B. Sarkar and Y. Wigfield, *Can. J. Biochem.*, 46 (1968) 601.
113 R. S. J. Phillips, *Chem. Rev.*, 66 (1966) 501.
114 M. M. T. Khan and A. E. Martell, *J. Am. Chem. Soc.*, 88 (1966) 668.
115 J. I. Watters and S. Matsumoto, *Inorg. Chem.*, 5 (1966) 361.
116 Y. Murakami and M. Takagi, *J. Phys. Chem.*, 72 (1968) 116.
117 G. L. Eichhorn, in *Advances in Chemistry Series*, No. 62, American Chemical Society, Washington, D.C., 1967, p. 378.
118 J. E. Falk, *Porphyrins and Metalloporphyrins*, Elsevier Publishing Co., Amsterdam, 1964.
119 P. S. Braterman, R. C. Davies and R. J. P. Williams, *Adv. Chem. Phys.*, 7 (1964) 359.
120 E. B. Fleischer, *Accounts Chem. Res.*, 3 (1970) 105.
121 J. N. Phillips, *Rev. Pure Appl. Chem.*, 10 (1960) 35.
122 A. B. P. Lever, *Adv. Inorg. Chem. Radiochem.*, 7 (1965) 27.
123 C. H. Kirksey, P. Hambright and C. B. Storm, *Inorg. Chem.*, 8 (1969) 2141.
124 C. H. Kirksey and P. Hambright, *Inorg. Chem.*, 9 (1970) 958.
125 J. R. Miller and G. D. Dorough, *J. Am. Chem. Soc.*, 74 (1952) 3977.
126 J. E. Falk, J. N. Phillips and E. A. Magnusson, *Nature*, 212 (1966) 1531; S. J. Cole, G. C. Curthoys and E. A. Magnuson, *J. Am. Chem. Soc.*, 92 (1970) 2991.
127 B. D. McLees and W. S. Caughey, *Biochemistry*, 7 (1968) 642.

Chapter 3

ELECTRONIC STRUCTURES OF IRON COMPLEXES

HARRY B. GRAY

Arthur Amos Noyes Laboratory of Chemical Physics, California Institute of Technology, Pasadena, Calif. 91109, U.S.A.

and

HARVEY J. SCHUGAR

Department of Chemistry, Rutgers University, New Brunswick, N.J. 08903, U.S.A.

INTRODUCTION

Investigations bearing on the important question of the electronic structures of transition metal ions in molecules of biological interest have been actively pursued for several years.

Although it is often difficult to obtain meaningful experimental data on metal ions in such highly dilute systems, in principle all the spectroscopic and magnetic techniques familiar to the inorganic chemist[1-3] are applicable. In a favorable situation, elucidation of both geometric and electronic structural features of coordination to the metal is possible.

In order to keep this chapter within reasonable bounds we have chosen to discuss a selected group of model complexes and related proteins in which iron (as Fe^{2+} or Fe^{3+}) is the central metal. The choice of iron is made because of its wide occurrence in biology. Our primary emphasis will be on models which relate to iron proteins which utilize nitrogen and oxygen donors as ligands.

LIGAND FIELD THEORY FOR OCTAHEDRAL COMPLEXES

Here we will be primarily concerned with the electronic structural features of Fe^{2+} and Fe^{3+} in complexes of octahedral stereochemistry. In discussions of magnetic and electronic spectral properties of these and other transition metal complexes, by far the most useful model is the ligand field (LF) theory[1]. Briefly, in an octahedral molecular environment the metal valence d orbitals split into two sublevels. The more energetic sublevel is called e_g and includes the two d orbitals ($d_{x^2-y^2}$ and d_{z^2}) which interact strongly in σ bonding with the ligands. Because of their spatial orientation, the three d orbitals in the t_{2g} sublevel (d_{xz}, d_{yz}, d_{xy}) can only participate in π interaction with the ligands and as a result are always lower in energy than the e_g orbitals. The splitting of e_g and t_{2g} (Fig. 1) is called the

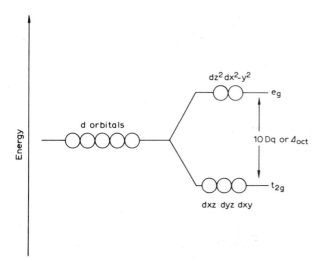

Fig. 1. Ligand field theory of d orbital splitting for an octahedral complex.

octahedral LF splitting and by general convention is designated $10\,Dq$ or Δ_{oct}.

Fe^{2+} complexes

The ground state configurations of octahedral metal complexes are constructed by placing the appropriate number of d valence electrons in the t_{2g} and e_g orbitals. Two cases are distinguished which may be illustrated with Fe^{2+} complexes: If $10\,Dq$ is relatively small, as in $Fe(H_2O)_6^{2+}$, the six d valence electrons spread over the t_{2g} and e_g sublevels to give the maximum spin multiplicity in accordance with Hund's rule. On the other hand, if $10\,Dq$ is larger than the energy required to pair electrons in the t_{2g} orbitals, as in $Fe(CN)_6^{4-}$, all six electrons will reside in the t_{2g} sublevel. To describe this difference in ground state spin multiplicity, $Fe(H_2O)_6^{2+}$ is referred to as a *high-spin* complex, whereas $Fe(CN)_6^{4-}$ is denoted *low-spin*. The LF descriptions of the ground state configurations of $Fe(H_2O)_6^{2+}$ and $Fe(CN)_6^{4-}$ are illustrated in Fig. 2.

Absorption bands arise in the near infra-red (NIR), visible (VIS) and ultra-violet (u.v.) regions due to electronic transitions involving the t_{2g} and e_g sublevels of octahedral complexes. For example, an electron can be promoted from t_{2g} to e_g in $Fe(H_2O)_6^{2+}$ to give an excited state configuration $(t_{2g})^3(e_g)^3$. In standard electronic state notation★ the transition is described as $^5T_{2g}(t_{2g})^4(e_g)^2 \rightarrow {}^5E_g(t_{2g})^3(e_g)^3$.

★ In notation such as $^5T_{2g}$ and ^5E_g the left-hand superscript refers to the spin multiplicity of the electronic state, which is equal to $2S + 1$. For 4 unpaired electrons $S = 2$ and $2S + 1 = 5$. T_{2g} (or E_g) refers to the symmetry of the many-electron spatial wavefunction. Several books[1] present introductory treatments of this notation.

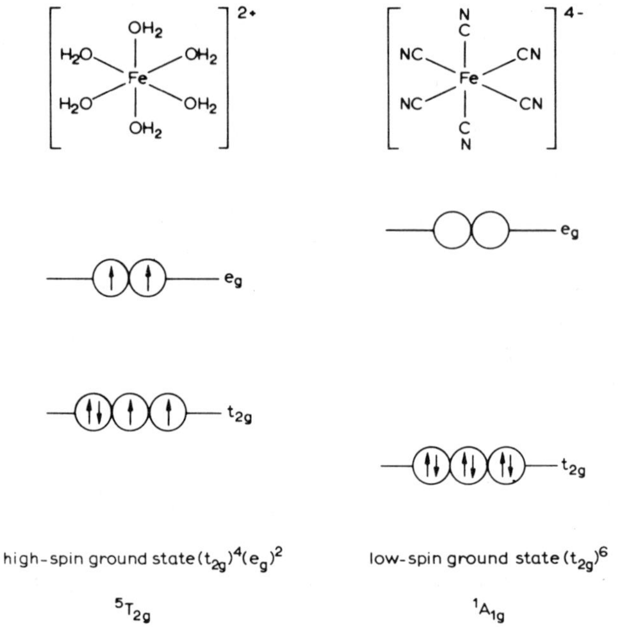

104

Fig. 2. Ground state electronic structures of $Fe(H_2O)_6^{2+}$ and $Fe(CN)_6^{4-}$ according to LF theory.

The absorption spectrum of $Fe(H_2O)_6^{2+}$ is typical of high-spin, octahedral d^6 complexes; the broad absorption band occurring in the near infra-red region at about 10,000 cm^{-1} is assigned as the $^5T_{2g} \rightarrow {}^5E_g$ transition[1]. This band usually shows some splitting, probably because the excited ^5E_g state is somewhat distorted from regular octahedral symmetry. Ground-state distortion may also play some role, particularly in the case of inherently less symmetrical complexes of the type $[Fe(II)L_4X_2]$.

The ground state of the low-spin complex $Fe(CN)_6^{4-}$ is the spin singlet $(S = 0; 2S + 1 = 1)^1A_{1g}(t_{2g})^6$. The LF transition $t_{2g} \rightarrow e_g$ gives rise to two excited spin-singlet states, $^1T_{1g}$ and $^1T_{2g}$, which are separated in energy by electron repulsion effects. Electron repulsion energies are generally given in terms of the Racah parameters B and C[1]. The calculated separation of the $^1T_{1g}$ and $^1T_{2g}$ states is 16B. The electronic absorption spectrum of $Fe(CN)_6^{4-}$ consists of two spin-allowed LF bands, at 31,000 and 37,000 cm^{-1}, which have been identified as the $^1A_{1g} \rightarrow {}^1T_{1g}$ and $^1A_{1g} \rightarrow {}^1T_{2g}$ transitions, respectively[4].

Fe^{3+} complexes

For the same set of ligands the octahedral LF splitting is larger for Fe^{3+} than for Fe^{2+}, so the tendency toward a low-spin ground-state structure is greater. Monomeric octahedral complexes of the type $[Fe(III)O_6]$, nevertheless, always have high-spin ground states.

The high-spin ground state of $3d^5$ is $^6A_{1g}(t_{2g})^3(e_g)^2$. The lowest electronic excited states in order of increasing energy are $^4T_{1g}$, $^4T_{2g}$, and a degenerate pair $(^4E_g, \, ^4A_{1g})$. A diagram of these and other important energy levels for an octahedral $^6A_{1g}$ complex is shown in Fig. 3.

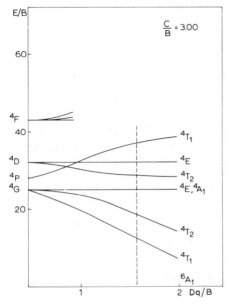

Fig. 3. Electronic energy levels calculated for a high-spin d^5 metal ion in an octahedral field assuming a Racah parameter ratio (C/B) of 3.00.

As is evident from Fig. 3 all the LF transitions in a $^6A_{1g}$ complex are spin-forbidden and thus only weak absorption bands are expected. The LF absorption spectrum of $Fe(H_2O)_6^{3+}$ is shown in the upper curve of Fig. 4[5]. The three lowest energy bands are assigned as $^6A_{1g} \rightarrow \, ^4T_{1g}$, $^6A_{1g} \rightarrow \, ^4T_{2g}$, and $^6A_{1g} \rightarrow (^4E_g, \, ^4A_{1g})$, respectively. The molar extinction coefficients of these bands are very low, as expected for spin-forbidden transitions. An analysis of this spectrum is summarized in Table I.

At this juncture it is important to point out the difference in LF spectra of octahedral $[Fe(III)O_6]$ and tetrahedral $[Fe(III)O_4]$ systems. Information on the latter system has been obtained from a sample of orthoclase feldspar ($KAlSi_3O_8$) containing a small percentage of Fe^{3+} in tetrahedral Al^{3+} sites[5]. The LF absorption spectrum of this sample consists of three bands, at 444, 418, and 377 nm, as shown in the lower curve of

TABLE I

ELECTRONIC SPECTRAL DATA FOR $Fe(H_2O)_6^{3+}$ IN $Fe_2(SO_4)_3 \cdot (NH_4)_2SO_4 \cdot 24H_2O^a$

$^6A_1 \rightarrow$	$\bar{v}(cm^{-1})^b$	ϵ
$^4T_1$	12,600	0.05
$^4T_2$	18,200	0.01
	24,200	
$(^4A_1, {}^4E)$	24,600	1.3
	25,400	
$^4T_2$	27,700	1

[a] From ref. 5.
[b] The spectrum may be fit satisfactorily using the following parameter values: $10\,Dq$ (oct) = 13,700 cm^{-1}; C/B = 3.13; B = 945 cm^{-1}.

Fig. 4. The LF theory for tetrahedral $[Fe(III)O_4]$ is expressed in the energy level diagram of Fig. 5. The fact that the first two LF bands in the tetrahedral case are at much higher energies than in the $[Fe(III)O_6]$ octahedral model complex is predicted by theory[6].

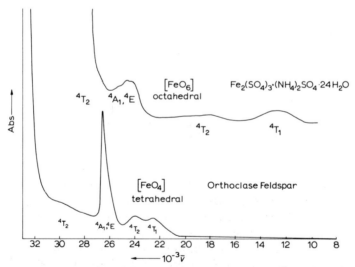

Fig. 4. Electronic absorption spectra of (a) $Fe(H_2O)_6^{3+}$ in ferric ammonium sulfate and (b) tetrahedral $[Fe(III)O_4]$ in orthoclase feldspar.

We assign the three observed bands in the $[Fe(III)O_4]$ to the transitions to $^4T_1$, $^4T_2$, and $(^4E, {}^4A_1)$ excited states, respectively. The assignments are summarized in Table II. Analysis of this spectrum gives $10\,Dq$ (tet) = 7350 cm^{-1} and a value of 540 cm^{-1} for B. The value of $10\,Dq$ is expected[1] to be smaller for [Fe(III)O$_4$] than for [Fe(III)O$_6$], as is found. It is also reasonable that the molar extinction coefficients of the LF bands for the noncentrosymmetric [Fe(III)O$_4$] are about a factor of 10 greater than the respective bands observed for the model [Fe(III)O$_6$] case.

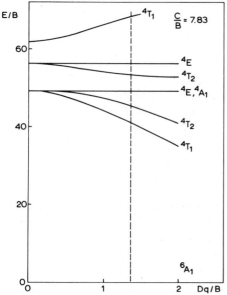

Fig. 5. Electronic energy levels calculated for a d^5 high-spin metal ion in a tetrahedral field assuming $C/B = 7.83$.

With N-donor and CN$^-$ ligands, low-spin ground states are obtained for Fe^{3+}. Thus both Fe(en)$_3^{3+}$ (en = ethylenediamine) and Fe(CN)$_6^{3-}$ have magnetic properties which establish a $^2T_{2g}(t_{2g})^5$ ground state. The lowest excited LF configuration $(t_{2g})^4(e_g)^1$ gives rise to a number of spin-doublet

TABLE II

ELECTRONIC SPECTRAL DATA FOR [Fe(III)O$_4$] IN ORTHOCLASE FELDSPAR[a]

$^6A_1 \rightarrow$	$\bar{v}(\text{cm}^{-1})$	ϵ	Calcd[b]
$^4T_1$	22,500	0.73	22,150
$^4T_2$	23,900	0.76	24,550
$(^4A_1, {}^4E)$	26,500	4.1	26,550
$^4T_2$	29,200	0.1	28,940

[a] From ref. 5.
[b] For 10 Dq (tet) = 7350, B = 540, C = 4230 cm^{-1}.

states and as a result the spectra of $Fe(en)_3^{3+}$ [7] and $Fe(CN)_6^{3-}$ [8] are complicated. The latter complex also exhibits low energy $\pi CN \to Fe^{3+}$ charge transfer absorption. However, assignments for several LF transitions in both $Fe(en)_3^{3+}$ [7] and $Fe(CN)_6^{3-}$ [8] have been proposed.

CHARGE TRANSFER SPECTRA

As was mentioned above, analysis of the LF spectra of Fe^{3+} complexes is often complicated by the presence of intense low-energy ligand $\to Fe^{3+}$ charge transfer bands. The origin of these bands is thought to be related to the oxidative nature of Fe^{3+}. Conversely, low-energy $Fe^{2+} \to$ ligand transitions are expected because of the reductive nature of Fe^{2+}. A classic example of ligand $\to Fe^{3+}$ charge transfer absorption is provided by the blood-red color exhibited by the high-spin $FeNCS^{2+}$ complex (presumably $Fe(H_2O)_5NCS^{2+}$). The electronic transition involved is believed to originate in the highest filled π level of NCS^- and terminate on the t_{2g} level of the Fe^{3+} ($\pi NCS \to t_{2g}$)[9].

A systematic study of ligand $\to Fe^{3+}$ charge transfer has been made in the case of low-spin $[Fe(III)(CN)_5X]^{3-}$ systems[10]. The parent complex, $Fe(CN)_6^{3-}$, exhibits a band at about 24 kcm^{-1} attributable to the charge transfer transition $\sigma CN \to t_{2g}$[8]. On substitution of a CN^- by N_3^-, or $NCSe^-$, an additional low-energy charge transfer absorption develops in the 16–19 kcm^{-1} region which has been assigned $\pi X \to t_{2g}$[10]. An order of $X \to t_{2g}$ charge transfer excitation energy of $NCSe^- < N_3^- < NCS^- < CN^-$ is found.

DIMERIC Fe^{3+} COMPLEXES

Oxobridged dimers

A characteristic reaction of aqueous Fe^{3+} complexes is polymerization via dimeric oxo and dihydroxo bridging structures. A number of model dimers have been isolated and characterized, and their spectral and magnetic properties have been elucidated[11].

The most extensively studied oxobridged system is the complex $(HEDTAFe)_2O^{2-}$ [11], which contains an approximately linear ($\sim 165°$) $Fe(III)–O–Fe(III)$ bridging unit. The magnetic moment of the ethylenediammonium (enH_2^{2+}) salt of this dimeric anion in the temperature range 10–300K is compared in Fig. 6 with the moment of a typical monomeric, high-spin complex $[Fe(picolinate)_2(OH_2)Cl]$. The decrease in μ_{eff} as the temperature is lowered has been interpreted as antiferromagnetic behavior arising from spin–spin coupling of two $S = 5/2$ Fe^{3+} units[11]. The results can be fitted by the standard spin–spin interaction model which assumes a Hamiltonian $\mathcal{H} = -2JS_1 \cdot S_2$ (J is a proportionality constant and is treated as a parameter in fitting data). A J value of -86 cm^{-1} allows an excellent fit of the experimental curve (Fig. 6).

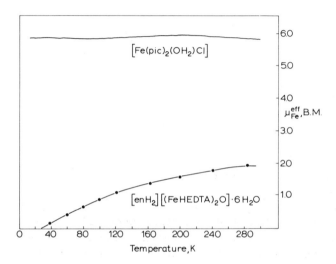

Fig. 6. Magnetic moment versus temperature for model monomeric and oxobridged dimeric Fe^{3+} complexes.

It is clear from several similar studies[11] that a J value in the range -95 ± 10 cm^{-1} is characteristic of an approximately linear Fe(III)–O–Fe(III) unit. Oxobridged dimers of ferric ion (and other transition metal ions) also show strong infra-red absorption at ~ 850 cm^{-1} due to the antisymmetric Fe–O–Fe stretching vibration[11,12].

The electronic spectrum of $(HEDTAFe)_2O^{2-}$ has been studied in detail[11]. It reveals a typical pattern of energies for an octahedral high-spin Fe^{3+} complex. The spectrum is shown in Fig. 7.

The first result worthy of note is the fact that the intensities of the LF bands are much greater ($\times 10^2$) than in an analogous monomeric case. This intensity enhancement is not unusual for spin-coupled systems and has been observed in a large number of Mn^{2+} salts (MnF_2 [13] and $KMnF_3$ [14]). Perhaps a more interesting observation is the appearance of a number of intense u.v. bands in the region between 25 and 42 kcm^{-1}. These bands are difficult to assign to *individual* Fe^{3+} centers and therefore have been interpreted as simultaneous pair electronic (*SPE*) excitations[11]. The two antiferromagnetically coupled Fe^{3+} ions can undergo a simultaneous LF

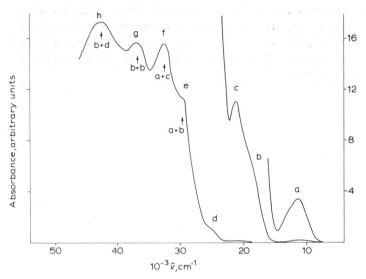

Fig. 7. Electronic absorption spectrum of $(HEDTAFe)_2O^{2-}$.

excitation by a single photon such that the transition energy is to a good approximation the sum of the two one-center energies. A detailed assignment of the electronic spectrum of $(HEDTAFe)_2O^{2-}$ is given in Table III.

TABLE III

ELECTRONIC SPECTRAL DATA FOR $enH_2[(HEDTAFe)_2O] \cdot 6H_2O$[a]

Band	$\bar{v}_{max}(km\ cm^{-1})$	Assignment
a	11.2	$^6A_1 \rightarrow {}^4T_1$
b	18.2	$^6A_1 \rightarrow {}^4T_2$
c	21.0	$^6A_1 \rightarrow ({}^4A_1, {}^4E)$
d	24.4	$^6A_1 \rightarrow {}^4T_2$
e	29.2	$a + b = 29.4$
f	32.5	$a + c = 32.2$
g	36.8	$b + b = 36.4$
h	42.6	$b + d = 42.6$

[a] From ref. 11.

Dihydroxobridged dimers

Dihydroxobridged ferric dimers have also been associated with the aqueous chemistry of ferric ion. They are representative of the hydrolysis and polymerization (or olation) reactions characteristic of "acidic" metal ions, *e.g.*,

$$Fe(H_2O)_6{}^{3+} \rightleftharpoons H^+ + Fe(H_2O)_5OH^{2+}$$

$$Fe(H_2O)_5OH^{2+} \underset{}{\overset{-H_2O}{\rightleftharpoons}} (H_2O)_4Fe \overset{OH}{\underset{OH}{<>}} Fe(H_2O)_4{}^{4+}$$

aquo dimer

While crystalline salts of the aquo dimer have never been isolated, its existence has been inferred from an analysis of titration data[15], the appearance of a u.v. band at ~ 335 nm[16], and a reduction in paramagnetism[17]. The magnetic properties of the aquo dimer, however, have not yet been resolved[18,19].

The only reasonably well documented case of dihydroxobridging in Fe^{3+} is the yellow-green crystalline material of probable structure

$$[(pic)_2Fe \overset{OH}{\underset{OH}{<>}} Fe(pic)_2]$$ (pic = picolinate)[20]. This complex displays a u.v. absorption peak at 342 nm and is weakly antiferromagnetic ($J = -8$ cm^{-1}). A number of dimeric Fe^{3+} complexes with dialkoxobridging units have been prepared which show similar magnetic properties[21].

ELECTRONIC STRUCTURES OF MODEL FERRIC CHELATES AND RELATED COMPOUNDS OF BIOLOGICAL INTEREST

Ferriporphyrin dimers

The paramagnetism of hemin in alkaline solution (*i.e.*, hemin hydroxide or polymerization products thereof) is known to be reduced[22]. Possibly related to this observation are studies based on polarography[23] and redox potentiometry[24] which indicate that dimeric hemin materials of unspecified structure exist in alkaline solutions. After [(phthalocyanato Mn)$_2$O] was characterized by X-ray diffraction[25], it was proposed[18] that oxobridged ferriporphyrin dimers could accommodate the above magnetic results. In connection with the electrochemical results, it is of interest that a two-electron reduction has been observed with fast-sweep techniques for $(FeHEDTA)_2O^{2-}$ (or $(FeEDTA)_2O^{4-}$) in labile equilibrium with $FeHEDTA(OH)^-$ (or $FeEDTA(OH)^{2-}$), the analogous high-spin monomer[26].

An oxobridged ferric porphyrin dimer was first synthesized and characterized by Caughey and coworkers[27], who assigned a structure of [deuteroporphyrin dimethylester Fe(III)]$_2$O on the basis of infra-red absorption at 840 cm^{-1} and molecular weight data. Additional measurements consistent with this formulation were later reported[28]. [TetraphenylporphineFe(III)]$_2$O has also been synthesized and fully characterized by X-ray diffraction studies[29]. In contrast to the related high-spin, monomeric chloro complex, the dimer is antiferromagnetic. The Fe–O–Fe angle of 168° and Fe–O bond distance of 1.76 Å found[29] for this dimer are similar to the corresponding values of 165° and 1.79 Å reported[30] for (FeHEDTA)$_2$O^{2-}.

The u.v. and visible spectral regions of the porphyrin dimers are dominated by intense bands that are generally considered to arise solely from intraligand and charge transfer excitations. There has been no real hope of locating the *normally* weak LF transitions localized on the ferric ion. However, as previously discussed, enhancements of LF bands by a factor of ~100 have been observed in (FeHEDTA)$_2$O^{2-} and related materials. Similar or possibly even greater enhancements could occur in the oxobridged Fe^{3+} porphyrin dimers. In addition, such dimers would be expected to display fairly intense u.v. bands arising from *SPE* transitions. For these reasons we suggest that a re-examination of the electronic spectra of dimeric ferriporphyrins is in order, paying particular attention to the special spectral features of the Fe$_2$O structural unit.

Ferrihemoprotein hydroxides

The magnetic properties of ferrihemoprotein hydroxides and ferriporphyrin hydroxides are similar in the sense that both sets of compounds display reduced paramagnetism. The situation with regard to the ferrihemoprotein hydroxides has been reviewed in detail in an important paper by George, Beetlestone, and Griffith[31]. These workers interpreted the observed magnetic behavior in terms of a thermal equilibrium of high-spin and low-spin Fe^{3+} states. The implication of this interpretation is that the LF splitting of OH$^-$ is larger than H$_2$O, because the aquo monomers are known to be high-spin. This is curious in view of the fact that the reverse order of LF splitting (H$_2$O > OH$^-$) is commonly observed in simple inorganic hydroxo and aquo complexes[1].

From steric considerations, the formation of oxobridged ferrihemoprotein dimers in alkaline solution does not appear feasible. Structural studies show that in crystals the ferriporphyrin moieties are surrounded by protein in metmyoglobin and methemoglobin[32]. It should be noted, however, that heme group exchange in solution is fairly rapid in ferrihemoglobin. The rate of exchange, which may necessitate the dissociation of the ferriporphyrin unit, increased 52% when the pH was increased from 6.4 to 7.7[33]. Thus it is not inconceivable that at pH 10 ferriporphyrin may be dissociated from the protein and dimerized (pK values for ferrihemoproteins fall in the range $8.2 \sim 11.3^{31}$). The possibility of oxobridged structures in cytochrome c oxidase has also been discussed[34].

The visible absorption spectrum reported[28] for [deuteroporphyrin-dimethylester Fe(III)]$_2$O closely resembles the spectra of ferrimyoglobin, ferrihemoglobin, and ferriperoxidase hydroxide[31]. In view of this fact and the above discussion we conclude that the relationship between solution magnetic properties and structure in the ferrihemoprotein hydroxides requires further clarification.

Hemerythrin and the problem of oxygen binding

An excellent example of a metal–protein structural problem that has been substantially elucidated by spectroscopic and magnetic studies is provided by the oxygen-transport hemerythrin system. The protein from the sipunculid *Golfingia gouldii* has been studied extensively by Klotz and coworkers (Chapter 11)[35]. Eight subunits, each containing two iron atoms capable of reversibly binding one oxygen molecule, associate in the full protein to give a molecular weight of about 108,000. Magnetic measurements[35] show that the deoxy form of the protein contains high-spin Fe(II). The electronic absorption spectrum of this form is relatively clear down to 300 nm; in the u.v., the peak at about 280 nm characteristic of tyrosine appears.

Upon oxidation to the met [Fe(III)] protein, absorption peaks appear in the visible region. Each subunit of the met protein binds ligands such as Cl$^-$, Br$^-$, and N$_3^-$ in a 1:2 Fe(III) complex, and the visible peak positions of these derivatives vary slightly[36]. The electronic absorption spectrum of metchlorohemerythrin is shown in Fig. 8. The bands at 668 and 500 nm may be assigned as LF bands in a $^6A_1$-type Fe(III) complex. Their moderate intensities[36,37] are consistent with a model in which two Fe(III) ions are spin-coupled in an oxobridged dimer of the type [Cl–Fe(III)–O–Fe(III)].

The bands at 384 and 331 nm in the metchloro derivative are also attributable to electronic transitions involving one or both of the Fe(III) ions. Their positions and intensities can be rationalized if they are assigned to *SPE* excitations in an oxobridged dimer. Alternatively, ligand → Fe(III) charge transfer transitions are also possible explanations. At wavelengths lower than 331 nm, the tyrosine absorption takes over and additional peaks due to Fe(III) are presumably buried.

The electronic absorption spectrum of oxyhemerythrin is shown in Fig. 9[37]. This spectrum closely resembles that of metchlorohemerythrin, except for the much greater intensity in the band at about 500 nm. The spectrum can be considered as reasonable proof of a suggestion first made by Klotz[38] that in the oxy form both irons have been oxidized to Fe(III)

114

and the O_2 reduced to O_2^{2-}. Large intensity at 501 nm can be easily understood in this model because a similar absorption band has been observed in
the complex formed between [Fe(III)–EDTA]$^-$ and H_2O_2 in basic solution[39].
Presumably, this band is attributable to an $O_2^{2-} \rightarrow$ Fe(III) or HOO$^- \rightarrow$ Fe(III)
charge transfer transition.

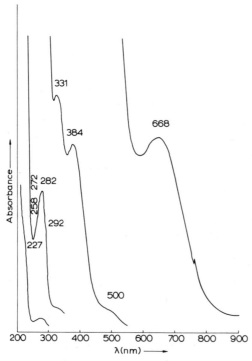

Fig. 8. Electronic absorption spectrum of metchlorohemerythrin.

The bands at 370 and 317 nm in oxyhemerythrin are similar to the
384 and 331 nm bands in the metchloro derivative and in all probability
analogous assignments are appropriate.

The spectral data for the metchloro and oxy proteins suggest coordination structures involving dimeric Fe(III) units. Recent magnetic susceptibility experiments on metaquo and oxyhemerythrin using a magnetometer
equipped with an ultrasensitive sensor (a superconducting quantum
mechanical device) have essentially established the nature of these dimeric
units[40]. Magnetic moment data in the 2–200K range for the oxy form (Fig.
10) conclusively establish antiferromagnetic behavior. Furthermore, the
data can be fit assuming spin coupling of a high-spin ($S = 5/2$) Fe(III) pair,
with $J = $ -77 cm^{-1}. The J value is so similar to that of the (HEDTAFe)$_2$O^{2-}

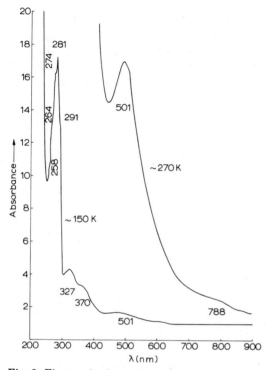

Fig. 9. Electronic absorption spectrum of oxyhemerythrin.

dimer that an oxobridged Fe(III) structure for oxyhemerythrin seems highly probable. Magnetic moment data shown in Fig. 11 show that the metaquo form is also antiferromagnetically coupled. The J value of -135 cm^{-1} could mean that the Fe(III)–O–Fe(III) unit is more nearly linear in this case.

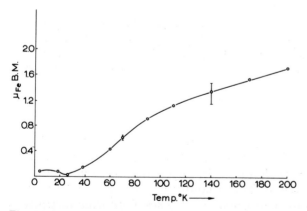

Fig. 10. Magnetic moment data for oxyhemerythrin in the range 2–200K.

References pp. 117–118

116

The spectroscopic and magnetic data leave little doubt that a type of oxidative addition mechanism is operative in O_2 binding by hemerythrin. The two electrons necessary to reduce O_2 are furnished by the two Fe(II) ions, which from the magnetic and spectral data are antiferromagnetically coupled Fe(III) ions in the $[O_2^{2-}-Fe(III)-O-Fe(III)]$ oxidative addition product. Two-electron reductive elimination of O_2 by this unit completes the reversible process.

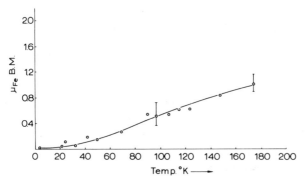

Fig. 11. Magnetic moment data for metaquohemerythrin in the range 2–200K.

The oxidative addition model for reversible O_2 binding by metal proteins has also been proposed for hemocyanin[6]. Hemocyanin is a copper protein which binds one O_2 molecule for every two copper atoms (Chapter 12). The deoxy Cu(I) form has no appreciable absorption in the visible region. When oxygenated, the protein is blue and exhibits a rich visible spectrum, with bands at 700 (ϵ 75), 570 (ϵ 500), 440 (ϵ 65), and 347 nm (ϵ 8900)[41]. The pattern of bands around 570 nm leaves little doubt that oxyhemocyanin contains Cu(II). The enhanced LF band intensities further suggest a dimeric Cu(II) complex[6]. The spectral data for oxyhemocyanin are thus entirely consistent with a hemerythrin-type oxidative addition O_2-binding model. Extension of this concept to oxyhemoglobin requires a seven-coordinate Fe(IV) structure. A discussion of this structure and other models of O_2-binding in hemoglobin has been presented elsewhere[6].

ACKNOWLEDGMENTS

The research from Caltech reported in this paper was supported by the National Science Foundation. The authors are particularly grateful to George Rossman, Dana Powers, John Webb, Jack Thibeault, and Vinnie Miskowski for helpful discussions. This is contribution No. 4512 from the Arthur Amos Noyes Laboratory of Chemical Physics, California Institute of Technology, Pasadena, California 91109, U.S.A.

REFERENCES

1 A. B. P. Lever, *Inorganic Electronic Spectroscopy*, Elsevier, New York, 1968;
   D. Sutton, *Electronic Spectra of Transition Metal Compounds*, McGraw-Hill,
   New York, 1968; H. L. Schläfer and G. Gliemann, *Basic Principles of Ligand
   Field Theory*, Wiley-Interscience, New York, 1969.
2 A. Earnshaw, *Introduction to Magnetochemistry*, Academic Press, New York,
   1968; B. N. Figgis and J. Lewis, in H. B. Jonassen and A. Weissberger, *Techniques
   of Inorganic Chemistry*, Vol. 4, Interscience, New York, 1965, pp. 137–248.
3 G. K. Wertheim, *Mössbauer Effect: Principles and Applications*, Academic Press,
   New York, 1964; R. H. Herber, in F. A. Cotton, *Progress in Inorganic Chemistry*,
   Vol. 8, Interscience, New York, 1967, pp. 1-41 and references therein.
4 H. B. Gray and N. A. Beach, *J. Am. Chem. Soc.*, 85 (1963) 2922.
5 G. R. Rossman, *Ph.D. Thesis*, California Institute of Technology, 1971.
6 H. B. Gray, *Adv. Chem. Ser.*, 100 (1971) 365.
7 G. A. Renovitch and W. A. Baker, Jr., *J. Am. Chem. Soc.*, 90 (1968) 3585.
8 J. J. Alexander and H. B. Gray, *J. Am. Chem. Soc.*, 90 (1968) 4260.
9 C. K. Jørgensen, *Absorption Spectra and Chemical Bonding*, Addison-Wesley,
   Reading, Mass., 1962.
10 D. F. Gutterman, *Ph.D. Thesis*, Columbia University, 1969.
11 H. J. Schugar, G. R. Rossman, C. G. Barraclough and H. B. Gray, *J. Am. Chem.
   Soc.*, 94 (1972) 2683 and references therein.
12 R. M. Wing and K. P. Callahan, *Inorg. Chem.*, 8 (1969) 871.
13 L. L. Lohr, Jr. and D. S. McClure, *J. Chem. Phys.*, 49 (1968) 3516.
14 J. Ferguson, H. J. Guggenheim and Y. Tanabe, *J. Phys. Soc. Japan*, 21 (1966)
   692.
15 B. O. A. Hedström, *Arkiv. Kemi*, 6 (1953) 1.
16 R. M. Milburn and W. C. Vosburgh, *J. Am. Chem. Soc.*, 77 (1955) 1352.
17 L. N. Mulay and P. W. Selwood, *J. Am. Chem. Soc.*, 77 (1955) 2693.
18 H. J. Schugar, C. Walling, R. B. Jones and H. B. Gray, *J. Am. Chem. Soc.*, 89
   (1967) 3712.
19 J. Máthé and E. Bakk-Máthé, *Revue Roumaine de Chemie*, 11 (1966) 225.
20 H. J. Schugar, G. R. Rossman and H. B. Gray, *J. Am. Chem. Soc.*, 91 (1969) 4564.
21 C. -H. Wu, G. R. Rossman, H. B. Gray, G. S. Hammond and H. J. Schugar, *J. Am.
   Chem. Soc.*, 11 (1972) 990.
22 W. A. Rawlinson and P. B. Scutt, *Aust. J. Sci. Res.*, A5 (1952) 173.
23 J. Jordan and T. M. Bednarski, *J. Am. Chem. Soc.*, 86 (1964) 5690.
24 J. Shack and W. M. Clark, *J. Biol. Chem.*, 171 (1947) 143.
25 L. H. Vogt, Jr., A. Zalkin and D. H. Templeton, *Inorg. Chem.*, 6 (1967) 1725.
26 H. J. Schugar, A. T. Hubbard, F. C. Anson and H. B. Gray, *J. Am. Chem. Soc.*, 91
   (1969) 71.
27 N. Sadasivan, H. I. Eberspaecher, W. H. Fuchsman and W. S. Caughey, *Biochemistry*,
   8 (1969) 534.
28 I. A. Cohen, *J. Am. Chem. Soc.*, 91 (1969) 1980.
29 E. B. Fleischer and T. S. Srivastava, *J. Am. Chem. Soc.*, 91 (1969) 2403.
30 S. J. Lippard, H. J. Schugar and C. Walling, *Inorg. Chem.*, 6 (1967) 1825.
31 P. George, J. Beetlestone and J. S. Griffith, *Rev. Mod. Phys.*, 36 (1964) 441.
32 R. E. Dickerson and I. Geis, *The Structure and Action of Proteins*, Harper and Row,
   New York, 1969.
33 H. F. Bunn and J. H. Jandl, *Proc. Natn. Acad. Sci. U.S.*, 56 (1966) 974.

118

34 W. S. Caughey, *Adv. Chem. Ser.*, 100 (1971) 248.
35 M. Y. Okamura, I. M. Klotz, C. E. Johnson, M. R. C. Winter and R. J. P. Williams, *Biochemistry*, 8 (1969) 1951; and references therein.
36 K. Garbett, D. W. Darnall, I. M. Klotz, R. J. P. Williams, *Arch. Biochem. Biophys.*, 103 (1969) 419.
37 S. Simon, B. Grube, G. R. Rossman and H. B. Gray, unpublished data.
38 I. M. Klotz, T. A. Klotz and H. A. Fiess, *Arch. Biochem. Biophys.*, 68 (1957) 284.
39 C. Walling, M. Kurz and H. J. Schugar, *Inorg. Chem.*, 9 (1970) 931.
40 J. W. Dawson, H. B. Gray, H. E. Hoenig, G. R. Rossman, J. M. Schredder and R.-H. Wang, *Biochemistry*, 11 (1972) 461.
41 K. E. Van Holde, *Biochemistry*, 6 (1967) 93.

PART II

INTERACTION OF METAL IONS WITH AMINO ACIDS,
PEPTIDES AND RELATED NATURAL CHELATORS,
AND PROTEINS

Chapter 4

METAL COMPLEXES OF AMINO ACIDS
AND PEPTIDES[*]

HANS C. FREEMAN

School of Chemistry, University of Sydney, Sydney 2006, Australia

The metal complexes of amino acids and peptides have provided both coordination chemists and biochemists with grist for their experimental mills for a long time. Solid derivatives of both Cu(II) and Pt(II) were prepared from alanine by Strecker in 1850[1], and even before this the production of a deep-red colour in the presence of Cu(II) salts and alkali had been used by Wiedemann (1847) to characterize the newly discovered compound biuret[2]. Wiedemann obtained a crystalline but impure product by evaporating his solution, and he noted the need to use an excess of biuret to produce the colour reaction if ammonium hydroxide was used as the alkali. It was more than a hundred years before the structure of his crystals was revealed by X-ray diffraction, and it took almost as long as this before modern coordination theory led to an understanding of the equilibrium between $Cu(Biu)_2^{2-}$ and $Cu(NH_3)_4^{2+}$ ions, and of the reason why the former are red and the latter are blue.

Since these early discoveries, the interactions between metals and amino acids and peptides have become of considerable interest as coordination phenomena, as models for metal–protein reactions, and as models for biological systems in which the properties of proteins are modified by the fact that metal atoms are attached to them. The literature relevant to this topic was first reviewed by Gurd and Wilcox in 1956[3], and a comprehensive account of the basic chemistry up to 1957 was given by Greenstein and Winitz[4]. Interesting examples of metal complexes involving amino acids

---

[*]*Abbreviations used in this chapter:* $HL^{\pm}$ or HL = zwitterion form of an amino acid or peptide; *e.g.*, HGly = glycine, $^+NH_3CH_2COO^-$; HSar = *N*-methylglycine, sarcosine; HAla = alanine; HAsp = aspartic acid; HCys = cysteine; HGlu = glutamic acid; HHis = histidine; HMet = methionine; HNle = norleucine; HSer = serine; HTyr = tyrosine; HGly-Gly = glycylglycine, $^+NH_3CH_2CONHCH_2COO^-$. LH = neutral form of amino acid or peptide; *e.g.*, GlyH = glycine, NH_2CH_2COOH; Gly-GlyH = glycylglycine, $NH_2CH_2CONHCH_2COOH$. $BiuH_2$ = biuret, $NH_2CONHCONH_2$; ImH = imidazole, $C_3N_2H_4$. The symbols M(II), M^{2+} are used to indicate the metal in oxidation state II, and the ionic species M^{2+}, respectively. K_1, K_2, K_3 = stepwise formation constants of complex species ML^{+n-1}, ML_2^{+n-2}, ML_3^{+n-3}, where free metal ion = M^{+n}. β_{pqr} = $[M_pH_qL_r^{+np+q-r}]/[M^{+n}]^p[H^+]^q[L^-]^r$, the overall stability constant of the species $M_pH_qL_r^{+np+q-r}$.

References pp. 162–166

and peptides as ligands occur throughout Martell and Calvin's book[5] and Nakamoto and McCarthy's *Spectroscopy and Structure of Metal Chelate Compounds*[6]. Areas to which recent reviewers have paid particular attention have included stereoselectivity and reactivity[7], the spectroscopic behaviour of copper in peptide complexes and proteins[8], crystal structure analysis[9], and the relation of model compounds to the understanding of enzyme activity[10]. A comprehensive survey of the recent literature up to the end of 1968 has been given by Gillard and Laurie in the first of a periodical series of specialist reports[11].

This chapter deals with the main types of potentially metal-binding atoms in amino acids and peptides, using as examples their interactions with a small number of metal ions. Before we discuss these interactions, let us note that metal-binding by functional groups in a protein may differ in two important respects from metal-binding by the same groups in smaller peptides.

Firstly, the functional groups through which a metal atom is bound to a protein may well be brought close together by the tertiary structure of the protein, but be separated by many amino acid residues along the protein chain. Such functional groups act relatively independently of one another. Each is subject to geometrical constraints imposed by the protein chain in its own immediate vicinity, and by contacts with other functional groups coordinated to the same metal atom. These constraints are among the properties which we wish to study, but they are difficult to reproduce in small model compounds, precisely because the ligands are small molecules. Even the simplest amino acid, glycine, has an NH_2-terminus and a COO-terminus separated by only a few Ångstrom units, and the simplest peptide, glycylglycine, has a peptide (-CO-NH-) group in addition to its two termini. Whenever a polyfunctional ligand acts in a multidentate fashion, one molecule of it displaces two or more aquo- or other unidentate ligands from the metal. The resulting chelate complex has additional thermodynamic stability because the entropy (measure of randomness) of the system has been increased. The "chelate effect" explains why there are few complexes in which amino acids or peptides act as unidentate or non-chelating ligands. It would therefore be risky to extrapolate unreservedly from metal–peptide to metal–protein interactions, since a metal-donor bond which is not stable when it alone is present may become stabilized if its formation leads to the closure of a five- or six-membered ring.

Secondly, there is increasing evidence that the active sites of enzymes — including metallo-enzymes — frequently lie in clefts or pockets in the protein structure which are lined with predominantly non-polar amino acid side-chains. Metal-binding at or near such active sites may therefore take place in what are really non-aqueous solutions, whose dielectric constants must be very different from the dielectric constants of the aqueous electrolyte solutions in which most metal–peptide complexes have been studied. While metal–peptide interactions in aqueous solutions may adequately represent the conditions at the interface between a protein and the surrounding medium, they may not be very good models for what happens in the interior of a protein molecule.

. A great deal of our information concerning metal-binding sites on amino acids is derived from crystal structure analyses. This does not mean that every interaction which is found in a particular structure persists when the complex is dissolved: some intermolecular interactions obviously do not persist, or dissolution would not occur. Neither is it implied that interactions which have not yet been confirmed by crystal structure analysis cannot take place in solution. The risk of placing too great an emphasis on crystal structural data is, however, no greater than the danger of drawing structural inferences from data which have been obtained by techniques which do not directly measure structural properties. At least one can be certain that any complex species whose existence is confirmed by crystal structure analysis was in the solution from which the crystals grew. It may not have been the only species, or even the most important species, in the solution but it is unlikely that new complexes are created as an artefact of the process of crystallization. The wisest course is to use the growing fund of crystal structural evidence both to suggest and to test models for the processes or properties deduced from other methods.

FUNCTIONAL GROUPS OF AMINO ACIDS AND PEPTIDES AS METAL-BINDING SITES

The participation of a particular functional group in metal-binding depends on two factors. How successfully does the functional group compete against others in the vicinity? And how successfully do the metal ions compete against protons for the potential donor atoms? Part of the answer to the first question lies in the acid dissociation constants of the functional groups. The lower the pK_a, the greater is the availability of the donor atom for the formation of a metal-ligand bond. According to this criterion, the order of metal-binding tendencies will be carboxyl > imidazole > amino ($pK_{COOH} \simeq 1.8$, $pK_{ImH_2^+} \simeq 6.5$, $pK_{NH_3^+} \simeq 9.0$). It is risky to use this criterion alone because the order of the pK values may not be the same as that of the enthalpy changes accompanying complex formation, which provide a measure of the relative thermodynamic stabilities of the metal-ligand and proton-ligand bonds. Finally (as already noted) bonds with low enthalpies of formation may nevertheless be stabilized by favourable entropy effects.

Terminal amino groups

Terminal amino groups are among the most common metal-binding loci, despite their high pK_a values. Coordination is favoured (i) by the

strong electron-donor (basic) character of N(amino) atoms, (ii) by the
relatively strong ligand-field effect of N(amino) compared with other
potential donor atoms in transition metal complexes, and (iii) by the fact
that an O(carboxyl) or O(peptide) atom capable of forming a chelate ring
is never more than three or four atoms away. The only geometrical constraint seems to be a requirement that the metal–N(amino)–C_α angle should
be near the tetrahedral value (109°, s.d. 1°, in α-amino acids; 110°, s.d. 0.4°,
in peptides; 113°, s.d. 2°, in β-amino acids).

Chelation through terminal amino groups has been found in the crystal
structures of all complexes in which amino acids or peptides function as
bidentate or higher-dentate ligands. Typical examples are $Zn(Gly)_2 \cdot H_2O$
and $Zn(Gly\text{-}Gly)_2 \cdot H_2O$ (I and II)[12,13]. For a number of transition metal

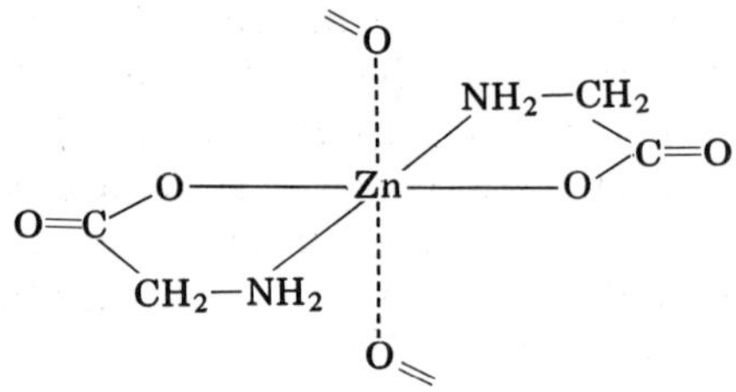

(I)

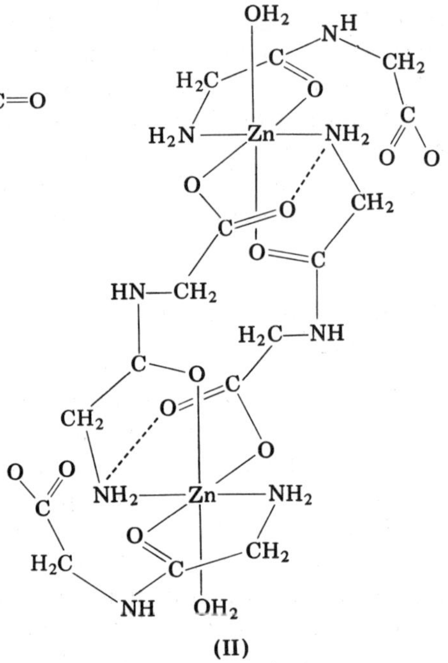

(II)

complexes of amino acids and peptides, the thermodynamic functions for
the complexation reactions $M^{2+} + L^- \rightleftharpoons ML^+$ and $ML^+ + L^- \rightleftharpoons ML_2$ have
been determined from the temperature gradients of the equilibrium con-
stants[14-16] and from calorimetric measurements[15,17-19]. The most important
negative contributions to the enthalpies of chelation come from the formation of the metal–N(amino) bonds[14,15]. The enthalpy changes accompanying
metal–O(carboxyl) bonding are actually *unfavourable*[15], and chelation
depends entirely on the increase in entropy which results from the release
of aquo-ligands and from the mutual neutralization of the metal and
carboxyl group charges.

It is much less common to find coordination through terminal amino groups without the formation of chelate rings. One metal where this occurs is Ag(I). The electronic configuration is d^{10}, and the coordination geometry is characteristically digonal (linear) or tetrahedral. The closure of five- or six-membered chelate rings is geometrically impossible in linear, and difficult in tetrahedral, complexes and the coordination of Ag(I) by multidentate ligands leads to the formation of complexes which are polynuclear or in which some of the functional groups are not used for metal-binding. Complexation of the first type is illustrated by the crystal structure of Ag(Gly), which has endless $-Ag-NH_2CH_2COO-Ag-NH_2CH_2COO-$ chains (III)[20]. The same complex also crystallizes as $Ag(Gly)\cdot\frac{1}{2}H_2O$, in which alternate Ag atoms are bonded to two N(amino) and two O(carboxyl) donors, respectively (IV)[20]. In both these structures the ligands must be bidentate because

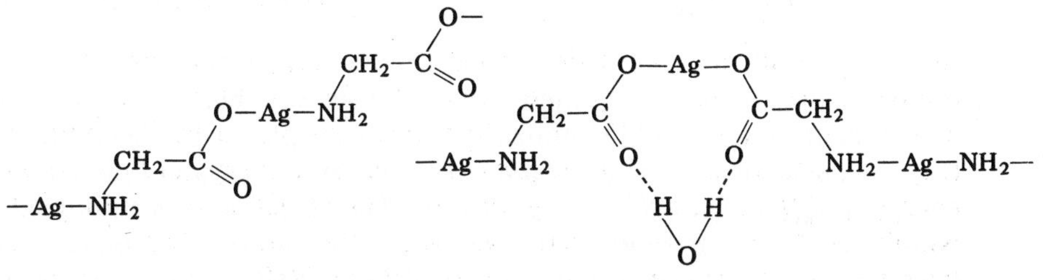

(III)   (IV)

there is only one ligand molecule per Ag atom. $Ag-OH_2$ bonds are known to be very weak, so that H_2O molecules cannot replace amino and carboxyl groups as a means of satisfying the requirement that every Ag atom shall be bonded to at least two donor atoms. In aqueous solutions and in solid complexes with metal:ligand ratios of 1:2, more donor atoms than necessary are available, and the bonds which are formed should be those with the largest enthalpies of formation. The necessary enthalpy data have not been reported, but for amino acids with non-functional side-chains the stability constants corresponding to the formation of AgL and AgL_2^- (log β_{101} = 3.5–4.0, log β_{102} = 6.5–7.5[21–23]) are similar to those of Ag–ammine complexes, and increase in the same order as the acid dissociation constants $K_{NH_3^+}$ of the amino acid ligands. It is concluded that Ag-binding occurs only through the N(amino) atoms[23,24]. A similar conclusion is reached from an examination of the infra-red spectrum of solid $Li[Ag(Nle)_2]$. The asymmetric carboxyl frequency of the complex is the same as that of free norleucine (HNle), suggesting that the carboxyl group is free[25] (although this line of argument is open to criticism in the absence of observations of the metal–ligand stretching frequencies[26]).

Another metal to which amino acids are frequently coordinated only through their N(amino) atoms is Pt(II). Among the complexes characterized either by the preparative pathways[27–29] or by the contact shifts of the proton resonances in the proton magnetic resonance (p.m.r.) spectra[30] are $Pt(Gly)_4^{2-}$, $Pt(Gly)_3^-$, $Pt(GlyH)_2Cl_2$ and $Pt(GlyH)_2(NH_3)_2^{2+}$ (V–VIII). The

$$
\begin{array}{cc}
{}^-OOCCH_2NH_2 \quad NH_2CH_2COO^- \\
\diagdown \quad\diagup \\
Pt \\
\diagup \quad \diagdown \\
{}^-OOCCH_2NH_2 \quad NH_2CH_2COO^-
\end{array}
$$

(V)

$[Pt(Gly)_4]^{2-}$

$$
\begin{array}{cc}
H_2C\text{---}NH_2 \quad NH_2CH_2COO^- \\
\quad\;\; \diagdown\;\diagup \\
\quad\;\; Pt \\
\quad\;\; \diagup\;\diagdown \\
O\text{=}C\text{---}O \quad\; NH_2CH_2COO^-
\end{array}
$$

(VI)

$[Pt(Gly)_3]^-$

$$
\begin{array}{cc}
Cl \quad NH_2CH_2COOH \\
\diagdown\quad\diagup \\
Pt \\
\diagup\quad\diagdown \\
Cl \quad NH_2CH_2COOH
\end{array}
$$

(VII)

$[Pt(GlyH)_2Cl_2]$

$$
\begin{array}{cc}
H_3N \quad NH_2CH_2COOH \\
\diagdown\quad\diagup \\
Pt \\
\diagup\quad\diagdown \\
H_3N \quad NH_2CH_2COOH
\end{array}
$$

(VIII)

$[Pt(GlyH)_2(NH_3)_2]^{2+}$

reason for unidentate coordination in these complexes is that nitrogen donors, being at the strong-field end of the spectrochemical series, cause considerable crystal-field stabilization in square-planar Pt(II) complexes (*i.e.* the enthalpies of bonding are enhanced by the crystal field stabilization energy (*CFSE*) of the d^8 configuration). The effect is so pronounced that N(amino) atoms can bind Pt(II) even when the carboxyl groups of the ligand molecules are protonated (as in **VII** and **VIII**), thus reversing the order deduced from the pK_a values.

Terminal and side-chain carboxyl groups

The tendency of carboxyl groups to occur in chelate rings has already been noted. With a metal having suitable coordination geometry, the ability of a carboxyl group to participate in chelation depends upon the presence of a second donor atom at the correct spacing for the completion of a five- or six-membered ring. This condition is satisfied in α- and β-amino acids by the availability of the terminal amino group, thus accounting for the chelate structures of the majority of amino acid complexes. In peptides, except in special circumstances (p. 153), the terminal carboxyl group is *not* favourably located with respect to a second coordinating group unless the COO-terminal residue happens to be histidine. Metal–O(carboxyl) interactions with peptides must then be non-chelating or absent. Further, at pH values where amino acids and peptides are in their zwitterion forms $^+NH_3CHR\text{-}(CONHCHR)_nCOO^-$ (and where basic side-chain groups such as imidazole are protonated), the negatively charged carboxyl groups are almost the only groups immediately available for metal-binding. The probability of metal–carboxyl binding under these conditions is highest for metal ions which are unaffected by *CFSE*. Those metal ions which form strong covalent bonds with good electron donors, or which have large *CFSE* values in the presence of strong-field ligands, tend to compete successfully against protons for the lone-pair electrons of the terminal N(amino) atoms, as we have seen in the Pt(II) complexes **VII** and **VIII**.

Experimental evidence for unidentate coordination through O(carboxyl) atoms is provided by the p.m.r. contact shifts (in D_2O solutions at low pD) for the histidine complex of Co(II)[31], and by the infra-red carboxylate stretching frequencies for 1:1 DL-alanine and L-histidine complexes of a series of first-row transition elements[32]. In addition, non-chelating metal–O(carboxyl) interactions have been found in a variety of crystal structures of complexes prepared at low pH. One might expect a metal–O(carboxyl) interaction to take the form of a single bond from the positive metal ion to a negatively charged O(carboxyl) atom, with a balance between ionic and covalent characters depending on the electronegativity of the metal. This expectation is too optimistic. Figure 1 shows five types of

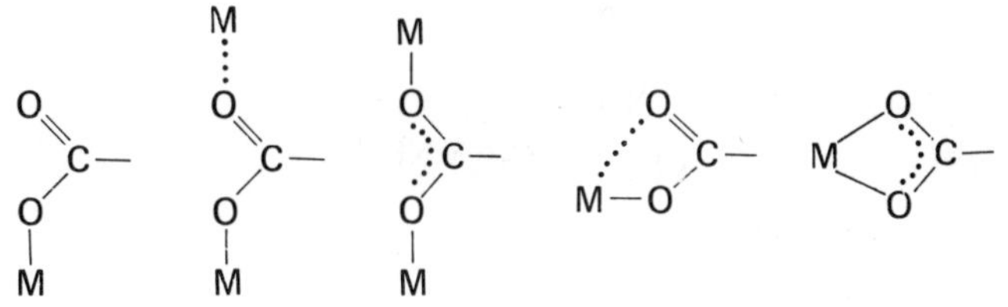

Fig. 1. O(carboxyl)–metal interactions.

metal–O(carboxyl) interactions which can be distinguished in crystal structures of amino acid and peptide complexes. In *type a* a single metal atom is bonded to one O(carboxyl) atom. In *type b* the "free" O(carboxyl) atom binds, albeit less strongly, a second metal atom. In *type c* the carboxyl group binds two metal atoms more or less equally, so that it provides a symmetrical bridge between them. (Differences between the infra-red spectra enable *types b* and *c* to be further sub-divided according to whether the disposition of the metal–oxygen bonds is *anti–anti*, *syn–anti* or *syn–syn*[33].) In *types d* and *e*, the carboxyl group itself acts as a bidentate ligand, forming an unsymmetrical four-membered chelate ring in *type d* and a symmetrical ring in *type e*.

Type a. The simplest type of unidentate coordination through carboxyl groups has been established in only one structure, $Fe(HGly)SO_4 \cdot 5H_2O$[34]. The complex is more correctly described by the formula $[Fe(OH_2)_6][Fe(HGly)_2$-$(OH_2)_4](SO_4)_2$. The glycine ligands in the octahedral $[Fe(HGly)_2(OH_2)_4]^{2+}$

ions are *trans* to each other, and are bonded to the metal only through single O(carboxyl) atoms (IX).

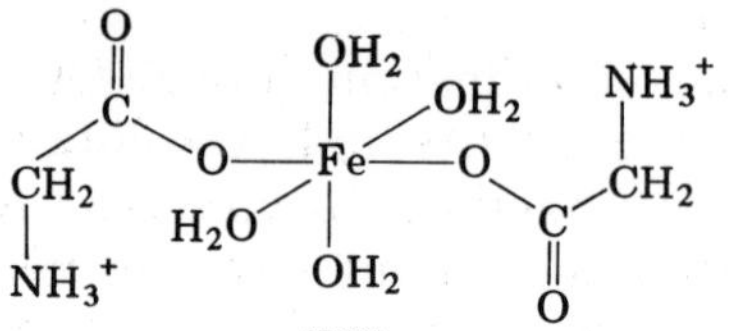

(IX)

Type c. The reaction of Ag^+ ions with amino acids and peptides at low pH yields complexes such as $Ag(HGly)NO_3$ and $Ag(HGly\text{-}Gly)NO_3$ (X and XI)[20,35]. The coordination of the Ag atoms is approximately digonal, and

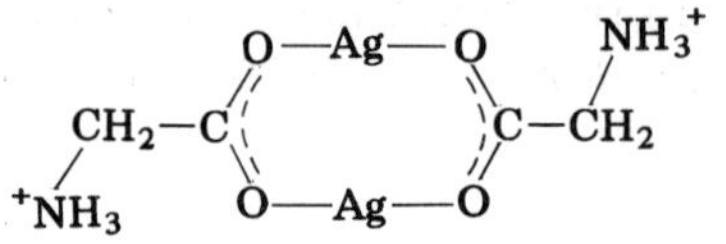

(X)

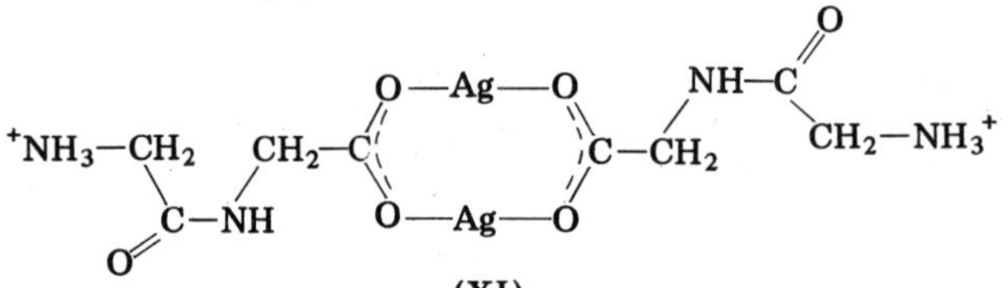

(XI)

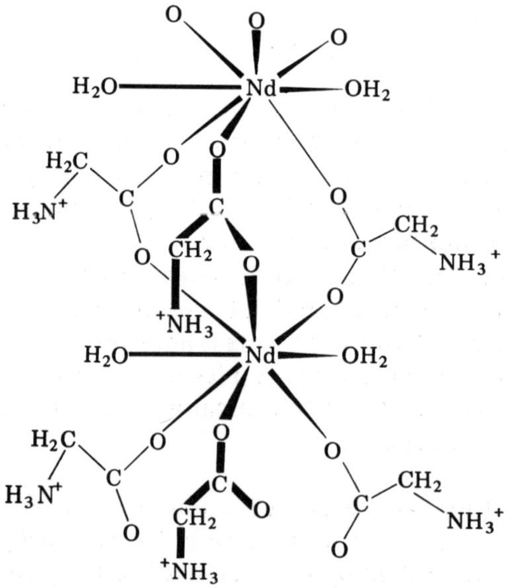

(XII)

the bond configuration about the carboxyl groups is *syn–syn*. The same type of metal–O(carboxyl) interaction, but with an *anti–anti* configuration, is found in $[Nd(HGly)_3(OH_2)_2]Cl_3 \cdot H_2O$ (XII)[36]. The large Nd(III) atoms are 8-coordinated. Six of the coordination positions are occupied by O(carboxyl) atoms belonging to six different glycine molecules. Neighbouring Nd(III) atoms are connected in the crystal by bridges consisting of the carboxyl groups of three glycines, all with the geometry of *type c*.

Type d. Unsymmetrical four-membered metal–carboxyl chelate rings occur in complexes where chelation through the carboxyl group and a second group is impossible or energetically unfavourable. Typically such interactions are found in crystal structures where Cu(II) and Zn(II) atoms are bound by the terminal carboxyl groups of peptides, or by the side-chain carboxyl groups of glutamic acid residues. For example, in $[Zn(Gly\text{-}Gly\text{-}Gly)](SO_4)_{\frac{1}{2}} \cdot 4H_2O$ (XIII)[37], the peptide molecule binds one Zn atom at its NH_2-terminal end and another at its COO-terminal end. The complexes form endless chains. Each Zn atom is coordinated by one peptide molecule through its terminal amino group and its first O(peptide) atom, and by another through both oxygen atoms of its carboxyl group. In $Zn(H_{-1}Glu) \cdot 2H_2O$ (XIV)[38], there

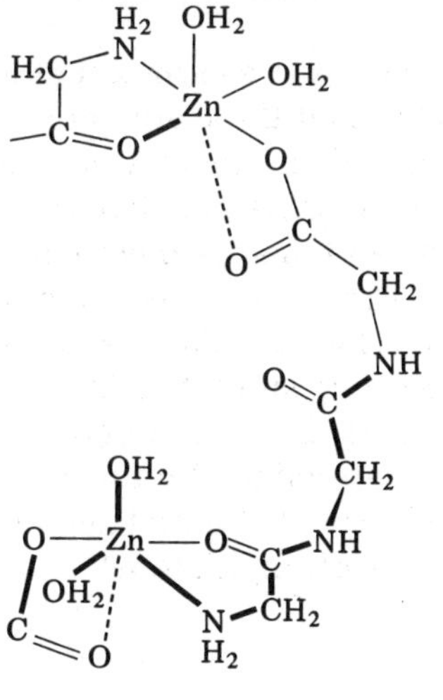

(XIII) 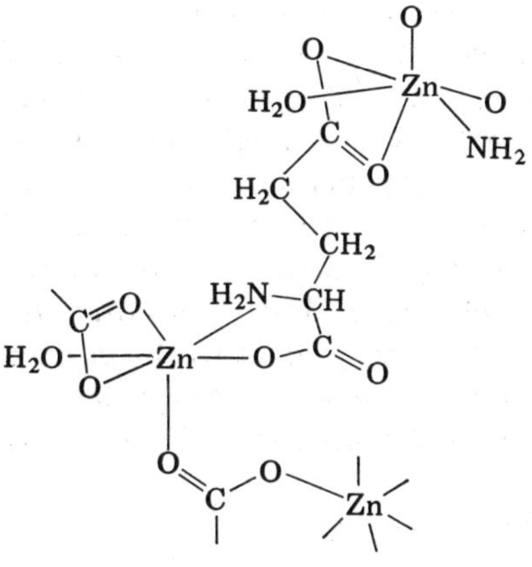   (XIV)

are *three different* Zn–O(carboxyl) interactions: one Zn–O(carboxyl) bond is in a normal amino acid chelate ring, one bond is formed with the "free" oxygen of a carboxyl group in a neighbouring chelate ring (*type b*), and two bonds are part of an irregular chelate ring with the side-chain carboxyl group of a third amino acid ligand (*type d*). In octahedral complexes this type of coordination causes distortions from the regular geometry, since the less strongly bonded O(carboxyl) atom inevitably occupies an irregular coordination position with respect to the metal. The displacement of two unidentate ligands by a single carboxyl group presumably compensates for the reduced enthalpy of formation caused by the less favourable bonding geometry.

Type e. A four-membered chelate ring with approximately equal metal–O(carboxyl) distances occurs in the crystals of [Ca(HGly-Gly-Gly)(OH$_2$)$_2$]-Cl$_2$·H$_2$O (XV)[39]. One of the O(carboxyl) atoms makes a contact with a second Ca ion. The Ca(II) atom is 7-coordinate. In addition to three Ca–O(carboxyl) bonds it forms two Ca–O(peptide) and two Ca–OH$_2$ bonds. Four different peptide molecules are involved in these contacts, and conversely each peptide molecule binds four different Ca(II) atoms.

Chelate rings (types a, b and c). The metal–O(carboxyl) interactions found in chelate rings are represented by *types a, b* and *c* in Fig. 1. Once one oxygen atom of a carboxyl group is involved in a chelate ring, the second O(carboxyl) atom can remain free to form hydrogen bonds or it may form a bond with a second metal atom in an adjacent complex (see, *e.g.*, I). Interactions of *types b* and *c* occur frequently in crystal structures, suggesting that they may also contribute to the cohesion of dimeric and polymeric complexes in solution.
    The relative bond-orders of the two C–O bonds in a carboxyl group can be deduced from the bond lengths, provided that these are known to sufficient precision. The bond lengths commonly lie in the ranges 1.25–1.30 Å and 1.22–1.27 Å, respectively. In a five-membered chelate ring the M–O(carboxyl)–C angle is always close to 114°, and in a six-membered chelate ring generally between 123° and 126°. There is no rule which forces the metal atom to lie in the plane of the carboxyl group, but the deviation is seldom greater than 0.5 Å in a five-membered ring and 0.8 Å in a six-membered ring. Where a second O(carboxyl)–metal bond is present, its direction with respect to the carboxyl plane is determined by the relative positions of adjacent complexes. Values of 113° to 125° have been observed for the M–O(carboxyl)–C angle at the second O(carboxyl) atom[9].

Peptide groups

    It is well known that peptide groups must remain — to a first approximation — planar in order to preserve the resonance between the canonical forms:

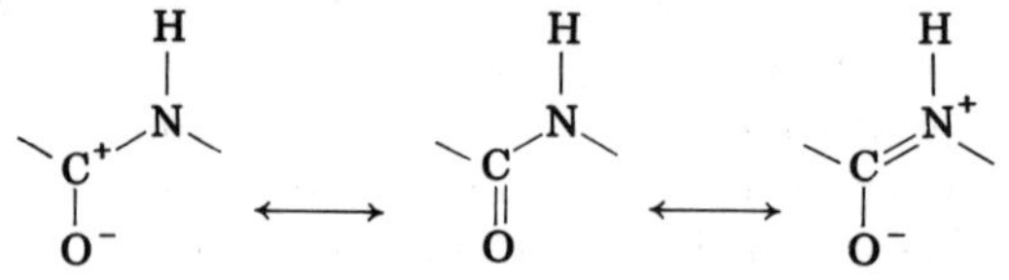

The O(peptide) atom is only weakly basic. Metal–binding is weak, and when it occurs it is often stabilized by the closure of a chelate ring with an adjacent terminal amino group. The N(peptide) atom binds a metal only when the process is accompanied by the dissociation of the peptide proton[40]. Metal-binding at the N(peptide) atom would otherwise imply the formation of a fourth bond, a change from trigonal sp^2 to tetrahedral sp^3 hydridization, and the loss of the peptide group resonance. Many structural formulae showing metal-binding at the nitrogens of protonated peptide groups appeared in the literature before this principle was understood. All such formulae should be disregarded.

O(peptide) atoms

There are, strictly speaking, only two complexes in which *non-chelating* interactions between metal ions and O(peptide) atoms have been found by crystal structure analysis. The first is 2HCys-Gly·NaI [or Na(HCys-Gly)$_2$I], in which the Na$^+$ ion lies 2.23 Å from each of two O(peptide) atoms and 2.16 Å from each of two O(carboxyl) atoms[41]. The second is [Ca(HGly-Gly-Gly)(OH$_2$)$_2$]Cl$_2$·H$_2$O (**XV**), which we have already

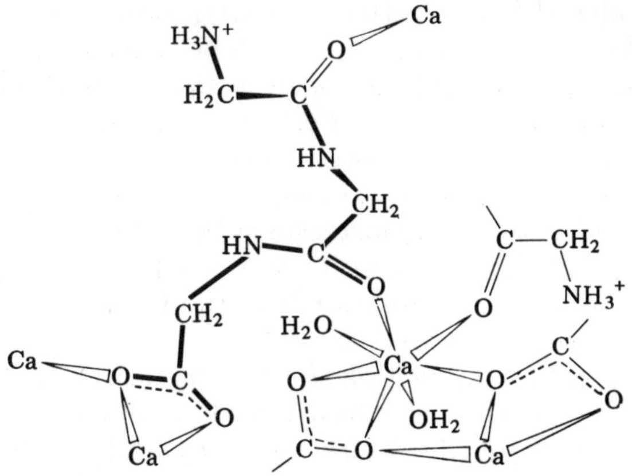

(XV)

described in connection with metal-O(carboxyl) binding (p. 130). In this complex there are two Ca–O(peptide) bonds per peptide molecule[39]. The peptide ligand has an unusual conformation, and the reported N-C$_\alpha$-C′ angle of the central glycyl residue (120°) is much larger than the average value in other peptides (111°).

The absence of structural data becomes slightly less acute if we admit complexes in which the ligands are not peptides but "peptide-like". Biuret, NH$_2$CONHCONH$_2$, acts as a unidentate ligand in the complexes Cd(BiuH$_2$)$_2$Cl$_2$ and Hg(BiuH$_2$)$_2$Cl$_2$ [42]. In both structures the biuret molecules form single metal–O(amide) bonds. The metal atoms are joined into infinite –metal–Cl_2–metal– ribbons by double chloro-bridges which complete the octahedral coordination. The formamide complex $Cd(HCONH_2)_2Cl_2$ has an analogous structure[43]. In the N-methylacetamide complex $Li(CH_3CONHCH_3)Cl$, all the amide C=O bonds point towards Li^+ ions and all the amide N–H bonds are directed at Cl^- ions[44]. In $Na(CH_3CONHCOCH_3)_2Br$, the diacetamide molecules pack with their O(amide) atoms at 2.3 Å from the Na^+ ions[45].

Interactions between peptide groups and alkali metal or alkaline earth ions obviously have a strongly ionic character, but there is evidence that they persist in solution. Nuclear magnetic resonance (n.m.r.) proton shifts show that metal–O(amide) interactions similar to those in the cited struc- ture exist in solutions of N-methylacetamide and Al^{3+}, Th^{4+}, Mg^{2+} and Li^+ ions; the metal–ligand bond strengths decrease in this order[46,47]. Without being specific about individual bonds, metal–O(carboxyl) and metal– O(peptide) interactions are also indicated by the fact that the solubilities of amino acids and peptides in water are altered in the presence of alkali and alkaline earth halides[48]. For instance, $[Ca(HGly-Gly-Gly)(OH_2)_2]Cl \cdot H_2O$ (XV) is only one of a series of stoichiometric complexes which the chlorides, bromides and iodides of Ca(II), Sr(II) and Ba(II) form with amino acids and peptides. Whenever such a complex can be isolated, it is also found that the solubility of the peptide is greater in a solution of the salt than in pure water[48]. Additional evidence for calcium–peptide inter- actions in solution comes from the converse observation that the solubility of calcium iodate in water is enhanced in the presence of glycylglycine and some other peptides and amino acids[49]. The increases in the solubilities of alkaline earth iodates have actually been used to derive stability con- stants for the metal–peptide complexes in solution[50]. Both the thermodyn- amic and kinetic stabilities of these complexes are low.

For *chelating* metal–O(peptide) interactions there is no dearth of supporting data. Among the complexes which have already been discussed, $Zn(Gly-Gly)_2 \cdot 2H_2O$ (II)[19] and $Zn(Gly-Gly-Gly)(SO_4)_{\frac{1}{2}} \cdot 4H_2O$ (XIII)[37] pro- vide examples of chelate rings where the donors are N(amino) in addition to O(peptide). The coordination of Cu(II) in $Cu(Gly-Gly-Gly)Cl \cdot 1\frac{1}{2}H_2O$ is similar to that of Zn(II) in XIII[51], and there is little doubt that this mode of chelation is a general phenomenon whenever a terminal amino group becomes involved in metal-binding — with the exception of complexes where N(peptide) atoms are deprotonated (see below). The peptide-like ligand biuret similarly acts as a bidentate chelator through both its O(amide) atoms in $Zn(BiuH_2)_2Cl_2$ (XVI)[52], as well as in the analogous complexes with Ni(II), Mn(II) (deduced from infra-red spectra[53,54]) and Cu(II) (shown by structure analysis[55]).

A variety of techniques have confirmed that the five-membered che- late rings with N(amino) and O(peptide) donors survive in metal–peptide solutions. The n.m.r. spectrum of glycylglycine in D_2O solution, for instance, has two proton resonances due to the two non-equivalent $-CH_2-$ groups. The addition of Cd^{2+} ions to the solution causes a greater chemical shift in one frequency than the other. The more sensitive signal must belong to the $-CH_2-$ group which is closest to the donor atoms, $i.e.$, the $-CH_2-$ between the NH_2- and peptide groups. It turns out that this signal is also the one which disappears first (due to selective paramagnetic line broadening) when small concentrations of Cu^{2+} ions are added to the solution, thus proving that the initial sites of chelation are the same for Cd^{2+} and Cu^{2+}. Up to this point the experiment merely identifies the protons to be associated with the n.m.r. frequencies, assuming that the donor groups are known. By extending the observations to Cd(II) complexes of amino acids and peptides with side-chains, it is possible to arrive at assignments which are independent of this assumption. The coordination sites in glycylglycine are thereby confirmed[56]. In the spectra of tri- and tetra-peptides at low pD, the line which disappears in the presence of Cu^{2+} ions inevitably belongs to the methylene protons of the NH_2-terminal amino acid residue, again leading to the result that chelation occurs through the N(amino) and first O(peptide) atoms[57].

The same conclusion is reached from the infra-red spectra of metal–peptide complexes in D_2O. When HL = glycylglycine, the peptide C=O frequency shifts from 1645 cm^{-1} in the free ligand to 1625 cm^{-1} in CuL^+. When HL = di- or tri-glycylglycine, CuL^+ contains free as well as coordinated O(peptide) atoms, and both the expected C=O frequencies are observed[58-60].

Finally, new light is thrown on the formation of these chelates by a careful calorimetric study of Cu(II)–peptide systems[18]. Some results from this study are quoted in Table I. Equilibrium constants are functions of both the enthalpies and entropies of reaction ($-RT \ln K = \Delta G = \Delta H - T\Delta S$). ΔH depends mainly upon the making and breaking of metal–ligand bonds, and to a smaller extent on changes in steric factors such as chelate-ring strain. The three complexes CuL^+ (L^- = Gly^-, Gly-Gly^- and Gly-Gly-Gly^-) have virtually the same enthalpy of formation, suggesting that the same type of complex is formed with all three ligands. This is in agreement with the structural evidence that 5-membered chelate rings are formed by both amino acids and peptides. As a corollary to Table I, the decrease in stepwise stability constants from $Cu(Gly)^+$ to $Cu(Gly$-Gly-$Gly)^+$ is seen to be entirely an entropy effect.

N(peptide) atoms

The labilization of peptide protons, and the formation of metal–N(peptide) instead of metal–O(peptide) bonds, are limited to metals which have a $CFSE$ which can be significantly increased by the substitution of strong-field (nitrogen) for weak-field (oxygen) donors. Those known at present are Co(III) (d^6), Co(II) (d^7), Ni(II), Pd(II), Pt(II) (all d^8), and Cu(II) (d^9). The ligand field effects of N(peptide) donors manifest themselves in different ways for different metal ions: by the stabilization of Co(III) with respect to Co(II) in the case of cobalt, by transitions from blue, paramagnetic, octahedral to yellow, diamagnetic, square-planar complex species in the case of Ni(II), by the very low pH values at which peptide protons are displaced by Pd(II) and Pt(II) (indicating the avidity with which both metals bind nitrogen donors), and by the characteristic colour change from blue through violet to pink (the "biuret reaction") when peptides are titrated with alkali in the presence of Cu(II).

TABLE I

THERMODYNAMIC FUNCTIONS FOR SOME REACTIONS OF Cu(II) WITH GLYCINE AND GLYCINE PEPTIDES (25°C, I = 0.1 M)[18]

[HG = HGly$^\pm$, HGG = HGly-Gly$^\pm$, HGGG = HGly-Gly-Gly$^\pm$. Numbers in parentheses are standard deviations right-adjusted to the least significant digit of the preceding quantity.]

Reaction	$\log_{10} K$	ΔG (kcal mole^{-1})	ΔH (kcal mole^{-1})	ΔS (cal deg^{-1} mole^{-1})
HGG$^\pm \rightleftharpoons$ GG$^-$ + H$^+$	−8.09(1)	+11.03(1)	+10.6(1)	−1.5(4)
HGGG$^\pm \rightleftharpoons$ GGG$^-$ + H$^+$	−7.87(1)	+10.73(1)	+10.1(1)	−2.1(4)
Cu^{2+} + G$^- \rightleftharpoons$ CuG$^+$	+8.62	−11.71(1)	−6.76(4)	+16.6(3)
Cu^{2+} + GG$^- \rightleftharpoons$ CuGG$^+$	+5.56(1)	−7.58(1)	−6.1(2)	+5.0(7)
Cu^{2+} + GGG$^- \rightleftharpoons$ CuGGG$^+$	+5.04(1)	−6.68(1)	−6.3(2)	+2(1)
CuGG$^+ \rightleftharpoons$ CuH$_{-1}$GG + H$^+$	−4.06(1)	+5.54(1)	+6.9(2)	+4.5(7)
CuGGG$^+ \rightleftharpoons$ CuH$_{-1}$GGG + H$^+$	−5.06(1)	+6.90(1)	+7.5(2)	+2.0(7)
CuH$_{-1}$GGG $\rightleftharpoons$ CuH$_{-2}$GGG$^-$ + H$^+$	−6.78(2)	+9.25(2)	+7.4(2)	−6.2(8)

The peptide complexes of Cu(II) and Ni(II) have been particularly amenable to investigation by a large range of experimental methods, because they are at the same time thermodynamically relatively stable and kinetically labile. Equilibria in Cu(II)- and Ni(II)-peptide systems are reached sufficiently rapidly (especially in the case of Cu(II)) for the effective use of potentiometric titrations and for the calorimetric measurement of thermodynamic parameters. The kinetics of reaction, however, have to be followed by stopped-flow or relaxation methods.

The results of a typical potentiometric study are shown in Fig. 2, which represents the titration of a tetrapeptide in the absence and presence of an equimolar concentration of Cu^{2+} ion[61]. The peptide is in its zwitterion form at pH 6, and only one equivalent of OH$^-$ is required to titrate a proton from the terminal –NH$_3{}^+$ group between pH 7 and pH 9. In the presence of Cu^{2+}, the –NH$_3{}^+$ proton is titrated at a much lower pH because Cu^{2+} competes with H$^+$ for the –NH$_2$ group. Three more protons are then titrated between pH 5 and pH 10. Clearly, if the species formed by the reaction of

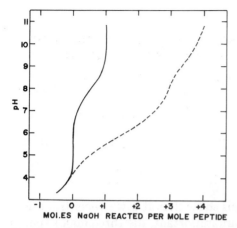

Fig. 2. Titration of a tetrapeptide, alanylglycylglycylglycine (———), and its Cu(II) complex (- - - - -) with sodium hydroxide. (Ionic strength = 0.16, T = 25.0°C)[61].

Cu^{2+} and L^- is denoted by CuL^+, then the dissociation of three protons leads successively to the formation of $CuH_{-1}L$, $CuH_{-2}L^-$, $CuH_{-3}L^{2-}$. The dissociation which yields $CuH_{-1}L$ does not occur if the second amino acid residue from the NH_2-terminus of the peptide is sarcosyl or prolyl, *i.e.*, if the first peptide group has no dissociable proton. Similarly, the formation of $CuH_{-2}L^-$ is prevented if one of the same residues occupies the third position along the peptide chain. The peptide groups are thereby established as the sites of the additional proton dissociations. Deprotonation makes the N(peptide) atoms available for metal-binding without loss of the peptide resonance energy. (In addition, the protons of coordinated H_2O molecules may be labilized after all the peptide groups capable of metal-binding have been deprotonated.)

Among metal–peptide systems, none has been investigated more intensively by potentiometric titrations than Cu(II)–diglycylglycine[62]. The titrations were made with a glass electrode in 3 M $NaClO_4$ solution for five metal:ligand ratios for each of five total metal concentrations (including zero). The data extended over the range $1.5 < pH < 11.0$ for total metal concentrations from 1 to 100 mM and total ligand concentrations from 2 to 250 mM. A second set of data was recorded for a series of solutions in which the ligand concentration was varied while the free hydrogen ion and total metal concentrations were kept constant. The free metal ion concentration was measured by means of a Cu–amalgam electrode. The data were processed by a combination of graphical and numerical methods[62].

As a result of this study, no less than thirteen complex species are revealed as contributing to the system. These are shown with their formation constants in Table II. The formation constants may be used to compute the composition of the equilibrium mixture present in solution under

TABLE II

Cu(II)-DIGLYCYLGLYCINE COMPLEX SPECIES $Cu_pH_qL_r^{+2p+q-r}$ AND THEIR LOG STABILITY CONSTANTS (L$^-$ = Gly-Gly-Gly$^-$, 25°C, 3 M NaClO$_4$)[62]

Species	log β_{pqr}	Species	log β_{pqr}	Species	log β_{pqr}
$CuHL^{2+}$	10.13	$CuH_2L_2^{2+}$	19.0	$Cu_2H_2L_2^{4+}$	21.0
CuL^+	5.66	CuL_2	10.17	$Cu_2HL_2^{3+}$	17.3
$CuH_{-1}L$	-0.13	$CuH_{-1}L_2^-$	3.91	$Cu_2L_2^{2+}$	13.12
$CuH_{-2}L^-$	-6.86	$CuH_{-2}L_2^{2-}$	-4.81	$Cu_2H_{-2}L_2$	1.44
				$Cu_2H_{-4}L_2^{2-}$	-13.4

any given conditions. Thus, calculations show that at low concentrations ($\sim 10^{-3}M$) only mononuclear species are important, the three major species being CuL^+, $CuH_{-1}L$ and $CuH_{-2}L^-$ (Table III). At higher concentrations

TABLE III

COMPOSITION OF Cu(II)-DIGLYCYLGLYCINE SOLUTIONS (CONCENTRATION $\sim 10^{-3}$ M)[62]

pH	Approximate percentage of Cu(II)[a] present as			
	Free Cu^{2+}	CuL^+	$CuH_{-1}L$	$CuH_{-2}L^-$
5	72	20	3	—
6	16	24	47	9
7	—	2	28	60
8	—	—	4	85
9	—	—	—	90

[a] Minor species: $CuHL^{2+}$ at pH < 6, CuL_2 at pH 6-7, $CuH_{-1}L_2^-$ at pH 7-9, and $CuH_{-2}L_2^{2-}$ at pH $\geqslant$ 9.

($\geqslant 10^{-2}M$) polynuclear species and species with a metal : ligand ratio of 1 : 2 become increasingly important. When $10^{-1}M$ solutions are titrated, 1 : 1 complexes never account for more than half the total Cu(II) at any pH (Table IV).

The chemical conclusions are clear. At low concentrations, the series

$$CuHL^{2+} \rightleftharpoons CuL^+ \rightleftharpoons CuH_{-1}L \rightleftharpoons CuH_{-2}L^-$$

accounts for most of the species involved, implying that the complexes undergo successive deprotonation reactions as the pH is increased. At higher concentrations of metal and ligand, there is evidence for

$$CuH_2L_2^{2+} \rightleftharpoons CuL_2 \rightleftharpoons CuH_{-1}L_2^- \rightleftharpoons CuH_{-2}L_2^{2-}$$

and

$$Cu_2HL_2{}^{3+} \rightleftharpoons Cu_2L_2{}^{2+} \rightleftharpoons Cu_2H_{-2}L_2 \rightleftharpoons Cu_2H_{-4}L_2{}^{2-}$$

It is convenient to arrange the complexes in series like these in order to discuss their structural relationships, but it must be remembered that all

TABLE IV

PREDOMINANT SPECIES PRESENT IN Cu(II)-DIGLYCYLGLYCINE SOLUTIONS (CONCENTRATION $\sim 10^{-1}M$)[62]

pH	Approximate percentage of Cu(II) present as			
	Free Cu^{2+}	1:1 complexes	1:2 complexes[a]	2:2 complexes
3	Cu^{2+} (60%)	$CuHL^{2+}$ (35%)	—	—
4	Cu^{2+} (15%)	$CuHL^{2+}$ (30%)	—	$Cu_2HL_2{}^{3+}$ (20%)
5	—	CuL^+ (20%)	—	$Cu_2L_2{}^{2+}$ (50%)
6	—	$CuH_{-1}L$ (20%)	CuL_2 (20%)	$Cu_2H_{-2}L_2$ (30%)
7	—	$CuH_{-2}L^-$ (30%)	$CuH_{-1}L_2$ (35%)	$Cu_2H_{-2}L_2$ (15%)
8	—	$CuH_{-2}L^-$ (45%)	$CuH_{-1}L_2$ (35%)	$Cu_2H_{-4}L_2{}^{2-}$ (10%)
9	—	$CuH_{-2}L^-$ (50%)	$CuH_{-2}L_2$ (30%)	$Cu_2H_{-4}L_2{}^{2-}$ (10%)

[a]Minor species: $CuH_2L_2{}^{2+}$ at pH 4. Most of the species listed also contribute significantly to the equilibria at pH values one unit above and one unit below those where they are the predominant species.

the 1:1, 1:2 and 2:2 complexes are in equilibrium with one another and with free ligand, metal and hydrogen ions.

Unfortunately such a detailed description requires the accumulation of massive equilibrium data, and our knowledge of other metal–peptide systems has often been based on much scantier experimental evidence. It has been a common practice to fit the stability constants for a postulated series of complex species to a small number of titration curves, or even to a single titration curve. There is no harm in this procedure, provided that the results are treated with the reservation that not all the species present in a given system may have been taken into account.

Some thermodynamic data for the deprotonation reactions of Cu(II)-di- and tri-peptides are included in Table I. The labilization of the peptide protons is seen to be largely an enthalpy effect. The deprotonations $(CuL^+ \rightarrow CuH_{-1}L \rightarrow CuH_{-2}L^-)$ are endothermic (ΔH positive), but appreciably less so than the dissociations of protons from the uncomplexed ligands $(HL \rightarrow L^-)$. The enthalpies of proton dissociation for CuL^+ and $CuH_{-1}L$ are approximately constant, leading to the surprising result that a more positive entropy contribution alone makes it easier to remove a proton from CuL^+ than from $CuH_{-1}L$. A partial explanation for the difference is that the deprotonation of CuL^+ yields a neutral complex, the positive entropy change being due mainly to the "chelate effect"; the loss of the next proton leads to a separation of charged ions (H^+ and $CuH_{-2}L^-$) to which water molecules become attracted in hydration shells with a reduction in their randomness[18].

The most direct evidence that the dissociation of peptide protons is accompanied by the formation of metal–N(peptide) bonds comes from X-ray crystal structure analyses. The structure of the species CuL^+, in which the metal is chelated only by the N(amino) and first O(peptide) atoms, has already been discussed, using $Cu(Gly\text{-}Gly\text{-}Gly)Cl \cdot 1\frac{1}{2}H_2O$ as an example (p. 132). Crystal structure analyses similarly provide illustrations of complex species in which Cu(II) is bonded to deprotonated peptide ligands. Examples are $CuH_{-1}L$ (**XXI, XXII**)[63,64], $CuH_{-2}L^-$ (in $Cu_2H_{-4}L_2{}^{2-}$, **XVII**)[65],

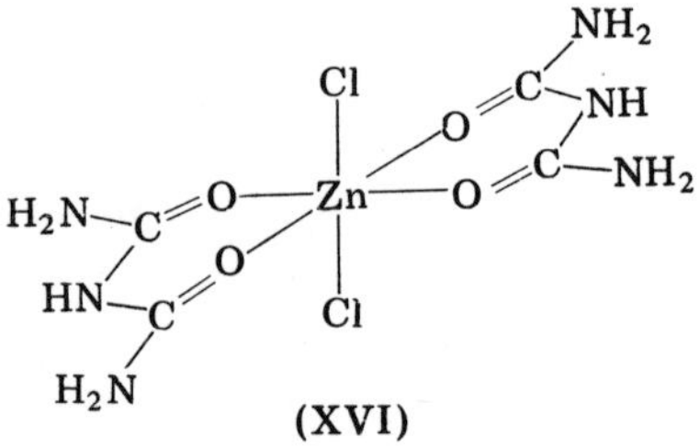

(XVI)

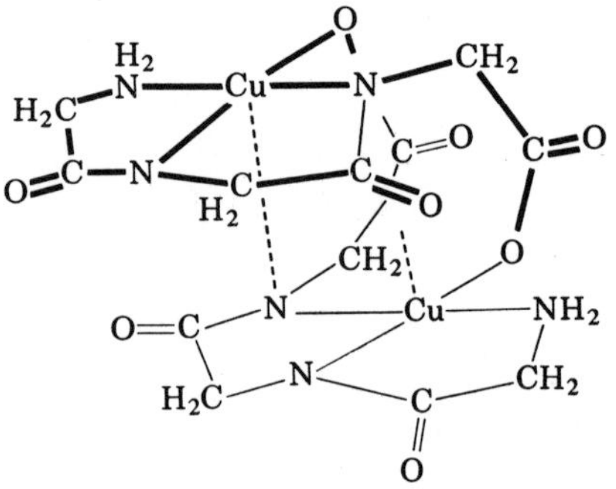

(XVII)

$[Cu(H_{-2}Gly\text{-}Gly\text{-}Gly)]_2{}^{2-}$
in $Na[Cu(H_{-2}Gly\text{-}Gly\text{-}Gly)] \cdot H_2O$

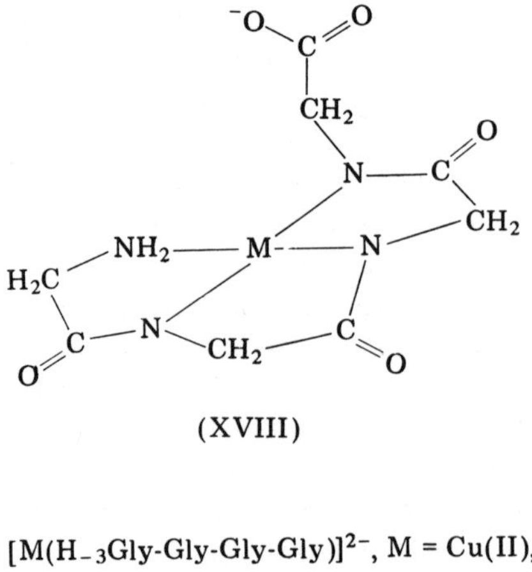

(XVIII)

$[M(H_{-3}Gly\text{-}Gly\text{-}Gly\text{-}Gly)]^{2-}$, M = Cu(II), Ni(II)

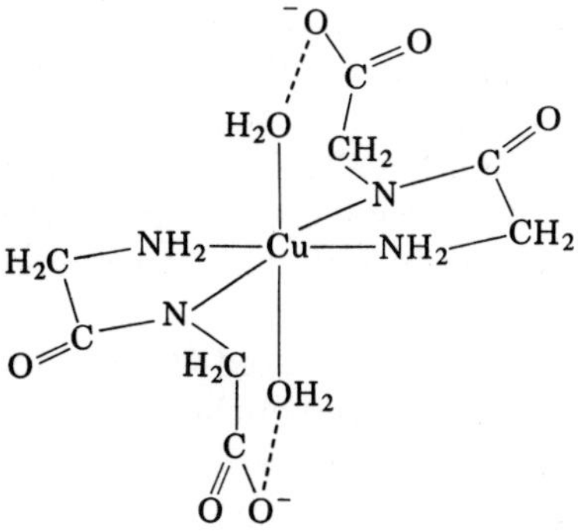

(XIX)

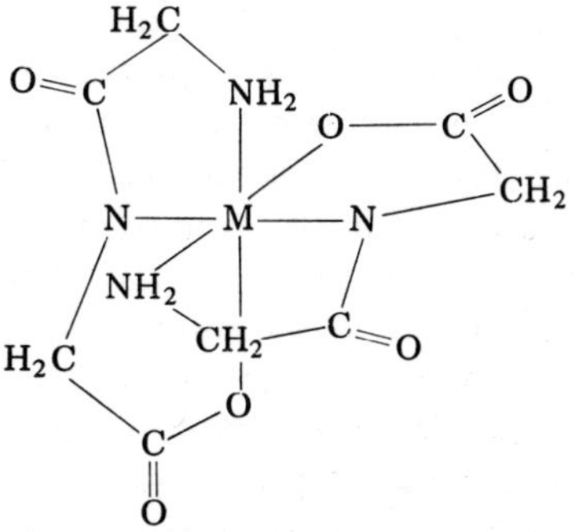

M = Ni(II), Co(III)

(XX)

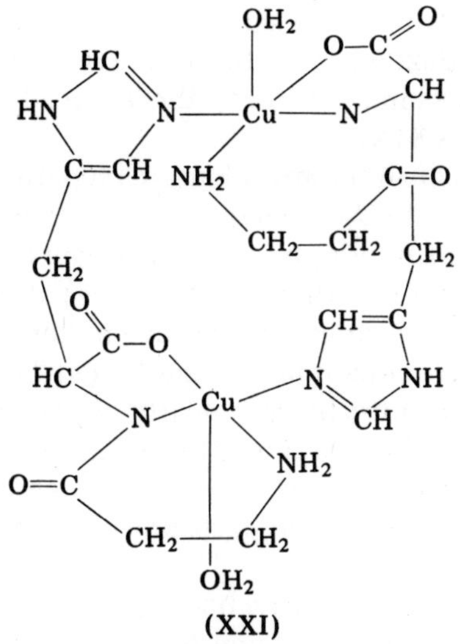

(XXI)

Dimeric complex in Cu(β-Ala-L-H$_{-1}$His) $\cdot$ 2H$_2$O

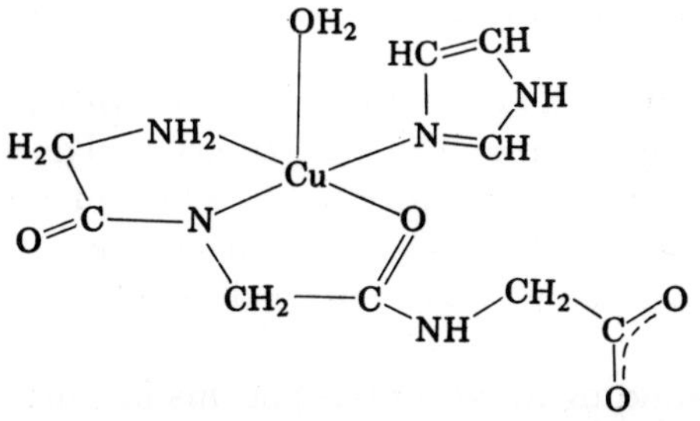

(XXII)

$CuH_{-3}L^{2-}$ (**XVIII**)[66], and $CuH_{-2}L_2^{2-}$ (**XIX**)[67]. A correlation exists between the structures and colours of these species. As the number of N(peptide) donors attached to a Cu(II) atom increases, the principal d–d transition of Cu(II) moves to shorter wavelengths. Simultaneously, there is a progressive reduction in the sum of the bond-orders of the two axial metal–ligand bonds until, in $CuH_{-3}L^{2-}$, the axial ligands are infinitely far away and Cu has coordination number 4[9].

While it is geometrically impossible for a coordinated O(peptide) atom to participate in more than one five- or six-membered chelate ring, coordination at a deprotonated N(peptide) atom assembles the ligand into an ideal conformation for the closure of two adjacent chelate rings having the metal–N(peptide) bond in common. There may thus be a thermodynamic bonus in the form of an increased chelate effect, since an additional unidentate H_2O molecule is displaced by the peptide ligand. The donors most likely to be in a chelating position adjacent to an N(peptide) are an N(amino) or N(peptide) atom on the NH_2-terminal side, and an N(peptide), O(peptide), O(carboxyl), histidyl N(imidazole) or methionyl S(thioether) on the COO-terminal side. Some examples in addition to those already given occur in formulae **XX, XXXVI, XXXIX**.

The participation of an N(peptide) atom in multidentate chelation is governed by two geometrical conditions. (1) Only small angular distortions from the normal trigonal configuration of bonds about the nitrogen are permitted. This condition accounts for the fact that an N(peptide) atom can be shared by two five-membered chelate rings, or by a five-membered and a six-membered ring, but not by two six-membered rings[63]. (2) The metal and the atoms of the peptide group must be very nearly coplanar. As a corollary, if an N(peptide) atom and the two donor atoms adjacent to it in a tridentate or tetradentate peptide are bonded to the same metal atom, then they must occupy coplanar coordination positions (as in **XVIII, XX, XXII**).

The dimensions of the peptide group are slightly, but significantly, changed by coordination at the N(peptide) atom[9]. When the metal is Cu(II), the C=O bond-length changes from 1.24 Å to 1.26 Å, and the C–N bond-length from 1.325 Å to 1.30 Å. This shows that the formation of the metal–N(peptide) bond decreases the contribution of the resonance structure $-\overset{|}{N}-\overset{|}{C}=O$ and increases that of $-\overset{|}{N}{}^+=\overset{|}{C}-O^-$, *i.e.* that the electron shift to the metal is smaller than that to the proton which the Cu^{2+} replaces. The same conclusion is reached from the decrease of the infra-red C=O frequency of Cu(II)-peptides in D_2O solution. As the pD is raised, the frequencies for free and coordinated C=O (1645, 1625 cm^{-1}) are replaced by a band at 1600–1610 cm^{-1} [58-60].

Our understanding of metal-binding at N(peptide) atoms in Cu(II)-peptide complexes is aided by the study of interactions between the same ligands and Ni(II). The existence of Ni(II) complexes in which the peptide groups are deprotonated has been demonstrated by preparative[68], potentiometric[60, 69-71], infra-red spectroscopic[60], and crystallographic[72] studies, which all produce results very similar to those for Cu(II)–peptides. The complex species of which crystal structure analyses have been made are $NiH_{-3}L^{2-}$ (XVIII) and $NiH_{-2}L_2^{2-}$ (XX)[72]. The thermodynamic stabilities of Ni(II)–peptides are somewhat lower than those of the corresponding Cu(II)–peptides. For instance, the formation constants of some Ni(II) complexes with HGly-Gly-Gly (followed by the values for Cu(II) from Table II) are $\log \beta_{101}$ = 3.7 (5.7), $\log \beta_{1-11}$ = -5.1 (-0.1), and $\log \beta_{1-21}$ = -12.8 (-6.9). Since the differences between successive constants represent the pK_a values of the peptide groups, it follows that the peptide protons are labilized less effectively by Ni(II) ($pK_a \sim 8$) than by Cu(II) ($pK_a \sim 6$).

The Ni(II)–peptides are, however, kinetically more stable than their Cu(II) counterparts. During potentiometric titrations it is necessary to allow five minutes for the equilibration of the solution after each addition of base[69]. Stopped-flow measurements have been made of the kinetics of the reverse reaction, namely the protonation of $MH_{-2}L^-$. The rate is 25 to 50 times slower for $Ni(H_{-2}Gly\text{-}Gly\text{-}Gly)^-$ than for $Cu(H_{-2}Gly\text{-}Gly\text{-}Gly)^-$ [71,73]. For both complexes the reaction proceeds much more slowly than a diffusion-controlled process, and the rate-limiting step is the rearrangement from metal–N(peptide) to metal–O(peptide) coordination.

At low pH, the solutions of Ni(II)–peptides are blue-green, showing that the complex species are octahedral and paramagnetic. The regular octahedral geometry enables Ni(II) to form complexes with metal:ligand ratios of 1:1, 1:2 and 1:3, so long as the peptides remain bidentate. With tripeptides or longer ligands, the dissociation of two or three protons as the pH is raised leads to the formation of yellow, square-planar, diamagnetic complexes in which the metal:ligand ratio is 1:1. A surprising consequence of the stability constants quoted for Ni(II)–diglycylglycine is that the second peptide proton is lost *more easily* than the first proton ($pK_{a2} = \beta_{1-11} - \beta_{1-21}$ = 7.7; $pK_{a1} = \beta_{101} - \beta_{1-11}$ = 8.8). In other words, the species $NiH_{-1}L$ (octahedral) is thermodynamically unstable with respect to both NiL^+ (octahedral) and $NiH_{-2}L^-$ (square-planar). The deprotonation is a (so called) cooperative reaction, and although $NiH_{-1}L$ must be an intermediate it never represents more than 5% of the Ni(II) in the solution[71].

The cooperative nature of the transition from NiL^+ to $NiH_{-2}L^-$ is a result of the change from octahedral to square-planar coordination, but at the same time the change in coordination geometry depends on the increase in the ligand field as N(peptide) atoms become bonded to the metal. Complexes of Ni(II) are either octahedral and paramagnetic, or square-planar and diamagnetic. There is not, as in the case of Cu(II), a gradation of geometries. The *CFSE* of square-planar Ni(II), expressed in units of $\Delta(10Dq)$, is greater than that of octahedral Ni(II). The stronger the ligand field is, the more the square-planar configuration is favoured by the increase in *CFSE*. Coordination by a single N(peptide) donor is insufficient to cause a change in geometry, but the accessibility of the square-planar configuration, with its higher *CFSE*, promotes the dissociation of *additional* peptide protons so as to make more strong-field donors available. In addition, the donor atoms in square-planar Ni(II)–peptide complexes are about 0.2 Å closer to the metal than in octahedral complexes[72], so that the ligand field effects are further enhanced. The entropy effect (release of axial ligands) works in the same direction.

The reaction between Ni(II) and glycylglycine is apparently anomalous. At pH $\sim$10, Ni(Cly-Gly)$_2$ (= NiL$_2$) loses two protons to yield the blue-green Ni(H$_{-1}$Gly-Gly)$_2^{2-}$ ion (= NiH$_{-2}$L$_2^{2-}$)[69]. The sodium salt of this ion has been crystallized in two hydrated modifications, whose crystal structures show that the complex is bis-tridentate and octahedral (**XX**). Yet, when HL = glycinamide, the species NiH$_{-2}$L$_2^{2-}$ is yellow and square-planar.

The d^8 elements below Ni in the Periodic Table are Pd and Pt. Like Ni(II), the divalent ions of these two metals induce the ionization of peptide hydrogens. They form square-planar complexes in which the deprotonated N(peptide) atoms are metal-binding sites. As one goes down the group, the crystal field stabilization by strong-field donors and therefore the effectiveness of the metal ions in labilizing peptide protons increase. In the presence of Pd(II), the peptide protons are titratable at pH 3.5[74] compared with pH 8–9 for Ni(II). When PtCl$_4^{2-}$ and a peptide are mixed in solution, deprotonation of the peptide groups occurs at an even lower pH. This is shown by the structure of Pt(Gly-L-Met)Cl·H$_2$O, a complex which crystallizes at pH 2.5 (formula **XXXIX**, see p. 157)[75]. The hydrogen atoms in this complex were located by neutron diffraction, so that there is no doubt that the peptide group is deprotonated while the carboxyl group is still neutral. [It does not follow from this observation that PtCl$_4^{2-}$–protein binding always involves deprotonated peptide groups, since the N(peptide) atoms in a protein are generally not as accessible as in a dissolved peptide molecule.]

Recently it has been reported that Co(II) also promotes the dissociation of peptide protons[76]. The reaction requires a pH around 10, and a metal:ligand ratio of 1:2. It yields an octahedral complex species in which the ligating atoms include two N(amino) and two N(peptide) atoms — four strong-field donors. In a strong ligand field the higher oxidation state of cobalt becomes favoured. Accordingly, solutions containing the species Co^{II}(H$_{-1}$Gly-Gly)$_2^{2-}$ have a strong tendency to undergo oxidation (via an oxygenated intermediate which is a reversible carrier of molecular oxygen) to Co^{III}(H$_{-1}$Gly-Gly)$_2^-$. The Co(III) complex species cannot be prepared directly by reacting Co^{III}(Gly-Gly)$_2^+$ with base, because Co(III) complexes characteristically resist reactions which involve the rupture of metal–ligand bonds.

The kinetic stability of Co(III)–ligand bonds can be exploited in experiments which are impossible with the more labile Ni(II)- and Cu(II)-

peptide systems. The structure of the $Co^{III}(H_{-1}Gly\text{-}Gly)_2^-$ ion in four different salts is the same as that of $Ni(H_{-1}Gly\text{-}Gly)_2^{2-}$ (formula **XX**)[77,78]. The only symmetry element of both ions is a two-fold symmetry axis, and they should exist as enantiomers. This can be proven for the Co(III) complex (but not for the Ni(II) complex) by resolving the enantiomers by starch-column chromatography[79]. When a strong acid is added to a solution of $Co(H_{-1}Gly\text{-}Gly)_2^-$, the anions are rapidly and reversibly protonated to give $Co(Gly\text{-}Gly)_2^{+}$[76,77,79]. A structure analysis of $Co(Gly\text{-}Gly)_2(ClO_4)$ has been used to prove that protonation leaves the metal–ligand bonds intact, and that the added protons become attached to the O(peptide) atoms[78]. The protonated cations with Co(III) still bonded to N(peptide) are thermodynamically unstable, and appear to undergo a slow reductive change to Co(II)[77].

N(imidazole) atoms

Histidyl side-chains are important metal-binding groups both in natural metalloproteins (*e.g.* carboxypeptidase (Chapter 15), myoglobin and haemoglobin (Chapter 25)) and in metal–protein complexes prepared in laboratory studies (Chapter 7). In many of these metal–protein interactions the functional groups attached to the metal belong to amino acid residues which are not adjacent to one another in the protein chain. The histidyl residues are therefore typically non-chelating.

Non-chelating metal–imidazole interactions

The simulation of these metal–protein interactions by simpler systems is made difficult by the fact that histidine and histidyl-peptides have a great tendency to act as chelating ligands. The only complex in which an imidazole side chain has been shown to act in a non-chelating fashion is $Cu(\beta\text{-Ala-L-H}_{-1}His)\cdot 2H_2O$ ("copper carnosine", **XXI**)[63]. What is particularly interesting about this complex is that the Cu(II) is bound to the N(3) atom of the imidazole ring (as in the Cu(II) complex of sperm whale metmyoglobin[80]), whereas in histidine and histidyl peptide chelates the imidazole N(1) atom is invariably involved. It is not yet proven that the structure found in the crystalline state persists in solution, but it leads to a satisfactory interpretation of the antecedent potentiometric titration data[63]. The existence of a dimeric species has been shown, independently of any structural model, from the line-broadening of the e.p.r. spectra of frozen solutions of the complex[81].

One device which has been used to circumvent the chelating tendency of histidine has been the preparation of mixed–ligand complexes in which histidyl side-chains are represented by imidazole molecules. A mixed complex of this type was first shown to exist in solutions at pH 5.8 to 7, as a result of the observation that the rate of imidazole-catalyzed hydrolysis of *p*-nitrophenylacetate is much reduced in the presence of Cu(II) and glycyl-glycine[82]. Mixed complexes of imidazole and glycine, glycylglycine or diglycylglycine with Cu(II), Ni(II) and Cd(II) have subsequently been iso-lated and characterized[64,83,84], and the crystal structures of a number of them have been solved[64]. An example is $[Cu(H_{-1}Gly\text{-}Gly\text{-}Gly)(ImH)(OH_2)]\cdot$ H_2O (**XXII**). Finally, there is much spectroscopic and magnetic evidence[85-87] concerning complexes which contain only imidazole and no amino acid or peptide ligands. Four complexes of this type whose structures have been determined are $[Cu(ImH)_4]I_2$ (cationic)[88], $[Zn(ImH)_6]Cl_2\cdot4H_2O$ (cationic)[89], $Zn(ImH)_2Cl_2$ (molecular complex linked by hydrogen bonds)[90], and $[Zn(Im)_2]_\infty$ (polymer)[91] (**XXIII–XXVI**).

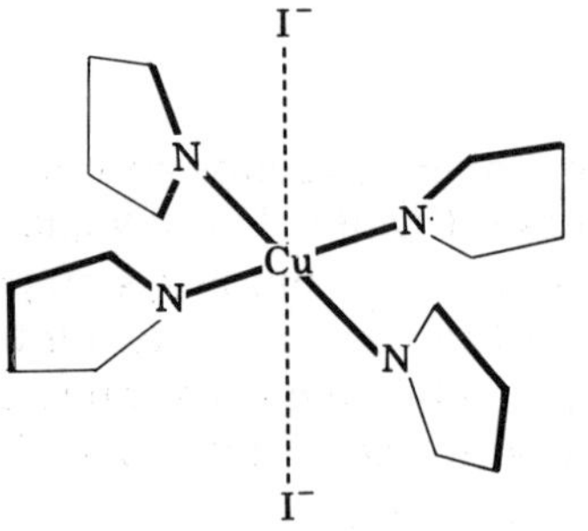

(**XXIII**)

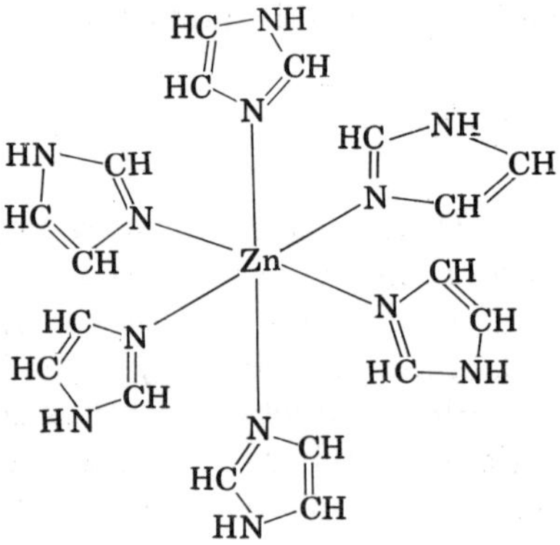

(**XXIV**)

$[Zn(ImH)_6]^{2+}$

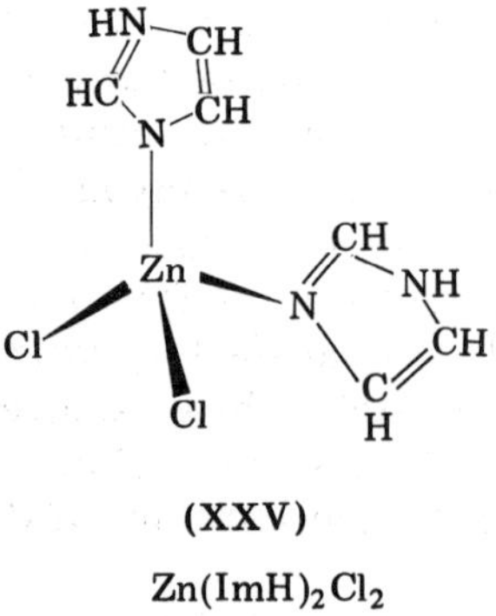

(**XXV**)

$Zn(ImH)_2Cl_2$

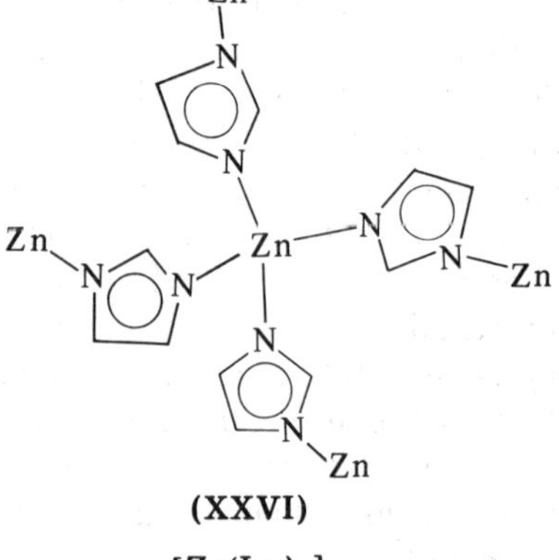

(**XXVI**)

$[Zn(Im)_2]_\infty$

The most useful information derived from the structural studies of such model compounds — mixed-ligand and pure imidazole complexes, as well as histidine chelates — has been that the imidazole group has great flexi-bility as a ligand. The metal–N(imidazole) bond can lie up to 30° from the imidazole plane (as it does in $Cd(L\text{-}His)_2\cdot5H_2O$[92]) and the M–N(imidazole)–C angles at the donor atoms range from $121°$ to $131°$. There is evidence from the structures of $[Cu(H_{-1}Gly\text{-}Gly)(ImH)(OH_2)]\cdot1\frac{1}{2}H_2O$ and $[Cu(H_{-1}Gly\text{-}Gly\text{-}Gly)(ImH)(OH_2)]\cdot H_2O$ (**XXII**)[64] that sterically unhindered imidazole rings tend to be coplanar with the M–N(imidazole) bond and three other metal–ligand bonds (and therefore, in the case of Cu, perpendicular to the singly occupied d_{z^2} orbital). Two imidazole rings occupying adjacent coordination positions in an octahedral or square-planar complex would get in each other's way if they were coplanar, and such rings are rotated about the M–N(imidazole) bonds far enough to make acceptable contacts[64]. On the whole, rotation about a M–N(imidazole) bond appears to be governed by the steric requirements of the rest of the histidyl residue, by contacts between the imidazole group and neighbouring ligands, and by hydrogen-bonding of the free ("pyrrole") N(imidazole) atom, rather than by electronic factors.

The flexibility of imidazole–metal bonds may be one reason for the effectiveness of histidine side-chains as metal-binding sites in proteins. Another reason may be the simple fact that imidazole groups are good enough electron donors to form strong covalent bonds. Calorimetric measurements show that metal–N(imidazole) bonds contribute only slightly less than metal–N(amino) bonds to the enthalpy changes for complexation reactions[116,157]. The enthalpy changes involved in complexing M^{2+} ions in solution to give $M(ImH)_4^{2+}$ are -23.0, -18.4 and -16.2 kcal mole^{-1} when M = Cu, Ni and Zn, respectively. The first-step enthalpy changes for the complexation of the same metal ions by one molecule of imidazole are -7.6, -5.8 and -3.8 kcal mole^{-1} [158]. The most important contributions to the metal–N(imidazole) bond energies come from σ-bonding. π-Bonding and *CFSE* effects are also possible. The evidence for $d_\pi\text{-}p_\pi$ bonding varies from complex to complex (see below), and where *CFSE* is present at all it plays a smaller part in stabilizing bonds with N(imidazole) than with N(amino) and N(peptide) atoms. The position of N(imidazole) in the spectrochemical series seems to be near H_2O (*i.e.*, below N(amino) and N(peptide)[9,93]).

Curiously, no correlation has been found between metal–imidazole bond-lengths and the rotations of the imidazole rings about the metal-nitrogen bonds[64], but evidence for π-electron delocalization is provided by chemical, magnetic and spectroscopic properties. (1) In $[Cu(ImH)_4]I_2$ (**XXIII**), the metal is stabilized in oxidation state II so that it does not undergo the usual reduction in the presence of I^- ions ($Cu^{2+} + I^- \rightarrow Cu^+ + \frac{1}{2}I_2$). Clearly the exceptional stability of Cu(II) in this complex is due to interaction with the four imidazole ligands. These lie in planes which are perpendicular to the plane of the four Cu–N bonds, and this orientation is favourable for $d_{xy}\text{-}p_\pi$ overlap. Appreciable delocalization of the unpaired electron from the Cu atom into the imidazole rings is indicated by copper hyperfine splitting and also hyperfine splitting due to nitrogen atoms in the electron spin resonance (e.s.r.) spectrum of the complex at $80°K$[88].

(2) A broad band in the ultra-violet spectrum of $Co(CO_3)(ImH)_2(OH_2)_2$, has likewise been attributed to π-electron delocalization. The complex is irregular octahedral, and each imidazole ligand lies in a plane roughly perpendicular to that of its own and three other metal-donor bonds[94]. (3) In the polymeric complex $[Co(Im)_2]_\infty$, the Co(II) atoms are tetrahedrally coordinated. The magnetic moment is abnormally low, and this is explained by superexchange coupling of the Co(II) atoms, i.e., interactions across the mobile π-electron systems of the imidazolato ligands which link the metal atoms[86]. (4) $[Cu(Im)_2]_\infty$ crystallizes in blue, green and brown polymeric modifications[95]. (The blue form is known to contain 4-coordinated Cu(II) atoms linked into a three-dimensional polymer by bidentate imidazolato ligands[96].) In the blue and green modifications there are strong interactions between Cu atoms through the imidazole rings, shown in each case by a large negative Weiss constant (the absolute temperature where the reciprocal of the magnetic susceptibility is zero)[95].

The third property of imidazole groups which accounts for their popularity as metal-binding sites is their availability at physiological pH. In a series of histidyl peptides, $pK_{ImH_2^+}$ lies between 5.5 and 7.2, three pK units below $pK_{NH_3^+}$[97]. A transition metal ion bonded to N(imidazole) does not gain as much *CFSE* as one bonded to N(amino), but it does not have to compete so hard against protons for the nitrogen electron-pair. In the case of histidine itself, the balance between the imidazole and amino groups as donors provides an intriguing problem in coordination chemistry.

Chelation by histidine

The partly protonated form of the ligand, H_2His^{2+} is represented in formula **XXVII**. The acid dissociation constants for the sequence $H_3His^{2+} \rightleftharpoons$

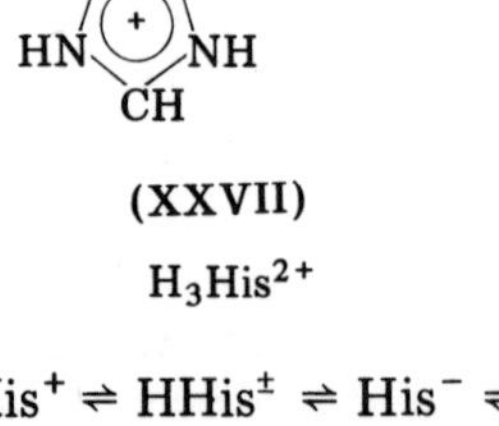

(XXVII)

H_3His^{2+}

$H_2His^+ \rightleftharpoons HHis^\pm \rightleftharpoons His^- \rightleftharpoons H_{-1}His^{2-}$ are $pK_{COOH} = 1.8$, $pK_{Imidazolium} = 6.0$, $pK_{NH_3^+} = 9.1$ and $pK_{Imidazole} = 14$. It is very tempting to assume that the sequence in which the protons are removed by titration is also the sequence in which potential donor atoms are used in metal-binding as the pH

increases. According to this sequence, the steps in metal-histidine complexation must be: (i) unidentate coordination via O(carboxyl) at low pH;
(ii) metal-binding at the N(imidazole), leading to the formation of a seven-
membered chelate ring; (iii) binding at the N(amino) atom, making histidine
a tridentate ligand; and possibly (iv) the dissociation of the second proton
from the imidazole group.

The published crystal structures of metal–histidine complexes do not
cast much light on the order of the proton dissociation and metal-binding
reactions. With one exception (p. 150) they deal with the pH range where
histidine acts as a tridentate chelator (step (iii), above). The Co(II)[98,99] and
Ni(II)[100] complexes of histidine are octahedral, as expected (**XXVIII**). The
Zn(II) complex is based on a tetrahedral structure with strong Zn–N(amino)
and Zn–N(imidazole) bonds; there are weaker Zn–O(carboxyl) bonds in
directions which emerge from two of the tetrahedron faces (**XXIX**). The

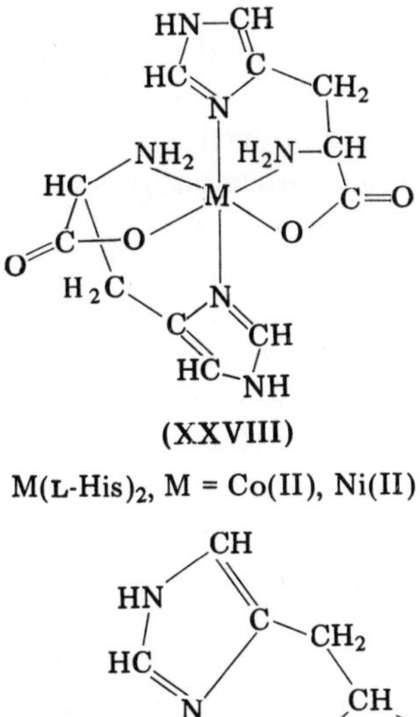

(**XXVIII**)

$M(\text{L-His})_2$, M = Co(II), Ni(II)

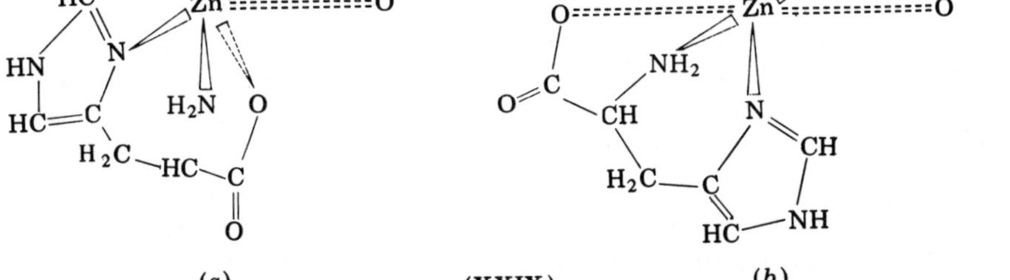

(a) (**XXIX**) (b)

$Zn(\text{L-His})_2$ in (a) $Zn(\text{L-His})_2 \cdot 2H_2O$ and (b) $Zn(\text{D-His})(\text{L-His}) \cdot 5H_2O$

Zn(II) derivative of DL-histidine crystallizes with a structure in which there are equal numbers of Zn(L-His)$_2$ and Zn(D-His)$_2$ complexes. The Zn(L-His)$_2$ complexes in Zn(L-His)$_2\cdot$2H$_2$O and Zn(D-His)(L-His)$\cdot$5H$_2$O, however, have different configurations[100a]. The significance of this fact in terms of "stereoselectivity" has been discussed elsewhere[9]. The histidine complexes of Cd(II) behave similarly to those of Zn(II), but have a less irregular geometry[92].

Evidence that a reaction sequence consistent with the order of pK_a values is followed in many cases has been adduced from infra-red spectra (although the observations, regrettably, have not yet been extended into the region where metal–ligand frequencies can be observed)[32]. At very low pH (pD = 2.2–3 in D$_2$O solution), complexation with 1:1 ratios of Mn(II), Fe(II), Co(II), Ni(II), Zn(II) and Cd(II) causes similar shifts in the infra-red antisymmetric carboxylate frequencies of both α-alanine (1615 cm^{-1}) and histidine (1625 cm^{-1}). This indicates unidentate coordination at the –COO$^-$ group. Above pD = 3, a band assigned to the ring-stretching vibration of imidazole (1490–1501 cm^{-1}) appears in the spectra of the histidine complexes. Between pD 4 and pD 7, another band assigned to the NC$_\alpha$-C(carboxyl) antisymmetric stretching vibration (1110 cm^{-1}) is replaced by a new band (1100 cm^{-1}) corresponding to coordinated N(amino). (For Cu(II), Zn(II) and Cd(II), precipitation prevents observations at pD values high enough for metal–N(amino) binding.)

The literature is replete with formation constants[101-104] and enthalpy-entropy data[17, 105, 106] for metal–histidine complexes, but most of these again refer only to the formation of the species M(His)$^+$ and M(His)$_2$ (step (iii)). The Co(II)–histidine system is one of the few which have been studied over a wide pH range and by a variety of experimental methods.

The interest in this system arises partly from the ability of Co^{II}(His)$_2$ to form an oxygenated derivative from which molecular oxygen can be recovered[76,107-109], but this is not a theme which will be explored here. Measurements of contact shifts in the p.m.r. spectra of Co(II)–histidine solutions[31] show that unidentate coordination at O(carboxyl) takes place at pH < 4, that octahedral 1:1 and 1:2 complexes in which histidine is tridentate are present at 4 < pH < 11, and that a tetrahedral complex is formed at even higher pH. The existence of the octahedral complex is confirmed by the crystal structure analyses of both Co(L-His)$_2\cdot$H$_2$O (**XXVIII**)[98] and Co(L-His)(D-His)$\cdot$2H$_2$O[99]. The crystalline Co(II) derivative of racemic histidine contains exclusively Co(L-His)(D-His) complexes[99], in agreement with the deduction from p.m.r. chemical shifts[31] and potentiometric titrations[110] that this species has a greater stability in solution than the optically pure species. The proposed tetrahedral structure of the blue high-pH complex Co(L-H$_{-1}$His)$_2{}^{2-}$ is supported by both potentiometric titrations and the assignment of u.v.–visible absorption bands[111]. This species is formed after the dissociation of the second ("pyrrole") protons from the imidazole groups. The histidine ligands are bidentate, the donors are N(amino) and N(imidazole), and the tetrahedral configuration is attributable to charge repulsion.

A second, particularly thorough p.m.r. spectroscopic study has dealt with 1:2 Pt(II)-L-histidine complexation[30]. Chemical shifts indicate that — as expected for Pt(II) — the coordination is square-planar, and that the donor atoms are N(amino) and N(imidazole) over the entire pH range $1 \leqslant pH \leqslant 12$. Large up-field shifts for C_α:H and C_2:H as the pH is raised clearly show the dissociation of protons from both the carboxyl and imidazolium groups. Histidine therefore chelates Pt(II) in the forms **XXX** and **XXXI** at low and high pH, respectively. The complexes exist as two sets of

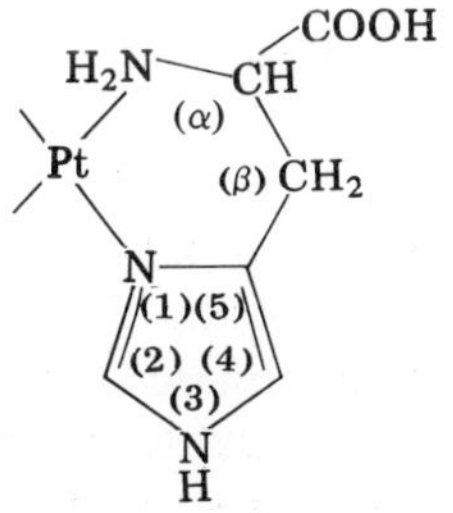

(XXX)

Coordination in $[Pt(HHis)_2]^{2+}$

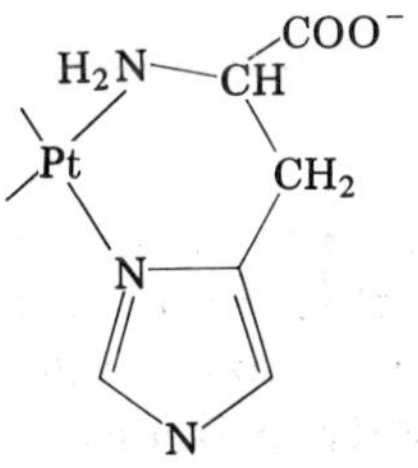

(XXXI)

Coordination in $[Pt(H_{-1}His)_2]^{2-}$

isomers, and complete analyses of the spectra have been made for both sets. The chemical shifts of C_2:H enable one series to be identified as *cis* and the other as *trans*. Finally, the effects of $^1H-^1H$ and $^{195}Pt-^1H$ coupling lead to the assignment of chelate ring conformations to all the species.

For the analogous Pd(II) complex, Pd(L-His)$_2$, the wavelength of the u.v. absorption maximum and the circular dichroism (c.d.) pattern (both compared with similar data for other amino acid complexes) indicate that the structure is square-planar, with Pd-N(amino) and Pd-N(imidazole) bonding[74].

Pd(II) and Pt(II) therefore provide examples where the order of metal-binding is not at all the order in which the potential donor atoms are deprotonated in the absence of metal ions. The reason, no doubt, is to be found in the large increases in *CFSE* when Pd(II) and Pt(II) are bonded to N(amino) and N(imidazole) instead of C(carboxyl) and N(imidazole) (see p. 142). This effect can influence the behaviour of only those metal ions which are sensitive to crystal-field stabilization, but there is a second reason why the order of the pK_a values may not always be followed in metal-binding. Step (ii) of the proposed sequence leads to the formation of a seven-membered chelate ring (**XXXII**), for which the entropy change is

(**XXXII**)

7-Membered chelate ring in proposed structure of $M(HHis)^{2+}$

known to be unfavourable. Seven-membered chelate rings are seldom found in other coordination compounds, and (despite the infra-red data[32]) it remains to be explained why they become stabilized when histidine and histidyl peptides chelate metals via O(carboxyl) and N(imidazole) atoms.

In the case of Cu(II)–histidine interaction, there is a direct and unresolved conflict between the infra-red data for the 1:1 complex in D_2O solution[32], and the crystal structure analysis of a 1:2 complex. The infra-red spectra indicate that Cu–O(carboxyl) binding at very low pH is followed at slightly higher pH by the formation of a seven-membered chelate ring in which N(imidazole) is the second donor. Temperature-jump and stopped-flow measurements of the reaction between Cu^{2+} and $HHis^{\pm}$ in the pH range 2.5 to 4.0 can be correlated with the formation of complexes $Cu(HHis)^{2+}$ and $Cu(HHis)_2^{2+}$, in which the rate-determining steps are the closures of seven-membered [O(carboxyl), N(imidazole)] chelate-rings. The authors of the data are unable to reproduce the observed kinetics if their calculations are based on the formation of five-membered [O(carboxyl), N(amino)] chelate-rings[112]. The kinetic calculations do not appear to take account of the existence of species other than $Cu(HHis)^{2+}$ and $Cu(HHis)_2^{2+}$. The most recent and extensive of many potentiometric titration studies of the Cu(II)–histidine system[104] shows that the species $Cu(His)^+$ and $Cu(HHis)(His)^+$ would represent up to 7% and 24%, respectively, of the total Cu under the conditions of some of the reaction rate measurements. In contradiction of the infra-red and kinetic evidence, the crystal structure analysis of a complex crystallized at pH 3.7, $[Cu(L\text{-}HHis)_2(OH_2)_2](NO_3)_2$, shows that the histidine molecules are coordinated via their N(amino) and O(carboxyl) atoms, that both nitrogens of each imidazole group are protonated, and that the N(imidazole) atoms are not involved in chelation (**XXXIII**)[113].

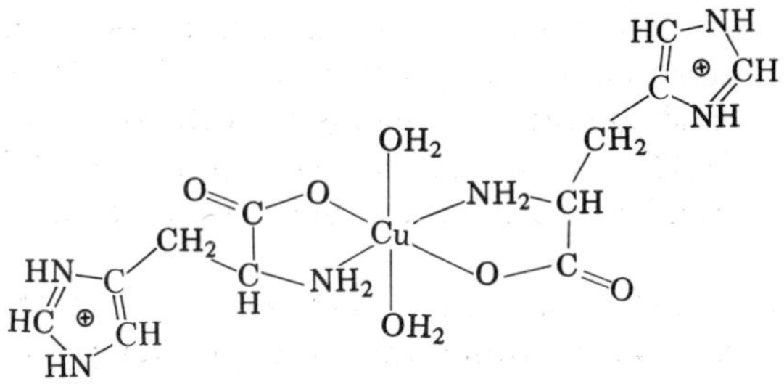

(XXXIII)

Whatever the structures of the Cu(II)–histidine complexes Cu(HHis)$^{2+}$ and Cu(HHis)$_2$$^{2+}$ may be at low pH, there is substantial agreement that the coordination in the species Cu(His)$^+$ and Cu(His)$_2$ at higher pH values is mainly through the N(amino) and N(imidazole) atoms. Involvement of the N(amino) atom in the range from pH 5 to pH 11 is shown by infra-red spectral data of the type mentioned earlier (p. 148)[32], and direct evidence of Cu-binding at the imidazole group is provided by the paramagnetic line-broadening of the n.m.r. signals of the C$_2$:H and C$_4$:H protons of the imidazole ring. The formation constants[115], enthalpies of formation[105,116], and degradabilities by H$_2$O$_2$[114] of the Cu(II)–histidine complexes are very similar to those of the Cu(II)–histamine complexes formed under the same conditions. (Histamine has no carboxyl group, so that the metal *must* be bonded to the N(amino) and N(imidazole) atoms.) A similar argument has been based on the o.r.d. spectrum of Cu(His)$_2$ at pH 8. This spectrum does not change dramatically if histidine is replaced by its methyl ester (showing that the carboxyl group cannot be strongly coordinated), but is quite different from the o.r.d. spectrum of the complex with l-methylhistidine in which coordination of the N(imidazole) is prevented[117]. These conclusions are supported by X-ray structural results. In the complex [Cu(L-His)(L-Thr)-(OH$_2$)]·H$_2$O which predominates in mixed-ligand solutions at pH 6–8, the histidinato ligand forms strong Cu–N(amino) and Cu–N(imidazole) bonds, with a much weaker Cu–O(carboxyl) bond in an irregular direction (XXXIV)[118]. There has, however, been a suggestion based on a recent study

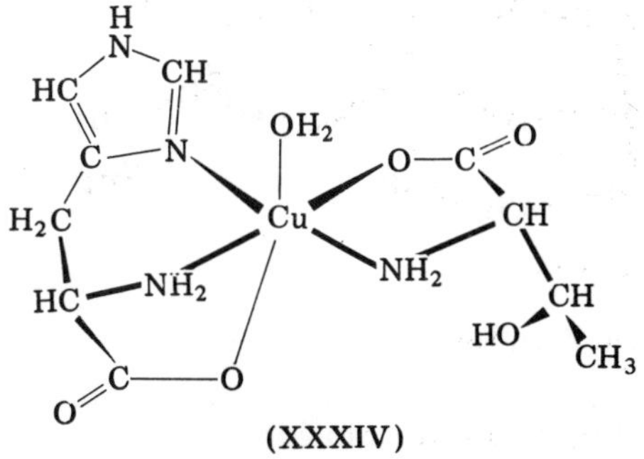

(XXXIV)

of the d–d o.r.d. and c.d. spectra of some mixed-ligand complexes[119], that at pH 8.3 Cu(His)$_2$ itself is a *mixed* complex in which one histidinato residue is tridentate as in **XXXIV**, and the other is chelated via the N(amino) and O(carboxyl) atoms as in **XXXIII**. At the time of writing, this work has been challenged[114,117]. Clearly, the last word has not yet been written on this interesting topic.

With the exception of the Cu(II) complex of carnosine[63,120-122], the metal chelates of histidyl peptides have aroused far less controversy and passion than those of histidine. The reader is referred to the original literature[93,97].

Side-chain O(hydroxyl) atoms

The side-chain hydroxyl groups of serine, threonine and tyrosine appear not to act as metal-binding sites. A p.m.r. study of serine in the presence of Cd^{2+} ions gives no sign of proton chemical shifts due to Cd–OH binding[56]. The enthalpy of formation of Cu(Ser)$_2$ at pH 9.1 to 10.1 (–12.6 to –13.4 kcal mole^{-1}) agrees well with the ΔH values for Cu(Gly)$_2$ and Cu(Ala)$_2$ (–12 to –13 and –11 to –12 kcal mole^{-1}, respectively) where there is no question of interaction with functional side-chains[106]. [At pH 12, ΔH for Cu(Ser)$_2$ rises abruptly to –23 kcal mole^{-1}, possibly owing to the formation of hydroxo-bridged polynuclear species[106].] For other first-row transition metals, the stability constants of serine and threonine complexes on the one hand, and α-alanine and α-aminobutyric acid complexes on the other hand, are so similar that the same type of chelation is presumed to be common to all of them[16]. The crystal structures of Ni(L-Ser)$_2$(OH$_2$)$_2$[123], Cu(L-Ser)$_2$[124], and Zn(L-Ser)$_2$[125] confirm that the coordination is simply through N(amino) and O(carboxyl). In Zn(L-Ser)$_2$ there is a non-chelating interaction between a serine side-chain in one complex and the Zn atom of a neighbouring complex (**XXXV**), but the quoted thermodynamic results

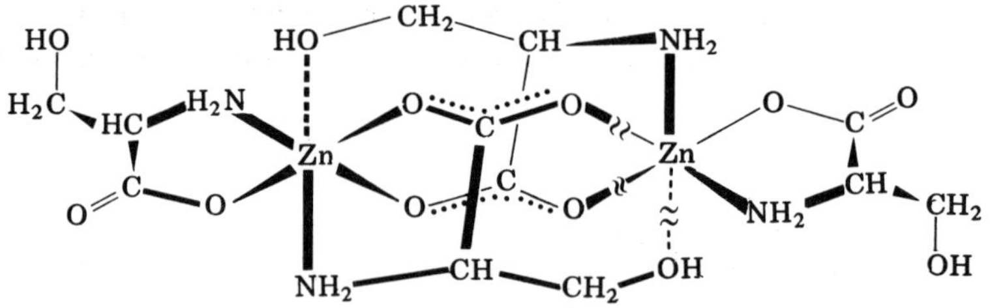

(XXXV)

Coordination in Zn(L-Ser)$_2$

suggest that this interaction is not important in solution.

It is appropriate to mention here an unexpected type of contact between Cu(II) atoms and tyrosyl side-chains, which has been discovered by the structure analysis of $Cu(Gly-L-Leu-L-H_{-1}Tyr) \cdot 4H_2O \cdot \frac{1}{2}Et_2O^{126}$. The coordination of the Cu by five ligand atoms is square-pyramidal (**XXXVI**).

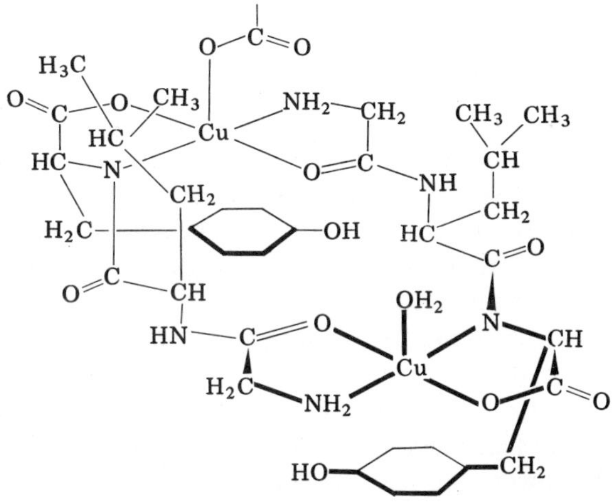

(XXXVI)

Dimer in $Cu(Gly\text{-}L\text{-}Leu\text{-}L\text{-}H_{-1}Tyr) \cdot 4H_2O \cdot \frac{1}{2}Et_2O$

The phenol ring lies roughly parallel to the base of the pyramid, and several aromatic carbon atoms make contacts with the Cu atom which are significantly smaller (3.2–3.3 Å) than the estimated sum of the van der Waals' radii. A second aspect of the structure is of special interest in connection with the deprotonation reaction of peptide groups (p. 135). The complex belongs to the type $(CuH_{-1}L)_2$, but the deprotonated peptide group of the ligand is adjacent to the carboxyl terminus instead of the amino terminus. The tripeptide forms five-membered chelate rings with *two* Cu atoms, and each Cu is coordinated by *two* ligand molecules, thus accounting for the dimeric formula unit.

S(sulphydryl) atoms★

The metals which have a pronounced preference for sulphur (compared with oxygen) donors occupy a triangular area near the centre of the Periodic Table, and have been labelled "class (b)"[127]. They are characterized generally by the ability to form not only strong σ-bonds with easily polarizable ligands, but also π-bonds by back-donation of electrons from metal

---

★The discussion of sulphur-containing ligands incorporates helpful suggestions from Professor S. E. Livingstone.

d_π to ligand d_π or p_π orbitals. The explanation of their affinities for sulphur ligands is three-fold[128].

(a) The electronegativity of sulphur is low (being smaller than that of Br and roughly equal to that of C), and its polarizability is high. A sulphur atom will become highly polarized in the field of a small metal ion with a high charge density, even if the metal ion has a d^{10} configuration (Cu(I), Ag(I), Hg(II)).

(b) Calculations show that for d^{10} ions neither polarization, nor the heat of formation of covalent single bonds, can account even for the orders of the stabilities of comparable metal–O and metal–S bonds[128]. The presence of metal→ligand d_π–d_π bonding is therefore inferred, though it must be admitted that there is no direct evidence for it at this time.

(c) It is possible in the cases of bonds with low-spin d^8 ions (Pd(II), Pt(II), Au(III)) that additional stabilization is caused by an increase in *CFSE*. This question is not yet resolved, since sulphur ligands seem to be distributed all along the spectrochemical series. Some occur towards the lower end near Br¯ ion, and others (*e.g.* thioether) in intermediate and high positions in the series[129].

The special affinity of cysteinyl side-chains for Ag(I) and Hg(II) is well known and has been extensively exploited in the preparation of heavy-atom protein derivatives. The stability of $Ag(H_{-1}Cys)_2{}^{3-}$ is approximately the same as that of $Ag(CN)_2{}^-$. In solutions of low pH the predominant species is $Ag(HCys)_2{}^+$, and complexation is not negligible even at pH 1.3[130]. Metal binding by the ionized thiolo groups of cysteine side-chains is not, however, confined to "class (b)" metal ions. The formation constants of the cysteinato complexes of Mn(II), Fe(II), Co(II), Ni(II), Zn(II) and Pb(II)[131] are all considerably greater than those of the glycinato[132-134] and histidinato[102,104] complexes of the same metals (Table V). A variety of other properties confirm that the enhanced stabilities in all these cases are accompanied by metal–sulphur bonding.

For example, the absorptions due to S–H are absent from the solid-state infra-red spectra of a series of cysteine complexes with Zn(II), Cd(II), Hg(II) and Pb(II)[108]. Two of these complexes are $Na_2[Zn(H_{-1}Cys)_2] \cdot 4H_2O$ and $[Zn_3(Cys)_4][ZnCl_4]$, which are obviously prepared at high and low pH, respectively. In the spectrum of the first, the appearance of NH_2 bands indicates metal–N(amino) bonding, and the carboxylate frequencies (1588 and 1404 cm⁻¹) are those of free –COO¯. A crystal structure analysis confirms that the coordination geometry is tetrahedral and that the donor atoms are indeed N(amino) and S(sulphydryl)[136]. The existence and structure of a polynuclear ion $[Zn\{Zn(Cys)_2\}_2]^{2+}$ in the second complex are still conjectural, but a doublet (1510, 1600 cm⁻¹) and a high asymmetric car-

TABLE V

LOG STABILITY CONSTANTS OF SOME COMPLEXES ML^+ AND ML_2 WITH GLYCINE[132-134], HISTIDINE[102,104] AND CYSTEINE[131]

L	Log β_{101}			Log β_{102}		
	Gly^-	His^-	Cys^-	Gly^-	His^-	Cys^-
Mn(II)	3.2	3.2	5.8	5.5	6.1	10.3
Fe(II)	4.1	—	7.7	7.6	—	14.2
Co(II)	4.8	6.7	9.4	8.6	12.1	17.4
Ni(II)	5.7	8.4	10.8	10.6	15.1	23.5
Cu(II)	8.2	10.1	—	15.2	18.1	—
Zn(II)	5.0	6.3	10.3	9.2	11.7	20.0
Pb(II)	5.3	6.0	13.8	8.6	(9.0)	18.1

boxylate stretching frequency (1642 cm^{-1}) are interpreted as evidence that the cysteine $-NH_3^+$ groups are free and the $-COO$ groups are coordinated. Accordingly, cysteine coordinates Zn(II) through N and S at high pH, and through O and S at low pH[135]. In the absence of crystal field effects this is exactly what one would expect.

Qualitative evidence for metal–sulphur binding is also obtained from comparisons between the u.v.–visible spectra of cysteinato and other aminoacidato complexes. The u.v. absorption band of $Pd(Cys)_2$ has an extinction ($\epsilon = 1300$ M^{-1} cm^{-1} at $\lambda_{max} = 337$ nm) which is three times as great as the extinctions of corresponding bands in the spectra of Pd(II) complexes in which amino acids are known to be coordinated through N(amino) and O(carboxyl) atoms[74]. Comparisons between the spectra of the Co(III) complexes of cysteine, $NH_2-CH_2-CH_2-SH$ and $HOOC-CH_2-CH_2-SH$ show that the coordination is through N(amino) and S(sulphydryl) in $K_3[Co^{III}-(H_{-1}Cys)_3] \cdot 3H_2O$ (green, high pH), and through O(carboxyl) and S(sulphydryl) in $[Co^{III}(Cys)_3] \cdot 3H_2O$ (red, low pH)[137]. The infra-red spectrum of the green complex confirms that the carboxyl groups are uncoordinated[138].

Cysteinato complexes of Fe(III) are especially important because of their possible relevance to the study of ferredoxin (Chap. 22). Fe^{3+} (like Cu^{2+}) catalyzes the oxidation of cysteine. Extremely labile complexes with inferred metal:ligand ratios of 1:1 (blue), 1:2 (red) and 1:3 (violet and green forms, thermally interconvertible) have been isolated at -78°C. Coordination is through O and S in the blue, red and violet forms, and through N and S in the green form. The evidence consists of the absorption, o.r.d. and c.d. spectra, and comparisons of the spectra with those of analogous thioglycollate complexes[139]. In the related complex $[Fe^{II}(L\text{-}Cys)_2-(CO)_2]$ the coordination is through N and S. The thiol group has a greater affinity for Ag^+ than for Fe^{2+}, and Ag^+ ions convert the complex into $[Fe(L\text{-}CysAg)_2(CO)_2]$ where the Fe(II) is now bound to the cysteinato ligand at N(amino) and O(carboxyl)[140].

Interactions between Cu(II) and cysteine similarly are complicated by the reduction of the metal ion. The amino acid is oxidized to cystine. The interactions between Cu(I) and mercaptides have been extensively investigated by titrations in water–acetonitrile solutions[141].

The complexes which have been mentioned so far provide examples of unidentate coordination through S alone, and bidentate chelation through N and O, N and S, and O and S. Tridentate chelation through all functional groups of cysteine is the basis of a detailed interpretation of the o.r.d. spectra of solutions containing Co(II) and cysteinato ions[142]. It has also been found in the crystal structure of $Na_2[Mo_2^V(L\text{-}H_{-1}Cys)_2O_4] \cdot 5H_2O$ (XXXVII)[143]. Other interesting features of the structure are $Mo^V\text{-}O\text{-}Mo^V$

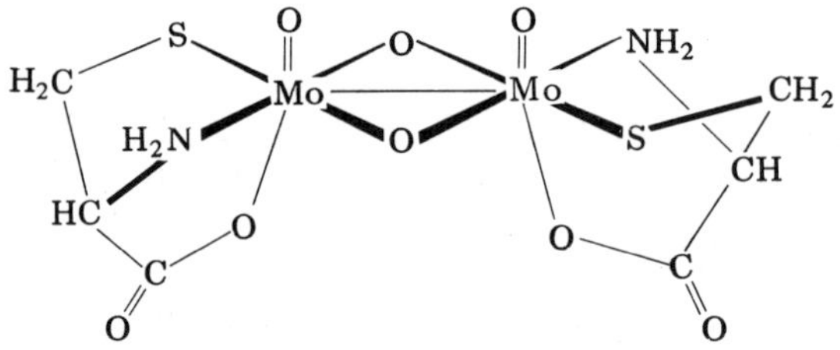

(XXXVII)

bridges, $Mo^V{=}O$ double bonds, and a $Mo^V\text{-}Mo^V$ metal–metal bond (2.57 Å). The Mo(V) atoms are 7-coordinate. The complex was originally reported as $Na_2[Mo_2O_4(L\text{-}H_{-1}Cys)_2(OH_2)_2] \cdot 3H_2O$, and an infra-red frequency at 1590 cm^{-1} was assigned as the asymmetric frequency of an uncoordinated $-COO^-$ group[144].

S(thioether) atoms

The preceding discussion of thiolo ($-S^-$) groups applies largely to thioether ($-SR$) groups as well. S(thioether) atoms have smaller polarizabilities and are weaker donors than S(sulphydryl) atoms, but they have fewer lone-pair electrons and should therefore be better d_π electron acceptors[128]. The only metal ions which are bound at methionine $-CH_2CH_2SCH_3$ side-chains are a small group of "class (b)" ions with d^8 and d^{10} configurations: Pd(II), Pt(II), Ag(I), Cu(I) and Hg(II).

The binding of Pt(II) by methionine side-chains has been established in crystal structure analyses of several protein derivatives, and $Pt^{II}Cl_4^{2-}$ has been proposed as a specific label for exposed methionine residues[145]. There are also four methionine complexes whose crystal structure analyses confirm that both Pd(II) and Pt(II) are coordinated by the $-SCH_3$ groups (as predicted from their chemical properties[146] and infra-red spectra)[147]. Six-membered chelate rings between N(amino) and S(thioether) atoms occur in Pd(DL-MetH)Cl$_2$ [148], Pt(DL-MetH)Cl$_2$ and Pt(L-MetH)Cl$_2$[75], and between N(peptide) and S(thioether) atoms in Pt(Gly-L-Met)Cl·H$_2$O[75] (**XXXVIII** and **XXXIX**). All four complexes are, incidentally, prepared at low pH.

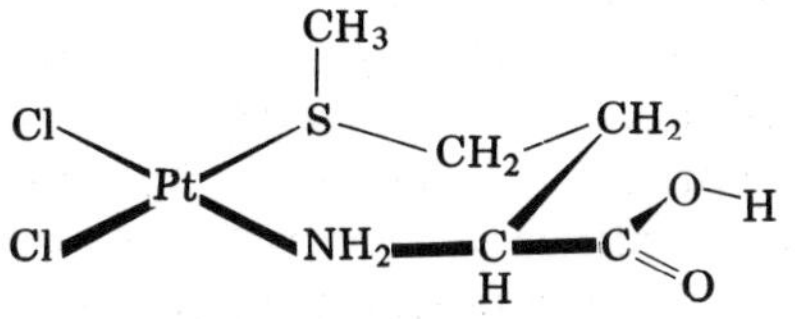

(XXXVIII)

M(L-MetH)Cl$_2$ in Pt(L-MetH)Cl$_2$,
Pt(DL-MetH)Cl$_2$ and Pd(DL-MetH)Cl$_2$

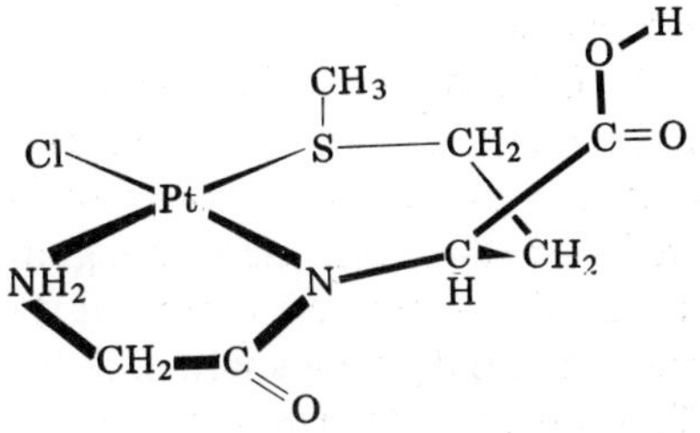

(XXXIX)

They provide further examples of coordination through amino (and, in **XXXIX**, peptide) groups while the carboxyl groups of the ligands are still protonated (see p. 126). Coordination of Pd(II) by methionine through N and S donors at a higher pH has been deduced from the high extinction coefficient of Pd(Met)$_2$ in the ultra-violet region (ϵ = 2220 at λ_{max} = 315 nm)[74].

At each of these methionyl side-chain metal-binding sites the donor S atom sits at the apex of a trigonal pyramid, whose base has the C(methylene), C(methyl) and Pt atoms at its corners. Coordination thus creates a new chiral (asymmetric) centre at the S(thioether) atom. Both chiralities are found in the crystal structures of Pt(II)–methionine chelates: for example, there are two crystallographically independent complexes in the crystal structure of Pt(L-MetH)Cl$_2$, and their sulphur donor atoms have configurations of opposite handedness[75]. This observation is consistent with the occurrence of two alternative PtCl$_4$-binding sites at a methionine residue in cytochrome-c, one on each side of the sulphur atom[145].

Evidence for Ag(I)–sulphur bonding is provided by the high stability constant of the Ag(Met)$_2$$^-$ ion[149], and by a difference of 100 cm^{-1} between the NH$_2$ stretching frequencies of Li[Ag(Met)$_2$] and Li[Ag(Nle)$_2$][25]. In Ag(Met)$_2$$^-$ only the methionine sulphur atoms are involved in coordination to Ag(I), leaving the amino and carboxyl groups free to form chelates with atoms of "class (a)" metals. Mixed-metal complexes can be made in this way with Cr(III), Fe(III), Co(II), Ni(II) and Cu(II)[25].

The formation of a Cu(I)–methionine complex can be demonstrated by the potentiometric titration of mixtures of methionine and

$Cu^I(CH_3CN)_4(ClO_4)$ in water–acetonitrile solution. Cu(I)–oxygen coordination is unknown, and the metal must be bound to methionine at the S(thioether) and possibly at the N(amino) atoms[141]. It is also important that Cu(II) — not a "class (b)" metal ion — does not interact with $-SCH_3$ groups. The crystal structure analysis of $Cu(DL\text{-}Met)_2$ shows that Cu(II) is chelated by methionine through the amino and carboxyl groups in the normal way[150].

Until recently, Hg(II) was thought to exhibit only "class (a)" behaviour in its reaction with methionine (see below). At very low pH, however, the addition of Hg^{2+} ions to a solution of methionine causes large down-field shifts of the n.m.r. proton signals of the protons closest to the sulphur atom. The shifts are maximal when the metal:ligand ratio is 1:2, so that the principal species which is formed has the composition $Hg(H_2Met)_2^{4+}$. The species is thermodynamically stable and kinetically labile[151].

The methionine $-SCH_3$ groups are, however, not bonded to the metal atom in $Hg(Met)_2$ or in any of the complexes $M(Met)_2$ [M = Mn(II), Co(II), Ni(II), Cu(II), Zn(II), Cd(II), Pb(II)] and $M'(Met)_3[M' = Al(III), Cr(III), Bi(III), Fe(III), Rh(III)]^{25}$. The infra-red spectroscopic evidence is indirect (i.e. it proves the presence of bonding through N and O rather than the absence of bonding through S) because the C–S stretching frequency is inaccessible under the reported experimental conditions. Proof that the $-SCH_3$ groups in complexes such as $Cr(Met)_3$ are not involved in chelation comes from their ability to bind Ag(I), leading to the formation of the same types of mixed-metal complexes as were mentioned earlier[25]. In the case of $Hg(Met)_2$, measurements of n.m.r. proton shifts in Hg(II)–methionine solutions at pH 9 support the conclusions drawn from the infra-red spectra[151].

Bridging S(disulphide) atoms

The –S–S– bridge of cystine is at the same time an important structural link in proteins, and a very reactive centre for some metals. Ag(I) in aqueous solution causes a disproportionation in which cysteine (stabilized as the Ag(I) complex) and a sulphinic acid are formed[152]: $2RSSR + 2H_2O \xrightarrow{Ag^+} 3RS^- + RSO_2H + 3H^+$. Cu(I), but not Cu(II), is bound by cystine. The complex may contain Cu in an indeterminate oxidation state. The equilibrium can be approached from both sides and, in the presence of Cu(II)-specific ligands, can be shifted further towards cysteine[141].

$$2Cu^+ + RSSR \; \rightleftharpoons \; Cu^+ \begin{array}{c} R \\ | \\ S \\ | \\ S \\ | \\ R \end{array} Cu^+ \; \longleftrightarrow \; Cu^{2+} \begin{array}{c} R \\ | \\ S \\ / \\ \\ S \\ | \\ R \end{array} Cu^{2+} \; \rightleftharpoons \; 2Cu^{2+} + 2RS^-$$

This explains why the reduction of Cu(II) to Cu(I) by cysteine is inhibited in the presence of histidine (a Cu(II)-specific ligand). A violet-black complex is stabilized in the solution[153]. Its colour is no doubt due to charge-transfer of the type indicated by the central formulae in the above equilibria.

THE USEFULNESS OF INFORMATION DERIVED FROM THE STUDY OF "MODEL COMPOUNDS"

The claim has often been made that complexes of the types which have been discussed in this chapter are "of biological interest". In a few cases it is true that a particular complex turns up in some biological system. In many more cases the phrase "of biological interest" expresses the hope that the systematic study of many complexes will contribute to the discovery of the rules which govern the interactions between metal ions and naturally occurring ligands. Many naturally occurring ligands are tremendously complicated molecules. The ligands whose complexes can be systematically studied are relatively simple. A metal–peptide complex may have some properties in common with a metal–protein complex. So far as these properties alone are concerned, the simple complex is a "model compound" for the biological interaction. But we must remember that it is a model — not a replica[154].

As an example of the usefulness and limitations of model compounds we shall examine the extent to which the crystal structures of simple Zn(II) complexes help us to understand one aspect — the geometry — of the Zn(II)-binding site of carboxypeptidase A. The example is chosen, because we can use the known crystal structure of the protein to check the correctness of our predictions. In addition, the model structures supply some, but not all, of the numerical details which the protein structure analysis alone cannot (yet) reveal with precision.

The active site of carboxypeptidase (Chapter 15) includes a Zn atom. The ligands are the residues His-69, His-196 and Glu-72, so that the known donor atoms include two N(imidazole) atoms, and one or both O(carboxyl) atoms of the glutamate side-chain carboxyl group. In amino acid and peptide complexes of Zn(II), the coordination number of the Zn atom is always at least four (tetrahedral coordination), but frequently six (octahedral coordination) or intermediate (irregular coordination). If, as indicated by the structure analysis of the complete protein[155,156], the coordination is tetrahedral, then an H_2O molecule (or possibly a halide or other ion) must occupy the fourth coordination position. During hydrolysis, a peptide substrate molecule is anchored to the Zn atom via the O(peptide) atom of the peptide group which is being hydrolyzed. The structures of model compounds do not lead to a prediction whether the substrate will form an additional bond with the Zn atom, or whether it will displace one of the existing ligands (*e.g.* the H_2O molecule). It is fairly certain that if the Zn–O(peptide) bond is not stabilized by chelation, it must be stabilized by other contacts between the enzyme and substrate molecules.

In order to describe the environment of the Zn atom we need two kinds of structural information: the lengths of the Zn–ligand bonds, and the most probable values of the angles between the bonds. The information which can be distilled from all the published structure analyses of relevant "model compounds" is presented in Table VI.

Table VI leads to the following conclusions: (a) the Zn–N(imidazole) bonds are 2.0 Å long, assuming that the coordination geometry is more tetrahedral than octahedral; (b) the bond-lengths of the Zn–O(carboxyl), Zn–OH$_2$ and/or Zn–O(peptide) bonds are 2.1–2.2 Å. We must keep in mind the possibility that the carboxyl group may coordinate the Zn through both of its O(carboxyl) atoms. If this is the case, then the second bond is about 2.6 Å long, and the angle between the two Zn–O(carboxyl) bonds is about 60°; (c) the only recorded examples of N(imidazole)–Zn– O(carboxyl) bond angles occur in two Zn(II)–histidine complexes, in which the Zn–O(carboxyl)

TABLE VI

SOME STRUCTURAL PARAMETERS OF Zn(II) COMPLEXES[a]

(a) Complexes included in survey

Complex	Formula No.	Coord. No of Zn(II)	Geometry	Donor atoms
Zn(Im)$_2$	XXVI	4	Tetrahedral	4 N(imidazole)
Zn(ImH)$_2$Cl$_2$	XXV	4		2 N(imidazole), 2Cl
Zn(L-His)$_2 \cdot 2H_2O$	XXIX	4 + 2	Tetrahedral plus two weaker bonds	2 N(amino), 2 N(imidazole), 2 O(carboxyl)
Zn(DL-His)$_2 \cdot 5H_2O$	XXIX	4 + 2		
Zn(L·Ser)$_2$	XXXV	5	Irregular square-pyramidal	2 N(amino), 3 O(carboxyl)
Zn(Gly-Gly-Gly) (SO$_4$)$_{\frac{1}{2}} \cdot 4H_2O$	XIII	6	Irregular octahedral	1 N(amino), 2 O(carboxyl), 1 O(peptide), 2H$_2$O
Zn(L-H$_{-1}$Glu)$\cdot 2H_2O$	XIV	6		1 N(amino), 4 O(carboxyl), 1H$_2$O
Zn(ImH)$_6$Cl$_2 \cdot 4H_2O$	XXIV	6		6 N(imidazole)
Zn(Gly-Gly)$_2 \cdot 2H_2O$	II	6	Octahedral	2 N(amino), 1 O(carboxyl), 2 O(peptide), 1H$_2$O
Zn(L-H$_{-1}$Asp)$\cdot 3H_2O$		6		1 N(amino), 3 O(carboxyl), 2H$_2$O
Zn(BiuH$_2$)$_2$Cl$_2$	XVI	6		4 O(amide), 2 Cl

TABLE VI—cont.

(b) Zn(II)-ligand bond-lengths[b]

Bond	Coordination	Length (Å)		No. of examples
		Mean	Range	
Zn–N(imidazole)	Tetrahedral	2.00	1.96–2.04	12
	Octahedral	2.20	2.15–2.26	6
Zn–O(carboxyl)	Irregular	2.8	2.6[b]–2.9	4
	Square pyramidal	2.08	1.96–2.16	3
	Octahedral	2.09	2.03–2.20	9
Zn–O(peptide)	Octahedral	2.18	2.18–2.19	2
Zn–O(amide)	Octahedral	2.04	2.03–2.05	2
Zn–OH_2	Octahedral	2.14	2.07–2.19	6

(c) Ligand-Zn-ligand bond-angles[c]

Atoms	Coordination	Angle (degrees)		No. of examples
		Mean	Range	
N(imidazole)–Zn–N(imidazole)	Tetrahedral	109	105–117	13
	Irregular	115	112–117	2
	Octahderal	90	87–93	12
N(imidazole)–Zn–O(carboxyl)	Irregular	—	77–164	2
O(carboxyl)–Zn–O(carboxyl)	Irregular	—	81–139	2
	Octahedral	94	89–103	6
	Octahedral	—	87[b]–149[b]	2
O(carboxyl)–Zn–OH_2	Octahedral	93	88–100	6
	Octahedral	—	77[b]	1

[a] Numerical data are taken from ref. 9, with the exception of values for Zn(L-Ser)$_2$[125]
and Zn(Gly-Gly-Gly)(SO$_4$)$_{\frac{1}{2}}$·4H$_2$O[37]
[b] Indicates that the parameter involves the irregular bond where *both* oxygen atoms of a carboxyl group interact with the same Zn atom
[c] *Not* including angles where both bonds are in the same chelate ring

bonds are weak and the O(carboxyl) atoms occupy irregular coordination positions. At best, these angles can be used to set limits, 77° and 164°, between which angles in other complexes probably lie; (d) we have even less information about bond-angles involving Zn–OH_2 and Zn–O(peptide) bonds, respectively; (e) the geometries of the metal-binding functional groups are easier to predict, since examples drawn from complexes with a variety of metals show that they remain fairly constant. Thus we can say with some confidence that the C–O–Zn bond-angle at a donor O(carboxyl) atom will be close to 104°, that the C–N–Zn angles at the donor N(imidazole) atoms will be in the range 121–131°, and so on.

162

The picture of the active site which finally emerges is shown in Fig. 3. This is what we should predict if, as is often the case, our information about the active site were limited to a knowledge of the ligating groups. The predicted geometry should be compared with the detailed interpretation of the protein structure itself (Chapter 15).

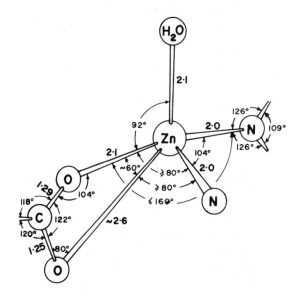

Fig. 3. Predicted configuration around a Zn(II) atom whose ligands are two histidyl and one glutamyl side-chains and a H_2O molecule.

The author thanks the following for helpful criticisms of the manuscript: Professor R. W. Green, Professor S. E. Livingstone, Dr. D. A. Langs, Dr. J. K. Fawcett, Miss Marcia Golomb, C. B. Acland, D. J. Fensom, and R. J. Flook. Support from the U.S. Public Health Service (Grant GM-10867) and the Australian Research Grants Committee (Grant 65-15552) is gratefully acknowledged.

REFERENCES

1 A. Strecker, *Annalen*, 75 (1850) 27.
2 G. Wiedemann, *J. Prakt. Chem.*, 42 (1847) 255; 43 (1848) 271.
3 F. R. N. Gurd and P. E. Wilcox, *Adv. Protein Chem.*, 11 (1956) 311.
4 J. P. Greenstein and M. Winitz, *The Chemistry of the Amino Acids*, Vol. 1, Wiley, New York, 1961, Chapter 6.
5 A. E. Martell and M. Calvin, *Chemistry of Metal Chelate Compounds*, Prentice-Hall, New York, 1962.

163

6 K. Nakamoto and P. J. McCarthy, *Spectroscopy and Structure of Metal Chelate Compounds*, Wiley, New York, 1968.
7 R. D. Gillard, *Inorg. Chem. Acta. Rev.*, 1 (1967) 69.
8 A. S. Brill, R. B. Martin and R. J. P. Williams, in B. Pullman (Ed.), *Electronic Aspects of Biochemistry*, Academic Press, New York, 1964, p. 519.
9 H. C. Freeman, *Adv. Protein Chem.*, 22 (1967) 257.
10 B. L. Vallee and R. J. P. Williams, *Proc. Natn. Acad. Sci. U.S.*, 59 (1968) 498.
11 R. D. Gillard and S. H. Laurie, in *Amino acids, Peptides and Proteins*, Vol. 1, The Chemical Society, London, 1969, p. 262.
12 B. W. Low, F. L. Hirshfeld and F. M. Richards, *J. Am. Chem. Soc.*, 81 (1959) 4412.
13 C. Sandmark and I. Lindqvist, personal communication, 1965, quoted in ref. 9.
14 V. S. Sharma and H. B. Mathur, *Indian J. Chem.*, 3 (1965) 475.
15 S. Boyd, J. R. Brannan, H. S. Dunsmore, G. H. Nancollas, *J. Chem. Eng. Data*, 12 (1967) 12.
16 E. V. Raju and H. B. Mathur, *J. Inorg. Nucl. Chem.*, 30 (1968) 2181.
17 W. F. Stack and H. A. Skinner, *Trans. Faraday Soc.*, 63 (1967) 1136.
18 A. P. Brunetti, M. C. Lim and G. H. Nancollas, *J. Am. Chem. Soc.*, 90 (1968) 5120.
19 J. E. Letter and J. E. Bauman, *J. Am. Chem. Soc.*, 92 (1969) 437; 92 (1969) 443.
20 C. B. Acland and H. C. Freeman, *Chem. Commun.*, (1971) 1016.
21 S. P. Datta and A. K. Grzybowski, *J. Chem. Soc.*, (1959) 1091.
22 D. J. Alner, *J. Chem. Soc.*, (1962) 3283.
23 Y. M. Azizov, A. K. Miftakhova and V. F. Toropova, *Russ. J. Inorg. Chem.*, 12 (1967) 345.
24 F. J. C. Rossotti, in J. Lewis and R. G. Wilkins, *Modern Coordination Chemistry*, Interscience, New York, 1960, p. 57.
25 C. A. McAuliffe, J. V. Quagliano and L. M. Vallarino, *Inorg. Chem.*, 5 (1966) 1996.
26 F. A. Cotton, in J. Lewis and R. G. Wilkins, *Modern Coordination Chemistry*, Interscience, New York, 1960, p. 387.
27 A. A. Grinberg, A. I. Stetsenko and E. N. In'kova, *Dokl. Akad. Nauk SSSR*, 136 (1961) 821.
28 L. M. Volshtein and I. O. Volodina, *Russ. J. Inorg. Chem.*, 5 (1960) 840.
29 L. M. Volshtein and G. G. Motyagina, *Russ. J. Inorg. Chem.*, 5 (1960) 949.
30 L. E. Erickson, J. W. McDonald, J. K. Howie and R. P. Clow, *J. Am. Chem. Soc.*, 90 (1968) 6371.
31 C. C. McDonald and W. D. Phillips, *J. Am. Chem. Soc.*, 85 (1963) 3736.
32 R. H. Carlson and T. L. Brown, *Inorg. Chem.* 5 (1966) 268.
33 K. Nakamoto, *Infra-Red Spectra of Inorganic and Coordination Compounds*, 2nd edn., Wiley–Interscience, New York, 1970, p. 222.
34 I. Lindqvist and R. Rosenstein, *Acta Chem. Scand.*, 14 (1960) 1228.
35 J. K. M. Rao and M. A. Viswamitra, *Acta Cryst.*, B28 (1972) 1484.
36 K. F. Belyaeva, M. A. Porai-Koshits, T. I. Malinovskii, L. A. Aslanov, L. S. Sukhanova and L. I. Martynenko, *Zh. Strukt. Khim.*, 10 (1969) 557; *J. Strukt. Chem.*, 10 (1969) 470.
37 D. Van der Helm and H. B. Nicholas, Abstracts, Am. Cryst. Assoc. Meeting, Tucson, Paper F5, 1968; *Acta Cryst.*, B26 (1970) 1858.
38 C. Gramaccioli, *Acta Cryst.*, 21 (1966) 600.
39 D. Van der Helm and T. V. Willoughby, *Acta Cryst.* B25 (1969) 2317.
40 B. R. Rabin, *Biochem. Soc. Symp.*, 15 (1958) 21.
41 H. B. Dyer, *Acta Cryst.*, 4 (1951) 42.
42 L. Cavalca, M. Nardelli and G. Fava, *Acta Cryst.*, 13 (1960) 594.

164

43 A. Mitschler, J. Fischer and R. Weiss, *Acta Cryst.*, 22 (1967) 236.
44 D. J. Haas, *Nature*, 201 (1964) 64.
45 J. F. Roux and J. C. A. Boeyens, *Acta Cryst.*, B25 (1969) 1700.
46 J. F. Hinton and E. S. Amis, *Chem. Commun.*, (1967) 100.
47 J. F. Hinton, E. S. Amis and W. Mettetal, *Spectrochim. Acta*, A25 (1969) 119.
48 P. Pfeiffer, *Organische Molekülverbindungen*, Enke, Stuttgart, 1927, quoted in ref. 4.
49 C. W. Davies and G. M. Waind, *J. Chem. Soc.*, (1950) 301.
50 C. B. Monk, *Trans. Faraday Soc.*, 47 (1951) 285, 292, 297, 1233.
51 H. C. Freeman, G. Robinson and J. C. Schoone, *Acta Cryst.*, 17 (1964) 719.
52 M. Nardelli, G. Fava and G. Giraldi, *Acta Cryst.*, 16 (1963) 343.
53 B. B. Kedzia, P. X. Armendarez and K. Nakamoto, *J. Inorg. Nucl. Chem.*, 30 (1968) 849.
54 R. H. Nuttall and G. A. Melson, *J. Inorg. Nucl. Chem.*, 31 (1969) 2979.
55 H. C. Freeman and J. E. W. L. Smith, *Acta Cryst.*, 20 (1966) 153.
56 N. C. Li, R. L. Scruggs and E. D. Becker, *J. Am. Chem. Soc.*, 84 (1962) 4650.
57 M. K. Kim and A. E. Martell, *J. Am. Chem. Soc.*, 91 (1969) 872.
58 M. K. Kim and A. E. Martell, *Biochemistry*, 3 (1964) 1169.
59 M. K. Kim and A. E. Martell, *J. Am. Chem. Soc.*, 88 (1966) 914.
60 A. E. Martell and M. K. Kim, personal communication.
61 C. R. Hartzell and F. R. N. Gurd, *J. Biol. Chem.*, 244 (1969) 147.
62 R. Österberg and B. Sjöberg, *J. Biol. Chem.*, 243 (1968) 3038.
63 H. C. Freeman and J. T. Szymanski, *Acta Cryst.*, 22 (1967) 406.
64 J. D. Bell, H. C. Freeman, A. M. Wood, R. Driver and W. R. Walker, *Chem. Commun.*, (1969) 1441.
65 H. C. Freeman, J. C. Schoone and J. G. Sime, *Acta Cryst.*, 18 (1965) 381.
66 H. C. Freeman and M. R. Taylor, *Acta Cryst.*, 18 (1965) 939.
67 A. Sugihara, T. Ashida, Y. Sasada and M. Kakudo, *Acta Cryst.*, B24 (1968) 203.
68 A. R. Manyak, C. B. Murphy and A. E. Martell, *Arch. Biochem. Biophys.*, 59 (1955) 373.
69 R. B. Martin, M. Chamberlin and J. T. Edsall, *J. Am. Chem. Soc.*, 82 (1960) 495.
70. M. K. Kim and A. E. Martell, *J. Am. Chem. Soc.*, 89 (1967) 5138.
71 E. J. Billo and D. W. Margerum, *J. Am. Chem. Soc.*, 92 (1970) 6811.
72 H. C. Freeman, J. M. Guss and R. L. Sinclair, *Chem. Commun.*, (1968) 485.
73 G. K. Pagenkopf and D. W. Margerum, *J. Am. Chem. Soc.*, 90 (1968) 501.
74 E. W. Wilson and R. B. Martin, *Inorg. Chem.* 9 (1970) 528.
75 H. C. Freeman and M. L. Golomb, *Chem. Commun.*, (1970) 1523.
76 M. S. Michailidis and R. B. Martin, *J. Am. Chem. Soc.*, 91 (1969) 4683.
77 R. D. Gillard, E. D. McKenzie, R. Mason and G. B. Robertson, *Nature*, 209 (1966) 1347.
78 M. T. Barnet, H. C. Freeman, D. A. Buckingham, I. N. Hsu, D. Van der Helm, *Chem. Commun.*, (1970) 367.
79 R. D. Gillard, P. M. Harrison and E. D. McKenzie, *J. Chem. Soc. (A)*, (1967) 618.
80 L. J. Banaszak, H. C. Watson and J. C. Kendrew, *J. Mol. Biol.*, 12 (1965) 130.
81 J. F. Boas, J. R. Pilbrow, C. R. Hartzell and T. D. Smith, *J. Chem. Soc. (A)*, (1969) 572.
82 W. L. Koltun, M. Fried and F. R. N. Gurd, *J. Am. Chem. Soc.*, 82 (1960) 233.
83 G. N. Rao and N. C. Li, *Can. J. Chem.*, 44 (1966) 1637.
84 R. Driver and W. R. Walker, *Aust. J. Chem.*, 21 (1968) 671.
85 J. T. Edsall, G. Felsenfeld, D. S. Goodman and F. R. N. Gurd, *J. Am. Chem. Soc.*, (1954) 3054.
86 W. J. Eilbeck, F. Holmes and A. E. Underhill, *J. Chem. Soc. (A)*, (1967) 757.

87 D. L. Goodgame, M. Goodgame, P. J. Hayward and G. W. Rayner-Canham. *Inorg. Chem.*, 7 (1968) 2447.
88 F. Akhtar, D. M. L. Goodgame, M. Goodgame, G. W. Rayner-Canham and A. C. Skapski, *Chem. Commun.*, (1968) 1389.
89 C. Sandmark, *Acta Chem. Scand.*, 21 (1967) 993.
90 B. K. S. Lundberg, *Acta Cryst.*, 21 (1967) 901.
91 B. Strandberg, B. Svensson and C. I. Bränden, personal communication in ref. 9.
92 R. Candlin and M. M. Harding, *J. Chem. Soc. (A)*, (1967) 421.
93 G. F. Bryce and F. R. N. Gurd, *J. Biol. Chem.*, 241 (1966) 122.
94 E. Baraniak, H. C. Freeman and J. M. James, *J. Chem. Soc. (A)*, (1970) 2558.
95 M. Inoue, M. Kishita and M. Kubo, *Bull. Chem. Soc. Japan*, 39 (1966) 1352.
96 J. A. J. Jarvis and A. F. Wells, *Acta Cryst.*, 13 (1960) 1028; see also ref. 9, p. 413.
97 G. F. Bryce, R. W. Roeske and F. R. N. Gurd, *J. Biol. Chem.*, 240 (1965) 3837.
98 M. M. Harding and H. A. Long, *J. Chem. Soc. (A)*, (1968) 2554.
99 R. Candlin and M. M. Harding, *J. Chem. Soc. (A)*, (1970) 384.
100 K. A. Fraser and M. M. Harding, *J. Chem. Soc. (A)*, (1967) 415.
101 S. Valladas-Dubois, *Compt. rend.*, 237 (1953) 1408; *Bull. Soc. Chim. Fr.* (1954) 831; *Compt. rend.*, 246 (1958) 2619.
101a M. M. Harding and S. J. Cole, *Acta Cryst.*, 16 (1963) 643; R. H. Kretsinger, R. F. Bryan and F. A. Cotton, *ibid.*, 651.
102 D. D. Perrin and V. S. Sharma, *J. Chem. Soc. (A)*, (1967) 724.
103 D. J. Perkins, *Biochem. J.*, 55 (1953) 649.
104 H. C. Freeman and R. P. Martin, *J. Biol. Chem.*, 244 (1969) 4823.
105 E. V. Raju and H. B. Mathur, *J. Inorg. Nucl. Chem.*, 31 (1969) 425.
106 A. C. R. Thornton and H. A. Skinner, *Trans. Faraday Soc.*, 65 (1969) 2044.
107 D. Burk, J. Z. Hearon, L. Caroline and A. L. Schade, *J. Biol. Chem.*, 165 (1946) 723.
108 A. Earnshaw and L. F. Larkworthy, *Nature*, 192 (1961) 1068.
109 J. Simplicio and R. G. Wilkins, *J. Am. Chem. Soc.*, 89 (1967) 6092.
110 P. J. Morris and B. R. Martin, *J. Inorg. Nucl. Chem.*, 32 (1970) 2891.
111 P. J. Morris and R. B. Martin, *J. Am. Chem. Soc.* 92 (1970) 1543.
112 A. F. Pearlmutter and J. E. Stuehr, *J. Am. Chem. Soc.*, 90 (1968) 858.
113 B. Evertsson, *Acta Cryst.*, B25 (1969) 30.
114 H. Sigel, R. Griesser and D. B. McCormick, *Arch. Biochem. Biophys.*, 134 (1969) 217.
115 M. A. Doran, S. Chaberek and A. E. Martell, *J. Am. Chem. Soc.*, 86 (1964) 2129.
116 J. L. Meyer and J. E. Bauman, *J. Am. Chem. Soc.*, 92 (1970) 4210.
117 H. Sigel, R. E. McKenzie and D. B. McCormick, *Biochim. Biophys. Acta*, 200 (1970) 411.
118 H. C. Freeman, J. M. Guss, M. J. Healy, R. P. Martin, C. E. Nockolds and B. Sarkar, *Chem. Commun.*, (1969) 225.
119 K. M. Wellman and B. K. Wong, *Proc. Nat. Acad. Sci. U.S.*, 64 (1969) 824.
120 R. B. Martin and J. T. Edsall, *J. Am. Chem. Soc.*, 82 (1960) 1107.
121 G. R. Lenz and A. E. Martell, *Biochemistry*, 3 (1964) 745.
122 K. Kustin and R. F. Pasternack, *J. Am. Chem. Soc.*, 90 (1968) 2295.
123 D. Van der Helm and M. B. Hossain, *Acta Cryst.*, B25 (1969) 457.
124 D. Van der Helm and W. A. Franks, *Acta Cryst.*, B25 (1969) 451.
125 D. Van der Helm, A. F. Nicholas and C. G. Fisher, *Acta Cryst.*, B26 (1970) 1172.
126 D. Van der Helm and W. A. Franks, *J. Am. Chem. Soc.*, 90 (1968) 5627; W. A. Franks and D. Van der Helm, *Acta Cryst.*, B27 (1970) 1299.
127 S. Ahrland, J. Chatt and N. R. Davies, *Quart. Rev.*, 12 (1958) 265.
128 S. E. Livingstone, *Quart. Rev.* 19 (1965) 386.

166

129 C. K. Jørgensen, *J. Inorg. Nucl. Chem.*, 24 (1962) 1571; *Inorg. Chim. Acta Rev.*, 2 (1968) 65.
130 S. Valladas-Dubois, *Compt. rend.*, 231 (1950) 53.
131 D. A. Doornbos, *Pharm. Weekblad*, 103 (1968) 1213.
132 A. Gergely, *Acta Chim. Acad. Sci. Hung.* 59 (1969) 309.
133 W. P. Evans and C. B. Monk, *Trans. Faraday Soc.*, 51 (1955) 1244.
134 A. Albert, *Biochem. J.*, 47 (1950) 531.
135 H. Shindo and T. L. Brown, *J. Am. Chem. Soc.*, 87 (1965) 1904.
136 R. P. K. Lum and K. Eriks, Abstracts, Am. Cryst. Assoc. Ottawa, Paper D4, 1970.
137 G. Gorin, J. E. Spessard, G. A. Wessler and J. P. Oliver, *J. Am. Chem. Soc.*, 81 (1959) 3193
138 B. J. McCormick and G. Gorin, *Inorg. Chem.*, 2 (1963) 928.
139 A. Tomita, H. Hirai and S. Makishima, *Inorg. Chem.*, 7 (1968) 760.
140 A. Tomita, H. Hirai and S. Makishima, *Inorg. Nucl. Chem. Lett*, 4 (1968) 715.
141 P. Hemmerich in J. Peisach, P. Aisen and W. E. Blumberg, *The Biochemistry of Copper*, Academic Press, New York, 1966, p. 15.
142 D. W. Urry, D. Miles, D. J. Caldwell and H. Eyring, *J. Phys. Chem.* 69 (1965) 1603.
143 J. R. Knox and C. K. Prout, *Acta Cryst.*, B25 (1969) 1857.
144 A. Kay and P. C. H. Mitchell, *Nature*, 219 (1968) 267.
145 R. E. Dickerson, D. Eisenberg, J. Varnum and M. L. Kopka, *J. Mol. Biol.*, 45 (1969) 77.
146 L. M. Volshtein and M. F. Mogilevkina, *Russ. J. Inorg. Chem.*, 8 (1963) 304.
147 C. A. McAuliffe, *J. Chem. Soc. (A)*, (1967) 641.
148 N. C. Stephenson, J. F. McConnell and R. Warren, *Inorg. Nucl. Chem. Lett.*, 3. (1967) 553; R. C. Warren, J. F. McConnell and N. C. Stephenson, *Acta Cryst.*, B26 (1970) 1402.
149 G. R. Lenz and A. E. Martell, *Biochemistry*, 3 (1964) 750.
150 M. V. Veidis and G. J. Palenik, *Chem. Commun.*, (1969) 1277.
151 D. F. S. Nattusch and L. J. Porter, *Chem. Commun.*, (1970) 596.
152 Ref. 4, p. 645.
153 P. S. Hallman, *Proc. XII. I.C.C.C.*, (1970) 127.
154 B. G. Malmström, Plenary Lecture, *XII Int. Conf. Coordination Chem., Sydney, 1969*, Butterworth, London, 1970.
155 W. N. Lipscomb, *Accounts Chem. Res.*, 3 (1970) 81.
156 W. N. Lipscomb, J. A. Hartsuck, F. A. Quiocho and G. N. Reeke, *Proc. Natn. Acad. Sci.*, 64 (1969) 28.
157 F. Holmes and D. R. Williams, *J. Chem. Soc. (A)*, (1967) 1702.
158 J. E. Bauman and J. C. Wang, *Inorg. Chem.*, 3 (1964) 370; see also D. R. Williams, *J. Chem. Soc. (A)*, (1968) 2965.

Chapter 5

MICROBIAL IRON TRANSPORT COMPOUNDS (SIDEROCHROMES)

J. B. NEILANDS

Biochemistry Department, University of California, Berkeley, California, U.S.A.

I. GENERAL REVIEW

A. *Introduction*

In the preceding chapter we have seen what functional groups on peptides are capable of complexing metal ions. In this chapter we shall discuss the structures of some specific peptides and other coordinating

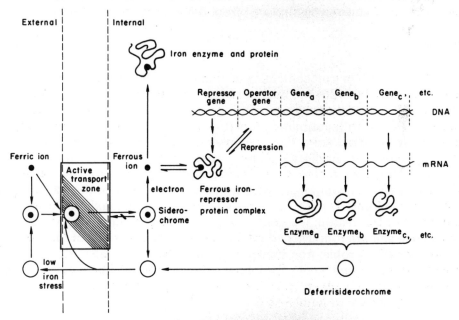

Fig. 1. Schematic illustration of the microbial iron transport system and its operon. In this scheme the information contained in the genes is transcribed into messenger RNA molecules and translated from the latter into the synthesis of the enzyme proteins. A series of genes, such as those designated (a), (b), and (c) may be controlled by an operator gene. The inhibition of this operator gene by a repressor thus prevents all of the proteins corresponding to all of the genes controlled by the operator gene from being synthesized.

agents that are produced by micro-organisms and appear to be specifically designed to complex iron.

In the course of evolution, the transition from an anaerobic to an aerobic atmosphere required the invention of powerful sequestering agents for the very insoluble Fe(III) ion (iron(III) hydroxide $K_{so} \cong 10^{-36}$). As a consequence, most aerobic and facultative aerobic micro-organisms probably contain an iron-gathering and transporting mechanism approximately equivalent to that shown in Fig. 1. The features of this scheme are the following: (a) biosynthesis of a Fe(III) ion-specific ligand by a sequence of enzymic reactions; (b) regulation by iron of the genetic apparatus which gives rise to the enzymes just described; (c) diffusion of the ligand to the membrane, cell surface or medium, complexation of Fe(III) ion and active transport of the metal ion, as its coordination derivative, into the cell; (d) retention of the iron(III) complex within the cell, followed by reduction and delivery of the iron atom at the point of demand by the biosynthetic machinery of the cell. Under conditions of low iron stress the overproduction of the ligand may cause a substantial amount of the metal-free molecules to be displaced from the cell into the medium.

B. Terminology

The red-brown iron-containing metabolites with a characteristic absorption band at 420–440 nm have been given the generic name sidero-chromes[1]. Those with growth-promoting activity were designated sidera-mines and those with antibiotic properties sideromycins. This terminology is not entirely satisfactory, since a number of the compounds are neither growth factors nor antibiotics. Further, some of the antibiotics in this class may serve as growth factors depending upon the test organism (see Table II). Also, a class of compounds with the same biofunction as the sidero-chromes, the iron(III) phenolates, possess an absorption maximum near 500 nm. For this reason, it seems desirable to abandon the designation sideramine and sideromycin and retain the term siderochrome or "irono-phore" for all microbial iron transport factors. Siderochrome (Gr.: *sidero* = iron; *chrome* = color) will be the designation used here but it is suggested that the workers in the field have a responsibility to devise a satisfactory system of nomenclature.

C. Classification

This review will focus on the chemistry of the microbial iron transport compounds, but enough biological information will be included to document the fact that organic "capsules" or "envelopes" are indeed the main role of these substances. Thus the segment of Fig. 1 marked "active transport zone" is too unchartered for review at this time. The preoccupation with structure is willful since whatever the biofunction the latter will be dependent on and dictated by the chemical constitution of the molecules.

At this time we recognize two very general types of siderochromes, *viz.*, the phenolates and the hydroxamates. Both of these ligands are weak acids ($pK_a \sim 9$), and only oxygen atoms are in the coordination sphere of the bound metal ion (Fig. 2). The structures shown in Fig. 2 are three-to-one complexes in which all six octahedrally deployed coordination sites

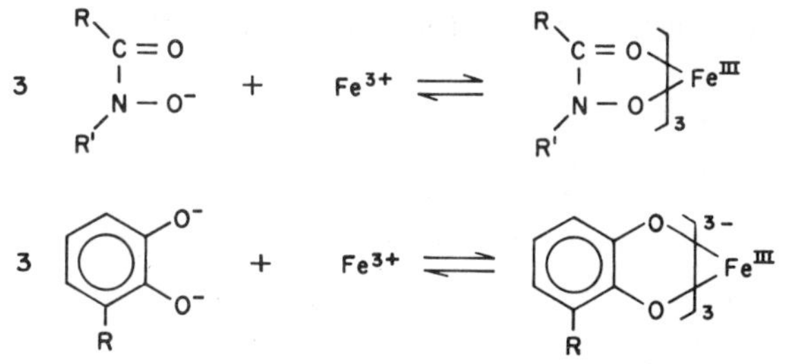

Fig. 2. Types of iron(III) phenolate (upper) and iron(III) hydroxamate (lower) complexes found in microbial iron transport compounds.

of the Fe(III) ion are occupied. The substances which are most intimately and unequivocally connected to iron metabolism have the six oxygens contained within the same molecule and are represented by enterobactin (I)[2] and ferrichrome (Va)[3] in the phenolate and hydroxamate series, respectively. A number of products believed to act in iron transport contain less than six binding atoms per molecule; such substances will also be described here, but the primary focus will be on the enterobactin–ferrichrome type systems.

The phenolate–hydroxamate classification is a tidy one but, unfortunately, it is incompatible with the diversity of structures represented in the microbial iron transport factors. Thus, the mycobacteria synthesize an array of compounds, collectively named mycobactins (XI)[4], which contain two hydroxamate bonds (*i.e.*, four oxygen atoms), a single phenolate group, and a residue of a substituted oxazoline, the latter contributing nitrogen as the sixth metal-binding atom. The mycobactins can hence be regarded as a hybrid of the pure trihydroxamate and triphenolate systems. Further, aerobactin (IX)[5], from *Aerobacter aerogenes*, and deferrischizokinen (Xa)[5a], from *Bacillus megaterium*, contain two hydroxamate groups and an internal residue of citric acid. The latter moiety provides a single free carboxyl and hydroxyl group, either of which might coordinate to a hydroxamate-bound Fe(III) ion (X). In general, it can be said that hydroxamates occur in fungi, yeast, and bacteria. Thus far a triphenolate has only been found in true bacteria.

D. Biosynthesis

Before launching a discussion of the structure of the iron transport compounds *per se*, it will be instructive to refer to some of the factors regulating their biosynthesis, the most important of which is iron. The role of iron in this event was first described by Garibaldi and Neilands, some 15 years ago, who demonstrated the massive accumulation of the metal-free ligand when the smut fungus, *Ustilago sphaerogena*, and other microbial species were cultured at low concentrations of iron[6]. Total elimination of the iron from the medium will, naturally, prevent all growth of the organism, and the data shown in Fig. 3 obviously must be

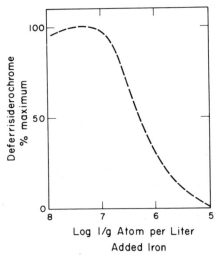

Fig. 3. A generalized curve showing repression by iron of deferrisiderochrome production as observed in a variety of microbial species.

related to cell yields. A slight stimulation by iron of deferrisiderochrome production may hence represent superior cell yields, or it may suggest participation of iron in the enzymology of the biosynthesis. At a concentration of about 10^{-5} M iron salts there is usually abundant cell growth but no excretion of the transport ligands. The fact that the regulating effect of the metal ion can be seen already at a concentration of about 10^{-7} M suggests the mechanism is that of repression, but more convincing evidence for this comes from experiments in which iron is added to a culture which is actively synthesizing hydroxamate — the synthesis continues for several hours, thus ruling out enzyme inhibition by the added iron. The manipulation of the iron nutrition of certain yeasts and fungi enables the production of hydroxamate up to a level of several g/l of medium[7]. Such large yields have never been seen for phenolates from bacteria even when the iron level is "optimum", although phenolate production is similarly iron-repressible. This reflects the differences in the operons and enzymes which direct the synthesis of the hydroxamates and phenolates. We assume, in analogy to the biosynthesis of small peptides like gramicidin, that the fragments of the deferrisiderochromes are assembled by a non-coded process. Garibaldi[7a] made the interesting discovery that a fluorescent pseudomonad, which had been isolated from chicken spoiled at 2°C, failed to form a hydroxamate-type siderochrome at temperatures over 28°C. For growth at higher temperatures the strain required more iron or a trace of the iron transport compound.

The origin of the complexing centers of the deferrisiderochromes is shown in Fig. 4. This illustrates the fact that the biosynthesis of the

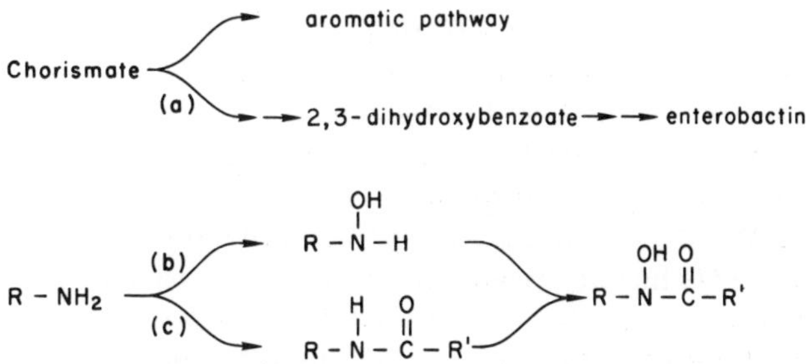

Chorismate —(a)
→ aromatic pathway
→ → 2,3-dihydroxybenzoate → → enterobactin

R — NH₂ —(b)(c)

R − N − H (OH)

H O
R − N − C − R'

OH O
R − N − C − R'

Fig. 4. Biosynthetic pathways for the origin of the metal-complexing atoms of the phenolate (upper) and hydroxamate (lower) deferrisiderochromes.

phenolate, enterobactin, is relatively well understood down to the enzyme level[8]; iron has not been implicated in this biosynthesis. Thus pathway (a) in Fig. 4 represents a spur on the route to the common aromatic compounds which occurs after chorismate and which terminates in enterobactin; the sole purpose of this spur is apparently the synthesis of a siderochrome (enterobactin, I). In the case of the hydroxamate siderochromes, hydroxylation of the N atom may occur either before (pathway b, Fig. 4) or after (pathway c, Fig. 4) acylation. The former reaction, which would be favored thermodynamically, holds for ferrichrome[9] and hadacidin[10] whereas with aspergillic acid[11] (and possibly with mycobactin[4]) the free hydroxylamino compound may not be an intermediate.

Knowledge of the iron repressibility of these biosyntheses has proven of utmost practical importance in obtaining sufficient quantities of the iron transport compounds for chemical and biological investigations.

Failure to find such compounds may mean: (a) the organism has an alternative iron incorporation mechanism, (b) the compounds are not excreted, (c) they are made in amounts below the detection limit, (d) the organisms require a special level of iron for synthesis, or (e) substances naturally present, or artifactually formed, in the medium may take over the transport function. Carbon substrates which can only be oxidized through the tricarboxylic acid (TCA) cycle should precipitate a maximum demand for iron[12].

Unfortunately, very little is known about the metabolism and degradation of the siderochromes. A soil organism, *Pseudomonas* Fc-1, is capable of growth on the ferrichromes as sole source of C and N. The initial attack is by way of a peptidase which cleaves the cyclic peptide at the carboxyl carbon atom of the δ-N-acyl-δ-N-hydroxyornithine tripeptide sequence[13]. Aluminum can substitute for iron, but the demetallocompounds are not split.

E. Biofunction

From the point of view of structure the ferrichrome compounds (V) are among the most intensely investigated in the entire realm of biochemistry. A complete crystallographic representation has been elaborated for ferrichrome A (Vb)[14], ferrichrome (Va)[15] has been synthesized, the substances have been examined by proton magnetic resonance (p.m.r.)[16] electron spin resonance (e.s.r.) and Mössbauer spectroscopy[17]. In the entire phenolate series, only enterobactin (I) has not been chemically synthesized. It is instructive to consider now how well these rather fully understood structures can be reconciled with the biological role illustrated in Fig. 1.

The original supposition that ferrichrome (Va)[18] and itoic acid (IV)[19] act as iron transport agents was based in large part on the observation that the iron(III) complexes are extremely stable (formation constant $\sim 10^{30}$) while the iron(II) analogues are relatively weak. This provides a release mechanism for the iron and, at the same time, it rules out an electron transfer role based on oxido-reduction of the central metal ion. It was recognized that augmented production in iron deficiency would have definite evolutionary advantages to the cell, and this was also compatible with the notion that both ferrichrome (Va) and itoic acid (IV) act as iron transport agents. A second line of evidence was derived from the observation that certain fungi and bacteria, mainly soil and dung inhabiting species, require siderochromes for growth (see Table II). Subsequently, it was found that synthetic chelating agents, at the proper concentration, could promote the growth of natural siderochrome auxotrophs[20]. Ferrichrome has been shown by Emery[21] to play a direct iron transport role in *U. sphaerogena*, a smut fungus notorious for over-production of siderochromes. *Bacillus megaterium*, which in the wild type produces schizokinen[22], can be mutated to siderochrome dependency[23]. The stable chromic complex of mycobactin competitively antagonizes the growth factor activity of mycobactin for certain mycobacteria[4]. This implies that a metal complex of mycobactin,

presumably iron(III) mycobactin (XI), is the physiologically active form
of the metabolite. Finally, the iron transport role of natural iron(III)
hydroxamates can be reconciled with experiments on their antagonism of
the toxicity of closely related antibiotics of the siderochrome class. Thus,
siderochromes with growth-factor activity block the permeation of such
antibiotics into the cells of *Staphylococcus aureus*[24]; presumably the
mechanism is one of competition for the same transport system. Once
inside the cell the siderochrome antibiotic may exert its toxicity in various
processes[25] (depending on its particular structure) which have no relation
to iron metabolism and which accordingly are not reversed by siderochrome
analogues.

Since the biological activities of the iron(III) hydroxamates and
iron(III) phenolates can be equated — the former characteristic of fungal
and yeast species and the latter functional (according to present knowledge)
only in the true bacteria — it has seemed reasonable to attack the biological
mechanism of action by use of organisms such as *Bacillus* sp. and, especially,
members of the Enterobacteriaceae such as *Escherichia coli* and *Salmonella
typhimurium*. It has not been possible, for example, to obtain ferrichrome
auxotrophs from *Ustilago* sp.; such mutants would be desirable in view of
the high concentration of siderochromes present in the wild type. Thus,
the substantial evidence for active transport of the siderochromes first
discovered by Peters and Warren[26] from studies with *Bacillus subtilis* was
promptly confirmed by others using *E. coli*[27] and *S. typhimurium*[28].

The isolation from *B. subtilis* cultures of the first phenolate deferrisiderochrome, itoic acid[19] (*I*ron *T*ransferring *O*rtho-phenol) and its
characterization as 2,3-dihydroxybenzoylglycine (IV) was followed by
the isolation, from other bacterial species, of the serine analogue (III)[29],
the bis-lysine derivative (II)[30] and enterobactin (I)[2]. With the exception of
the bis-lysine compound, all of the other phenolate deferrisiderochromes
have been shown to transport iron in the source organisms, *viz.*, *B. subtilis*,
E. coli and *S. typhimurium*, respectively.

As shown in Fig. 5, Peters and Warren[26] found itoic acid to stimulate
iron uptake in mutant strains of *B. subtilis* partly blocked in the biosynthesis of this deferrisiderochrome. The uptake was both temperature and
energy dependent and it appeared to be magnified in iron-starved cells.
Similar data were obtained by Wang and Newton[27] who, working with *E.
coli*, found iron uptake to require an active transport system as well as
2,3-dihydroxybenzoic acid or equivalent; enterobactin was unknown at
the time these studies were performed.

Investigation of the biochemical genetics of *S. typhimurium* demonstrates convincingly that siderochromes are involved in iron transport. This
organism yields a series of mutants which are able to grow on minimal,
citrate-containing media only upon the addition of iron salts, enterobactin,
or other synthetic or naturally occurring complexing agents[28]. On minimum

medium, citrate, apparently ties up the iron and makes it unavailable to the cells. Enterobactin (I) at low concentrations facilitates the uptake of

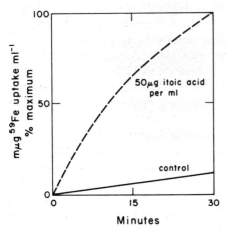

Fig. 5. Iron uptake by *Bacillus subtilis* strain dhb-4 with and without 50 μg of itoic acid per ml. (Data of Peters and Warren, *Biochim. Biophys Acta*, 165 (1968) 225.)

[55]Fe by cell suspensions of *S. typhimurium* auxotrophs and is 100 times as active, on a molar basis, in this capacity as III. The mutants fall into two classes, *viz.*, those blocked in the biosynthesis of 2,3-dihydroxybenzoic acid and those unable to conjugate this phenolic acid with serine. Transduction experiments with Phage P22 suggest that the genes for enterobactin biosynthesis are clustered in an "iron operon". *S. typhimurium* LT2 grows well on complex, natural media and is, in general, sensitive to albomycin (Fig. 6); strains resistant to the antibiotic were suspected to lack a siderochrome transport system, a hypothesis confirmed by the use of tritiated

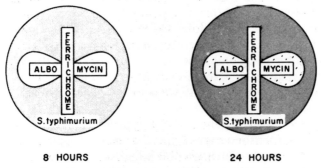

8 HOURS   24 HOURS

Fig. 6. Illustration of ferrichrome antagonism of albomycin and the technique of selection of mutants of *Salmonella typhimurium* defective in ferrichrome transport.

ferrichrome (Fig. 7)[31]. Thus *S. typhimurium*, even though it does not
appear to synthesize hydroxamates, is equipped with a transport system for

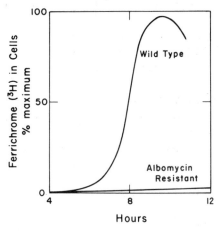

Fig. 7. Uptake of tritiated ferrichrome by cells of *Salmonella typhimurium* (J. R. Pollack, Doctoral dissertation, University of California, Berkeley, 1970).

a type of siderochrome widely distributed in nature. Since some entero-
bacteria are able to make both phenolate and hydroxamate siderochromes[5],
the capacity of *S. typhimurium* to form the latter may have been lost in
the process of evolution.

F. Fitness of siderochrome structure

Having defined the function of siderochromes as that of iron transport,
it may be instructive to review the chemical structures and properties of a
typical iron(III) phenolate (iron(III) enterobactin, **Ia**), and iron(III) hydroxamate (ferrichrome, **Va**) in the light of the declared biological role.

Enterobactin itself is soluble in ether, but the iron(III) complex will
have a triple negative charge and will only dissolve in polar solvents. Ferrichrome is sparingly soluble in methanol and insoluble in nearly all other
non-polar media. Thus, water-solubility of the iron complex appears to be
a constant feature of these systems. On the other hand, the mycobactins,
with or without a metal ion, are only soluble in lipophylic media, and this
may account for the fact that siderochromes other than mycobactin are
inactive in auxotrophic mycobacteria.

In ferrichrome three hydroxamate anions combine with the Fe(III)
ion to form a neutral complex; in enterobactin the iron(III) complex will
have a net charge of − 3 (like ferrichrome A, due to the dicarboxylic acid
moiety in the acyl substituent of the hydroxamic acid function). All lower

phenolates, such as Fe^{III}. (itoic acid)$_3$, similarly, will have a large negative charge. Rhodotorulic acid (VI)[32], however, a deferrisiderochrome commonly found in certain yeasts, will have a unit net positive charge in the 1:1 iron complex form. Thus, net charge *per se* does not seem to be an important determinant for siderochrome function.

It has already been noted that the exclusive presence of oxygen as the coordinating atoms in all of the deferrisiderochromes will assure a high iron(III)/iron(II) stability ratio. The iron in the earth's crust, having been oxidized by O_2 evolved in photosynthesis, will be mainly in the iron(III) state and efficient complexing will require a ligand with high affinity for the +3 state. Examination of a limited range of metal ions for their complexing ability with synthetic and natural hydroxamic acids showed Fe(III) ion to bind most firmly (log K_S deferrioxamine B, 30.6) (Table I)[33]. Such high

TABLE I

STABILITY CONSTANTS OF IRON(III) TRIHYDROXAMATE SIDEROCHROMES[a]

Ligand	log K	Increment over triacethydroxamate
Deferriferrichrome	29.1	0.8
Deferriferrichrysin	30.0	1.7
Deferriferrioxamine B	30.5	2.2
Deferriferrioxamine D$_1$	30.8	2.5
Deferriferrichrome A	~32.0	3.7
Deferriferrioxamine E	32.4	4.1

[a] G. Anderegg et al., Helv. Chim. Acta, 46 (1963) 1409; J. B. Neilands, Experentia Suppl., IX (1964) 22.

binding constants rival those reported for the best synthetic iron(III) coordination reagents used in inorganic analytical chemistry. The iron of ferrichrome is "ionic", high spin[34] and fairly rapidly exchangeable[35]. The iron(III) complex forms readily and is stable at high values of pH; only in dilute alkali can it be removed as the hydroxide. Ferrichrome A, with its cluster of three negative charges, holds the iron even more firmly than in ferrichrome and in a milieu substantially more resistant to attack by alkali[36]. Less is known about the properties of iron phenolate sidero-chromes, except that the metal is observed to enter the complex more slowly under conditions where iron(III) hydroxamates form rapidly[37]. Iron(III) phenolates derived from catechols are known to be stable and have, in fact, served as the basis for the commerical manufacture of ink.

As shown in Fig. 1 under normal conditions of iron nutrition $[(Fe) \cong > 10^{-7} M]$ the transport ligands are presumed to be passively excreted to the cell surface where they can snatch iron from the medium. The cell wall may act as a matrix to limit diffusion of such molecules;

unrestrained excretion to the medium would be improvident for water-dwelling organisms, but might be nutritionally useful for soil-inhabiting species. At low iron stress, however, the disconnection of the regulatory control over the biosynthesis of the deferrisiderochrome will cause its overproduction and hence accumulation in the medium. This may serve to make very small amounts of iron in the environment available to the cell, and it may also be very important ecologically for the growth of mixed populations in the soil.

The profound conformational change in the ligand which follows complexation in aqueous media has been investigated in the ferrichrome series with the aid of p.m.r.[16]. A similar conformational change can be predicted for enterobactin. This may furnish a selective device at the cell surface for exclusion of the deferrisiderochromes. Interestingly, the ionophores for monovalent alkali metal cations display a similar conformational differential between the free and complexed states[38].

Unfortunately, we still know too little about the chemistry of microbial membranes to predict a detailed mechanism for the incorporation of siderochromes. Active transport[26,27], however, is the general process by which the siderochromes are taken into the cell. Resistance to the antibiotic siderochrome albomycin has been used to select mutants defective in siderochrome transport, but it has not been established if these mutants lack specific binding proteins[31]. The transport of iron, like that of other essential nutrilites, may very well be a complicated process and involve multiple pathways.

In the hydroxamate siderochromes it seems obvious that the complex cannot back-diffuse through the membrane. Thus, ferrichrome was originally isolated[39] from the cells of U. sphaerogena, and experiments using a radioactively labelled ligand suggest that the deferricompound does not exit the cell until the iron is depleted by fixation in the iron enzymes and iron proteins[21]. Again, we may invoke a conformational differential between free and complexed species and/or a requirement for an intact energy metabolism system to account for retention of the chelate compound. High levels of cobalt in the growth medium cause the accumulation of siderochromes in the cells or mycelium of fungi[40]. The mechanism of this effect of cobalt is obscure, but it may represent competition with Fe(II) iron for a repressor protein, the cobalt–protein–complex being unable to attach to the operator gene.

A flavoprotein preparation from U. sphaerogena catalyzes the reduction of ferrichrome iron by means of TPNH or DPNH, with some preference for the former[41]. The specificity of this reaction is questionable and, in any case, it would not be of great benefit to the cell to liberate iron from the siderochrome prior to demand for the metal. To do so would only be to make the iron susceptible to hydroxylation and precipitation. Thus a much more attractive hypothesis is one which visualizes direct donation of the iron to an acceptor such as porphyrinogen or a porphyrinogen–apoenzyme complex. In ageing low iron cultures certain polyfunctional siderochromes of both the phenolate and hydroxamate type suffer degradation, but it has not been established that such a "destruct" mechanism is actually required for release of the iron. The intense potency of the siderochromes would imply that the carrier molecules survive and catalytically cycle iron into the cell.

Ferrochelatase is known to be lipid-associated and ferrichrome is an effective donor of iron for this enzyme[42]. Ferrichrome is universally active for siderochrome auxotrophs (with the exception of mycobacteria), an activity which may be explained by the ability of this particular transport agent to dissolve in non-equeous solvents such as alcohols. In the case of *Arthrobacter* JG-9, siderochrome dependency can be entirely replaced by synthetic, lipid-soluble complexing agents, when these are furnished at an appropriate concentration[20]. All of this suggests that, at least in *Arthrobacter*, the siderochromes are not specific "co-enzymes" in the internal iron metabolism of the cell but simply serve to deliver the metal ion to a point where it can be converted to a physiologically functional derivative.

There is one aspect of the iron *transfer* reaction which requires special mention. The deferriferrichromes are routinely prepared by low-iron fermentation and crystallized by addition of a Fe(III) or Fe(II) (with aeration) salt. As a consequence of the unique spatial disposition of the three hydroxamic acid bonds, the six oxygen atoms of the ligand assume a single, specific arrangement around the metal ion[43]. A Cotton effect is seen in the region of the wavelength maximum of the iron(III) trihydroxamate ($\sim$ 420–440 nm)[3], and the crystallographic formulation of ferrichrome A shows that the absolute configuration around the iron is that of a left-handed propeller[14]. Interestingly enough, construction of iron(III) enterobactin with space-filling molecular models shows that an identical absolute configuration is also feasible in this siderochrome[2].

G. Distribution

Probably all aerobic and facultative aerobic microbial cells produce or require siderochromes. A survey of a very large number of yeasts and yeast-like organisms showed that rhodotorulic acid (**VI**) is a common deferrisiderochrome in these species and disclosed the formation, by *Crytococcus melibiosum*, of a new alanine-containing ferrichrome (component C, **Vg**)[7]. In general, siderochromes derived from δ-*N*-acyl-L-δ-*N*-hydroxyornithine are commonly encountered in fungal species from the Ascomycetes, Basidiomycetes and the Fungi Imperfecti classes[13]. One actinomycete, *Streptomyces griseus*, also forms a ferrichrome-type antibiotic (albomycin, grisein, **Vi–k**), and *Pseudomonas fluorescens* elaborates a cyclic decapeptide, ferribactin[44], which contains both D and L forms of δ-*N*-hydroxyornithine.

The next higher homologue of the latter amino acid, L-ε-N-hydroxylysine, is found in a family of deferrisiderochromes from mycobacteria (mycobactins, **XI**)[4], and in aerobactin (**IX**)[5] from *Aerobacter aerogenes*. Siderochromes of the actinomycetes characteristically contain bases which can be regarded as decarboxylation products of the α-amino-ω-hydroxy-amino acids, *viz.*, α-amino-ω-hydroxyaminopentane or hexane; such products, designated ferrioxamines or ferrimycins (**XV, XVI**) have been found in all thoroughly investigated actinomycetes[45].

The ferric phenolate siderochromes are similarly widely distributed in the microbial world, with a bias toward the true bacteria. The isolated and characterized deferrisiderochromes include 2,3-dihydroxybenzoylglycine (itoic acid, **IV**)[19], from *B. subtilis*, 2,3-dihydroxybenzoylserine (**III**) from *E. coli*[29] and *A. aerogenes*[46], bis-2,3-dihydroxybenzoyllysine (**II**)[30] from *Azotobacter vinelandii* and enterobactin (**I**)[2] from *S. typhimurium*. The specific catechol of these deferrisiderochromes, 2,3-dihydroxybenzoic acid, may be found in actinomycetes, fungi and higher plants[8].

Siderochrome activity can be detected in soil and in dung, and it must be concluded that such compounds are of general significance in microbial iron metabolism. An index of their importance is indicated by the existence of both natural and induced auxotrophs for such metabolites (Table II). Hemin is active for native auxotrophs; doubtless such strains have adapted to the use of several natural iron-containing compounds. The fact that siderochrome antibiotics such as albomycin (**Vh**) and the ferrimycins (**XVe**) are "broad spectrum" speaks for a wide distribution of transport mechanisms for these substances[47]. Fe(II) iron may be sufficiently soluble in the culture media of anaerobic species to eliminate the need for siderochrome-like carriers, but this point needs to be verified experimentally. Specific transport systems for metal ions other than iron may not be required since such ions are generally much more soluble than iron and moreover, they are needed in smaller amounts.

H. Summary

There are features of the scheme shown in Fig. 1 which resemble mammalian iron metabolism, even though higher organisms do not appear to contain siderochromes[48]. Thus, there is no mechanism in animals for excretion of iron. The element is admitted through the intestinal wall according to the demand. This much the microbes and animals have in common; however, probably the latter alone have the intricate device for salvaging iron and storing it in ferritin, although the latter protein apparently occurs in certain *Phycomyces* sp.[49]. Siderochromes will be present in the gut of higher animals but the significance of these products to the iron metabolism of the host is unclear. The ability to obtain iron seems to be a key to virulence and pathogenicity in microbial infestions[50].

180

TABLE II

SIDEROCHROME DEPENDENT AUXOTROPHIC MICRO-ORGANISMS

Source	Active factors[a]	Reference
A. Natural		
Pilobolus kleinii	Dung, extract, coprogen, ferrichrome, hemin	C. W. Hesseltine *et al.*, *Mycologia*, 45 (1953) 7.
Arthrobacter terregens	Terregens factor, hemin, coprogen, ferrichrome, aspergillic acid, soil and liver extracts, synthetic chelating agents, mycobactin	M. O. Burton and A. G. Lochhead, *Can. J. Bot.*, 31 (1953) 145; N. E. Morrison and E. E. Dewbrey, *J. Bact.*, 92 (1966) 1848.
Arthrobacter JG-9 (*A. flavescens* ATCC 25091)	Terregens factor, ferrioxamine B, ferrichrome, aspergillic acid, coprogen, grisein, mycobactin, hemin, higher fusarinines, rhodotorulic acid, acethydroxamic acid and synthetic chelating agents.	B. F. Burnham and J. B. Neilands, *J. Biol. Chem.*, 236 (1961) 554; ref. 20.
Microbacterium lacticum ATCC 8181	Ferrichrome, coprogen, terregens factor, heme compounds, glucosyl-glycine	D. Hendlin and A. L. Demain, *Nature*, 184 (1959) 1894.
Mycobacterium johnei	Mycobactins	Ref. 4
B. Induced		
Bacillus megaterium ATCC 19213 (Sk mutants)	Schizokinen, ferrioxamine B, ferrimycin A, ferrichrome, mycobactin, 2,3-dihydroxybenzoic acid, acethydroxamic acid, natural and synthetic chelating agents	Ref. 23
Escherichia coli (Chr2 mutant)	2,3-Dihydroxybenzoylserine, citrate	Ref. 27
Salmonella typhimurium LT2 (Enb mutants)	Dihydroxybenzoylserine, enterobactin, ferrichrome, schizokinen, deferriferrioxamine E, rhodotorulic acid, ascorbate and EDTA	Ref. 28

[a] Deferrisiderochromes will generally be as active as the iron complexes unless steps are taken to remove contaminating iron from the medium. Most siderochrome auxotrophs have the ability to grow in the presence of high concentrations of inorganic iron.

The role of pacifarin[51], a siderochrome of the phenolate class[52], in this process is of considerable interest from the medical point of view. In spite of some interesting leads, *viz.*, a heme requirement for certain nematodes[53], no one has yet made a systematic search for siderochromes throughout the animal kingdom. Animal tissues have strictly limited ability to transform aromatic compounds, and their siderochromes, if they exist, might hence be expected to belong to the hydroxamate group. Interestingly enough, certain animal membranes are believed to contain low molecular weight iron chelating substances[54].

Deferriferrioxamine B mesylate is marketed by the Ciba Pharmaceutical Cc. under the trade name "Desferal" and has had important uses in removing pathological deposits of iron and in neutralizing accidentally ingested quantities of iron[55]. Hydroxyurea, a synthetic hydroxamic acid, is used in cancer chemotherapy; it controls DNA synthesis apparently through binding to the iron in protein B_2 of the ribotide reductase complex[56].

Hydroxamic acids and/or siderochromes have been found in algae[57] and some higher plants[58], but their function in these life forms has not been defined. Higher plants grown at low iron levels excrete substances resembling deferrisiderochromes[59]. Synthetic chelating agents make iron available to plants without incorporation of the ligand[60]; however, a systematic study of plant metabolism of microbial siderochromes known to commonly occur in the soil would be of interest. Synthetic hydroxamic acids have been applied in agriculture to increase iron availability and such substances may, through inhibition of urease, prevent loss of ammonia from the soil[61].

In summary, the siderochromes play a role in microbial iron transport which in some ways parallels the membrane-active "ionophores" for the alkali metal cations[38]. Apparently, the slight decrease in diffusional mobility arising from the order of magnitude increase in molecular weight of the complexed ion is more than offset by the advantages of the organic cloak. The synthesis of the latter can be precisely regulated and, moreover, a specific transport system for the complex can be localized in the cell membrane. By increasing the internal density of deferrisiderochrome molecules, which happens at low iron concentrations, the organism can excrete more ligand and can clutch more metal ions from the medium. In severe iron deficiency, the ligand is excreted into the medium in massive amounts and serves to dissolve insoluble deposits of iron. Meanwhile, the transport system protects the cell against invasion by toxic quantities of inorganic iron and rejects the ligand unless it is laden with its specific metal ion.

There have been a number of reviews in recent years dealing with various aspects of siderochromes. The reader is referred to articles which emphasize the ferrichromes[3,13], ferrioxamines[1], siderochrome antibiotics[47], and mycobactins[4]. This chapter will stress in particular the phenolate siderochromes, since this information has not hitherto been collected in the literature.

II. STRUCTURE AND PROPERTIES

A. *Methodology*

Characterization of the siderochromes is a fairly straightforward exercise in natural product chemistry. The first objective, however, is to boost production and this can generally be achieved by proper adjustment of the iron concentration of the medium. Occasionally it may be necessary, as in *Rhodotorula pilimanae*, to add more C and N to the medium in order to secure maximum yields[32]. It is best to isolate the substance without the iron but, if this is not feasible, the iron can always be removed from the complex by various techniques[3].

The known phenolate deferrisiderochromes are all soluble in organic solvents at a pH $\sim$ 2 units below the pK_a of the carboxyl group. Even though enterobactin (**I**) has no free acid groups it will be present in the medium complexed to ions and must be separated from these by acidification prior to extraction. The hydrolytic fragments obtained with $6N$ HCl, 2,3-dihydroxybenzoic acid and an α-amino acid, can be detected by the Arnow[62] and ninhydrin reagents, respectively. Furthermore, the polyhydric phenolic acids display intense and characteristic fluorescence under ultraviolet illumination[19].

The hydroxamate siderochromes are more diverse in structure but here again some common characterization steps can be followed. The substances can generally be extracted into either chloroform–phenol (1:1, w/w) or benzyl alcohol[3]. From these organic solvents the active principle can be returned to water by addition of many volumes of ether. Hydrolysis of the iron-free molecules with $6N$ HCl yields tetrazolium-positive hydroxylamino moieties from the hydroxamate functions; hydrolysis with $6N$ HI converts the hydroxylamino groups to primary amino groups[4]. Hydrolysis with HCl in the presence of iron will produce artifacts in excellent yield. Periodate is the reagent of choice for oxidation of the hydroxamic acid bonds since this treatment will leave intact any ester or amide linkages in the molecule[3].

Hydroxamate siderochromes are universally of the "secondary" type ($-CON(OH)R$) in which the hydroxylamino N is attached to an alkyl group. All such hydroxamates are smoothly converted on periodate scission to the easily detected *cis*-nitrosoalkane dimer (a_{mM} at 267 nm $\cong$ 10)[3]. Carboxyl groups arising after periodate oxidation can be presumed to be attached to hydroxylamino groups in hydroxamic acid bonds in the original molecule. Changes in the iron(III) hydroxamate spectrum *vs.* pH can reveal details of the structure of the iron-binding center.

Both the dihydroxybenzoyl nucleus and the free hydroxyamino groups are unstable at pH levels alkaline to their respective pK_a values.

B. Phenolates

1. Enterobactin

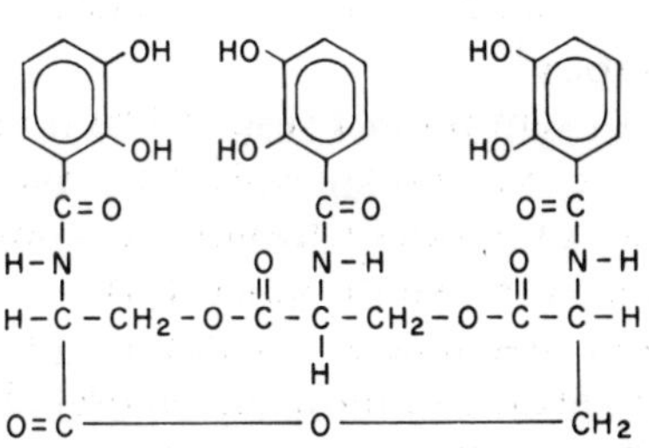

I Enterobactin 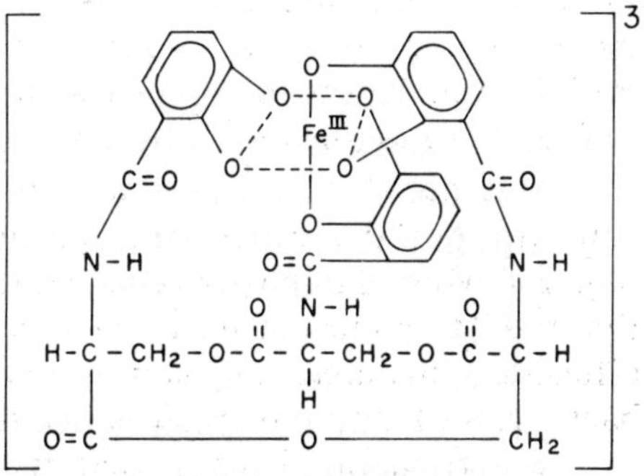   Ia Iron(III) enterobactin

Enterobactin (I) was obtained from low iron cultures of *Salmonella typhimurium* LT2 to which had been added relatively high levels of manganese ion[2*]. The compound crystallized from ethanol–water, m.p. 202–203°C.

Enterobactin is essentially insoluble in water but dissolves readily in acetone, dioxane, dimethylsulfoxide or methanol. Its water solubility is increased at pH values between 7 and 8 where the 2-hydroxyl group begins to ionize. It is neutral on paper electrophoresis at pH 5 and exhibits the bright blue fluorescence under ultra-violet light characteristic of the 2,3-dihydroxybenzoyl nucleus. Hydrolysis in 6N HCl affords L-serine and 2,3-dihydroxybenzoic acid as the sole identifiable fragments; elementary analyses for C, H, and N suggest an anhydride of these two fragments as the structure of enterobactin.

The infra-red and p.m.r. spectra of enterobactin resemble those of 2,3-dihydroxybenzoylserine. A strong absorption at 1760 cm^{-1} can be assigned to a strained ester group carbonyl stretch. The p.m.r. spectrum at 220 MHz in deuterodimethylsulfoxide with tetramethylsilane as internal standard gave these data: δ = 8.90, doublet (amide); δ = 7.30, doublet (aromatic); δ = 6.96, doublet (aromatic); δ = 6.71, triplet (aromatic); δ = 4.91, complex (alpha); δ = 4.64, complex (methylene); and δ = 4.41, complex (methylene). In the mass spectrometer a major ion occurred at

---

*See note added in proof, p. 200.

m/e of 223, corresponding to the monomer; methylation of enterobactin with diazomethane gave ions with mass equivalent to a dimer, or greater. Sedimentation equilibria in water and D_2O showed enterobactin to be a trimer, molecular weight 669, with the structure shown for I. Enterobactin is thus a cyclic triester of L-serine in which all three amino groups have been acylated with residues of 2,3-dihydroxybenzoic acid.

Iron(III) enterobactin (Ia) behaves on paper electrophoresis at pH 6.5 as a trivalent anion; the structure is assumed to be as shown in Ia. The three catechol groups are located on one side of the plane of the 12-membered cyclic polyester and assume a left-hand propeller around the Fe(III) iron. This is the same absolute arrangement, the D configuration, as found in ferrichrome A by crystallography[14].

Mutants of S. typhimurium unable to synthesize I respond to it at a concentration two orders of magnitude less, on a molar basis, than that required when 2,3-dihydroxybenzolserine is the siderochrome[2]. It appears that the latter substance can be used for iron transport, albeit less efficiently, in Salmonella and probably in other species. However, enterobactin is probably the major active siderochrome in these micro-organisms★.

Enterobactin displays pacifarin activity in the mouse[62b] and it is apparently identical to the inhibitor of colicin B which is found in enteric bacteria[62c].

2. 2-N,6-N-Di(2,3-dihydroxybenzoyl)-L-lysine (II)

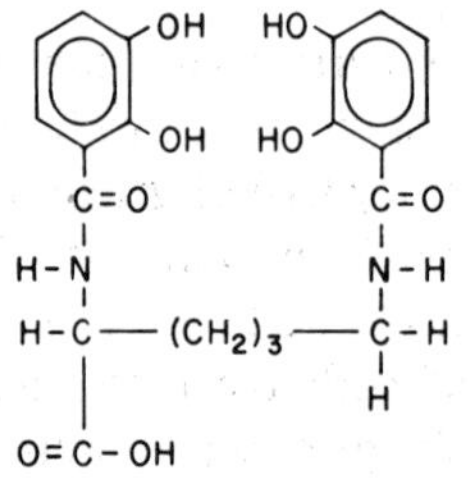

II

This compound was obtained as an amorphous white solid in yields of approximately 90 mg/l by growth of *Azotobacter vinelandii* O on low-iron nitrate medium[30]. It was indistinguishable from a compound chemically synthesized by acylation of both N atoms of L-lysine with 2,3-dihydroxybenzoic acid. Although no studies have been performed on the iron

----

★ Mutations affecting iron transport in *Escherichia coli* are comparable to those in *Salmonella typhimurium*, with the expected displacement along the chromosome. However, unlike *Salmonella*, the former organism apparently has an inducible uptake system for iron(III) citrate (G. B. Cox, F. Gibson, R. K. J. Luke, N. A. Newton, I. G. O'Brien and H. Rosenberg, *J. Bact.*, 104 (1970) 219.)

complex, it is of interest to note that exactly the same number of atoms, namely 5, separate the two acylated amino groups in enterobactin. This would suggest that **II** is involved in iron transport in *A. vinelandii*.

3. 2,3-dihydroxy-N-benzoyl-L-serine (III)

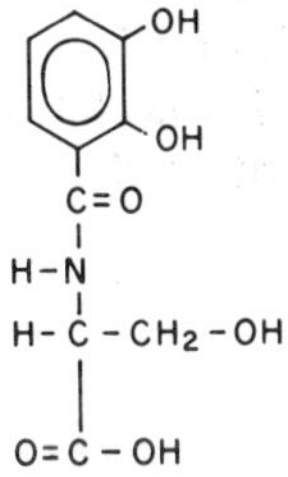

OH
OH
C=O
H–N
H–C–CH₂–OH
O=C–OH

III

Brot *et al.*[29] found that growth of a methionine–vitamin B_{12} auxo-troph of *E. coli* K_{12} on glucose salts medium without added iron gave a substance in yields of 75 mg/l, isolated as the Fe(III) complex. A solution of the compound was decolorized at pH 2 or in the presence of 0.1 *M* EDTA. The ligand was characterized as 2,3-dihydroxy-*N*-benzoyl-serine (**III**), a structure confirmed by O'Brien *et al.*[46] through synthesis of the D and DL configurations. The configuration of the natural compound is believed to be L, although the synthetic product containing D-serine has comparable activity for a mutant strain of *Escherichia coli*. Iron acts as a powerful repressor for both the enzyme conjugating 2,3-dihydroxybenzoic acid with L-serine in *E. coli*[63] and the entire enzyme system converting chorismate to 2,3-dihydroxybenzoate in *A. aerogenes*[8]. **III** has been obtained from the latter organism[46] and from *S. typhimurium*[50,31], and may be correlated with the virulence of *S. typhimurium* via its ability to obtain iron from human sera[50].

4. 2,3-Dihydroxybenzoylglycine (itoic acid) (IV)

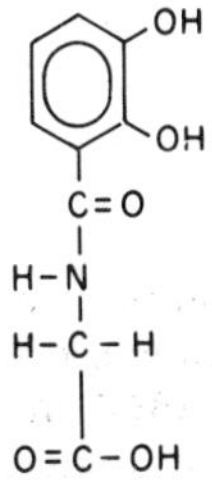

OH
OH
C=O
H–N
H–C–H
O=C–OH

IV Itoic acid

Itoic acid (**IV**), the glycine conjugate of 2,3-dihydroxybenzoic acid, was the first phenolate siderochrome to be described[19]. It was isolated from low-iron cultures of *Bacillus subtilis* in amounts of about 50 mg/l. The compound was synthesized chemically by coupling 2,3-dihydroxybenzoic acid with glycine ester, followed by saponification. Production was shown to be completely repressed in *B. subtilis* at an iron concentration of 3×10^{-6} g atoms/l[64] and iron(III) schizokinen was found to be more active in this capacity than inorganic iron[65]. Phenolic acids were found to facilitate iron transport in *B. subtilis*[26], a finding which confirms that such "monomeric" siderochromes as **IV** can play a role equivalent to that of enterobactin.

C. Hydroxamates

1. Ferrichrome family

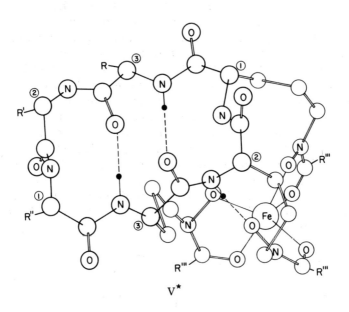

V*

---

* Crystal structure after Zalkin *et al.*[14] for ferrichrome A. The number on the α-carbons refer to the assignment of Zalkin *et al.*[14]. For a discussion of the solution conformation of the ferrichrome compounds see: Emery, *Biochemistry*, 6 (1967) 3858; M. Llinás, M. P. Klein and J. B. Neilands, *J. Mol. Biol.*, 52 (1970) 399; and M. Llinás, Doctoral Dissertation, University of California Berkeley, 1971.

Va Ferrichrome: R = R′ = R″ = H; R‴ = CH₃—

Vb Ferrichrome A: R = R′ = HOCH₂—;

$$R'' = H; R''' = \quad \begin{array}{c} CH_3 \\ \diagdown \diagup \\ CH_2CO_2H \end{array} \quad \text{(trans)}$$

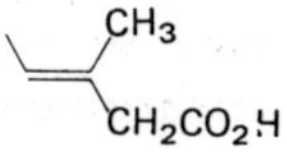

Vc Ferrichrysin: R = R′ = HOCH₂—; R″ = H; R‴ = CH₃—

Vd Ferricrocin: R = R″ = H; R′ = HOCH₂—; R‴ = CH₃—*

Ve Ferrirubin: R = R′ = HOCH₂—;

$$R'' = H; R''' = \quad \begin{array}{c} CH_2 \\ \diagdown \diagup \\ CH_2CH_2OH \end{array} \quad \text{(trans)}$$

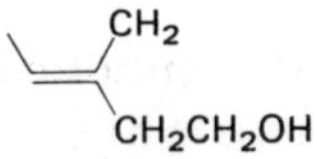

Vf Ferrirhodin: R = R′ = HOCH₂—;

$$R'' = H; R''' = \quad \begin{array}{c} CH_2CH_2OH \\ \diagdown \diagup \\ CH_3 \end{array} \quad \text{(cis)}$$

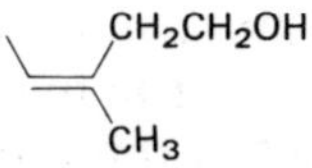

Vg Ferrichrome C: R = R″ = H; R′ = R‴ = CH₃-*

Vh Sake Colorant A: R = HOCH₂-; R′ = R‴ = CH₃-; R″ = H*

Vi Albomycin δ₂: R = Acyl—N=
$$\begin{array}{c} CH_3 \quad O \\ N \\ \diagdown \\ N-SO_2-O-CH_2-; \end{array}$$

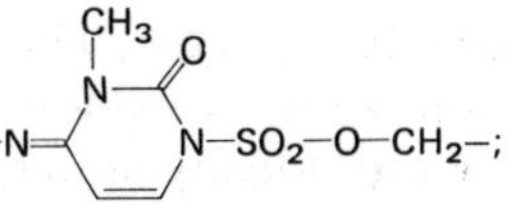

R′ = R″ = HOCH₂—; R‴ = CH₃—

Vj Albomycin ε: R = H—N=
$$\begin{array}{c} CH_3 \quad O \\ N \\ \diagdown \\ N-SO_2-O-CH_2-; \end{array}$$

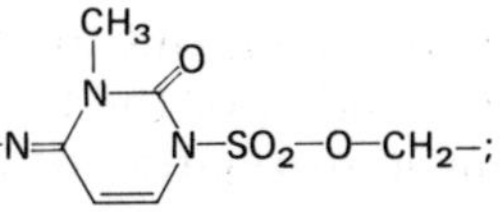

R′ = R″ = HOCH₂—; R‴ = CH₃—

Vk Albomycin δ₁: R = O=
$$\begin{array}{c} CH_3 \quad O \\ N \\ \diagdown \\ N-SO_2-O-CH_2-; \end{array}$$

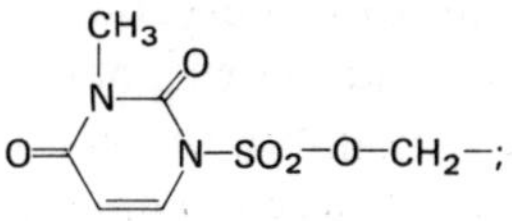

R′ = R″ = HOCH₂—; R‴ = CH₃—

Grisein: This antibiotic may be identical with one of the components of albomycin.

Ferribactin: A cyclic decapeptide containing both D and L forms of N^δ-hydroxyornithine.

---

* Amino acid sequence determined by p.m.r. (M. Llinás, personal communication).

References pp. 200–202

The characteristic structural feature of the ferrichrome family is a cyclic hexapeptide containing a tripeptide sequence of δ-N-acyl-L-δ-N-hydroxyornithine and a tripeptide composed of small, neutral amino acids. The latter may be glycine, serine or alanine alone or in some particular combination. Thus ferrichrome (**Va**) contains triglycine while albomycin (**Vi–k**) is comprise of a run of triserine. The sequence serylserylglycine is common to ferrichrome A (**Vb**), ferrichrysin (**Vc**), ferrirubin (**Ve**) and ferrirhodin (**Vf**). Sake colorant A[66] and ferrichrome C[7] contain alanine. The conformation of the albomycins could thus be unique since they are the only members of the series not bearing a glycyl residue at position R″ in **V**.

The ferrichrome compounds form stable, neutral complexes with 6-coordinate metal ions which prefer octahedral geometry. Among the bio-elements Fe(III) ion is the most efficiently complexed; the log stability constants for ferric triacethydroxamate, ferrichrome and ferrichrome A are 28.3, 29.1, and 32, respectively (see Table I). In spite of the tight binding, inorganic iron exchanges into ferrichrome at neutral pH with a half time of only a few minutes[35]. In the ferrichrome series both the position of the absorption maximum and the millimolar absorbancy coefficients will depend upon such factors as number of ferric ions bound, pH and nature of the acyl substituent. As a rule, at neutral pH the a_{mM} will be 3 to 4 at 420 to 440 nm[3]. Iron(II) hydroxamate complexes are colorless and relatively unstable. The properties of iron(III) hydroxamates are incompatible with an oxido-reduction mechanism for the iron but are, on the other hand, ideally suited for the biological role proposed, *i.e.*, metal transport.

The complete crystallographic structure of ferrichrome A (**Vb**) worked out by Zalkin *et al.*[14] confirms the formulations based on degradative and synthetic studies of the ferrichromes and, in addition, provides evidence for the conformation of the molecule. The peptide moiety is an approximately planar, anti-parallel ("cross-β") reactangle with two hydroxamate-bearing residues (and hence the metal ion) situated at one end of the molecule. In the crystal only a single transannular H bond was located, furnished by residue orn[3], and a second H bond was identified between orn[2] and the hydroxamate oxygen of the same residue. In solution the complexes possess four slowly exchanging hydrogens and these have been assigned by p.m.r. analysis of alumichrome to orn[1] ("internal"), orn[2] (H bonded), orn[3] (H bonded) and gly[3] (H bonded or "internal") (ref. 16 and references therein). The iron-free peptides, which have never been crystallized, exhibit no slowly exchanging protons in aqueous media; in dimethylsulfoxide two protons in the demetallo molecules exhibit decreased temperature dependence of chemical shift. These protons were identified by p.m.r. as gly[3] and orn[3]. Thus in non-aqueous media the demetallo peptide moiety appears to have a conformation corresponding to that of a Schwyzer-type cyclohexapeptide with two transannular H bridges.

Polycrystalline ferrichrome A was found to be paramagnetic down to at least 10K, and g values, energy eigen values and eigen functions have been computed. The Mössbauer parameters of both ferrichrome A[17] and synthetic iron(III) hydroxamates have been measured[67]. In the three iron(III) hydroxamate rings the hydroxylamino-oxygen bond lengths are 1.97, 1.96 and 2.00 Å; in the same three rings the carbonyl oxygen bond distances are 2.02, 2.03 and 2.06 Å, respectively[14]. These measurements indicate that all three hydroxamates can penetrate to the central metal ion and hence confirm the stability of the complex (see Table I).

The ferrichromes are presumed to arise by oxidation of ornithine on the δ-N atom followed by acylation with a carbon fragment related to mevalonic acid, the latter furnished as the coenzyme A derivative. The ferrichromes are commonly produced by the Ascomycetes, Basidiomycetes and Fungi Imperfecti, representatives of V having been isolated from cultures of *Aspergillus, Penicillium, Neurospora, Ustilago* and other smuts[3]. In addition, one yeast, *Crypotococcus melibiosus* forms ferrichrome C (**Vg**)[7] and albomycin (**Vi–k**) is produced by *Actinomyces griseus*.

Regardless of the test organism, ferrichrome is efficient in reversing the toxicity of albomycin and is usually active as a growth factor at levels down to ng/ml. Ferrichrome A, on the other hand, is generally devoid of biological activity and it may thus represent a specialized siderochrome specific for the source organism.

2. *Rhodotorulic acid family*

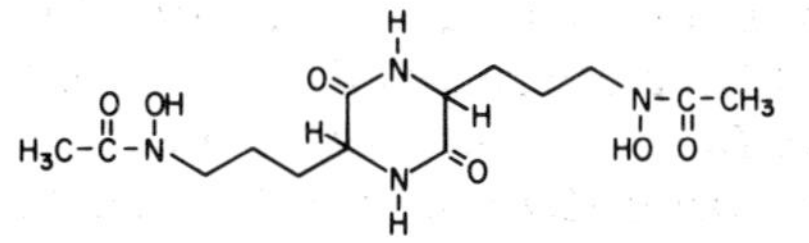

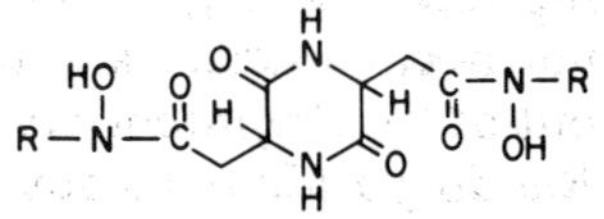

VI Rhodotorulic acid

VII Retro-rhodotorulic acids
VIIa L, R = H
VIIb L, R = CH₃
VIIc D, R = CH₃
VIId L, R = CH₂CH₂CH₃

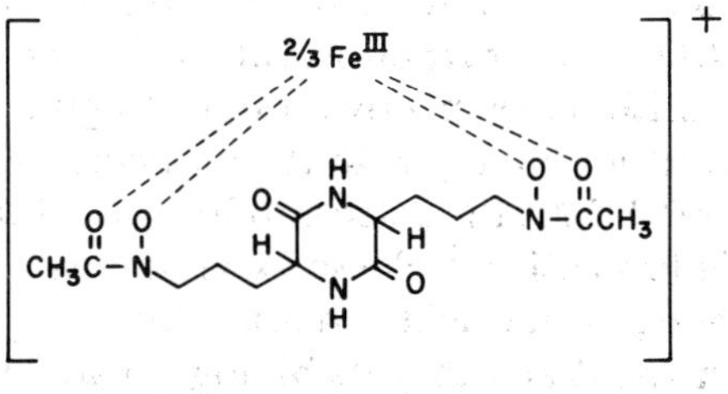

VIII Iron(III) rhodotorulate

A cyclic dipeptide or diketopiperazine of δ-N-acetyl-L-(S)-δ-N-hydroxy-ornithine, chemically defined as N,N'3,6-dioxopiperazine-2(S), 5(S)-di(tri-methylene)-bis-acetohydroxamic acid or rhodotorulic acid (**VI**), has been isolated from iron-starved cultures of *Rhodotorula pilimanae* and related yeasts[32]. The compound has been characterized by reduction to the known diketopiperazine of δ-N-acetyl-L-ornithine and by such physical analyses as infra-red, nuclear magnetic and mass spectrometry. The *cis*-peptide bonds in the ring of **VI** gave, as expected, infra-red spectra with lack of either "amide II" bands or N–H stretch bands above 3250 cm^{-1} and a prominent in-plane bending near 1450 cm^{-1}. Similarly, comparison of the characteristic p.m.r. spectra of ring associated protons in **VI** with those in model diketopiperazines confirms the postulated structure, and examination of the optical rotatory dispersion in the deep ultra-violet proved the configuration to be L,L.

While there is no doubt about the structure, rhodotorulic acid itself has not been chemically synthesized. However, a number of L and D analogues of the type illustrated by **VII**, all with shorter side chains than **VI**, have been prepared[68]. These are termed "retro-rhodotorulic acids" inasmuch as the relative positions of the diketopiperazine nucleus and acyl substituent in **VI** have been reversed about the hydroxamic acid linkage.

From the absorption spectrum of iron(III) rhodotorulate at low pH it can be concluded that both hydroxamates can bind to a centrally coordinated metal ion (**VIII**). Solutions containing $\frac{2}{3}$ mole of Fe(III) salt/mole of **VI** behave as polynuclear complexes with molecular weights of several thousand. Rhodotorulic acid appears to follow the biosynthetic route established for ferrichrome. A survey of 142 different cultures representing 19 genera of yeasts showed hydroxamate production by 2 unclassified strains and by 52 strains among the genera *Aessosporon* (3 out of 3), *Cryptococcus* (1 out of 43), *Leucosporidium* (3 out of 11), *Rhodosporidium* (4 out of 14), *Rhodotorula* (27 out of 39), *Sporidiobolus* (2 out of 2) and *Sporobolomyces* (12 out of 13). With the exception of the single *Cryptococcus* strain, which makes ferrichrome C, most of the other yeasts were observed to form rhodotorulic acid in crystallizable amounts[7].

While rhodotorulic acid is a potent growth factor for *Arthrobacter* JG-9 (half maximum growth at $\sim 3 \times 10^{-9}$ M), it cannot reverse the toxicity of albomycin for *E. coli* or *B. subtilis*[32]. Deoxyrhodotorulic acid, the diketopiperazine of δ -N-acetyl-L-ornithine, is inactive. Interestingly, analogues **VIIb** and **VIIc** show identical potency with *Arthrobacter* JG-9 and give half maximum growth at $\sim 10^{-5}$ M; the retro-rhodotorulic form **VIId** is only an order of magnitude less active[68]. These experiments demonstrate unequivocally that rhodotorulic acid cannot serve in *Arthrobacter* sp. as an acyl transfer agent and that its biological activity must be related to the iron-binding capacity.

Dimerum acid (**VIIIa**), the diketopiperazine of the *trans* isomer of fusarinine, has been obtained at the end of the fermentation period in cultures of *Fusarium dimerum*[68a]. In this substance the acyl moiety of ferrirubin (**Ve**) has been substituted in rhodotorulic acid in place of acetic acid. However, such a molecule does not appear to be a precursor of rhodotorulic acid[68b].

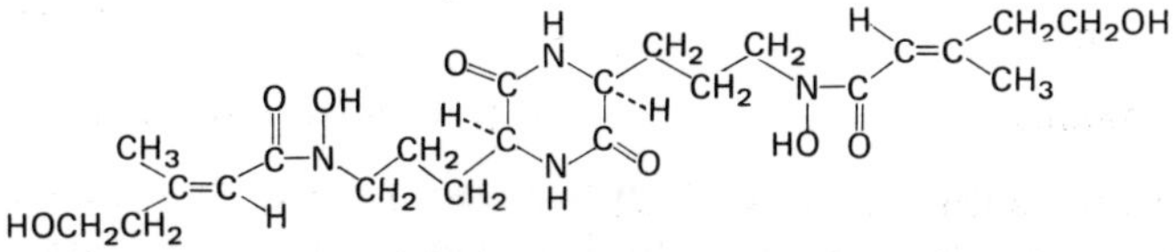

VIIIa Dimerum acid

Coprogen (**VIIIb**), one of the first iron transport agents to be isolated and a product of *Neurospora* and other fungal species, has been characterized recently as the iron(III) complex of a molecule of dimerum acid esterified on one of the primary hydroxyl groups by a residue of *N*-acetyl fusarinine[68c]; coprogen B is the deacetyl analogue[68a].

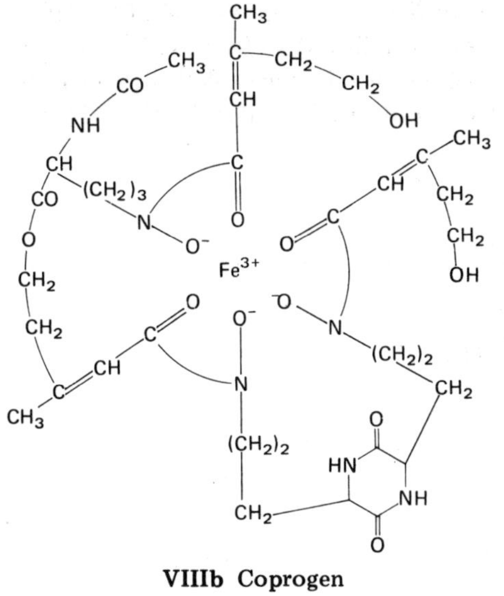

VIIIb Coprogen

3. Aerobactin

Aerobacter aerogenes 62-1 grown 20 hours on glucose mineral salts medium without added iron formed crude aerobactin (**IX**) in yields of 1 g/l[5]. Reductive hydrolysis with HI gave lysine while degradation with HCl afforded citric acid and two residues of the amino acid found in mycobactin, L-ϵ-*N*-hydroxylysine. P.m.r. analysis, electrometric titration, periodate oxidation, the infra-red spectrum and the spectrum of the

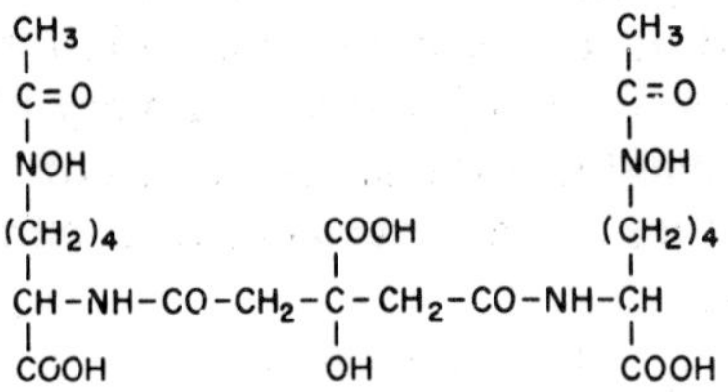

IX Aerobactin

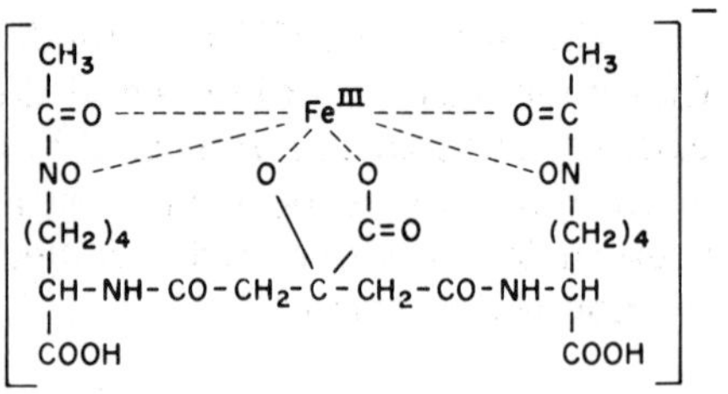

X Iron(III) aerobactin

ferric complex *vs.* pH support the structure shown for IX. The substance probably plays a role in iron transport in *A. aerogenes* 62-1, a strain which also forms 2,3-dihydroxybenzoyl compounds. Other strains of *Aerobacter* sp. do not produce either hydroxamates or phenolates, an observation which reinforces the concept of multiple pathways of microbial iron transport. In any event, inspection of a molecular model of iron(III) aerobactin indicates that the free carboxyl and hydroxyl groups of the citric acid unit, as well as the two hydroxamate groups, can bind to the Fe(III) ion. However, the structure illustrated in X needs to be verified experimentally.

The ligand of schizokinen (Xa) is believed to be a close relative of aerobactin in which the central residue of citric acid is linked symmetrically via substituted amide bonds to two moles of 1-amino-3-acethydroxamido-propane[5a].

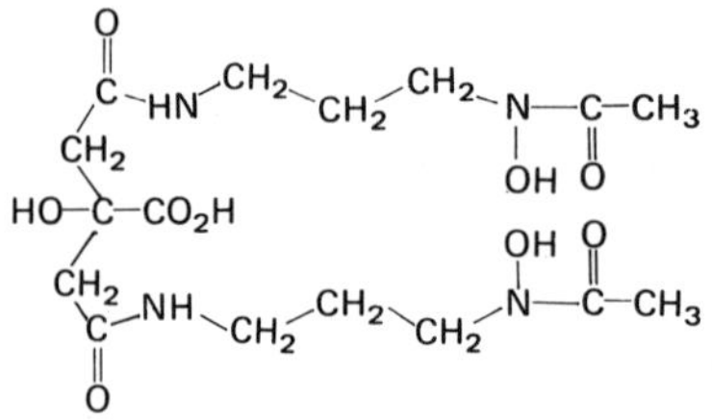

Xa Deferrischizokinen

4. Mycobactin family

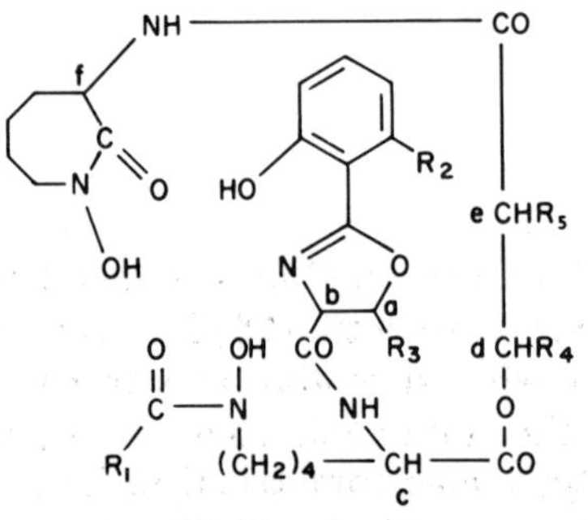

XI Mycobactin

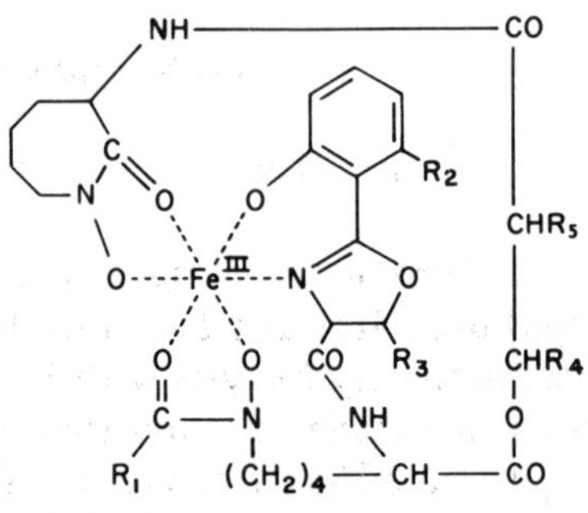

XII Iron(III) mycobactin

Mycobactin	R_1	R_2	R_3	R_4	R_5	a	b	c	d	e	f
A^b(XIIa)	13Δ	CH_3	H	CH_3	H						
F^a(XIIb)	17, 15, 13, 11, 9Δ	H	CH_3	CH_3	H	Threo			S	( )	L
H (XIIc)	19, 17Δ	CH_3	CH_3	CH_3	H	R	L	L	S	( )	L
M (XIId)	1		H	CH_3	18, 17, 16, $15^{\underline{c}}$	CH_3				R^d	S^d
N^b(XIIe)	2		H	CH_3	18, 17, 16, 15^c	CH_3					
P (XIIf)	19, 17, 15 $cis\Delta^1 n$	CH_3	H	C_2H_5	CH_3	( )	L	L	S	R	L
R (XIIg)	19Δ	H	H	C_2H_5	CH_3	( )	L	L	R	S	L
S (XIIh)	19, 17, 15, $13cis\Delta$	H	H	CH_3	H	( )	L	L	S	( )	L
T (XIIi)	20, 19, 18, 17 ⎫ 20, 19, 18, 17Δ ⎭	H	H	CH_3	H	( )			R	( )	L

General structure of the iron(III) mycobactins. Side chains R_1 are alkyl groups having the number of C atoms shown; double bonds are indicated where they are known. Figures show the main types of side chain; some occur in greater abundance in each myobactin. Asymmetric centers are labeled a–f. Blanks indicate that the configuration has not been determined. Key: ( ), lack of asymmetric center; [a], structure inferred from mixture with XIIc; [b], tentative structures; [c], saturated alkyl groups having the number of C atoms shown; [d], relative configuration at d and e is erythro—absolute configuration is uncertain. (G. A. Snow, *Bact. Rev.*, 34 (1970) 100.)

The mycobactins have been the subject of a recent review by Snow[4]. In 1911 Twort and Ingram showed that *Mycobacterium johnei* required for growth a factor found in organic solvent extracts of other mycobacteria. More than 30 years later interest in the factor, subsequently named mycobactin, was revived in connection with efforts to secure a specific anti-metabolite for *Mycobacterium tuberculosis.* A combination of proper iron nutrition and growth time was found to give maximum yield of the mycobactins, the most thoroughly studied of which is component P(XIIf) from *Mycobacterium phlei.* Unlike the other hydroxamates, the mycobactins and their iron complexes are lipid-soluble and bound to the cells; soaking in ethanol is the preferred method of extraction. Yields of up to 2% of the dry weight of the cells have been reported, depending on the source organism.

A second unusual feature of the mycobactins is the microheterogeneity arising from the presence of homologous fatty acids in the side chains. Each individual mycobactin has been distinguished by letters followed by a number equivalent to the carbons in the side chain. For several mycobactins the side chain length, abundance of each side chain, presence or absence of unsaturation, etc., have all been established but thus far only component P has been fully characterized; here the main species has the n-*cis*-octadec-2-enoyl side chain and is designated mycobactin P 18-*cis*-Δ_2 (c.f. **XIIf**).

Two general classes of mycobactins are recognized and are exemplified by types M(**XIId**) and P(**XIIf**). The P type mycobactins contain, in the mycobactic acid unit, either salicylic acid or 6-methyl salicylic acid, serine or threonine, a mixture of long-chain fatty acids (usually with a double bond adjacent to the carboxyl group, as in ferrichrome A), and L-ε-N-hydroxylysine. The cobactin unit is comprised of a β-hydroxy acid (either 3-hydroxybutyric or 3-hydroxy-2-methylpentanoic acid) and a second residue of L-ε-N-hydroxylysine. Variations within the P-type series arise by substitutions in the benzene ring, the oxazoline ring or the β-hydroxy acid fragment, or the presence of different optical isomers in the latter unit [c.f. mycobactins S(XIIh) and T(XIIi)]. The M type myco-bactins have only a short side chain at R in XI, *i.e.*, acetyl and propionyl in mycobactins M(XIId) and N(XIIe), respectively. The configuration of the hydroxy acid is again *erythro*; however, this unit bears a long alkyl side chain in mycobactins M.

The mycobactins form neutral, lipid-soluble ferric complexes, of the type shown for XII, which exceed the stability of ferrioxamine B. The prime difference between the iron(III) trihydroxamates and the iron(III) mycobactins is that in the latter one hydroxamate is replaced with a bidentate ligand containing a phenolic hydroxyl group and the N atom of the oxazoline ring. Examination of molecular models indicates that the hexadentate structure shown in XII is feasible but the exact disposition of the atoms around the iron has not been established by crystallography. Probably only a few of the 16 isomeric arrangements are possible. The aluminum, gallium and chromium complexes of mycobactin P have been crystallized; the first-named of these is sufficiently volatile for mass spectrometry, a process which reveals the presence of homologues. The chromic complex, which dissociates very slowly, was observed to anta-gonize the growth-promoting effect of mycobactin for *M. johnei*; presumably, therefore, the physiologically significant form of mycobactin is the iron(III) complex. In spite of the obvious structural analogies between the mycobactins and the trihydroxamate siderochromes, none of the latter have the slightest activity for mycobactin-dependent strains. Fastidious structural demands by *M. johnei* and/or the unique solubility properties of mycobactin and its iron(III) derivative may explain this specificity.

Little is yet known about the biosynthesis of mycobactins. All strains of mycobacteria that could be propagated on synthetic media were found to produce iron-chelating substances related to the ligand of XI and the distribution of mycobactins among various strains has proven to be a useful taxonomic tool.

5. *Fusarinine family*

Fusarium roseum ATCC 12822 and other species of *Fusarium* cultured at low iron concentrations ($< \sim 10^{-7} M$) produced a series of basic, ninhydrin-positive hydroxamates termed fusarinines (XIII)[69]. Accumula-

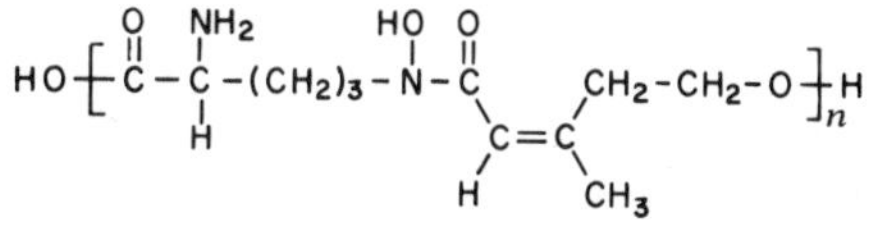

XIII Fusarinines

XIIIa Fusarinine: $n = 1$
XIIIb Fusarinine A: $n = 2$
XIIIc Fusarinine B: $n = 3$
XIIId Fusarinine C (Deferrifusigen): $n = 3$, cyclo

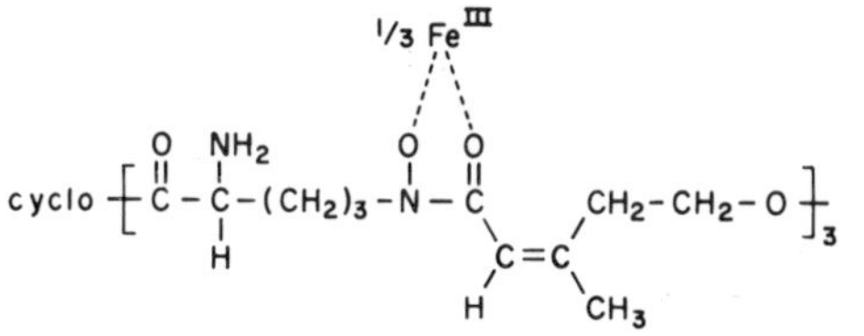

XIV Fusigen

tion of fusarinine, the monomer (**XIIIa**), reached a peak concentration of ~ 1 mM after 5 days of growth and thereafter rapidly disappeared from the medium. Only 25–30 mg of component A (**XIIIb**) and even less of component B (**XIIIc**) could be isolated per liter of medium. Fusarinine C was shown to be a cyclic compound identical to deferrifusigen[70]. All of these substances consist of units of L-δ-N-hydroxyornithine and cis-5-hydroxy-3-methylpent-2-enoic acid esterified head-to-tail. The acyl moiety is thus identical to that found in ferrirhodin.

The ferric fusarinines displayed the expected behavior on acidification, i.e., the spectrum of iron(III) **XIIIc** is rather insensitive to this treatment. Fusigen, as isolated from *Fusarium cubense*, showed about the same activity in the antagonism test with *Staphylococcus aureus* as coprogen, i.e., it was about 1% as active as ferrioxamine B[70]. Although fusarinine cannot support the growth of *Arthrobacter* JG-9, fusarinines A, B and C display slight activity with this organism.

6. Ferrioxamine family

Siderochromes of the ferrioxamine class contain units of acetate, succinate and α-amino-ω-hydroxyaminoalkane and fall into two types, the linear (**XV**) and the cyclic (**XVI**)[45]. Certain derivatives of the ferrioxamines display antibiotic activity and have been designated as ferrimycins. There is no structural relation between these antibiotics and albomycin except that in both cases the toxic agent is a further elaboration of a complete and functional iron transport agent. Thus if the substituted uracil is split off from albomycin the compound acts as a growth factor

References pp. 200–202

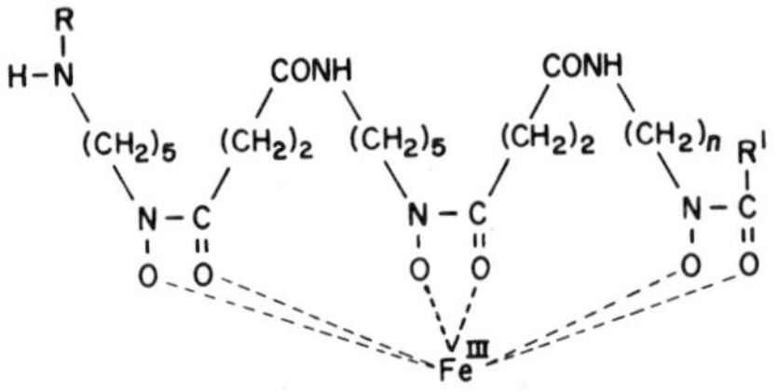

XV   Ferrioxamines (linear)

	R	n	R'
XVa Ferrioxamine B	H	5	CH_3-
XVb Ferrioxamine D$_1$	CH_3CO-	5	CH_3-
XVc Ferrioxamine G	H	5	$HO_2C(CH_2)_2-$
XVd Ferrioxamine A$_1$	H	4	$HO_2C(CH_2)_2-$
XVe Ferrimycin A$_1$	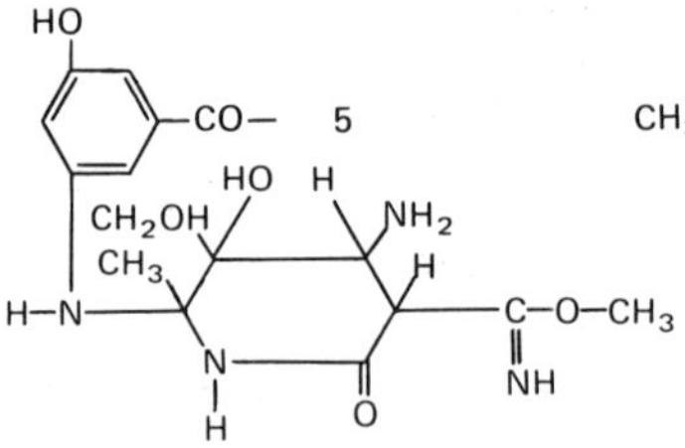	5	CH_3-

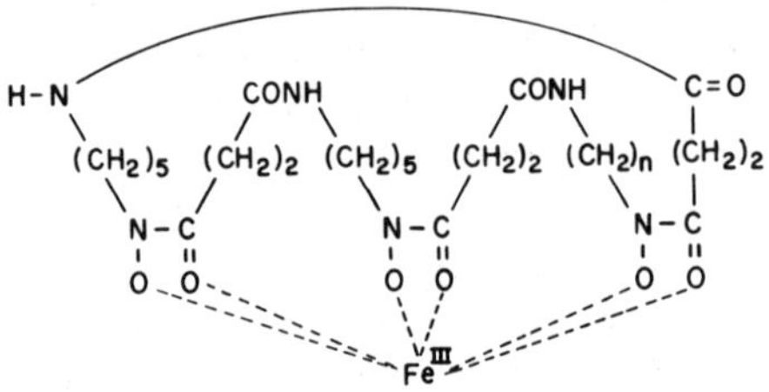

XVI   Ferrioxamines (cyclic)

	n
XVIa Ferrioxamine E	5
XVIb Ferrioxamine D$_2$	4

Metabolite "C" A derivative of ferrioxamine B in which the terminal amino group is oxidized to COOH.

Ferrioxamine B may be identical to terregens factor.

Ferrioxamine A$_2$ contains one residue of acetic acid, two of succinic acid, one of 1-amino-5-hydroxyaminopentane and two of 1-amino-4-hydroxyaminobutane. Components C and F are basic ferrioxamines.

for siderochrome auxotrophs. The antagonism test (Fig. 6) has been used by the Swiss workers to isolate the long series of ferrioxamines, both growth factors and antibiotics[45].

The ferrioxamines are relatively simple molecules which are devoid of optical activity in the ligand. Accordingly, a number have been prepared by chemical synthesis. The complexation constants are in the same range as those reported for ferrichrome (Table I). Ferrioxamines have been found in all actinomycetes which have been investigated for their presence.

D. Other compounds

A number of siderochromes remain uncharacterized, the most prominent of which are probably ferribactin[44], from *P. fluorescens*, and some of the iron-containing antibiotics. Several other microbial products probably contain one or more hydroxamate linkages but the relevance of these, if any, to iron transport and metabolism is unknown. Some, such as succinimycin[71], danomycin[72] and ASK[73], are antibiotics.

The aspergillic acid family[11] of antibiotics is of historical interest as the first microbial metabolite shown to contain the hydroxamic acid bond. Aspergillic acid itself (**XVIIa**) acts as a growth factor for *Arthrobacter* JG-9 but the formation by *Aspergillus sclerotiorum* of at least some members of the series (**XVIIb**, **XVIIe**) is not regulated by iron[68]. The single hydroxamic acid bond in these compounds arises by oxidation of the precursor cyclic amide, flavacol, in a pathway which is the reverse of that noted for the ferrichromes.

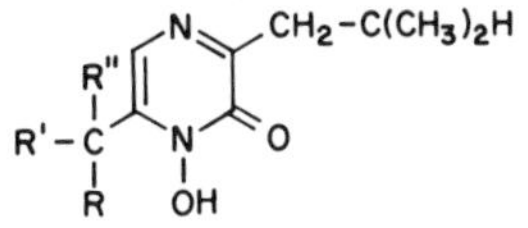

XVII Aspergillic acids

XVIIa Aspergillic acid: R = H, R′ = C_2H_5, R″ = CH_3
XVIIb Neo-aspergillic acid: R = R″ = H, R′ = $C(CH_3)_2H$
XVIIc Muta-aspergillic acid: R = OH, R′ = R″ = CH_3
XVIId Hydroxy-aspergillic acid: R = OH, R′ = C_2H_5, R″ = CH_3
XVIIe Neo-hydroxy-aspergillic acid: R = OH, R″ = $C(CH_3)_2H$, R‴ = H

Mycelianamide (**XVIII**)[74], a diketopiperazine with both ring nitrogens hydroxylated, may be slightly over-produced at low iron levels; otherwise there is nothing to connect it to iron metabolism.

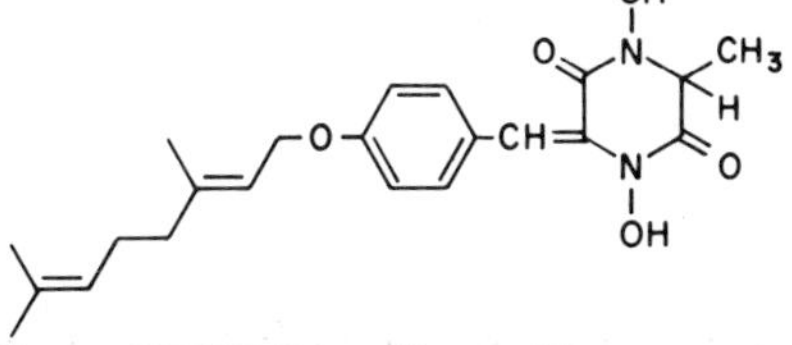

XVIII Mycelianamide

A third type of diketopiperazine-like substance, pulcherriminic acid (**XIX**)[75], is formed both intra- and extra-cellularly in high iron media where it occurs as the very insoluble ferric complex, pulcherrimin. The substance is synthesized by *Candida pulcherrima* and related yeasts. It has no activity for siderochrome-dependent *Arthrobacter* sp. and is hence probably unrelated to iron metabolism[76]. In **XIX** the ring is further oxidized over that in mycelianamide and although a structure can be written with a single hydroxamic acid bond, the conjugated form of **XIX** is probably the dominant species. It is derived biosynthetically from L-leucine diketo-piperazine.

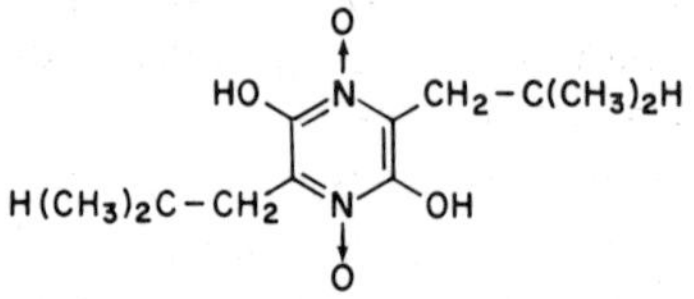

XIX Pulcherriminic acid

The anti-tumor agent *N*-formyl-*N*-hydroxyglycine, or hadacidin (**XX**) is identical with asymmetrin, a plant growth inhibitor of fungal origin. In *Penicillium aurantio-violaceum* F 4070b hadacidin[10] arises by oxidation of glycine, followed by formylation, and again iron seems not to repress the synthesis[68].

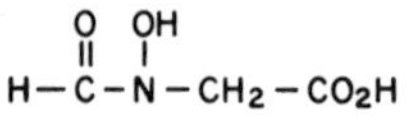

XX Hadacidin

Actinonin (**XXI**), a primary hydroxamic acid (H rather than R on the nitrogen), completes the list of microbial products containing an acylated hydroxylamino group. *O*-Benzylactinonin was found to be toxic, so the importance of the hydroxamic linkage for antibiotic activity is questionable[77].

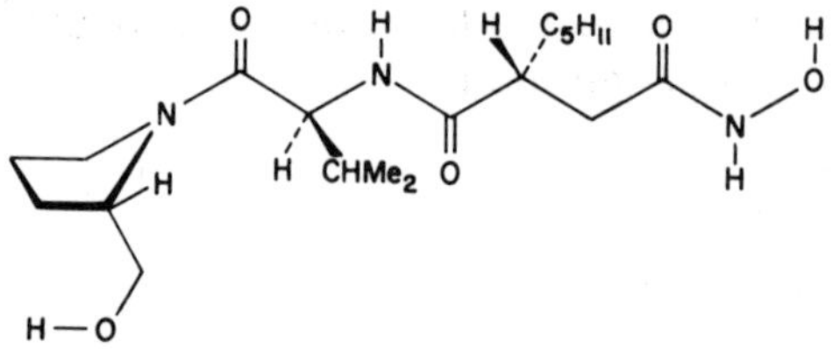

XXI Actinonin

XXII, a cyclic hydroxamic acid from higher plants[58], has no known biological role. However, bioassay with *Arthrobacter* JG-9 has demonstrated the presence of siderochrome activity in aqueous extracts of leaves of tomato, lettuce, cauliflower, leek and cabbage, and in extracts of carrot roots[59]. Ferrioxamine B is translocated in tomato plants growing in nutrient solution[78].

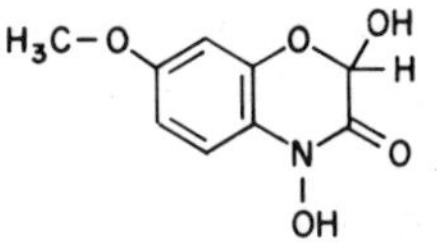

XXII 2,4-Dihydroxy-7-methoxy-1,4-benzoxazin-3-one (DIMBOA)

Finally, two Fe(II) ion-binding agents produced by micro-organisms are worthy of note even though these substances cannot now be associated with iron metabolism. Ferroverdin (**XXIII**), a green pigment from the mycelium of *Streptomyces* sp., has been characterized crystallographically (**XIIIa**)[79]. Each iron atom is bonded to a phenolate oxygen and the nitrogen of the nitroso group in the nitrosophenyl ligand. *Pseudomonas* GH and *P. roseus fluorescens* form pyrimine (**XXIV**)[80,81], a pyridine ring substituted at the 2-position with the γ-carboxyl of L-glutamic acid. At all reasonable values of pH pyrimine exists in the imine form and, as such, forms a stable, magenta-colored complex with Fe(II) iron (**XXV**).

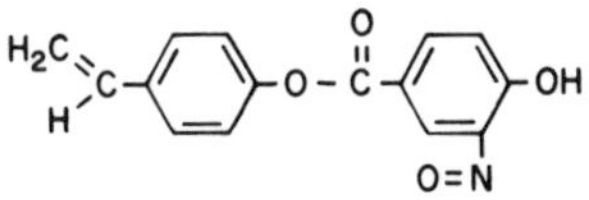

XXIII Deferroferroverdin

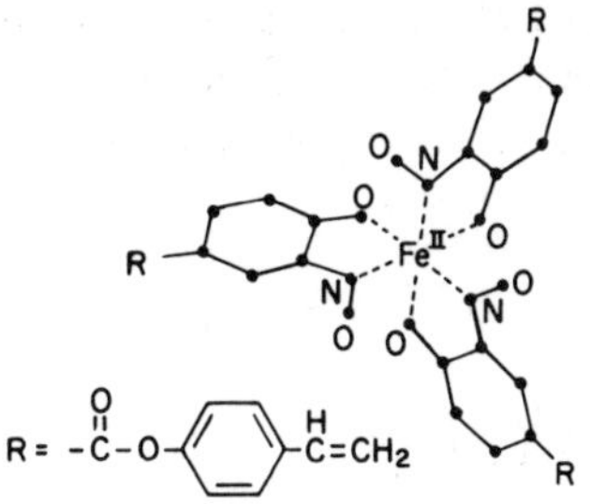

XXIIIa Ferroverdin, crystal structure[79]

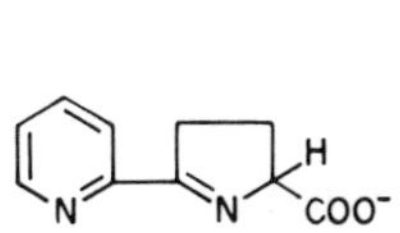

XXIV Pyrimine

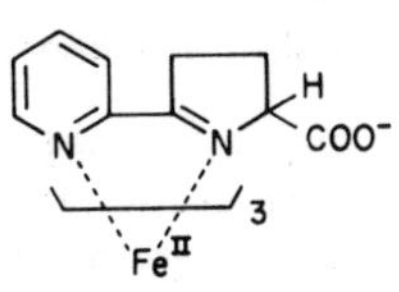

XXV Ferropyrimine

III. CONCLUSIONS

Practically all aerobic and facultative aerobic micro-organisms which have been investigated, whether bacteria, yeasts or fungi, contain or require a transport system for iron. The path of iron in microbial metabolism has been outlined by examination of mutants of *Salmonella typhimurium* and other enteric bacteria. The system is comprised of (a) low molecular weight, Fe(III)-ion specific carriers (siderochromes) of the phenolate, *i.e.*, enterobactin, or the hydroxamate, *i.e.*, ferrichrome type, and (b) a cell-bound, active transport mechanism for the uptake of the specific siderochrome.

While little is yet known about the precise mechanism of uptake of siderochromes into the cell, a catalogue of such transport compounds has been assembled which at once reveals certain unifying principles. Their biosynthesis is strongly repressed by iron, they show high Fe^{III}/Fe^{II} binding specificity, they occur and are active either as single units or as dimers or trimers linked through amide or ester bonds, and they show a profound conformational change on complexation of the Fe(III) ion.

It is concluded that the siderochromes fulfill a unique and essential function in regulating the iron nutrition of the microbial cell.

NOTE ADDED IN PROOF (see page 183)

An identical substance, named enterochelin, was isolated from the broth of *Escherichia coli* and other enteric bacteria by Gibson and co-workers[62a]. *S. typhimurium* made only *ca.* 6 mg/10 l of medium, but Gibson's *Aerobacter aerogenes* strain 62-1 could be induced to excrete larger amounts.

REFERENCES

1 W. Keller-Schierlein, V. Prelog and H. Zähner, *Fortsch. Chem. Org. Natur.*, 22 (1964) 279.
2 J. R. Pollack and J. C. Neilands, *Biochem. Biophys. Res. Commun.*, 38 (1970) 989.
3 J. B. Neilands, *Structure and Bonding*, 1 (1966) 59.
4 G. A. Snow, *Bact. Rev.*, 34 (1970) 100.
5 F. Gibson and D. I. Magrath, *Biochim. Biophys. Acta*, 192 (1969) 175.
5a K. B. Mullis, J. R. Pollack and J. B. Neilands, *Biochemistry*, 10 (1971) 4894.
6 J. A. Garibaldi and J. B. Neilands, *Nature*, 177 (1956) 526.
7 C. L. Atkin, J. B. Neilands and H. Phaff, *J. Bact.*, 103 (1970) 722.
7a J. A. Garibaldi, *J. Bact.*, 105 (1971) 1036.
8 F. Gibson and J. Pittard, *Bact. Rev.*, 32 (1968) 465.
9 T. Emergy, *Biochemistry*, 5 (1966) 3694.
10 T. Emery, in D. Gottlieb and P. D. Shaw, *Antibiotics*, Vol. 2, Springer, Berlin, 1967, p. 439.
11 J. C. MacDonald, *ibid.*, pp. 43–51.
12 J. A. Garibaldi, personal communication.

13 J. B. Neilands, *Science*, 156 (1967) 1443.
14 A. Zalkin, J. D. Forrester and D. H. Templeton, *J. Am. Chem. Soc.*, 88 (1966) 1810.
15 W. Keller-Schierlein and B. Maurer, *Helv. Chim. Acta*, 52 (1969) 603.
16 M. Llinás, M. P. Klein and J. B. Neilands, *J. Mol. Biol.*, 52 (1970) 399.
17 H. H. Wickman, M. P. Klein and D. A. Shirley, *J. Chem. Phys.*, 42 (1965) 2113.
18 J. B. Neilands, *Bact. Rev.*, 21 (1957) 101.
19 T. Ito and J. B. Neilands, *J. Am. Chem. Soc.*, 80 (1958) 4645.
20 N. E. Morrison, A. D. Antoine and E. E. Dewbrey, *J. Bact.*, 89 (1965) 1630.
21 T. Emery, Abstr. 158th Am. Chem. Soc. Meeting, New York, 1969, Sept. 8-12.
22 B. R. Byers, M. V. Powell and C. E. Lankford, *J. Bact.*, 93 (1967) 286.
23 J. L. Arceneaux and C. E. Lankford, *Biochem. Biophys. Res. Commun.*, 24 (1966) 370.
24 F. Knüsel, B. Schiess and W. Zimmerman, *Arch. Mikrobiol.*, 68 (1969) 99.
25 W. Zimmerman and F. Knüsel, *ibid.*, p. 107.
26 W. J. Peters and R. A. J. Warren, *Biochim. Biophys. Acta*, 165 (1968) 225.
27 C. C. Wang and A. Newton, *J. Bact.*, 98 (1969) 1135, 1142.
28 J. R. Pollack, B. N. Ames and J. B. Neilands, *J. Bact.*, 104 (1970) 635.
29 N. Brot, J. Goodwin and H. Fales, *Biochem. Biophys. Res. Commun.*, 25 (1966) 454.
30 J. L. Corbin and W. A. Bulen, *Biochemistry*, 8 (1969) 757.
31 J. R. Pollack, B. N. Ames and J. B. Neilands, *Abstr. Fed. Proc.*, 29 (1970) 3132.
32 C. L. Atkin and J. B. Neilands, *Biochemistry*, 7 (1968) 3734.
33 G. Anderegg, F. L'Eplattenier and G. Schwarzenbach, *Helv. Chim. Acta*, 46 (1963) 1409.
34 A. Ehrenberg, *Nature*, 178 (1956) 379.
35 W. Lovenberg, B. B. Buchanan and J. C. Rabionwitz, *J. Biol. Chem.*, 238 (1963) 3899.
36 T. Emery and J. B. Neilands, *J. Am. Chem. Soc.*, 82 (1960) 3658.
37 J. B. Neilands in J. E. Falk, R. Lemberg and R. K. Morton, *Haematin Enzymes*, Vol. 1, Pergamon Press, Oxford and New York, 1961, p. 194.
38 M. M. Shemyakin, Y. A. Ovchinnikov, V. T. Ivanov, V. K. Antonov, *et al.*, *J. Membrane Biol.*, 1 (1969) 402.
39 J. B. Neilands, *J. Am. Chem. Soc.*, 74 (1952) 4846.
40 H. Komai and J. B. Neilands, *Science*, 153 (1966) 751.
41 H. Komai, Doctoral dissertation, University of California, Berkeley, 1968.
42 O. T. G. Jones, personal communication.
43 R. D. Gillard and P. R. Mitchell, *Structure and Bonding*, 7 (1970) 46.
44 B. Maurer, A. Muller, W. Keller-Schierlein and H. Zähner, *Arch. Mikrobiol.*, 60 (1968) 326.
45 V. Prelog, *Pure Appl. Chem.*, 6 (1963) 327.
46 I. G. O'Brien, G. B. Cox and F. Gibson, *Biochim. Biophys. Acta*, 177 (1969) 321.
47 J. Nüesch and F. Knüsel, in D. Gottlieb and P. D. Shaw, *Antibiotics*, Vol. 1, Springer, Berlin, 1967, p. 499.
48 T. G. Spiro and P. Saltman, *Structure and Bonding*, 6 (1969) 116.
49 K. Bergman, P. V. Burke, E. Cerdá-Olmedo, C. N. David, *et al.*, *Bact. Rev.*, 33 (1969) 99.
50 T. D. Wilkins and C. E. Lankford, *J. Infect. Dis.*, 121 (1970) 129.
51 H. A. Schneider, *Science*, 158 (1967) 597.
52 E. J. Wawszkiewicz and H. A. Schneider, *Abstr. Fed. Proc.*, 29 (1970) 3256.
53 E. F. Hieb, E. L. R. Stokstad and M. Rothstein, *Science*, 168 (1970) 143.
54 E. C. Larkin, L. R. Weintraub and W. H. Crosby, *Am. J. Physiol.*, 218 (1970) 7.

202

55 L. Hedenberg, *Scand. J. Haemat. Suppl.*, 6 (1969) 86pp; see also *Clin. Pharmacol. Ther.*, 10 (4) (1969) 595.
56 I. H. Krakoff, N. C. Brown and P. Reichard, *Cancer Res.*, 28 (1968) 1559.
57 H. Zähner, E. Bachmann, R. Hutter and J. Nüesch, *Path. Mikrobiol.*, 25 (1962) 708.
58 C. L. Tipton, J. A. Klun, R. R. Husted and M. D. Pierson, *Biochemistry*, 6 (1967) 2866 (see also ref. 3).
59 E. R. Page, *Biochem. J.*, 100 (1966) 34P.
60 C. A. Price, *A. Rev. Plant Physiol.*, 19 (1968) 239.
61 K. B. Pugh and J. S. Waid, *Soil Biol. Chem. Biochem.*, 1 (1969) 195, 207.
62 L. E. Arnow, *J. Biol. Chem.*, 118 (1937) 531.
62a I. G. O'Brien and F. Gibson, *Biochim. Biophys. Acta*, 215 (1970) 393.
62b E. J. Wawszkiewicz, H. A. Schneider, B. Starcher, J. R. Pollack and J. B. Neilands, *Proc. Natn. Acad. Sci. U.S.*, 68 (1971) 2870.
62c Sonia K. Guterman, *Biophys. Res. Commun.*, 44 (1971) 1149.
63 N. Brot and J. Goodwin, *J. Biol. Chem.*, 243 (1968) 510.
64 W. J. Peters and R. A. J. Warren, *J. Bact.*, 95 (1968) 360.
65 D. N. Downer, W. B. Davis and B. R. Byers, *J. Bact.*, 101 (1970) 181.
66 N. Tadenuma and S. Sato, *Agric. Biol. Chem.*, 31 (1967) 1482.
67 L. M. Epstein and D. R. Straub, *Inorg. Chem.*, 8 (1969) 453.
68 C. L. Atkin, Doctoral dissertation, University of California, Berkeley, 1970.
68a H. Diekmann, *Arch. Mikrobiol.*, 73 (1970) 65.
68b H. Akers, M. Llinás and J. B. Neilands, Abstr., Am. Soc. Biol. Chem. Meeting, San Francisco, 1971.
68c W. Keller-Schierlein and H. Diekmann, *Helv. Chim. Acta*, 53 (1970) 2035.
69 J. M. Sayer and T. F. Emery, *Biochemistry*, 7 (1968) 184.
70 H. Diekmann and H. Zähner, *Eur. J. Biochem.*, 3 (1967) 213.
71 T. H. Haskell, R. H. Bunge, J. C. French and Q. R. Bartz, *J. Antibiot.*, 16 (1963) 67.
72 H. Tsukiura, M. Okanishi, T. Ohmori, H. Koshiyama, T. Miyaki, H. Kitazima and H. Kawaguchi, *ibid.*, 17 (1964) 39.
73 I. R. Shimi, G. M. Imam and B. M. Haroun, *J. Antibiot.*, 22 (1969) 106.
74 R. B. Bates, J. H. Schauble and M. Soucek, *Tetrahedron Lett.*, (1963) 1683.
75 J. C. MacDonald, *Biochem. J.*, 96 (1965) 533.
76 M. O. Burton, F. J. Sowden and A. G. Lochhead, *Can. J. Biochem. Physiol.*, 32 (1954) 400.
77 M. M. Attwood, *J. Gen. Microbiol.*, 55 (1969) 209.
78 E. Stutz, *Experientia*, 20 (1964) 430.
79 S. Candeloro, D. Grdenić, N. Taylor, B. Thompson, M. Viswamitra and D. Crowfoot-Hodgkin, *Nature*, 224 (1969) 589.
80 R. Shiman and J. B. Neilands, *Biochemistry*, 4 (1965) 2233.
81 M. Pouteau-Thouvenot, M. Choussy, A. Gaudemer and M. Barbier, *Bull. Soc. Chim. Biol.*, 52 (1970) 51.

Chapter 6

ALKALI METAL CHELATORS—THE IONOPHORES

B. C. PRESSMAN

Department of Pharmacology, University of Miami School of Medicine, Miami, Florida, U.S.A.

The *ionophores* (*i.e., ion bearers*[1]) are a group of compounds which possess a molecular architecture especially designed for carrying small ions across lipid barriers. They have generated considerable interest for their ability to alter the cation permeability of both natural and artificial membrane systems. On the one hand they provide a means of altering the permeability of, and gradients across, biological membranes thereby perturbing membrane-linked metabolic processes. On the other hand they can confer on artificial membranes selective permeability and electrical properties characteristic of biological membranes suggesting that the ionophores are analogous to naturally occurring membrane components.

The known ionophores are relatively stable both chemically and metabolically. They are macrocyclic, or capable of forming macrocyclic rings by means of hydrogen bonds, and range in molecular weight from 200 to 2000. They contain high proportions of oxygen which is periodically distributed along the molecular backbone. A distinguishing property of ionophores is their ability to form lipid-soluble cation complexes which are intimately involved in their transport function.

The ionophores may be divided into two subgroups. Members of the neutral ionophore group have no dissociable protons and the cation complexes they form accordingly have a net positive charge. The backbone of the carboxylic group of ionophores is constructed of a chain of covalent carbon-to-carbon bonds. They have a single terminal carboxyl which is involved in head-to-tail ring cloture via hydrogen bonding. As will be seen, the presence or absence of this single carboxyl group has a profound effect on the equilibria catalyzed by ionophores.

HISTORICAL BACKGROUND

The first compound to be recognized as an ionophore is valinomycin, which was originally isolated[2] and characterized[3] by Brockmann. It was subsequently synthesized and identified as a cyclic dodecadepsipeptide by Shemyakin *et al.*[4] who determined its antibiotic spectrum along with those of several of its synthetic analogues[5,6]. McMurray and Begg discovered valinomycin to be a powerful uncoupler of oxidative phosphorylation in mitochondria[7], a property found by Moore and Pressman to be strongly dependent on the presence of K^+, but not Na^+ [8]. By means of ion selective electrodes it was established that the primary effect of valinomycin is on mitochondrial K^+ *transport* and its apparent uncoupling activity results from its induction of an energy-dissipating cyclic movement of K^+ [8,9]. A detailed account of the sequence of observations leading to recognition of the ionophorous properties of valinomycin has been published[10] as well as more general reviews of the transport properties of ionophores[11-13].

From the critical structure–activity relationship it was incorrectly inferred, at an early stage, that valinomycin interacts with a pre-existing highly selective transport receptor of mitochondria[8]. The demonstration that the valinomycin group of ionophores increases the cation permeability of artificial lipid bilayers, however, established that transport activity and ion selectivity are intrinsic to the valinomycin molecule *per se*[14,15]. On the basis of a detailed examination of its established alkali ion complexing ability and its transport activity in bulk phases, it was ultimately concluded that the ionophores owe their properties to their ability to serve as mobile ion carriers[1,16-19].

The capacity of nigericin to form lipid-soluable alkali salts was first recognized by Harned *et al.*[20]. This compound, along with the related dianemycin, were found to inhibit the respiration and phosphorylation of intact, but not disrupted, mitochondria[21]. From its ability to reverse the valinomycin-induced uptake of K^+ by mitochondria[22,23], Lardy and co-workers concluded that the nigericin group of ionophores blocks the ion transport mechanism of mitochondria[16]. However, from its transport properties in bulk oil–water systems and erythrocytes, and its alkali ion complexing properties, Pressman *et al.* concluded that nigericin, and the other carboxylic ionophores related to it, act as mobile ion carriers, just as do the neutral group. The differences in behavior of the ionophore sub-classes stems from the presence or absence of a net charge on the cation complex[1,18].

Attempts were made to account for the selective complexation of the valinomycin type ionophores on the basis of a molecular form fit between representations of their known formal structures as planer rings, and the radius of the complexed cation[14,24]. In order to explain the complexing equivalency of the large-ringed valinomycin (36 atoms) and the smaller-ringed enniatins (18 atoms) it was suggested that the complexed ion could adjust to the ionophore ring by undergoing variable degrees of desolvation[14]. Establishment of the structure of the K^+ complex of the ionophore non-actin (32 atom ring) by X-ray crystallographic analysis[25,26] indicated that its ring is not planer but forms a remarkable convoluted cage about the unsolvated cation. The general principle of the complexing molecule enveloping the unsolvated form of the complexed cation extends to all known ionophores.

GENERAL ARCHITECTURAL FEATURES OF IONOPHORES

The ionophores form multiple ion-dipole ligands with the complexed cation via a series of oxygen atoms, strategically deployed by their periodic insertion within the cyclic molecular backbone. The neutral ionophores are typically covalently cyclic, containing rings ranging from 18 to 40 atoms. The rings usually contain repeating subunits with regular alternations of optical asymmetry. Valinomycin and the enniatins are depsipeptides[27], their rings being held together by alternating amide and ester groups. The repeating units of the macrotetralide actins[28] are held together exclusively by ester linkages. Several compounds with ionophore-like activity are held together exclusively by peptide bonds although the gramicidins are not covalently cyclic[29] and do not form highly lipophylic cation complexes. The synthetic crown polyethers[30] are even more inert with only stable ether oxygens within the molecular backbone.

The known carboxylic ionophores are formally linear chains containing heterocyclic rings with a carboxyl at one end of the molecule and one or two hydroxyl groups at the other end. In the complexes the carboxyl is deprotonated and forms a ring by head-to-tail hydrogen bonding with the opposing hydroxyl groups. Because of the deprotonation requirement, complexation by the carboxylic ionophores is pH dependent, being favored by higher pH[1]. Such complexes are electrically neutral zwitterions in contradistinction to the "neutral" ionophore complexes which actually form charged complexes which are pH insensitive.

SPECIFIC IONOPHORE FAMILIES

Macrotetralide actins

Four homologous macrotetralide actins containing 18 ring atoms (Fig. 1A) are produced by the same strain of actinomyces. The lowest homolog, nonactin, consists of a symmetrical alteration of four D and L enantiomorphs (no other diastereoisomers have been observed) of nonactinic acid, a hydroxycarboxylic acid containing an ether oxygen in a tetrahydrofuran ring about midway in its chain. In the higher homologs, monactin, dinactin and trinactin, the nonactinic acid residues are replaced respectively by one, two or three residues of the monomethylated nonactinic acid, homononactinic acid[28]. Although various racemic mixtures are possible if the sequence of nonactinic and homoanactinic residues are randomized, the natural higher homologues of nonactin are all optically active. Still higher homologues of the actin series are known containing dimethylated

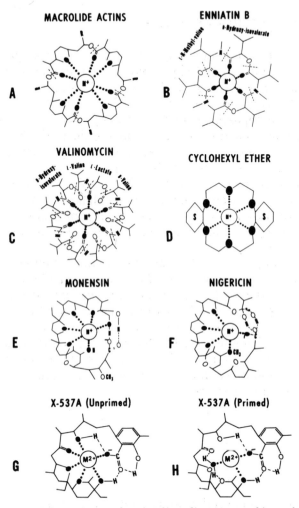

MACROLIDE ACTINS

ENNIATIN B

A

B

VALINOMYCIN

CYCLOHEXYL ETHER

C

D

MONENSIN

NIGERICIN

E

F

X-537A (Unprimed)

X-537A (Primed)

G

H

Fig. 1. Structures and cation liganding atoms of ionophores of known structure.

nonactinic acid (bis-homononactinic acid); however, little has been reported about the conformation and ion selectivity of these compounds[31].

The complete three dimensional structure of nonactin was obtained by X-ray crystallography of the thiocyanate salt of the K^+ complex. Ion-dipole interaction with the complexed cation occurs with the four ether and four carbonyl oxygens arranged at the apices of a cube: the unhydrated

K^+-to-O center-to-center distance is 2.7–2.9 Å. The position of the thiocyanate groups is not clearly revealed by X-ray crystallography implying
that they are somewhat loosely oriented within the unit crystal cell about
the spherical cation complex[26].

The rigorous establishment of the conformation of this complex, the
first obtained for an ionophore, offered considerable insight into the molecular behavior of these compounds. The precise configurational arrangement of the component residues appears to be necessary for folding the
molecular backbone into the conformation required for complexation.
This conformation has been described picturesquely, " . . . as resembling
the seam of a tennis ball"[25]. The unsolvated state of the cation indicates
that considerable free energy must be expanded for its desolvation when
the complex forms in a polar medium. In order for such a complex to form
in the presence of water, a condition which obtains at a biological membrane interface, this free energy must be recouped through ion-dipole
interaction between the complexed ion and the liganding oxygens. The
high degree of ion selectivity arises from the free energy difference between
dehydration and complexation, which in turn depends on the flexibility
limitations of the ionophore backbone. The complete encagement of the
cation indicates that the ionophore must have been in some other more
open conformation before complexation in order for the cation to enter.
Crystallographic studies show that free nonactin forms a different unit
cell and does have a conformation different from that of the K^+ complex[26];
however, the detailed structure of the free ionophore has not been reported.

The above conclusions depend on the assumption that the conformation of the ionophore in the crystalline state also obtains in solution.
This can be confirmed through nuclear magnetic resonance (n.m.r.) studies
which reveal different conformations in solution for the K^+, Na^+ and Cs^+
complexes as well as the free ionophore[32,33]. Other applications in n.m.r.
to the study of ionophores will be described under valinomycin. The ion
selectivity of the macrolide actins (mixed sample of nonactin and monactin),
first observed as the ion dependency of stimulation of mitochondrial respiration is $K^+ > Rb^+ > Cs^+ > Na^+ > Li^+$ [9]. By the criterion of ATPase
induction, another function of transport-linked energy dissipation by mitochondria, the same series was obtained[34,35]. As the homologous ionophore
series is ascended, potency in inducing K^+-dependent ATPase increases but
the K^+: Na^+ selectivity ratios decrease. Electrometric measurements across
bulk phases also yielded the same sequence with the lower homologues[36];
however, with the lipid bilayer an inversion of K^+ and Rb^+ selectivities
(*i.e.*, $Rb^+ > K^+$) was reported[14].

The stability constants obtained in methanol by vapor pressure
osmometry[12,36] and in aqueous acetone by n.m.r.[33], indicate a marked
K^+: Na^+ preference; however, in the less polar dry acetone, the stability
constants for the K^+, Na^+ and Cs^+ complexes are all similar (Table I). Thus

208

in dry acetone where the differential free energies of ion desolvation do not make strong contributions to the complexation equilibria, the intrinsic affinity of nonactin for Na^+, K^+ and Cs^+ are about equal indicating considerable flexibility in the ionophore backbone in accommodating ions of

TABLE I

INFLUENCE OF SOLVENT ON SELECTIVITY OF MACROLIDE ACTINS

Ionophore	Solvent	Anion	Relative	Selectivity		Reference
			Na^+	K^+	Cs^+	
Nonactin	Acetone	ClO_4^-	1	1	0.20	33
	Acetone–H_2O	ClO_4^-	0.012	1	0.024	33
	CH_3OH	SCN^-	0.025	1		44
	BuOH–toluene	SCN^-	0.004	1	0.77	51
	CH_2Cl_2	Picrate	0.017	1	0.061	76
Monactin	CH_3OH	SCN^-	0.004	1		44
	BuOH–toluene	SCN^-	0.010	1	0.42	51
	CH_2Cl_2	Picrate	0.009	1	0.029	76
Dinactin	BuOH–toluene	SCN^-	0.022	1	0.21	51
	CH_2Cl_2	Picrate	0.013	1	0.023	76
Trinactin	BuOH–toluene	SCN^-	0.042	1	0.087	51
	CH_2Cl_2	Picrate	0.011	1	0.019	76

Methods used in determining affinity constants in ref. 33, n.m.r.; ref. 44, vapor pressure osmometry; refs. 51 and 76, two-phase partition studies.

differing radii. In the more polar solvents the same selectivities observed in the multiphase systems are obtained supporting the role of solvation factors in ion selectivity. The smaller Na^+ ion requires more energy than K^+ for desolvation while the larger Cs^+, although requiring less energy than K^+ for desolvation, may keep the ionophore cage more open maintaining both cation and ionophore more accessible for solvation. Systematic determination of the complexation affinities in solvents of moderate polarity show a decrease in $K^+:Na^+$ selectivities with the higher macrotetrolide actin homologues (Table I) (c.f. also mitochondrial criteria in above paragraph.)

Enniatin

The enniatins are a family of cyclic hexadepsipeptides, containing 18 ring atoms, obtained from certain strains of *Fusarium*[37,38]. The hydroxyacid is D-hydroxyisovaleric acid which alternates with an *N*-methyl-L-amino acid. In enniatin A, B and C the amino acids are isoleucine, valine and leucine, respectively. The structure of enniatin B is shown in Fig. 1B. Still another variant of enniatin is beauvaricin in which the aminoacid is *N*-methylphenylalanine[39].

Because of the small ring size few conformational options are available to the enniatin complexes. Space-filling models indicate a close fit with K^+ when the carbonyls are oriented towards the interior and this structure for the KI complex has been confirmed by X-ray crystallography[40]. The six liganding oxygens, 2.6–2.8 Å from the K^+, lie in two triads, 1.5 Å above and below the mean plane of the ring. The top and bottom of the disc-shaped complex is more exposed than are the complexes of the macro-tetralide actins or valinomycin, but shielded enough to impede close contact between the cation and its counteranion. The anions are accordingly not discretely localized within the crystal lattice.

The conformation of the enniatins in solution has been studied in considerable detail by o.r.d., c.d. and n.m.r.[41,42]. Enniatin B gradually unfolds with the carbonyls pushed progressively outward as the complexed ion increases in size. In this respect it shows less selectivity than the more rigid valinomycin molecule, although the rank order of selectivities, $K^+ > Rb^+ > Cs^+ > Na^+ > Li^+$ by various criteria, resembles that of other typical neutral ionophores[43,44]. One interesting variant of the enniatin series is the synthetic enantiomorph of enniatin B. This compound has the same activity towards bacteria and mitochondria as does the natural compound[43,45]. This indicates that a selective receptor site could not be involved in the interaction of these ionophores with membrane systems. Naturally occurring receptors would presumably contain optically active protein components which could be expected discriminate between a competent trigger and its enantiomorph.

Valinomycin

Brockmann correctly identified the residues of valinomycin as D and L-valine, L-lactate and D-isovalerate; however, he concluded that the sequence repeats only twice[3]. Shemyakin *et al.* subsequently showed that the repeating sequence occurs three times as a dodecadepsipeptide[4] (Fig. 1C) and established this by synthesis[5]. An extensive series of valinomycin analogues was synthesized by Shemyakin and co-workers which showed that very little could be changed in the molecule without loss of biological activity as measured by mitochondrial or antibiotic assays[5,6,9]. Both the addition or deletion of a repeating four-residue sequence abolished activity, as did inverting any optically active centers. A single optically inactive glycine could be substituted for a valine, or all optically active centers could be inverted and activity retained. Opening the chain destroyed activity, even if the end groups were covered by formyl (free amino end) or ethanolamide (free carboxyl end) groups.

The chemical shifts observed during n.m.r. studies established that a unique valinomycin conformation exists for each cation species complexed, the resonance perturbations being greatest in the region of the ring backbone[42,46-50]. By establishing the relationship of the coupling constant between the amide protons and the valine α-protons to the dihedral angle about the amide linkages, a three-dimensional conformation for complexed valinomycin was arrived at. Infra-red spectroscopy indicated that the ester carbonyls are hydrogen bonded to the amide protons leaving the amide carbonyls free for liganding to cations. The hydrogen bonds create an ordered secondary structure in which the valinomycin backbone loops up and down to form a bracelet 4 Å high and 8 Å in diameter[48]. The alkyl sidechains project from the edges of the bracelet which is rendered asymmetric by the grouping of all the lactate residues at one edge of the braclet. In the complexed structure, the amide carbonyls turn inward, three at each bracelet edge, to complete a cage about the entrapped cation. In the uncomplexed form the carbonyls opposite the lactate edge turn outward permitting a pathway for cations to the interior of the cage. To some extent the hydrogen bonds of the free valinomycin may break resulting in a small fraction of an even more open structure[42,45]. The conclusions were confirmed by subsequent n.m.r. studies at higher resolution (220 MHz); the evidence for hydrogen bonding was strengthened by the low, temperature dependent chemical shift of the amide protons and their slow rate of exchange with D_2O[49,50]. Further n.m.r. studies of valinomycin as well as the macrolide actins enniatin B, and dicyclohexyl 18-crown-6 are contained in the thesis of Haynes[51].

Virtually the same cage structure was obtained independently from X-ray crystallographic analysis of the chloroaurate of the K^+ complex of valinomycin[52]. This again supports the validity of extending the solid state crystalline conformations to the solution state. The position of the anion within the lattice is clearly evident hence the anion is highly oriented with respect to the ionophore complex (c.f. nonactin). The principal liganding carbonyl oxygens are 2.7-2.8 Å, center-to-center, from the cation while the hydrogen bonded carbonyl oxygens are somewhat further away.

Nuclear magnetic resonance also yields information about the dynamics of complexation. The lactate methyl doublet of valinomycin is particularly sharp and ideal for measuring the shifts and shape alterations during cation titration from which cation-ionophore exchange rates may be obtained. In a low dielectric solvent, $DCDl_3$, the exchange rate is barely measurable, indicating that the complex is not likely to dissociate within the low dielectric environment of the interior of a lipid membrane. Raising the dielectric constant by the progressive addition of CD_3OD leads to measurable increases of exchange rate consistent with the ready dissociation of the complex at the high dielectric region of the membrane interphase[46]. Because of the rigidity conferred by the intramolecular hydrogen bonding, the valinomycin cage, specifically tailored for K^+, has difficulty in contracting to the size of Na^+. Consequently the K^+:Na^+ selectivity by all criteria is the greatest of all known ionophores[8-15,41-44].

Gramicidin

The gramicidins are a family of linear pentadecapeptides produced by a strain of *B. brevis*[53]. Structural options are the choice of valine or iso-leucine at the beginning of the chain and phenylalanine, tryosine or trypto-phane as the eleventh residue of the fifteen residue chain[29]. The alternation of D and L configurations, typical of many ionophores, is present and the end groups are blocked with formyl (R–$\overset{\text{H}}{\text{N}}$–CH$_2$OH) and ethanolamide (R–$\overset{\text{O}}{\text{C}}$–NH–CH$_2$–CH$_2$OH) groups. The exact conformation of the molecule is not known although its tendency to dimerize is recognized[29]. Despite the similarity of gramicidin to the above ionophores in structure and effects on membranes gramicidin does not form readily demonstrable lipid-soluble complexes and may not function as a true ionophorous mobile carrier. The gramicidins have been synthesized[29].

Confusion has occasionally arisen between the above ("Dubos") gramicidins[53] and gramicidin S[54], which is a cyclic basic decapeptide pro-duced by another strain of *B. brevis*. Gramicidin S, which is also structur-ally related to the tyrocidins[55], uncouples oxidative phosphorylation in mitochondria as can the other ionophores although this effect is not ion dependent. Gramicidin S forms multiple hydrogen bonds across the interior of the molecule so that no ring opening is available for multiligand com-plexation[50]. It also contains basic ornithine residues which would render the molecule positively charged thereby discouraging cation complexation due to electrostatic repulsion. The uncoupling properties of this molecule therefore likely arise from its surface activity. Tyrocidins A, B, C, purified by counter-current distribution, uncouple but do not induce mitochondrial ion uptake. Commercial tyrocidin, however, does have ionophore activity presumably owing to contamination by gramicidin which is coproduced by the same organism. This may explain the reports of tyrocidin-induced membrane transport[56].

Polycyclic ethers

This series of ionophores are obtained by Pederson who condensed epoxyethylene with resorcinol and its derivatives by refluxing with alkali[30, 57, 58]. The most readily available member of this series, dibenzo-18-crown-6, is obtained in high yield if the condensation is directed by KOH; if NaOH is used instead, a wider variety of products results. The aromatic crown ethers are not too soluble in either water or organic solvents and their effectiveness as ionophores is improved by ring saturation. This can produce a variety of geometric isomers two of which constitute the bulk of unresolved dicyclohexyl-18-crown-6 (Fig. 1D). The most interesting aspect of the crown polyethers is that variants have been synthesized in which the ring size is systematically altered. Ion selectivity can thus be correlated with the size of the ring opening (Table II). As might be expected, compounds with smaller rings (14–15 members) prefer Na^+;

TABLE II

EFFECT OF SIZE OF RING OPENING ON POLYETHER ION SELECTIVITY

Polyether	Size of ring hole (Å)	Log K_A for K^+	Relative affinities		
			Na($D = 1.9$ Å)	K($D = 2.66$ Å)	Cs($D = 2.34$ Å)
Dicyclohexyl-14-Crown-4	1.5	1.30	7.6	1	0.16
Cyclohexyl-15-Crown-5	2.2	2.18	1.3	1	0.16
Dicyclohexyl-18-Crown-6 (Isomer A)	3.2	6.01	0.016	1	0.040
Dibenzo-18-Crown-6	3.2	5.00	0.23	1	0.036
Dibenzo-21-Crown-7	4.3	4.30	0.012	1	0.80
Dibenzo-24-Crown-8		3.49		1	0.19
Dibenzo-30-Crown-10		4.60	0.002	1	
18-Crown-6	3.2	6.10	0.021	1	0.033
21-Crown-7	4.3	4.41		1	4.0
24-Crown-8		3.48		1	4.7

Selectivity constants determined in methanol by means of cation selective electrodes[59], ring holes measured on models built on Fisher–Hirschfelder–Taylor atoms.

preference for K^+ is greatest with 19-membered rings while Cs^+ preference is greatest with rings of 21–24 members. The increase in association constant (K_A) for K^+ in the dibenzo series when going from a 24 to 30 membered ring is presumably due to facilitation of molecular folding into a more spherical conformation[59].

Dicyclohexyl-18-crown-6 is soluble enough in water to permit complexing to be followed either by ion selective electrodes[59] or calorimetrically[60]. Under comparable conditions complexing is 10,000-fold stronger in methanol than in water, apparently reflecting the difficulty encountered by the ionophore in dislodging water from the hydration sphere of the cation[60]. Dicyclohexyl-18-crown-6 forms 2:1 and 3:2 complexes with Cs^+ and Rb^+ [58].

X-ray crystallographic studies have been carried out on a complex crystal containing the Rb^+ and Na^+ complexes of dibenzo-18-crown-6, as well as free ionophore. The cations are held coplaner with the oxygen atoms (c.f. enniatin above). The Rb–O distances are 2.86–2.93 Å and the Na–O distances are 2.74–2.89 Å. The open structure of these ionophore complexes is evident from the fact that the anion, SCN⁻, is in van der Waals contact with Rb (Rb–N distance 2.94 Å) i.e. forms an ion pair, while the Na–N distance is appreciably further, 3.32 Å. The conformation of the free ionophore within the lattice is less symmetrical than that of the complexed ionophore[61,62]. In solution the n.m.r. proton spectrum, at least at 60 MHz, is not sufficiently well resolved to lend itself to simple conformational analysis[51].

The *dentates* are related synthetic polyethers containing nitrogen, *e.g.* N(-C-C-O-C-C-O=C-C)₃N[63]. They are relatively wide spectrum complexing agents, however, since their polarity is too great to facilitate entry into lipids, they are not fully functional ionophores. X-ray crystallography of the RbSCN complex of the above compound indicates that the Rb–O distances are 2.90 Å and that the nitrogen atoms are also in contact with the encaged cation (Rb–N, 3.00 Å)[64].

Miscellaneous

The cyclic decapeptide antamanide, obtained from the mushroom *Amanita phalloides*[65] acts as an antidote against the accompanying toxic oligopeptide phaloidin. Antamanide forms lipid soluble alkali cation complexes (Na⁺ favored over K⁺) and promotes active K⁺ transport in mitochondria[11,66]. It contains all L configurations; however, the four proline residues may efficiently contort the ring for optimum focusing of the carbonyls for complexation thereby obviating the need for D configurations.

Another all L containing cyclic polypeptide is alamethicin, produced by *Trichoderma viride*[67]. Seven of its seventeen residues consist of the unusual aminoacid α-aminoisobutyric acid. Seventeen residues form the ring while one glutamine with a free carboxyl hangs pendant from the ring[68]. Although it behaves like a typical ionophore in stimulating mitochondrial ion uptake and forming lipid-soluble cation complexes[18], it has a distinguishing electrometric property. Alamethecin, as do the other valinomycin type antibiotics, induces lipid bilayers to carry a cationic current. In this case, however, the initial membrane resistance drops to a lower value when the potential exceeds a critical threshold level, a phenomenon analogous to the behavior of excitable membranes such as those of nerve and muscle[69,70]. Monazomycin exhibits similar behavior but here membrane resistance *rises* with increasing transmembrane potential[69,70]. Monazomycin also stimulates mitochondrial ion transport[16]; however, it fails to solubilize alkali ions in toluene[71] and therefore its classification as a true ionophore is not established. Monazomycin has a molecular weight of 1050–1200 and contains twenty oxygen atoms and a single basic nitrogen; its exact struc-

ture is presently unknown, but possibly is related to the polyene antibiotics[72]. The antibiotic, LL-491 (mol. wt. 1500)[72a] resembles monazomycin in activity but has only about 20% of its activity towards
mitochondria[71].

The monamycins[73] are a family of cyclic octadecapeptides with alternating D and L configurations in the ring. Several of the amino acids are
piperidazine or chloropiperidazine derivatives[74]. They form K^+ complexes[75]
and conduct ions across lipid bilayers hence appear to be bonafide ionophores. As in the case of alamethicin and monazomycin, the current voltage
characteristics of lipid bilayer conductance is non-linear and displays
potential-dependent switching[75].

Cationic selectivities

Several criteria have been used to evaluate the cationic selectivities of
the valinomycin type ionophores, *e.g.*: (1) cationic dependency of the
uncoupling of mitochondrial oxidative phosphorylation; (2) conductivity
of the lipid bilayer; (3) bionic potentials developed across the lipid bilayer
when separating different cationic species; (4) the ability to form lipid-
soluble complexes in organic solvent–water, two phase systems; (5) direct
titrations of the ionophore in a single solvent using some physical measurement as an indicator. As pointed out previously complex formation is difficult to detect for the gramicidins and monazomycin and so criterion (4) is
not met by these compounds. Criterion (4) may be evaluated by two alternative techniques, direct measurement of the migration of complexed
cations into the organic phase[1,10,18,46] and indirectly by measurement of
the anions accompanying the cation, *e.g.* SCN^{-}[10], or picrate[76]. Some of the
techniques used for titrating ionophores with cations in single phases
include ion selective electrodes[59], vapor pressure osmometry[44], conductivity[41,42,51] and n.m.r.[33,42,51]. The ion selectivity of valinomycin by several
criteria is indicated in Fig. 2. It can be observed that very similar ion discrimination patterns result by all methods.

The neutral ionophores show the usual preference series $K^+ \simeq Rb^+ >$
$Cs^+ > Na^+ > Li^+$ by all criteria. The degree of preference is extremely
variable; compared as the biologically relevant $K^+:Na^+$ ratios they range
from 10,000:1 for valinomycin, 25:1 for the actins and enniatins and close
to unity for the gramicidins (criterion 1)[43]. Appreciable variations occur
within each series of analogs which are solvent dependent.

Since the carboxylic ionophores are usually electrically silent, criteria
(2) and (3) above cannot be used to evaluate them. They evoke a release of
endogenous K^+ in exchange for H^+ in mitochondria, which can be used to
gauge their dynamic capacity to carry K^+ across membranes; however, such
measurements cannot be simply extended to other ions. Instead of uncoupling mitochondria, as do the valinomycin type ionophores, the carboxylic

ionophores inhibit mitochondrial respiration, apparently by indirectly preventing substrate entry. This inhibition can be prevented by providing high

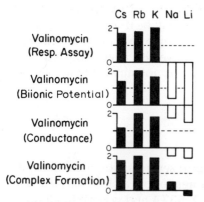

Fig. 2. Selectivity of valinomycin by various criteria[46]. The scales in Figs. 2 and 3 are logarithmic, with the most favored ion assigned a value of 2, so that a difference of one unit between two cations represents a selectivity of 10.

extra-mitochondrial levels of K^+ [23]. The complexation equilibria for these ionophores can, however, be measured by two-phase ion distribution techniques. The results of such a study (Fig. 3) show that despite their similarity of structure, the carboxylic ionophores exhibit a considerably wider range of selectivity than do the closely related valinomycin type[10,18,46].

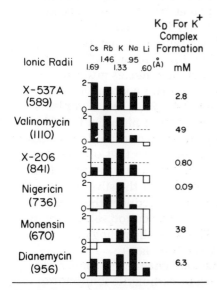

Fig. 3. Selectivity of various ionophores by two-phase partition studies[18].

References pp. 223–226

Ionophores can complex other ions of comparable diameter and charge density although quantitative data concerning them are fragmentary. By the criterion of mitochondrial interaction, NH_4^+ is complexed slightly less effectively than Cs^+ by valinomycin and the enniatins, and, somewhat more effectively than K^+ by the actins[76]. This may be due to the availability of the other bridge oxygens for hydrogen bonding to NH_4^+, an interaction not available to metal ions. Tl^+ [76] and Ag^+ [59] have also been reported to complex with the actins and polyethers, which is not surprising in view of their atomic radii and charge density.

CARBOXYLATE IONOPHORES

Although the neutral ionophores fall into several distinct families, the known structures of the carboxylic ionophores follow a close pattern. The covalent structures of monensin, nigericin, X-537A and grisorixin all feature chains containing ether oxygens within five and six membered heterocyclic rings; additional oxygens occur along the backbone. The single carboxyl at one end of the chain forms a ring by hydrogen bonding to one or two hydroxyls at the opposite end of the molecule.

The first carboxylic ionophore whose structure was established is monensin (Fig. 1E) isolated from a strain of *Streptomyces*[77]. X-ray crystallography of its silver salt shows six cation–oxygen ion-dipole ligands ranging from 2.40 to 2.68 Å. The carboxylate oxygen however does not form a strong ligand as it is 3.8 Å from the cation. The 22-atom ring about the cation is irregularly puckered but roughly planer[78,79]. Monensin shows a moderate degree of selectivity favoring Na^+ over K^+ by 10:1[10,18].

The silver salt of nigericin[20] (Fig. 1F) contains a 24-membered ring. Two of the oxygens corresponding to the ion-dipole ligands of monensin are displaced by the carboxylate oxygen which now moves into the coordination sphere of the cation (Ag–O, 2.25 Å) and forms a strong ionic bond. Together with four ion-dipole ligands (Ag–O, 2.45–2.69 Å) it completes a five-membered coordination shell[80,81]. The affinity of nigericin for K^+ is 400 fold stronger than monensin, although they are known to be almost identical in structure. The structure of grisorixin[82] is identical to that of nigericin except for omission of a single hydroxyl on the terminal ring opposite that containing the carboxyl. The coordinating oxygens of the silver salt are the same as in nigericin; however, the missing hydroxyl reduces the number of head-to-tail hydrogen bonds from two to one[83]. The cation selectivity of this ionophore has not yet been reported. The Ag–O liganding distances range from 2.4–2.7 Å (ion-dipole) to 2.2 Å (carboxylate).

The greater number of hydrogens which are chemically similar but non-identical make the proton n.m.r. spectra of carboxylic ionophores more complicated to interpret than those of neutral ionophores. The fields

217

arising from the zwitterionic structure of the carboxylic ionophores can be
examined by another application of n.m.r. however. The nuclide ^{23}Na, by
virtue of its $\frac{3}{2}$ nuclear spin, senses the asymmetry of its surrounding field as
a broadening of its n.m.r. band. Thus the greater asymmetry determined
for the Na$^+$ complex of nigericin, as compared to monensin, correlates with
the closer approach of the negative carboxyl oxygen, to the cation[47] predicted by the crystalline structures of the silver salts[79,80].

The most significant structural departure of X-537A (Hoffmann La
Roche designation)[84] from the other carboxylate ionophores is the aromatic ring to which the terminal carboxyl is attached. The ring also contains a phenolic hydroxyl which renders it readily amenable to chemical
alteration. A second unique feature is the ketonic carbonyl on the straight
chain region of the antibiotic. Since this group appears to be involved in
ion-dipole bonding, the complexing properties of this molecule ought to
be strongly sensitive to reagents forming ketonic adducts.

X-537A is the only carboxylic ionophore so far determined to have a
preference for Cs$^+$[10,18]. Its relatively small degree of selectivity, roughly
following the lyotropic series, indicates perhaps a greater flexibility in the
complexing ring compared to the antibiotics of sharper discrimination.

The structure of X-537A has been worked out by X-ray crystallographic analysis of its barium salt[85]. The divalent ion complex consists of
an electrically neutral zwitterion containing two ionophore anions. One
(*unprimed*) coordinated to the cation via five ion-dipole oxygens (Ba–O,
2.7–3.1 Å) in addition to the charged carboxylate group (Ba–O, 2.8 Å)
(Fig. 1G). The other (*primed*), although sharing almost the same conformation as the *unprimed* ionophore molecule, coordinates only via the carboxylate (Ba–O, 2.6 Å) and terminal hydroxyl (Ba–O, 2.8 Å) (Fig. 1H). The
primed ionophore molecule also hydrogen bonds to a single water which in
turns bonds to the barium (Ba–O, 2.7 Å) rendering the cation 9-coordinate.

In solution we have been able to demonstrate that X-537A complexes
with Sr^{2+}, Ca^{2+} and Mg^{2+} as well as Ba^{2+} and the alkali ions. The binding
isotherms for Ba^{2+} are consistent with an (ionophore)$_2$ Ba^{2+} complex[71].
X-537A acts as a divalent ion ionophore; we have confirmed its ability to
carry ^{133}Ba^{2+} and ^{90}Sr^{2+} across bulk phases of toluene butanol[71].

Structural variants of monensin (monensin B and C) have been
reported[77] although not enough material has been available to work out
their complexing selectivity. Preliminary evidence is that dianemycin (Na$^+$
preferring) is also closely related to this series[80], as is presumably X-206[84]
(K$^+$ preferring). Dianemycin resembles X-537A in that its alkali ion ionic
discrimination spectrum is relatively broad[10,18]. It may therefore be significant that dianemycin also shows a tendency to complex with divalent
cations[71].

The oxygen-cation liganding distances of those ionophore complexes
studied by X-ray crystallography have been summarized in Table III.

TABLE III

MOLECULAR PROPERTIES OF IONOPHORES

Ionophore	Salt	Ligand distances (Å)		Complexation selectivity
		Ion dipole	Carboxyl	
Nonactin	KSCN	2.7-2.9[26]	—	$NH_4 > K > Rb > Cs >$ $Na > Li$[44]
Enniatin B	KI	2.6-2.8[40]	—	$K > Rb > Cs > Na > Li$[51]
Valinomycin	KAuCl$_3$	2.7-2.8[52]	—	$K > Rb > Cs > Na > Li$[18]
Dibenzo-18-Crown 6	RbSCN	2.7-2.8[62]	—	$Ag > K > Na > NH_4 >$ $Cs > Li$[59]
Dentate	RbSCN	2.9[64]	—	
Monensin	Ag$^+$	2.4-2.7[79]	3.8	$Na > K > Rb > Li > Cs$[18]
Nigericin	Ag$^+$	2.4-2.7[80]	2.2	$K > Rb > Na > Cs > Li$[18]
X-537A	Ba^{2+}	2.7[85]	2.8	$Cs > Rb = K > Na > Li$[18]
Grisorexin	Ag$^+$	2.4-2.7[83]	2.2	

DYNAMICS OF IONOPHORE COMPLEX FORMATION

Although a naive determination of closeness of fit by manipulation of molecular models may suggest the ion selectivity of a given ionophore, the equilibria of complex formation depend on additional factors. The alkali cations show large differences in solvation energy in the free ionic state. When complex formation takes place in the presence of polar solvents the desolvation process accordingly is itself strongly dependent on the cation species. The free energy of complex formation is, therefore, a function of the difference between the ion desolvation and ion–ionophore association processes. Thus the observed selectivity of a given ionophore may vary as the solvent system is changed. The ion-dipole interaction between the ionophore and ion is the same type of interaction involved in ion solvation; during complexation the ionophore in effect solvates the complexed ion. The complexation process may in turn be resolved into two energetic components, the enthalpic arising from the interaction between ion and ionophore, and the entropic involved in altering the ionophore conformation, from that favored in free state to that in the complex required for focusing the appropriate oxygen atoms about the cations.

The following sequence of events is inferred for the dynamic ionophore catalysed transport of ions across biological membranes and other lipid barriers. The membrane may be considered as a thin lipid phase inserted in an aqueous continuum, the uncomplexed cation concentrated primarily within the aqueous phase and the hydrophobic ionophore concentrated within the lipid phase. The mutual repulsive forces within the electronegative oxygen system would force the uncomplexed ionophore into a conformation different from that in the complexed state. Thus the uncomplexed ionophore presumably senses the presence of a cation in the right spot before commencing rearrangement to the complexed conformation. This implies a weak transient ionophore–cation complex at the

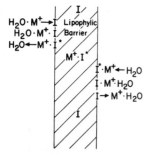

Fig. 4. Sequence of events taking place during neutral ionophore-induced transport. Starting at the top, the ionophore, I, diffuses to the lipid barrier interface where it encounters the hydrated cation, $M^+ \cdot H_2O$. A weak complex, $M^+ \cdot H_2O \cdot I$, is formed at the interphase. I now undergoes a conformational change to I^* as it replaces the solvation shell of M^+ to form the lipid soluble complex $M^+ \cdot I^*$ which diffuses freely through the membrane. At the other interphase the complex is attacked by solvent, rehydrating the cation and freeing I to diffuse back to the opposite interphase and begin another catalytic transport cycle.

interphase. The ionophore can then change its conformation enabling its oxygen system to displace the solvation water of the cation. This completes formation of the stable lipophylic ionophore complex which diffuses across the membrane whereupon attack of the complex by water dissociates the complex. The ionophore is now free to diffuse back to the other side of the membrane and undergo another catalytic cycle of cation transport. This reaction sequence is illustrated in Fig. 4.

TRANSMEMBRANE EQUILIBRIA CATALYZED BY IONOPHORES

The neutral ionophores carry ions across lipid barriers as a charged species, the uncharged free ionophore returning to complete the catalytic cycle. Thus these ionophores are capable of the net translocation of charges and the expression for the equilibrium catalysed by these ionophores contains an electrical term; equilibrium is attained when the electrochemical activity of the transported ion on both sides of the membrane becomes equal. The equilibrium condition is described by the Nernst equation for the diffusion potential:

$$\Delta E = \frac{RT}{nF} \ln \frac{[M^+]_1}{[M^+]_2} = 59 \ m.v. \ . \log \frac{[M^+]_1}{[M^+]_2} \tag{1}$$

References pp. 223–226

where $[M^+]_1$ and $[M^+]_2$ are concentrations of cation species M^+ on side 1 and side 2 of the barrier, respectively.

Where more than one cation species present can complex with the ionophore the Goldman equation applies:

$$\Delta E = 59 \; m.v. \; \log \frac{[M^+]_1 + P_{mn}[N^+]_1 + P_{mo}[O^+]_1 \cdots}{[M^+]_2 + P_{mn}[N^+]_2 + P_{mo}[O^+]_2 \cdots} \qquad (2)$$

The factors P_{mn} and P_{mo} represent the relative permeabilities of ion species M and N, and M and O, respectively. These factors are functions of the ionophore complex concentrations at the membrane interphase, $[IM^+]$ which, in the absence of high, saturating cation concentrations, depends on the equilibrium expression:

$$K_D = \frac{[I].[M^+]}{[IM^+]} \qquad (3)$$

Hence the biionic potential set up across a membrane or lipid barrier between solutions of two complexing ion species expresses the relative affinities of the opposing ion species for the ionophore and the mobilities of the complexes across the membrane. A second function of these same permeability coefficients is the relative membrane conductances for equal concentrations of various ions. The close correlations of membrane conductances and complexing equilibria for the series of alkali ions are given in Fig. 2 along with corresponding selectivity values based on mitochrondrial uncoupling.

In order to facilitate migration of cations from an aqueous medium to a lipophylic phase containing ionophore, it is necessary to provide a lipid compatible anion to migrate with the cation. This prevents the buildup of enough interfacial potential (c.f. eqn. 1) to equilibrate the electrochemical potential of the cation in both phases before a measurable cation migration can occur. At fixed concentrations of such anions as SCN^- and at higher concentrations of cation, cation migration approaches as ideal Langmuir saturation isotherm from which the ionophore affinity for a specific cation may be determined:

$$\frac{1}{[MI^+]_{org}} = \frac{1}{[M^+]_{H_2O}} \cdot \frac{K_D}{[I]_{org}} + \frac{1}{[I]_{org}} \qquad (4)$$

where $[MI^+]_{org}$ is measured as the amount of cation in the organic phase (less the no ionophore control), $[M^+]_{H_2O}$ is the concentration of M^+ added to water and $[I]_{org}$ is the concentration of ionophore added to the organic phase. The apparent dissociation constant, K_D, is a function of the lipid compatible anion concentration but does not vary appreciably with pH (c.f. carboxylic ionophores below).

The carboxylic ionophores tend to enter lipid phases either as the electrically neutral zwitterionic complexes or as the uncharged, undissociated carboxylic acid. At neutral pH little ionophore is available to traverse a lipid barrier as a charged species, hence this class of ionophores is virtually electrically silent towards a membrane. At higher pH the uncomplexed carboxylate form of these ionophores would presumably become prevalent enough to carry current and produce electrometrically detectable effects on membranes. Although no electrical term is included in the transmembrane equilibration expression, the prevailing proton gradient does appear:

$$\frac{[H^+]_1}{[H^+]_2} = \frac{[M^+]_2}{[M^+]_1} = \frac{[N^+]_2}{[N^+]_1} = \frac{[O^+]_2}{[O^+]_1} \cdots \tag{5}$$

Identical gradients are ultimately reached for each ionic species capable of crossing the membrane along with the ionophore regardless of its affinity for the ionophore.

In two phase distribution measurements, however, the cation species competes with the protons present for ionophore:

$$I.H + M^+ \rightleftharpoons I^-.M^+ + H^+ \tag{6}$$

Thus, at a fixed pH, an ideal Langmuir saturation isotherm can be obtained as a function of cation concentration and affinity. Since no net transfer of charge takes place, no interphase potential is produced and this equilibrate is not affected by the anion species present. Just as the apparent cation affinity of a carboxylic ionophore is influenced by pH, the apparent dissociation constant of the carboxyl group, the pK_a, is affected by the cation species and concentration present. The equilibria describing the interaction between carboxylic ionophores and divalent cations are somewhat more complicated.

APPLICATION OF IONOPHORES

Although most of the ionophores were regarded as antibiotics upon isolation, their effectiveness in disrupting mitochondrial energy conservation in higher organisms[8,9] limits their therapeutic application. The antibiotic activity of ionophores also appears to arise from their disruption of the energy and ion metabolism of micro-organisms[86-88]. Gramicidin although systemically toxic has maintained limited medical use as a topical antibiotic. The toxicity of ionophores to the brine shrimp *Artemia salina*, has been used as a screening procedure for their detection[89].

Some of the carboxylic antibiotics have limited ability to cross the poultry gut and monensin has been reported to be effective in destroying coccids, intestinal parasites which infect fowl[90]. Valinomycin and the actins have been utilized in ion selective electrodes (c.f. eqns. (1) and (2)[91-93]. Although to date exploitation of these electrodes have been limited the extremely high $K^+:Na^+$ selectivity of the valinomycin electrode might well find application in the monitoring of K^+ in the presence of high concentrations of Na^+ such as obtained in blood or urine. This would make it possible to monitor serum K^+ conveniently during the treatment of cardiac crisis or diabetic coma.

The ionophores have been utilized extensively as tools for experimentally perturbing the energy and ion metabolism of subcellular organelles, *i.e.* mitochondria, both isolated[1,7-11,13,16,18,21,24,34,35,46] and *in situ* (ascites tumor)[94-96] as well as chloroplasts[97-100] and chromatophores[100-105]. They have also been used to increase selectively the permeability of other biological membrane systems, *e.g.* erythrocytes[106-108] and bladder[109].

One of the most striking properties of ionophores is their ability to induce in artificial lipid bilayers permselective conductance mimicking that of natural membranes. Electrometrically this permselectivity also parallels that of biological membranes. Lipid membranes treated with some of the ionophores, *e.g.* alamethicin, offer a means of studying the molecular origin of electroexcitability which characterizes the membranes of nerve and muscle[69,70].

Since all the properties of ionophores described appear to derive from their ability to function as mobile ion carriers, attention has been focused upon them as model membrane-bound carrier systems. Indeed their substrate specificity and turnover numbers (*e.g.* valinomycin, 200/sec, nigericin 500/sec in the mitochondrial membrane) are well within the range of more *orthodox* enzymes, although the ability of valinomycin to be sublimed without decomposition is unusual among biological catalysts. The small size of ionophores has facilitated study of the molecular architecture and conformational transformations responsible for their properties by a variety of sophisticated physical methods. The same general architectural feature, cation liganded in the polar interior of a cyclic molecule with lipophylic exterior arc shared by the membrane permeable iron complexing agents of biological origin (c.f. Chapter 5) consistent with a reoccurring solution in nature to related membrane transport problems.

Natural membrane-bound carriers would be expected to require additional properties beyond those shown by ionophores. They must remain anchored within a given membrane so that they would equilibrate with all cellular membrane systems. This could be accomplished by having an ionophore-like structure attached to a larger bulkier protein molecule. In order to construct an active transport system it would be necessary to have a part of the natural carrier capable of reacting with an energy source such as ATP. One could imagine the binding and subsequent hydrolysis of ATP to exert an *allosteric* effect on the conformation of the active ionophore region of a carrier which in turn reversibly alters the affinity for substrate to be transported. Thus it would be possible to force a carrier in a high affinity conformation to bind a substrate in a region of low electrochemical activity. As the carrier would bring the substrate to a region of high substrate chemical activity, the affinity would be lowered allosterically by appropriate binding or hydrolysis of ATP effecting discharge of the substrate. Such a system for cyclic alteration of a substrate affinity site by an extrinsic energy system resembles the transport scheme proposed by Willbrandt and Rosenberg[110].

The ionophores have become established both as tools for altering membrane function and models for understanding membrane behavior. In the future we might anticipate the isolation of their natural counterparts from membranes hopefully aided by our studies in depth of their simpler prototypes.

NOTE ADDED IN PROOF

Recently, the structures and conformation of two additional carboxylic ionophores, X-206[111] and dianemycin[112], have been established by X-ray crystallography.

REFERENCES

1 B. C. Pressman, E. J. Harris, W. S. Jagger and J. H. Johnson, *Proc. Natn. Acad. Sci.*, 58 (1967) 1949.
2 H. Brockmann and G. Schmidt-Kastner, *Chem. Ber.*, 88 (1955) 57.
3 H. Brockmann, M. Springorum, G. Träxler and I. Höfer, *Naturwiss.*, 50 (1963) 689.
4 M. M. Shemyakin, N. A. Aldanova, E. I. Vinogradova and M. Yu. Feigina, *Tetrahedron Lett.* (1963) 1921.
5 M. M. Shemyakin, E. I. Vinogradova, M. Yu. Feigina, N. A. Aldanova, N. F. Loginova, N. F. Ryabovd and I. D. Pavlenko, *Experientia*, 21 (1965) 548.
6 M. M. Shemyakin, E. I. Vinogradova, M. Yu. Feigina, N. A. Aldanova, Yu. B. Shvetsov, and L. A. Fonina, *Zh. Obshch. Khim. (USSR)*, 36 (1966) 1391.
7 W. McMurray and R. W. Begg, *Arch. Biochem. Biophys.*, 84 (1959) 546.
8 C. Moore and B. C. Pressman, *Biochem. Biophys. Res. Commun.*, 15 (1964) 562.
9 B. C. Pressman, *Proc. Natn. Acad. Sci.*, 53 (1965) 1076.
10 B. C. Pressman, in *Mitochondria, Structure and Function, 5th Meeting, Fed. Europ. Biochem. Soc.*, Academic Press, New York, 1969, p. 315.
11 B. C. Pressman, in E. Racker, *Membranes of Mitochondria and Cholorplasts*, Reinhold Van Nostrand, New York, 1970, p. 213.
12 W. Simon, L. A. R. Pioda and H. K. Wipf, in T. Bücher and H. Sies, *Inhibitors, Tools in Cell Research*, Springer-Verlag, New York, 1969, p. 356.
13 B. C. Pressman, *Antimicrobial Agents and Chemotherapy-1969*, (1970) 28.
14 P. Mueller and D. O. Rudin, *Biochem, Biophys. Res. Commun.*, 26 (1967) 398.
15 A. A. Lev and E. P. Buzhinsky, *Cytology (USSR)*, 9 (1967) 102.
16 H. A. Lardy, S. N. Graven and S. Estrada-O, *Federation Proc.*, 26 (1967) 1355.

224

17 G. Eisenman, *Federation Proc.*, 27 (1968) 1249.
18 B. C. Pressman, *Federation Proc.*, 27 (1968) 1283.
19 H. K. Wipf, W. Pache, P. Jordan, H. Zähner, W. Keller, Schierlein and W. Simon, *Biochem. Biophys. Res. Commun.*, 36 (1969) 387.
20 R. L. Harned, P. H. Hidy, C. J. Corum and K. L. Jones, *Antibiot. Chemother.*, 1 (1951) 594.
21 H. A. Lardy, D. Johnson and W. C. McMurray, *Arch. Biochem. Biochys.*, 18 (1958) 587.
22 S. N. Graven, S. Estrada-O and H. A. Lardy, *Proc. Natn. Acad. Sci.*, 53 (1965) 1076.
23 S. N. Graven, S. Estrada-O and H. A. Lardy, *Proc. Natn. Acad. Sci.*, 56 (1965) 654.
24 J. B. Chappell and A. R. Crofts, *Biochem. J.*, 95 (1965) 393.
25 B. T. Kilbourn, J. D. Dunitz, L. A. R. Pioda and W. Simon, *J. Mol. Biol.*, 30 (1967) 559.
26 M. Dobler, J. D. Dunitz and B. T. Kilbourn, *Helv. Chim. Acta*, 52 (1969) 2573.
27 M. M. Shemyakin, *Angew. Chem.*, 72 (1960) 342.
28 J. Beck, H. Gerlach, V. Prelog and W. Voser, *Helv. Chim. Acta*, 45 (1962) 621.
29 R. Serger and B. Witkop, *J. Am. Chem. Soc.*, 87 (1965) 2011.
30 C. J. Pederson, *J. Am. Chem. Soc.*, 89 (1967) 7017.
31 W. Keller-Schierlein, in T. Bücher and H. Sies, *Inhibitors Tools in Cell Research*, Springer-Verlag, New York, 1969, p. 365.
32 J. H. Prestegrad and S. I. Chan, *Biochemistry*, 8 (1969) 3921.
33 J. H. Prestegard and S. I. Chan, *J. Am. Chem. Soc.*, 92 (1970) 4440.
34 S. N. Graven, H. A. Lardy, D. Johnson and A. Rutter, *Biochemistry*, 5 (1966).
35 S. N. Graven, H. A. Lardy and A. Rutter, *Biochemistry*, 5 (1966) 1735.
36 L. A. R. Pioda, H. A. Wachter, R. E. Dohner and W. Simon, *Helv. Chim. Acta*, 50 (1967) 1373.
37 P. A. Plattner, K. Vogler, R. O. Studer, P. Quitt and W. Keller-Schierlein, *Helv. Chim. Acta*, 46 (1963) 927.
38 P. Quitt, R. O. Studer and K. Vogler, *Helv. Chim. Acta*, 46 (1963) 1715.
39 R. L. Hamill, C. E. Higgens, H. E. Boaz and M. Gorman, *Tetrahedron Lett.*, (1969) 4255.
40 M. Dobler, J. D. Dunitz and J. Krajewski, *J. Biol. Biol.*, 42 (1969) 603.
41 M. M. Shemyakin, Yu. A. Ovchinnikov, V. T. Ivanov, V. K. Antonov, *et al.*, *Biochem. Biophys. Res. Commun.*, 29 (1967) 834.
42 M. M. Shemyakin, Yu. A. Ovchinnikov, V. T. Ivanov, V. K. Antonov, *et al.*, *J. Memb. Biol.*, 1 (1969) 402.
43 B. C. Pressman and E. J. Harris, Abstracts of Seventh International Congress of Biochemistry, Tokyo, August 1967, Vol. V, p. 900.
44 L. Pioda, Dissertation, Eidenossischen Technische Hochschule, Zurich, 1969.
45 M. M. Shemyakin, Yu. A. Ovchinnikov, V. T. Ivanos and A. V. Evstratov, *Nature*, 213 (1967) 412.
46 B. C. Pressman and D. H. Haynes, in D. C. Tosteson, *The Molecular Basis of Membrane Function*, Prentice-Hall Inc., Englewood Cliffs, N.J., 1969, p. 221.
47 D. H. Haynes, A. Kowalsky and B. C. Pressman, *J. Biol. Chem.*, 244 (1969) 502.
48 V. T. Ivanov, I. A. Laine, N. D. Abdullaev, L. B. Senyavina, *et al.*, *Biochem. Biophys. Res. Commun.*, 34 (1969).
49 M. Ohnishi and D. W. Urry, *Biochem. Biophys. Res. Commun.*, 36 (1969) 164.
50 D. W. Urry and M. Ohnishi, in D. W. Urry, *Spectroscopic Approaches to Biomolecular Conformation*, American Medical Association, Chicago, Ill. 1970, p. 263.

51 D. H. Haynes, Dissertation Thesis, University of Pennsylvania, 1970.
52 M. Pinkerton, L. K. Steinrauf and P. Dawkins, *Biochem. Biophys. Res. Commun.*, 35 (1969) 512.
53 R. Dubos and R. Hotchkiss, *J. Exp. Med.*, 73 (1941) 629.
54 A. Stern, W. Gibbons and L. C. Craig, *Proc. Natn. Acad. Sci.*, 61 (1968) 734.
55 R. Hotchkiss, *Adv. Enzymol.*, 4 (1944) 153.
56 M. C. Goodall, *Biochim. Biophys. Acta*, 203 (1970) 28.
57 C. J. Pederson, *Federation Proc.*, 27 (1968) 1305.
58 C. J. Pederson, *J. Am. Chem. Soc.*, 92 (1970) 389, 391.
59 H. K. Frensdorff, *J. Am. Chem. Soc.*, 93 (1971) 600.
60 R. M. Izaat, J. H. Rytting, D. P. Nelson, B. L. Haymore and J. J. Christensen, *Science*, 164 (1969) 443.
61 D. Bright and M. R. Truter, *Nature*, 225 (1970) 176.
62 D. Bright and M. R. Truter, *J. Chem. Soc. (B)*, (1970) 1544.
63 B. Dietrich, J. M. Lehn and J. P. Sauvage, *Tetrahedron Lett.*, 34 (1969) 2889.
64 B. Metz, D. Moras and R. Weiss, *Chem. Commun.*, (1970) 217.
65 T. Wieland, G. Luben, H. Ottenheyn, J. Faesel, *et al.*, *Angew. Chem.*, 80 (1968) 209.
66 T. Wieland, H. Faulstich, W. Burgermeister, W. Otting, *et al.*, *FEBS Letters*, 9 (1970) 89.
67 C. E. Meyer and F. Reusser, *Experientia*, 23 (1967) 85.
68 J. W. Payne, R. Jakes and B. S. Hartley, *Biochem. J.*, 117 (1970) 757.
69 P. Mueller and D. O. Rudin, *Nature*, 217 (1968) 713.
70 P. Mueller and D. O. Rudin, *J. Theor. Biol.*, 18 (1968) 222.
71 B. C. Pressman, unpublished observations.
72 K. Akasakim, K. Karasawa, M. Watanabe, H. Yonehara and H. Umezawa, *J. Antibiot. Ser. A.*, 16 (1963) 127.
72a L. A. Mitcher, A. J. Shayand and N. Bohonos, *Appl. Microbiol.*, 15 (1967) 1002.
73 C. H. Hassall and K. E. Magnus, *Nature*, 184 (1959) 1223.
74 K. Bevan, J. S. Davies, M. J. Hall, C. H. Hassall, *et al.*, *Experientia*, 26 (1970) 122.
75 M. C. Goodall, *Nature*, 225 (1970) 1258.
76 G. Eisenman, S. Ciani and G. Szabo, *J. Memb. Biol.*, 1 (1969) 294.
77 M. E. Haney and M. M. Hoehn, *Antimicrobial Agents and Chemotherapy-1967* (1968) 349.
78 A. Agtarap, J. W. Chamberlin, M. Pinkerton and L. K. Steinrauf, *J. Am. Chem. Soc.*, 89 (1967) 5737.
79 M. Pinkerton and L. K. Steinrauf, *J. Mol. Biol.*, 49 (1970).
80 L. K. Steinrauf and M. Pinkerton, *Biochem. Biophys. Res. Commun.*, 33 (1968) 29.
81 T. Kubota, S. Matsutani, M. Shiro and H. Koyama, *Chem. Commun.*, (1968) 1541.
82 P. Gachon, A. Kergomard, H. Veschambre, C. Esteve and T. Staron, *J. Chem. Soc. D.*, (1970) 1421.
83 A. Alleaume and D. Hickel, *J. Chem. Soc. D.*, (1970) 1422.
84 J. Berger, A. I. Rachlin, W. E. Scott, L. H. Sternbach and M. W. Goldberg, *J. Am. Chem. Soc.*, 73 (1951) 5295.
85 S. M. Johnson, J. Herrin, S. J. Liu and I. C. Paul, *J. Am. Chem. Soc.*, 92 (1970) 428.
86 B. C. Pressman, in *Wirkungsmechanismen von Fungiziden und Antibiotika* (Biologische Gesellschaft der D.D.R., Sektion Microbiologie) Academie Verlag, Berlin, 1967, p. 3.
87 F. M. Harold and J. Baarda, *Bacteriology*, 94 (1967) 53.
88 F. M. Harold, *Bact. Rev.*, (1969) 45.

89 C. E. Higgens, J. E. Westhead, M. Gorman and R. L. Hamill, *Antimicrobial Agents and Chemotherapy-1968* (1969).

90 R. F. Shumard and M. E. Callender, *Antimicrobial Agents and Chemotherapy-1967* (1968) 369.

91 Z. Stafanac and W. Simon, *Microchem. J.*, 12 (1967) 125.

92 L. A. R. Pioda and W. Simon, *Chimia (Switzerland)*, 23 (1969) 72.

93 M. S. Frant and J. W. Ross, *Science*, 167 (1970) 987.

94 C. E. Wenner, E. J. Harris and B. C. Pressman, *J. Biol. Chem.*, 242 (1967) 3454.

95 E. E. Gordon, K. Nordenbrand and L. Ernster, *Nature*, 213 (1967) 82.

96 E. E. Gordon and J. Bernstein, *Biochim. Biophys. Acta*, 205 (1970) 464.

97 L. Packer, J. M. Allen and M. Starks, *Arch. Biochem. Biophys.* 128 (1968) 142.

98 N. Shavit and A. San Pietro, *Progress in Photosynthesis Research*, 3 (1969) 1392.

99 S. J. D. Karlish and M. Avron, *FEBS Letters*, 1 (1968) 21.

100 S. J. D. Karlisch and M. Avron, *Eur. J. Biochem.*, 9 (1969) 291.

101 N. Shavit and A. San Pietro, *Biochem. Biophy. Res. Commun.*, 28 (1967) 277.

102 N. Shavit, A. Thore, D. L. Keister and A. San Pietro, *Proc. Natn. Acad. Sci. U.S.*, 59 (1968) 917.

103 J. B. Jackson, A. R. Crofts and L. V. Von Stediggk, *Eur. J. Biochem.*, 6 (1968) 3499.

104 W. Junge and H. T. Witt, *Z. Naturforsch.*, 23b (1968) 244.

105 M. Nishimura and B. C. Pressman, *Biochemistry*, 8 (1969) 1360.

106 J. B. Chappell and A. R. Crofts, J. M. Tager *et al.*, (Eds.), *Regulation of Metabolic Processes in Motochondria, Biochim. Biophys. Acta Library*, Amsterdam. Vol. 7, 1966, p. 293.

107 E. J. Harris and B. C. Pressman, *Nature*, 216 (1967) 918.

108 D. C. Tosteson, T. E. Andreoli, E. Tieffenberg and P. Cook., *J. Gen. Physiol.*, 51 (1968) 373S.

109 H. R. Wyssbrod, *Biochim. Biophys. Acta*, 193 (1969) 361.

110 W. Willbrandt and T. Rosenberg, *Pharmacol. Rev.*, 13 (1961) 109.

111 J. F. Blount and J. W. Westley, *Chem. Commun.*, (1971) 927.

112 E. W. Czerwinsky and L. K. Steinrauf, *Biochem. Biophys. Res. Commun.*, 45 (1971) 1284.

Chapter 7

METAL–PROTEIN COMPLEXES

ESTHER BRESLOW

Department of Biochemistry, Cornell University Medical College, New York, U.S.A.

I. INTRODUCTION

Metal ion–protein interactions necessarily differ from those of metal ions with amino acids and small peptides to the extent that the α-NH$_2$ and α-COOH of long polypeptide chains are separated covalently by a number of intervening residues. These interactions also differ because of the influence of peptide chain conformation which may block reaction at potential metal-binding sites or which may place amino acid side-chains which are distant in sequence in suitable position to form chelates. The suitable juxtaposition of ligand side-chains to allow strong chelation with specific metals finds its most significant expression in metalloproteins and metalloenzymes where the tight interaction between metal and protein plays a critical and specific biological role. Metalloproteins and metalloenzymes will be covered elsewhere. This chapter will deal in general with the *in vitro* behavior of proteins in the presence of metal ions with which they have not necessarily been designed by nature to react. By and large, the protein–metal complexes represented here have no known biologic function, with the possible exceptions of the cupric ion–albumin and zinc–insulin complexes for which transport and storage functions have been postulated respectively.

Historically interest arose in the study of non-specific metal–protein interactions because of the use of metal ions in protein purification, the known inhibitory effects of metal ions on certain enzymes, and the possibility that non-specific metal ion–protein complexes would serve as metalloprotein models. With the growing probability that the environment of the metal ion in metalloproteins is uniquely strained[1], present studies of non-specific protein–metal ion interactions have been concerned chiefly with mechanisms of metal ion–mediated protein denaturation and enzyme inhibition and with the use of metal ions as structural probes by which the accessibility and reactivity of potential ligand groups on proteins can be gauged. With respect to the latter, it is clear that heightened reactivity of a potential coordinating atom on a protein relative to its individual behavior in model systems suggests the proximity of a second electron donor not found in the models used. Conversely, a relatively diminished reactivity of a potential binding site is an indication of steric hindrance and/or a conformational change associated with binding. For those proteins which, as isolated, are in their lowest free energy state, all such conformational changes will be associated with positive free energy contributions to binding, and a lower apparent binding constant will be observed than would be found in the absence of conformational change.

II. CONSIDERATIONS AND APPROACHES IN THE STUDY OF METAL-PROTEIN COMPLEXES

The approach used in the study of any metal–protein interaction necessarily depends on the specific system and the information which is being sought. In general, however, the following questions are asked: How many binding sites are there for a particular metal ion? How strong are these sites? And what are the coordinating side-chain or backbone atoms?

Determination of the numbers of sites and their affinities can be achieved by direct binding techniques such as equilibrium dialysis[2] or gel-filtration chromatography[3]. In all such studies, the number of metal ions bound are determined directly by analysis of the metal–protein complex as a function of the equilibrium concentration of free metal ion. pH is particularly critical in metal-binding studies, both for obtaining meaningful data and in the interpretation of data (see section III) and, depending on the study, it may be controlled or allowed to change. The use of most buffers is, if possible, to be avoided because of competition between protein and buffer for metal and because of the distinct possibility that ternary complexes involving metal, protein and buffer will be formed[4]. For translation of raw binding data into the numbers and affinities of different sites, the reader is referred to standard texts[5].

Similarly, an estimate of the numbers and relative affinities of sites present can be made from indirect binding studies in which binding is observed by a change in some property of the system when metal–protein coordination occurs. This approach, which is typified by potentiometric titration (see section III) and a variety of spectroscopic techniques, demands some known quantitative correlation between the magnitude of the effect and the degree of binding. Both its advantages and limitations depend on the extent to which the properties of metal-complexes formed at different binding sites differ. For example, such an approach can allow clear identification of a weak binding site whose presence might go unobserved by direct binding studies, if the properties of the site are sufficiently distinctive.

Approaches used in the identification of side chains or atoms to which metal ions are bound include direct observation of the complex by

X-ray crystallography (see myoglobin studies) and nuclear magnetic resonance (n.m.r.) studies (see metal ion–RNase studies) as well as a variety of indirect approaches. The latter include the effect of specific protein side-chain modification on metal-binding, titrimetric determination of the pK_a of the ligand groups involved (see section III), and, where possible, correlation of protein–metal ion spectra with those of appropriate model systems. The last, although hazardous in metalloproteins, has been particularly fruitful in the study of non-specific metal ion–protein complexes. The effects of protein modification on metal-binding, while clearly useful, must be interpreted allowing for the possibility that modification of side-chains which do not participate in binding may alter binding by changing the net charge on the protein or its conformation.

III. pH AND OTHER FACTORS AFFECTING METAL ION–LIGAND AFFINITY

Potential metal ion-binding sites on proteins are principally the ionizable side-chains and the –NH and –C=O groups of the peptide chain backbone. A contribution to binding in specific instances by the hetero-atoms of the side-chains of serine, threonine, cystine and methionine cannot be discounted (one of the ligand atoms to iron in cytochrome c appears to be a methionine sulfur), but involvement of these groups in non-specific complexes has not yet been demonstrated. Within this broad framework, a number of factors other than protein conformation determine the probability that a given metal ion will interact at a given site. Foremost among these are the relative intrinsic affinity of the metal ion for the different coordination atoms available and pH. The reader is referred elsewhere for a discussion of the former[6]. The interactions between H^+ ion and metal ions on binding to proteins, however, have profound implications both with respect to the availability to metal ions of different binding sites at different pH values and to an interpretation of titration curves of metal–protein complexes. Accordingly, it is worth while to summarize the important aspects of these interactions.

Consider a single binding site, L^-, which can react either with a single H^+ ion or with a metal ion, M^{2+}. Intrinsic association constants with H^+ ion and with metal ion are defined respectively by the relations:

$$L^- + H^+ = LH \qquad\qquad K_h = \frac{(LH)}{(L^-)(H^+)} \qquad\qquad (1)$$

$$L^- + M^{2+} = LM^+ \qquad\qquad K_m = \frac{(LM^+)}{(L^-)(M^{2+})} \qquad\qquad (2)$$

For a set of n such independent sites on a macromolecule, in equilibrium with both H^+ and M^{2+}, the above equilibria can be reformulated most generally as:

$$K_h = \frac{\bar{v}_h \, e^{+2w\bar{Z}}}{(n - \bar{v}_h - \bar{v}m)(H^+)} \tag{3}$$

$$K_m = \frac{\bar{v}_m \, e^{+2w\bar{Z}(Zm)}}{(n - \bar{v}_m - \bar{v}_h)(M^{2+})} \tag{4}$$

where $\bar{v}_h$ and $\bar{v}_m$ are the number of sites occupied by H^+ ion and metal ion respectively, and K_h and K_m are the intrinsic affinities of each site for H^+ and M^{2+} respectively. The terms $e^{+2w\bar{Z}}$ and $e^{+2w\bar{Z}(Zm)}$ are electrostatic interaction factors to allow for charge interactions on the macromolecule[7] where $\bar{Z}$ is the net protein charge on the metal ion and w is an electrostatic factor which varies inversely with ionic strength and molecular size. By appropriate combination of eqns. (1) and (2) or eqns. (3) and (4), an apparent pH-dependent constant, K'_m can be defined which relates the number of binding sites carrying metal to those not carrying metal as:

$$K'_m = \frac{(LM^+)}{(L^- + LH)(M^{2+})} = \frac{K_m}{1 + K_h(H^+)} = \frac{K_m \, e^{-2w\bar{Z}(Zm)}}{1 + K_h(H^+) \, e^{-2w\bar{Z}}} \tag{5}$$

Equation (5) has several important implications. First, the apparent affinity for metal ion will be less than the intrinsic affinity until the pH is approximatily two or more units above the pK_a of the coordinating atom. For binding sites with very high pK_a values, such as the side-chains of arginine, this eliminates any significant contribution on their part to binding as *single participants* at neutral pH. Secondly, for a binding site containing a single ionizable group, the change in apparent metal ion affinity with change in pH can be used to determine the pK_a of that group. And last, for proteins, which contain several potential classes of sites with different H^+ ion and metal ion-affinities, the *relative* distribution of metal ions among these sites can be expected to be pH-dependent. For example, the intrinsic affinities of carboxyl groups for H^+ ion and Cu(II) respectively are $10^{4.5}$ and $10^{1.8}$; the intrinsic affinities of imidazole side-chains for H^+ and Cu(II) are 10^7 and 10^4. By eqn. (5), the apparent affinity of carboxyls for Cu(II) is slightly greater than that of imidazoles at pH 4.0, while the reverse is true at pH 7. It is to be expected that the predominant loci of bound Cu(II) ions will shift from carboxyl groups at low pH to imidazole groups as the pH is raised to neutrality, as has been observed[4].

A second consequence of competition between H^+ and metal ion is that binding of metal by a site which is normally protonated in the absence of metal at the pH of binding leads to displacement of protons;

at pH values well above the pK_a no protons are displaced (except in certain instances for a small displacement due to a change in net protein charge with a concurrent electrostatic effect on acidity.) It can readily be shown that, for any single coordinating group, the difference in H^+ ion bound by the uncomplexed group relative to the fully complexed one is given by its H^+ ion titration curve. This has practical consequences in that, in binding situations where the binding site does not change with pH, the pK_a of the coordinating group (and therefore its probable identity) can be assessed by measuring the decreasing displacement of protons as a function of pH. However, where a change in binding site with change in pH does occur, the titrimetric relationships are more complex. For example, the change with increasing pH to a site of higher pK_a (relative to the pH) will lead to an

TABLE I

IONIZABLE GROUPS ON PROTEINS AND THEIR "NORMAL" ASSOCIATION CONSTANTS FOR H^+, Cu(II) AND Zn(II) AT 25°C[a]

Group	$\log K_h$ (pK_a)	$\log K_m$	
		Cu(II)	Zn(II)
α-COOH	3.1–3.5	—	—
β, γ-COOH (Asp, Glu)	4.4–4.7	1.8 (acetate)[55]	1.0 (acetate)[55]
Imidazole (His)	6.4–7.2	3.1–4.4[10]	2.5 (imidazole)[55]
α-NH$_2$	7.8–8	4.9–5.5[9]	
ϵ-NH$_2$ (Lys)	9.6–10.5	4.3 (NH$_3$)[55]	2.6 (NH$_3$)[55]
Phenolic-OH (Tyr)	9.6–9.8	—	—
Sulfhydryl (Cys)	9	—	7[6]
Guanidinium (Arg)	>12	—	—

[a] Values of $\log K_h$ are representative normal values found in proteins. Values of $\log K_m$ are based on 1:1 model complexes indicated in parentheses and/or in references noted.

increased displacement of protons as a function of pH. Such situations will be illustrated by specific example. For reference, representative association constants of typical protein coordinating groups for H^+, Cu(II), and Zn(II) are shown in Table I.

IV. SELECTED SYSTEMS

The vast majority of non-specific metal protein interactions studied have been concerned with the interactions of Cu(II) and Zn(II) and such interactions will be the main focus here. Those properties of Cu(II) and

Zn(II) which are of special interest when considering their interactions with proteins may be summarized as follows: (1) complexes of Zn are generally tetrahedral or octahedral while those of Cu(II) are square-planar or distorted octahedral; (2) the coordinating atom preference of each is $S > N > O$ but for any individual atom, the interaction of Cu(II) is approximately 10-fold stronger than that of Zn(II); (3) the association of Cu(II) [but not typically of Zn(II)] with small peptides is characterized by a tendency to coordinate with the nitrogen atoms of the peptide bond with concurrent displacement of the peptide bond proton[8]. At high pH this tendency is manifest by formation of biuret-type complexes in which Cu(II) is coordinated with peptide bond nitrogen atoms alone. In the pH region 5–9, it is manifest by coordination of Cu(II) with the α-NH$_2$ and/or with intra-chain imidazoles and successive adjacent peptide bond nitrogens as in the 1:1 complexes of Cu(II) with Gly-Gly-Gly-Gly-Gly and acetyl-Gly-Gly-Gly-His respectively[9,10]. Thus Cu(II) coordinates with the α-NH$_2$ of Gly-Gly-Gly-Gly-Gly and displaces protons from three adjacent peptide bonds with apparent pK values of 5.85, 6.95 and 8.1. With acetyl-Gly-Gly-Gly-His, Cu(II) coordinates with the N$_1$ of the imidazole and displaces peptide bond protons with apparent pK values of 6.5, 7.35 and 8.8.

It is also relevant to note the useful relationship between the field strength of atoms bound to Cu(II) and the position of the principal visible absorption bands of Cu(II)-complexes[8]. These bands shift to shorter wavelengths as the number of strong field electron donors to Cu(II) increases, varying from above 700 nm for complexes such as the 1:1 complex with imidazole to 515 nm for complexes in which the electron donors are an α-NH$_2$ and 3 peptide bond nitrogens. In general, coordinating groups may be classified according to field strength as –NH$_2$ > peptide bond N > imidazole > oxygen.

A. Zinc ions and serum albumin

Serum albumin is a protein of approximately 69,000 molecular weight which contains (among a full complement of the other amino acids) 16 histidine residues. Early equilibrium dialysis studies of Zn(II)-binding and of the effect of this binding on the H$^+$ equilibria of serum albumin were prompted by the use of Zn(II) in the fractionation of blood proteins and led to the conclusion that the principal mode of Zn(II)-binding between pH 5.5 and 7.5 is a 1:1 interaction between Zn(II) and each of the 16 imidazole side-chains[2]. The arguments in support of such an interaction were: (1) large pH decreases accompanying Zn(II) binding indicate that Zn(II) binds to sites normally protonated at pH 5.5. Because carboxyls and lysines can be excluded as the most significant binding sites (the pK_a of the former is too low and guanidination of the latter does not effect binding), imidazoles are the most probable sites; (2) values of $\bar{v}_m$ and (H^+) at equilibrium are best interpreted to indicate that the number of imidazoles $(n - \bar{v}_m)$ participating in the H^+ ion equilibrium in eqn. (3) is reduced by one for every mole of Zn(II) bound. And finally, the assumption of a 1:1 Zn(II)–imidazole interaction leads to values of K_m, derived from the data via the relationships in eqns. (3) and (4), which agree well with the known first association constant of Zn(II) and imidazole under the same conditions.

These conclusions have been questioned by Ral and Lal[11,12], who point out that the original value of K_m was based on an erroneous value of K_h and that corrected average values of K_m are too high for 1:1 Zn(II)–imidazole interaction. Moreover, their own electrophoretic and polarographic studies[11,12] indicate that approximately two of the Zn(II)-binding sites are stronger than the others and that only the affinities of the weaker sites are compatible with 1:1 Zn(II)–imidazole interaction. On the basis of the observed release of one proton by each of the first two Zn(II) bound at pH 6.6, these authors suggest that the strongest two Zn(II)-binding sites each contain an imidazole together with a carboxylate side-chain or a peptide bond carbonyl oxygen. It is relevant here that the strongest two Zn(II)-binding sites are probably *not* identical with the strongest two Cu(II)-binding sites (see section IVE) since the proton displacement by Cu(II) and Zn(II) from their strongest sites differs[12] and high concentrations of Zn(II) do not block interaction of Cu(II) with its strongest site.

Other aspects of Zn(II)–serum albumin interaction also remain unclarified. Key among these are the disputed effects of carboxyl esterification on Zn(II)-binding[12,13] and the effects of temperature. Incubation of serum mercaptalbumin at 37°C leads to an increase in the number of Zn(II)-binding sites such that the number of sites exceeds the number of imidazoles[14]. The new sites have not been identified.

B. *Interaction of zinc and insulin*

The hormone insulin is isolated from biological materials in association with Zn(II) and can be crystallized as a zinc–insulin complex containing a minimum of two Zn(II) per unit cell of 36,000 daltons[15]. Zinc-free insulin is biologically active and, tentatively, it does not appear that reassociation with zinc *in vivo* is necessary for this activity.

The insulin monomer (molecular weight 6000) contains two dissimilar polypeptide chains, A and B, linked by disulfide bridges; there are two histidines per monomer, at positions B_5 and B_{10} respectively. In solution, Zn(II)-free insulin exists as a mixture of molecular weight species of which the dimer appears to be the basic unit[16]. Addition of 0.7 Zn(II) per dimer leads to a monodisperse hexamer[16] in agreement with the crystal structure. Although more than two Zn(II) can be bound per hexamer, the

234

first two Zn(II) are unique in that they represent the minimum necessary for crystallization and cannot be removed by dialysis[17].

Early titration studies by Tanford and Epstein[18] of insulin and Zn(II)–insulin showed that Zn(II)–insulin exhibited two types of titration behavior —a "direct" titration curve reversible between pH 3.5 and pH 8 and a metastable "reversed" titration curve formed after exposure to pH values below 2 or above 11, as shown in Fig. 1. The "direct" curve which represents titration of the crystalline material is the most relevant to present discussion, and was shown to have two obvious implications[18]. First, two

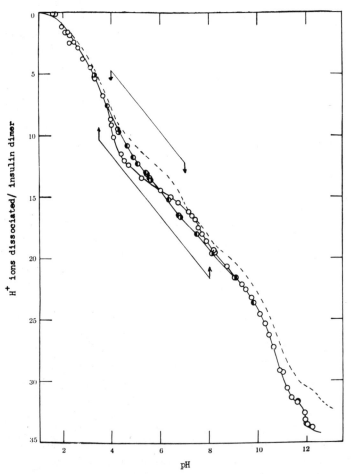

Fig. 1. Hydrogen ion titration curves of zinc insulin at 25°C, ionic strength 0.075; o, direct curve; ◑, acid-reversed; ◕, base-reversed. The dashed line is the curve for zinc-free insulin. Regions of partial insolubility in the direct and reversed curves are indicated, respectively, by the lower and upper arrows. (From Tanford and Epstein[18].)

additional protons per dimer are titrated in Zn(II)–insulin than in insulin between pH 8 and 12 and were assigned to dissociation of water in the coordination sphere of bound Zn(II). Secondly, two side-chains per dimer which titrate in insulin near pH 6.5 (identified as imidazoles) are deprotonated in the presence of Zn(II) at pH 4. This can be seen most directly by examination of the difference in H^+ bound between insulin and crystalline Zn(II)–insulin over the pH interval 4.5 to 7.5. At pH 4.5, binding of zinc leads to the release of two protons from insulin; the number of protons released diminishes to zero over the pH interval 5.5 to 7.5, corresponding to the expected pH region of imidazole titration in the absence of zinc. A zinc content of 0.7 Zn(II) per dimer thus permits the conclusion that each Zn(II) is coordinated with three imidazoles and three water molecules[19]. (It is relevant to note that the original assignment of two imidazoles per zinc was based on an erroneous zinc content of one zinc per dimer[18,19].)

Identification of the groups coordinated to zinc as imidazoles was questioned by Marcker and coworkers[20,21] who instead proposed a hexameric model in which each of two zincs coordinates with the α-NH_2 from the B-chain of each of three dimers. Their model was based both on available X-ray data which suggested that each unit cell contained three dimers related by a 3-fold rotation axis[21] and on chemical data which indicated that modification of the α-NH_2 of the B chains blocked molecular weight changes associated with Zn(II)-binding[20]. Brill and Venable[19], however, in a subsequent re-analysis of Tanford's direct titration data, pointed out that the α-NH_2 could not be a binding site for Zn(II) because, at pH 7.3 (a pH where the α-NH_2 groups are largely protonated in the absence of zinc) no fewer protons are bound by crystalline Zn(II)–insulin than by insulin itself (Fig. 1). In addition, they showed that the direct titration curve of Zn(II)–insulin was in good agreement with a model in which each of the two zincs per hexamer was bound to three imidazoles with a binding constant of the second zinc (10^{13}) which is greater than that of the first (10^9).

Recent X-ray data at 2.8 Å resolution[22] are in good agreement with the chemical data. These data show that each unit cell of Zn(II)–insulin contains three insulin dimers arranged as a hexamer around a three-fold symmetry axis and connected by two zinc ions. Each Zn(II) is coordinated with three B_{10} imidazoles (one from each dimer) and with three additional groups tentatively identified as water molecules; the arrangement of coordinating atoms does not yield a regular octahedron. The basis of the effect through which the α-NH_2 groups of the B chains were erroneously identified as ligands to zinc[20] is now also clear. Thus it appears that in the hexamer the dimers are in very close contact (almost "hooked" together) at the 1 position of the B chains; such contact would be hindered by modification of the α-NH_2.

References pp. 247–249

Finally, it is instructive to examine the derived Zn(II)–insulin binding constants in the light of the X-ray data. The cooperativity of binding is readily explained by the fact that the entropy decrease associated with bringing the randomly aggregated dimers of Zn(II)–free insulin into the hexameric configuration is associated with the binding of the first Zn(II)[19]; in effect the second Zn(II) binds to a pre-constrained tris-imidazole site. Although there is no suitable model for the second Zn(II)-site, the observed binding constant of 10^{13} can be reasonably related to a value of 10^{10} found for the binding of Zn(II) to a bis-imidazole ligand[23].

C. Cupric ion complexes of oxytocin and vasopressin

Oxytocin and vasopressin are structurally related polypeptide hormones (Fig. 2) which are found in the hypothalamus and posterior pituitary glands

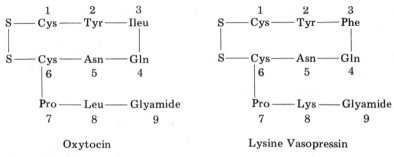

Fig. 2. The primary structures of oxytocin and lysine vasopressin.

of mammals and some lower organisms. While it may be debated whether the hormones are indeed large peptides or small proteins, in their reactivity towards Cu(II) they exhibit complexities analogous to those anticipated for proteins.

Both hormones bind a single Cu(II) with high affinity[24,25] yielding a pink complex at pH 8 with a single visible absorption band located at 525 nm. Titration of the Cu(II) complexes of both oxytocin (Fig. 3) and vasopressin[25] show that binding of cupric ion is accompanied by the release of four protons — one from the α-NH$_2$ (position 1) and three from additional groups normally protonated at pH 8. Participation of the α-NH$_2$ in binding has been confirmed by showing that no Cu(II) complex is formed by deamino-oxytocin in which the α-NH$_2$ is replaced by hydrogen[24] or by N-acetyl lysine vasopressin in which the α-NH$_2$ is acetylated[25]. Lack of participation of the phenolic hydroxyl in binding has been demonstrated by showing that replacement of tyrosine with phenylalanine is without effect on complex formation (Fig. 3). The three additional groups deprotonated by Cu(II) have been identified as peptide bond (or amide) nitrogen atoms because of the spectral similarity of the hormone complexes with that formed by Cu(II) and Gly-Gly-Gly-Gly in which Cu(II) is coordinated to the α-NH$_2$ and 3 adjacent peptide bond nitrogens[24]. For the vasopressin-Cu(II) complex a specific structure has been proposed involving coordination of Cu(II) with the α-NH$_2$ and three sequentially adjacent peptide bonds, in strict analogy with the Cu(II) complex of Gly-Gly-Gly-Gly[25].

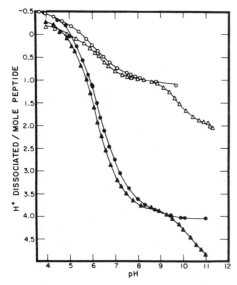

Fig. 3. Continuous H$^+$ titration curves of oxytocin and 2-phenylalanine oxytocin (in which the tyrosine in position 2 is replaced by phenylalanine) and their Cu(II) complexes. △-△, oxytocin; ▲-▲, oxytocin + CuCl$_2$,; o-o, 2-phenylalanine oxytocin; ●-●, 2-phenylalanine oxytocin + CuCl$_2$; 25°C; 0.16 ionic strength. (From Breslow[24].)

It is, however, unlikely that the Cu(II) complexes of the hormones are exactly analogous to those of simple linear peptides. Apparent pK values for dissociation of the third peptide bond proton in Cu(II)–vasopressin and Cu(II)–oxytocin are 6.8 and 7.4 respectively[24,25] (see Fig. 3), appreciably below the value of 8.1 observed with pentapeptides such as tetra-glycylglycine[9]. Moreover, the steepness of the titration curves of both Cu(II)–oxytocin (Fig. 3) and Cu(II)–vasopressin[25] near pH 6.5 indicates a cooperative involvement of the three peptide bond nitrogens, in contrast to the step-wise involvement seen in linear peptides[9]. A predominantly all or none development of the hormone complexes is in fact suggested by the spectral data which indicate that the exclusive species formed at all pH values above 6 is spectrally almost identical to the final complex[24,25]. The probable explanation of these data is that the ring structure of the hormones constrains one or more peptide bond (or

References pp. 247–249

amide) nitrogens in more favorable juxtaposition to the α-NH$_2$ than found in linear peptides. Such an effect of conformation means that the peptide bond nitrogen atoms involved together with the α-NH$_2$ are not necessarily sequentially adjacent to it. Thus, recent n.m.r. studies suggest that the Cu(II) complex of oxytocin is transannular, involving coordination of Cu(II) with the α-NH$_2$, and the peptide bond nitrogens of the cysteine in position 6, the tyrosine in position 2 and the asparagine in position 5[26].

D. Interaction of sperm whale metmyoglobin with cupric and zinc ions

Sperm whale metmyoglobin (metMb) was the first protein whose three-dimensional structure was resolved by X-ray crystallography[27], a feat which quickly led to its study as a model for the interpretation of protein–metal interactions. Several aspects of myoglobin chemistry are particularly relevant to its metal ion interactions.

The single polypeptide chain of metMb (Fig. 4) is a highly ordered structure of molecular weight 17,000 containing 75% α-helix which is readily and reversibly unfolded by H$^+$ ion near pH 4.5. The heme is bound through coordination of iron to a histidine (F$_8$), ionic interactions between the carboxyl side-chains of heme and positively charged groups on the protein, and non-polar interactions between the protoporphyrin ring of the heme and non-polar amino acid side-chains. Denaturation is accompanied by a change in heme spectrum and an exposure to solvent of 4–6 histidine side-chains which are unreactive to either H$^+$ ions or to carboxymethylation in the native state[28].

It was quickly apparent that metmyoglobin undergoes a reversible denaturation in the presence of Cu(II) and Zn(II), as evidenced by a change in heme spectrum[29], a marked insolubility of the protein–metal complex[29], the unmasking of buried imidazoles[29], and, finally, a decrease in α-helical content[30]. The driving force for the denaturation is clearly that the denatured form has a higher metal ion-affinity than the native structure[29]. However, denaturation is only manifest at levels of $\bar{v}_m$ above one[29] and the first metal ion appears to bind to a site on the native structure[29,31].

The binding sites of the first Cu(II) and Zn(II) to the native structure at pH 6 were determined crystallographically[31]. Interestingly, although binding of Cu(II) and Zn(II) are competitive[21], the first Cu(II) and Zn(II) bind to overlapping but non-identical sites (Fig. 4). Thus Zn(II) is clearly bound to the imidazole of His-GH$_1$ and in close proximity to (but not clearly bonded to) Lys-A$_{14}$ and Asn-GH$_4$. Cu(II), alternatively, is bound to His A$_{10}$ and is also adjacent to Lys-A$_{14}$ and Asn-GH$_4$, the differences between the two metals presumably arising from their different stereochemical requirements. Although the X-ray data are ambiguous with respect to the coordination of groups other than imidazole, the involvement of

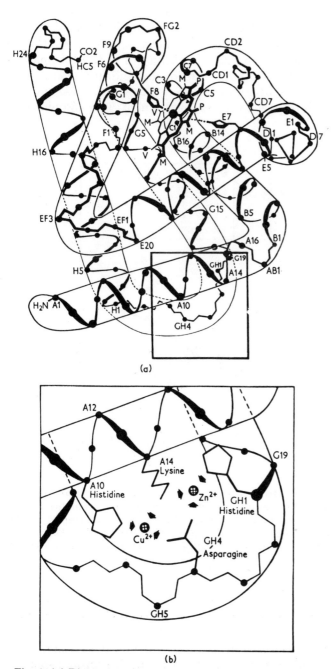

(a)

(b)

Fig. 4. (a) Diagrammatic representation of the myoglobin molecule after Dickerson[56]. (b) Enlarged sketch of the part of the myoglobin molecule enclosed on the rectangular area in (a) showing the relationship between Cu(II) and Zn(II) ions and potential binding groups. (From Banaszak, Watson and Kendrew[31].)

other electron donors (at least to cupric ion) is necessitated by the binding data[29]. At pH 6.8, these can be interpreted to indicate a value of K'_m for the first Cu(II) of 2.5×10^5, whereas the intrinsic affinity of Cu(II) for imidazole is only 10^4. It is unlikely that all the additional stabilization can arise from the carbonyl oxygen of asparagine, but titration data (see insert, Fig. 5) are ambiguous with respect to lysine involvement. This ambiguity is

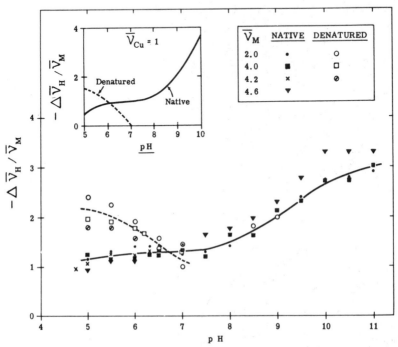

Fig. 5. The number of protons displaced ($\Delta \bar{v}_h$) per bound Cu(II) as a function of pH at different values of $\bar{v}_m$ (where $\bar{v}_m$ = moles cupric ion bound per mole protein). Data for native metMb represent net proton displacements observed upon addition of Cu(II) to the native protein. Data for denatured metMb represent observed proton displacements by Cu(II) from acid-denatured metMb. Conditions: 25°C, ionic strength 0.16 (From Breslow and Gurd[29].)

increased because binding sites for the first Cu(II) alter above pH 7 as evidenced by the increased displacement of protons by the first Cu(II) and also by e.p.r. data[32].

Titration studies have been particularly useful in elucidating the nature of the Cu(II)-binding sites in the denatured state[29]. Proton displacement by Cu(II) from acid-denatured metmyoglobin is shown in Fig. 5. The data indicate that for each of at least 4 bound Cu(II), two protons are displaced at pH 5 and only 1 proton at pH 7. Thus two groups which are normally protonated at pH 5 bind to each of the first 4 Cu(II); one of these titrates between pH 5 and 7 and can therefore be identified as an imidazole, a conclusion which is supported by the decrease in Cu(II)-binding attendant to histidine modification[33]. The identity of the second group deprotonated by Cu(II) at pH 5 is discussed below. It is relevant to note here that titration and binding studies indicate that most Cu(II) has the same binding sites on metmyoglobin both before and after acid denaturation. Thus the lesser proton displacement by Cu(II) from native metMb at pH 5 (Fig. 5) can be explained by the binding of each Cu(II) to two groups normally protonated at pH 5 together with the usual proton uptake accompanying unmasking by denaturation of basic imidazoles[29].

The second group from which each Cu(II) displaces a proton at pH 5-6 appears to be a peptide bond N although the apparent pK of this dissociation in metMb complexes is lower than seen in model complexes[34]. Lysines can be discounted as the proton source because guanidination has no effect on Cu(II)-binding[29]. Involvement of peptide bond nitrogens is supported in particular by the visible absorption spectra of Cu(II)-apomyoglobin complexes which indicate that, at pH 5.5, at least one strong-field ligand atom must be involved together with each imidazole[35]. In addition, both titration data (Fig. 5) and spectral data[35] indicate that the number of peptide bonds involved with each Cu(II) increases as the pH is raised above 8. It should also be noted, however, that at least at pH 11, one of the first 4 Cu(II)-binding sites appears to contain the α-NH$_2$ terminus[34].

The probable binding of Cu(II) to imidazoles and their peptide bond nitrogens at neutral pH suggests one mechanism for the Cu(II)-induced denaturation of metMb since it has been shown that the structural requirements of the α-helix are incompatible with those of similar Cu(II) complexes[36]. However, Zn(II), which reputedly has little tendency to coordinate with peptide bond N atoms also denatures metMb with a change in both heme spectra[37] and α-helix content[30]. In addition, studies of modified myoglobins suggest that changes in heme spectra and loss of α-helical content may not be inevitably associated[33]. In this respect it is relevant that binding of Zn(II) and Cu(II) ion to the heme-linked F$_8$ imidazole has been implicated as a critical step in myoglobin denaturation[33,37].

Other aspects of myoglobin–metal interactions are also unresolved. Among these is the relationship between the apparent maximum number of binding sites at neutral pH (approximately 7 for both copper and zinc) and the presence of 12 histidines in the protein. Thus, some imidazoles appear to be unavailable for reaction with metal ion despite the fact that the protein is denatured, or at least they are of very low reactivity compared to the others. In this instance, such a division of reactivity into two classes is a probable reflection of local amino acid sequence[38].

E. Cupric ion complexes of serum albumin

The optical properties of Cu(II) have allowed a more precise definition of Cu(II)–serum albumin interaction than obtained for Zn(II)–serum
albumin interaction, including an apparently complete picture of the
principal Cu(II)-binding site.

Following early studies of Cu(II)–serum albumin interaction[4,39], it
became clear that the first Cu(II) binds more tightly than do subsequent
Cu(II)[40]. A postulated involvement of the α-NH$_2$ at the strong Cu(II)-
binding site[40] was given support by the demonstration that one equivalent
Cu(II) blocked reaction of the α-NH$_2$ with 2,4-dinitrofluorobenzene[41] and
by subsequent titrimetric studies which indicated that one of the
coordinating groups at the strongest binding site had a pK_a near 8[35]. In
addition, the 525 nm absorption band of the 1:1 Cu(II)–serum albumin
complex and the displacement of two protons by the first Cu(II) at pH 9
indicated that two peptide bond N's and a possible fourth coordinating
atom participated along with the α-NH$_2$ at the strong site[35]. Localization
of the strong Cu(II)-binding site within the first 24 residues and demonstration of the sequence Asp-Thre-His-Lys at the N-terminus led to subsequent
identification of the fourth ligand group as an imidazole[42]. Thus it has
been suggested that, in analogy with the spectrally identical Cu(II) complex
of Gly-Gly-His, the strong Cu(II)-binding site of serum albumin contains
the imidazole of His-3, the α-NH$_2$, and the two intervening peptide bond
N's[42]. A model of the proposed strong Cu(II)-binding site in serum albumin
is shown in Fig. 6. That the assignment of coordinating groups is correct
seems assured by the demonstration that the first Cu(II) bound protects
both the α-NH$_2$ and His-3 from alkylation[43] and that the 1:1 Cu(II)
complex of serum albumin has the same optical and rotatory properties
as that of the peptide Asp-Thre-His-Lys[43]. There is no evidence to support
involvement of the epsilon-NH$_2$ of Lys-4 in the complex.

Less is known about the binding of subsequent Cu(II) ions to serum
albumin near neutral pH. Above pH 10 all bound Cu(II) show properties
consistent with multiple peptide bond N involvement[35]. However, titrimetric and spectral changes associated with binding of the second to
fourth Cu(II), while consistent with the involvement of imidazoles as
binding sites suggest that fewer peptide bond nitrogens are involved at
pH 8-9 than found in model histidine-containing peptides under comparable conditions[35]. The results suggest that either access to peptide
bond N sites is hindered in serum albumin or that there are secondary
chelate sites present which are intermediate in stability between the
strongest site and histidine–peptide bond N sites. The latter alternative
is supported for the second binding site by electrophoretic binding
studies[12].

One of the weaker Cu(II)-binding sites of particular interest is that observed by Klotz and coworkers[44,45] in the presence of four or more equivalents Cu(II). This site shows a strong absorption maximum at 375 nm which weakens in intensity with time. Its formation is blocked

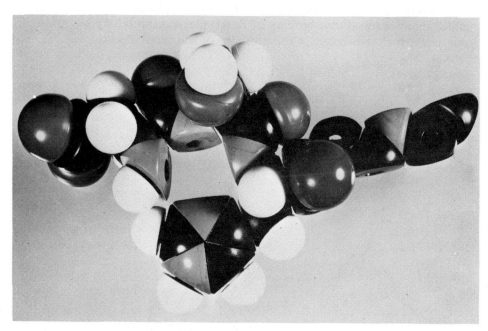

Fig. 6. Model of binding site for the first Cu(II) in bovine serum albumin: the terminal sequence, Asp-Thre-His- is shown in the conformation postulated for its Cu(II) complex, but with the metal ion deleted. (From Peters and Blumenstock[42].)

by reagents which react with sulfhydryls and its disappearance is facilitated by reagents which react with disulfide bonds. The postulated structure contains Cu(II) bound to a sulfhydryl in close proximity to a disulfide bond.

F. Ribonuclease–metal ion interactions

Pancreatic ribonuclease (RNase) catalyzes the hydrolysis of ribonucleic acid through cleavage of the 3′,5′-phosphodiester bonds which join adjacent nucleotides. The reaction is believed to proceed via formation of a 2′,3′-cyclic nucleoside phosphate intermediate which is then hydrolyzed to the 3′-phosphate. Both 2′- and 3′-pyrimidine nucleotides are potent inhibitors of the enzyme.

Structurally, RNase consists of a single polypeptide chain of 13,683 molecular weight with four disulfides which help to stabilize the native conformation. On the basis of both chemical and X-ray crystallographic

244

data, it has been shown that two of the four histidines, His-12 and -119, are spatially close and necessary for enzymatic activity. Lys-41 also appears to be catalytically important.

RNase activity is inhibited by both Cu(II) and Zn(II)[46]. Ross, Mathias and Rabin[47] showed that binding of Zn(II) alone is inhibitory but relatively weak. However, in the presence of the product cytidine

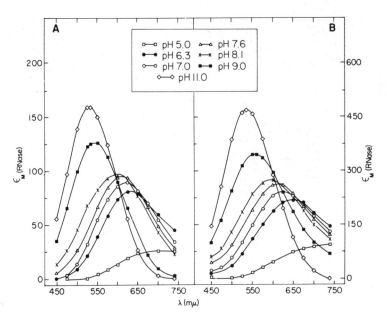

Fig. 7. Spectra of RNase in the presence of $CuCl_2$. Protein concentration = 3–4%; ionic strength 0.16. A, molar ratio of $CuCl_2$ to RNase = 1; B, molar ratio of $CuCl_2$ to RNase = 3. Results are reported as the extinction per mole of RNase. The oridinate in B is 3 times the ordinate in A. (From Breslow and Girotti[49].)

3'-phosphate (3'-CMP), inhibition is observed at very low Zn(II) concentrations and was attributed to the formation of a strong "ternary" complex between Zn(II), RNase and 3'-CMP. For Cu(II), it was initially suggested that inhibition might be due to chelation of a single Cu(II) between His-12 and His-119[48]. More recent studies, however, have shown that Cu(II) also forms a "ternary" complex with RNase and 3'-CMP[49] (see below) and that Cu(II)-binding in the absence of nucleotides is more complex than initially envisioned. From the results of several studies, the following conclusions about the interaction of Cu(II) with RNase *in the absence of nucleotides* appear probable.

(1) In the pH region 5–7, RNase has five sites of overlapping affinities for Cu(II)

A multiplicity of sites of similar affinities was first suggested on the basis of spectrophotometric and titration studies[49] which showed that, under conditions of essentially complete binding, the addition of each of four molar equivalents of Cu(II) to RNase produced essentially identical changes in visible absorption spectra (Fig. 7) and proton equilibria. Gel-filtration binding studies[3] now indicate five significant sites whose affinities for Cu(II) increase sharply between pH 5.5 and 7. At both pH 5.5 and 7, the affinity of one site for free cupric ion is approximately 10-fold greater than each of the other four[3,52]. Of the three Cu(II)-binding sites which have been observed by proton relaxation rate studies at pH 5.1, one site is 11-fold stronger than each of the other two[50].

(2) Three sites contain imidazole side-chains and one site contains the α-NH₂; peptide bond nitrogens are involved at all four of these sites, the extent of involvement increasing with pH.

The α-NH$_2$ is the strongest site at pH 7, but an active site imidazole is the strongest site at pH 5.5. There is no evidence to support chelation of a single Cu(II) between His-12 and His-119.

Participation of a single imidazole together with its adjacent peptide bond nitrogens at each of four sites was first suggested by the observation that both the changes in visible absorption spectra of Cu(II)–RNase complexes with pH and the titrimetric properties of Cu(II)–RNase complexes (Fig. 8) were analogous to those of the Cu(II)–complexes of acetyl-Gly-Gly-Gly-His[49]. The subsequent demonstration (see below) that the α-NH$_2$ also contributes to binding indicates that this interpretation of the titration data needs refinement. However, additional evidence for Cu(II)–imidazole interaction has been obtained from n.m.r. studies[51] which indicate that, at pH 5.5, Cu(II) broadens the C-2 proton resonance of His-12, -105 and -119. No evidence for participation of His-48 in Cu(II)-binding has yet been seen. The lack of simultaneous coordination of a single Cu(II) between His-12 and -119 has been argued from titration data[49] and is also evident from n.m.r. studies which indicate that His-12 is perturbed by lower Cu(II)-concentrations than His-119[51].

Participation of the α-NH$_2$ in Cu(II)-binding was demonstrated by showing that deamination of the α-NH$_2$ leads to loss of the strongest Cu(II)-binding site at pH 7 and to spectral changes in the Cu(II)–protein complex which suggest deletion of a site containing the α-NH$_2$ and peptide bond nitrogen atoms[52]. At pH 5.5, however, deamination leads to only minor changes in Cu(II)-binding[52], while interaction of RNase with 2'-CMP (which binds to the active site) appears to block binding at the strongest site[53]. Since carboxymethylation of His-12 also affects the strong Cu(II)-binding site at pH 5.1[50] the data suggest that His-12 may be the strongest Cu(II)-binding site at pH 5.5

Other aspects of Cu(II)-binding to RNase in the absence of nucleotides which are of interest include the ready formation of biuret-type

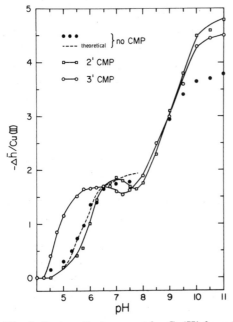

Fig. 8. Proton displacement by Cu(II) from RNase in the presence and absence of CMP. Conditions: 25°C, ionic strength 0.16, protein concentration = 1%. Experimental: •, protons displaced ($-\Delta\bar{h}$) from RNase by 1 eq. of Cu(II); ○, protons displaced from the 3'-CMP-RNase complex by 1 eq. of Cu(II); □, protons displaced from the 2'-CMP-RNase complex by 1 eq. of Cu(II). ---, theoretical curve calculated for the distribution of Cu(II) among four independent imidazole sites, each containing two peptide bond nitrogens below pH 8. (From Breslow and Girotti[49].)

complexes above pH 8 (as evidenced in part by the large proton displacement by Cu(II) at high pH in Fig. 8), interactions among the binding sites[49,50,52], and the binding by RNase of both free cupric ion and the cupric ion monoacetate complex[3]. In addition it has been shown that RNase undergoes extensive aggregation in the presence of Cu(II) near pH 7[54]. The conditions under which aggregation occurs suggest that it may result from the formation of intermolecular bridges between protein-bound –CuOH groups[54].

In the presence of 3'-CMP, binding of Cu(II) to RNase changes remarkably. This was first evidenced by a large shift to longer wavelengths of the absorbance of Cu(II)–RNase complexes in the presence of 3'-CMP[49], together with a change in titration behavior (Fig. 8) which indicated that 3'-CMP increased the strength of Cu(II)–RNase interaction[49]. Interestingly, the effect of 3'-CMP involves the facilitated interaction of Cu(II) at two cooperatively interacting binding sites as shown in particular by gel-filtration studies[53]. The nature of the binding sites on the 3'-CMP–RNase complex is not known. However, it is significant that the ternary complex of Zn(II) with RNase and 3'-CMP, whose stereochemistry may be expected to differ from that of the Cu(II) complex, also involves the facilitated binding of two metal ions[49,54]. It is also relevant that 2'-CMP does not form a similar ternary complex with Cu(II) and RNase (see Fig. 8), but instead competes with Cu(II)[49,53]. Differences between 2'-CMP and 3'-CMP in this respect suggest that the phosphate of the nucleotide may participate in binding to the bound metal ion.

V. CONCLUSIONS

It is evident from the above that identification of metal-binding sites is most straightforward when these sites all fall into the same class and/or when they are clearly analogous to sites in known model systems. New approaches are obviously needed to identify those sites which may be unique to a particular protein. It is also evident that even a full knowledge of the three-dimensional structure of a protein (consider myoglobin) does not enable us at present to predict in detail either the nature of the binding sites or the effect of binding on the properties of the protein. An increased predictability in this area undoubtedly awaits a more quantitative under-standing of the contributions of potential ligand atoms to protein conforma-tion.

REFERENCES

1 B. L. Vallee and R. J. P. Williams, *Proc. Natn. Acad. Sci., U.S.*, 59 (1968) 498.
2 F. R. N. Gurd and D. S. Goodman, *J. Am. Chem. Soc.*, 74 (1952) 670.
3 A. W. Girott and E. Breslow, *J. Biol. Chem.*, 243 (1968) 216.
4 I. M. Klotz and H. A. Fiess, *J. Phys. Coll. Chem.*, 55 (1951) 101.
5 J. T. Edsall and J. Wyman, *Biophysical Chemistry*, Vol. 1, Academic Press, New York, 1958.
6 F. R. N. Gurd and P. E. Wilcox, *Adv. Protein Chem.*, 11 (1956) 311.
7 K. Linderstrom-Lang, *Compt. rend. trav. lab. Carlsberg*, 15 (1924) No. 7.
8 H. C. Freeman, *Adv. Protein Chem.*, 22 (1967) 257.
9 C. R. Hartzell and F. R. N. Gurd, *J. Biol. Chem.*, 244 (1969) 147.
10 G. F. Bryce, R. W. Roeske and F. R. N. Gurd, *J. Biol. Chem.*, 240 (1965) 3837.
11 M. S. N. Rao and H. Lal, *J. Am. Chem. Soc.*, 80 (1958) 3222.
12 M. S. N. Rao and H. Lal, *J. Am. Chem. Soc.*, 80 (1958) 3226.

248

13 D. J. Perkins, *Biochem. J.*, 80 (1961) 668.
14 F. R. N. Gurd, *J. Phys. Chem.*, 54 (1954) 788.
15 J. Schlichtkrull, *Acta Chem. Scand.*, 10 (1956) 1455.
16 K. Marcker, *Acta Chem. Scand.*, 14 (1960) 194.
17 L. W. Cunningham, R. L. Fischer and C. S. Vestling, *J. Am. Chem. Soc.*, 77 (1955) 5703.
18 C. Tanford and J. Epstein, *J. Am. Chem. Soc.*, 76 (1954) 2170.
19 A. S. Brill and J. H. Venable, Jr., *J. Am. Chem. Soc.*, 89 (1967) 3622.
20 K. Marcker, *Acta Chem. Scand.*, 14 (1960) 2071.
21 K. Marcker and J. Grae, *Acta Chem. Scand.*, 16 (1962) 41.
22 M. J. Adams, T. L. Blundell, E. J. Dodson, G. G. Dodson, *et al.*, *Nature*, 224 (1969) 491.
23 C. W. C. Drey and J. S. Fruton, *Biochemistry*, 4 (1965) 1258.
24 E. Breslow, *Biochim. Biophys. Acta*, 53 (1961) 606.
25 B. J. Campbell, F. S. Chu and S. Hubbard, *Biochemistry*, 2 (1963) 764.
26 R. Walter, R. T. Havran, I. L. Schwartz and L. F. Johnson, in E. Scoffane, *Proc. 10th European Peptide Symposium*, North Holland Publishing Co., Amsterdam, (in press).
27 J. C. Kendrew, H. C. Watson, B. E. Strandberg, R. E. Dickerson, D. C. Phillips and V. C. Shore, *Nature*, 190 (1961) 666.
28 L. J. Banaszak, P. A. Andrews, J. W. Brugner, E. H. Eylar and F. R. N. Gurd, *J. Biol. Chem.*, 238 (1963) 3307.
29 E. Breslow and F. R. N. Gurd, *J. Biol. Chem.*, 238 (1963) 1332.
30 K. D. Hardman, *Ph.D. Thesis*, Indiana Univ., 1965.
31 L. Banaszak, H. C. Watson and J. C. Kendrew, *J. Mol. Biol.*, 12 (1965) 130.
32 F. R. N. Gurd, K. E. Falk, B. G. Malmstrom and T. Vanngard, *J. Biol. Chem.*, 242 (1967) 5724.
33 C. R. Hartzell, K. D. Hardman, J. M. Gillespie and F. R. N. Gurd, *J. Biol. Chem.*, 242 (1967) 47.
34 G. F. Bryce, R. W. Roeske and F. R. N. Gurd, *J. Biol. Chem.*, 241 (1966) 1072.
35 E. Breslow, *J. Biol. Chem.*, 239 (1964) 3252.
36 W. T. Shearer, R. K. Brown, G. F. Bryce and F. R. N. Gurd, *J. Biol. Chem.*, 241 (1966) 2665.
37 J. R. Cann, *Biochemistry*, 3 (1964) 714.
38 F. R. N. Gurd and G. F. Bryce, in J. Peisach, P. Aisen and W. E. Blumberg, *The Biochemistry of Copper*, Academic Press, New York, 1966.
39 I. M. Klotz and H. G. Curme, *J. Am. Chem. Soc.*, 70 (1948) 939.
40 I. M. Kolthoff and B. R. Willeford, Jr., *J. Am. Chem. Soc.*, 79 (1957) 2656.
41 T. Peters, Jr., *Biochim. Biophys. Acta*, 39 (1960) 546.
42 T. Peters and F. A. Blumenstock, *J. Biol. Chem.*, 242 (1967) 1574.
43 R. A. Bradshaw, W. T. Shearer and F. R. N. Gurd, *J. Biol. Chem.*, 243 (1968) 3817.
44 I. M. Klotz, J. M. Urquhart and H. A. Fiess, *J. Am. Chem. Soc.*, 74 (1952) 5537.
45 I. M. Klotz, J. M. Urquhart, T. A. Klotz and J. Ayers, *J. Am. Chem. Soc.*, 77 (1955) 1919.
46 F. F. Davis and F. W. Allen, *J. Biol. Chem.*, 217 (1955) 13.
47 C. A. Ross, A. P. Mathias and B. R. Rabin, *Biochem. J.*, 85 (1962) 145.
48 A. M. Crestfield, W. H. Stein and S. Moore, *J. Biol. Chem.*, 238 (1963) 2421.
49 E. Breslow and A. W. Girotti, *J. Biol. Chem.*, 241 (1966) 5651.
50 B. K. Joyce and M. Cohn, *J. Biol. Chem.*, 244 (1969) 811.
51 M. Ihnat and R. Bersohn, personal communication. Similar data were cited in ref. 50 as a communication from G. Roberts and O. Jardetzky.

52 A. W. Girotti and E. Breslow, *J. Biol. Chem.*, 245 (1970) 3066.
53 E. Breslow and A. W. Girotti, *J. Biol. Chem.*, 245 (1970) 1527.
54 E. Breslow and A. W. Girotti, unpublished observations.
55 A. E. Martell and L. G. Sillen, *Stability Constants*, Special Publication No. 17, The Chemical Society, London, 1964.
56 R. E. Dickerson, in H. Neurath, *The Proteins*, Vol. 2, Academic Press, New York, 1964.

PART III

METALLOPROTEINS IN STORAGE AND TRANSFER

Chapter 8

FERRITIN

PAULINE M. HARRISON AND T. G. HOY

Department of Biochemistry, University of Sheffield, Sheffield, Gt. Britain

THE FUNCTION OF FERRITIN AS AN IRON-STORAGE COMPOUND

Ferritin is a red-brown, water-soluble, iron-containing protein, which is very widely distributed in both animal and plant kingdoms. It has been found in vertebrates[1-3] and invertebrates[4-8], in flowering plants[9-11] and even in fungi[12,13]. Its widespread occurrence suggests an ancient evolutionary origin and almost universal functional importance in multicellular organisms. Ferritin was first isolated from horse spleen by Laufberger[14] in 1937. Its role as an iron-storage compound in animals was established by Granick[15,16] and it seems to have a similar role in plants.

In humans it is well established that most of the iron present in the body is continuously recycled, very little is absorbed from the diet and very little is excreted[17]. About 70% of body iron is present as haemoglobin compared with 20–25% as storage iron. Under normal conditions most of the storage iron occurs as ferritin, the remainder being in the form of iron-rich, insoluble granules known as haemosiderin[16,18]. When haemoglobin is broken down its iron may be stored temporarily as ferritin and then re-utilized for further haemoglobin production[17]. Some of the body iron may be stored as ferritin or haemosiderin for longer periods. It can be mobilized after blood loss, if the diet is low in iron, or to meet the demands of the growing foetus in pregnancy.

The role of ferritin[16] is to sequester Fe(III) iron in a form which is non-toxic, soluble and readily available. Fe(III) iron has a very low solubility at neutral pH and is toxic at quite low concentrations. Ferritin is stored mainly in the liver, spleen and bone-marrow. In animals fed an iron-rich diet, or injected with large doses of iron, ferritin is found in the gastro-intestinal mucosa[19]. It was originally thought that the presence of ferritin in the mucosa blocked the intake of iron from the gut[16,19]. Although it is no longer believed that there is an absolute "mucosal block" to iron absorption, the formation of ferritin in the mucosa may play a role, directly or indirectly, in the regulation of iron absorption[20].

What sort of a molecule would be expected if the functions of ferritin outlined above are to be fulfilled? The major part of this chapter will be devoted to describing what is known about the structure of ferritin in relation to its function.

For ferritin to be an efficient storage compound a high iron/protein ratio would be expected with protein surrounding the iron to make it soluble. If it is important to prevent accumulation of free iron, the protein might be expected to play an active part in iron uptake, normally to be present in the cell in a state which is unsaturated with respect to iron, and its synthesis to be iron-dependent. Iron uptake must be reversible, unless it is to be released only on turnover.

PREPARATION AND CHARACTERIZATION OF FERRITIN

Granick's[21] procedure, which depends on the relative stability of ferritin to heat, is still widely followed. An aqueous extract of minced tissue is heated to $80°C$ on a water bath. After cooling and removal of denatured proteins, ferritin is precipitated with ammonium sulphate and isolated by centrifugation. It can be purified by repeated crystallization from 5% $CdSO_4$. Dialysis at pH 4.6 prior to addition of $CdSO_4$ sometimes causes the precipitation of insoluble material. This and the brown mother liquor surrounding ferritin crystals is sometimes referred to as "non-crystallizable" ferritin. Ferritin can also be prepared without the heating step[22,23] and can be crystallized in the presence of a variety of divalent cations[14].

Horse spleen ferritin has a variable iron content, usually about 17–23% of its dry weight[1], with Fe/N about 2.0[24]. Ferritins from other organs and species may have less iron, 12–20% being common. This variability suggests heterogeneity and this is confirmed by its ultracentrifuge appearance[25] as shown in Fig. 1. In it can be seen a slow iron-free peak, about 18S, representing about 20–25% of the preparation and a broad, approximately 60S peak, representing species of varying iron contents. The major component of the iron-free protein obtained when iron is removed (see below) has a molecular weight of about 450,000. Isopycnic centrifugation separates ferritin into species whose buoyant densities vary continuously from 1.25 to 1.81[26]. Ferritin contains a maximum of about 4500 iron atoms per molecule, molecular weight about 900,000[26]. The ultra-violet and visible absorption spectra of ferritin are dominated by the iron[16]. Thus at 280 nm absorption due to the protein accounts for only about 7% of the total and for less than 1% at 310 nm.

Granick[27] showed that ferritin iron occurred in "micelles" of a hydrated iron oxide/phosphate complex of approximate composition $(FeOOH)_8(FeO:OPO_3H_2)$. The iron micelles can be largely freed from protein by treatment with strongly alkaline solutions[28], 5% sodium hypochlorite[29], hydrogen peroxide[30], sodium dodecyl sulphate (SDS)[31] or

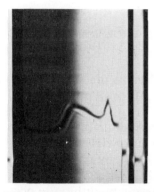

Fig. 1. Ultracentrifuge schlieren pattern of horse spleen ferritin in 0.1 M sodium acetate buffer, pH 4.4. Picture taken 20 min after reaching speed of 33,450 rev/min. Sedimentation right to left. Note the colourless iron-free peak (about 18S) and the broad, iron-containing peak (roughly 62S) containing ferritin molecules of varying iron contents.

acetic acid[32]. On liberation the micelles lose much of their phosphate[28] and aggregate to give a dark red-brown precipitate. They can be seen in the electron microscope as electron-dense particles 55–60 Å across[33]. Each "micelle" forms a central core inside a protein shell. This has been shown by electron microscopy of negatively stained preparations[34] (Fig. 2), by low-angle scattering from ferritin molecules in solution[26] and by X-ray diffraction of ferritin and apoferritin crystals[35,36] and of denatured ferritin[37]. The core diameter estimated by X-ray methods is 73 Å[26,36,37].

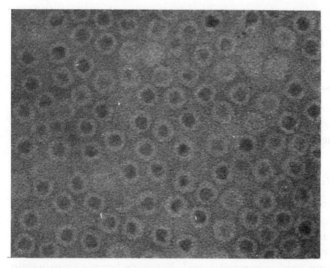

Fig. 2. Electron micrograph of ferritin negatively stained with sodium phosphotungstate, kindly supplied by G. H. Haggis. The ferritin iron cores are surrounded by protein shells, which appear light against a background of negative stain. Magnification x 500,000.

256

REMOVAL OF FERRITIN IRON : PREPARATION AND PROPERTIES OF THE IRON-FREE PROTEIN, APOFERRITIN

The procedures used to isolate the iron cores lead to the degradation of the protein. To obtain the protein moiety intact, the iron may be removed by the action of reducing agents.

Apoferritin can be conveniently prepared using sodium dithionite in sodium acetate buffer at pH 4.6–5.0[38]. The protein and Fe(II) iron can be separated by chelation of the latter with α,α'-bipyridyl[38], by precipitation of the protein with ammonium sulphate[39] or by gel filtration[22]. Reduction is slow and may take several hours. Very little reduction by dithionite occurs at pH 7.0. Like ferritin, apoferritin can be purified by repeated crystallization from 5% CdSO$_4$ solutions. Protein with a very low iron content can be separated from ferritin preparations by high speed centrifugation.

Estimates of the molecular weight of horse apoferritin range from 430,000 to 480,000[25,36,40-42]. X-ray measurements of crystals and solutions show that the molecule approximates to a smooth spherical shell with average inner and outer diameters 73 Å and 122 Å respectively[26,36,40]. The inner diameter exactly matches that derived from the iron core.

Evidence for the presence of subunits in apoferritin was obtained from the symmetry of X-ray differaction patterns[35,36] and confirmed by disaggregation to an approximately 2S subunit[32,43]. The subunit molecular weight has been estimated by several methods as about 23,000 corresponding to 20 subunits per molecule[42,44,45]. A very recent estimation based on migration in polyacrylamide gels containing SDS gives a value of only 18,000 ± 1100, or 24 subunits per 450,000 molecular weight[47]. The existing chemical evidence strongly suggests that the subunits are identical[2,48]. The amino acid composition of horse spleen apoferritin has been determined by several workers[22,44,48-50] and is shown in Table I.

The simple picture is complicated by heterogeneity in the protein as well as in the iron content. Minor components seen in the ultracentrifuge[22,23,25] and on gel electrophoresis[51-53] have been identified as monomer, dimer, trimer, etc., with the monomer accounting for 80–90% of the preparation[22,23,41,49]. The biological significance of the polymers is obscure. Heterogeneity in ferritin and apoferritin has been observed by isoelectric focusing[54,55]. This does not seem to be associated with iron or polymer content and might be due to variations in acetyl, amide, ionized carboxyl or other attached groups or ions. The average isoelectric point of horse ferritin is 4.4. Ferritins from different organs of the same animal have also been shown recently to have different electrophoretic mobilities[56-58] and amino acid compositions[59]. Ferritins from carcinogenic and non-carcinogenic tissues also have different electrophoretic mobilities[52,54].

TABLE I

AMINO ACID COMPOSITION OF APOFERRITIN FROM HORSE SPLEEN

Amino acid	Residues per 22,500 molecular weight	
	Ref. 48	Ref. 49
Cysteine	a	2.39
Aspartic acid	21.05	21.03
Threonine	6.67	6.70
Serine	10.91	11.55
Glutamic acid	29.01	29.19
Proline	3.43	2.51
Glycine	12.01	11.92
Alanine	16.99	17.26
Valine	8.43	8.59
Methionine	3.36	3.59
Isoleucine	4.25	4.56
Leucine	30.35	30.52
Tyrosine	6.07	6.33
Phenylalanine	8.92	8.87
Histidine	7.03	7.14
Lysine	10.62	10.90
Arginine	11.49	11.98
Tryptophan	1.06[b]	0.98[c]

[a] Not determined
[b] Ref. 45
[c] Ref. 44

QUATERNARY STRUCTURE OF APOFERRITIN

The structure of ferritin is formally analogous to that of a small spherical virus, both being multisubunit proteins with a non-protein "core", although ferritin is much smaller in size than any of the viruses and has fewer protein subunits. Nevertheless some resemblance might be expected in its basic architecture to that of the icosahedral viruses.

Two different crystalline forms have been obtained for horse ferritin (cubic and orthorhombic) and two for apoferritin (cubic and tetragonal)[35,36]. The cubic forms, growing as octahedra, are isomorphous. This enables the effect of iron on the diffraction patterns to be observed directly. The molecules are close-packed with intermolecular contact distance 130 Å. The space group, F432, would formally indicate that each molecule contains 24 n identical subunits but the arrangement in orthorhombic crystals suggests lower molecular symmetry. The diffraction patterns of both cubic and orthorhombic crystals indicate that the molecules may have pseudo-icosahedral symmetry. A quaternary structure proposed to account for this symmetry and for most of the chemical evidence consists of twenty protein subunits situated at the twenty faces of an icosahedron[36, 60]. This structure is shown in Fig. 3(a). If this arrangement is correct, the symmetry of the cubic crystals would have to arise statistically from several molecular orientations. A model proposed earlier[35], which has octahedral not icosahedral symmetry, is shown in Fig. 3(b). This consists of twenty four subunits situated at the apices of a snub cube. This model accounts for many features of the diffraction patterns of both cubic and tetragonal crystals

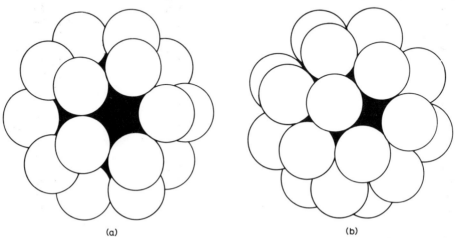

(a)   (b)

Fig. 3. Alternative models of the quaternary structure of apoferritin. (a) Twenty subunits at the apices of a pentagonal dodecahedron; (b) twenty four subunits at the apices of a snub cube. Subunits are represented as small spheres in contact. Note that both models have intersubunit spaces which would allow access of iron atoms to the centre of the molecule. The size of these spaces would depend on the true shapes of the subunits which are unlikely to be exactly spherical.

and of their Patterson vector maps. An exact solution to the quaternary structure of the protein depends on the success of a more detailed analysis of the crystal structure which is now being attempted. While a complete understanding of the protein's function depends ultimately on a knowledge of the fine details of its structure, much can be understood about the nature of the iron core, its relationship to the protein and the biosynthesis of the ferritin molecule without this knowledge. For the remaining discussion it is only necessary to know that the protein shell is a regular assembly of subunits covering the iron core and that this type of structure allows the passage of iron in and out of the molecule by way of intersubunit channels.

MORPHOLOGY AND ATOMIC STRUCTURE OF FERRITIN CORES

Results reported above indicate that apoferritin contains a central cavity about 73 Å in diameter, while in ferritin varying numbers of iron atoms are present up to about 4500. In molecules of maximum iron content the space inside the protein must be rather well filled. In low-iron molecules the iron component could either be distributed randomly throughout the cavity, or occur in isolated regions as depicted in Fig. 4.

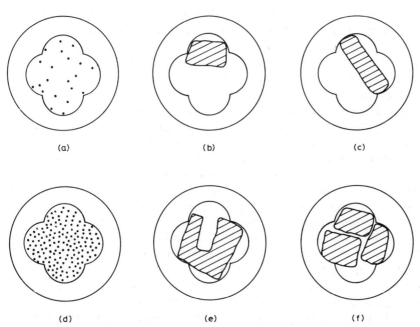

Fig. 4. Diagrammatic representation of ways in which the iron component may be distributed in molecules of different iron contents. (a) The protein is represented as an outer, nearly spherical shell. (a), (b) and (c) are molecules of low iron content, while (d), (e) and (f) are molecules of fairly high iron content. If the iron component is amorphous and distributed randomly throughout the space inside the protein, cores of different iron contents would have the same shape but different densities as in (a) and (d). If the iron occurs in localized regions as in (b), (c), (e) or (f), a variety of shapes might be found, depending on the number, arrangement and shapes of the various particles within the protein. (b) and (c) contain single particles, (e) contains a single particle but it has a cleavage line running part-way through it, while (f) contains three separate particles. Electron microscope and X-ray evidence support structures such as (b), (c), (e) and (f) and indicate that the hatched regions are small crystallites.

In Figs. 4(a) and (d) the material might be expected to be amorphous; in Figs. 4(b), (c), (e) and (f), crystalline. If the iron component is randomly distributed, then the cores, while differing in density, should always have a uniform appearance and the same shape irrespective of iron content. If it is crystalline, the core might sometimes be subdivided as in Fig. 4(f) and have

260

a variety of shapes as in Figs. 4(b), (c), (e) and (f). The nature of the core material has been investigated by several techniques.

(a) Electron microscopy

Early electron micrographs indicated that ferritin iron cores were divided into a number of discrete particles, "tetrads" being seen in some[33]. Several alternative symmetrical arrangements of four or six particles were proposed[61-64]. Huxley (quoted in ref. 2) and Haggis[65] found that cores viewed on very thin support films gave the "classical" subunit appearance only if grossly underfocused. Haggis[65] observed that the cores had a very variable appearance in different molecules, sometimes looking uniform in density and polyhedral, but with a variety of shapes (triangles, diamonds, squares, pentagons and hexagons), sometimes separated by cleavage lines into two or more dense regions. Occasionally a cleavage line might run partly across a core, forming a U-shape. Haggis concluded that "the iron is present in the form of small crystallites, possibly in some molecules a single small crystal, contained within an approximately spherical shell". This was borne out by further studies of ferritin fractions of high and low iron contents[66], as shown in Figs. 5(a) and (b). It can be seen that in low-iron molecules the cores appear much smaller than those of full ferritin. Haydon[67,68] has recently questioned Haggis's conclusions and attributes the appearance of crystallites in the latter's micrographs to artifacts due to background granularity. Haydon's failure to observe the types of image obtained by Haggis may be due to the use of too thick a supporting film[69]. Pictures obtained by the very recently developed technique of high-resolution scanning-transmission electron microscopy[70], such as that shown in Fig. 5(c), appear to support Haggis's view. Some cores, especially those of fairly low iron content, show evidence of substructure, others may be polyhedral.

(b) Low angle X-ray scattering

X-ray diffraction photographs of wet single crystals of unfractionated ferritin (dominated by molecules of high iron content) indicated that ferritin iron cores are not composed of a *regular* arrangement of four, six or eight discrete subunits[36]. On average the core appears to be a uniform polyhedron.

A more detailed investigation of ferritin cores was made by Fischbach and Anderegg[26] using low angle X-ray scattering. They examined a series of ferritin fractions, each having a narrow range of iron contents, in 53% sucrose solution. In these solutions scattering from ferritin protein is negligible, since it has the same electron density as the medium, while scattering

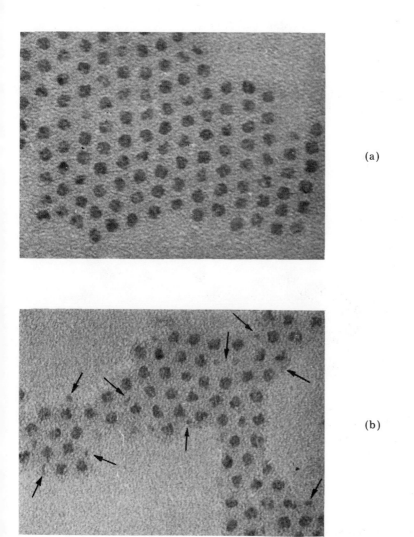

(a)

(b)

Fig. 5. Electron micrographs of unstained ferritin. (a) is a high density fraction (density > 1.82 g cm^{-3}) containing about 4000–4500 iron atoms per molecule ("full" ferritin). Note variations in iron core shape. (b) is a low density fraction (density 1.38 g cm^{-3}) containing about 450 iron atoms per molecule, mixed with the fraction shown in (a). Small particles can be seen (indicated by arrows) interspersed with large particles. These must represent the iron-cores of low-density fractions. (a) and (b) were taken with an AEI EM6B electron microscope at 60 kV and kingly supplied by G. H. Haggis. Magnification x 500,000.

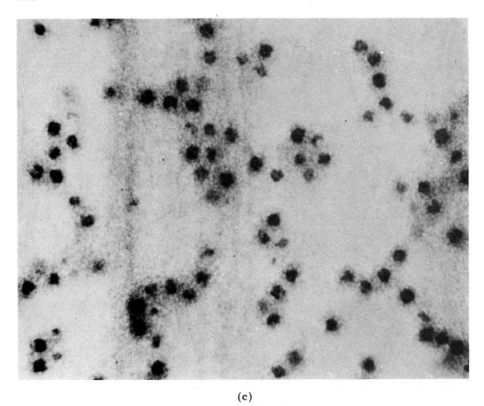

(c)

Fig. 5. (c) is a sample of unfractionated ferritin taken with a high resolution scanning electron microscope. Variations in iron-core shape, structure and density can be seen. Photograph by kind permission of A. V. Crewe and J. Wall x 480,000.

from the electron-dense cores is relatively large. Observed scattering was compared with curves calculated for several models. No collection of small, identical, roughly spherical particles gave scattering distributions which fitted the experimental results as well as a single particle. "Full" ferritin (density greater than 1.80) was approximated by a uniform sphere of diameter 73 Å and molecular weight 418,000. The average dimensions of the scattering objects became smaller the lower the iron content. Models which gave the best fit were single parallelipipeds or cylinders, one dimension of which always bridged the hole in apoferritin. The fraction of lowest density studied, which contained about 800 iron atoms, was approximated by a parallelipiped 17 x 34 x 68 Å. The theoretical curves were for single objects and did not allow for the possibility of variations in shape or size in core material in molecules of the same density. Nevertheless the models probably represent an average particle in the core. In the fraction containing roughly 450 iron atoms shown as an electron micrograph in Fig. 5(b), many of the cores appear rather isometric although some are elongated. The differences may represent actual differences in shape or different views of a similar shape. The greater particle asymmetry suggested by the low angle scattering results may be due to the larger number of iron atoms present in the fraction.

(c) High angle X-ray and electron diffraction studies on ferritin iron cores and iron-core analogues

Debye–Scherrer diagrams obtained with electrons or X-rays show clearly that ferritin iron cores are composed of crystalline material of small crystallite size[29,65,71]. Similar X-ray patterns are given by both wet and air-dried material[71] and by isolated cores and intact molecules[71,72] (although extra lines have been reported for unwashed cores released by alkali treatment[72]). The atomic structure of the cores seems to be independent of the presence of phosphate[71]. The latter is released by alkali treatment[28]. It is probably surface bound and may be involved in binding the core to the protein.

The iron core diffraction patterns are different from those of the well known ferric oxides or oxyhydroxides (α- or γ-Fe_2O_3, α-, β-, γ- or δ-FeOOH)[29,65,71]. Recent studies, however, of "ferric oxide-hydrates" prepared by hydrolysis of Fe(III) salts, which were previously reported to be amorphous, have now been shown to give Debye–Scherrer patterns, which are remarkably similar to those of ferritin[29,73,74]. Powder lines given by hydrolysates obtained by the addition of ammonia to iron(III) nitrate[73] and by heating an iron(III) nitrate solution to 85°C[29] are shown in Table II, together with those given by ferritin[71]. The patterns suggest these three materials have closely similar atomic structures.

The hydrolysates have the formula FeOOH · n H_2O with n equal to about 0.3 or 0.4[29,74]. The product obtained by ammonia addition measured 20–30 Å across in the electron microscope and was estimated to contain about 1000 iron atoms (molecular weight about 100,000)[74], while that produced by heating had a molecular weight 200,000–250,000[29]. An iron "polymer" obtained by addition of $KHCO_3$ to aqueous $Fe(NO_3)_3$ has also been studied by Spiro, Saltman and coworkers[75-78]. This polymer was estimated to have the composition $FeO_{0.75}OH(NO_3)_{0.5}(H_2O)_{0.36}$ and to contain about 1200 iron atoms. Its ultracentrifuge molecular weight was 150,000 and its diameter, measured from electron micrographs, 70 Å. These authors noted the similarity between this diameter and that of ferritin cores. The number of iron atoms present, however, is much smaller than that in well-filled ferritin molecules.

TABLE II

DIFFRACTION DATA FOR FERRITIN IRON CORES AND IRON CORE ANALOGUES

Spacings of observed lines[a] (Å)	Ferritin (electron diffraction)[65]	Ferritin (X-ray diffraction)[71]	Iron(III) oxide hydrate (X-rays)[73]	Iron(III) Nitrate hydrolysate (X-rays)[29]
4.70				Uncertain (U)
4.20				U (broad)
3.30				U (broad)
2.55 }	S	41 }	100	S } unresolved
2.47 }		73 }		M }
2.35				U
2.24	S	33	20	M-S
1.98	M	18	10	M (broad)
1.73	W	27	15	W
1.50	VW	40		M
			60	
1.47	S	100		S
1.40				U
1.34	VW			VW
1.25 }	W	}	7	VW
1.23 }		}		
1.18	W	} broad medium		U
1.12	VW	} broad weak	10	U
1.06	M			
0.96			2	
0.92	W			
0.88	VW		1	
0.85	M		3	

[a] Mean values

(d) Magnetic susceptibility and Mössbauer spectroscopy of ferritin and Fe(III) iron hydrolysates

High spin Fe(III) iron with five unpaired electrons is characterized by a magnetic moment of 5.9 μ_B. Free Fe^{3+} in strongly acid solution has a moment of 5.8 μ_B[79]. Magnetic susceptibility measurements made on ferritin gave a value of 3.8 μ_B for the magnetic moment[27,28,80]. Similar values have been reported for solutions of a polymeric iron(III) hydrolysate[77] and of a hydroxy bridged dimer[81] $Fe_2(OH)_2^{4+}$ (which may have contained polymer)[78]. To account for the low magnetic moment in ferritin, Michaelis, Coryell and Granick[27] suggested that the iron atoms might be in the rather unusual state having only three unpaired electrons with oxygen ligands in a square planar arrangement. Alternatively ferritin might contain a mixture of Fe^{3+} ions with five and one unpaired electron. Recently Blaise et al.[82,83] have found a marked field strength dependence of the magnetic susceptibility of ferritin and obtained a magnetic moment of 5.08 μ_B at high fields. They have also shown that the fine grain particles of ferritin cores behave as superparamagnets. Evidence of superparamagnetism in iron(III) oxide-hydrates has also been obtained[74,77].

A study of the Mössbauer spectra of ferritin has been carried out at several temperatures[77,82,84,85]. At 77K, or above, a two line spectrum is obtained, while at 4K six lines are observed. Between about 20 and 45K there is a gradual transition, with the six line spectrum gradually giving way to two lines as the temperature is raised. Such behaviour is indicative of superparamagnetic particles of variable particle size under 100 Å. The six line spectrum results from splitting of the energy levels due to a magnetic field at the iron nucleus. The hyperfine spectrum and the magnetic measurements indicate that the iron cores are antiferromagnetic[82,84]. The observed magnetic moment must be due to uncompensated spins resulting from statistical deviations in the number of sites occupied by the two magnetic sublattices in the very small particles[86]. Quadrupole splitting leading to the two line spectrum is indicative of an electrical field gradient caused by an unsymmetrical distribution of oxygen atoms around the iron nuclei.

If the iron nuclei are present in two sufficiently different sites their Mössbauer spectra may be broadened, unsymmetrical and sometimes resolvable into two sets of lines, as seen in γ-Fe_2O_3[87] and δ-$FeOOH$[88], which contain both octahedrally and tetrahedrally co-ordinated iron atoms. The spectra of ferritin iron cores look rather symmetrical and do not suggest the presence of more than one site unless they have very similar field strengths and isomer shifts. Values[82,84] for the quadrupole splitting 0.60 ± 0.10 to 0.74 ± 0.04 mm/sec, isomer shift relative to stainless steel 0.47 ± 0.05 to 0.50 ± 0.05 mm/sec and internal magnetic field 493 ± 10 kOe are similar to those observed in a iron(III) hydrolysate[74,77] and in iron(III) oxides and oxhydroxides[89,90]. The isomer shift values are rather closer to those for iron in octahedral sites, while the internal magnetic field is rather closer to values found for the iron in tetrahedral sites.

Mössbauer spectra of ferritin cores which have been largely freed from protein are shown in Fig. 6.

ATOMIC STRUCTURE OF FERRITIN CORES

The atomic structures of ferritin mineral and the iron(III) hydrolysates are probably the same or closely similar. The hydrated iron oxide exhibits an infra-red absorption band at 1620–1630 cm^{-1} attributed to water, which disappears almost completely on drying[74]. A strong band at 3400 cm^{-1} remains and is probably due to OH groups within the lattice. This is also suggested by nuclear magnetic resonance (n.m.r.) spectra[74].

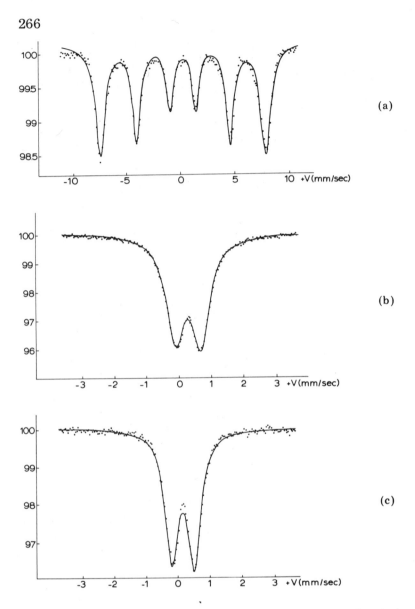

Fig. 6. Mössbauer absorption spectra of ferritin iron cores at various temperatures with computer calculated fits superimposed. (J. M. Williams and P. M. Harrison, to be published.) The iron cores were isolated by disaggregation in 67% acetic acid followed by digestion of the protein subunits by pepsin in 1 M acetic acid pH 2.4.
(a) Temperature 10K. Isomer shift (relative to Fe in Pd) = 0.30 ± 0.04 mm/sec. Internal magnetic field = 473 ± 10 KOe.
(b) Temperature 77K. Isomer shift = 0.29 ± 0.04 mm/sec. Quadrupole splitting (2ϵ) = 0.79 ± 0.04 mm/sec.
(c) Temperature 295K. Isomer shift = 0.18 ± 0.03 mm/sec. Quadrupole splitting = 0.71 ± 0.04 mm/sec.

A number of different unit cells and atomic arrangements have been proposed to account for the observed diffraction pattern[29,71-73]. The strong lines at spaces 2.54 and 1.47 Å, like those from δ-FeOOH, suggest that the structure is based on a system of close-packed oxygen atoms, 2.94 Å apart.

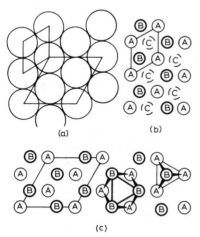

Fig. 7. (a) represents a single layer of close-packed oxygen atoms with the outlines of two alternative unit cells proposed for ferritin iron-cores drawn in. The small unit cell[71] has a side a = 2.94 Å, the inter-oxygen distance, while the larger unit cell[29] has a = 5.08 Å (2.94 x (3)$\frac{1}{2}$). A third suggested unit cell[72] has a = 11.79 Å (2.94 x 4). To maintain close-packing in three dimensions successive oxygen layers must be placed over spaces between oxygen atoms in the layers above or below. There are three possible positions for oxygen layers A, B and C, but several different sequences of layers are possible. Two of these suggested as alternative arrangements for ferritin cores are shown in (b) and (c). In (c) the unit cell proposed by Towe and Bradley[29], the stacking is hexagonal ABAB . . . with c = 9.40 Å corresponding to four oxygen layers. The large unit cell results from the positioning of the iron atoms. In (b) the arrangement suggested by Harrison et al.[71], the oxygen layers are arranged in sequence ABAC . . . giving a four layer sequence, c = 9.40 Å. Iron atoms can be placed in spaces between oxygen layers so that they are either octahedrally or tetrahedrally co-ordinated by oxygen atoms. Two such sites are indicated on the right of (C), the iron atoms being situated at the centres of the octa-hedron shown. Towe and Bradley's arrangement has all iron atoms in octahedral sites, while the structure of Harrison et al. both octahedral and tetrahedral sites can be occupied at random.

The actual unit cell adopted by the mineral depends both on the way suc-cessive layers of oxygen atoms are stacked and on the positions of the metal ions in the interstices. As shown in Fig. 7 oxygens can be placed in positions A, B or C. Stacking ABAB . . . gives hexagonal close-packing and ABCABC . . . cubic close-packing. More complex arrangements are possible, e.g. ABACABAC . . ., as in brookite, TiO_2[91]. For ferritin Harrison et al.[71] have proposed a simple hexagonal unit cell with A = 2.94 Å and c = 9.40 Å while

Girardet and Lawrence[72] suggest a larger one with a = 11.79 Å and c = 9.90 Å. Towe and Bradley[29] propose a related hexagonal cell for the iron(III) nitrate hydrolysate, with a = 5.08 Å and c = 9.40 Å while van der Giessen[73] has suggested his iron oxide hydrate is cubic, a = 8.37 Å. The atomic arrangement proposed by Harrison et al.[71] has oxygen layers ABACABAC . . . with iron atoms randomly distributed amongst all the available octahedral and tetrahedral sites. Towe and Bradley's[29] arrangement is ABAB . . . with all the iron atoms in octahedral sites, as in haematite, but with a lower site occupancy giving rise to a four-layer unit. For the cubic cell no arrangement of iron atoms could be found which accounted for the observed intensities[74]. No structure has been proposed for the very large hexagonal cell, although it was concluded that iron atoms must occupy both octahedral and tetrahedral sites[85].

A quite different type of atomic structure has been suggested by Brady et al.[77] based on the radial distribution function of an iron "polymer" obtained by addition of $KHCO_3$ to aqueous $Fe(NO_3)_3$. The iron atoms are all tetrahedrally coordinated to O^{2-} or OH^- and linked through them into a flat ribbon. The ribbon may be folded to form a spherical particle.

A survey of iron and oxygen arrangements is being carried out by T. G. Hoy with the aid of a computer.

RELATIONSHIP BETWEEN PROTEIN SHELL AND ITS IRON-OXIDE HYDRATE CORE

The external geometrical form of a macroscopic crystal depends on its atomic structure. Well developed faces usually have high atomic population densities although crystals growing under constraints may show different face developments from those growing freely. Macroscopic crystals are normally a mosaic of much smaller crystallites aligned nearly parallel. Conditions of crystallization may affect the size and shape of both the macroscopic crystal and the microscopic crystallites and the regularity of their arrangement in the mosaic.

If ferritin cores are crystalline, then their external form should be geometrically related to their atomic structure. If the core crystallites grow as regular polyhedra fitting exactly into the space inside the protein, then the atomic structure of the core would always lie in a specific orientation with respect to that of the protein. Such a situation would be expected if the cores formed independently of the protein and acted as templates for subunit aggregation. If the iron-oxide hydrate grew at specific sites inside a preformed protein shell, then the direction of crystal growth might be determined by the protein. Alternatively crystallization within the protein might occur randomly in many different orientations.

Electron microscope studies suggest that the core morphology is variable, different molecules having different numbers and arrangements of crystallites. Partly filled molecules sometimes show an empty space at the centre, suggesting that the crystallites are attached to the protein; well filled molecules tend to have a more uniform appearance[65,66]. X-ray diffraction at low angles from ferritin crystals also shows that the cores are variable[92]. Patterns from low density fractions are consistent with variations of shape, position and orientation of crystallites within the cores. In high

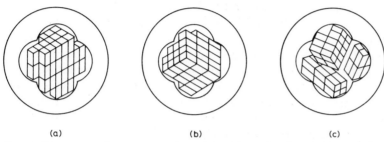

(a)   (b)   (c)

Fig. 8. Diagrammatic representation of ferritin molecules which contain nearly their full complement of iron, summarizing results of electron microscopy and X-ray diffraction. The core material is crystalline. Some molecules may contain a single crystallite as in (a) and (b), others may contain a few crystallites as in (c). The core crystallites grow in different orientations with respect to the protein in different molecules (a) and (b), or within the same molecule (c). Ferritin of high iron content contains all species (a), (b) and (c), but (a) and (b) may predominate.

density fractions a more regular shape is indicated. This probably represents the protein cavity largely or completed filled with iron oxide hydrate. Diffraction at higher angles shows both the single crystal pattern, owing to the regular lattice of protein molecules, and the Debye–Scherrer pattern of the cores. This implies that the crystal structure of the core material is not regularly arranged with respect to the crystal structure of the protein[92].

The size of core crystallites estimated from Debye–Scherrer line breadths for ferritin is 75 ± 30 Å[71] or 68 Å[85]. Since the size of the hole is 73 Å the core material is present either as a single crystallite, or as a small number of crystallites in some molecules[93]. Full ferritin thus appears to be a mixture of molecules in which the core has grown to fill the cavity starting from different sites and growing in different orientations with respect to the protein in different molecules, as depicted in Fig. 8.

EFFECT OF THE IRON COMPONENT ON THE PROPERTIES AND STRUCTURE OF THE PROTEIN

Ferritin is more resistant to attack by some proteolytic enzymes than apoferritin[24,48]. At pH 8.5 apoferritin is digested faster and to a greater extent by trypsin, chymotrypsin and subtilisin and at pH 7.0 and 5.0 by papain[94]. Pepsin hydrolyses apoferritin and ferritin at the same rate at pH 3.0, but digests apoferritin 2-3 times faster than ferritin at pH 2.5[94]. The effect of cathepsin D on ferritin and apoferritin is the same at both pH values[94]. Crichton[48] attributes the difference in tryptic digestion of ferritin and apoferritin to conformational differences in the protein. Alternatively, the iron may inhibit proteolysis by inhibiting the unfolding of the protein.

Optical rotatory dispersion studies on ferritin and apoferritin in aqueous solutions indicate a comparatively high helix content, 40-50% in ferritin, 50-60% in apoferritin[95]. The mean residue rotations at 233 nm for ferritins containing from 100 to 4000 iron atoms, separated from native ferritin by density gradient centrifugation, were all the same, but lower than that of apoferritin derived from ferritin by dithionite reduction[95]. The removal of the iron atoms apparently causes a configurational change in the molecules, unless some other effect can be attributed to the dithionite. It is perhaps rather surprising that an increased degree of helix content should be associated with greater ease of attack by proteolytic enzymes. Evidence for differences between ferritin and apoferritin either in surface conformation or ion binding is provided by DEAE cellulose chromatography. With this technique ferritin is found to be heterogeneous whereas apoferritin gives a single peak[22]. Dithionite treatment also tends to cause polymerization of ferritin monomers[22,49]. On the other hand, ferritin and apoferritin have the same electrophoretic mobilities and serological reactions[96].

X-ray evidence indicates that ferritins of different iron contents and dithionite reduced apoferritin have essentially the same protein conformations in the crystals[92]. At resolutions between about 2.5 and 10 Å the diffraction patterns depend very little on the presence of iron. It seems highly unlikely, at least in the crystalline proteins, that the helix contents of ferritin and apoferritin would differ by as much as 10%. Relatively small surface differences such as the dispositions or chemical modifications of some side chains could occur without detection.

BIOSYNTHESIS OF FERRITIN AND ITS RECONSTITUTION in vitro

Many years ago it was demonstrated that ferritin appeared in mucosal cells in response to iron feeding and that this involved the de novo synthesis of ferritin protein[19]. This has been confirmed for other types of tissue and tracer studies have shown that the first molecules to be formed contain little or no iron[31,97-106]. They gradually accumulate iron over a 72-hour period[103] as shown in Fig. 9.

The mechanism by which iron stimulates the synthesis of ferritin protein has not been elucidated. The failure of actinomycin D to inhibit synthesis found in most experiments indicates a post-transcriptional effect[103,105,107]. Iron might act by repressing a translational inhibitor[107], by promoting the assembly of protein subunits[103], or in some other way. It has been suggested that the increased net biosynthesis results from stabilization of iron-rich molecules to degradation by intracellular enzymes[103], but turn-over is too slow for this to be the sole cause[31,107], while *in vitro* cathepsin D attacks ferritin and apoferritin at the same rate[94].

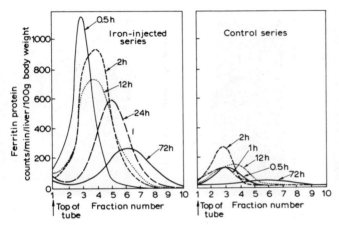

Fig. 9. Distribution of radioactive ferritin isolated from rat liver on a sucrose density gradient at various times after injection with 5 μC of leucine-^{14}C per 100 g of body weight. In the experiments illustrated on the left each rat received an injected dose of 400 μg of iron per 100 g body weight two hours before the labelled amino acid. In the control series, on the right, each rat received an injection of 0.9% NaCl instead of iron. Note that at the earliest time interval in both groups the maximum amount of label has been taken up by fractions of least iron content (lowest density) but at later times maximum labelling is found in ferritins of increasing iron content. A comparison of the two groups shows clearly the inductive effect of iron on the synthesis of ferritin protein. Reproduced by kind permission of J. W. Drysdale and H. N. Munro[103] from *J. Biol. Chem.*, 241 (1966) 3630.

The possibility that iron in the form of intracellular hydrous ferric oxide polymers, 70 Å in diameter, promotes the assembly of free subunits was suggested by Saltman and co-workers[108]. This mechanism can only be reconciled with *in vivo* formation of apoferritin if the latter disaggregates and reforms around the core[109]. The variations in size and shape of the ferritin iron cores make it rather unlikely that such irregular objects would control the assembly of the protein shell[109]. If this was the only mechanism by which the biosynthesis was stimulated it would also require a considerable prior accumulation of iron in the cell.

In vitro reassembly experiments with apoferritin subunits suggest that the presence of iron, whether polynuclear or mononuclear, is not obligatory for formation of completed shells[32]. Molecules disaggregated in 67% acetic acid and reassembled by gradually raising the pH in the presence of thiols, give molecules which are similar to native apoferritin[32] in electron microscopic appearance, sedimentation coefficient, electrophoretic mobility and immunological reactions and which give similar crystals from $CdSO_4$ solutions. Ferritin molecules disaggregated with acetic acid or SDS reassembled to give apoferritin not ferritin while the cores aggregated with each other[32,110].

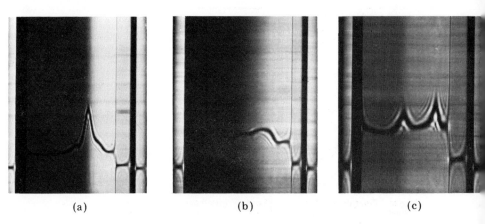

(a)   (b)   (c)

Fig. 10. Ultracentrifuge schlieren patterns of "ferritins" reconstituted from Fe^{2+} and apoferritin under oxidizing conditions. Pictures taken 20 min after reaching speed of 33,450 rev/min. Sedimentation from right to left. Solutions in 0.1 M NaCl. (a) Reconstituted in bicarbonate, pH 6.9–7.9 with O_2 as oxidizing agent. The major, iron-containing, component has S about 45. The small slow shoulder is apoferritin and the small fast component represents reconstituted molecular dimers. (b) Reconstituted in Tris buffer, pH 6.7–7.3 with O_2 as oxidizing agent. The product has a broad range of iron contents with the fast shoulder from 65S. (c) Reconstituted in unbuffered solution, pH 5.8–7.4, with KIO_3 as oxidizing agent. The colourless, iron-free 18S component is apoferritin. The iron-containing component has S about 59. The apoferritin used in (c) was isolated from ferritin by high speed centrifugation. A similar result has been obtained using dithionite reduced ferritin. The apoferritin used in (a) and (b) was obtained by dithionite reduction.

Completed apoferritin shells can accumulate iron *in vitro* to give a product resembling ferritin. Reconstitution was first achieved by Bielig and Bayer[111] from iron(II) ammonium sulphate and apoferritin in bicarbonate buffer at pH 7.6 in the presence of atmospheric oxygen. Loewus and Fineberg[112] confirmed this result and also showed that the uptake of an *iron(III)* salt by apoferritin could be mediated by a boiled extract of liver.

"Ferritin" reconstituted by Bielig and Bayer's method does not seem to accumulate its full complement of iron[71,108,111]. The product has a sedimentation coefficient averaging about 45S (Fig. 10(a)) and a diffraction pattern and crystallite size like those of low iron ferritin. With imidazole or Tris buffer at pH 6.7–7.3 instead of bicarbonate a broader range of iron contents is obtained with some molecules nearly full (Fig. 10(b)). A third reconstitution from apoferritin and iron(II) ammonium sulphate in unbuffered solution, pH 5.8–7.4, uses KIO_3 as the oxidizing agent in the presence

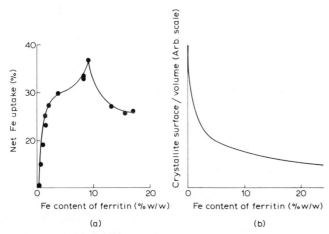

(a)   (b)

Fig. 11. (a) The net uptake of iron from $^{59}Fe_2$ transferrin by ferritins of various iron contents (obtained from ferritin by reduction with $Na_2S_2O_4$ for varying times). In these experiments 1 ml 0.2% part iron saturated ferritin was dialyzed against 10 ml 0.2% $^{59}Fe_2$-transferrin, 1.0 mM ascorbic acid, 1.0 mM AMP in 0.1 M Tris buffer, ph 7.4. Reproduced by kind permission of J. P. G. Miller and D. J. Perkins[113] from *Eur. J. Biochem.*, 10 (1969) 149. (b) Plot of surface/volume for ferritins of various iron contents, calculated on the assumption that all the iron is present in each ferritin molecule as a single cubic crystallite. Compare with (a). Note, however, that the total surface area of each crystallite may not be available for addition of iron atoms since the crystallite probably adheres to the protein. Available surface area may decrease when the protein cavity is more than half filled with iron component.

of $Na_2S_4O_6$ (procedure first used by Orme-Johnson, Gilchrist and Collins)[71]. The product is shown in Fig. 10(c). In this case reconstitution seems to be almost an "all-or-none" process. The fast iron-containing peak has S about 60. It gives diffraction patterns like native ferritin indicating similarity of structure and crystallite size despite the absence of phosphate[71]. In the absence of apoferritin different iron compounds are obtained[71]. Oxidation of Fe^{2+} by O_2 in bicarbonate gives α-FeOOH, while oxidation with KIO_3 in $Na_2S_4O_6$ solution gives γ-FeOOH. Clearly the protein influences the product. While in the first method some iron may be precipitated outside the protein, in the third, virtually all the iron finds its way into the protein shell. The three methods started with Fe^{2+} and gave a product resembling iron(III) oxide hydrate inside the protein. Attempts to reconstitute ferritin by hydrolyzing Fe^{3+} in the presence of apoferritin have so far been unsuccessful.

Serum iron attached to transferrin is known to be transferred to ferritin *in vivo*. Miller and Perkins[113] have studied this transfer *in vitro*. Ferritin, in a dialysis sac, was immersed in a solution of transferrin labelled with Fe^{59} in Tris buffer at pH 7.4. The effect of various mediators on the transfer iron to the ferritin compartment was tested. Both ATP and ascorbic acid were found to mediate transfer, confirming previous results[114,115]. ATP acts not as a source of metabolic energy but as a chelator[113,114]. It can be replaced by AMP, ADP, lactate, glucose or other chelating agents but strong chelating agents like EDTA or nitrilotriacetic acid remove iron from transferrin but do not transfer it to ferritin. In the presence of ATP increasing amounts of ascorbic acid increase transfer. Very little iron is transferred in the absence of a reducing agent. The transfer process involves atmospheric oxygen and is complete when the reducing agent is exhausted. The role of the chelator is presumably to remove iron from transferrin, since the reconstitution experiments show that it is not required for the uptake of iron by ferritin.

Miller and Perkins[113] showed that iron was taken up by ferritin against an iron concentration gradient. Using ferritins from which varying amounts of iron had been removed by reduction they found the dependence of iron uptake on iron content shown in Fig. 11(a). It can be seen that very little iron is taken up by ferritins of low iron content compared with those containing 3% of iron or more.

MECHANISM OF FERRITIN FORMATION FROM APOFERRITIN

The structural evidence and reconstitution experiments suggest that ferritin is probably produced from apoferritin and Fe^{2+} by a process involving oxidation of the iron followed by hydrolysis and crystallization. The iron enters the protein through intersubunit spaces. At what point it becomes oxidized is uncertain. Perhaps the protein plays a role in removing electrons from Fe^{2+}, possibly via iron atoms chelated to it. Mazur, Baez and Shorr[116,117] suggested that a small amount of Fe^{2+} is chelated to surface thiol groups. In *in vitro* experiments, however, prior reaction with sulphydryl reagents affected neither the uptake nor the release of iron. Hydroxyls have a higher affinity for Fe^{3+} than for Fe^{2+} at neutral pH. If the iron has entered the molecule as a iron(II) chelate the chelating agent may be displaced by hydroxyls on oxidation to Fe^{3+}. Protons are produced in the hydrolysis of Fe(III) ions. The protein may promote hydrolysis by combining with these protons. This might be important in the initial stages of nucleation.

In solutions of Fe(III) ions there is a concentration below which polynuclear complexes are unstable and the iron occurs only in mononuclear
form. For Fe(III) ion at pH 7 this "mononuclear barrier" occurs at concentrations less than $10^{-3.7} M$[118]. In the apoferritin molecule this concentration
is already exceeded by the presence of a single iron atom within its central
cavity. The addition of further Fe(III) ions would therefore be expected to
yield polynuclear complexes, which would soon become too large to escape
through the intersubunit spaces. Since the polynuclear complex is crystalline, its formation may be influenced by factors normally affecting
nucleation and crystal growth. The number of nuclei, the rate of growth
and the number and size of crystals ultimately formed are affected by the
degree of supersaturation, temperature and the presence of nucleus-forming
centres. In ferritin increase in pH and iron concentration might cause multi-
plication of nuclei and rapid deposition of oxide-hydrate, perhaps resulting
in a less well-filled core. Differences in pH may be one of the factors
accounting for the differences in reconstitution illustrated in Figs. 10(a),
(b) and (c). A decrease in temperature from 25° to 0°C led to a poorer
reconstitution by Bielig and Bayer's method. A crystal nucleus has a large
surface/volume ratio and large surface energy. Unless it can grow above a
critical size it will tend to dissolve. Once above this size it is energetically
more favourable to add more atoms to this nucleus than to start the formation of another nucleus. Such considerations may explain the results
shown in Figs. 11(a) and 10(c). Under the experimental conditions of
Fig. 11(a) a critical stage is reached at about 2–3% iron (200–300 Fe per
molecule). A comparison of Fig. 11(a) with a plot of surface/volume ratio
against iron content, calculated on the assumption that all the iron is
present as a single cubic crystallite (Fig. 11(b)) suggests the crystallite surface energy may be an important factor in iron binding. Fig. 11(a) shows
that molecules containing above about 10% iron take up a decreased
amount of iron. This may be because the iron component is enclosed in
protein and once the protein is half full the surface available for the
addition of new ions decreases. Fig. 10(c) also suggests that, under the
conditions of this reconstitution, addition of iron atoms to partly filled
molecules is more favourable than the formation of new nuclei inside new
apoferritin molecules. Theories of crystal growth show that an atom added
to an existing layer will generally be held more strongly than one forming a
new layer. Lateral crystal growth is suggested by the low angle scattering
results of Fischbach and Anderegg[26] on low iron ferritin. Iron which is
deposited slowly in ferritin may lead to small numbers of crystallites, perhaps only one in many of the molecules of full ferritin[93]. Rapid accumulation may lead to multiple nucleation, perhaps giving less well-filled molecules. Iron might be more easily removed from such molecules. It might
be biologically advantageous for ferritin to consist of a mixture of molecules
from some of which iron can be more rapidly mobilized than from others.

MOBILIZATION OF FERRITIN IRON

The mechanism by which ferritin iron is mobilized *in vivo* is still controversial. It may be released from the intact molecules, from the cores after degradation of the protein, or both. It may be released by reduction to the more soluble ferrous ions, or by formation of soluble ferric chelates, or both.

It has been suggested that haemosiderin consists mainly of ferritin cores stripped of protein by proteolysis[30]. Haemosiderin gives a similar diffraction pattern to that of ferritin cores, although its crystallite size is rather smaller on average[42]. Tracer studies indicate, however, that although some of its iron may come from ferritin, in conditions of iron overload it is formed independently[119].

The ability of the reducing agents, cysteine, ascorbic acid and glutathione to release ferritin iron at pH 7.4 suggests the possibility of a reductive mechanism *in vivo*[116,120]. Experiments by Green and Mazur[121] show that ferritin iron may act as an electron acceptor for reduced xanthine oxidase and that this may bring about the release of ferritin iron to the plasma under conditions of low oxygen tension.

An alternative mechanism involving chelators has been proposed although the biological chelator has not been identified[78,122]. Chelating agents which have a high affinity for iron slowly liberate it from ferritin. Chelators which have been used successfully are desferrioxamine[123], 1,10-phenanthroline[124] and nitriloacetic acid[125]. EDTA removed little iron under comparable conditions[125]. Iron release by chelators is a slow process requiring several weeks. A chelator with a very high binding constant is required to compete with hydroxyls. It seems rather doubtful whether this could be the sole mechanism for iron release *in vivo*, although possibly reduction and chelation are combined. Further work is required to settle this problem.

NOTE ADDED IN PROOF

Further evidence supporting a twenty four subunit structure for ferritin protein has come from subunit molecular weight determination by sedimentation equilibrium and by gel filtration in 6 M guanidine hydrochloride[126,127].

The partial purification from mammalian liver of a ferrireductase enzyme capable of catalysing the reduction of bound ferritin Fe(III) to free Fe(II) has recently been announced[128]. The properties of this enzyme suggest that it is not xanthine oxidase.

REFERENCES

1 L. Michaelis, *Adv. Protein Chem.*, 3 (1947) 53.
2 P. M. Harrison, *Iron Metabolism: An International Symposium*, Springer-Verlag, Berlin, 1964, p. 40.
3 T. Kato, *Seikagaku*, 41 (1969) 61.
4 J. Roche, M. Bessis and J. Breton-Gorius, *Compt. rend. Acad. Sci.*, 252 (1961) 3886.
5 K. M. Towe and H. A. Lowenstam, *J. Ultrastruct. Res.*, 17 (1967) 1.
6 K. M. Towe, H. A. Lowenstam and M. H. Nesson, *Science*, 142 (1963) 63.
7 I. F. Heneine, G. Gazzinelli and W. L. Tafuri, *Comp. Biochem. Physiol.*, 28 (1969) 391.
8 A. Baba, *J. Biochem.*, 65 (1969) 915.
9 B. B. Hyde, A. J. Hodge, A. Kahn and M. L. Birnstiel, *J. Ultrastruct. Res.*, 9 (1963) 248.
10 A. W. Robards and P. G. Humpherson, *Planta*, 76 (1967) 169.
11 J. Seckbach, *J. Ultrastruct. Res.*, 22 (1968) 413.
12 A. Peat and G. H. Banbury, *Planta*, 79 (1968) 268.
13 C. N. David, *Dissert. Abstr.*, 29 (1969) 2733.
14 V. Laufberger, *Bull. Soc. Chim. Biol.*, 19 (1937) 1575.
15 P. Hahn, S. Granick, W. Bale and L. Michaelis, *J. Biol. Chem.*, 150 (1943) 407.
16 S. Granick, *Chem. Rev.*, 38 (1946) 379.
17 T. H. Bothwell and C. A. Finch, *Iron Metabolism*, Little, Brown & Company, Boston, 1962.
18 S. F. Cook, *J. Biol. Chem.*, 82 (1929) 595.
19 S. Granick, *J. Biol. Chem.*, 164 (1946) 737.
20 M. E. Conrad and W. H. Crosby, *Blood*, 22 (1963) 406.
21 S. Granick, *J. Biol. Chem.*, 146 (1942) 451.
22 A. A. Suran and H. Tarver, *Arch. Biochem. Biophys.*, 111 (1965) 399.
23 P. M. Harrison and D. W. Gregory, *J. Mol. Biol.*, 14 (1965) 626.
24 A. Mazur, I. Litt and E. Shorr, *J. Biol. Chem.*, 187 (1950) 473.
25 A. Rothen, *J. Biol. Chem.*, 152 (1944) 679.
26 F. A. Fischbach and J. W. Anderegg, *J. Mol. Biol.*, 14 (1965) 458.
27 L. Michaelis, C. D. Coryell and S. Granick, *J. Biol. Chem.*, 148 (1943) 463.
28 S. Granick and P. F. Hahn, *J. Biol. Chem.*, 155 (1944) 661.
29 K. M. Towe and W. F. Bradley, *J. Colloid Interface Sci.*, 24 (1967) 384.
30 G. T. Matioli and R. F. Baker, *J. Ultrastruct. Res.*, 8 (1963) 477.
31 J. W. Drysdale, in *Regulatory Mechanisms for Protein Synthesis in Mammalian Cells*, Academic Press, New York, 1968, p. 431.
32 P. M. Harrison and D. W. Gregory, *Nature*, 220 (1968) 578.
33 J. L. Farrant, *Biochim. Biophys. Acta*, 13 (1954) 569.
34 G. W. Richter, *J. Biophys. Biochem. Cytol.*, 6 (1959) 531.
35 P. M. Harrison, *J. Mol. Biol.*, 1 (1959) 69.
36 P. M. Harrison, *J. Mol. Biol.*, 6 (1963) 404.
37 V. Kleinwachter, *Arch. Biochem.*, 105 (1964) 352.
38 S. Granick and L. Michaelis, *J. Biol. Chem.*, 147 (1943) 91.
39 M. Behrens and M. Taubert, *Hoppe-Seyler's Z. Physiol. Chem.*, 290 (1952) 156.
40 H-J. Bielig, O. Kratky, G. Rohns and H. Wawra, *Biochim. Biophys. Acta*, 112 (1966) 110.
41 G. W. Richter and G. F. Walker, *Biochemistry*, 6 (1967) 2871.

278

42 P. M. Harrison, in *Iron Metabolism and Anemia*, Pan American Health Organization, Scientific Publications No. 184, 1969, p. 2.
43 T. Hofmann and P. M. Harrison, *J. Mol. Biol.*, 6 (1963) 256.
44 P. M. Harrison, T. Hofmann and W. I. P. Mainwaring, *J. Mol. Biol.*, 4 (1962) 251.
45 J. R. Spies, *Anal. Chem.*, 37 (1967) 1412.
46 P. M. Harrison and T. Hofmann, *J. Mol. Biol.*, 4 (1962) 239.
47 R. R. Crichton and C. F. A. Bryce, *Febs. Letters*, 6 (1970) 121.
48 R. R. Crichton, *Biochim. Biophys. Acta*, 194 (1969) 34.
49 M. A. Williams and P. M. Harrison, *Biochem. J.*, 110 (1968) 265.
50 S. Shinjyo, *Seikagaku*, 39 (1967) 23.
51 R. Kopp, A. Vogt and G. Maass, *Nature*, 198 (1963) 892.
52 G. W. Richter, *Lab. Invest.*, 12 (1963) 1026.
53 J. J. Theron, A. O. Hawtrey and V. Schirren, *Clin. Chim. Acta*, 8 (1963) 165.
54 Y. Makino and K. Konno, *J. Biochem.*, 65 (1969) 471.
55 P. G. Righetti, J. W. Drysdale and R. A. Malt, *Federation Proc.*, 29 (1970) 813 Abs.
56 C. P. Alfrey, E. C. Lynch and C. C. Whitley, *J. Lab. Clin. Med.*, 70 (1967) 419.
57 T. G. Gabuzda and F. H. Gardner, *Blood*, 29 (1967) 770.
58 T. G. Gabuzda and J. Pearson, *Biochim. Biophys. Acta*, 194 (1969) 50.
59 R. R. Crichton, J. A. Millar and R. L. C. Cumming, *Biochem. J.*, 117 (1970) 35P.
60 T. G. Hoy and P. M. Harrison, *Acta Cryst. Congress. Suppl.*, (1969) S261 Abstr.
61 G. W. Richter, *J. Exp. Med.*, 109 (1959) 197.
62 M. Bessis and J. Breton-Gorius, *Compt. rend. Acad. Sci.*, 250 (1960) 1360.
63 A. R. Muir, *Quart. J. Exp. Physiol.*, 45 (1960) 192.
64 E. F. J. van Bruggen, E. H. Wiebenga and M. Gruber, *J. Mol. Biol.*, 2 (1960) 81.
65 G. H. Haggis, *J. Mol. Biol.*, 14 (1965) 598.
66 G. H. Haggis, *Sixth International Congress for Electron Microscopy*, Maruzen Co., Ltd., Kyoto, 1966, p. 127.
67 G. B. Haydon, *J. Microsc.*, 89 (1969) 251.
68 G. B. Haydon, *J. Microsc.*, 91 (1970) 65.
69 G. H. Haggis, *J. Microsc.*, (1970) in press.
70 A. V. Crewe and J. Wall, *J. Mol. Biol.*, 48 (1970) 375.
71 P. M. Harrison, F. A. Fischbach, T. G. Hoy and G. H. Haggis, *Nature*, 216 (1967) 1188.
72 J-L. Girardet and J. J. Lawrence, *Bull. Soc. Fr. Mineral. Cristallog.*, 91 (1968) 440.
73 A. A. van der Giessen, *J. Inorg. Nucl. Chem.*, 28 (1966) 2155.
74 A. A. van der Giessen, *Thesis*, Technical University, Eindhoven, 1968.
75 T. G. Spiro, S. E. Allerton, J. Renner, A. Terzis, R. Bills and P. Saltman, *J. Am. Chem. Soc.*, 88 (1966) 2721.
76 T. G. Spiro, L. Pape and P. Saltman, *J. Am. Chem. Soc.*, 89 (1967) 5555.
77 G. W. Brady, C. R. Kurkjian, E. F. X. Lyden, M. B. Robin, P. Saltman, T. Spiro and A. Terzis, *Biochemistry*, 7 (1968) 2185.
78 T. G. Spiro and P. Saltman, *Structure and Bonding*, 6 (1969) 116.
79 L. N. Mulay and P. W. Selwood, *J. Am. Chem. Soc.*, 77 (1955) 2693.
80 G. Schoffa, *Z. Naturforsch.* 20b (1965) 167.
81 H. Schugar, C. Walling, R. B. Jones and H. B. Gray, *J. Am. Chem. Soc.*, 89 (1967) 3712.
82 A. Blaise, J. Chappert and J. L. Girardet, *Compt. rend. Acad. Sci.*, 261 (1965) 2310.
83 A. Blaise, J. Feron, J. L. Girardet and J. J. Lawrence, *Compt. rend. Acad. Sci.*, 265B (1967) 1077.
84 J. F. Boas and B. Window, *Aust. J. Phys.*, 19 (1966) 573.

85 J-L. Girardet, *Thesis*, University of Grenoble, (1969).

86 L. Neel, *J. Phys. Soc. Japan*, 17 (1962) 676.

87 R. J. Armstrong, A. H. Merrish and G. A. Sawatzky, *Phys. Lett.*, 23 (1966) 414.

88 I. Dezsi, L. Keszthelyi, D. Kulgawizuk, B. Molnar and N. A. Eissa, *Phys. Stat. Sol.*, 22 (1967) 617.

89 W. Kundig, H. Bommel, G. Constabaris and R. H. Lindquist, *Phys. Rev.*, 142 (1965) 327.

90 M. J. Rossiter and A. E. M. Hodgson, *J. Inorg. Nucl. Chem.*, 27 (1965) 63.

91 L. Pauling and J. H. Sturdivant, *Z. Krist.*, 68 (1928) 239.

92 F. A. Fischbach, P. M. Harrison and T. G. Hoy, *J. Mol. Biol.*, 39 (1969) 235.

93 P. M. Harrison and T. G. Hoy, *J. Microsc.*, 91 (1970) 61.

94 R. R. Crichton, *Biochim. Biophys. Acta*, 229 (1971) 75.

95 I. Listowsky, J. J. Betheil and S. England, *Biochemistry*, 6 (1967) 1341.

96 A. Mazur and E. Shorr, *J. Biol. Chem.*, 182 (1950) 607.

97 R. A. Fineberg and D. M. Greenberg, *J. Biol. Chem.*, 214 (1955) 97.

98 R. A. Fineberg and D. M. Greenberg, *J. Biol. Chem.*, 214 (1955) 107.

99 R. B. Loftfield and E. A. Eigner, *J. Biol. Chem.*, 231 (1958) 925.

100 G. W. Richter, *Nature*, 190 (1961) 413.

101 G. T. Matioli and G. H. Eylar, *Proc. Natn. Acad. Sci.*, 52 (1964) 508.

102 R. Saddi and A. von der Decken, *Biochim. Biophys. Acta*, 90 (1964) 196.

103 J. W. Drysdale and H. N. Munro, *J. Biol. Chem.*, 241 (1966) 3630.

104 A. O. A. Miller, *Biochim. Biophys. Acta*, 155 (1968) 262.

105 J. A. Smith, J. W. Drysdale, A. Goldberg and H. N. Munro, *Br. J. Haemat.*, 14 (1968) 79.

106 Y. Yoshino, J. Manis and D. Schachter, *J. Biol. Chem.*, 243 (1968) 2911.

107 L. L. H. Chu and R. A. Fineberg, *J. Biol. Chem.*, 244 (1969) 3847.

108 L. Pape, J. S. Multani, C. Stitt and P. Saltman, *Biochemistry*, 7 (1968) 606.

109 J. W. Drysdale, G. H. Haggis and P. M. Harrison, *Nature*, 219 (1968) 1045.

110 H. Smith-Johannsen and J. W. Drysdale, *Biochim. Biophys. Acta*, 194 (1969) 43.

111 H-J. Bielig and E. Bayer, *Naturwiss.*, 42 (1955) 125.

112 M. W. Loewus and R. A. Fineberg, *Biochim. Biophys. Acta*, 26 (1957) 441.

113 J. P. G. Miller and D. J. Perkins, *Eur. J. Biochem.*, 10 (1969) 146.

114 A. Mazur, S. Green and A. Carleton, *J. Biol. Chem.*, 235 (1960) 595.

115 A. Mazur, A. Carleton and A. Carlsen, *J. Biol. Chem.*, 236 (1961) 1109.

116 A. Mazur, S. Baez and E. Shorr, *J. Biol. Chem.*, 213 (1955) 147.

117 A. Mazur, in *Essays in Biochemistry*, John Wiley & Sons, New York, 1956, p. 198.

118 J. Schubert in *Iron Metabolism: An International Symposium*, Springer-Verlag, Berlin, 1964, p. 466.

119 P. Sturgeon and A. Shoden, in *Iron Metabolism: An International Symposium*, Springer-Verlag, Berlin, 1964, p. 121.

120 H-J. Bielig and E. Bayer, *Naturwiss.*, 42 (1955) 466.

121 S. Green and A. Mazur, *J. Biol. Chem.*, 227 (1957) 653.

122 P. Saltman, *J. Chem. Educ.*, 42 (1965) 682.

123 F. Wöhler, *Acta Haemat.*, 30 (1963) 65.

124 M. M. Jones and D. O. Johnson, *Nature*, 216 (1967) 509.

125 L. Pape, J. S. Multani, C. Stitt and P. Saltman, *Biochemistry*, 7 (1968) 613.

126 C. F. A. Bryce and R. R. Crichton, *J. Biol. Chem.*, 246 (1971) 4198.

127 I. Bjork and W. W. Fish, *Biochemistry*, 10 (1971) 2844.

128 S. Osaki and S. Sirivech, *Federation Proc.*, 30 (1971) 1292 Abstr.

Chapter 9

THE TRANSFERRINS (SIDEROPHILINS)

P. AISEN

Department of Biophysics, Albert Einstein College of Medicine, New York, U.S.A.

I. HISTORY AND NOMENCLATURE

Although the presence in blood serum of a non-heme fraction which was neither ultrafiltrable nor dialyzable was recognized as far back as the 1920s[1], the impetus to blood protein fractionation research provided by the second World War was required before the precise physico-chemical nature of this iron-bearing component of blood could be established. Holmberg and Laurell in Sweden established the existence of an iron-binding protein in blood[2] which they punningly named transferrin[3], while Schade and Caroline in the U.S.A. demonstrated that the bacteriostatic activity of blood serum was in part due to a protein with an enormous avidity for iron which was therefore named siderophilin[4,5]. Schade and Caroline further established the remarkable similarity of the iron-binding properties of the blood protein to those of conalbumin, a protein present in egg white[6]. Earlier, in 1939, an iron-containing protein had been recognized in bovine milk and accurately if prosaically called "milk red protein"[7]. The similarity of this protein to transferrin and conalbumin was subsequently appreciated. Since then, there has been considerable contention and remarkably little agreement about the proper choice of names for these iron-binding proteins, which have been shown to be present in a wide variety of bodily fluids and tissues.

The term transferrin has gained wide acceptance in the clinical literature of the English speaking countries as well as elsewhere. Because it accurately describes what is probably the most important biological function of the serum protein the retention of this designation seems both reasonable and useful. The protein isolated from egg white appears to be identical in its primary polypeptide structure with that isolated from the blood serum of the same species[8], so that the designation used by Feeney and his coworkers, ovotransferrin, seems justified; nevertheless objections to this term have been voiced on the grounds that the iron transporting function of this protein has yet to be established[9]. Accordingly, the term conalbumin will be retained in this review, while recognizing its close similarity to the serum protein. Except for their metal-binding sites, on the other hand, the iron-binding protein of milk and other secreted fluids seem to be structurely quite different from the protein of blood serum[10,11], so that a distinctive name for this class of iron-loving proteins is probably desirable. Because the first and most carefully studied source of this protein has been milk, the name lactoferrin has been proposed and has gained considerable acceptance; the term milk red protein has also been used. To stress the common properties of the iron-binding protein, Feeney and coworkers have introduced the designation lactotransferrin for the milk protein[9]. This introduces some confusion, however, because in addition to the protein peculiar to the external secretions the milk of some species also contains an iron-binding protein almost indistinguishable and perhaps derived from that of blood[12]. Furthermore, in rabbit milk the predominant and perhaps only iron-binding protein is identical, except for its content of sialic acid, to the serum protein, although probably secreted by the mammary gland[13]. It therefore seems wise to retain the new term lacto-ferrin for the protein characteristic of milk and external secretions, and designate the protein found in milk, which in its structure closely resembles serum transferrin, as milk transferrin. Recently Schade and his coworkers have proposed the term ekkrinosiderophilin for protein of the lactoferrin type isolated from fluids of external secretion[14]. The report that an identical protein has been identified in a variety of tissues[15], however, would render this suggestion somewhat inappropriate, and lactoferrin will be used in this review.

Thus, in terms of primary protein structure there are two broad classes of iron-bearing proteins: serum transferrin and egg white conalbumin on the one hand, and the lactoferrins of milk, tears, gastric juice, bronchial secretions and tissues on the other. The common property which distinguishes these proteins is their unique metal-binding powers so that the generic designation siderophilin for all proteins with this property appears to be the most accurately descriptive and appropriate. However, the use of the term transferrin has gained wide currency, and serves to relate all members of this class to the most carefully studied prototype protein, serum transferrin. Until further studies and discussion lead to general agreement among workers in the field, it would seem that both terms must be retained to facilitate communication.

II. ISOLATION AND PURIFICATION

A. Serum transferrin

The first preparations of serum transferrin were achieved with the methods of Cohn[16] and by ammonium sulfate fractionation combined with precipitation by ethanol[1]. Subsequently, ion-exchange chromatographic procedures have been introduced which are simpler and probably

282

gentler than the methods involving alcohol[17]. This latter point is probably of some importance since Schade has pointed out that some preparations of transferrin may be inactivated with respect to their physiological functions, although their iron-binding properties remain intact[18]. The greatest problem in the purification of transferrin has generally been separation from the heme-binding protein of serum, hemopexin. Crystallization of serum transferrin has been effected by gradual addition of alcohol to concentrated solutions of the protein[19], by simultaneously diffusing alcohol into a solution of the protein while concentrating it on an ultrafiltration apparatus[20] and from concentrated solutions at low ionic strength and pH[13].

B. Conalbumin

Conalbumin has been prepared from hen egg white by ammonium sulfate fractionation[21], ethanol fractionation[22], and ion-exchange chromatography using CM–cellulose or CM–Sephadex[23]. A simplied procedure for the preparation of relatively large amounts of conalbumin, which employs both ammonium sulfate precipitation and ion-exchange chromatography, has been given by Feeney and Komatsu[9]. Both apoconalbumin and iron-conalbumin have been crystallized from ethanol–water solutions[21].

It should be noted that commercially available preparations of conalbumin and transferrin are usually heterogeneous and apt to contain varying quantities of low molecular weight substances, particularly chelating agents, used in their preparations[24].

C. Lactoferrin

Groves, in 1959, reported in detail a procedure for the isolation of lactoferrin from the acid-precipitated casein fraction of bovine milk[25]. The method utilized both ammonium sulfate fractionation and DEAE cellulose chromatography, and yielded about 1 g of protein from 15 gallons of milk. A much simpler procedure for the isolation of lactoferrin from human milk has been given by Johansson[26], which entails direct adsorption of the protein on CM–Sephadex added batch-wise to the milk. Yields of 0.5 to 1 g of protein per liter of milk were reported; this increase over that obtained from bovine milk probably reflects both procedural differences and the greater concentration of lactoferrin present in human milk. Lactoferrin may be crystallized without the use of alcohol from dilute phosphate buffer[26].

III. CHEMICAL COMPOSITION

A. Amino acid composition

In general, the amino acid composition and N-terminal amino acid analyses of transferrin and conalbumin reported from a number of laboratories have been in good agreement with each other (Table I), and show no

TABLE I

AMINO ACID COMPOSITION OF THE SIDEROPHILINS

Amino acid	Human serum transferrin/ 76,700 g[27]	Human lacto-ferrin residues/ 77,000 g[a 28]	Bovine lacto-ferrin residues/ 77,100 g[29]	Conalbumin residues/76,600 g (adapted from ref. 8)
Lysine	52	45	42	62
Histidine	19	10	10	13
Arginine	24	43	32	33
Aspartic acid	83	66	71	74
Threonine	31	28	39	35
Serine	41	42	45	42
Glutamic acid	59	71	73	70
Proline	31	34	31	31
Glycine	55	53	43	58
Alanine	60	58	59	52
Cysteine ($\frac{1}{2}$ cystine)	22	13	28	22
Valine	44	40	43	46
Methionine	4	5	4	11
Isoleucine	14	16	17	24
Leucine	58	53	61	48
Tyrosine	24	23	19	20
Phenylalanine	29	30	25	25
Tryptophan	9	1 or 0	9	18

[a] Molecular weight of 77,000 assumed for purposes of comparison. The only reported value for the molecular weight of human lactoferrin is 95,000[9].

unusual features. Since conalbumin and fowl serum transferrin have been reported to have identical protein moieties[8], the difference in amino acid composition between human serum transferrin and conalbumin probably reflects species variations. Both transferrin and conalbumin appear to contain only one N-terminal amino acid residue per molecule; small amounts of a second amino acid residue which have been reported on occasion are probably due to impurities in the protein preparation being analyzed, or to partial hydrolysis of susceptible peptide bonds[30].

The amino acid compositions of human and bovine lactoferrin differ significantly from those of the corresponding serum proteins. Human lactoferrin appears to contain little or no tryptophan, a property which may prove to be of value in spectrophotometric studies (see below). One N-terminal alanine residue has been found in bovine lactoferrin[29]. No explanation is as yet available for the reported absence of an amino acid residue with a free α-amino group in human lactoferrin[31]★.

B. Peptide mapping

Further evidence for the difference in structure between transferrin and lactoferrin has been obtained from peptide-mapping studies of proteolytic digests of the these proteins. Interestingly, much greater resistance to proteolytic digestion is observed when metal is bound to the protein than when it is absent[32]. Studies of the human protein have shown that fewer peptides are observed in the 2-dimensional map of tryptic digests than would be expected on the basis of number of lysine and arginine residues known to be present[27]. The implications of this observation will be discussed in the section on the subunit structure of transferrin.

C. Carbohydrate content and structure

All the known siderophilins are properly classified as glycoproteins. Transferrin contains about 5% carbohydrate, disposed in two branched chains, identical in composition, terminating in sialic (N-acetyl neuraminic) acid residues, and probably joined to the polypeptide chain by asparaginyl linkages[33]. The terminal sialic acid residues may be completely removed by treatment with neuraminidase[34]. Transferrin deficient in neuraminic acid and possibly other sugars has been reported to be present in umbilical cord blood and cerebrospinal fluid[35]. After removal of sialic acid from transferrin by neurominidase, the protein serves as a substrate for galactose oxidase, indicating that neuraminic acid is bound terminally to galactose in the carbohydrate chains[36].

Although conalbumin and serum transferrin from the domestic fowl appear to be identical in their amino acid composition, immunochemical reactivities, and peptide patterns, distinct differences have been noted in their carbohydrate components. No sialic acid or galactose was found in the glycopeptide of the egg protein. In contrast to human serum protein the carbohydrate appeared to be arranged in a single chain in both proteins from the fowl[37].

★ Recent studies by P. Masson indicate that human lactoferrin contains 10 tryptophan residues per molecule (personal communication).

Human lactoferrin has been reported to contain about 6.6% carbohydrate, arranged in three chains each terminating with a sialic acid residue and composed of one or two residues of fucose, six residues of hexose and four of N-acetylglucosamine[38]. Evidence has also been presented that the carbohydrate chains are joined to the protein moiety by a glycosidyl linkage to the hydroxyl group of threonine. The carbohydrate content of bovine lactoferrin appears similar to that of the human protein, except that only one sialic acid residue per molecule has been found[25].

The functional or structural role of the carbohydrate portions of the siderophilins is not known. Removal of sialic acid from transferrin does not seem to affect its metal-binding properties or its capacity to serve as an iron donor for the reticulocyte[39].

D. *Immunochemistry*

All siderophilins are antigenic. Conalbumin and transferrin obtained from egg white and serum, respectively, of the domestic fowl are immunochemically identical, although each is synthesized at a different site[8]. In contrast, there is no cross immunoreactivity between human serum transferrin and human lactoferrin[40], almost certainly demonstrating that the proteins are not structurally related to each other except for their metal-binding abilities and that the milk protein is not derived from the blood protein.

IV. GENERAL PHYSICAL PROPERTIES

A. *Molecular weight*

Although earlier determinations of the molecular weight of serum transferrin based on ultracentrifugal measurements have yielded values ranging from 68,000 to 93,000, more recent studies have generated results clustering about a molecular weight of 75,000 to 80,000 daltons[27]. No tendencies to aggregation or dissociation have been observed in native transferrin solutions[27]. The molecular weight is the same whether the transferrin is saturated with iron, half-saturated or devoid or iron[41].

Conalbumin and lactoferrin have been less extensively studied than transferrin. The best estimate of the molecular weight of conalbumin is about 76,000 daltons, as determined from its iron-binding capacity[42]. Sedimentation equilibrium measurements of bovine lactoferrin show a molecular weight of 77,100[29], while that of human lactoferrin is about 80,000 from iron-binding studies[43], and 95,000[11] or 82,000[14] by sedimentation-velocity ultracentrifugation[11].

286

B. Optical properties

The extinction coefficients of transferrin complexes in the ultra-violet and visible regions are shown in Table II. Of particular interest is the increase in ultra-violet absorbancy when metal ions are bound[44]; the origin of this increase will be considered below.

TABLE II

EXTINCTION COEFFICIENTS OF HUMAN SERUM TRANSFERRIN

Protein	λ	$E_{1\ cm}^{1\%}$	Ref.
Human serum apotransferrin	280	11.4	9
Human serum Fe^{3+}-transferrin	280	14.3	44
	470	0.57	9
Human serum Cr^{3+}-transferrin	280	12.75	44
	440	0.22	45
	615	0.18	45
Human serum Co^{3+}-transferrin	280	14.3 (45)	
	405	1.2	9,44
Human serum Mn^{3+}-transferrin	280	14.3	44
	330	1.3	45
	430	1.2	45

Quenching of tryptophan fluorescence in transferrin has been observed when metal ions are bound[46].

C. Hydrodynamic properties

Based on measurement of dielectric dispersion and intrinsic viscosity, the transferrin molecule in aqueous solution approximates a prolate ellipsoid of revolution with an axial asymmetry ratio of 3.0 to 4.5[27,47]. Since the binding of metal ions causes slight, if any, change in intrinsic viscosity and no change in dielectric behavior[28,48] the likelihood is that the geometry and hydration of the molecule are not substantially affected.

Measurements of the frequency dependence of proton relaxation rate in transferrin solutions have led to the conclusion that only about 13, or 2%, of the water molecules in the first hydration sphere of the protein are relatively tightly (irrotationally) bound; this number is increased by two when the protein is saturated with metal[49,50].

On the basis that hydrogen exchange rates proceed more rapidly in apoconalbumin than in the metal complexes of the protein, Ulmer has suggested that the metal-saturated protein has a more compact conformation[51].

D. Subunit structure

A protein with two specific sites and two identical carbohydrate chains might well be expected to have a paired subunit structure. This point has provoked many experiments in a number of laboratories. Jeppsson reported reduced, alkylated transferrin, when dried and re-dissolved, dissociated into subunits of molecular weight 39,000 to 42,000 as measured by an approach-to-equilibrium centrifugation method[52]. Independent studies in the laboratories of Bezkorovainy[53] and Feeney[54] of reduced-alkylated transferrin failed to reveal any evidence of dissociation into subunits. Subsequently, studies by Bezkorovainy of reduced-alkylated and sulfitolyzed transferrin and conalbumin again showed no evidence of subunit structure[55]. Indeed, viscosity measurements of the treated transferrin in the presence of 8 M urea showed a tendency to aggregation[55]. Recently, a careful examination of the question of subunit structure of transferrin utilizing the techniques of sedimentation equilibrium centrifugation, intrinsic viscosity measurements, and gel filtration in 6 M guanidine hydrochloride further corroborated the single chain nature of human serum transferrin[27]. If subunits do indeed exist, they must be joined by unusual and as yet unknown linkages. In the light of present knowledge, the suggestion that the structural gene for transferrin underwent duplication and subsequent fusion in its evolutionary history[27,54] is an attractive one.

V. THE METAL-BINDING SITES

A. Is there an interaction between binding sites?

With the isolation of relatively pure transferrin and conalbumin came the demonstration that each protein was capable of specifically and tightly binding two atoms of iron[1]. For the development of the characteristic red color, carbon dioxide or one of its ionized forms was also shown to be required[5]. Schade and coworkers were able to show that for each ion atom bound one bicarbonate ion was required for the full development of color and two to three protons were released into solution[5].

A quantitative study of the mechanism of iron complex formation with conalbumin was undertaken by Warner and Weber[42] who found that the binding of each iron atom at pH values between 7.5 and 9.5 was accompanied by the release of three protons into solution, while the binding of copper entailed the release of two protons. Although it is difficult to be certain that protons are not released as a consequence of hydrolysis of the metal ion undergoing binding, the likelihood is that they are actually displaced from the proteins during binding, particularly as the analysis of equilibrium binding studies at somewhat varying pH is consistent with this view[56]. Since two protons are released for the binding of the divalent cation, and three for that of the trivalent ion, the net charge of the protein would not be affected if no other charged species participated in binding. Measurement of free electrophoretic mobilities showed, however, that there was an increase in the anionic mobility corresponding to a gain in net negative charge of the protein of one unit for each ion atom bound[42]. On the basis of this observation, as well as the experiments of Schade and his co-authors, it was concluded that a bicarbonate ion was the form in which carbon dioxide had participated in the metal-combining function of transferrin and conalbumin. Subsequently, Masson and Heremans showed that similar events accompanied the binding of iron and copper to lactoferrin[43].

Thus, the binding of iron to the siderophilins may be represented by the following equations for the first and second atoms bound.

$$Fe^{3+} + H_6 Tfn + HCO_3^- \rightleftarrows [Fe^{3+} H_3 Tfn \cdot HCO_3] + 3H^+ \qquad (1)$$

$$Fe^{3+} + [Fe^{3+} H_3 Tfn \cdot HCO_3] + HCO_3^- \rightleftarrows [(Fe^{3+})_2 Tfn \cdot$$

$$(HCO_3)_2] + 3H^+ \qquad (2)$$

This formulation has led to one of the more interesting and lively controversies in the chemistry and physiology of the siderophilins. Using citrate as a competing iron-complexing agent and measuring the color developed in the solutions of conalbumin in which the thermodynamic activity of the iron was varied by varying citrate concentration, Warner and Weber concluded that the binding of iron to conalbumin occurred in pairwise fashion with the second stability constant much greater than the first[42]. Schade proposed the same interpretation of the mode of iron release from transferrin based on non-thermodynamic observations of rate of iron uptake from the protein by liver cells[57]. Finally, Woodworth advanced arguments based on measurements of iron-binding kinetics to substantiate further the pairwise mechanism of iron-binding in conalbumin[58]. Until quite recently, therefore, the prevalent view has been that iron atoms go on and off the siderophilin molecule two at a time, so that the only stable species in solution would have 0 or 2 iron atoms per molecule.

In 1963, Malmström and his associates published a series of careful experiments using the classical thermodynamic method of equilibrium dialysis, supplemented by electron paramagnetic resonance (e.p.r.) spectroscopy studies[56]. Taking particular care to show that equilibrium had indeed been achieved, and subjecting their data to a novel graphical analysis, these investigators concluded that the binding of iron to transferrin could be described as involving equivalent and noninteracting sites. If so, this necessarily leads to the conclusion that in solutions of transferrin less than fully saturated with iron, three species of protein molecules must co-exist: those with two iron atoms bound, those with only one bound, and those devoid of metal[59]. The existence of three kinds of molecules in conalbumin solutions incompletely saturated with iron was first tentatively recognized from electrophoresis patterns by Warner and Weber[21], who did not, however, reconcile this observation with their conclusions from equilibrium studies. Later, it was confirmed in detail for transferrin using moving-boundary electrophoresis[59], and finally for conalbumin by isoelectric focusing experiments[60]. Although it has been suggested that the species of intermediate mobility in the electrophoresis and isoelectric focusing experiments may have represented a dimer of apoprotein and iron-saturated protein[61], subsequent measurements proved this suggestion untenable[41].

The bulk of currently available evidence, then, supports the view that the thermodynamic constants which describe the binding of iron to transferrin are intrinsically identical. A re-examination of iron-binding to conalbumin by the method of equilibrium dialysis also points to a close similarity in the metal-binding behavior of this protein to that of transferrin[62]. It was suggested that earlier data of Warner and Weber may have been recorded before equilibrium was reached, since only 18 hours elapsed between the addition of the iron to the apoconalbumin solutions and the time of observation; a time-dependence study later indicated that at least several days were required for true equilibrium to be attained[56]. The physiological studies of Schade are not thermodynamic measurements, and so are not necessarily in conflict with the foregoing. Quantitative studies of iron-binding to lactoferrin are not available at this time.

B. The meaning of the binding constants

Some confusion has existed as to the meaning of the numbers obtained from equilibrium dialysis measurements of metal-binding to the siderophilins. The binding constants, K_1 and K_2, reflect the chemical events given in eqns. (1) and (2) above, as follows[55]:

$$K_1 = \frac{[Fe^{3+} H_3 Tfn \cdot HCO_3] \cdot [H^+]^3}{[Fe^{3+}] [H_6 Tfn] [HCO_3^-]} \tag{3}$$

$$K_2 = \frac{[(Fe^{3+})_2 Tfn(HCO_3)_2] [H^+]^3}{[Fe^{3+}] [Fe^{3+} H_3 Tfn \cdot HCO_3] [HCO_3^-]} \tag{4}$$

Since three protons are released and one bicarbonate ion bound for each iron atom bound, the true thermodynamic constants are each directly proportional to the third power of the hydrogen iron concentration and inversely proportional to the bicarbonate ion concentration in the solutions in which measurements are made. If the two specific sites are equivalent and independent, statistical considerations indicate that a ratio of K_1 to K_2 must be four.

At fixed conditions of pH and bicarbonate ion activity, the effects of these terms in the binding equations may be incorporated into apparent equilibrium constants as follows[56]:

$$K_1'(\text{const. pH, pCO}_2) = \frac{K_1 \cdot [\text{HCO}_3^-]}{[\text{H}]^3} = \frac{K_1 \cdot \text{pCO}_2 \cdot K_{\text{CO}_2}}{[\text{H}]^4} \tag{5}$$

where K_{CO_2} is the appropriate constant for the hydration of CO_2 and ionization of carbonic acid. Because bicarbonate ion activity depends inversely on pH, the effect is to create a fourth power dependence of the apparent constant on the pH at which equilibrium studies are made, and a direct dependence on the carbon dioxide tension of the atmosphere with which they are in equilibrium. Under the relatively fixed conditions of pH and pCO_2 which prevail in the human circulation, the apparent K_1 is near $10^{24} M^{-1}$. From this, it can be shown that an iron atom will require something of the order of 10,000 years to separate spontaneously from a transferrin molecule in the blood[63]. Of the quantities used to describe quantitatively metal binding to transferrin, the apparent stability constant probably carries the greatest physiological relevance.

When the pK values of the protein's proton-releasing groups are taken into consideration then the intrinsic chelation constant for metal binding is

$$K_{1\,\text{chel}} = K_1 \cdot 10^{\Sigma pK_i} \tag{6}$$

Three protons are released at pH values at least as high as 9.5, so that the pK values of the releasing groups must exceed this value. Accordingly, the intrinsic chelation constant for the binding of iron to transferrin must exceed 10^{32}, and may be as high as $10^{36,56}$.

The identity of the binding constants immediately leads to the question whether the binding sites themselves are identical. This problem is all the more provocative in a protein composed of a single polypeptide chain. Physiological and physical studies have both shed light on this still incompletely resolved point, which will be returned to below.

C. Kinds and state of metal ions specifically bound to transferrin

In addition to the well studied iron and copper complexes of transferrin the protein will also accept a variety of multivalent cations. Zinc, chromium, cobalt, manganese, cadmium, nickel and gallium are all bound with the same stoichometry as iron[9,44,64,65]. The binding of chromium, manganese, cobalt and copper is also accompanied by the binding of a bicarbonate ion[45]. Participation of bicarbonate in the binding of other metal ions has not yet been studied. Indium is also capable of combining with transferrin, but the involvement of the same specific sites of the protein has not been clearly demonstrated[66].

Presumably because of hydrolysis problems, the original studies on the binding of iron to siderophilins was carried out with Fe(II) ions. The question of the valence state of the bound iron was not settled until room temperature measurements of static magnetic susceptibility by Ehrenberg and Laurell showed that the specifically bound metal was in the high spin (d^5) Fe(III) state[67]. The paramagnetic susceptibility was somewhat greater than would be predicted for non-interacting Fe(III) ions so that the suggestion was advanced that a weak interaction between metal ions bound to the same molecule was present. Subsequent electronic paramagnetic resonance analyses[57] and static susceptibility studies at low temperature[45] failed to confirm the presence of such interaction, with reasonably good conformance to Curie's law observed at temperatures between 1.5 and 77K.

Each of the metal complexes of transferrin shows a distinctive color and absorption bands in the visible region. Observing that the characteristic color formation in the manganese and cobalt complexes was accelerated by the addition of hydrogen peroxide, Inman suggested that these metal ions were probably oxidized to the trivalent state when bound[9]. This was confirmed by static susceptibility measurements at low temperatures, which also indicated that cobalt was bound in the low spin (d^6) diamagnetic form[45].

It is of considerable physiological interest to know whether the binding sites of transferrin can accommodate ferrous as well as Fe(III) ions, or whether the former must be oxidized to the latter before binding will take place. Unpublished studies by Gaber in the author's laboratory, based on difference spectrophotometry and isotope exchange experiments, indicate that Fe^{2+} is bound, if at all, only very weakly[68].

D. The role of bicarbonate

Because the characteristic red and yellow colors of the iron and copper complexes of the siderophilins did not develop in the absence of bicarbonate, it was clear that this ion played a central role in metal–protein complex formation[5]. Direct measurement of the amount of carbon dioxide released on acid denaturation of the iron[42], copper[69], chromium, manganese and cobalt[45] complexes confirmed the earlier suggestion of Schade[5] that one bicarbonate ion was bound for each metal ion bound. The facultative nature of bicarbonate binding was demonstrated by a series of electron paramagnetic resonance spectroscopic studies of metal binding to transferrin, in which it was shown that specific binding of iron and copper, at least, could take place even in the absence of bicarbonate[70]. The complexes so formed were colorless, and hence had presumably eluded detection before the availability of e.p.r. techniques. In the absence of bicarbonate, the iron–protein bond was evidently very much weaker than in its presence, since hydrolysis of the iron with formation of insoluble iron(III) hydroxide was observed as a bicarbonate-free iron transferrin complex was allowed to stand at neutral or higher pH. A possible physiological function of this effect will be discussed in the section on biological functions of siderophilins.

The bicarbonate binding site will also accommodate a variety of other anions, including ethylenediaminetetraacetate, nitrilotriacetate, and oxalate[70,71]. The common property of these anions is the presence of two or more carboxyl groups; it is not yet known whether this is essential for occupancy of the anion binding site by ions other than bicarbonate. The relationship and proximity of the locus of bicarbonate-binding to that of metal-binding is also unknown, although they must be very intimate since the anion-binding site is occupied only when the metal-binding site is. Furthermore, occupancy of the bicarbonate site has been shown to impede the accessibility of solvent water to specifically complexed metal ion[72]. When bound, bicarbonate cannot be removed by exhaustive evacuation of the protein solution, nor can it be replaced by other anions capable of inhabiting its site. From these considerations, it seems likely that the bicarbonate is coordinated not only to the metal but in some way bound to one or more groups on the protein as well.

E. The kinetics of iron binding

Studies of the kinetics of metal complex formation with transferrin are complicated by two factors: the velocity at which binding occurs may be so rapid as to exceed the resolution capabilities of ordinary instruments, and the presence of side reactions, including those of hydrolysis and those of ligand displacement when the metal is presented to the protein in chelated form, must be taken into account. In studying this problem, Saltman and his coworkers have separated out two phases of the binding process when the Fe^{3+} is presented to the protein as the nitrilotriacetate complex: an early fast phase, which they could not measure, and a second slower phase which obeyed first-order kinetics with respect to the concentration of the iron complex[24]. The latter observation is consistent with the thermodynamics studies indicating the independent binding of Fe(III) ions[55]. These studies are at variance with those reported by Woodworth, who found a simple second order dependence of the rate of complex formation on the concentration of iron, again present as the iron(III) nitrilotriacetate complex[58]. The time limitations of the manual methods employed, and of the response of the spectrometer, are such that a definitive resolution of this controversy must await use of fast reaction techniques, particularly stopped flow spectrophotometry.

F. Spectroscopic studies

Optical studies

Most of the metal complexes of the siderophilins show intense absorption bands in the visible regions, which account for their characteristic colors (Table II). Detailed quantum mechanical analyses have not yet been undertaken for the chromophoric centers, so that the origin of the absorption bands in the visible region is still not understood. Simple charge transfer is probably excluded[73]. It seems reasonable to suppose that they result from transitions among the primary 3-d orbitals of the central metal ion, allowed because of the admixture of 4p and 4s orbitals resulting from the strained geometry imposed by the protein on its coordinating ligand groups. This has been shown to be the case for the copper–protein ceruloplasmin[74]. Other noteworthy features of the visible absorption bands in the metal siderophilin complexes are worth mentioning here: in the presence of bicarbonate, the copper complex shows an unusual deep yellow color due to the absorption maximum for 430 nm, while two distinct bands are present in the visible region in the chromium–transferrin-bicarbonate complex.

Changes best demonstrated by difference spectrophotometry occur in the ultra-violet absorption region when metal ions are bound to siderophilins[45,64,70]. Compared to reference solutions of apoprotein, the metal-protein complexes show absorption maxima near 245 nm and 295 nm[64]. There is some disagreement in the literature about the sharpness and intensity of the difference band at 295 nm[46]; This may be due to differences in concentration of the solutions being studied with consequent variations in the contributions of instrumental error owing to light scattering or the general difficulties in measuring small differences between large absorbancies. The differential absorption at 295 nm probably arises from the perturbation of tyrosyl residues by bound metal ion[76]. Tryptophanyl residues may also contribute to the difference spectra[63]. If so, studies of human lactoferrin should show variation from similar studies in serum transferrin since the latter protein contains little or no tryptophan.

Spectrophotometric titrations studies of the iron and copper complexes of siderophilin have also incriminated tyrosyl residues in the metal-binding function of the protein[75,76]. Ultra-violet difference spectra of proteins at alkaline vs. neutral pH show absorption bands at 295 nm which are very similar to those obtained with n-acetyl-1-tyrosine at pH 10.8 vs. pH 7, that is, with the ionized $vs.$ the unionized phenolic OH-group[64]. The difference spectra in proteins are therefore attributed to tyrosine ionization. The extinction coefficient of the difference band may be obtained from the absorption of the fully alkaline denatured protein compared to native protein, taking into account the number of tryosyl residues in the molecule, or from measurements on a model compound. There is some discrepancy in the values obtained by different methods, as well as values reported for different proteins and it is difficult to remove this uncertainty[64,76]. Furthermore, the molar absorbancies reported from alkaline denaturation studies may not be simply applicable to measurements on native protein[76]. There does seem to be good agreement, however, that the observed changes are due to perturbation of tyrosines on binding of metal to siderophilin, despite the disagreement as to the number of residues actually involved. Tan and Woodworth interpreted their data as indicating the participation of two tyrosines for the binding of iron and one for the binding of copper[64]; Aasa and Aisen reported 1.3 residues affected by the binding of copper[76], and Wishnia, Weber and Warner suggested three were involved in iron binding[75]. It might be well to point out that a difference in the number of tyrosine residues spectrophotometrically titrated in the presence and absence of complexed metal does not necessarily indicate their direct coordination to the metal. It may be, for instance, that the observed differences result from pK values altered as a result of conformational changes or electrostatic effects induced by metal-binding.

Optical rotatory dispersion and circular dichroism studies

Both transferrin and conalbumin exhibit Cotton effects in the visible region which depend on the presence of specifically bound metal ion[77]. The 470 nm absorption band of the iron complexes generates a negative Cotton effect, while the 429 nm band of the manganese complexes is associated with a positive effect. In contrast, the copper complexes of conalbumin and transferrin which absorb maximally at 430 nm do not show anomalous rotatory dispersions in this region. Optical activity associated with this band was identified by circular dichroism studies, however[78].

The extraordinary capacity of the metal-binding sites of transferrin to accommodate both copper and iron ions merits comments here. When copper is bound, the environment about the metal ion exhibits very nearly axial symmetry[76]. The slight departure from axiality evident in the e.p.r. spectrum taken at 35 Gc is consistent with the optical activity observed by Nagy and Lehrer[78]. In contrast, the e.p.r. spectrum[55] and the optical activity of the iron complex indicate a much lower symmetry at the metal-binding site. Evidently the ligands provided by the protein have a high degree of flexibility, accounting for the accommodating nature of the sites.

G. *Magnetic resonance spectroscopy*

Probably the most revealing information about the nature of the metal-binding sites in the siderophilins has been obtained from the magnetic resonance studies, particularly electron paramagnetic resonance (e.p.r.) spectroscopy. The e.p.r. spectrum of the native Fe–transferrin–HCO_3 complex is characteristic of high-spin Fe^{3+} in a rhombically distorted octahedral environment[56,79]. The sharp signal at $g = 4.14$ almost certainly originates from a single kind of Fe^{3+}, with no difference in line shape observed as a function of degree of saturation of the protein with iron[56]. A detailed theoretical treatment of this spectrum has recently been given by Aasa[80], who has analyzed the experimental results in terms of a spin Hamiltonian of the type:

$$\mathcal{H} = \tfrac{1}{3}D\,[3S_z^2 - S(S+1)] + E(S_x^2 - S_y^2) + g\beta\vec{B}\cdot\vec{S}.$$

The major lines near $g = 4$ arise from the middle of the 3 Kramer's doublets of the spin = 5/2 manifold. The experimental spectra, being obtained from frozen solutions, are interpreted as powder spectra composed of a super-position of spectra from randomly oriented crystallites. The experimentally observed lines were shown to arise from transitions in principal planes as well as along principal axes of the crystal field tensor, as shown earlier in somewhat different form by Blumberg[81] and by Dowsing and Gibson[82]. Values for the parameters of the spin Hamiltonian, as deduced by Aasa are given in Table III, along with the parameters of the copper and chromium complexes.

TABLE III

SPIN HAMILTONIAN PARAMETERS OF TRANSFERRIN COMPLEXES

Complex	$\lvert D\rvert$ (cm^{-1})	$\lvert E/D\rvert$	$g_{\parallel}$	$g_{\perp}$	$A_{\parallel}$ (gauss)	^{14}N Superhyperfine splitting	No. of N ligands observed
Fe^{3+}–bicarbonate	0.27	0.31–0.32					
Cr^{3+}–bicarbonate							
Type 1	0.55	0.32–0.33					
Type 2	0.55	0.27					
Cu^{2+}–bicarbonate			2.312	$g_x = 2.042$ $g_y = 2.059$	156	12	1
Cu^{2+} (bicarbonate-free)			2.205	2.05	210	9.5	4

In the absence of bound bicarbonate, the e.p.r. spectrum of Fe-transferrin is considerably altered, but is still indicative of specifically bound metal[70]. A change is observed in the spectrum on going from neutral to acid pH, with the appearance of new component indicating a difference between binding sites under such conditions[76]. When bicarbonate is absent,

References pp. 303–305

ternary complexes of iron, transferrin, and a variety of metal chelating agents may be revealed by their distinctive e.p.r. spectra[70].

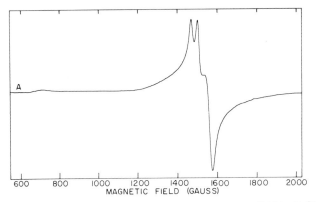

Fig. 1a. E.p.r. spectrum of transferrin, 1.3×10^{-3} M in Fe^{3+}. Temperature, 77K; microwave frequency, 9090 MHz; receiver gain, 1.5×10^4.

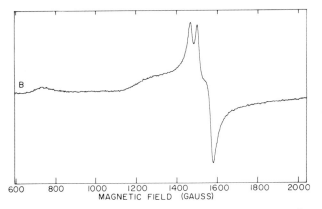

Fig. 1b. E.p.r. spectrum of bovine lactoferrin, 6.9×10^{-4} M in Fe^{3+}. Temperature, 77K; microwave frequency, 9090 MHz; receiver gain, 10×10^4.

Although the main features the e.p.r. spectrum of the iron–conalbumin–bicarbonate complex are identical to those of transferrin, another component in the low-field region of the lines near $g' = 4.3$ is also present[56]. A similar component has been observed by the author in lactoferrin (Fig. 1). Detailed studies have not yet been done, but it seems likely that the additional component must arise from a Fe(III) ion complexed in a slightly different way from that in transferrin. Thus, there may well be some difference between the binding sites in these proteins.

The e.p.r. spectra of the copper complexes of transferrin are also
dependent on bicarbonate binding[75]. In both cases, the spectra are
characteristic of copper in an axial environment. Nitrogen superhyperfine
structure, indicative of interaction with one nitrogen nucleus is observed
in the low-field hyperfine line of yellow copper–transferrin–bicarbonate
complex. In the absence of bicarbonate the colorless binary complex of
copper and transferrin shows a superhyperfine structure best interpreted
as indicating an interaction of the unpaired spin of the copper with four
nitrogen nuclei (Fig. 2). It is not yet understood why the change occurs
as bicarbonate is bound, but it is clear that the metal binding sites of
transferrin must have four nitrogen ligands available. This direct demonstration of the participation of nitrogen ligands in the coordination of
metal ions by transferrin corroborates and supplements earlier inferences
based on hydrogen ion titration studies, which pointed to the involvement
of two histidine imidazole groups for each ion bound[83].

Studies of the e.p.r. spectra of the specific chromium complexes of
transferrin have also been informative. When obtained at physiological
pH, the spectrum reveals the presence of two similar but distinguishable
chromium ions, which are displaced at different rates by Fe^{3+} [48]. This
may be the spectroscopic counterpart of the difference between the iron-
donating capacity of the binding sites of transferrin in physiological
studies which is demonstrable despite the random nature of iron
binding[84,85].

Further evidence that tryrosyl residues participate in the coordination of metal irons to the siderophilins has been provided by high
resolution nuclear magnetic resonance (n.m.r.) studies[65]. Specific binding
of gallium (Ga^{3+}) resulted in a shift to higher field of a portion of the
aromatic region of the spectrum, while binding of Fe^{3+} decreased intensity
in this region as a consequence either of paramagnetic broadening or shift.
The introduction of this technique would appear to have opened a fertile
field for the study of metal-binding to siderophilin.

H. Chemical modification studies

Additional information about the metal-binding sites of the siderophilins has been provided by studying the effects of chemical modifications
of the protein on its metal-combining functions. Modification of less than
50% of the free amino groups of metal-free siderophilins did not substantially affect their iron-binding properties, while more extensive modification produced a gradual loss of iron-binding capacity[86]. In contrast, alkylation of histidine residues is accompanied by loss of iron-binding activities
when 14 of the 17 histidines in the molecule are modified[87]. When only 10
histidines are altered, there is virtually no loss of iron-binding capacity.
This data has been interpreted as indicating that two histidines particpate

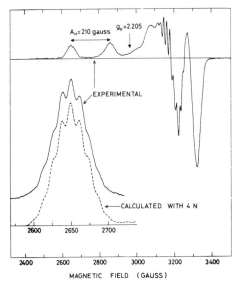

Fig. 2a. E.p.r. spectrum obtained at 77K of ^{65}Cu–transferrin, bicarbonate-free. Microwave frequency, 9149 MHz. The broken curve is a computer-simulation of the low-field
hyperfine line, using Gaussian line shape (width, 10 gauss); four equivalent nitrogen
nuclei with 12 gauss superhyperfine splitting; and copper parameters $g_\parallel$ = 2.205,
$g_\perp$ = 2.05, $A_\parallel$ = 210 gauss, and $A_\perp$ = 21 gauss. Reprinted with permission from ref. 76.

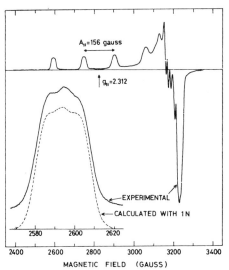

Fig. 2b. E.p.r. spectrum obtained at 77K of ^{65}Cu–transferrin–bicarbonate. Microwave
frequency, 9151 MHz. The broken curve is a computer simulation assuming Gaussian line
shape (width, 9.5 gauss); one nitrogen nucleus (splitting 9.5 gauss); and the parameters
$g_\parallel$ = 2.312, $g_\perp$ = 2.05, $A_\parallel$ = 156 gauss and $A_\perp$ = 16 gauss. Reprinted with permission
from ref. 76.

in the binding of each iron atom, but this suggestion may be open to question since carboxymethylation of the histidines shows no evidence of cooperativity, but rather seems to go in a random fashion while metal-binding function is abruptly lost when more than 10 residues have been changed. It would seem that the absence of cooperativity in histidine modifications should be reflected in a gradual but accelerating loss of iron-binding activity with the number of histidines changed. Possibly, then, the loss of iron-binding reflect changes in protein tertiary structure as increasing numbers of histidines are modified.

Several studies have been published of the effects of tyrosine modifications on the metal-binding sites of the siderophilins. On comparing the susceptibility of metal-free and iron-saturated proteins to chemical modification of tyrosyl residues, it was found that approximately six fewer tyrosines underwent modification when iron was bound than when the protein was metal-free. This was true when the modification procedure involved acetylation of the hydroxyl groups by N-acetylimidazole[88], and then modification consisted of nitration of tetranitromethane[87]. The phenolic residues of metal-bearing protein also showed greater resistance to iodination[88]. The agreement of these results with the spectrophotometric titrations from Warner's laboratory offers strong evidence of the participation of tyrosyl residues in metal-complex formation. But a caveat is necessary however, as pointed out by Buttkus, Clark and Feeney[86]: "As the proper orientation of the metal-binding sites is most probably stabilized by hydrogen bonds and charge interactions, the entire conformation of the molecule is important. Upon chemical modifications, these inter-molecular interactions are altered and the resulting changes may lead to variations in the stability of the chromogenetic complex, spectral shifts, and eventually to loss of metal-binding capacity of the protein."

I. Summary of current status of knowledge of the metal-binding sites of the siderophilins

At physiological pH and in the presence of bicarbonate the two iron-binding sites of transferrin are identical with respect to their e.p.r. parameters. At lower pH and in the absence of bicarbonate, however, a difference between the sites is evident. Four nitrogen ligands are available at each site, but not all of these may become liganded to bound metal under all circumstances. Two of these nitrogen atoms may reside in imidazole rings of histidine, but the source of the other two is not known. Two or three tyrosyl residues probably participate directly in iron-binding, and one or two residues in copper-binding. Although very likely, the direct coordination of the phenolic oxygen functions to the metal iron has not been demonstrated. A water molecule is directly coordinated to bound copper, and probably to bound iron was well.

The stability constants for the two sites are identical, or nearly so, for transferrin and conalbumin. Nevertheless, differences between the sites have been recognized in bicarbonate-free Fe^{3+}-transferrin at low pH, or when the bound metal is Cr^{3+}. A difference in the physiological iron-donating ability of the two sites of transferrin has also been observed.

Despite their usually differing stereochemical requirements, both Fe^{3+} and Cu^{2+} are accommodated by the metal-binding sites of siderophilins. A possible explanation is that there is considerable flexibility of the polypeptide side chains in the region of the binding sites, so that each ion may choose an environment of the ligands and symmetry it most prefers.

VI. BIOLOGIC FUNCTIONS OF THE SIDEROPHILINS

A. The need for a physiological iron carrier

Elementary thermodynamic calculations show that under physiological conditions of pH and oxygen tension the Fe(III) state of iron in solution at equilibrium will be favored. Taking the solubility product of iron(III) hydroxide as $10^{-36} M^4$, it follows that the equilibrium concentration of free ferric ion in blood cannot exceed $10^{-14} M$. The indispensible requirement for an iron carrier is clear.

Possibly an even more important function than transport is carried out by transferrin, however. The protein is capable of specifically recognizing the hemoglobin-synthesizing reticulocyte, and this recognition function insures delivery of its iron only to those cells which have specific need for it[89]. Furthermore, the reticulocyte itself will accept iron efficiently only from specific carriers[90], of which transferrin is probably the best. Thus, the recognition function is a dual one, involving this specific interaction of a protein with a cell.

B. The transferrin–reticulocyte interaction

Specific receptors on the surface of reticulocyte are required if it is to interact with and accept iron from transferrin. Treatment of the cell surface with trypsin, which is known to hydrolyze a sialoglycopeptide from the cell membrane[91], largely abolishes its ability to take up iron from the protein[89].

Morgan suggested that the interaction between transferrin and reticulocytes takes place in four steps[92]:

1. Adsorption of transferrin to receptor sites on the surface of the reticulocyte.

2. Formation of a firmer union between transferrin and the reticulocyte. This step may involve actual penetration of the protein into the interior of the cell[93].
3. Transfer of iron from protein to cell.
4. Release of protein.

Transferrin with Fe^{3+} or Cr^{3+} bound has a greater affinity for reticulocyte receptors than does apotransferrin[94,95]. This effect in part depends on the particular metal ion complexed, and appears to be due to a higher dissociation rate constant for the reaction between reticulocytes and transferrin for protein devoid of metal. Thus, molecules bearing Cr^{3+} or Fe^{3+} have a longer mean residence time on the reticulocyte, the average residence time of a protein molecule on the cell surface probably being between 5 and 10 min[4]. Manganese, copper and zinc transferrins behave like apotransferrin[93]. Jandl and Katz[96], and Kornfeld[94] have calculated that there are about 50,000 receptor sites on a reticulocyte surface, so that when saturated about 2% of the surface area of the cell is occupied by transferrin. Baker and Morgan have estimated that as many as 500,000 protein molecules may be bound to the reticulocyte[97].

The rate of iron uptake by reticulocytes is independent of the iron-saturation of transferrin with which they are incubated provided the total concentration of Fe–transferrin is constant, indicating that apotransferrin does not significantly impede the iron transfer process[57,98]. Discrepant reports concerning the effects of transferrin concentration on iron uptake by reticulocytes exist in the literature[57,96,98,99]. It has been suggested that some of the disagreement may be attributed to differences in the physiological integrity of the transferrin preparations being used[18,99]. This writer would also emphasize that studies of the effects of transferrin concentration and saturation must take into account the observation that, although exchange among transferrin molecules does not occur spontaneously because of the enormous stability of the metal binding to the protein, it may be promoted by the presence of iron complexing agents such as citrate[62].

The effects of structural modifications on the biological activity of transferrin have been studied by Kornfeld[100]. These studies appear able to differentiate between the effects of modification on the binding of the protein to the reticulocyte cell surface, and the subsequent specific uptake of iron by the cell. Enzymatic removal of most of the carbohydrate bound to transferrin caused no significant alteration in either protein binding or iron transfer.

The mechanism by which iron is transferred from the transferrin molecule to the hemoglobin synthesizing apparatus of the reticulocyte is not known. Because of the enormous stability of the iron–protein complex, it cannot involve simple dissociation. Iron uptake *in vitro* is not affected by chelating agents in the incubating medium. Since the stability of the iron–transferrin complex is greatly weakened in the absence of bicarbonate, the suggestion has been made that the interaction may involve the bicarbonate-binding site of the protein[101]. With the demonstration that the transferrin molecule is capable of penetrating to the interior of the reticulocyte[93], it may also be that local pH within the cell is reduced sufficiently to permit release of ion from the protein.

The concentration of transferrin in the blood (total iron-binding capacity) does not affect the absorption of iron from the gut, although the subsequent distribution of absorbed iron depends on whether unsaturated transferrin is available as a carrier[102].

C. Transferrin as an iron buffer

Because circulating transferrin is only about 1/3 saturated with iron, it has the capacity to accommodate fluctuating iron requirements so that it is an effective iron buffer as well as iron supplier. The importance of this function is probably best revealed when it is absent. Under conditions of chronic iron overload so that transferrin is saturated with iron, or when there is a congenital deficiency of the protein[103], iron deposition in tissues to the point of toxicity is prone to occur.

D. The biologic functions of conalbumin and lactoferrin

A specific role in iron transport has not been demonstrated for either conalbumin or lactoferrin. The ability of conalbumin to combine with iron and make it unavailable to micro organisms for growth may well serve a biologic function in protecting the egg against microbial infection[6]. The ubiquitous distribution of lactoferrin in the secretions of the respiratory, digestive, urinary, and lacrimal tracts may be related to a bacteriostatic activity of this protein[15]. The antimicrobial properties of conalbumin and lactoferrin are lost when they are saturated with iron[6].

Lactoferrin may play a role in the transport of iron from the blood to the milk of the nursing mother, but this has not been clearly shown. There has also been some speculation about the role of this protein in regulating the absorption of iron from the gastrointestinal track[104].

E. Genetic heterogeneity of transferrins

At least 18 genetic variants of human transferrin have been recognized[105]. The biologic significance, if any, of this variability is not known.

RESUME

The transferrins (siderophilins) are a group of proteins from diverse sources characterized by the ability to bind, specifically, tightly and reversibly, Fe(III) and other transition metal ions. Serum transferrin, which has been most extensively studied, functions to transport ion in blood and to direct it to reticulocytes requiring it for the biosynthesis of hemoglobin. The transferrin–reticulocyte system affords a system, so far unique, for studying the interaction of a metal with a protein and a protein with a cell.

NOTE ADDED IN PROOF

Gibson and Price (*Biochem. Biophys. Res. Commun.*, 46 (1972) 646) have recently reported evidence that specific binding of Fe^{3+} to transferrin does not occur in the absence of a suitable anion, thus calling into doubt the existence of a binary Fe-transferrin complex. Previous results obtained by Aasa and Aisen[76] and Aisen et al.[70] have been attributed to the persistence of citrate, used for the preparation of apotransferrin in the samples studied by these investigators.

REFERENCES

1   C. -B. Laurell and B. Ingelman, *Acta Chem. Scand.*, 1 (1947) 770.
2   C. G. Holmberg and C. -B. Laurell, *Acta Physiol. Scand.*, 10 (1945) 307.
3   C. G. Holmberg and C. -B. Laurell, *Acta Chem. Scand.*, 1 (1947) 944.
4   A. L. Schade and L. Caroline, *Science*, 104 (1946) 3420.
5   A. L. Schade, R. W. Reinhart and H. Levy, *Arch. Biochem.*, 20 (1949) 170.
6   A. L. Schade and L. Caroline, *Science*, 100 (1944) 14.
7   M. Sörenson and S. P. L. Sörenson, *Compt. Rend. Trav. Lab. Carlsberg*, 23 (1939) 55.
8   J. Williams, *Biochem. J.*, 83 (1962) 355.
9   R. E. Feeney and St. K. Komatsu, *Structure and Bonding*, 1 (1966) 149.
10 C. Bron, B. Blanc and H. Isliker, *Helv. Physiol. Acta*, 25 (1967) 337.
11 J. Montreuil, J. Tonnelat and S. Mullet, *Biochim. Biophys. Acta*, 45 (1960) 413.
12 M. L. Groves, *Biochim. Biophys. Acta*, 100 (1965) 154.
13 E. Baker, D. C. Shaw and E. H. Morgan, *Biochemistry*, 7 (1968) 1371.
14 A. L. Schade, C. Pallavicini and U. Wiesmann, *Protides of the Biological Fluids, Proc. Colloq. Bruges*, 16 (1968) 619.
15 P. L. Masson, J. F. Heremans, E. Schonne and P. A. Crabbe, *Protides of the Biological Fluids, Proc. Colloq. Bruges*, 16 (1968) 633.
16 B. A. Koechlin, *J. Am. Chem. Soc.*, 74 (1952) 2649.
17 B. Gelotte, P. Flodin and J. Killander, *Arch. Biochem. Biophys.*, *Supp.*, 1 (1962) 319.

18 A. L. Schade, *Sonderdruck Behringwerk-Mitteilung.*, 39 (1961).

19 C. -B. Laurell, *Acta Chem. Scand.*, 7 (1953) 1407.

20 A. Leibman and P. Aisen, *Arch. Biochem. Biophys.*, 121 (1967) 717.

21 R. C. Warner and I. Weber, *J. Biol. Chem.*, 191 (1951) 173.

22 J. A. Bain and H. F. Deutsch, *J. Biol. Chem.*, 172 (1948) 547.

23 R. C. Woodworth and A. L. Schade, *Arch. Biochem. Biophys.*, 82 (1959) 78.

24 G. W. Bates, C. Billups and P. Saltman, *J. Biol. Chem.*, 242 (1967) 2810.

25 M. L. Groves, *J. Am. Chem. Soc.*, 82 (1960) 3345.

26 B. G. Johansson, *Acta Chem. Scand.*, 23 (1969) 683.

27 K. G. Mann, W. W. Fish, A. C. Cox and C. Tanford, *Biochemistry*, 9 (1970) 1348.

28 B. Blanc, E. Bujard and J. Mauron, *Experientia*, 19 (1963) 299.

29 F. J. Castellino and K. G. Mann, submitted to *Biochemistry*.

30 J. -O. Jeppsson, *Ph.D. Dissertation*, Department of Microbiology, University of Umeå, Umeå, Sweden, 1967, p. 18.

31 J. Montreuil, G. Biserte, S. Mullet, M. Spik and N. Leroy, *Compt. Rend. Acad. Sci., Paris*, 252 (1961) 4065.

32 P. R. Azari and R. E. Feeney, *J. Biol. Chem.*, 232 (1958) 293.

33 G. A. Jamieson, *J. Biol. Chem.*, 240 (1965) 2914.

34 G. A. Jamieson, *Biochim. Biophys. Acta*, 121 (1966) 326.

35 A. G. Bearn and W. C. Parker, in F. Gross, *Iron Metabolism: An International Symposium*, Springer-Verlag, Berlin, 1964, p. 67.

36 J. C. Robinson and J. E. Pierce, *Arch. Biochem. Biophys.*, 106 (1964) 348.

37 J. Williams, *Biochem. J.*, 108 (1968) 57.

38 R. Got, Y. Goussault and J. Font, *Carbohydr. Res.*, 3 (1966) 157.

39 E. H. Morgan, G. Marsaglia, E. R. Giblett and C. A. Finch, *J. Lab. Clin. Med.*, 69 (1967) 370.

40 J. Montreuil, J. Tonnelat and S. Mullet, *Biochim. Biophys. Acta*, 45 (1960) 413.

41 P. Aisen, S. H. Koenig, W. Schillinger, I. H. Scheinberg, K. G. Mann and W. Fish, *Nature*, in press.

42 R. Warner and I. Weber, *J. Am. Chem. Soc.*, 75 (1953) 5094.

43 P. L. Masson and J. F. Heremans, *Eur. J. Biochem.*, 6 (1968) 579.

44 D. J. Perkins, *Protides of the Biological Fluids, Proc. Colloq. Bruges*, 14 (1966) 85.

45 P. Aisen, R. Aasa and A. G. Redfield, *J. Biol. Chem.*, 244 (1969) 4628.

46 S. S. Lehrer, *J. Biol. Chem.*, 244 (1969) 3613.

47 A. Bezkorovainy and M. E. Rafelson, Jr., *Arch. Biochem. Biophys.*, 107 (1964) 302.

48 R. Verbruggen, V. Blaton, M. Y. Rosenau-Motreff and H. Pecters, *Protides of the Biological Fluids, Proc. Colloq. Bruges*, 16 (1968) 101.

49 S. H. Koenig and W. Schillinger, *J. Biol. Chem.*, 244 (1969) 3283.

50 S. H. Koenig and W. Schillinger, *J. Biol. Chem.*, 244 (1969) 6520.

51 D. D. Ulmer, *Biochim. Biophys. Acta*, 181 (1969) 305.

52 J. -O. Jeppsson, *Acta Chem. Scand.*, 21 (1967) 1686.

53 A. Bezkorovainy and D. Grohlich, *Biochim. Biophys. Acta*, 147 (1967) 497.

54 F. C. Greene and R. E. Feeney, *Biochemistry*, 7 (1968) 1366.

55 A. Bezkorovainy, D. Grohlich and C. M. Gerbeck, *Biochem. J.*, 110 (1968) 765.

56 R. Aasa, B. G. Malmström, P. Saltman, and T. Vänngård, *Biochim. Biophys. Acta*, 75 (1963) 203.

57 A. L. Schade, *Farmaco*, 19 (1964) 185.

58 R. Woodworth, *Protides of the Biological Fluids, Proc. Colloq. Bruges*, 14 (1966) 37.

59 P. Aisen, A. Leibman and H. A. Reich, *J. Biol. Chem.*, 241 (1966) 1666.

60 R. V. Wenn and J. Williams, *Biochem. J.*, 108 (1968) 69.

61 R. C. Woodworth, A. T. Tan and L. R. Virkaitis, *Nature*, 223 (1969) 833.

62 P. Aisen and A. Leibman, *Biochem. Biophys. Res. Commun.*, 30 (1968) 407.

63 P. Aisen and A. Leibman, *Biochem. Biophys. Res. Commun.*, 32 (1968) 220.

64 A. T. Tan and R. Woodworth, *Biochemistry*, 8 (1969) 3711.

65 R. C. Woodworth, K. G. Morallee and R. J. P. William, *Biochemistry*, 9 (1970) 839.
66 F. Hosain, P. A. McIntyre, K. Poulose, H. S. Stern and H. N. Wagner, Jr., *Clin. Chim. Acta*, 24 (1969) 69.
67 A. Ehrenberg and C. -B. Laurell, *Acta Chem. Scand.*, 9 (1955) 68.
68 B. Gaber and P. Aisen, unpublished observation.
69 A. L. Schade and R. W. Reinhart, *Protides of the Biological Fluids, Proc. Colloq. Bruges*, 14 (1966) 75.
70 P. Aisen, R. Aasa, B. G. Malmström and T. Vänngård, *J. Biol. Chem.*, 242 (1967) 2484.
71 J. W. Young and D. J. Perkins, *Eur. J. Biochem.*, 4 (1968) 385.
72 B. Gaber, S. H. Koenig, W. Schillinger and P. Aisen, *J. Biol. Chem.*, in press.
73 W. G. Blumber, J. Eisinger, P. Aisen, A. G. Morell and I. H. Scheinberg, *J. Biol. Chem.*, 238 (1963) 1675.
74 A. S. Brill and G. F. Bryce, *J. Chem. Phys.*, 48 (1968) 4398.
75 A. Wishnia, I. Weber and R. C. Warner, *J. Am. Chem. Soc.*, 83 (1961) 2071.
76 R. Aasa and P. Aisen, *J. Biol. Chem.*, 243 (1968) 2399.
77 D. Ulmer and B. L. Vallee, *Biochemistry*, 2 (1963) 1335.
78 B. Nagy and S. S. Lehrer, *Abstracts, Third International Congress of Biophysics*, Cambridge, Mass., 1969, p. 11.
79 J. J. Windle, A. K. Wiersema, J. R. Clark and R. E. Feeney, *Biochemistry*, 2 (1963) 1341.
80 R. Aasa, *J. Chem. Phys.*, in press.
81 W. E. Blumberg, in A. Ehrenberg, B. G. Malmström and T. Vänngård, *Magnetic Resonance in Biological Systems*, Pergamon Press, Oxford, 1967, p. 119.
82 R. D. Dowsing and J. F. Gibson, *J. Chem. Phys.*, 50 (1969) 294.
83 E. E. Hazen, Jr., *Ph.D. Thesis*, Harvard University, 1962.
84 R. J. Dern, A. Monti and M. F. Glynn, *J. Lab. Clin. Med.*, 61 (1963) 280.
85 J. Fletcher and E. R. Huehns, *Nature*, 215 (1967) 584.
86 H. Buttkus, J. R. Clark and R. E. Feeney, *Biochemistry*, 4 (1965) 998.
87 W. F. Line, D. Grohlich and A. Bezkorovainy, *Biochemistry*, 6 (1967) 3393.
88 S. K. Kornatsu and R. E. Feeney, *Biochemistry*, 6 (1967) 1136.
89 J. H. Jandl, J. K. Inman, R. L. Simmons and D. W. Allen, *J. Clin. Invest.*, 38 (1959) 161.
90 J. V. Princiotto, M. Rubin, G. C. Shashaty and E. J. Zapolski, *J. Clin. Invest.*, 43 (1964) 825.
91 G. M. W. Cook, D. H. Heard and G. V. F. Seaman, *Nature*, 188 (1960) 1011.
92 E. H. Morgan, *Br. J. Haematol.*, 10 (1964) 442.
93 E. H. Morgan and T. C. Appleton, *Nature*, 223 (1969) 1371.
94 S. Kornfeld, *Biochim. Biophys. Acta*, 194 (1969) 25.
95 E. Baker and E. H. Morgan, *Biochemistry*, 8 (1969) 2954.
96 J. H. Jandl and J. H. Katz, *J. Clin. Invest.*, 42 (1963) 314.
97 E. Baker and E. H. Morgan, *Biochemistry*, 8 (1969) 1133.
98 J. H. Katz and J. H. Jandl, in F. Gross, *Iron Metabolism*, Springer-Verlag, Berlin, 1964, p. 112.
99 E. H. Morgan and C. -B. Laurell, *Br. J. Haematol*, 9 (1963) 471.
100 S. Kornfeld, *Biochemistry*, 7 (1968) 945.
101 P. Aisen, Mount Sinai J. Med., in press.
102 M. S. Wheby and G. Umpierre, *New Engl. J. Med.*, 271 (1964) 1391.
103 L. Heilmeyer, in F. Gross, *Iron Metabolism, An International Symposium*, Springer-Verlag, Berline, 1964, p. 201.
104 P. DeLacy, P. L. Masson and J. F. Heremans, *Protides of the Biological Fluids, Proc. Colloq. Bruges*, 16 (1968) 633.
105 E. R. Giblett, *Genetic Markers in Human Blood*, Blackwell Scientific Publications, Oxford, 1969, p. 138.

Chapter 10

CERULOPLASMIN *

I. H. SCHEINBERG AND A. G. MORELL

Department of Medicine, Albert Einstein College of Medicine, New York, U.S.A.

INTRODUCTION

 Copper, like iron, possesses the dual qualities of essentiality and toxicity for many biological systems including man[1-4]. Ewes grazing on copper-
deficient soil drop lambs unable to myelinate their developing neurons
unless copper salts are placed in the pastures. Uptake of body iron stores by
transferrin, necessary before iron can be incorporated into hemoglobin, is
markedly impaired in a state of copper deficiency[5,6]. Cytochrome oxidase
and tyrosinase, among other proteins, contain copper as an essential structural component[7,8]. Recent evidence suggests that normal taste acuity[9] and
normal development of bone[10] and collagen[11,12] are dependent on copper.
Nonetheless, physiologic consequences of copper deficiency are seen almost
only in animals. Human copper deficiency is almost unknown since a great
excess of the metal is generally present in all diets.
 Despite the large intake of dietary copper, the concentrations of the
metal in the tissues of a normal human being remain quite constant throughout life: the net long-term balance is zero. Since copper is a small diffusible
ion which is rather readily absorbed from the gastrointestinal tract, some
mechanism must exist which prevents a steady accumulation of copper in
tissues by limiting the absorption or promoting the excretion of copper
under physiologic conditions. A faulty mechanism of this sort, or no mechanism at all, is inherited by about four humans in a million in whom toxic
effects of the excessive deposits of tissue copper are ultimately manifest[13].
The earliest sign of this disorder — known as Wilson's disease — is a deficiency
of ceruloplasmin, a copper-containing glycoprotein of plasma which has been
more intensively studied than any other biologic compound of copper.

PHYSICAL AND CHEMICAL PROPERTIES OF CERULOPLASMIN

 Ceruloplasmin has been demonstrated in the plasma of man, monkey,
pig, horse, donkey, cow, deer, goat, dog, cat, guinea pig, rabbit, mouse, rat
and chicken[14-18]. In the frog it is found only after metamorphosis[19]. The

---

characteristic blue color and enzymatic activity of the protein are both due to its copper content and are similar, but not identical, in all species. Human ceruloplasmin is an intensely blue, α_2-globulin containing 0.3% copper and about 8% carbohydrate. It possesses oxidase activity toward a number of polyamines, polyphenols and the Fe(II) ion. Grammes of it may be readily prepared in good yield from pooled human plasma, which normally contains 20–40 mg of ceruloplasmin per 100 ml, or from ceruloplasmin-rich plasma fractions such as Cohn's Fraction IV. Several preparatory methods, such as chromatography on DEAE or hydroxylapatite, or precipitation by ammonium sulfate, ethanol, or ethanol–chloroform, have been devised[20-27]. The degree of purification is generally proportional to the ratio of light absorption at 610 nm to that at 280 nm, which is 0.046 in the best preparations. From the latter, the human protein can be crystallized as tetragonal prisms[28] (see Fig. 1).

Fig. 1. Crystals of human ceruloplasmin (× 75) (photomicrograph by Milton Kurtz and Irmin Sternlieb).

The amino acid composition of human ceruloplasmin is shown in Table I. No work related to sequence studies has been reported. The secondary structure appears to involve β and random configuration with little

TABLE I

AMINO ACID COMPOSITION OF HUMAN CERULOPLASMIN

Amino acid	Residues per 151,000 M.W.[a][79]
Aspartic acid	143
Threonine	89
Serine	73
Glutamic acid	132
Proline	60
Glycine	89
Alanine	58
$\frac{1}{2}$-Cystine	10
Cysteine	—
Valine	73
Methionine	19
Isoleucine	61
Leucine	80
Tyrosine	57
Phenylalanine	66
Lysine	77
Histidine	50
Arginine	44
Tryptophane	23

N-terminal amino acids per molecule: valine, 0.9 and lysine 0.25[57].

-SH groups detected per molecule: in ceruloplasmin   1[45,57,80]
in apoceruloplasmin, 5[45]
in apoceruloplasmin, 4[81]

[a] Assuming a total molecular weight of 151,000 of which approximately 9% is assumed to be carbohydrate and copper, this gives a molecular weight due to the amino acid residues of between 137,000 and 138,000.

or no α-helix. From its hydrodynamic properties (Table II), the tertiary structure appears to be that of a fairly compact molecule. Its quarternary structure has recently been investigated by Simons and Bearn[29] who propose a tetrameric model with two polypeptide chains of molecular weight 15,900, with N-terminal valines, and two chains of molecular weight 59,000, with N-terminal lysines.

Some optical, electrophoretic and crystallographic properties of human ceruloplasmin are presented in Tables III and IV.

TABLE II

PHYSICAL PROPERTIES OF HUMAN CERULOPLASMIN

Molecular weight	151,000 (sedimentation-diffusion)[34]
	160,000 (sedimentation-diffusion)[33]
	155,000 (approach to equilibrium)[33]
	143,000 (approach to equilibrium)[82]
	132,000 (X-ray crystallography)[32]

Sedimentation coefficient (S)	7.08[33]
Diffusion coefficient ($cm^2 sec^{-1}$)	3.76×10^{-7} [33]
Partial specific volume ($cm^3 g^{-1}$)	0.714[28]
	0.713[33]
Axial ratio (a/b)	3.6[45]
Viscosity ($cm^{-1} g sec^{-1}$)	4.1[45]
Isoelectric point	4.4[34]
Unit cell dimensions of crystal (space group, 14)	$a = b = 268$ Å [32]
	$c = 129$ Å [32]

TABLE III

OPTICAL PROPERTIES OF HUMAN CERULOPLASMIN

Wavelength (nm)	Gram-atomic absorption coefficient[a]	$\epsilon_{1 cm}^{1\%}$
794	850[83]	
610	4400[83]	0.68[23]
459	460[83]	0.71[45]
332	1600[83]	
280		14.6[23]
		15.4[45]

[a] Assuming that all Cu^{2+}, and only Cu^{2+}, ions are responsible for all absorption

TABLE IV

ELECTROPHORETIC PROPERTIES OF HUMAN CERULOPLASMIN[a]

Buffer	$\gamma/2$	pH	Mobility ($cm^2 V^{-1} sec^{-1} \times 10^5$)
Barbital	0.1	8.60	−5.20[33]
Tris	0.1	8.41	−5.26[33]
Phosphate	0.1	7.01	−4.72[33]
Acetate	0.1	5.48	−2.74[33]
Acetate	0.2	5.40	−3.50[44]

[a] Moving boundary electrophoresis. Two or more oxidase-active components can be separated by starch gel-[84], paper-[85] or microimmuno-electrophoresis[86].

COPPER IN CERULOPLASMIN

Copper atoms are an integral part of the ceruloplasmin structure, and remain attached to the protein during its entire life span in the circulation[30]. There is no exchange of ceruloplasmin copper with injected inorganic radioactive copper, *in vivo*. Consequently, hypotheses of ceruloplasmin's physiological function which postulate reversible ionization of copper from the protein (thereby, for example, creating a concentration gradient to regulate absorption of dietary copper[31]) are improbable.

The great stability of the copper–protein bonds in ceruloplasmin depends on the integrity of the entire protein molecule. Dissociation of the molecule into subunits, or its proteolytic degradation, results in rupture or marked weakening of these bonds. Their unique character can only be inferred from studies of the intact protein, or of a preparation of its apo-protein which can revert to the holoprotein upon addition of copper.

Number of copper atoms in ceruloplasmin

The number of copper atoms in each molecule of ceruloplasmin is uncertain because of lack of accurate knowledge of its molecular weight and, until recently, of its copper content. The molecular weight of the protein is probably between 150,000 and 160,000, but one X-ray study suggests it is as low as 132,000[32]. Early reports indicated the copper content of the protein to be 0.32-0.34%[33,34] but more recent analyses of crystalline preparations show it to be closer to 0.28%[35]. From these recent data, the number of copper atoms calculated per molecule is either six or seven.

Properties of ceruloplasmin due to copper

The oxidase activity of ceruloplasmin is a function of the reversible reduction of cupric ions by substrate and oxygen[36]. Whether all of the cupric ions in the protein are enzymatically active is unknown. The intense blue color of ceruloplasmin is also due to cupric ion and can be completely, and reversibly, abolished by reducing agents. Again, it is uncertain whether all of the protein's cupric ions contribute to the blue color[37].

Valence state of copper in ceruloplasmin

The valences of the copper atoms in ceruloplasmin were first determined by chemical methods, utilizing reagents specific for cuprous ions[38] and, later, by polarography and paramagnetic studies of cupric ions. The results are reasonably consistent and indicate that approximately half of the protein's copper is cupric (Table V). The cupric ions are not all equivalent since the disappearance of blue color, consequent to their progressive reduction by ascorbate, is related non-linearly to changes in the proton relaxation effects produced by cupric ions[39]. The cuprous ions of ceruloplasmin cannot be oxidized without denaturing the protein, although the cupric ions can be reversibly reduced. Reoxidation of the latter can be affected only by molecular oxygen, and not by such oxidizing ions as Cu^{2+}, Fe^{3+}, or ferricyanide[34]. Like cupric ions, the cuprous ions are also heterogeneous since only two of the three or four per molecule are bound to sulfur. There is no coordinate binding of oxygen to either cupric or cuprous ions in native, blue ceruloplasmin[40].

TABLE V

CHARACTERISTICS OF COPPER IN CERULOPLASMIN

Atoms/molecule:	8[33,34]	
	7[35,87,88]	
	6[32]	
Cu^{2+}: total Cu:		
Electron paramagnetic resonance	0.52[89]	
	0.29[83]	
	0.43, 0.48[90]	
	0.40[91]	
Magnetic susceptibility	0.40[92]	
	0.44[87]	
Polarography	0.54[93]	
$g_\perp$	2.048[94]	
$g_\parallel$	2.214[94]	
$A_\parallel$	0.008 cm^{-1}[94]	

Exchange of ceruloplasmin copper with ionic copper

The protein's copper atoms may be made to exchange with inorganic radioactive cuprous ions, though only *in vitro*, provided that all the copper atoms in the protein are first reduced★[31,41]. The rate of exchange varies widely for the various copper atoms of the protein. No exchange occurs in the presence of cysteine, and the ubiquitous distribution of cysteine residues *in vivo* may be at least part of the reason that the copper of ^{64}Cu- or ^{67}Cu-labelled ceruloplasmin seems to be irreversibly bound *in vivo*[30].

★In the original report of this phenomenon[31], only half of the copper atoms were exchanged. With longer incubation and an increase in the concentration of added cuprous ion, all atoms undergo exchange.

APOCERULOPLASMIN

When copper is dissociated from ceruloplasmin by acidification below its isoelectric point, or by treatment with cyanide[42], a colorless protein is produced from which it is impossible to regenerate native ceruloplasmin by the addition of copper ions. Reversible removal of copper from ceruloplasmin can, however, be achieved by treatment of the protein with diethyl-dithiocarbamate (DTC)[43]. Prior reduction of the protein's copper by cysteine, at a pH of 5.0, and high ionic strength are required to form a colloidal complex of ceruloplasmin copper and DTC which is readily removed by ultracentrifugation. The apoprotein may be purified by precipitation from the supernatant with ammonium sulfate, re-solution, and dialysis. In solution, it is stable for several months in the frozen state and ceruloplasmin can be reconstituted from it by the addition of cuprous ions and reoxidation by air. The resultant blue protein is indistinguishable in all of its properties from native ceruloplasmin. No species of the protein with a number of copper atoms intermediate between 0 and its full complement has ever been detected during interconversions of ceruloplasmin and its apoprotein[44].

Recent studies in this laboratory reveal that the reversible removal of copper from ceruloplasmin results in moderate conformational changes in the protein, and that two of the cuprous ions in the native protein are bound to sulfur[45].

Apoceruloplasmin injected into experimental animals appears to be rapidly lost from the circulation.

CERULOPLASMIN AS AN OXIDASE

The enzymatic properties of ceruloplasmin were first described by Holmberg and Laurell[46] who showed it to be the only polyamine and polyphenol oxidase of plasma. Catalysis of the oxidation of paraphenylene-diamine, and similar compounds, by ceruloplasmin is the common basis of the quantitative estimation of the latter in solution[31,47,48]. Substrates of biological interest include adrenalin, noradrenalin, serotonin, catecholamine and dopamine, but no physiologic significance of their oxidation by ceruloplasmin has been demonstrated[49].

There has been considerable controversy regarding the enzymatic properties of ceruloplasmin, particularly with respect to ascorbic acid as a possible substrate[25,50]. A current concept of the protein's enzymatic role assumes its cupric ions are cyclically reduced and oxidized by Fe(II) ion and molecular oxygen[51]. Several compounds, including DOPA and ascorbic acid, may be oxidized through this couple. Osaki and co-workers have termed the cycle itself "ferroxidase activity" and have suggested that ceruloplasmin thus oxidizes hepatic stores of Fe(II) iron to Fe(III) iron which is then bound to transferrin[52]. The resulting mobilization of iron from the liver has been demonstrated *in vivo*, utilizing copper-deficient pigs[53], though physiologic, or pathologic consequences related to this phenomenon are still unproven. Thus, patients with Wilson's disease, particularly during treatment, may exhibit marked deficiency, if not complete absence of ceruloplasmin for periods of years. Nevertheless, though such patients may have mild anemias when treatment is begun, in almost all of them hemoglobin and hematocrit levels soon rise to normal, and no abnormality of iron metabolism is demonstrable. Two recent findings may resolve this paradoxical situation. Topham and Frieden[54] have reported the identification and purification of a non-ceruloplasmin, copper-containing protein of human serum which also has ferroxidase activity but appears to be distinct from ceruloplasmin. And it has also been proposed that citrate may be a third component of human serum with ferroxidase activity[55].

HETEROGENEITY AND GENETIC POLYMORPHISM OF CERULOPLASMIN

Broman[56] first showed human ceruloplasmin to consist of two species, which were not artifacts, and which exhibited different chromatographic properties on hydroxylapatite[57, 58], and electrophoretic mobilities[59]. The meaning of this heterogeneity, except for its possible relationship to inherited isozymes, discussed below, was unclear. The copper, amino acid and carbohydrate composition, enzymatic activity, and spectral characteristics were identical in both chromatographic fractions. Richterich[60] observed differences in the relative concentrations of the two isozymes in neonates, adults and patients with Wilson's disease, but these could not be confirmed by Hirschman[61] or Broman[58]. Holtzman prepared peptide maps from crystalline ceruloplasmin derived from normal individuals and from patients with Wilson's disease and could demonstrate no difference between the two[62]. Unpublished work in our laboratory indicated these isozymes were synthesized at the same rate and from the same copper pool.

Shreffler and his co-workers described true genetic polymorphism of ceruloplasmin manifested by five phenotypes: CpA, CpB (the common form), CpAB, CpAC and CpBC[63]. The gene frequencies calculated for American Caucasians were: CpA = 0.006, CpB = 0.994, and CpC = 0; and for American Negroes: CpA = 0.053, CpB = 0.944, and CpC = 0.003. The data were interpreted as consistent with three co-dominant alleles at an autosomal locus each of which directed the synthesis of an isozyme with differing electrophoretic characteristics.

CERULOPLASMIN AS A GLYCOPROTEIN

Jamieson's data on the carbohydrate composition of ceruloplasmin are given in Table VI. There are nine or ten heterosaccharide chains, each of which contains six or seven carbohydrate residues, in a ceruloplasmin molecule. The chains are very similar but probably not identical in composition[64]. N-Acetylneuraminic acid (NANA) (or, rarely, fucose) is found at the non-reducing end of each chain and at the other end an N-acetyl-D-glucosamine is probably linked to the amide N of asparagine of the protein.

TABLE VI

CARBOHYDRATE COMPOSITION OF HUMAN CERULOPLASMIN[64]

% Carbohydrate	7–8
Heterosaccharide chains (per molecule)	9 or 10
Sialic acid residues (per molecule)	9
N-acetylglucosamine residues (per molecule)	18
Fucose residues (per molecule)	2
Mannose + galactose residues (per molecule)	36
Mannose:galactose residues (per molecule)	3:2 / 1:1[62]

Sialic acid may be removed from ceruloplasmin by mild acid hydrolysis or enzymatically. Sequential treatment of enzymatically produced asialoceruloplasmin with β-galactosidase and hexosaminidase permits the removal of D-galactose followed by that of N-acetyl-D-glucosamine. Removal of sialic acid produces an asialoceruloplasmin with its immunochemical properties, as well as its copper content, spectral and enzymatic characteristics, unchanged from the native protein[35]. Of course, the loss of the negatively charged sialic acid residues results in modification of the protein's electrophoretic mobility. Asialoceruloplasmin can be crystallized from 0.10 M sodium acetate, pH 5.45, and 1% NaCl at 4°C in the form of blue hexagonal prisms terminating at both ends in shallow pyramids, and differing from the crystal form of ceruloplasmin[65] (Fig. 2).

Employing completely desialylated ceruloplasmin as an acceptor, Hickman and co-workers[66] were able to restore up to 75% of sialic acid with a particulate sialyl transferase from rat liver and CMP-NANA-1-[14]C. Interestingly, desialylated apoceruloplasmin was found to be a better acceptor than desialylated ceruloplasmin for NANA, suggesting that in the biosynthetic process the addition of carbohydrate to the polypeptide may precede that of copper.

Tritium can be incorporated into the terminal carbinol of asialoceruloplasmin by sequential treatment of this protein with galactose oxidase and

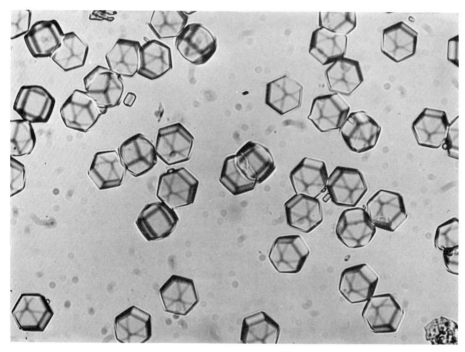

Fig. 2. Crystals of human desialylated ceruloplasmin[65] (× 400) (photomicrograph by Irmin Sternlieb).

tritiated borohydride. Like ceruloplasmin, tritiated asialoceruloplasmin may also be labelled with radioactive copper by the exchange reaction or by reconstitution of the apoprotein[67].

Radioactive homologous asialoceruloplasmin, when injected into rabbits, disappears from the circulation within minutes, although the physiologic half-life of native rabbit ceruloplasmin is 55 hours. Asialoceruloplasmin is taken up by the parenchymal cells of the liver and not by cells of the reticuloendothelial system[67]. This rapid transfer of asialoceruloplasmin from plasma to liver occurs only if intact galactose residues are present terminally in two or more of the polysaccharide chains[68]. If these residues are converted to aldehyde derivatives by galactose oxidase, removed by β-galactosidase, or covered by sialic acid again, survival of the protein in the circulation approximates that of the native protein[67].

Following uptake by the liver, catabolism of asialoceruloplasmin occurs rapidly, and principally in lysosomes[69]. About half of the protein-bound copper is split from the molecule within 20 min of hepatic uptake, and the cleavage of galactose is even faster. Seven other glycoproteins (orosomucoid, thyroglobulin, α_2-macroglobulin, haptoglobin, lactoferrin, fetuin and human gonadotrophic hormones)[70] are also rapidly taken up by the liver when freed of sialic acid. The survival of asialotransferrin in the circulation is, in singular contrast, identical to that of native transferrin. Whether these observations are an indication of the normal routes, sites and rates of catabolism of ceruloplasmin, as well as of other glycoproteins, remains to be determined.

BIOSYNTHESIS OF CERULOPLASMIN

Ceruloplasmin, like many plasma proteins, is synthesized in the liver[71], and the rate appears to be under hormonal control[72]. The concentration of ceruloplasmin in plasma is remarkably stable, though it is low in the neonate. Ceruloplasmin is also found in spinal fluid[73], lymph, joint fluid, lacrimal, nasal and gastrointestinal secretions[74]. In humans a low rate of synthesis is seen only in Wilson's disease and in copper deficiency[3].

Incorporation of copper occurs during biosynthesis only. The intracellular copper pool utilized for biosynthesis of ceruloplasmin is very small and probably microsomal[75]. The very low rate of incorporation of radioactive copper into ceruloplasmin in patients with Wilson's disease probably reflects both an increased size of this copper pool in the liver, and a low intrinsic rate of synthesis of the protein. Heterozygous carriers of one of the genes responsible for Wilson's disease generally have normal rates of ceruloplasmin synthesis but a less than normal incorporation of radioactive copper into the protein, probably due to modest increases in the size of the hepatic copper pool.

Biosynthesis of ceruloplasmin involves, in probable sequence, synthesis of several polypeptide chains, their assembly, glycosylation and copper addition. Apoceruloplasmin has recently been reported to be present in the plasma of copper-deficient rats[76] and has been detected, immunochemically, in normal human serum and in serum from hypoceruloplasminemic patients with Wilson's disease[77,78]. The precise nature of this copper-free, white protein, and its physiologic significance, if any, are unknown at present.

ACKNOWLEDGEMENTS

The work described herein, which was performed in the authors' laboratory, was supported, in part, by a grant from the National Institute of Arthritis and Metabolic Diseases (AM 01059) and The National Genetic Foundation, Inc.

REFERENCES

1 C. A. Elvehjem, *Physiol. Rev.*, 15 (1935) 471.
2 H. R. Marston, *Physiol. Rev.*, 32 (1952) 66.
3 I. H. Scheinberg and I. Sternlieb, *Pharmacol. Rev.*, 12 (1960) 355.
4 E. J. Underwood, *Trace Elements in Human and Animal Nutrition*, 2nd edn., Academic Press, New York, 1962.
5 G. R. Lee, G. E. Cartwright and M. M. Wintrobe, *Proc. Soc. Exp. Biol. Med.*, 127 (1968) 977.
6 G. R. Lee, S. Nacht, J. N. Lukens and G. E. Cartwright, *J. Clin. Invest.*, 47 (1968) 2058.
7 J. Peisach, P. Aisen and W. E. Blumberg (eds.), *The Biochemistry of Copper*, Academic Press, New York, 1966.
8 B. G. Malmström and J. B. Neilands, *A. Rev. Biochem.*, 33 (1964) 331.
9 R. I. Henkin, H. R. Keiser, I. A. Jaffe, I. Sternlieb and I. H. Scheinberg, *Lancet*, ii (1967) 1268.
10 R. I. Henkin, H. R. Keiser and M. Kare, *Proc. Soc. Exp. Biol. Med.*, 129 (1968) 516.
11 M. E. Nimni, *Biochim. Biophys. Acta*, 111 (1965) 576.
12 I. A. Jaffe, P. Merriman and D. Jacobus, *Science*, 161 (1968) 1016.
13 I. H. Scheinberg and I. Sternlieb, *A. Rev. Med.*, 16 (1965) 119.
14 S. Garattini, A. Giachetti and L. Pieri, *Arch. Biochem. Biophys.*, 91 (1960) 83.
15 U. S. Seal, *Comp. Biochem. Physiol.*, 13 (1964) 143.
16 C. J. A. Van Den Hamer, G. Buyze and M. C. M. van der Heyden, in H. Peeters *Protides of the Biological Fluids*, Vol. 11, Elsevier, Amsterdam, 1964, p. 382.
17 G. M. Martin, M. A. Derr and E. P. Benditt, *Lab. Invest.*, 13 (1964) 282.
18 J. B. Bingley and A. T. Dick, *Clin. Chem. Acta*, 25 (1969) 480.
19 T. Inaba and E. Frieden, *J. Biol. Chem.*, 242 (1967) 4789.
20 B. E. Sanders, O. P. Miller and M. N. Richard, *Arch. Biochem. Biophys.*, 84 (1959) 60.
21 G. Curzon and L. Vallet, *Nature*, 183 (1959) 751.
22 G. Curzon and L. Vallet, *Biochem. J.*, 74 (1960) 279.
23 H. F. Deutsch, *Arch. Biochem. Biophys.*, 89 (1960) 225.
24 H. F. Deutsch, C. B. Kasper and D. Welsch, *Arch. Biochem. Biophys.*, 99 (1962) 132.
25 A. G. Morell, P. Aisen and I. H. Scheinberg, *J. Biol. Chem.*, 237 (1962) 3455.
26 J. T. Sgouris, F. C. Coryell, H. Gallick, R. W. Storey, K. B. McCall and H. D. Anderson, *Vox Sang.*, 7 (1962) 394.
27 L. Broman and K. Kjellin, *Biochim. Biophys. Acta*, 82 (1964) 101.
28 A. G. Morell, C. J. A. Van Der Hamer and I. H. Scheinberg, *J. Biol. Chem.*, 244 (1969) 3494.
29 K. Simons and A. G. Bearn, *Biochim. Biophys. Acta*, 175 (1969) 260.
30 I. Sternlieb, A. G. Morell, W. D. Tucker, M. W. Greene and I. H. Scheinberg, *J. Clin. Invest.*, 40 (1961) 1834.
31 I. H. Scheinberg and A. G. Morell, *J. Clin. Invest.*, 36 (1957) 1193.
32 B. Magdoff-Fairchild, F. M. Lovell and B. W. Low, *J. Biol. Chem.*, 244 (1969) 3497.
33 C. B. Kasper and H. F. Deutsch, *J. Biol. Chem.*, 238 (1963) 2325.
34 C. G. Holmberg and C. -B. Laurell, *Acta Chem. Scand.*, 2 (1948) 550.
35 A. G. Morell, C. J. A. Van Den Hamer, I. H. Scheinberg and G. Ashwell, *J. Biol. Chem.*, 241 (1966) 3745.
36 L. Broman, B. G. Malmström, R. Aasa and T. Vänngard, *Biochim. Biophys. Acta*, 75 (1963) 365.

318

37 L. -E. Andreasson and T. Vänngard, *Biochim. Biophys. Acta*, 200 (1970) 247.
38 G. Felsenfeld, *Arch. Biochem. Biophys.*, 87 (1960) 247.
39 A. G. Morell, P. Aisen, I. H. Scheinberg, W. E. Blumberg and J. Eisinger, *Federation Proc.*, 22 (1963) 595.
40 A. G. Morell, P. Aisen, W. E. Blumberg and I. H. Scheinberg, *J. Biol. Chem.*, 239 (1964) 1042.
41 J. Marriott and D. J. Perkins, *Biochim. Biophys. Acta*, 117 (1966) 387.
42 C. B. Kasper and H. F. Deutsch, *J. Biol. Chem.*, 238 (1963) 2343.
43 A. G. Morell and I. H. Scheinberg, *Science*, 127 (1958) 588.
44 P. Aisen and A. G. Morell, *J. Biol. Chem.*, 240 (1965) 1974.
45 C. J. A. Van Den Hamer, A. G. Morell and I. H. Scheinberg, unpublished results.
46 C. G. Holmberg and C. -B. Laurell, *Acta Chem. Scand.*, 5 (1951) 476.
47 E. W. Rice, *J. Lab. Clin. Med.*, 55 (1960) 325.
48 H. A. Ravin, *Lancet*, i (1956) 726.
49 C. -B. Laurell, in F. W. Putnam, *The Plasma Proteins*, Vol. 1, Academic Press, New York, 1960, p. 349.
50 S. Osaki, J. A. McDermott and E. Frieden, *J. Biol. Chem.*, 239 (1964) 3570.
51 G. Curzon and S. O'Reilly, *Biochem. Biophys. Res. Commun.*, 2 (1960) 284.
52 S. Osaki, D. A. Johnson and E. Frieden, *J. Biol. Chem.*, 241 (1966) 2746.
53 H. A. Ragan, S. Nacht, G. R. Lee, C. R. Bishop and G. E. Cartwright, *Am. J. Physiol.*, 217 (1969) 1320.
54 R. W. Topham and E. Frieden, *J. Biol. Chem.*, 245 (1970) 6698.
55 G. R. Lee, S. Nacht, D. Christensen, S. Hansen and G. E. Cartwright, *Proc. Soc. Exp. Biol. Med.*, 131 (1969) 918.
56 L. Broman, *Nature*, 182 (1958) 1655.
57 H. F. Deutsch and G. B. Fisher, *J. Biol. Chem.*, 239 (1964) 3325.
58 L. Broman, *Acta Soc. Med. Upsal. Suppl.*, 69 (1964) 7.
59 A. G. Morell and I. H. Scheinberg, *Science*, 131 (1960) 930.
60 R. Richterich, E. Gautier, H. Stillhart and E. Rossi, *Helv. Paediat. Acta*, 15 (1960) 424.
61 S. Z. Hirschman, A. G. Morell and I. H. Scheinberg, *Ann. N.Y. Acad. Sci.*, 94 (1961) 960.
62 N. A. Holtzman, M. N. Naughton, F. L. Iber and B. M. Gaumnitz, *J. Clin. Invest.*, 46 (1967) 993.
63 D. C. Shreffler, G. J. Brewer, J. C. Gall and M. S. Honeyman, *Biochem. Genet.*, 1 (1967) 101.
64 G. A. Jamieson, *J. Biol. Chem.*, 240 (1965) 2019.
65 A. G. Morell, I. Sternlieb and I. H. Scheinberg, *Science*, 166 (1969) 1293.
66 J. Hickman, G. Ashwell, A. G. Morell, C. J. A. Van Den Hamer and I. H. Scheinberg, *J. Biol. Chem.*, 245 (1970) 759.
67 A. G. Morell, R. A. Irvine, I. Sternlieb, I. H. Scheinberg and G. Ashwell, *J. Biol. Chem.*, 243 (1968) 155.
68 C. J. A. Van Den Hamer, A. G. Morell, I. H. Scheinberg, J. Hickman and G. Ashwell, *J. Biol. Chem.*, 245 (1970) 4397.
69 G. Gregoriadis, A. G. Morell, I. Sternlieb and I. H. Scheinberg, *J. Biol. Chem.*, 245 (1970) 5833.
70 A. G. Morell, G. Gregoriadis, I. H. Scheinberg, J. Hickman and G. Ashwell, *J. Biol. Chem.*, 246 (1971) 1461.
71 C. A. Owen, Jr., and J. B. Hazelrig, *Am. J. Physiol.*, 210 (1966) 1059.
72 G. W. Evans, N. F. Cornatzer and W. E. Cornatzer, *Am. J. Physiol.*, 218 (1970) 613.
73 K. Jensen, *Acta Neurol. Scand.*, 39 (1963) 237.
74 J. F. Wilson, D. C. Heiner and M. E. Lahey, *J. Pediatrics*, 60 (1962) 787.
75 I. Sternlieb and I. H. Scheinberg, *Trans. Ass. Am. Phys.*, 75 (1962) 228.

76 N. A. Holtzman and B. M. Gaumnitz, *J. Biol. Chem.*, 245 (1970) 2350.
77 R. J. Carrico, H. F. Deutsch, H. Beinert and W. H. Orme-Johnson, *J. Biol. Chem.*, 244 (1969) 4141.
78 R. J. Carrico and H. F. Deutsch, *Biochem. Med.*, 3 (1969) 117.
79 J. T. Edsall and P. -F. Spahr, personal communication, 1960.
80 J. Witwicki and K. Zakrzewski, *Eur. J. Biochem.*, 10 (1969) 284.
81 J. O. Erickson, R. D. Gray and E. Frieden, *Proc. Soc. Exp. Biol. Med.*, 134 (1970) 117.
82 W. N. Poillon and A. G. Bearn, *Biochim. Biophys. Acta*, 127 (1966) 407.
83 W. E. Blumberg, J. Eisinger, P. Aisen, A. G. Morell and I. H. Scheinberg, *J. Biol. Chem.*, 238 (1963) 1675.
84 A. G. Morell and I. H. Scheinberg, *Science*, 131 (1960) 930.
85 D. V. Siva Sankar, *Federation Proc.* 18 (1959) 441.
86 J. Uriel, H. Gotz and P. Grabar, *J. Suisse Med.*, 87 (1957) 431.
87 P. Aisen, S. H. Koenig and H. R. Lilienthal, *J. Mol. Biol.*, 28 (1967) 225.
88 C. B. Kasper, *Biochemistry*, 6 (1967) 3185.
89 T. Vänngård, in A. Ehrenberg, B. G. Malmström and T. Vänngård, *Magnetic Resonance in Biological Systems*, Pergamon, Oxford, 1966, p. 213.
90 L. Broman, B. G. Malmström, R. Aasa and T. Vänngård, *J. Mol. Biol.*, 5 (1962) 301.
91 C. B. Kasper, H. F. Deutsch and H. Beinert, *J. Biol. Chem.*, 238 (1963) 2338.
92 A. Ehrenberg, B. G. Malmström, L. Broman and R. Mosbach, *J. Mol. Biol.*, 5 (1962) 450.
93 V. Kalons and H. Hobzova, in H. Peeters, *Protides of the Biological Fluids*, Vol. 14, Elsevier, Amsterdam, 1966, p. 173.
94 T. Vänngård and R. Aasa, in W. Low, *Paramagnetic Resonance*, Academic Press, New York, 1963, p. 509.

Chapter 11

HEMERYTHRIN*

M. Y. OKAMURA AND I. M. KLOTZ

Biochemistry Division, Department of Chemistry, Northwestern University, Evanston, Illinois 60201, U.S.A.

INTRODUCTION

   Among invertebrate animals oxygen-carrying proteins are often the non-heme pigments, hemerythrin and hemocyanin, in place of the familiar hemoglobin. Despite their functional resemblances there are great differences among these three proteins both in regard to their macromolecular structure and with respect to the molecular nature of the active oxygen-carrying site. Some of the contrasts between these proteins are summarized in Table I. Both hemoglobin and hemerythrin have iron as the oxygen-carrying metal, but despite the appearance of "heme" in the name hemerythrin (as well as hemocyanin) it does not contain a heme group. Also

TABLE I

COMPARISON OF SOME PROPERTIES OF OXYGEN-CARRYING PROTEINS

	Hemoglobin	Hemerythrin	Hemocyanin
Metal	Fe	Fe	Cu
Oxidation state of metal in deoxy protein	II	II	I
Metal: O_2	Fe:O_2	2Fe:O_2	2Cu:O_2
Color oxygenated	Red	Violet–pink	Blue
Color deoxygenated	Red-purple	Colorless	Colorless
Coordination of Fe	Porphyrin ring	Protein side chains	Protein side chains
Molecular weight	65,000	108,000	400,000–20,000,000
Number of subunits	4	8	Many

*Many of the studies described here were supported by a grant (No. HE-08299) from the National Heart Institute, U.S. Public Health Service. Abbreviations used; salen, N,N'-bissalicylideneethylenediamine; phen, 1,10-phenanthroline; bipy, 2,2'-bipyridine; terpy, 2,2',2''-terpyridine; Ac, acetate; EDTA, ethylenediaminetetraacetate; (-)PDTA, *levo*-propylenediaminetetraacetate; HEDTA, N-hydroxyethylethylenediaminetriacetate; B, 2,13-dimethyl-3,6,9,12,18-pentaazabicyclo 12.3.1 octadeca-1,[23] 2,12,14,16-pentaene; en, ethylenediamine.

hemerythrin combines with molecular oxygen in a ratio $2Fe:O_2$ rather than in the ratio $Fe:O_2$ found in hemoglobin. There are also substantial differences in molecular weight as well as in the number of subunits within a given macromolecule in its native state. In spite of these differences both hemerythrin and hemoglobin function effectively as oxygen carriers. It would obviously be of interest to learn to what extent a molecular structure can be varied and still perform the same function, in this case carrying oxygen.

DISTRIBUTION AND PREPARATION

Hemerythrin has been found in representatives of four different invertebrate phyla: sipunculids, polychaetes, priapulids, and bachiopods[1]. Most of the work reviewed here has been carried out with hemerythrin from the sipunculid *Golfingia gouldii.*

The protein from *Golfingia gouldii* can be isolated from the coelomic fluid of these marine worms. The coelomic cavities of about 100–200 worms are opened and the fluid drained. After coagulation seems completed, the fluid is filtered through glass wool. The erythrocytes are then washed several times with sea water (or with 2.5% NaCl) and centrifuged down after each wash. Some non-pigmented cells appear as a layer on top of the red cells and are removed by suction. A volume of distilled water equal to about one-half the initial volume of coelomic fluid is added to lyse the red cells, during about 12 hours. The cell debris is then removed by high speed centrifugation. The dark red solution of hemerythrin is then dialyzed extensively against a 0.4% NaCl solution. It can then be crystallized by dialysis against a 20% solution of ethanol containing 0.4% NaCl. The whole procedure is carried out in the cold (4°C) and takes 3–4 days. The resulting crystals of oxyhemerythrin are reasonably stable if stored at 4°C but are slowly converted to the met or iron(III) form of the protein[2].

CHEMICAL FORMS OF HEMERYTHRIN

Some of the different forms of hemerythrin are indicated schematically in Fig. 1. The physiologically significant reaction of this protein is its reversible combination with molecular oxygen. In the presence of suitable reagents other changes take place. The addition of an oxidizing agent, such as $K_3Fe(CN)_6$, to oxygenated hemerythrin produces metaquohemerythrin in which the iron is in the +3 oxidation state. Methemerythrin, in turn, can combine with a variety of small ligands to form derivatives exhibiting characteristic optical absorption spectra[3]. Native hemerythrin is an octa-

322

meric molecule but it may be dissociated into its monomeric form by
addition of reagents that react with SH groups[4].

Interconversions of chemical forms of hemerythrin

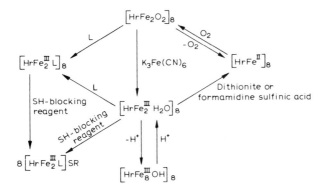

Fig. 1. Interconversions between chemical forms of hemerythrin.

Oxygen equilibrium

   The physiologically significant reaction of hemerythrin is its reversible
combination with molecular oxygen. The oxygen binding curve of *Sipunculus nudus* hemerythrin shows, in contrast to hemoglobin, only a slight
sigmoidal character and lacks a Bohr effect (*i.e.* oxygenation is independent
of pH)*[5]. Hemoglobin curves on the other hand are strongly sigmoidal
especially at lower pH and exhibit a marked Bohr effect. The difference
between the oxygen equilibria in these two proteins can be rationalized in
terms of their physiological function[6]. Hemerythrin is generally found in
the main body cavity of the organism and is thought to serve mainly in
oxygen storage whereas hemoglobin is found in the circulatory system and
serves chiefly in oxygen transport. It is in oxygen transport that the sigmoidal shape and the pH dependence of the oxygen equilibrium curves are
functionally valuable characteristics[7].
   Another important feature of the oxygen equilibrium of hemerythrin
is its high oxygen affinity, characteristic of the oxygen-binding pigments
found in the main body cavity of the organisms. This high affinity for
oxygen has been associated with a large favorable change in enthalpy in the
oxygen-binding process, the heats of reaction ranging from –12 to –20
kcal/mole[6].

---

*Hemerythrin from *Lingula*, however, does show a Bohr effect[6].

Perhaps the most significant difference between hemerythrin and hemoglobin is the iron to oxygen stoichiometry. For hemerythrin there are two iron atoms required for each oxygen molecule bound[2], whereas in hemoglobin one iron atom binds one oxygen molecule.

The kinetics of oxygenation and deoxygenation of hemerythrin from *Sipunculus nudus* have also been studied[5]. Both processes are fast but the system exhibits more than one relaxation time. Furthermore, there is a difference in the rate constant for deoxygenation obtained from temperature-jump and stopped-flow deoxygenating with dithionite. These results indicate that there may be conformational changes occurring in addition to the simple bimolecular process of combination of protein with oxygen.

Oxidation and reactions with small ligands

If oxyhemerythrin is allowed to stand even at $4°C$ it changes slowly to methemerythrin and loses its capacity to bind oxygen. Alternatively, addition of an oxidizing agent, $K_3Fe(CN)_6$, to oxyhemerythrin causes rapid oxidation and production of methemerythrin. Conversion of oxyhemerythrin to methemerythrin is also induced when relatively high concentrations of a ligand ($\sim 0.2\ M$) are present in a solution of oxyhemerythrin[3]. Hydrogen peroxide is detected as one of the reaction products★[8]. This reaction is first order in ligand concentration, is acid catalyzed in the pH range 6-9, and probably occurs by a displacement of peroxide ion (or hydroperoxide ion):

$$HL + HrFe_2O_2 \rightarrow HrFe_2^{3+}L + HO_2^-$$

The anions F^-, NO_2^-, N_3^-, and HCO_2^- are very effective catalysts, SCN^- less so, and SO_4^{2-}, $C_2H_3O_2^-$ and Cl^- are ineffective. The catalytic activity appears to decrease with increasing acidity of the anion, and maybe due to a general acid catalysis.

When hemerythrin is oxidized at neutral pH in the absence of ligand, metaquohemerythrin results. As the pH is raised above pH 9 metaquohemerythrin undergoes a characteristic spectral change as it is converted into methydroxyhemerythrin★★. The apparent pK_a for this transformation has been found to be markedly increased by specific anions[9], such as ClO_4^- and NO_3^-, which are bound to the protein but are not directly coordinated to the iron. In addition to affecting conversion of metaquohemerythrin

---

★When F^- and NO_2^- are the ligands the reaction appears to be pseudo first order. However when N_3^- is the ligand there is an initial fast reaction followed by a biphasic reaction.

★★The terms "metaquo" and "methydroxy" refer to the oxidized forms of hemerythrin at pH values near 7 and 9, respectively. It is possible that a hydroxyl ion instead of a water molecule is bound to iron in metaquohemerythrin, as has been proposed for hemoglobin[8a].

324

these anions affect other properties of hemerythrin such as reactivity of the SH groups on the protein[9].

The iron in methemerythrin can also form complexes with a variety of anionic ligands (e.g. N_3^-, SCN^-, Cl^-, F^-). In the cases where the stoichiometry has been established (e.g. N_3^- and SCN^-) it has been found that one ligand molecule is bound by two Fe atoms[3,4]. However, on the basis of the optical and circular dichroic spectra there is reason to believe that the stoichiometry of the F^-, H_2O, and OH^- derivatives of methemerythrin is $1 Fe : 1 L$[10].

SUBUNIT FORMATION

The hemerythrin molecule consists of eight subunits and can be dissociated into monomers by a variety of means. The gentlest of these is the reaction of hemerythrin with mercaptan blocking reagents, such as organic mercurials or cystine[4,11]. Various lines of evidence indicate that the monomer thus formed differs in conformation from the native monomer. The ultra-violet and circular dichroic spectra of the monomer cysteine derivative are different from the corresponding spectra of the octamer as well as from those of the monomer prepared by simple dilution of the protein[12].

The monomer prepared by blocking the SH group also differs from native monomer or octamer in its ability to bind a variety of bidentate chelating agents such as o-phenanthroline or tiron, to form non-dialyzable complexes[13]. o-Phenanthroline complexes with the Fe(II) form of hemerythrin iron, whereas tiron complexes with the Fe(III) form. The latter complex, with the SH-blocked monomer, is associated with an electron spin resonance (e.s.r.) signal at $g = 4.2$, characteristic of iron(III).

PROTEIN LIGANDS TO IRON

Since hemerythrin does not contain a porphyrin moiety, the iron must be coordinated, in part, to side chains provided by the protein matrix. The primary structure of hemerythrin has been established[14] and is shown in Fig. 2.

Of the 113 residues shown in Fig. 2, a large number are unlikely to form coordination linkages to iron. These include all the apolar side chains (such as Phe, Ile, Val, etc.) as well as those of threonine, serine, asparagine, glutamine and probably arginine. In contrast, cysteine, lysine, histidine, tyrosine, methionine, aspartic acid and glutamic acid are all potential candidates, and these total 43 (see Table II), more than enough to provide 8–10 ligands needed for 2 Fe atoms per subunit.★ With the possible exception of the sulfurs of cysteine and methionine all of the possible protein ligands are non-polarizable, "hard" ligands[15].

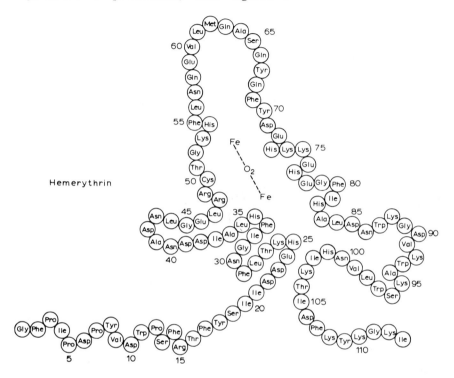

Fig. 2. Primary sequence of hemerythrin.

The sulfhydryl group of Cys, long thought to be a likely ligand, has been ruled out by experiments showing that its blockage by a mercurial did not produce any change in the absorption spectrum of the iron[4]. Similarly the ε-amino groups of lysine, as well as the α-amino terminal group can be ruled out as possible ligands since they can be blocked, for example with trinitrobenzene sulfonate, without a concomitant change in the spectrum[16].

On the other hand the reactivity of four of the histidines with diazonium-1H-tetrazole is attenuated in the protein; this could result from linkage to the iron[16]. Also the reaction of two tyrosines with tetranitromethane results in the release of one iron atom per subunit; this indicates

---

★One of the difficulties in determining the protein ligands lies in the fact that it has not been possible to remove the iron from hemerythrin in a reversible manner.

References pp. 342–343

that tyrosine may be involved in iron binding[17]. Of the other possible ligands to iron, including the carboxyl groups of glutamic or aspartic acid and the thioether group of methionine (Table II), no experimental evidence one way or the other has yet been provided.

TABLE II

POSSIBLE IRON LIGANDS

Group	Amino acid residue	Number in protein	Conclusions from chemical modification
SH	Cys	1	not involved
NH_2	Lys NH$_2$ terminal	11 1	not involved
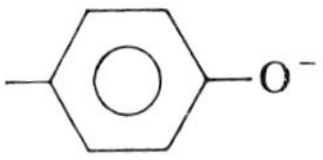 —O$^-$	Tyr	5	possibly 2 involved
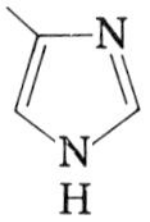	His	7	possibly 4 involved
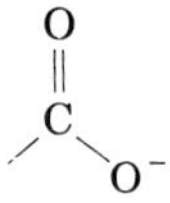	Glu Asp	6 11	— —
-S-CH$_3$	Met	1	—

MAGNETIC AND SPECTROSCOPIC PROPERTIES OF ACTIVE SITE

Since the objective of studying the electromagnetic properties of the iron in hemerythrin is to elucidate the structure of the active site, the properties of model compounds whose structures are known must also be considered. Furthermore, in view of the presence of two iron atoms per subunit in hemerythrin it is appropriate to examine binuclear complexes of iron as well as the mononuclear species.

Model compounds

Fe(III) iron has a strong tendency to polymerize in neutral or basic solution. An increasing number of dimeric as well as polymeric inorganic and biological iron(III) complexes have been characterized[8,19]. Binuclear iron(III) complexes exhibit anomalous magnetic behavior which is due to an antiferromagnetic exchange interaction, $-2JS_1 . S_2$, between the two iron atoms (J negative). For coupled high-spin iron(III) this results in a diamagnetic singlet ($S = 0$) ground state and a manifold of thermally accessible excited states, $S = 1, 2, 3, 4, 5$. Since the observed magnetic properties are due to excited states which are reduced in population at lower temperature, these magnetic properties at low temperature will approach those of a diamagnetic complex. This is observed experimentally as a reduction of the apparent magnetic moment[18,20] and a decrease in the intensity of the e.s.r. signal with decreasing temperature[21,22]. In addition, at low temperature, the Mössbauer spectrum is characteristic of a diamagnetic species and lacks magnetic hyperfine broadening even in the presence of high magnetic fields[23-25].

The optical absorption spectra of antiferromagnetically coupled high-spin iron(III) complexes exhibit an enhancement of the intensity of the long wavelength (800-900 nm) d-d bands. This can be explained at least in part by the removal of the "spin forbiddeness" of the transition[26].

In the binuclear bridged iron(III) complexes the antiferromagnetic exchange interaction arises from a superexchange mechanism and is determined by the overlap between the orbitals of the iron atoms and the bridg-

TABLE III

STRUCTURAL AND MAGNETIC PROPERTIES OF IRON(III) BINUCLEAR COMPLEXES[a]

Structure		$-J$ (cm^{-1})	μ_{eff} (room tem.) per Fe atom
Fe—O—Fe	(I)	80-100	1.28-2.0
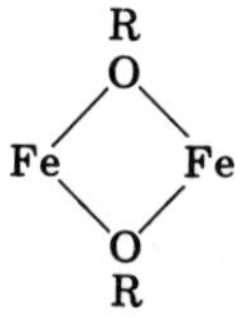 Fe $\diamond$ Fe (with R—O top and O—R bottom)	(II)[b]	8	5.3
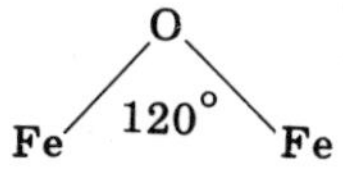 Fe $\overset{O}{\underset{120°}{\diagup\diagdown}}$ Fe	(III)	30	3.6[c]
Fe^{3+} low spin		0	2.0-2.5
Fe^{3+} high spin		0	5.7-6.0

[a] See ref. 10 and references therein
[b] R = H, salen
[c] Calculated for $g = 2$; no temperature independent paramagnetism

References pp. 342-343

ing ligand. The magnetic properties of several bridging groups are listed in Table III.

The linear oxo-bridge, (I) in Table III, found in many iron dimers exhibits a strong exchange interaction which is associated with a short FeO★ bond, indicating an appreciable amount of π-bonding.[27]. Another commonly found structural unit is the double hydroxy or alkoxy bridge, (II) in Table III. In these compounds the exchange interaction is much smaller partly due to weaker FeO single bonds and partly due to a weaker exchange interaction through a bridging angle of 90°.★★ A third situation occurs in trimeric iron(III) acetate; (III) in Table III; here the exchange interaction is intermediate between (I) and (II).

Hemerythrin

1. Absorption spectra and circular dichroism

The optical absorption spectra and circular dichroism data of hemerythrin have recently been reviewed[10]. The spectra of hemerythrin compare well with those of model oxo-bridged dimers as is shown in Table IV.

The optical absorption and circular dichroic spectra of all hemerythrin species, except aquo, hydroxy and fluoride, are independent of pH. Of particular interest are the bands in the region of wavelengths from 600–700 nm which have been assigned to d–d transitions. The extinction coefficients of these bands ($\epsilon \simeq 20$–50 per Fe) are much larger than expected for "spin forbidden" d–d transitions in monomeric high-spin iron(III) complexes. The intensification of the bands has been attributed to an exchange interaction coupling the two iron atoms.

In addition there are in hemerythrin charge transfer bands (from ligand to iron) which correlate well with charge transfer bands in simple $Fe^{3+}(H_2O)_5 L$ complexes, as well as other charge transfer bands at about 480 nm, 380 nm, and 330 nm.

The circular dichroic spectra of the met complexes with Cl^-, Br^-, NCO^-, CN^-, SCN^- and N_3^- are very similar except for differences due to the ligand-to-iron charge transfer band. In contrast the circular dichroic spectra of the met complexes of H_2O and OH^- are very different from those of the previously listed group of complexes. Fluoride complexes appear to be even more variable and exhibit circular dichroism similar to either the former or latter groups of ligands, depending upon the experimental conditions. These spectra have been interpreted by assigning a

---

★The Fe–O bond distance is about 1.8 Å for the oxo bridge, compared to 2 Å for a typical length[26a].

★★The weaker exchange in the 90° Fe–O–Fe bridge is expected since spin transfer through π overlap is much less than through σ overlap for FeO single bonds[27a].

TABLE IV

THE ABSORPTION SPECTRA OF HEMERYTHRIN COMPLEXES AND OF MODEL OXO-BRIDGED COMPLEXES[d]

Complex	Ligand[a]	Stoichiometry[b]	Absorption spectra[c]														
			λ(nm)	εm	λ(nm)	εm	λ(nm)	εm	λ(nm)	εm	λ(nm)	εm	λ(nm)	εm	λ(nm)	εm	λ(nm)	εm
Oxyhemerythrin	O_2	1	750sh	200	—	—	500	2200	360sh	5450	330	6800	—	—	—	—	—	—
	SH^-	1(?)	(not obsd)	—	—	—	510	~4000	~370sh	~5000	~330	~7000	—	—	—	—	—	—
	N_3^-	1	680	190	—	—	446	3700	380sh	4300	326	6750	—	—	—	—	—	—
	NCS^-	1	674	200	—	—	452	5100	370sh	4900	327	7200	—	—	—	—	—	—
	NCO^-	1	650	166	—	—	480sh	700	377	6500	334	6550	—	—	—	—	—	—
	Br^-	1	677	165	—	—	505sh	950	387	5400	331	6500	—	—	—	—	—	—
	Cl^-	1	656	180	—	—	490sh	750	380	6000	329	6600	—	—	—	—	—	—
	CN^-	1	695	140	—	—	493	770	374	5300	330	6400	—	—	—	—	—	—
	F^-	1 or 2	595sh	200	—	—	480sh	400	362	5000	317	5600	—	—	—	—	—	—
	OH^-	2	597	160	—	—	480sh	550	362	5900	320	6800	—	—	—	—	—	—
	H_2O	2	580sh	200	—	—	480sh	600	—	—	355	6400	—	—	—	—	—	—
	H_2O/I^-	—	550sh	800	—	—	480sh	1200	—	—	340	6300	—	—	—	—	—	—
$[Fe_3O(Ac)_6(H_2O)_3]^{2-}$	—	—	973	15.7	515sh	124	464	199	406sh	630	335	5010	—	—	—	—	—	—
$[Fe(H_2OB)_2O]^{4+}$	—	—	876	7.2	495	160	465	172	—	—	364	9360	—	—	—	—	—	—
$[FeEDTA)_2O]^{4-}$	—	—	878	6.6	540sh	113	475	199	405sh	810	335	10700	303	14100	274	15756	—	—
$[Fe(—)pDTA)_2O]^{4-}$	—	—	880	8.4	540sh	115	475	197	400sh	920	335	10600	305	13200	267	14100	240	17200
$[FeHEDTA)_2O]^{2-}$	—	—	875	6.5	540sh	104	474	196	405sh	830	335sh	11400	305	15400	267	14700	240	16500
$[Fe(phen)_2)_2O]^{4+}$	—	—	~970	—	580sh	—	—	—	—	—	352sh	—	310sh	—	270	16400	237	17600
$[Fe(bipy)_2)_2O]^{4+}$	—	—	960	—	575sh	—	525sh	—	—	—	345sh	—	310sh	—	273	—	225	—
$[Fe terpy)_2O]^{4+}$	—	—	~950	—	590sh	—	525	—	410	—	—	—	—	—	281	—	234	—
$[Fe salen)_2O]$	—	—	~1200	—	—	—	500sh	—	—	—	377	—	290sh	—	256sh	—	226	—
$[Fe (salen)Cl)_2]$	—	—	~1250	—	—	—	525	—	435	—	—	—	322	—	265	—	225	—

[a] Ligands forming complexes with hemerythrin. Iodide does not co-ordinate to the iron but binds nearby.

[b] The stoichiometries indicated have been proved only in the case of oxygen, azide, and thiocyanate. Those of cyanate, chloride, and hydroxide have been inferred from the interaction constant.

[c] The extinction coefficients are calculated per dimeric iron unit. Those bands indicated in italics are charge transfer bands arising from ligands other than the oxo bridge.

[d] Taken from ref. 10. See references therein.

sh signifies a "shoulder".

ligand to iron stoichiometry of 1:2 to the first situation and a stoichiometry of 1:1 to the latter behavior[10].

It is significant that the absorption spectrum as well as the circular dichroic spectrum of oxyhemerythrin is very similar to that of methemerythrin as was first pointed out by Nagy and Klotz[3], the similarity suggesting that the iron in hemerythrin is in the +3 valence state.

2. Magnetic susceptibility

The magnetic susceptibility at room temperature of various forms of hemerythrin have been measured in a number of laboratories using different techniques[25, 28, 29]. The observations are qualitatively in agreement. The deoxygenated form of hemerythrin contains iron(II) in the high-spin state. Oxygenated hemerythrin, as well as all derivatives of methemerythrin studied (N_3^-, SCN^-, F^-, CN^-, H_2O and CNO^-) have a lower magnetic susceptibility. Although an accurate diamagnetic correction has not been made the difference in the square of the effective magnetic moment between deoxyhemerythrin and the oxy- and met-hemerythrin is about $\Delta\mu_{eff}^2 = 18 \pm 2 (B.M.)^2$ per Fe.

In principle, the lower magnetic susceptibility at room temperature of oxyhemerythrin and methemerythrin could be attributed to spin pairing due to a large crystal field splitting of the d orbitals or to a strong antiferromagnetic interaction between adjacent iron atoms (see (I) in Table III).

The magnetic susceptibility at low temperatures and the Mössbauer spectra indicate that the latter is the case. The magnetic susceptibility of oxyhemerythrin and of methemerythrin is reduced at low temperatures[30, 31]. In the range 1.4–4.2°K the iron is diamagnetic[31]. This proves definitely (at least in methemerythrin) that the two iron atoms are antiferromagnetically coupled, since there is no way that a diamagnetic state can be achieved with an odd number of electrons as would be present in isolated iron(III) atoms[★7].

3. Mössbauer spectra

The Mössbauer spectra of hemerythrin derivatives[25, 32] give information about the electron density at the nucleus and the electric field gradient, which reflect the electronic state of the iron. The observed Mössbauer parameters are listed in Table V along with corresponding data for some model binuclear bridged iron(III) complexes.

★In oxyhemerythrin it is conceivable that a diamagnetic ground state is achieved by having the two Fe atoms as low-spin iron(II). However, it seems more likely that the reason for the low magnetic moment is the same as in methemerythrin.

TABLE V

COMPARISON OF MÖSSBAUER PROPERTIES OF HEMERYTHRIN
COMPLEXES WITH THOSE OF MODEL SYSTEMS

Complex	Ligand	Mössbauer parameters[a]	
		Isomer shift δ (mm/sec)	Quadrupole splitting ΔE_Q (mm/sec)
Oxyhemerythrin	O_2	0.46 / 0.47	1.87 / 0.94
Methemerythrin	NCS	0.55	-1.92
	Cl	0.50	2.04
	F	0.55	1.93
	N_3	0.50	1.91
	H_2O	0.46	1.57
$[Fe_3O(Ac)_6(H_2O)_3]^{2-}$	—	0.43	0.43
$[(Fe(H_2O)B)_2O]^{4+}$	—	0.54	0.85
$[(FeEDTA)_2O]^{4-}$	—	0.46	-1.44
$[(Fe(-)PDTA)_2O]^{4-}$	—	0.44	1.74
$[(FeHEDTA)_2O]^{2-}$	—	0.46	-1.68
$[(Fephen_2)_2O]^{4+}$	—	0.49	1.68
$[(Fe\ bipy_2)_2O]^{4+}$	—	0.48	1.51
$[(Fe\ terpy)_2O]^{4+}$	—	0.59	2.35
$[(Fe\ salen)_2O]$	—	0.46	+0.78
$[(Fe(salen)Cl)_2]$	—	0.51	-1.40
Deoxyhemerythrin	H_2O	1.15	2.86

[a] Mössbauer parameters at 77K, relative to iron foil. Sign of ΔE_a indicated where known. Taken from (13).

(a) Deoxyhemerythrin. The spectrum of deoxyhemerythrin consists of a single quadrupole doublet (Fig. 3). The large quadrupole splitting and high isomer shift (Table V) are characteristic of high-spin iron(II), in agreement with the magnetic susceptibility results. The similarity in Mössbauer spectrum of the two Fe nuclei in each subunit indicates that they are in similar environments.

(b) Methemerythrin. The Mössbauer spectra of the aquo, thiocyanate, fluoride, chloride and azide derivatives of methemerythrin exhibit a single quadrupole doublet (Fig. 4) at temperatures down to 4.2°K. The isomer shifts of all the derivatives (Table V) are in the range 0.45 to 0.55 mm/sec (relative to natural iron) characteristic of high-spin iron(III). The quadrupole splittings (Table V) are all very similar and are large compared to those of simple high-spin iron(III) salts. However, such large quadrupole splittings are not unusual for model binuclear high-spin iron(III) bridged complexes

332

or for monomeric high-spin iron(III) complexes of salen (Table V). The Mössbauer parameters are consistent with antiferromagnetically coupled high-spin iron(III).

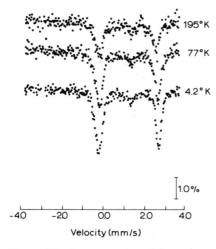

Fig. 3. Mössbauer spectra of deoxyhemerythrin at temperatures indicated.

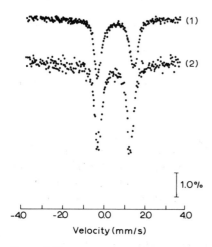

Fig. 4. Mössbauer spectra of (1) thiocyanatohemerythrin and (2) aquohemerythrin at 77K.

(c) Oxyhemerythrin. The Mössbauer spectrum of oxyhemerythrin (Fig. 5) consists of two quadrupole doublets both of which are clearly different from the spectrum of iron(II) in deoxyhemerythrin. This indicates that both iron atoms are involved in the oxygen binding and that the environments of the two iron nuclei are different. An interpretation

of the two doublets as an equilibrium between two electronic states or two geometrical isomers is unlikely since the relative intensities of the two doublets is unchanged at 4.2°K. Both doublets have isomer shifts very similar to those of met derivatives of hemerythrin. The outer pair of lines of oxyhemerythrin has the same large quadrupole splitting as the met derivatives while the inner pair exhibits a smaller quadrupole splitting. The similarity between the isomer shifts suggests that both iron atoms in oxyhemerythrin are in the +3 valence state.

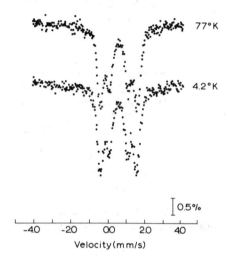

Fig. 5. Mössbauer spectra of oxyhemerythrin at temperatures indicated.

(d) Effect of magnetic fields. Paramagnetic states of iron can be characterized by the magnetic hyperfine splitting of their Mössbauer spectra[23,24]. However, since this splitting is proportional to an average $\langle\mu\rangle$ of the electron magnetic moment, μ, taken over the time-scale of the ^{57}Fe excited state, about 10^{-7} sec, it is not observed if relaxation processes (either spin lattice or spin–spin) average this moment to zero. The hyperfine field is of the order of several hundred kilogauss and can be induced by low temperatures and high magnetic fields.

The effect of a magnetic field at low temperatures is markedly different for a paramagnetic compound (a) as compared to antiferromagnetically coupled dimers (b) as is illustrated in Fig. 6. The splitting of the spectrum in (b) can be accounted for by the applied magnetic field. The Mössbauer spectra of oxy-, metaquo-, and metthiocyanato-hemerythrin are unchanged by a 5 kilogauss field at 4.2°K; such behavior is indicative of a diamagnetic state for the iron. The splitting of the Mössbauer spectrum of methiocyanatohemerythrin at 4.2°K in a 30 kilogauss magnetic field[32] can be accounted for entirely by the applied magnetic field (Fig. 7). The lack of

334

an induced hyperfine field under these conditions is a clear indication of a diamagnetic ground state as would be produced by an antiferromagnetic exchange interaction between the iron atoms.

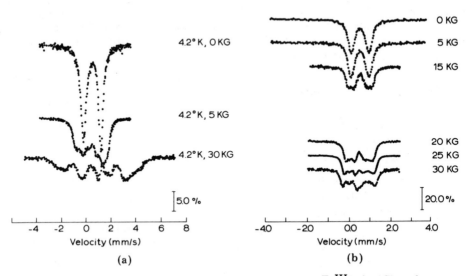

Fig. 6. Mössbauer spectra of (a) model iron(III) monomeric Fe^{III}(salen)Cl, and (b) dimeric [Fe^{III}(salen)]$_2$O at 4.2°K and the indicated magnetic fields. (The monomer (a) probably contains some dimer.)

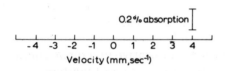

Fig. 7. Mössbauer spectrum of thiocyanatohemerythrin at 4.2°K and in 30 kilogauss magnetic field.

4. Electron spin resonance

The electron spin resonance properties of hemerythrin have been studied at a range of temperatures[25, 31]. There has been no report of an e.s.r. signal which can definitely be assigned to the iron. There is a signal at $g = 2$ in some samples of oxy- and met-hemerythrin which is probably due to a minor component in the protein sample. In addition there is a signal at $g = 1.94$ when oxyhemerythrin is reduced with sodium dithionite. This signal, however, does not appear when deoxyhemerythrin is prepared by deoxygenating oxyhemerythrin[29].

Electron spin resonance spectra of antiferromagnetically coupled iron(III) in model binuclear complexes have been observed[21,22]. The spectrum decreases in intensity with decreasing temperature, characteristic of a thermally excited state. However, other similar complexes have no e.s.r. spectrum, perhaps due to large zero field splittings or rapid relaxation[33]. Thus although an e.s.r. spectrum, due to an antiferromagnetically coupled iron(III) dimer in biological materials should be easily characterized by its temperature dependence, it is not always expected to be observable.

STRUCTURE OF THE ACTIVE SITE

Until the structure of hemerythrin is determined unequivocally by crystallographic methods the nature of the binding of iron at the active site can only be inferred from indirect measurements, such as absorption spectroscopy, Mössbauer spectroscopy and magnetic susceptibility. The interpretation of these measurements in terms of structural features is not always straightforward. Nevertheless these measurements provide criteria that the actual structure must meet, but they do not usually uniquely determine the structure as does X-ray diffraction.

The properties that the structure of hemerythrin must account for are:

1. The low magnetic susceptibility for oxy- and met-hemerythrin, and diamagnetism in the ground state at low temperatures.
2. Non-equivalence of the iron environments in oxy-hemerythrin and the equivalence of the iron environments in met- and in deoxyhemerythrin, respectively, indicated by Mössbauer spectroscopy.
3. The absence of a Bohr effect, or proton release upon oxygenation.
4. The lack of an e.s.r. spectrum from the iron in oxy- and met-hemerythrin.
5. The optical absorption and circular dichroic spectra.

Methemerythrin

The Mössbauer spectra at low temperature in a high magnetic field and the low temperature magnetic susceptibility measurements indicate that the iron in methemerythrin must be antiferromagnetically coupled. The low magnetic susceptibility at room temperature and the isomer shifts in the Mössbauer spectra, characteristic of high-spin iron(III), can best be reconciled by a strong ($|J| \geqslant 30$ cm^{-1}) antiferromagnetic coupling between two iron(III) atoms in a weak ligand field. This antiferromagnetic exchange interaction probably results from a superexchange coupling through a bridging ligand.

Since the cysteine residue has been shown not to be liganded to iron, this exchange interaction probably does not occur through a bridging sulfur atom (although the possibility of a sulfur bridge from methionine has not yet been eliminated). Bridging through the oxygen of an alkoxy group (such as the phenolic oxygen of tyrosine) or hydroxy group has precedent in inorganic iron(III) binuclear complexes. However, the exchange interaction through such bridging ligands would probably be too small to account for the low magnetic susceptibility of methemerythrin at room temperature.

The most plausible structural feature explaining both the Mössbauer and magnetic susceptibility data is an oxo-bridge, Fe–O–Fe.

Garbett and coworkers[10] have explained the 1:2 ligand to iron stoichiometry of hemerythrin complexes with azide and with thiocyanate and the similarity between the iron environments indicated by the Mössbauer spectra in terms of the symmetrical doubly bridged structure I.

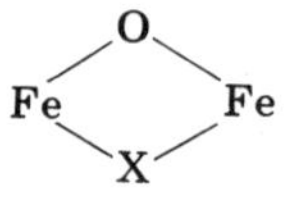

$$Fe \overset{\displaystyle O}{\underset{\displaystyle X}{<>}} Fe$$

(I)

$X = Cl^-, Br^-, F^-$

$= \ {>}N{=}C{=}S, \ {>}N{=}N{=}N, \ {>}C{=}N, \ {>}N{=}C{=}O$

The complexes of methemerythrin listed above have similar circular dichroic spectra.

Garbett *et al.* have also deduced from the differences between the circular dichroic spectra of the known 1:2 complexes and the circular dichroic spectra of aquo and hydroxy complexes, as well as the fluoride complex at high concentrations of fluoride, that the ligand:iron stoichiometry of these latter complexes is 2:2. This is represented by II.

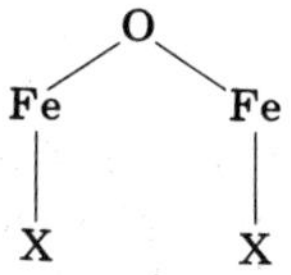

```
        O
      /   \
   Fe       Fe
    |         |
    X         X
     (II)
```

$X = H_2O, HO^-, F^-$

Oxo-bridged structures such as I, where the Fe–O–Fe bond angle must approach $90°$ due to van der Waals repulsion between O and X, have not yet been observed in inorganic iron(III) complexes. Thus it is difficult to assess the magnitude of the expected exchange interaction. Bond angles close to $90°$ have only been observed where the oxygen is bound to a third group such as a proton. This additional bonding is accompanied by a weakening of the FeO bond and a substantial reduction in the exchange interaction. It is possible that the protonation of the bent oxo-bridge in methemerythrin may be prevented by the protein environment or that the bent oxo-bridge may be only weakly hydrogen bonded to the protein. If this were so then the bent oxo-bridged structure I could retain the short FeO bond lengths and strong antiferromagnetic coupling usually associated with a single approximately linear, oxo-bridge.

However the strong Fe–O bond found in linear oxo bridges may be considerably weakened by large deviations from linearity as was pointed out by Dunitz and Orgel[27]. In addition it has been proposed that an exchange interaction through an Fe–O–Fe bond angle of $90°$ should be much weaker than an exchange interaction through a bridging angle of $180°$ [18,34]. If these proposals are correct, then the alternative structure III containing a single, approximately linear, oxo-bridge may be required to explain the magnetic susceptibility data.

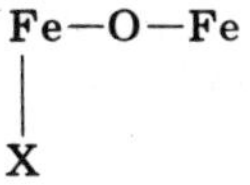

```
Fe—O—Fe
 |
 X
     (III)
```

Although structure III seems to conflict with the identical environments indicated by the Mössbauer data it is possible that the effect of the ligand X on one iron is just balanced by a ligand from the protein on the other iron. It should be noted that the Mössbauer spectra of the methemerythrins are relatively insensitive to the bound ligand. This might be expected since the isomer shift for high-spin iron(III) complexes is relatively constant. The quadrupole splitting may be dominated by the effect of the oxo bridge and may not be sensitive to other ligands unless they are particularly polarizable to distort the structure of the complex.

References pp. 342-343

Oxyhemerythrin

   The Mössbauer spectrum of oxyhemerythrin is entirely different from that of deoxyhemerythrin and consists of two doublets. This indicates that both iron atoms are affected by the binding of the oxygen molecule. However, the bonding of the two irons to the oxygen molecule is different. The alternative assignment of the two doublets either to an equilibrium between electronic states or geometric isomers is improbable since the invariance of the relative intensities of the two doublets over a wide temperature would require a fortuitous equivalence between the energies of the two electronic states or geometrical isomers.

   Although it is possible to explain the magnetic properties of oxyhemerythrin by having isolated iron atoms in the low-spin iron(II) state, the similarity between the isomer shifts of iron in oxyhemerythrin and methemerythrin suggest that in both cases the iron atoms are iron(III), strongly antiferromagnetically coupled. Thus the bound oxygen molecule should resemble the peroxide ion.

   From the proposed structure I for methemerythrin and in analogy with a model cobalt–oxygen complex, the structure IV has been proposed for oxyhemerythrin by Garbett and coworkers[10]:

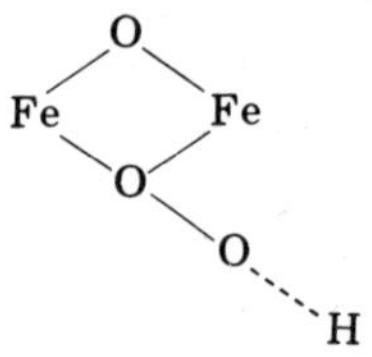

$$
\begin{array}{c}
\text{O} \\
\text{Fe} \diagup \diagdown \text{Fe} \\
\diagdown \text{O} \diagup \\
\text{O} \\
\cdots \text{H}
\end{array}
$$

(IV)

   The difference between the quadrupole splitting in the two iron atoms is attributed to the asymmetry in the bonding of the bridging peroxide ion which is presumed to be hydrogen bonded to a protein group.

   In analogy with the proposed structure (II) for methemerythrin it is possible that the bound oxygen (as peroxide ion) is bridged between the two iron atoms as shown in V:

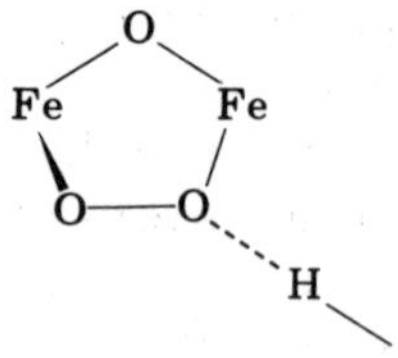

$$
\begin{array}{c}
\text{O} \\
\text{Fe} \qquad \text{Fe} \\
\text{O} - \text{O} \\
\cdots \text{H}
\end{array}
$$

(V)

The difference between the quadrupole splitting in the two iron atoms may be due to a difference in the strength of their bonds to the peroxide ion.* This difference may arise from hydrogen bonding to the protein.

Finally, in the oxyhemerythrin structure corresponding to structure III for methemerythrin the peroxide ion may be bound to a single iron atom either with the Griffith structure (VI) or the Pauling structure (VII).

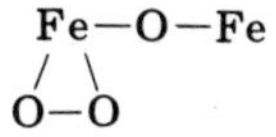

```
Fe—O—Fe
/ \
O—O
```
   (VI)

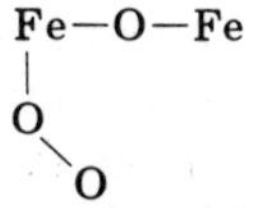

```
Fe—O—Fe
|
O
 \
  O
```
   (VII)

In **VI** the π-bonding of the oxygen molecule to the iron is significantly different from the bonding of other ligands, such as chloride or azide, to methemerythrin. In addition, the protein ligands must be distorted away from the bonded oxygen molecule. Both of these effects should have a large influence on the quadrupole splitting of the bonding iron. In **VII** the difference in the quadrupole splitting may be attributed to the greater polarizability of the peroxide ion. In both of these cases the environment of the non-bonding iron atom is the same as that in methemerythrin accounting for the similarity between the Mössbauer parameters of one iron in oxyhemerythrin and methemerythrin.

Deoxyhemerythrin

Much less can be said about the structure of deoxyhemerythrin. The two iron atoms are apparently in similar environments and the magnetic susceptibility is typical of a high-spin iron(II) compound. However this does not rule out bridging groups since the exchange interaction for iron(II) might be small.

One thing that must be accounted for in hemerythrin is the absence of a Bohr effect, that is the oxygen binding curves are not a function of pH[5,27]. Furthermore pH measurements indicate that there is no proton uptake or release when deoxyhemerythrin is oxygenated[8]. Thus if an oxo-bridge exists in the oxygenated form of hemerythrin then an oxo bridge or a dihydroxy bridge must either be present in deoxyhemerythrin or the

---

*Such non-equivalent bonding to oxygen has been observed in the recently determined structure of a synthetic cobalt oxygen carrier[34a].

protons released in forming the oxo bridge must be transferred to the protein or to the bonded O_2.

Thus it is apparent that the detailed atomic arrangement at the active site is still not unequivocally established.

STRUCTURE AND FUNCTION

One of the main goals of biochemistry in general and inorganic bio-chemistry in particular is to understand the function of biological molecules in terms of their structural and electronic properties. In explaining how hemerythrin functions as an oxygen carrier we must account for the stability (and reversibility) of the oxyhemerythrin complex with respect to molecu-lar oxygen and deoxyhemerythrin and also for the stability of oxyhemery-thrin with respect to methemerythrin and hydrogen peroxide[35]. The stabil-ity in the former respect is thermodynamic and must be explained in terms of the chemical bonding in the oxygenated complex. The stability in the latter respect is probably kinetic since irreversible oxidation does proceed slowly, and must be accounted for by the activation energy for the decomposition process (Fig. 8).

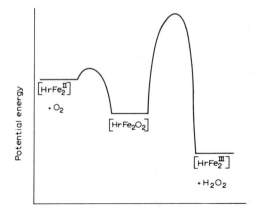

Fig. 8. Potential energy along the reaction coordinates in the oxygenation and in the oxidation of hemerythrin.

In all the oxygen carriers studied the strength of oxygen binding increases when the ligands are better electron donors. This is true for the oxygen adduct of Vaska's compound (p. 673) when iodide replaces chloride[36], the 1:1 oxygen adduct of Schiff base Co(II) complexes when substituted pyridines are used as bases[37], as well as hemoglobin derivatives when the side chains on the heme group are varied[38,39]. It is somewhat surprising, therefore, that hemerythrin which contains no good electron donor (soft) ligands such as the porphyrin group, can bind oxygen. The absence of such ligands may be the reason that there are two iron atoms in hemerythrin.

The strong binding of oxygen in hemerythrin can be explained if the two iron atoms transfer two electrons to antibonding oribitals on O_2 as has been previously proposed[40]. The bonding of the oxygen molecule as the peroxide ion, should be much stronger than bonding of a superoxide ion formed from a one electron transfer. The double coordination of two Fe to one O_2 molecule may also contribute appreciably to the strength of the binding.

The stability of oxyhemerythrin with respect to oxidation in turn is probably attributable in large measure to the nature of the protein environment. The iron in hemerythrin is probably buried in the interior of the protein, if one may judge from the difficulty of reversible removal of the metal, as well as from its inaccessibility to large chelating ligands in the native state[12]. The factors which may slow down the rate of decomposition of oxyhemerythrin may be the low dielectric and non-acidic environment in the vicinity of the iron oxygen complex which may prevent charge separation or acid-catalyzed oxidation of hemerythrin[2,41]. Another factor which may be of importance is the specific geometry imposed on the iron-oxygen complex by the protein, prohibiting a rearrangement of the decomposition products. This may be thought of as a manifestation of the "entatic" effect[42].

ACKNOWLEDGEMENTS

It is a pleasure to acknowledge Dr. Keith Garbett and Dr. Dennis Darnall for many stimulating discussions and Dr. R. J. P. Williams, Dr. C. E. Johnson and Dr. M. R. C. Winter for collaboration on the Mössbauer experiments.

NOTE ADDED IN PROOF

Since the completion of this manuscript, there have appeared a number of relevant articles on hemerythrin dealing principally with further studies of the Mössbauer spectra of hemerythrin[43,44] and magnetic susceptibility studies of hemerythrin at low temperatures[45,46] (p. 115). There appears to be a general agreement that, at least in oxy- and methemerythrin, the iron atoms are intimately coupled. However, the detailed structure of the active site remains unresolved. It is clear that additional insights are needed before an understanding of the function of hemerythrin in terms of its molecular structure can be considered soundly established.

342

REFERENCES

1 F. Ghirett, in O. Hayaishi, *Oxygenases*, Academic Press, New York, 1962, pp. 517–553.
2 I. M. Klotz, T. A. Klotz and H. A. Fiess, *Arch. Biochem. Biophys.*, 68 (1957) 284.
3 S. Keresztes-Nagy and I. M. Klotz, *Biochemistry*, 4 (1965) 919.
4 S. Keresztes-Nagy and I. M. Klotz, *Biochemistry*, 2 (1963) 923.
5 G. Bates, M. Brunori, G. Amiconi, E. Antonini and J. Wyman, *Biochemistry*, 7 (1968) 3016.
6 C. Manwell, *A. Rev. Physiol.*, 22 (1960) 191–244.
7 G. Wald and D. W. Allen, *J. Gen. Physiol.*, 40 (1957) 593–608.
8 M. Y. Okamura and I. M. Klotz, unpublished observations.
8a W. Caughey, in B. Chance, R. Estabrook and T. Yonetani, *Hemes and Hemoproteins*, Academic Press, New York, 1966, p. 276.
9 D. W. Darnall, K. Garbett and I. M. Klotz, *Biochem. Biophys. Res. Commun.*, 32 (1968) 264.
10 K. Garbett, D. W. Darnall, I. M. Klotz and R. J. P. Williams, *Arch. Biochem. Biophys.*, 103 (1969) 419.
11 I. M. Klotz and S. Keresztes-Nagy, *Biochemistry*, 2 (1963) 455.
12 D. Darnall and I. M. Klotz, unpublished observations.
13 K. Garbett, D. W. Darnall and I. M. Klotz, in preparation.
14 G. L. Klippenstein, J. W. Holleman and I. M. Klotz, *Biochemistry*, 7 (1968) 3868.
15 R. G. Pearson, *Survey of Progress in Chemistry*, Vol. 5, Academic Press, New York, 1969.
16 C. C. Fan and J. L. York, *Biochem. Biophys. Res. Commun.*, 36 (1969) 365.
17 R. L. Rill and I. M. Klotz, *Arch. Biochem. Biophys.*, 133 (1969) 103.
18 P. W. Ball, *Coord. Chem. Rev.*, 4 (1969) 361.
19 T. G. Spiro and P. Saltman, in P. Hemmerich *et al.*, *Structure and Bonding*, Vol. 6, Springer-Verlag, New York, 1969, 116–156.
20 K. Kambe, *J. Phys. Soc.*, Japan, 5 (1950) 48.
21 L. N. Mulay and N. L. Hofmann, *Inorg. Nucl. Chem. Lett.*, 2 (1966) 189.
22 M. Y. Okamura and B. M. Hoffman, *J. Chem. Phys.*, 51 (1969) 3128.
23 C. E. Johnson, in A. Ehrenberg, *Magnetic Resonance in Biological Systems*, Pergamon, New York, 1966, p. 405.
24 C. E. Johnson and D. O. Hall, *Nature*, 217 (1968) 217.
25 M. Y. Okamura, I. M. Klotz, C. E. Johnson, M. R. C. Winter and R. J. P. Williams, *Biochemistry*, 8 (1969) 1951.
26 W. M. Reiff, G. J. Long and W. A. Baker, *J. Am. Chem. Soc.*, 90 (1968) 6347.
26a S. J. Lippard, H. Schugar and C. Walling, *Inorg. Chem.*, 6 (1967) 1825.
27 J. D. Dunitz and L. E. Orgel, *J. Chem. Soc.*, (1953) 2593.
27a J. Owen and J. H. M. Thoruley, *Rep. Progr. Phys.*, 29(2) (1966) 675.
28 M. Kubo, *Bull. Chem. Soc., Japan*, 26 (1953) 246.
29 L. York, personal communication.
30 H. B. Gray, personal communication.
31 T. Moss, personal communication.
32 K. Garbett, C. E. Johnson, I. M. Klotz, M. Y. Okamura and R. J. P. Williams (1969), in preparation.
33 L. Marchant and B. M. Hoffman, personal communication.
34 M. A. Gilleo, *Phys. Rev.*, 109 (1958) 777.
34a B. Wang and W. P. Schaefer, 166 (1969) 1404.

35 F. Basolo and R. G. Pearson, *Mechanism of Inorganic Reactions*, 2nd edn., John Wiley and Sons, Inc., New York, 1968, pp. 641–648.
36 J. A. McGinnety, R. J. Doedens and J. A. Ibers, *Science*, 155 (1967) 709.
37 A. L. Crumbliss and F. Basolo, *J. Am. Chem. Soc.*, 92 (1970) 55.
38 A. Rossi-Fanelli and E. Antonini, *Arch. Biochem. Biophys.*, 80 (1959) 299, 308.
39 A. Rossi-Fanelli, E. Antonini and A. Cupato, *Arch. Biochem. Biophys.*, 85 (1959) 37.
40. I. M. Klotz and T. A. Klotz, *Science*, 121 (1955) 477.
41 J. H. Wang, *Accounts Chem. Res.* 3 (1970) 90.
42 B. L. Vallee and R. J. P., Williams, *Proc. Natn. Acad. Sci. U.S.*, 59 (1968) 498–505.
43 J. L. York and A. J. Bearden, *Biochemistry*, 9 (1970) 4549.
44 K. Garbett, C. E. Johnson, I. M. Klotz, M. Y. Okamura and R. J. P. Williams, *Arch. Biochem. Biophys.*, 142 (1971) 574.
45 T. H. Moss, C. Moleski and J. L. York, *Biochemistry*, 10 (1971) 840.
46 J. W. Dawson, H. B. Gray, H. E. Hoenig, G. R. Rossman, J. M. Schredder and R. Wang, *Biochemistry*, 11 (1972) 461.

Chapter 12

HEMOCYANIN

R. LONTIE AND R. WITTERS

Laboratorium voor Biochemie, Katholieke Universiteit te Leuven, B-3000, Louvain, Belgium

INTRODUCTION

Harless[1] in 1847 reported the presence of Cu in the blue blood of the roman snail *Helix pomatia*. Fredericq[2] in 1878 described a colorless extracellular protein in the blood of *Octopus vulgaris*, that gives an unstable deep-blue compound with oxygen. He coins the name hemocyanin from αἱμά, blood, and κύανος, blue. This term, of course, does not implicate the presence of a heme group.

Several review articles deal already with the occurrence, the composition, the molecular weight and dissociation, the oxygen binding and spectra, and with the immunological properties of hemocyanin[3-14].

OCCURRENCE

Hemocyanins are found in Arthropoda and Mollusca and show a copper content characteristic for each phylum (Table I).

Typical representatives for the Arthropoda are the horseshoe crab in the class Merostomata, crawfish, lobster and several crabs, in the class Crustacea, order Decapoda, and for the Mollusca the prosobranch limpets and whelks and the pulmonate snails in the class Gastropoda, squids and cuttle-fish in the class Cephalopoda, order Decapoda, and the octopus in the order Octopoda (see Table I for refs.). Hemocyanins have also been described in the Mollusca, class Amphineura, in the chitons[33-35].

In the Arthropoda hemocyanins are found, moreover, in the class Arachnida in scorpions[33,36] (see also *Androctonus australis* in Table I) and in spiders[17,37,38], in the class Crustacea, order Isopoda[39-41] and order Amphipoda[39,41], and in the class Chilopoda in centipedes[42].

MOLECULAR WEIGHT AND QUATERNARY STRUCTURE

A copper content of 0.166–0.180 for the Arthropoda and of 0.245–0.260 for the Mollusca corresponds to an average minimum molecular weight of 36,700 and of 25,100 respectively (Table I, see also ref. 16).

TABLE I

SEDIMENTATION COEFFICIENT (s_{20}) OF THE MAIN COMPONENT, COPPER
CONTENT AND MINIMUM MOLECULAR WEIGHT PER COPPER OF TYPICAL
HEMOCYANINS

Species	s_{20}	Ref.	% Cu	Ref.	M/Cu
Arthropoda:					
Merostomata					
Limulus polyphemus	56.6	21	0.170	16	37,400
Arachnida					
Androctonus australis	33.0	17	0.180	18	35,300
Crustacea					
Jasus lalandii	17.1	19	0.176	20	36,100
Palinurus vulgaris	16.4	21	0.170	16	37,400
Homarus americanus	24.5	22	0.175	23	36,300
H. vulgaris	22.6	21	0.169	16	37,600
Eriphia spinifrons	24.0	24	0.167	16	38,000
Cancer magister	25.0	25	0.166	25	38,300
Carcinus maenas	23.3	21	0.173	7	36,700
Maja squinado	23.4	26	0.182	7	34,900
				Mean	36,700
Mollusca:					
Gastropoda					
Busycon canaliculatum	102.0	15	0.245	27	25,900
Helix pomatia	104.3	28	0.250	7	25,400
Murex trunculus	102.7	29	0.257	16	24,700
Cephalopoda					
Sepia officinalis	55.9	21	0.256	7	24,800
Loligo pealei	59.0	30	0.260	27, 31	24,400
Octopus vulgaris	49.3	21	0.250	16, 32	25,400
Eledone moschata	49.1	21	0.252	16	25,200
				Mean	25,100

In the ultracentrifuge the hemocyanins of Arthropoda show sedimentation coefficients of 16, 24, 33, and 57 S (Table I; refs. 21 and 43). With *Limulus polyphemus e.g.* these four components are observed together in the stability region[21]. These components have also been studied in the electron microscope after negative staining[44-46]. They have been interpreted as monomers, dimers, tetramers and octamers[17,47].

The monomer (16 S, 10 x 10 nm in the electron microscope) is observed *e.g.* in *Palinurus vulgaris*[45] and in the isopod *Oniscus asellus*[46], the dimer (24 S, 22 x 9 nm) *e.g.* in *Cancer pagurus*[45] and in *Homarus vulgaris*[44,46,48], the tetramer (33 S, 22 x 20 nm) in the arachnids *Nemesia cementaria*[17,46,47] and *Avicularia metallica*[45], the octamer (57 S, 21 x 24 nm) in *Limulus polyphemus*[45,47].

As the dimer in *Homarus vulgaris* hemocyanin has a molecular weight of 825,000[22], the monomer seems to be constituted of 12 subunits of

34,400, corresponding to the molecular weight of the non-functional sub-unit obtained after succinylation[23].

A tentative model of the monomer is presented in Fig. 1. A distorted octahedral arrangement of six spheres is able to account for all the contours of the monomer seen in the electron microscope: regular hexagon, square, rectangle and rhomb. As the functional subunit has a molecular weight of 68,000–70,000[23], the 12 smallest subunits of Levin's octahedral model[44] have to be arranged in six pairs. This raises the question if one type of polypeptide, giving one subunit, is sterically able to yield the proposed quaternary structure.

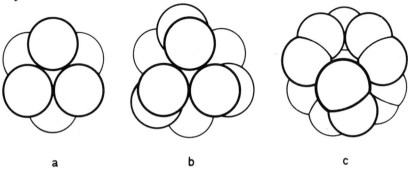

a   b   c

Fig. 1. Tentative model of the 16 S monomer of the hemocyanins of the Arthropoda. A, Distorted octahedral arrangement of six functional subunits. B, Top view of 12 subunits arranged in pairs. C, The same, side view.

The hemocyanins of Mollusca have also been studied in the ultracentrifuge[21,33,43,49] and in the electron microscope[44,45,50].

The shapes, as observed with the latter technique after negative staining, are outlined on Fig. 2. The rectangles and circles seen can be interpreted as cylinders with a fivefold axis.

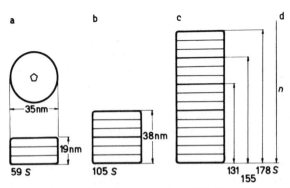

Fig. 2. Shape of the hemocyanins of Mollusca as seen in the electron microscope after negative staining and the corresponding sedimentation coefficients.

Two classes are observed in the ultracentrifuge: 59 *S, Loligo pealei*
e.g., and 105 *S, Helix pomatia e.g.* (Figs. 2a and b, see Table I for refs.).
The same cylindrical shape has been observed in the electron microscope
for *Cymbium neptuni*[46] and *Buccinum undatum*[51]. *Busycon canaliculatum*
shows, moreover, a tendency to linear polymerization[45] (Fig. 2c). The
sedimentation coefficients from 59 to 178 *S* have been shown to be in
accordance with the proposed models by a hydrodynamic treatment[52]. The
hemocyanin of *Kelletia kelletia* yields long linear polymers at the isoelectric
point[53] (Fig. 2d).

Some of the 105 *S* hemocyanins, like the α-component of *Helix
pomatia*, dissociate into halves in the stability region in the presence of
1 *M* alkaline chlorides[28, 54, 55], while the β-component does not. At the alkaline side of the stability region they dissociate, by a pH increase, over
halves into tenths and twentieths[28, 56]. As shown by light scattering[57], the
stability region of the purified hemocyanin of *Helix pomatia* is extended
from pH 8 to 10 by the addition of small amounts of Ca^{2+} as present in
the blood[58].

The twentieths seem to be the smallest functional subunit of *Helix
pomatia* hemocyanin obtainable by alkaline dissociation[56] or by succinylation[59], while the smallest subunit observed at pH 12.5 after succinylation,
reduction and alkylation has a molecular weight of the order of 22,300[60].

ABSORPTION SPECTRA

The blue color of the blood of several invertebrates, which had already
struck the earliest investigators[1,2], should not be confused with the color of
the true *blue* copper proteins, like ceruloplasmin (λ_{max} = 610 nm). The
blue color of *Helix pomatia* hemocyanin *e.g.* in the stability region, where
a molecular weight of 9×10^6 causes an intense scattering of light, results
from scattering plus absorption. A concentrated solution looks purplish in
transmission.

The oxyhemocyanins show, besides the usual protein band at 280 nm,
a copper band at 346 and at 580 nm[4] (Table II). The extinction coefficients
of the deoxygenated hemocyanin follow an inverse fourth power of the
wavelength (in the visible[5], in the visible and in the near ultra-violet[63]).
Especially with the hemocyanins of the gastropods with the highest molecular weights the contribution of scattering to the extinction coefficients
is important[55]. The values in Table II for *Helix pomatia* and *Octopus vulgaris*
have been corrected for scattering, those for *Loligo pealei* and *Omma(to)-
strephes sloani pacificus* for scattering and residual absorbance by subtrac-
tion of the value for apohemocyanin and deoxygenated hemocyanin
respectively.

TABLE II

EXTINCTION COEFFICIENTS (L/MOLE CM) EXPRESSED PER Cu FOR THE TWO COPPER ABSORPTION BANDS OF MOLLUSCAN HEMOCYANINS

Species	λ_{max}	ϵ_{max}/Cu	Ref.
Helix pomatia β	346	8,800	28
	580	540	
Loligo pealei	345	8,900	30
	580	370	
Omma(to)strephes sloani pacificus	347	6,100	61
	580	390	
Octopus vulgaris	347	8,900	62
	580	500	

At –195°C two new bands appear reversibly in the visible spectrum with a maximum at 420 and at 652 nm for *Cancer magister* hemocyanin[64], the maximum of the original band shifts from 585 to 566 nm. Two similar bands are revealed at the same temperature by a shoulder at 440 and at 700 nm for *Octopus vulgaris* hemocyanin[62] with a shift of the maximum of the original band from 580 to 570 nm. The spectrum at room temperature of the hemocyanin of *Omma(to)strephes sloani pacificus* shows a shoulder at 800 nm[61].

The two major copper bands, observed at room temperature, are optically active. Measurements of optical rotatory dispersion have shown the associated negative Cotton effects[30,65]. Circular dichroism measurements have revealed two negative and two positive bands (Table III). The striking similarity of the circular dichroism spectrum of the molluscan hemocyanins with that of the Cu(II) complex of Ac-Gly-Gly-His[67] at 490 and 590 nm has been pointed out by Van Holde[62].

TABLE III

CIRCULAR DICHROISM BANDS OF MOLLUSCAN HEMOCYANINS

Species	λ_{max}(nm)				Ref.
	–	+	–	+	
Loligo pealei	347	440	570	700	62
Octopus vulgaris	347	440	570	n.d.	62
	350	450	570	> 700	66
Helix pomatia	346	485	578	710	a

[a] Unpublished.

In contrast with the *blue* copper proteins the copper absorption bands of hemocyanins are linked to the presence of oxygen. A resonance between the following structures has been proposed[12, 68]:

$$Cu^+O_2Cu^+ \leftrightarrow Cu^{2+}O_2^-Cu^+ \leftrightarrow Cu^+O_2^-Cu^{2+} \leftrightarrow Cu^{2+}O_2^{2-}Cu^{2+}$$

for the oxygenated copper group, the last structure being less probable.

The two major absorption bands at 346 and 580 nm have been interpreted as charge transfer transitions[69]. The similarity of the near ultra-violet absorption band of hemocyanin (λ_{max} = 346 nm, log ϵ = 3.9) with the so-called peroxy-specific band of the decammine-μ-peroxocobalt(III)–cobalt(IV) complex (λ_{max} = 341 nm, log ϵ = 3.5)[70] has been stressed[71]. The O–O distance in this Co complex and in the oxygen-carrying bis(3-fluorosalicyl-aldehyde)ethylenediimine Co(II) complex is 1.31 Å, close to the O–O distance in the superoxide ion O_2^- (1.28 Å)[72].

The visible spectrum of the hemocyanins, when the two supplementary absorption bands found at –195°C or by circular dichroism are considered, is rather similar to the spectra of Cu(II) proteins like ceruloplasmin and laccase[62, 73]. There are small differences in λ_{max} and large differences in ϵ_{max}. These absorption bands could correspond to d–d transitions of Cu(II)[62].

Fresh oxyhemocyanins at room temperature, however, give no electron spin resonance spectrum at a g-value of 2.0: *Cancer magister*[74], *Limulus polyphemus*[75], *Jasus lalandii*[76]; *Octopus vulgaris*[77], *Busycon canaliculatum*[75], *Helix pomatia*[28]. The *Jasus lalandii* hemocyanin, measured at 77K, presented a signal corresponding to about 10% of the copper present[76], and very fresh *Homarus americanus* hemocyanin, at 1.5K, gave a Cu(II) signal corresponding to one quarter of the total copper[78].

THE BINDING OF OXYGEN

The relation between the blue color of hemocyanin and the binding of oxygen has already been noted by the earliest investigators[1,2]. The respiratory function was demonstrated for *Sepia officinalis*[79] and for *Octopus vulgaris*[80]. This physiological role has sometimes been questioned, but was definitely proven by Redmond[81].

Most hemocyanins show a cooperative (positive) oxygen binding under the conditions prevailing in the blood[5,75,82] with n values in the Hill equation up to 3.5–4.0.

As O_2 is bound by 2Cu[83-85], the monomer of the hemocyanins of the Arthropoda contains 6 oxygen-binding sites (Fig. 1), the hemocyanins of the Mollusca 90–100 (Fig. 2a) and 180–200 (Fig. 2b). So that hemocyanins must fit a general theory of allosteric transitions beyond the so-called heme–heme interaction[86].

With *Helix pomatia* hemocyanin *e.g.* the cooperativity is lost on dialysis[87], the same hyperbolic oxygenation curve observed from pH 4.0 to 9.0 indicating independent oxygen-binding sites. With fresh hemocyanin the sigmoidal oxygenation curve can be restored[88] at pH 7.8–8.2, when the solution is made 6 mM in Ca^{2+}, the concentration of the blood. With *Homarus americanus* hemocyanin the cooperativity of the oxygen binding was lost on dissociation into subunits with one oxygen binding site, by removal of Ca^{2+} in alkaline solutions of low ionic strength[23].

Sigmoidal oxygen equilibrium curves are liable to operate through conformational changes of the functional subunits. An interesting influence of oxygenation on the quaternary structure of *Loligo pealei* hemocyanin has been reported[89]: while deoxygenated hemocyanin shows a sedimentation coefficient of 59 S from pH 6 to 10 in the presence of 10 mM Mg^{2+}, a dissociation into fifths and into tenths occurs at pH 7 on partial oxygenation. Small differences are also seen in the circular dichroism spectrum between oxy- and deoxy-hemocyanin from 250 to 280 nm.

A new approach to the study of oxygen binding is also provided by the important enhancement of fluorescence on deoxygenation, as observed on the hemocyanin of the mollusc *Levantina hierosolima*[90].

THE BINDING OF OTHER LIGANDS

Hemocyanins yield reversibly colorless compounds (no absorption in the visible, nor in the near ultra-violet) with CO and, at least with molluscan hemocyanins, with CN$^-$, ethylisocyanide, SCN$^-$, and thiourea (for short treatments).

Limulus hemocyanin binds CO in the same proportion as O$_2$, but with a relative affinity of only one-twentieth[91]. *Octopus* hemocyanin binds 1CO/2Cu, the compound is split by CN^{-32}. No CO compound apparently was observed with *Palinurus vulgaris* hemocyanin[85]. The binding of CO by *Octopus vulgaris* hemocyanin was established by the inhibition of the pseudo-catalase and -peroxidase activity and by direct determination of the CO liberated by the addition of CN^{-92}. The compound with *Cancer magister* hemocyanin was studied with ^{14}CO, a half-saturation pressure of 134 mm was found[93]. With *Helix pomatia* hemocyanin a spectrophotometric determination of the competition between CO and O$_2$ gave an equilibrium constant of 1.00 at pH 9.2[94]. There is general agreement that 1CO is taken up per 2Cu, but no direct proof that a bond is formed with both atoms.

The reaction of *Limulus* hemocyanin with small concentrations of CN$^-$ has been investigated and seems to be partly reversible[95]. It is difficult to interpret the recovery in the presence of oxygen of the blue color of *Octopus vulgaris* hemocyanin treated with cyanide[96]. Experiments in progress at the laboratory show the reaction of *Helix pomatia* hemocyanin to be reversible for the low CN⁻ concentrations studied up to 0.2 mM.

Ethyl isocyanide also gives a reversible reaction with the hemocyanin of the whelk *Murex trunculus*[97].

SCN⁻ expels oxygen reversibly from *Helix pomatia* hemocyanin[98]. At constant pH the reaction can be described by the following equation:

$$(k'_a/k_a) - 1 = K([SCN^-]/p_{O_2})^{\frac{1}{2}}$$

wherein k_a means the extinction at 346 nm due to the oxygenated copper groups for a protein concentration of 1 g/l and a path length of 1 cm, and k'_a the maximum value of k_a when all the copper groups are oxygenated. The data indicate that 1SCN⁻ expels 1O_2 but decolorizes two oxygen binding sites[99].

Thiourea expels O_2 reversibly in a fast reaction. In a slow secondary reaction, accompanied by an O_2 absorption, the remaining copper bands are slowly destroyed in a process resembling aging[100].

By the action of NO *Helix pomatia* hemocyanin gives a green compound (no reaction was observed with *Homarus* hemocyanin)[101]. A similar compound has been mentioned by Mason (see ref. 93, discussion p. 472). This interesting derivative, as it shows a detailed electron paramagnetic resonance (e.p.r.) spectrum, merits a further investigation.

AGEING AND REGENERATION

The copper bands of several hemocyanin preparations seem variable: *Helix pomatia*[55,63], *Eriphia spinifrons* and *Octopus vulgaris*[102]. The copper bands decrease regularly as a function of time, and for *Helix pomatia* hemocyanin e.g. especially, when kept as an ammonium sulfate precipitate[103].

In contrast with the fresh solutions, the aged preparations give an electron spin resonance signal, characteristic for Cu(II) (*Jasus lalandii*[76], *Octopus vulgaris*[77], *Helix pomatia*[28]). No signal, however, was observed with aged preparations of *Limulus* and *Busycon* hemocyanin[78].

The copper bands of aged *Helix pomatia* hemocyanin can be regenerated by treatment with cysteine[104], H_2O_2 and hydroxylamine[104,105].

When deoxygenated hemocyanin from another mollusc *Busycon canaliculatum* is treated with small amounts of H_2O_2, the copper bands are partly destroyed and Cu(I) is accordingly oxidized to Cu(II). With larger amounts of H_2O_2 the copper bands are regenerated. With the deoxygenated hemocyanin of an arthropod, *Limulus polyphemus*, however, the copper bands are destroyed irreversibly by stoichiometric amounts of H_2O_2, they could not be regenerated by reducing agents[106]. Both hemocyanins are protected by O_2 from the attack by H_2O_2. The methemocyanin from *Cancer magister*, obtained by the action of H_2O_2, gives an electron

352

spin resonance spectrum[74]. A similar radiation inactivation is observed for *Limulus* hemocyanin, and inactivation and reactivation for *Busycotypus* hemocyanin, due to the intervention of peroxides[107]. Hemocyanins, moreover, show a catalase activity, that is inhibited by O_2. The activity is much larger for molluscan than for crustacean hemocyanins (relative activity: *Octopus vulgaris*, 100; *Helix ligata*, 41; *Maja squinado*, 3; *Palinurus vulgaris*, 3)[108].

A thermal equilibrium[109] could provide the simplest explanation for the *ageing* of the molluscan hemocyanins:

$$Cu^+O_2Cu^+ + 2H^+ \rightleftharpoons Cu^{2+}Cu^{2+} + H_2O_2$$

As an electron paramagnetic resonance signal is observed, there should, however, occur no spin coupling between all the Cu(II), implying a sufficient distance. The slow formation of metmyoglobin from oxygenated myoglobin has been ascribed to a similar formation of superoxide O_2^{-}[110].

The *catalase* activity of molluscan hemocyanins can also be explained by a simultaneous valence shift of both Cu atoms:

$$Cu^+O_2{}^{2-}Cu^+ + 4H^+ \rightarrow Cu^{2+}Cu^{2+} + 2H_2O$$

(reaction inhibited by O_2 by the formation of $Cu^+O_2Cu^+$, see p. 349).

$$Cu^{2+}O_2{}^{2-}Cu^{2+} \rightarrow Cu^+O_2Cu^+$$

The *regeneration* with H_2O_2 of aged molluscan hemocyanin is also given by the second equation, and can thus easily be written for 2 vicinal Cu^{2+} (it would be cumbersome with $1Cu^{2+}$ and $1Cu^+$). We suppose the regeneration by cysteine to be also mediated by slowly produced H_2O_2[111].

The presented mechanism would also explain why the ageing reaches a finite level, as the peroxide formed could be caught by the aged groups and regenerate them.

Hemocyanins do not resist lyophilization, but can be protected largely by the addition of non-reducing sugars, like sucrose[105,112,113]. A storage under CO prevents the ageing and preserves the sigmoidal oxygenation curve[94]. On ageing, namely, the partial oxidation of copper groups interrupts the progressive change in conformation, *i.e.* the cooperation of the oxygen binding sites is lost.

In order to characterize a preparation it is to be recommended to state ϵ_{max} at 346 nm or at least $A_{346\,nm}/A_{280\,nm}$.

APOHEMOCYANIN AND RECONSTITUTION

Cu is liberated from *Octopus vulgaris* hemocyanin on acidification with HCl[114], but can only be removed by CN^- under conditions, that do not disrupt the conformation of the protein, as shown already in Kubowitz[32].

It seems easier to remove Cu quantitatively by CN⁻ from arthropod than from molluscan hemocyanin (100 against 85% removal[16]; see also refs. 32 and 105).

A partial reconstitution of *Octopus vulgaris* hemocyanin has been achieved by an anaerobic treatment with Cu(I), working with the chloride complex (reconstitution of 50%[32], and 60%[113]), the ammine complex (65%[115]) and Cu_2O (80%[115]).

With the Cu(I) acetonitrile complex[116] a quantitative reconstitution of functional hemocyanin has been obtained[105,117], as indicated by the recovery of the copper bands and of the reversible oxygen binding. The recovery of the copper bands in partly reconstituted hemocyanin follows the square of the amount Cu bound, confirming that the oxygen binding sites are composed of pairs of equivalent and independent Cu atoms[105].

The fixation of Cu constitutes also an interesting problem in the biosynthesis of hemocyanin. The hepatopancreas, particularly in the cephalopods, is rich in Cu[118]. The reconstitution of functional hemocyanin was also effected with the apohemocyanin and the supernatant of the homogenate of the hepatopancreas of the spiny lobster, *Panulirus interruptus*[119].

THE COPPER BINDING SITES

The uniformity of the copper groups and of the copper binding sites, the number and nature of the ligands, and their geometry are the salient questions still to be answered (for a general discussion see refs. 109 and 120). Is the heterogeneity, often seen in the ultracentrifuge[43,54,121] and in zone electrophoresis[122-124] due to a difference in sequence of the polypeptide chain, that might even alter the Cu binding site? Different subunits have anyhow often to be postulated in order to explain the quaternary structure. The number of peptides on the fingerprints of *Helix pomatia* hemocyanin *e.g.* indicates at least two different smallest subunits with a molecular weight of 22,300[125].

Stability constants of the order of 10^{19} for the binding of Cu by *Limulus polyphemus* hemocyanin were found[126]. These values correspond to the stability constant of the Cu(I)–cysteine complex.

In the native protein the binding of copper is, moreover, stabilized by the conformation of the protein, as Cu cannot be liberated by most of the classical Cu reagents, except CN⁻ [127]. Cu becomes available to dithiocarbamate on denaturation by urea[128], and is also set free on partial hydrolysis[129,130].

As possible ligands have been considered: thiol groups[131], imidazole groups of histidine[132], amino groups[133], and disulfide bridges[134]. A Cu complex of arginine incidentally shows a definite catalase activity[135].

Thiol groups seem less likely, as only 1SH is liberated per 4Cu from *Cancer magister*[25] and from *Octopus vulgaris*[16,136] hemocyanin on anaerobic removal of Cu by CN⁻. It was still possible to reconstitute *Murex trunculus* hemocyanin after blocking the thiol groups of the apohemocyanin[137]. With *Jasus lalandii* hemocyanin[20], however, the reconstitution was partly inhibited on treating the apohemocyanin with Ag⁺.

Imidazole groups remain a possible candidate in function of the already mentioned similarity of the circular dichroism spectra of oxyhemocyanin and of the Cu(II) complex of Ac–Gly–Gly–His[62,67]. The electron paramagnetic resonance spectra[76] are not in opposition with this assumption, neither are the results of photooxidation, whereby the copper bands are destroyed at the same rate as the histidines[138,139].

CONCLUSIONS

Hemocyanins fall into two classes, according to the phyla Arthropoda and Mollusca. The average Cu contents of 0.173 and 0.253% respectively indicate minimum molecular weights of 36,700 and 25,100, corresponding to the smallest subunits found.

Deoxygenated hemocyanins contain Cu(I), as shown by the easy reconstitution with the Cu(I) acetonitrile complex, and, for molluscan hemocyanins, by the regeneration of the aged preparations with H_2O_2.

Measurements at $-195°C$ and by circular dichroism reveal three absorption bands in the visible and the near infra-red instead of the classical band at 580 nm.

Hemocyanins yield reversibly a colorless compound with CO, which is very useful for the preservation of the preparations. A green compound is obtained with NO. Molluscan hemocyanins form reversibly colorless compounds with cyanide, ethylisocyanide, SCN⁻ and thiourea.

Hemocyanins play a physiological role in the oxygen transport. Under the conditions that prevail in the blood, most hemocyanins show a cooperative oxygen binding.

The conformation of the protein stabilizes the Cu binding sites. The binding of Cu by thiol groups alone appears unlikely, while the imidazole group of histidine remains a serious candidate as a ligand.

NOTE ADDED IN PROOF

The electron spin resonance signal observed for aged preparations corresponds only to a few percent of the aged copper groups, implicating a spin coupling in most of the Cu^{2+} pairs. The electron paramagnetic resonance spectrum of NO-treated hemocyanin is due to the oxidation of Cu^+ to Cu^{2+}; the NO derivative of Cu^{2+} hemocyanin shows no definite signal.

REFERENCES

1 E. Harless, *Arch. Anat. Physiol.*, (1847) 148.
2 L. Fredericq, *Compt. Rend. Acad. Sci. Paris*, 87 (1878) 996.
3 G. Quagliariello, in H. Winterstein, *Handbuch der vergleichenden Physiologie*, Vol. I/1, G. Fischer, Jena, 1925, p. 597.
4 C. Dhéré, *Rev. Suisse Zool.*, 35 (1928) 277.
5 A. C. Redfield, *Biol. Rev.*, 9 (1934) 175.
6 F. Haurowitz, in C. Oppenheimer and L. Pincussen, *Tabulae Biologicae Periodicae*, Vol. IV, W. Junk, Berlin–Den Haag, 1935, p. 18.
7 J. Roche, *Essai sur la biochimie générale et comparée des pigments respiratoires*, Masson, Paris, 1936.
8 C. R. Dawson and M. F. Mallette, *Adv. Protein Chem.*, 2 (1945) 179.
9 A. C. Redfield, in W. D. McElroy and B. Glass, *Symposium on Copper Metabolism*, Johns Hopkins Press, Baltimore, Maryland, 1950, p. 174.
10 F. Haurowitz and R. L. Hardin, in H. Neurath and K. Bailey, *The Proteins*, Vol IIA, Academic Press, New York, 1954, p. 336.
11 H. J. Bielig and E. Bayer, in K. Lang and E. Lehnartz, *Hoppe-Seyler/Thierfelder, Handbuch der physiologisch- und pathologisch-chemischen Analyse*, Vol. IV/1, Springer, Berlin–Göttingen–Heidelberg, 10, 1960, p. 748.
12 C. Manwell, *A. Rev. Physiol.*, 22 (1960) 191.
13 F. Ghiretti, in O. Hayaishi, *Oxygenases*, Academic Press, New York and London, 1962, p. 517.
14 F. Ghiretti, in K. M. Wilbur and C. M. Yonge, *Physiology of Mollusca*, Vol. II, Academic Press, New York and London, 1966, p. 233.
15 W. M. Stanley and T. F. Anderson, *J. Biol. Chem.*, 146 (1942) 25.
16 A. Ghiretti-Magaldi, C. Nuzzolo and F. Ghiretti, *Biochemistry*, 5 (1966) 1943.
17 E. Feytmans, M. Wibo and J. Berthet, *Arch. Int. Physiol. Biochim.*, 74 (1966) 917.
18 M. Goyffon, *Compt. Rend. Soc. Biol.*, 162 (1968) 1123.
19 F. J. Joubert, *Biochim. Biophys. Acta*, 14 (1954) 127.
20 C. H. Moore, R. W. Henderson and L. W. Nichol, *Biochemistry*, 7 (1968) 4075.
21 I. B. Eriksson-Quensel and T. Svedberg, *Biol. Bull.*, 71 (1936) 498.
22 M. A. Lauffer and L. G. Swaby, *Biol. Bull.*, 108 (1955) 290.
23 S. M. Pickett, A. F. Riggs and J. L. Larimer, *Science*, 151 (1966) 1005.
24 L. Di Giamberardino, *Arch. Biochem. Biophys.*, 118 (1967) 273.
25 L. C. G. Thomson, M. Hines and H. S. Mason, *Arch. Biochem. Biophys.*, 83 (1959) 88.
26 T. Svedberg, *J. Biol. Chem.*, 103 (1933) 311.
27 F. Hernler and E. Philippi, *Z. Physiol. Chem.*, 216 (1933) 110.
28 R. Lontie and R. Witters, in J. Peisach, P. Aisen and W. E. Blumberg, *The Biochemistry of Copper*, Academic Press, New York and London, 1966, p. 455.
29 E. J. Wood, W. H. Bannister, C. J. Oliver, R. Lontie and R. Witters, *Comp. Biochem. Physiol.*, 40B (1971) 19.
30 L. B. Cohen and K. E. Van Holde, *Biochemistry*, 3 (1964) 1809.
31 K. E. Van Holde and L. B. Cohen, *Biochemistry*, 3 (1964) 1803.
32 F. Kubowitz, *Biochem. Z.*, 299 (1938) 32.
33 T. Svedberg and A. Hedenius, *Biol. Bull.*, 66 (1934) 191.
34 C. Manwell, *J. Cell. Comp. Physiol.*, 52 (1958) 341.
35 J. R. Redmond, *Physiol. Zool.*, 35 (1962) 304.
36 B. Padmanabhanaidu, *Comp. Biochem. Physiol.*, 17 (1966) 167.

356

37 W. C. Boyd, *Biol. Bull.*, 73 (1937) 181.
38 J. L. Cloudsley-Thompson, *J. Linn. Soc.*, 43 (1957) 134.
39 J. Berthet and P. Berthet, *Arch. Int. Physiol. Biochim.*, 71 (1963) 124.
40 J. Berthet, P. Baudhuin and M. Wibo, *Arch. Int. Physiol. Biochim.*, 72 (1964) 676.
41 W. Wieser, *Naturwiss.*, 52 (1965) 133.
42 R. G. Sundara, *Current Sci. (India)*, 38 (1969) 168.
43 T. Svedberg and K. O. Pedersen, *The Ultracentrifuge*, Clarendon Press, Oxford, 1940, p. 363.
44 Ö. Levin, *Acta Univ. Upsal.*, 24, 1963.
45 E. F. J. van Bruggen, in F. Ghiretti, *Physiology and Biochemistry of Haemocyanins*, Academic Press, London, 1968, p. 37.
46 P. Baudhuin, *Rev. Questions Sci.*, (1964) 77.
47 M. Wibo, P. Baudhuin and J. Berthet, *Arch. Int. Physiol. Biochim.*, 74 (1966) 945.
48 E. F. J. van Bruggen, V. Schuiten, E. H. Wiebenga and M. Gruber, *J. Mol. Biol.*, 7 (1963) 249.
49 S. Brohult, *J. Phys. Colloid Chem.*, 51 (1947) 206.
50 H. Fernández-Morán, E. F. J. van Bruggen and M. Ohtsuki, *J. Mol. Biol.*, 16 (1966) 191.
51 T. Eskeland, *J. Ultrastruct. Res.*, 17 (1967) 544.
52 V. A. Bloomfield, *Science*, 161 (1968) 1212.
53 R. M. Condie and R. B. Langer, *Science*, 144 (1964) 1138.
54 S. Brohult and K. Borgman, in *The Svedberg*, Almqvist and Wiksell, Uppsala-Stockholm, 1944, p. 429.
55 K. Heirwegh, H. Borginon and R. Lontie, *Biochim. Biophys. Acta*, 48 (1961) 517.
56 W. M. Konings, R. J. Siezen and M. Gruber, *Biochim. Biophys. Acta.* 194 (1969) 376.
57 P. Putzeys and J. Brosteaux, *Trans. Faraday Soc.*, 31 (1935) 1314.
58 J. Brosteaux, *Naturwiss.*, 16 (1937) 249.
59 W. N. Konings, J. Dijk, T. Wichertjes, E. C. Beuvery and M. Gruber, *Biochim. Biophys. Acta*, 188 (1969) 43.
60 J. Cox, R. Witters and R. Lontie, *Int. J. Biochem.*, 3 (1972) in the press.
61 T. Omura, T. Fujita, F. Yamada and S. Yamamoto, *J. Biochem. (Tokyo)*, 50 (1961) 400.
62 K. E. Van Holde, *Biochemistry*, 6 (1967) 93.
63 I. Moring Claesson, *Arkiv. Kemi*, 10 (1956) 1.
64 H. S. Mason, in F. Dickens and E. Neil, *Oxygen in the Animal Organism*, Macmillan, New York, 1964, p. 117 (discussion).
65 J. G. Foss, *Biochim. Biophys. Acta*, 79 (1964) 41.
66 H. Takesada and K. Hamaguchi, *J. Biochem. (Tokyo)*, 63 (1968) 725.
67 G. F. Bryce and F. R. N. Gurd, *J. Biol. Chem.*, 241 (1966) 1439.
68 L. E. Orgel, in E. M. Crook, *Metals and Enzyme Activity*, Cambridge University Press, London, 1958, p. 19.
69 E. Frieden, S. Osaki and H. Kobayashi, *J. Gen. Physiol.*, 49 (1965) 213.
70 S. Yamada, Y. Shimura and R. Tsuchida, *Bull. Chem. Soc. Japan*, 26 (1953) 72.
71 W. H. Bannister and E. J. Wood, *Nature*, 223 (1969) 53.
72 B.-C. Wang and W. P. Schaefer, *Science*, 166 (1969) 1404.
73 T. Tsuchida, *Biophysics*, 1 (1961) 42.
74 T. Nakamura and H. S. Mason, *Biochem. Biophys. Res. Commun.*, 3 (1960) 297.
75 C. Manwell, in F. Dickens and E. Neil, *Oxygen in the Animal Organism*, Macmillan, New York, 1964, p. 49.
76 J. F. Boas, J. R. Pilbrow, G. J. Troup, C. Moore and T. D. Smith, *J. Chem. Soc.*, (1969) 965.

357

77 E. Bayer and H. Fiedler, *Ann.*, 653 (1962) 149.
78 W. E. Blumberg, in J. Peisach, P. Aisen and W. E. Blumberg, *The Biochemistry of Copper*, Academic Press, New York and London, 1966, p. 472.
79 P. Bert, *Compt. Rend. Acad. Sci., Paris*, 65 (1867) 300.
80 H. Winterstein, *Biochem. Z.*, 19 (1909) 384.
81 J. R. Redmond, in F. Ghiretti, *Physiology and Biochemistry of Haemocyanins*, Academic Press, London and New York, 1968, p. 5.
82 H. P. Wolvekamp and T. H. Waterman, in T. H. Waterman, *The Physiology of Crustacea*, Vol. I, Academic Press, New York and London, 1960, p. 35.
83 A. C. Redfield, T. Coolidge and H. Montgomery, *J. Biol. Chem.*, 76 (1928) 197.
84 R. Guillemet and G. Gosselin, *Compt. Rend. Soc. Biol.*, 111 (1932) 733.
85 W. A. Rawlinson, *Aust. J. Exp. Biol. Med. Sci.*, 18 (1940) 131.
86 J. Wyman, *J. Am. Chem. Soc.*, 89 (1967) 2202.
87 E. Stedman and E. Stedman, *Biochem. J.*, 22 (1928) 889.
88 R. Lontie, *Meded. Vlaam. Chem. Ver.*, 16 (1954) 110.
89 H. A. DePhillips, K. W. Nickerson, M. Johnson and K. E. Van Holde, *Biochemistry*, 8 (1969) 3665.
90 N. Shaklai and E. Daniel, *Biochemistry*, 9 (1970) 564.
91 R. W. Root, *J. Biol. Chem.*, 104 (1934) 239.
92 E. Rocca and F. Ghiretti, *Boll. Soc. Ital. Biol. Sper.*, 39 (1963) 2075.
93 W. Vanneste and H. S. Mason, in J. Peisach, P. Aisen and W. E. Blumberg, *The Biochemistry of Copper*, Academic Press, New York and London, 1966, p. 465.
94 M. De Ley and R. Lontie, *FEBS Lett.*, 6 (1970) 125.
95 O. H. Pearson, *J. Biol. Chem.*, 115 (1936) 171.
96 A. Craifaleanu, *Boll. Soc. Nat. Napoli*, 32 (1919) 141.
97 E. J. Wood and W. H. Bannister, *Nature*, 215 (1967) 1091.
98 W. Rombauts and R. Lontie, *Arch. Int. Physiol. Biochim.*, 68 (1960) 695.
99 M. De Ley and R. Lontie, *Arch. Int. Physiol. Biochim.*, 76 (1968) 175.
100 W. Rombauts and R. Lontie, *Arch. Int. Physiol. Biochim.*, 68 (1960) 230.
101 C. Dhéré and A. Schneider, *Compt. Rend. Soc. Biol.*, 82 (1919) 1041.
102 A. Ghiretti-Magaldi, F. Ghiretti, G. Nardi, V. Parisi and R. Zito, *Boll. Soc. Ital. Biol. Sper.*, 38 (1962) 1851.
103 K. Heirwegh and R. Lontie, *Nature*, 185 (1960) 854.
104 K. Heirwegh, V. Blaton and R. Lontie, *Arch. Int. Physiol. Biochim.*, 73 (1965) 149.
105 W. N. Konings, R. van Driel, E. F. J. van Bruggen and M. Gruber, *Biochim. Biophys. Acta*, 194 (1969) 55.
106 G. Felsenfeld and M. P. Printz, *J. Am. Chem. Soc.*, 81 (1959) 6259.
107 J. Schubert and E. R. White, *Science*, 155 (1967) 1000.
108 F. Ghiretti, *Arch. Biochem. Biophys.*, 63 (1956) 165.
109 G. Morpurgo and R. J. P. Williams, in F. Ghiretti, *Physiology and Biochemistry of Haemocyanins*, Academic Press, London, 1968, p. 113.
110 P. George and C. J. Stratmann, *Biochem. J.*, 57 (1954) 568.
111 E. C. Slater, *Nature*, 170 (1952) 970.
112 K. Heirwegh, H. Borginon and R. Lontie, *Arch. Int. Physiol. Biochim.*, 67 (1959) 514.
113 A. Ghiretti-Magaldi, G. Nardi, F. Ghiretti and R. Zito, *Boll. Soc. Ital. Biol. Sper.*, 38 (1962) 1839.
114 M. Henze, *Z. Physiol. Chem.*, 33 (1901) 370.
115 A. Ghiretti-Magaldi and G. Nardi, in H. Peeters, *Protides Biol. Fluids, Proc. 11th Colloquium, Bruges, 1963*, Elsevier, Amsterdam, 1964, p. 507.
116 P. Hemmerich and C. Sigwart, *Experientia*, 19 (1963) 488.

358

117 R. Lontie, V. Blaton, M. Albert and B. Peeters, *Arch. Int. Physiol. Biochim.*, 73 (1965) 150.
118 R. Rocca, *Comp. Biochem. Physiol.*, 28 (1969) 67.
119 W. Johnston and A. A. Barber, *Comp. Biochem. Physiol.*, 28 (1969) 1259.
120 E. Bayer and P. Schretzmann, in C. K. Jørgensen *et al.*, *Structure and Bonding*, Vol. II, Springer, Berlin, 1967, p. 236.
121 R. Witters and R. Lontie, in F. Ghiretti, *Physiology and Biochemistry of Haemocyanins*, Academic Press, London, 1968, p. 61.
122 K. R. Woods, E. C. Paulsen, R. L. Engle, Jr. and J. H. Pert, *Science*, 127 (1958) 519.
123 F. G. Elliott and J. Hoebeke, in F. Ghiretti, *Physiology and Biochemistry of Haemocyanins*, Academic Press, London, 1968, p. 81.
124 E. J. Wood, C. M. Salisbury, N. Formosa and W. H. Bannister, *Comp. Biochem. Physiol.*, 26 (1968) 345.
125 M. Gruber, in F. Ghiretti, *Physiology and Biochemistry of Haemocyanins*, Academic Press, London, 1968, p. 49.
126 G. Felsenfeld, *J. Cell. Comp. Physiol.*, 43 (1954) 23.
127 E. Zuckerkandl, *Bull. Soc. Chim. Biol.* 41 (1959) 1629.
128 M. Griffé and R. Lontie, *Arch. Int. Physiol. Biochim.*, 69 (1961) 594.
129 G. Brauns and R. Lontie, *Arch. Int. Physiol. Biochim.*, 68 (1960) 211.
130 G. Nardi, A. Ghiretti-Magaldi, G. Caserta, R. Zito and F. Ghiretti, *Boll. Soc. Ital. Biol. Sper.*, 38 (1962) 1845.
131 I. M. Klotz and T. A. Klotz, *Science*, 121 (1955) 477.
132 R. Lontie, *Clin. Chim. Acta*, 3 (1958) 68.
133 R. Zito, G. Nardi, G. Caserta and A. Ghiretti-Magaldi, *Boll. Soc. Ital. Biol. Sper.*, 38 (1962) 1855.
134 P. Hemmerich, in J. Peisach, P. Aisen and W. E. Blumberg, *The Biochemistry of Copper*, Academic Press, New York and London, 1966, p. 15.
135 E. Rocca, *Boll. Soc. Ital. Biol. Sper.*, 39 (1963) 2077.
136 A. Ghiretti-Magaldi and C. Nuzzolo, *Comp. Biochem. Physiol.*, 16 (1965) 249.
137 E. J. Wood, C. M. Salisbury and W. H. Bannister, *Biochem. J.*, 108 (1968) 26P.
138 E. J. Wood and W. H. Bannister, *Biochim. Biophys. Acta*, 154 (1968) 10.
139 Y. Engelborghs, R. Witters and R. Lontie, *Arch. Int. Physiol. Biochim.*, 76 (1968) 372.

PART IV

METAL INDUCED LIGAND REACTIONS
AND METAL ENZYMES

Chapter 13

METAL INDUCED LIGAND REACTIONS INVOLVING SMALL MOLECULES

M. M. JONES

Department of Chemistry, Vanderbilt University, Nashville, Tennesse 37203, U.S.A.

AND

J. E. HIX, Jr.

Department of Chemistry, East Carolina University, Greenville, North Carolina 37834, U.S.A.

1. INTRODUCTION

When a small molecule is coordinated to a metal ion or other Lewis acid, it undergoes several changes which can dramatically alter its reactivity. First, it is changed from an independent species, with at least one free pair of valence shell electrons to a subordinate part of a larger molecule, the bond to which ties up at least one of its previous free pairs of electrons. Secondly, it is changed from a neutral or negative species to one which is bonded directly to a positive charge. Subsequent to coordination it must undergo reaction attached to a large, and generally unreactive, group of atoms, and in close proximity to a positive charge. That at least some of the reactions of the ligand would be altered by the coordination process seems obvious. The more detailed consideration of how a given ligand reaction will be altered by coordination is the subject of this chapter. As will be seen, there is a fair amount of experimental evidence which can serve as a guide to the correlation of such processes and also for purposes of making predictions. The goal of our inquiry includes both a deeper understanding of natural processes in which metal ions act as catalysts and the ability to design more effective catalysts for a variety of metal catalyzed ligand reactions. It should be noted at this point that metal ions may also activate certain enzymes by processes dependent primarily upon the assistance which coordination may furnish to the achievement of a particular conformation of the protein chain, rather than by furnishing coordination sites upon which the particular enzymatic reaction may occur. Thus the metal ion may be quite far removed from the active site of the enzyme and yet essential for its proper activity. This is in addition to possible activation processes involving enzymes in which the metal ion is an integral part of an active site.

2. GENERAL EFFECTS OF COORDINATION ON LIGANDS

The immediate result of attaching a ligand to a positively charged metal ion is to polarize the electron density of the ligand toward the metal ion. This is essentially the expected result from molecular orbital theory for complexes in which π-bonding is not important. It is thus a conclusion which can be applied with greatest assurance to the complexes of hard bases with hard acids[1]. Where π-bonding is important, this conclusion will be modified, the effect will be diminished if back donation from metal to ligand is important, or accentuated, if the π-bonding involves donation from ligand to metal.

For hard base–hard acid interactions (see Chapter 2) we can expect a typical set of consequences from this ligand polarization. These include:

1. an increase in the acidity of the hydrogen atoms on the donor atom;
2. an increase in the ease of attack by nucleophiles on the ligand and a decrease in the ease of attack by electrophiles;
3. a masking of certain of the reactions of the free ligand that are dependent upon a stereochemical or electronic interaction which is hindered or prevented by coordination.

These are in addition to some other general consequences of coordination which include a change in the redox properties of the ligand and, for tautomeric ligands, a shift in the tautomeric equilibrium to favor that tautomer which forms the most stable complexes. Most of these general features were first clearly enunciated by Meerwein in the late 1920s[2].

The increase of the acidity of hydrogen atoms on the donor atom was first noted by Werner[3] who found that the acidic properties of coordinated water or ammonia increased as the central metal ion was changed in the sequence $Co^{3+} < Cr^{3+} < Ru^{4+} < Pt^{4+}$. Studies on numerous complexes have shown this phenomenon to be generally applicable. A comparison of free and coordinated water can be made using the data on $Cu(H_2O)_4^{2+}$. An equilibrium constant of 1.07×10^{-8} was found[4] for the reaction

$$Cu(H_2O)_4^{2+} \rightleftharpoons Cu(H_2O)_3(OH)^+ + H^+$$

an increase about 10^7 times that of water itself. This kind of increase in acidity is also found for organic ligands which have hydrogen atoms attached to their donor atoms[5,6].

The increase in the ease with which a nucleophile can attack a species after it is coordinated was used to explain the role of metal ions in certain enzymatic reactions in the late 1940s and early 1950s[7], though such a role had been suspected earlier by Hellerman[8]. Reactions of this sort involving the attack of water are especially numerous and include the hydrolysis of simple species, such as $S_2O_7^{2-}$ [9], and condensed phosphates, as well as more complex species[10] such as phosphate esters, amino acid derivatives and peptides, which are discussed in more detail subsequently.

The masking of ligand reactions was one of the first properties of complexes to attract attention. The fact that coordinated organic ligands are less susceptible to attack by typical oxidizing agents was noted by 1857[11]. There are numerous instances of this masking, many of them involving donor atoms participating in the oxidation–reduction. Examples include the masking of coordinated oxalate against the action of bromine[12], of coordinated EDTA against the action of permanganate[13], and of coordinated HEDTA against vanadate[14]. It was also noted very early that coordinated anions do not give their characteristic precipitation tests or exhibit their usual reactions as bases (e.g. NH_3 vs. $Co(NH_3)_6^{3+}$). All in all, masking of certain reactions is a characteristic feature of coordination which can be anticipated to have significant biological effects.

An additional effect of coordination, which is especially important for larger, multidentate donors, results from the fact that the donor atoms are held in a relatively fixed position with respect to each other. This imposition of a specific ligand conformation on coordinated species usually has the result that reactions which require some other conformation are either difficult or impossible, while those which require the conformation which is found are enormously facilitated.

The last effect of coordination which must be mentioned is that it can result in the activation of certain small molecules for reaction with appropriate reagents. While this effect frequently involves nucleophilic attack, it can also facilitate certain redox processes.

3. CATALYSIS OF NUCLEOPHILIC ATTACK

The catalysis of nucleophilic attack via coordination of a substrate to a positively charged metal ion is one of the most general consequences of coordination. Its basis lies in the more favorable electrostatic interactions which are found in the complex. Following Laidler[15] we can estimate the electrostatic contribution to the entropy of activation for a model involving an activated complex consisting of two spherical reactants as:

$$\Delta S^{\ddagger}_{\text{e.s.}} = -10 Z_A Z_B \text{ (for water as solvent)}$$

If Z_A is the charge on the attacking reagent, Z_{B1} that on the ligand and Z_{B2} that on the metal ion prior to coordination, then

$$\Delta S^{\ddagger}_{\text{e.s.}}, \text{ligand} = -10 Z_A Z_{B1}$$

$$\Delta S^{\ddagger}_{\text{e.s.}}, \text{complex} = -10 Z_A (Z_{B1} + Z_{B2})$$

and the change in going to the complex is then

$$\Delta\Delta S^{\ddagger}_{\text{e.s.}} = -10 Z_A Z_{B2}$$

An alternative expression based on the single sphere activated complex can also be developed, but its use requires a knowledge of the radii of the activated complex as well as the reactants. The change in the entropy of activation predicted depends on the solvent and the charge on the species attacking the substrate. If the solvent is water and the attacking species is positively charged, coordination will make the entropy values more negative and the attainment of the transition state will therefore be more difficult. If the attacking species is negatively charged, the opposite will be true.

The same models may be used to obtain the ratio of the rate constants for the ligand and the complex. For the simple two-sphere model the result may be given as:

$$\ln \frac{k_c}{k_f} = - \frac{Z_M Z_B e^2}{\epsilon d_{AB} kT}$$

where k_c is the rate of reaction for the complex, k_f is that of the free ligand, Z_M is the charge on the metal complex excluding the ligand consideration, Z_B is the charge on the attacking species, ϵ is the dielectric constant of the reaction medium, d_{AB} is the separation distance of the substrate and attacking species in the activated complex (assumed to be the same in the reaction of both the complex and free ligand), k is Boltzmann's constant, and T is the absolute temperature. An analogous equation can be derived for an ion dipole reaction.

From such a model we would always expect nucleophilic attack to be accelerated by coordination to cations and we would expect this acceleration to be more pronounced with more highly charged cations and in solvents of lower dielectric constant. A great shortcoming of this model is that it does not incorporate the hard or soft nature of the interacting species and thus does not adequately reflect the specificities so characteristic of such catalytic processes. The derivation of a more refined equation is possible but it requires that the transition state be specified in greater detail.

A listing of the types of such reactions, all examples of metal ion catalyzed nucleophilic attack, largely drawn from Bender's review[10] illustrates the scope of such a generalization. In such a list one finds *hydrolysis* reactions of acid anhydrides, carboxylic acid esters and amides, phosphate esters, and Schiff bases, *carboxylation* and *decarboxylation*, *transaminations*, hydrations, hydrogenations, the splitting of ethers, and a variety of proceses which are analogous to the more thoroughly studied reactions in which the proton catalyzes the attack of a nucleophile other than OH^-. The great advantage of metal ions over the proton for many reactions of this type lies in the fact that they can be effective as catalysts in solutions of much higher pH than can be used otherwise. For this reason, such metal ions have been called "super-acids"[16]. This meaning is literally true in some cases where the metal ion chelates to the ligand and is much more effective than the proton as a catalyst[17]. In other cases where the metal ion coordinates to only one site on a ligand, its effective polarizing power, and catalytic effect is usually *not* greater than that of a proton; though it can exert a "super-acid" effect by functioning at a pH where the proton cannot.

4. SOME IMPORTANT HYDROLYTIC PROCESSES

As an example of how coordination aids nucleophilic attack of a ligand, let us consider several hydrolytic processes which have served as "model" systems for various enzymatic reactions. The systems most thoroughly understood include catalysis of the hydrolysis of amino acid esters, peptide linkages, and phosphate linkages.

In all of these reactions the thermodynamic stability of the various complexes involved plays a major role in the apparent catalysis. If no complex is formed between the metal ion and the ligand molecule, then only the reactivity of the free ligand will be detected experimentally. In aqueous solution, the metal ion must compete with hydrogen for coordination to the ligand. Thus as the acidity of the reaction medium is increased, not only is less hydroxide present to provide nucleophilic attack (water can also serve as a nucleophile, but usually at a much slower rate), but also less complex will be formed. On the other hand, as the basicity of the system is increased, hydrolysis of the metal ion itself occurs, forming insoluble, and therefore inactive, metal hydroxides. In most cases, then, one is quite restricted in the pH limitations required for catalysis to occur.

In relating observed rates of hydrolysis to the actual rates of hydrolysis of the various ligand-containing species, the overall composition of the reaction solution must be considered. Each ligand-containing species undergoes alkaline hydrolysis at its own characteristic rate, and the observed overall rate is a composite rate constructed from specific rates by summing over all ligand-containing species present in the reaction system:

$$\text{Rate}_{\text{observed}} = \text{Rate}_{\text{free ligand}} + \text{Rate}_{\text{complex}_1} + \text{Rate}_{\text{complex}_2} + \dots$$

Let us now confine our interest to the catalytic reactivity of specific complexes in the hydrolysis of the various ligand classes.

(a) Amino acid esters

That transition metal ions catalyze the rates of alkaline hydrolysis of amino acid esters has been known since first reported by Kroll in 1952[18]. Since that time many investigators have studied these catalyzed reactions, and the role of the metal ion in accelerating the hydrolysis is fairly well understood[19].

One would expect coordination of an amino acid ester to a positive metal ion or complex to increase the rate of alkaline hydrolysis from an electrostatic point of view since this reaction involves nucleophilic attack of the ligand. As a first approximation, the rate of catalysis would be expected to be determined essentially by the electrostatic charge of the reactive complex. As an approximation, this is usually quite valid since most trivalent complexes catalyze the rate of hydrolysis in solutions as acidic as pH 3, divalent complexes catalyze the rate of hydrolysis in solutions as acidic as pH 5–7, and nonovalent complexes catalyze the rate of hydrolysis in solutions of pH more basic than 7[17,20,21].

To approximate the effect of the metal ion on the hydrolysis of the ester more accurately, one must consider the interaction of the ligand with the metal ion in more detail. In most instances, the ester coordinates to the metal using only the amine nitrogen as a donor atom. The ester carbonyl is not nearly as nucleophilic as is the carboxyl group of the corresponding amino acid and therefore shows little tendency to chelate. When more electrophilic metal ions or complexes are used as catalysts, however, a weak interaction results between the metal and the ester carbonyl, causing direct polarization of the carbonyl bond. This effect of chelation can be seen below:

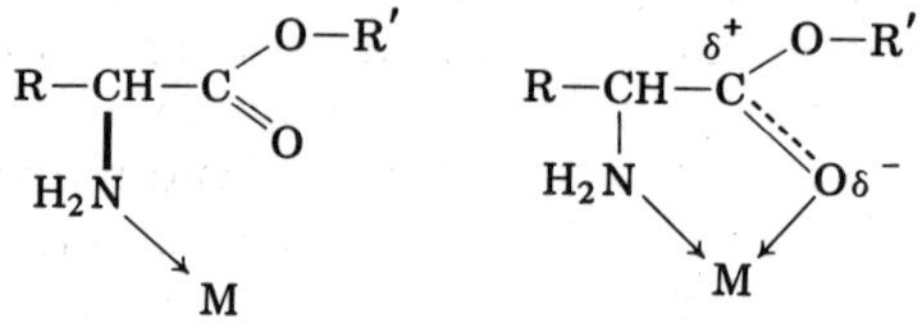

$$
\begin{array}{cc}
\text{R--CH--C} \overset{\displaystyle \text{O--R}'}{\underset{\displaystyle \text{O}}{\Big\langle}} & \text{R--CH--C} \overset{\displaystyle \delta^+ \; \text{O--R}'}{\underset{\displaystyle \text{O}\delta^-}{\Big\langle}} \\[2ex]
\underset{\displaystyle \searrow}{\text{H}_2\text{N}} \qquad\qquad & \underset{\displaystyle \searrow}{\text{H}_2\text{N}} \qquad\qquad \\[1ex]
\qquad\quad \text{M} & \qquad\quad \text{M}
\end{array}
$$

Note that chelation polarizes the carbonyl group so that the carbon atom is much more electrophilic. When this happens, not only is the complex much more susceptible to nucleophilic attack, but the reaction site on the ligand becomes the site of nucleophilic attack. In the case of copper(II) catalyzed hydrolysis of ethyl glycinate, chelation enhances the rate of hydrolysis by a factor of almost 200 times that predicted from electrostatic principles alone[22]. Busch has likewise shown this to be true for cobalt(III) catalyzed hydrolysis of coordinated amino acid esters[20].

Owing to the asymmetry present at the α-carbon in all amino acids and their derivatives (except glycine), stereoselectivity can result in the structures and reactions of complexes formed from these ligands. Stereoselectivity of this nature has only been found in cases where the ligand possesses at least two primary donor atoms other than the potentially coordinating ester group, as in histidine[23]. When esters of histidine are hydrolyzed in the presence of the D- and L-histidinato nickel(II) isomers, the ester and histidinate present in the catalyst must have opposite configurations for the direct interaction of the carbonyl group with the metal ion. When the ester and histidinate have the same configuration the ester group will be pointed away from the metal ion so that no direct interaction can occur.

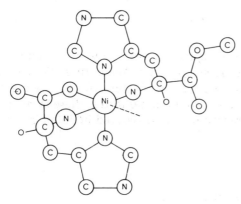

Same configuration

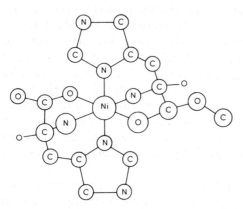

Opposite configuration

(b) Peptide linkages

Polypeptides and amino acid amides have also been shown to undergo catalyzed hydrolysis. The catalysis is quite often not as pronounced as observed in the case of the esters. Tetravalent metals such as thorium(IV) and cerium(III) and a number of heavier trivalent metals have been utilized extensively[24].

The catalysis of the hydrolysis of peptide linkages by cis-hydroxyaquo-triethylenetetraminecobalt(III) has been shown to be selective for the N-terminal amino acid[25]. Complexes of this type coordinate to the terminal amine group, and by induction help to cleave the peptide bond to the next amino acid residue. These complexes have been used extensively in the analysis of peptide fragments to determine the ordering of the amino acids present.

(c) Phosphate linkages

Catalysis of the cleavage of phosphate esters[26], acyl phosphates[27], and polyphosphates[28] has also been shown to occur. Again, these linkages are not as easily hydrolyzed as those of amino acid esters, but by the use of highly charged complexes and/or elevated temperatures, these systems hydrolyze quite readily. These reactions are considered in detail in Chapter 17.

5. REDOX PROPERTIES OF LIGANDS

The redox behavior of ligands is invariably altered by coordination, though the fact that the metal ion may be an oxidant or reductant may alter the general pattern. When the behavior of species such as the halide ions is examined, it is found that coordination generally makes it much more difficult to oxidize the ligand *by the use of an external oxidizing* agent; *e.g.* the oxidation of the I^- of HgI_4^{2-} or of the Br^- of $HgBr_4^{2-}$. To a considerable extent this phenomenon can be attributed to the decreased availability of the ligand electron pairs to an external reagent. When the metal ion itself is an oxidant one would expect coordination to *facilitate* ligand oxidation, as is indeed suggested by evidence available from multistep oxidants.

Thus in oxidations involving MnO_4^- or CrO_4^{2-}, the initial step is frequently slower than subsequent steps which proceed through lower oxidation states of Mn or Cr which are capable of coordination to a ligand undergoing oxidation. The fact that the coordination process itself will draw electron density toward the Lewis acid should make the action of such a species as an oxidant more rapid than it would otherwise be. It is very difficult to test this hypothesis, but in the case of copper(II), thallium(III), iron(III) and similar oxidants, inner sphere mechanisms for oxidation processes are commonly found and provide evidence favoring this view.

The systems of greatest interest are those in which a ligand is coordinated to readily reduced metal ions. In such cases the metal is reduced and, in most cases, a reactive free radical derived from the ligand is formed. This free radical can then undergo subsequent reaction to give further products.

Some such processes give useful products, as for example the copper(II) oxidation of α-hydroxyketones to α-diketones:

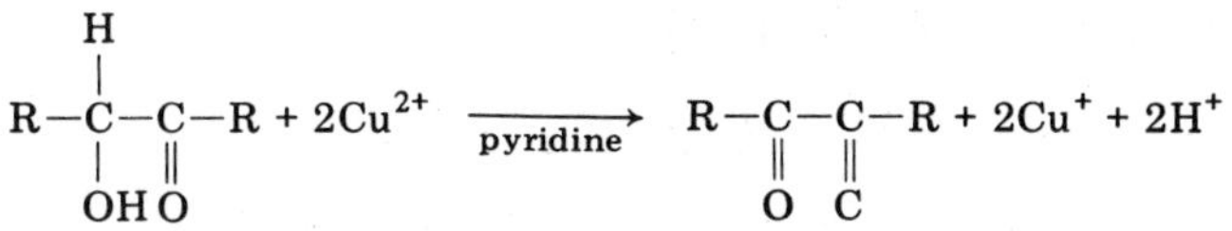

$$\underset{\underset{\text{OH O}}{|\;\;||}}{R-C-C-R} + 2Cu^{2+} \xrightarrow[\text{pyridine}]{} \underset{\underset{\text{O C}}{||\;\;||}}{R-C-C-R} + 2Cu^+ + 2H^+$$

which can be used to make benzil from benzoin[29] and furil from furoin[30]. In the presence of oxygen these oxidations proceed further and, by reoxidation of the copper(I), catalytic processes can be set up, as can be seen in the synthesis for nitriles developed by Brockmann[31]. Here the overall reaction are

$$RCHO + NH_3 + O_2 \rightarrow RCN + H_2O + H_2O_2$$

$$H_2O_2 + \text{substrates} \xrightarrow{Cu^{II}} H_2O + \text{Products}$$

The steps are as follows:

$$RCH_2OH \xrightarrow{Cu^{II} \text{ complex}} RCHO + H_2O$$

$$RCHO + NH_3 \rightarrow RCH{=}NH + H_2O$$

$$RCH{=}NH + Cu^{II} \text{ complex} \xrightarrow{\text{slow}} \text{imine radical} + H^+ + Cu^I \text{ complex}$$

$$\text{Imine radical} + Q_2 \rightarrow RC{\equiv}N + HO_2.$$

$$HO_2{\cdot} + H^+ \rightarrow H_2O_2$$

The copper(II) complex $[Cu(NH_3)_4(OCH_2R)]^+$ is postulated as the active species and, when the reaction is in the absence of oxygen, stoichiometric amounts of complex must be used, as would be expected if an intermediate complex with the Schiff base had to be formed.

There are a large number of autocatalytic processes of this sort which depend for their success on the easier oxidation of the substrate by a metal ion in a complex and the fact that the metal ion can be reoxidized by atmospheric oxygen. The copper(II) catalyzed autoxidation of ascorbic acid is another example of this same general type[32] (see Chapter 20).

A very striking use of coordination to facilitate an oxidation process is seen in the oxidation of ethylene to acetaldehyde by tetrachloropalladate[33]. The process is turned into a catalytic one by the use of a medium which contains copper(II) salts, hydrochloric acid and oxygen. The overall reaction is

$$C_2H_4 + PdCl_4{}^{2-} + 3H_2O \rightarrow CH_3CHO + Pd + 2H_3O^+ + 4Cl^-$$

This process involves the initial coordination and oxidation of the olefin, as follows:

$$PdCl_4{}^{2-} + C_2H_4 \rightleftharpoons [PdCl_3(C_2H_4)]^- + Cl^-$$

$$PdCl_3(C_2H_4)^- + H_2O \rightleftharpoons PdCl_2(H_2O)(C_2H_4) + Cl^-$$

$$PdCl_2(H_2O)(C_2H_4) + H_2O \rightleftharpoons PdCl_2(OH)(C_2H_4)^- + H_3O^+$$

$$PdCl_2(OH)(C_2H_4)^- \rightarrow CH_3CHO + Pd^\circ + 2Cl^-$$

followed by

$$Pd^\circ + 2CuCl_2 + 2Cl^- \rightarrow PdCl_4{}^{2-} + 2CuCl$$

and

$$4CuCl + 4HCl + O_2 \rightarrow 4CuCl_2 + 2H_2O.$$

Reduction reactions of ligands have been studied most extensively for complexes in which unsaturated molecules are coordinated to Pt(II). In such complexes the olefinic ligands are much more easily and rapidly reduced than the as free ligands[34]. Evidence on other types of ligands is quite scarce.

In summary, one can always expect the redox properties of a ligand to be altered by coordination. When the metal ion itself will not undergo a redox process, it usually reduces the ease of oxidation of the ligand by external reagents.

6. MASKING OF LIGAND REACTIVITY IN COMPLEXES

The masking of ligand reactivity in complexes refers to the retardation or complete elimination of a typical ligand reaction as a result of coordination. It has always been recognized that coordination causes the suppression of precipitation reactions of the halides. The prediction of the consequences of coordination in such simple cases follows directly from its effect on the solubility product. For more complex reactions predictions of masking are more difficult.

The following are the ways in which masking may arise.

1. The formation of the coordinate bond involves an electron pair which is essential to the reaction.
2. Coordination imposes a conformation on the ligand which is unfavorable for the reaction.
3. The oxidation potential of the ligand is altered so drastically by coordination that the reaction is no longer spontaneous.

4. The charge on the reactive site is changed by coordination in such a fashion as to make the reaction more difficult.
5. The character of the ligand is altered by coordination to favor a form (*e.g.* tautomer or derived ion) of the ligand which does not undergoes the reaction or else reacts much less readily.

In any specific case masking may arise as a combination of two or more of these factors, though it is customary to search for masking processes by the considered use of the particular factors which are most easily manipulated in the specific case.

One of the most useful applications of masking is in shielding the carboxyl and its α-amino group in amino acid reactions. The formation of a chelate with copper(II) very effectively masks both the amino group and its neighboring carboxylate group from most reagents with which they typically react. This allows other functional groups which are not masked to be the sole sites of reaction. Subsequently the copper(II) can be removed from the amino acid by conversion to its insoluble sulfide by the use of hydrogen sulfide or sodium sulfide. In outline these reactions are all of the type:

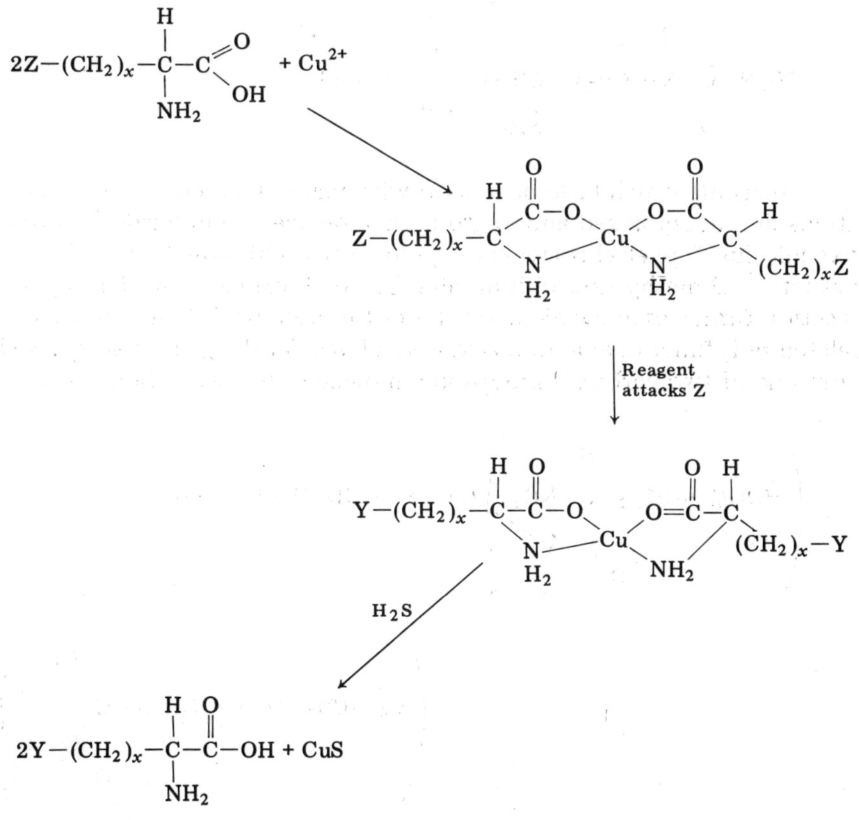

The synthetic applications have been reviewed[35,36], and include mostly reactions of terminal, unprotected amino or carboxyl groups. An example is the conversion of citrulline to ornithine[37]:

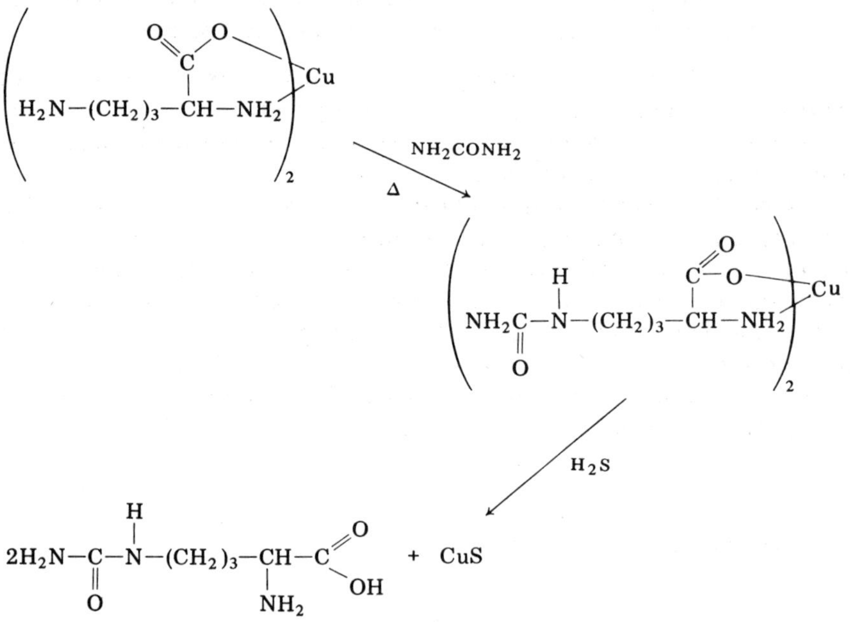

$$\left(\begin{array}{c} O \diagdown \diagup O \\ C \\ | \\ H_2N-(CH_2)_3-CH-NH_2 \end{array} \diagdown Cu \right)_2 \xrightarrow[\Delta]{NH_2CONH_2}$$

$$\left(\begin{array}{c} H \quad\quad C \diagdown O \\ | \quad\quad\quad | \\ NH_2C-N-(CH_2)_3-CH-NH_2 \\ \| \\ O \end{array} \diagdown Cu \right)_2 \xrightarrow{H_2S}$$

$$2H_2N-C-N-(CH_2)_3-CH-C \diagup^{O}_{\diagdown OH} \;+\; CuS$$

with H above the second C, $\|$ and O below the first C, and NH_2 below the CH.

Reagents which have been used with masked amino acid complexes to attack an uncomplexed amino group include acetic anhydride[38], carbobenzoxychloride[39], phenyl isocyanate[40], biotin acid chloride[41], potassium cyanate[42], O-methylisouronium salts[40], acid chlorides[40] etc. This type of reaction furnishes convenient syntheses for arginine[43], N-acetyllysine[36], and related polyfunctional amino acid derivatives. Masking can also be used to alter one of two carboxyl groups in a molecule such as glutamic acid[44]:

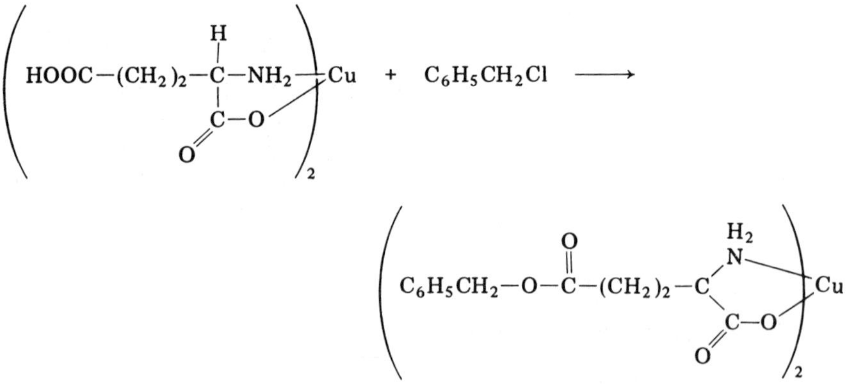

$$\left(\begin{array}{c} H \\ | \\ HOOC-(CH_2)_2-C-NH_2 \\ | \\ C-O \\ \| \\ O \end{array} Cu \right)_2 + C_6H_5CH_2Cl \longrightarrow$$

$$\left(\begin{array}{c} O \quad\quad\quad H_2 \\ \| \quad\quad\quad N \\ C_6H_5CH_2-O-C-(CH_2)_2-C \diagdown \\ C-O \\ \| \\ O \end{array} Cu \right)_2$$

The masking of amino groups by coordination can also be used in cases where there are no free functional groups, but a reagent is utilized to attack a less reactive part of a molecule than is otherwise possible. A good example is seen in the use of the Knoevenagel condensation on copper(II) complexes of amino acids[45]. In one example bis(glycinato)copper(II) is condensed with a basic solution of acetaldehyde to give bis(threoninato)-copper(II):

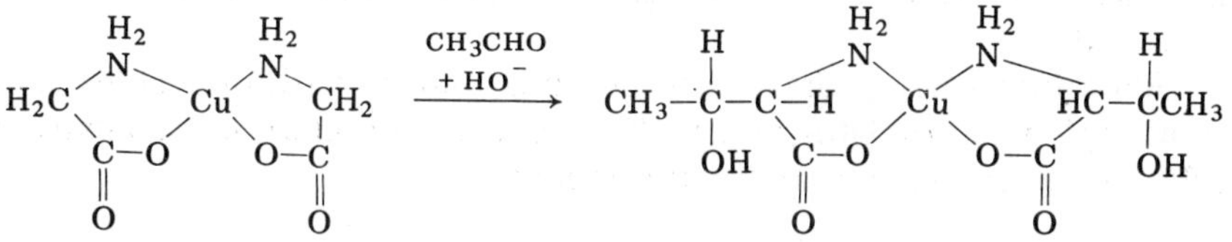

Other reactions of this type can be used to prepare β-hydroxyleucine[46] serine and phenylserine[47].

The terminal methylene groups in coordinated dipeptides in cobalt(III) complexes are activated towards proton exchange reactions while the protons in the amino group are not[48]. This behavior has been established for complexes with glycylglycine, alanyglycine, and glycylalanine and is consistent with the apparent patterns of reactivity established for such molecules in complexes. The exchange time for protons in these groups is less than the half life for conformational shifts.

Another type of masking appears to arise in the case of complexes of certain ligands which bear uncoordinated hydroxy groups. In some of these the hydroxy groups are unusually resistant to esterification, though it is not apparent why. Thus bis(2-hydroxyethyliminodiacetato)chromium(III):

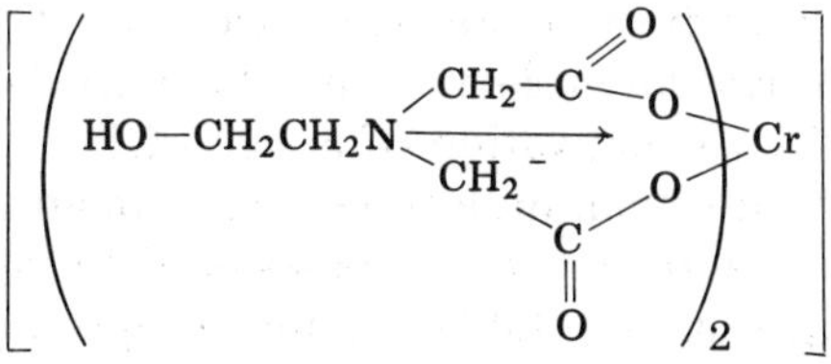

is very resistant to the action of typical acetylating agents, though it can be acetylated in low yield in refluxing acetic acid[49]. The origin of this difficulty is not known, though it may be related to factors which cause some alcohols to undergo esterification only with difficulty.

A final aspect of masking which may be noted is the use of coordination to reduce the acid strengths of some very strong Lewis acids to facilitate their use. Thus the SO_3-pyridine adduct will provide a small amount of SO_3 by dissociation.

References pp. 379-380

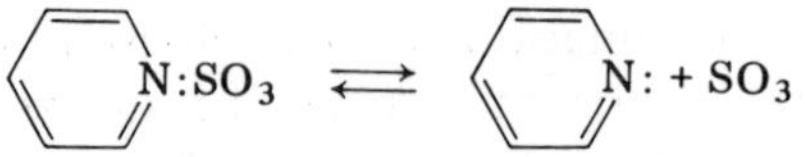

$$\langle N:SO_3 \rightleftharpoons \langle N: + SO_3$$

This type of equilibrium can be used to moderate the vigor of reactions in which Lewis acids (including metal ions) are used as catalysts under circumstances in which the introduction of the pure acid would lead to an uncontrolled reaction or undesirable side products. It eliminates any problems which would otherwise arise because of high localized concentration of the Lewis acid.

Masking has its greatest synthetic possibilities with molecules containing multiple reaction sites, some of which can be tied up in a chelate ring while others remain essentially unaffected.

7. THE ACTIVATION OF SMALL MOLECULES VIA COORDINATION[50-55]

A considerable number of diatomic and some slightly larger molecules can be activated by coordination to the point at which they enter reactions in a controllable fashion which is otherwise very difficult to achieve. A recently discovered example of this phenomenon is provided by the oxygen complexes formed by the isonitrile complexes of nickel(II) and palladium(II) which function as catalysts for the process[56,57]:

$$RNC + O_2 \xrightarrow[Ni(CNR)_4]{} RCNO$$

Coordination can be used to activate N_2, H_2, O_2, Cl_2, Br_2, $ClCN$, H_2O, $CH_2=CH_2$, CO, $NOCl$, NO_2Cl, and H_2O_2 as well as larger molecules. Many of these molecules are capable of participating in insertion reactions of low valent transition metal complexes, though there are also a wide variety of other processes by which specific individual molecules may be activated. In this section we will examine primarily the activation of H_2 and CO.

The use of coordination to activate small molecules is seen in an impressive fashion in the oxo reaction and related processes used to incorporate carbon monoxide and hydrogen into suitable organic compounds. The reaction is catalyzed by cobalt salts which are transformed into $HCo(CO)_4$ and $Co_2(CO)_8$ in the reaction mixture, which consists of unsaturated hydrocarbons or alcohols, carbon monoxide and hydrogen under pressure and a suitable cobalt salt. A variety of reactions can be carried out in such a system, the more important of which are[58]

(1) Hydroformylation:

$$CH_2=CH_2 + CO + H_2 \xrightarrow[\substack{100\text{-}200\,^\circ C \\ 100\text{-}400\ atm}]{HCo(CO)_4} CH_3CH_2CHO$$

(2) Hydrogenation

$$CH_2 = CH_2 + H_2 \longrightarrow CH_3CH_3$$

(3) Homologation

$$RCH_2OH + CO + 2H_2 \longrightarrow RCH_2CH_2OH + H_2O$$

(4) Hydrogenolysis

$$R_2CHOH + H_2 \longrightarrow R_2CH_2 + H_2O$$

A process basic to many of these systems is the *insertion reaction* in which a carbon monoxide or olefin molecule is inserted between the cobalt atom and an alkyl group or hydrogen atom bonded to it:

$$HCo(CO)_4 + C_2H_4 \longrightarrow H{-}C_2H_4{-}Co(CO)_4$$
$$CH_3Co(CO)_4 + CO \longrightarrow CH_3\overset{\|}{\underset{O}{C}}Co(CO)_4$$

An important aspect of these reactions appears to be the ability of the cobalt to form active intermediates in which it has a lower coordination number, and then to react with carbon monoxide or olefins. Other metal carbonyls can catalyze similar reactions, presumably via analogous mechanisms[59].

In a large number of cases square planar d^8 complexes of the transition elements undergo oxidative–addition reactions with small molecules which are split into two parts and may be activated for further reaction[60]. These reactions are often of the type:

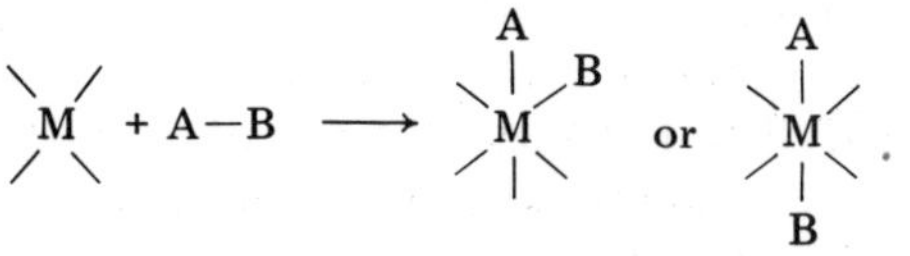

$$\backslash M / + A{-}B \longrightarrow \text{(complex A-M-B)} \text{ or } \text{(complex)} .$$

where A–B can be a halogen, pseudohalogen, alkylhalide, acyl halide, O_2, H_2, acetylenes, hydrogen halides, olefins, metal halides, non-metal hydrides or protonic acids (see also Chapter 20). Examples of such reactions include:

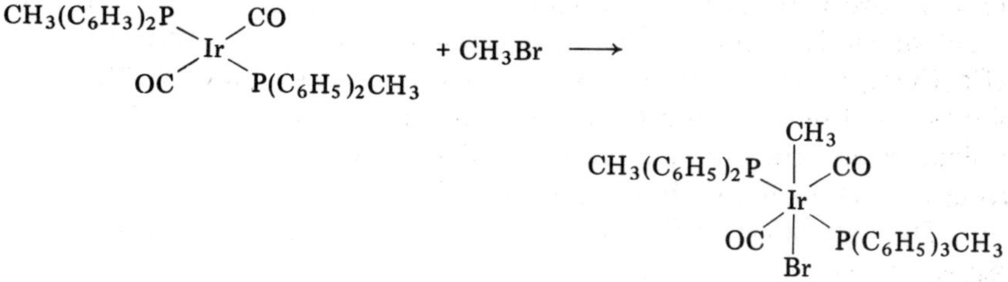

$$CH_3(C_6H_3)_2P, OC \text{ Ir } CO, P(C_6H_5)_2CH_3 + CH_3Br \longrightarrow$$

$$CH_3(C_6H_5)_2P, OC \text{ Ir } \overset{CH_3}{\underset{Br}{}} CO, P(C_6H_5)_3CH_3$$

and

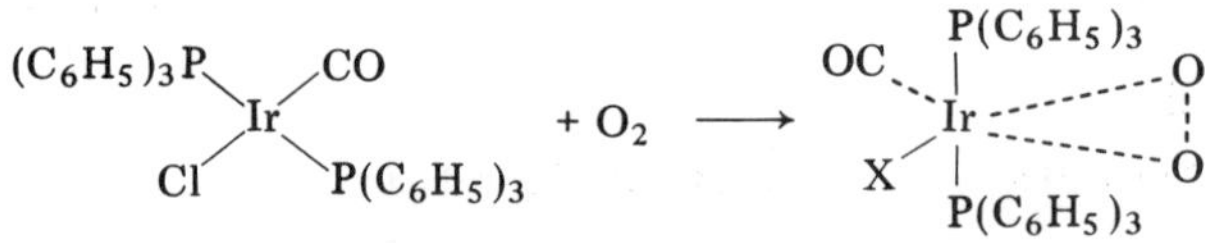

$$
(C_6H_5)_3P \diagdown \overset{CO}{\diagup} \quad \underset{Cl \diagup}{Ir} \diagdown P(C_6H_5)_3 \quad + O_2 \longrightarrow \quad \overset{P(C_6H_5)_3}{\underset{P(C_6H_5)_3}{OC \diagdown \underset{X}{Ir} \diagdown \overset{O}{\underset{O}{}}}}
$$

Complexes with a d^8 configuration can be very effective catalysts for hydrogenation (*e.g.* [$(C_6H_5)_3P$)_3 RhCl$], hydroformylation ($Ru(CO)_3$-($P(C_6H_5)_3)_2$), decarbonylation and ethylene dimerization[60].

The activation of molecular hydrogen, especially for hydrogenations, proceeds quite effectively with a large number of transition metal complexes. Following Halpern[61], we can examine the three mechanisms by which molecular hydrogen forms hydride complexes which are the active reducing agents. These are:

Heterolytic splitting

$$Ru^{III}Cl_6^{3-} + H_2 \rightleftharpoons Ru^{III}HCl_5^{3-} + H^+ + Cl^-$$

Homolytic splitting

$$2Co^{II}(CN)_5^{3-} + H_2 \rightleftharpoons 2Co^{III}H(CN)_5^{3-}$$

Insertion (dihydride formation)

$$Ir^I Cl(CO)(P(C_6H_5)_3)_2 + H_2 \rightleftharpoons Ir^{III}H_2Cl(CO)(P(C_6H_5)_3)_2.$$

These complexes appear to catalyze hydrogenation reactions by processes in which they first form a coordinate bond to the olefin, then transfer hydrogen to it and finally split it off. The formation of the hydride complex may precede or follow the formation of the bond to the olefin. A very large number of low valent complexes of iron, cobalt, nickel, and the platinum metals are capable of catalyzing hydrogenations homogeneously; the organic substrates which have been hydrogenated include alkenes, alkynes, aldehydes and aromatic compounds. Some of these materials are very effective, *e.g.* $RhCl(P(C_6H_5)_3)_3$ which catalyzes the hydrogenation of olefins at room temperature with hydrogen pressures of the order of one atmosphere. Other effective catalysts include $PtCl(P(C_6H_5)_3)_2 SnCl_3^-$, $RuCl_2(P(C_6H_5)_3)_2$, $HCo(CO)_4$, $IrH(CO)IP(C_6H_5)_3)_3$ and related species. Many of these complexes are closely related to the complexes used to activate other small molecules and they are generally capable of coordinating the olefins or other molecules whose hydrogenation they catalyze.

The activation of O_2 and N_2 is discussed in Chapters 20 and 23 respectively.

8. MISCELLANEOUS REACTIONS

Among the miscellaneous reactions may be included some general processes such as: (1) the interaction of Lewis acids with the π-electron systems of aromatic systems; and (2) the photochemical reactions of complexes. There are also a wide variety of processes which arise in special instances where the unique or unusual properties of one or more ligands are matched with a cation of just the right size or oxidation potential or stereochemistry to produce an unusual reaction.

The interaction of Lewis acids with aromatic systems is well established, and a large number of reasonably stable adducts have been isolated and characterized[62]. These include compounds with such compositions as $C_6H_6 \cdot Al_2Br_6$, $(CH_3)_3C_6H_3 \cdot GaBr_3$ $C_6H_6 \cdot 3TiCl_4$, $C_6H_6 \cdot 2SbCl_3$, and $(CH_3)_2C_6H_4 \cdot AlBr_3$. Bronsted acids form similar adducts with aromatic systems[62]. These are compounds whose structures are largely unknown. For the few of known structure, the bonding is difficult to explain. $C_6H_6Al_2Br_6$ seems to have the benzene bonded to the aluminum bromide dimers via the bromines[63]. In $C_6H_6CuAlCl_4$, the benzene is bonded to the copper(I) which has as its three other nearest neighbors chlorides of the $AlCl_4^-$ groups[64]. Here each copper(I) is bonded to but a single aromatic ring in contrast to the silver(I) ions in $C_6H_6 \cdot AgClO_4$, which are bonded to two benzene rings[65]. These compounds are related to those involved in the catalytic reactions of aromatic rings in non-aqueous solvents[62]. Their formation reactions occur quite readily in many cases, and solid adducts are easily isolated. The aromatic systems do not undergo any permanent change and can generally be recovered by treating the adducts with water.

The photochemical reactions of complexes are usually dependent upon the metal ion for the absorption of light. They include a large number of electron transfer processes[66,67] such as:

$$Co(NH_3)_5I^{2+} + h\nu \rightarrow [Co(NH_3)_5^{2+}] + I\cdot$$

$$Ce(H_2O)_x^{4+} + h\nu \rightarrow Ce^{3+}(aq) + H^+ + \cdot OH$$

$$FeSCN^{2+}(aq) + h\nu \rightarrow Fe^{2+}(aq) + \cdot SCN$$

as well as reactions in which a hydrated electron is produced

$$M^x(H_2O)_y + h\nu \rightarrow M^{x+1}(H_2O)_y + e(H_2O)_x$$

and subsequently reacts, often to form $H\cdot$ by reaction with H^+ [68].

Another use of photochemistry which may be mentioned is the photochemical isomerization of complexed olefins as in

$$1\text{-octene} \xrightarrow[h\nu]{Fe_3(CO)_{12}} \text{isomers}[69]$$

References pp. 379–380

Reactions in which a certain combination of ligands and metal ion produces an unusual reaction path are quite striking in providing a glimpse of the enormous potentialities of metal ion catalyses. One such example[70] involves the use of the zinc complex of the pyridinecarboxaldoxime anion (**I**) as a catalyst for the hydrolysis of 8-acetoxyquinoline-5-sulfonic acid (**II**).

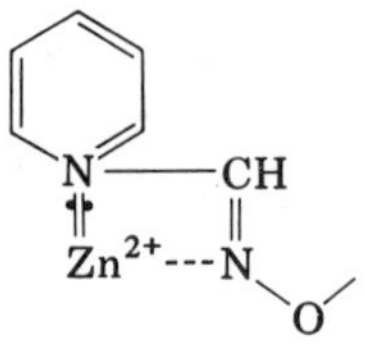

(I) 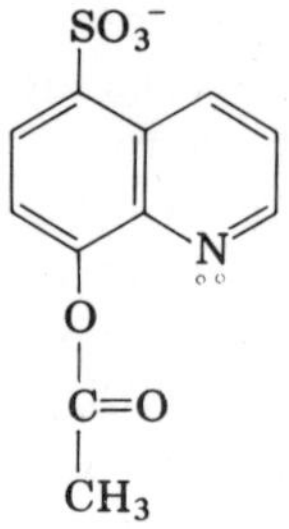   (II)

In this process a mixed complex is formed between **I** and **II** and the acetyl group is transferred rapidly to the oxygen of the oxime. It is then hydrolytically split off the oxime. The overall process, even though it occurs in two steps, is considerably more rapid than the uncatalyzed hydrolysis.

Another process of a sort more directly related to enzymatic processes involves reactions in which a metal complex is prepared near a hydrophobic cavity and then proceeds to accelerate hydrolytic splitting of a substrate molecule bound in the cavity[71]. In this case cycloamylose, a toroidal polysaccharide which bonds hydrocarbon groups in a hydrophobic center was the starting material. A hydroxy group near the opening of the toroid was esterified with pyridine-2,5-dicarboxylic acid which was then converted to a nickel(II) chelate. A mixed complex was then formed with pyridinecarboxaldoxime. When *p*-nitrophenyl acetate is added to this solution the phenyl group is accommodated in the hydrophobic center and the metal complex group at the opening of the toroid catalyzes the removal of the acetyl group. This type of interaction uses the metal ion only to catalyze a process and not to bind a substrate to an enzyme. It is probably more nearly a model of some enzymatic activities than systems which use the metal ion for both purposes.

9. SUMMARY

The principal effects of coordination on the reactions of ligands are based on the fact that the formation of new bonds alters the reactive patterns of the old ones. When one considers how the properties of coordination centers vary, from simple ions such as magnesium(II) or calcium(II)

379

to metal ions whose properties are drastically altered by the presence of ligands already in the coordination sphere (as iron in hemoglobin, cobalt in vitamin B_{12} and magnesium in chlorophyll), it becomes apparent that coordination centers can potentially interact with ligand orbitals ranging from those which are filled and low in energy to those which are empty and high in energy. The practical consequences of such bonding vary from the facilitation of nucleophilic attack, when ligands are coordinated to cations, to the more complex interactions which occur in the enzymatic fixation of nitrogen via iron and molybdenum complexes. The number and type of such reactions is limited only by the ingenuity of man and nature in designing novel coordination sites for special types of ligands.

REFERENCES

1 R. G. Pearson, *J. Am. Chem. Soc.*, 85 (1963) 3533; *Science*, 151 (1966) 172; *Chem. Brit.*, 3 (1967) 103.
2 H. Meerwein, *Ann. Chem. (Liebigs)*, 455 (1927) 227–253; *Schriften Konigsberger Gelehrten Ges. Naturiw. Kl.*, 3 (1927) 129–166.
3 A. Werner, *New Ideas on Inorganic Chemistry*, Longmans-Green, New York, 1911, p. 201.
4 K. J. Pedersen, *Kgl. Danske Vetensk. Akad. Math. -Phys. Medd.*, 20 (7) (1943) 20.
5 S. P. Bag, Q. Fernando and H. Freiser, *Inorg. Chem.*, 1 (1962) 887.
6 G. I. H. Hanania and D. H. Irving, *J. Chem. Soc.*, (1960) 2745, 2750.
7 E. Smith, in W. D. McElroy and B. Glass, *The Mechanism of Enzyme Action*, The Johns Hopkins University Press, Baltimore, Md, 1954, p. 309.
8 L. Hellerman and M. E. Perkins, *J. Biol. Chem.*, 112 (1935) 175; L. Hellerman and C. C. Stock, *J. Biol. Chem.*, 125 (1938) 771; L. Hellerman, *Physiol Rev.*, 17 (1937) 454.
9 W. Wieker and E. Thilo, *Z. Anorg. Allgem. Chem.* 306 (1960) 48; 313 (1961) 296; E. Thilo and F. Von Lampe, *Z. Anorg. Allgem. Chem.*, 349 (1967) 1.
10 M. Bender, *Adv. Chem. Ser.*, 37 (1963) 19.
11 W. Gibbs and F. M. Genth, *Am. J. Sci.* 23 (2) (1857) 241; 24 (2) (1857) 84.
12 S. Zsindely and E. Pungor, *Mikrochim. Acta*, 2 (1963) 209.
13 M. T. Beck and O. Kling, *Acta. Chem. Scand.*, 15 (1963) 453.
14 M. M. Jones, D. O. Johnston and C. J. Barnett, *J. Inorg. Nucl. Chem.*, 28 (1966) 1927.
15 K. J. Laidler, *Reaction Kinetics*, Vol. 2, Pergamon Press, Oxford, 1962, p. 6.
16 F. H. Westheimer, *Trans. N. Y. Acad. Sci.*, 18 (1955) 15.
17 H. L. Conley, Jr., and R. B. Martin, *J. Phys. Chem.*, 69 (1965) 2914.
18 H. Kroll, *J. Am. Chem. Soc.*, 74 (1952) 2036.
19 M. M. Jones, *Ligand Reactivity and Catalysis*, Academic Press, New York, 1968, pp. 34 et seq.
20 D. H. Busch and M. D. Alexander, *J. Am. Chem. Soc.*, 88 (1966) 1130.
21 J. E. Hix, Jr., and M. M. Jones, *Inorg. Chem.*, 5 (1966) 1863.
22 C. G. Regardh, *Acta Pharm. Suedica*, 3 (1966) 101.
23 J. E. Hix, Jr., and M. M. Jones, *J. Am. Chem. Soc.*, 90 (1968) 1723.
24 E. Bamann, J. G. Haas and H. Trapmann, *Arch. Pharm.*, 294 (1961) 569.
25 J. P. Collman and D. A. Buckingham, *J. Am. Chem. Soc.*, 85 (1963) 3039.
26 E. Bamann and W. P. Mutterlein, *Chem. Ber.*, 91 (1958) 471, 1322.

380

27 C. H. Oestreich and M. M. Jones, *Biochemistry*, 5 (1966) 2926.
28 R. Hofstetter and A. E. Martell, *J. Am. Chem. Soc.*, 81 (1959) 4461.
29 H. T. Clark and E. E. Dreger, *Org. Syntheses*, 1 (1941) 87.
30 W. W. Hartman and J. B. Dickey, *J. Am. Chem. Soc.*, 55 (1933) 1228.
31 W. Brockmann and P. J. Smit, *Rec. Trav. chim.*, 82 (1963) 757.
32 V. S. Butt and M. Hallaway, *Arch. Biochem. Biophys.*, 92 (1961) 24.
33 P. M. Henry, *Adv. Chem. Ser.*, 70 (1968) 127; also contains a discussion of the previous literature.
34 J. H. Flynn and H. M. Hulburt, *J. Am. Chem. Soc.*, 76 (1954) 3393, 3396.
35 R. A. Boisonnas, *Adv. Org. Chem.*, 3 (1963) 159.
36 J. F. McOmie, *Adv. Org. Chem.*, 3 (1963) 191.
37 A. C. Kurtz, *J. Biol. Chem.*, 122 (1938) 477; *J. Biol. Chem.*, 140 (1941) 705.
38 A. Neuberger and F. Sanger, *Biochem. J.*, 37 (1943) 515.
39 R. L. M. Synge, *Biochem. J.*, 42 (1948) 99.
40 A. C. Kurtz, *J. Biol. Chem.*, 180 (1949) 1253.
41 D. E. Wolf, J. Valliant, R. L. Peck and K. Folkers, *J. Am. Chem. Soc.*, 74 (1952) 2002.
42 L. H. Smith, *J. Am. Chem. Soc.*, 77 (1955) 6691.
43 F. Turba and K. H. Schuster, *Z. Physiol. Chem.*, 283 (1948) 27.
44 W. E. Hanby, S. G. Wales and J. Watson, *J. Chem. Soc.*, (1950) 3241.
45 Y. Ikutani, T. Okuda, M. Sato and S. Akabori, *Bull. Chem. Soc. Japan*, 32 (1959) 203.
46 Y. Ikutani, T. Okuda and S. Akabori, *Bull. Chem. Soc. Japan*, 33 (1960) 582.
47 K. Okawa and M. Sato, Brit. Pat. 814,063, 1959; *Chem. Abstr.*, 54 328g.
48 R. D. Gillard, P. R. Mitchell and N. C. Payne, *Chem. Commun.*, (1968) 1150.
49 R. A. Krause and S. D. Goldby, *Adv. Chem. Ser.*, 37 (1963) 143.
50 L. Vaska, *Accounts Chem. Res.*, 1 (1968) 335.
51 J. P. Collman, *Accounts Chem. Res.*, 1 (1968) 136.
52 A. D. Allen and F. Bottomley, *Accounts Chem. Res.*, 1 (1968) 360.
53 J. A. McGinnety and M. J. Mays, *Rep. Progr. Chem.*, 65A (1969) 353.
54 *Homogeneous Catalysis, Adv. Chem. Ser.*, 70 (1968)
55 *Homogeneous Catalysis with Special Reference to Hydrogenation and Oxidation, Discuss. Faraday Soc.*, 46 (1968).
56 S. Otsuka, A. Nakamura and Y. Tatsuno, *J. Am. Chem. Soc.*, 91 (1969) 6994.
57 S. Otsuka, A. Nakamura and T. Yoshida, *J. Am. Chem. Soc.*, 91 (1969) 7198.
58 F. Basolo and R. G. Pearson, *Mechanisms of Inorganic Reactions*, John Wiley and Sons, New York, 2nd edn., 1967, pp. 587–596.
59 I. Wender and P. Pino, *Organic Syntheses via Metal Carbonyls*, Vol. I, Interscience, New York, 1968.
60 J. P. Collman and W. R. Roper, *Adv. Organometall. Chem.*, 7 (1968) 53–94.
61 J. Halpern, *Adv. Chem. Ser.*, 70 (1968) 7.
62 G. A. Olah and M. W. Meyer, in G. A. Olah, *Friedel-Crafts and Related Reactions*, Vol. 1, John Wiley and Sons (Interscience), New York, 1963, p. 710.
63 D. D. Eley, J. H. Taylor and S. C. Wallwork, *J. Chem. Soc.*, (1961) 3867.
64 R. W. Turner and E. L. Amma, *J. Am. Chem. Soc.*, 85 (1963) 4046.
65 H. G. Smith and R. E. Rundle, *J. Am. Chem. Soc.*, 80 (1958) 5075.
66 A. W. Adamson, *Coord. Chem. Rev.*, 3 (1968) 169.
67 A. W. Adamson, W. L. Waltz, E. Zinato, D. W. Watts, P. D. Fleischauer and R. D. Lindholm, *Chem. Rev.*, 68 (1968) 541.
68 E. J. Hart, *Solvated Electron Adv. Chem. Ser.*, 50 (1965).
69 M. D. Carr, V. V. Kane and M. C. Whiting, *Proc. Chem. Soc.*, (1964) 408.
70 R. Breslow and D. Chipman, *J. Am. Chem. Soc.*, 87 (1965) 4196.
71 R. Breslow and L. Overman, *J. Am. Chem. Soc.*, 92 (1970) 1076.

Chapter 14

METAL ENZYMES

M. C. SCRUTTON*

Department of Biochemistry, Rutgers Medical School, Rutgers University, The State University of New Jersey, New Brunswick, New Jersey 08903, U.S.A.

1. INTRODUCTION

A. *The roles of metal ions in the catalytic mechanism of enzymes*

In 1950 Lehninger[1] estimated that approximately one-third of the enzymes described at that time either required addition of a metal ion as a cofactor in order to exhibit maximum activity or contained a tightly bound metal ion which appeared to be involved in the catalytic process. The percentage of enzymes recognized as metal enzymes may now have increased since the introduction and application of simpler methods for metal analysis in biological systems has extended the range of enzymes which are recognized as metallo-proteins. In addition enzymic functions have been described for some metallo-proteins, which had previously been purified on the basis of metal content rather than catalytic function[2].

It is the purpose of this chapter to examine the various roles proposed for the involvement of metal ions in enzymic catalysis; to describe and evaluate some of the methods which are applicable to the study of such catalytic roles with special reference to the types of data obtained; and to consider the present status of our knowledge of the role of metal ions in the catalytic mechanism of dehydrogenases, carboxylases and decarboxylases, isomerases, lyases, aldolases, transferases and synthetases. Succeeding chapters in this section will consider the role of metal ions in enzymic catalysis by peptidases (Chapter 15), carbonic anhydrase (Chapter 16), kinases (Chapter 17) and phosphatases (Chapter 18).

No distinction will be drawn here between metallo-enzymes (enzymes containing one or more tightly bound metal ions) and metal-activated enzymes (enzymes which require addition of exogeneous metal ion to exhibit maximum catalytic activity). Malmström and Rosenberg[3] have emphasized that the catalytic role of a metal ion is not correlated with its affinity for the apoenzyme in any instance examined thus far. Hence

*The unpublished data to which reference is made in this article as well as the preparation of the article were aided by USPHS grant No. AM-11712.

382

emphasis on differences in the affinity of enzymes for metal ions may obscure a unity existing at the functional level[4,5]. For example, an enzyme with a broad specificity for metal activation may behave as a metallo-protein in the presence of a metal ion which exhibits high affinity for the apoenzyme but as a metal-activated system in the presence of a more weakly bound metal ion[6]. Changes in pH may also cause alterations in the affinity of the given metal ion. In some cases, *e.g.* β-methylaspartase[7], such an alteration in pH converts a metallo-enzyme into a metal-activated system. It is, however, improbable that a fundamental alteration in catalytic mechanism occurs under either of these conditions. It has been suggested that a stability constant in the range 10^7–10^8 M^{-1} for the enzyme–metal ion complex may serve as an operational dividing line between the two situations[8].

Roles which have been proposed for metal ions in enzymic catalysis are illustrated in Fig. 1 in which L(ligand) may represent a substrate, an activator or an inhibitor of the enzyme.

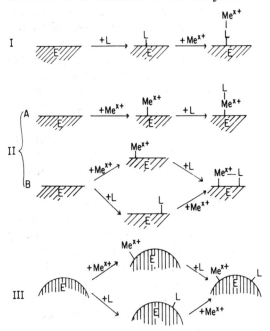

Fig. 1. Various coordination schemes for the interaction of an enzyme with metal ion (Me^{x+}) and ligand (L).

B. *The ligand bridge complex*

Figure 1 (I) represents the "ligand bridge" complex. In this complex the metal ion is coordinated solely with the ligand to give a species which is catalytically competent, and does not interact directly with the enzyme. Ligand bridge complexes are characteristic of many enzymes which utilize Me–NDP$^-$* and Me–NTP^{2-}* complexes *e.g.* creatine kinase[9,10], arginine kinase[11], adenylate kinase[12] and tetrahydrofolate synthetase[13]. The studies on these and other kinases are discussed in Chapter 18. The binary complexes of Me^{2+}* with ATP and ADP are species in which both the phosphate groups and the nitrogen atoms of the purine ring are coordinated to the metal ion[14,15] (cf. Chapter 33). In most instances ADP^{3-} (or ATP^{4-}) also interacts with an affinity approximating that observed for the active Me^{2+}-nucleotide complex, but acts as an inhibitor of the reaction[16]. The role of the metal ion in such substrate bridge complexes may, therefore, be confined to activation of the phosphorus atom on which attack occurs. An exception to this generalization may occur in the case of tetrahydrofolate synthetase[13] (cf. Section XII B). Examples of inhibitor (activator) bridge complexes are less common although the metal-dependent effects of *d*–TTP (inhibitor) and *d*–CTP (activator) on *d*–CMP deaminase may be due to formation of ligand bridge complexes since these effects are expressed in the presence of Ca^{2+} as well as of Mg^{2+} and Mn^{2+}[17] (cf. Section II D).

C. The metal bridge complex

A second role for metal ions in enzymic catalysis involves the formation of a bridge complex in which the metal ion provides all (Fig. 1, II A) or part (Fig. 1, II B) of the binding site for the ligand. The formation of an enzyme–metal–substrate bridge complex was first proposed by Hellerman[18] to describe data obtained for the activation of arginase by metal ions[19], and was also invoked to explain the slow metal-dependent activation of certain peptidases[20]. The metal bridge concept has subsequently been invoked to explain the metal-dependent activation of many enzymes, but until recently any evidence for the existence of a metal bridge complex has been indirect. However, nuclear magnetic resonance (n.m.r.) techniques which permit examination of events occurring in the first coordination sphere of certain paramagnetic metal ions, *e.g.* Mn^{2+}, have now been applied to the study of these systems[21]. The first study, in which the nuclear magnetic relaxation rates of substrate resonance were examined, permitted the unequivocal demonstration that interaction of Mn^{2+} and fluorophosphate with pyruvate kinase from rabbit muscle results in formation of a pyruvate kinase–Mn^{2+}–

---

*Abbreviations used are: PRR, the proton relaxation rate (the longitudinal nuclear magnetic relaxation rate of protons of water); PEP, phosphoenolpyruvate; Me^{2+}, divalent metal ion; Me$^+$, metal ion; E, enzyme; NMP(NDP)(NTP) any nucleoside monophosphate (diphosphate)(triphosphate); TDP, thiamine pyrophosphate; FDP, fructose-1,6-diphosphate; *t*RNA or transfer RNA, the RNA species which carries the anti-codons.

References pp. 428–437

fluorophosphate bridge complex which has properties consistent with its participation in catalysis[22] (cf. Chapter 18). Application of this technique has subsequently resulted in the detection of enzyme–metal–substrate, –product, or –inhibitor complexes for seven other enzymes[8]. These studies will be discussed in greater detail in the later sections on specific enzymes (Sections III–X) and in Chapter 18.

Metal bridge complexes have also been demonstrated by X-ray crystallography (carboxypeptidase)[23] (cf. Chapter 15) and by infra-red spectroscopy (carbonic anhydrase)[24] (cf. Chapter 16). Demonstration of the existence of a bridge complex by these latter techniques does not, however, provide an evaluation of the catalytic relevance of such a complex, which must be established by comparison of the kinetic and thermodynamic properties of the bridge complex with similar parameters obtained in studies on the catalytic process itself. In this respect the n.m.r. technique may have distinct advantages over other procedures (cf. Section II C). The importance of the application of these criteria cannot be overemphasized since formation of a catalytically irrelevant metal bridge complex appears to have been observed in one system[25].

In all enzyme–metal–ligand bridge complexes the unique coordination properties of the metal ion permit it to play an important role in mediating the protein–ligand interaction. However, an additional catalytic role for the metal ion is usually inferred where enzyme–metal–substrate bridge complexes are observed or suggested[8]. The proposal of a catalytic role is often derived from observation of catalysis by metal ions in a model system which bears some resemblance to the biological reaction[8,26,27]. Although the proposal of catalytic involvement of the metal ion is attractive for the biological system, convincing evidence for such an additional role has been obtained in few cases. Thus in biological systems the contributions of the metal ion to binding and to catalysis are not readily separated, and model studies have provided little valid insight.

For example, the metal ion, which is the most effective catalyst in the model system, is commonly found to be either inactive or the least effective in the biological reaction. This situation is well documented for oxaloacetate decarboxylation[28,29] and peptide bond hydrolysis[30]. Furthermore, although catalysis in biological systems must obey fundamental chemical and physical laws, the presence of the organized tertiary and quaternary structure of the protein may stabilize complexes and permit effects, e.g. directed proton transfer[31], which cannot readily be simulated in model systems. These considerations are illustrated in the mechanisms proposed by Wang and his coworkers for chymotrypsin[32] and carbonic anhydrase[33]. The divergence in viewpoints regarding the possible role of the enzyme protein is documented by the discussion between Wang and Caplow which follows the article on carbonic anhydrase[33]. Thus although model studies have provided useful information in many instances caution must be exercised in the extrapolation of such data to infer a catalytic role for the metal ion in metal bridge complexes.

Additionally Vallee and Williams have suggested that the environment of the metal ion in an enzyme may give rise to an "entatic" state in the absence of substrate[34]. The "entatic state" is defined as "the existence in the enzyme of an area with energy closer to that of a unimolecular transition state than to that of a conventional stable molecule, thereby constituting an energetically posed domain"[34]. For metallo-enzymes this postulate is based on the observation of absorption, electron paramagnetic resonance (e.p.r.) and/or optical rotary dispersion (o.r.d.) spectra and redox potentials for certain copper, iron and cobalt enzymes which do not appear to be explicable in the context of the properties either of typical complexes of these metal ions with small molecules or of metal–protein complexes which do not exhibit catalytic activity (cf. references in paper by Vallee and Williams[34]). The "entatic" absorption spectra obtained for several cobalt-metallo-enzymes do, however, resemble in many respects the spectra observed in complexes containing pentacoordinate Co^{2+} [37a,b]. No absolute correlation exists between abnormal spectral or redox properties, which may be attributed to asymmetric or distorted coordination[34], and the observation of catalytic activity. Thus for example, the Co^{2+} complexes of pyruvate kinase[35] and pyruvate carboxylase[36] are catalytically active but exhibit spectra typical of octahedral Co^{2+} complexes. Determination of ligand exchange rates on protein-bound metal ions have also failed to provide convincing evidence for the unusual kinetic properties which might characterize an activated state[8,37]. However, no kinetic studies have yet been reported for a metallo-enzyme which exhibits spectral or redox properties characteristic of the "entatic" state.

At least two pathways exist for formation of metal bridge complexes in instances where the metal ion is not tightly bound to the enzyme (Fig. 1):

$$E + Me^{2+} \rightleftharpoons E\text{--}Me^{2+} + \text{ligand} \rightleftharpoons E\text{--}Me^{2+}\text{--ligand} \qquad (1)$$

$$E + \text{ligand} \rightleftharpoons E\text{--ligand} + Me^{2+} \rightleftharpoons E\text{--}Me^{2+}\text{--ligand} \qquad (2)$$

and a third pathway is also feasible if the unbound metal ion exhibits significant affinity for the ligand:

$$Me^{2+} + \text{ligand} \rightleftharpoons Me^{2+}\text{--ligand} + E \rightleftharpoons E\text{--}Me^{2+}\text{--ligand} \qquad (3)$$

The pathway which is primarily responsible for formation of a bridge complex may be defined by a combination of initial rate, rapid reaction and binding studies if a kinetically significant enzyme–metal–ligand bridge complex is known to be formed, although few systems have been examined in sufficient detail to provide such data. For pyruvate kinase from muscle[35,38,39] random operation of all three pathways best describes the mechanism of formation of the metal bridge complexes involving PEP and ADP (cf. Chapter 18). This conclusion is not consistent with the proposal[16] that all complexes of nucleotides with enzymes which involve the participation of divalent metal ions are formed by interaction of the enzyme with the metal-ligand complex (eqn. 3).

D. *The enzyme bridge complex*

Metal ions may also interact with an enzyme at a different site from that occupied by the ligand and cause alterations in the properties of the catalytic site (or ligand binding site). In such a complex the enzyme acts as the bridge between the ligand and the metal sites (Fig. 1, III). In contrast to the ligand and metal bridge complexes (Sections I B and C) little attention has been paid to the mechanism(s) involved although many metal ion effects may be explicable on the basis of enzyme bridge complexes. For example, the effects of monovalent cations on the catalytic properties of many enzymes have been attributed to stabilization of a catalytically active state by formation of an enzyme bridge complex[40-42] although other explanations have also been proposed[43]. Similarly, the reactivation of glutamine synthetase from *Escherichia coli* by certain divalent metal ions may be due to stabilization of an active conformation of this enzyme[44]. The formation of enzyme bridge complexes are often accompanied by alterations in parameters which act as indicators of enzyme conformation[44-46]. However, these effects are not characteristic of the formation of enzyme bridge complexes since the metal–enzyme interaction, which occurs during formation of metal bridge complexes, often also causes similar alterations in parameters reflecting protein conformation[45].

It should be noted that many enzymes exhibit multiple metal ion requirements which may reflect the participation of more than one of these various types of bridge complexes in the catalytic mechanism, *e.g.* the triple metal ion requirement exhibited by acetyl-CoA synthetase[47,48].

II. EXPERIMENTAL APPROACHES TO THE STUDY OF METAL ENZYMES

A. *General considerations*

Many of the general approaches to the study of the mechanism of action of enzymes are also applicable to the examination of the role of metal ions in enzymic catalysis. The coordination scheme describing the interaction of enzyme, metal and ligand may be examined by the methods which are applicable to the determination of the stoichiometry and affinity of binding of any small molecule by a protein. Such methods include gel filtration either in the presence or absence of the small molecule[49], the dialysis rate method[50], ultrafiltration, the Hayes–Velick ultracentrifugation procedure[52], and equilibrium dialysis[53] as well as methods which measure only the affinity of interaction[54-58]. Determination of the coordination scheme for the interaction of metal ions and ligands with enzymes by these procedures is feasible only if other factors do not intervene. For example if an E-ligand-Me^{2+} complex is formed the enzyme should exhibit significant affinity for the metal ion only in the presence of ligand. Conversely where an E-Me^{2+}-ligand complex is formed no significant ligand binding should occur in the absence of the metal ion. In practice, however, the enzyme often exhibits significant affinity for both components of the complex regardless of the final coordination scheme. Evaluation of the data utilizing the criteria of stoichiometry and kinetic significance is therefore important. Significant complexes, especially of the E-ligand-Me^{2+} and E-Me^{2+}-ligand types, usually contain the three components in equimolar amounts. Furthermore, if the observed complex is catalytically significant, the dissociation constant(s) measured in the binding studies should approximate the constants determined in initial rate studies of the overall reaction, although differences in the protein concentrations employed may cause difficulties in interpretation in some cases. The required constants may be obtained from initial rate studies for the metal ion (as K_A) and for the ligand if this is an inhibitor (as K_i). However, dissociation constants for enzyme-substrate and enzyme-product complexes are not readily accessible unless assumptions are made regarding either the rate-limiting step of the reaction and/or the order of substrate addition (or product release). Confirmation of the kinetic significance of those complexes which are of most interest in the study of the mechanism of action of metal enzymes therefore presents the greatest difficulties.

Additionally studies of the variation of the initial rate of the reaction as a function of substrate and of metal ion concentration have been employed as a probe for the role of metal ions in enzymic catalysis. Although Malmström and Rosenberg[3] have shown that the rate equations for formation of E-Me^{2+}-substrate and E-substrate-Me^{2+} complexes in a single substrate system have the same form, definition of the order of addition of metal ion and substrate in metal-activated systems may be obtained for multi-substrate systems by application of the methods described by Cleland[59,60] if suitable corrections are applied for metal-substrate interaction.* Several examples of the use of this approach will be discussed in later sections (cf. also Chapter 18). In this context it should be

---

*No good treatise on all aspects of enzyme kinetics is available at present. A simple treatment of the kinetics of single substrate systems which includes definitions of terms such as K_M (Michaelis constant) and K_i (inhibitor constant) may be found in a programmed text[60a]. The kinetics of multi-substrate systems are best described in the articles by Cleland[59,60] and Dalziel[60b].

noted that the practice of maintaining a constant and saturating concentration of metal ion while investigating the order of substrate addition and/or product release in kinetic studies may often be unsatisfactory. In many systems uncomplexed Me^{2+} acts as an activator or an inhibitor of the reaction. Furthermore, for weak complexes, *e.g.* MgADP⁻ the relative concentrations of the various species present is a function of the total concentration of both components of the complex[16].

Finally, the coordination scheme for the interaction of enzymes with metal ions and ligands may be examined directly by X-ray crystallography although thus far only carboxypeptidase has been studied in sufficient detail[23]. The advantages of this technique are partly offset by several less desirable features. First, only static binding can be examined and the kinetic significance of the observed complex cannot be evaluated in the same experiment. Second, the present X-ray crystallographic studies are limited to the complexes of enzymes with poor substrates or inhibitors. Although the structures of these complexes are of considerable interest, conclusions regarding normal catalytic function which are based on observations of complexes formed by poor substrates may be unwarranted since the diminution in catalytic activity may reflect defective complex formation. Similar considerations apply to the use of infra-red spectroscopy to detect interaction of ligands with enzyme-bound metal ion[24].

All of the methods discussed thus far are applicable to the study of the mechanism of any enzyme whether or not a metal ion is involved. However, three techniques which are dependent on the unique properties of the metal ion(s) have found increasing application in studies on metal-enzymes. These techniques which will be considered in greater detail are: (i) e.p.r.; (ii) the measurement of paramagnetic contributions to the nuclear magnetic relaxation rates of magnetic nuclei (*e.g.* protons) of ligands; and (iii) metal–metal replacement studies.

B. *Electron paramagnetic resonance*

The theoretical basis of e.p.r. which probes the environment of paramagnetic metal ions by measurement of the ligand field affecting the unpaired electrons has been discussed elsewhere[62]. Several different applications of e.p.r. to the study of metal enzymes have been described.

First, the hyperfine structure in e.p.r. spectra which exhibit a high degree of resolution, such as those obtained from Fe^{3+} and Cu^{2+} complexes, may provide tentative identification of the nature and number of the ligands provided by the protein[63]. The effectiveness of this method may be improved by incorporation of magnetic nuclei into the suspected liganding groups of the protein. The e.p.r. spectra observed on substitution of ^{33}S and ^{77}Se for ^{32}S in several non-heme iron proteins[64,65] illustrates the potential of such isotopic substitution. The hyperfine splitting which results from such spin–spin interaction provides a more definitive indication of the nature of the ligands since the required coupling can only occur through chemical bonds[66]. A similar approach may be employed to confirm the existence of a E–Me^{2+}–ligand complex if a magnetic nucleus is present in the ligand. Direct coordination of ^{19}F$^-$ by iron in metymoglobin fluoride has been demonstrated using this procedure[67]. Conversely the failure to observe hyperfine splitting of the e.p.r. signal tends to exclude the presence of metal–ligand interaction in the complex with the enzyme[68]. The extension of this approach to enzymes containing other metals, e.g. Mn^{2+}, has been thus far been frustrated by the complexity of the e.p.r. spectra and the tendency to a loss of resolution on complex formation. However, recently a well-resolved e.p.r. spectrum has been observed for the Mn^{2+} complex of concanavalin A[69] indicating the possibility of obtaining useful spectra for this metal ion in macromolecular complexes.

Second, the sensitivity of the amplitude of the e.p.r. spectrum of Mn^{2+} to its environment may be employed to investigate the coordination scheme. For example, the amplitude observed for a Mn^{2+}–ligand complex should decrease on addition of the enzyme if an E–Mn^{2+}–ligand complex is formed. If the amplitude of the Mn^{2+}–ligand e.p.r. spectrum is unaffected by addition of the enzyme and other evidence indicates formation of a ternary complex, formation of an E–ligand–Mn^{2+} complex is indicated. This situation has been observed for the interaction of MnADP$^-$ with creatine kinase[9].

Third, the introduction of stable organic molecules containing free radicals ("spin labels") in the study of biological systems[71] has extended the use of e.p.r. to the study of spin-labelled systems. The "spin-label" may be attached to an amino acid residue in or near the active site of the enzyme[72] or may be incorporated into a substrate analog[73]. In either case the extent of immobilization of the protein-bound "spin-label" may be estimated by comparison of the e.p.r. spectra observed for the free and bound states, and the effects of various agents, such as diamagnetic metal ions, on the environment of the "spin-label" may be investigated[72]. In such experiments the spin label acts as a "reporter group"[74] and differs only in that e.p.r. provides the method of detection. However, in the presence of a paramagnetic metal ion, e.g. Mn^{2+}, Co^{2+}, spin–spin interaction dominates the relaxation process and causes a marked decrease in the amplitude of the e.p.r. signal due to the spin label if the spins behave as though they are immobilized in a rigid lattice[74a]. In this instance the distance between the spins may be calculated from this effect if a value is assigned for the correlation time for the dipolar interaction[72,74a]. The use of "spin labels" with specificities for different amino acid residues or for different locations within the active site may permit the use of this approach to map the active sites of metal enzymes[74b].

References pp. 428–437

C. Paramagnetic effects on the nuclear magnetic relaxation rates of ligand nuclei

(a) General considerations
    This complex technique will not be discussed in detail here but certain aspects require examination in the present context. Extensive discussions of the technique may be found in reviews by Kowalsky and Cohn[75], Mildvan and Cohn[76] and Sheard and Bradbury[76a] as well as in Chapter 18.
    In the presence of certain paramagnetic ions, *e.g.* Mn^{2+}, the paramagnetic contribution to the longitudinal ($1/T_{1p}$) and transverse ($1/T_{2p}$) nuclear magnetic relaxation rates of the nucleus under observation, *e.g.* water protons, is given by[77,78]:

$$\frac{1}{T_{1p}} = \frac{pq}{T_{1M} + \tau_M} + \frac{1}{T_1{}^{(os)}} \tag{4}$$

$$\frac{1}{T_{2p}} = \frac{pq}{T_{2M} + \tau_M} + \frac{1}{T_2{}^{(os)}} \tag{5}$$

In eqns. (4) and (5) p is the ratio of the concentration of paramagnetic ion to the concentration of ligand; q, the coordination number for the ligand; $1/\tau_M$, the rate of ligand exchange on the paramagnetic ion; $1/T_{1M}(1/T_{2M})$, the relaxation rates of coordinated ligand nuclei; and $1/T_1{}^{(os)}(1/T_2{}^{(os)})$, the contribution to the relaxation rates due to dipolar interaction of the paramagnetic ion with nuclei outside the first coordination sphere (outer sphere relaxation). It should be noted that eqn. (5) holds for Mn^{2+} but is not applicable to all paramagnetic ions[77]. Equations (4) and (5) contain several important parameters for the analysis of the catalytic role of metal ions if these and other equations are applicable to macromolecular complexes. First, comparison of q for water protons in $E-Me^{x+}$ and $E-Me^{x+}$-ligand complexes determines the coordination scheme for these interactions. Second, comparison of $1/\tau_M$ of substrate exchange in a $E-Me^{x+}$-substrate complex with the turnover number of the enzyme indicates if the observed complex forms and dissociates rapidly enough to participate in catalysis. Third, r, the distance between the ligand protons and the paramagnetic ion which may be calculated from $1/T_{1M}$ if a value is assigned for the correlation time (τ_c) describing this dipolar interaction[79,80], may be used to determine if a $E-Me^{x+}$-ligand complex is formed. And finally, the coupling constant for the hyperfine (scalar) interaction may be obtained from $1/T_{2M}$ and $1/T_{1M}$ if a value is assigned for the correlation time (τ_e) describing this hyperfine interaction[79,81]. This latter parameter is of particular importance to the proof of direct coordination since hyperfine interaction occurs only through chemical bonds[66]. Since $1/T_{1p}$ and $1/T_{2p}$ can be measured simultaneously it is possible in principle to obtain both kinetic and structural data in the same experiment provided that suitable processes dominate the relaxation rates. Extraction of this information is, however, often difficult since the process(es) which dominate(s) the relaxation rates in the system under study must first be identified. For example, $1/T_{1p}$ may be dominated by $1/T_{1M}$, $1/\tau_M$ or $1/T_1^{(os)}$ and in turn $1/T_{1M}$ may contain contributions from $1/\tau_s$ (the electron spin relaxation time); from $1/\tau_r$ (the rotational correlation time); or from $1/\tau_M$ [81]. Where only one process contributes to the relaxation rate it is usually possible to identify this process by examination of the frequency and temperature dependence of the observed relaxation rate although the narrow range of temperature stability characteristic of most biological systems limits the usefulness of this latter variable. The behavior expected for a series of situations has been summarized in tabular form by Mildvan and Cohn [76]. However, the analysis described above becomes impossible if the overall relaxation rate contains contributions from more than one process.

In the context of current research it is especially important to note three points. First, *both* a negative temperature dependence of $1/T_{1p}$ *and* equality of $1/T_{1p}$ with $1/T_{2p}{}^*$ are required to identify τ_M as the process which dominates the overall relaxation rate. Domination of $1/T_{1p}$ or $1/T_{2p}$ by τ_s can *also* give rise to a negative temperature dependence for the relaxation rate but in this latter case $1/T_{2p} > 1/T_{1p}$ [76,82]. Second, even if τ_M dominates $1/T_{1p}$ or $1/T_{2p}$ for water protons in a series of E–Me^{x+}–ligand complexes, valid relative hydration numbers cannot be determined by comparison of $1/T_{1p}$–$(1/T_{2p})$ at a given temperature *unless* similar activation energies are observed for τ_M [82]. Third, in most instances the correlation times for dipolar interaction (τ_c) between the metal ion and the magnetic nucleus of the ligand in a E–Me^{x+}–ligand complex must be estimated [22]. If protons are employed as the magnetic nucleus the ranges obtained for the metal–ligand distance are usually not adequate to permit identification of the liganding group(s) although they may be consistent with direct coordination. This problem may be solved by a direct determination of τ_c from the frequency dependence of $1/T_{1p}$ and/or $1/T_{2p}$ but determination of the paramagnetic effects on the relaxation rates of magnetic nuclei (^{13}C, ^{17}O) which are closer to the metal ion, may provide a more satisfying approach to this problem. For ^{13}C such an approach will be facilitated by the large chemical shifts observed for this nucleus.

(b) The enhancement phenomenon and its application

In the nuclear magnetic resonance (n.m.r.) studies performed thus far on biological systems much use has been made of the enhancement

---

*Since domination of $1/T_{2p}$ by τ_M may persist after this process has ceased to dominate $1/T_{1p}$, this criterion is more accurately stated as a requirement that $1/T_{2p}$ and $1/T_{1p}$ are either equal or approach equality at lower temperature. It should also be noted that this analysis only applies when the metal ion exhibits a long τ_s, e.g. Mn^{2+}.

phenomenon, which was discovered by Eisinger *et al.*[81] for the complexes of DNA with certain paramagnetic ions, *e.g.* Mn^{2+}, Cr^{3+}, Cu^{2+} and by Cohn and Leigh[9,21] for complexes of various enzymes with Mn^{2+} in the presence or absence of substrates. The enhancement factor for $1/T_1(\epsilon_1)$ is defined by[81]:

$$\epsilon_1 = \frac{1/T_{1p}^{\star}}{1/T_{1p}} = \frac{1/T_1^{\star} - 1/T_{1(0)}^{\star}}{1/T_1 - 1/T_{1(0)}} \tag{6}$$

In equation (6) $1/T_{1(0)}$ is the relaxation rate observed in the absence of the paramagnetic ion and the star designates relaxation rates measured in the presence of a macromolecule. The enhancement factor for $1/T_2(\epsilon_2)$ is defined similarly. When $1/T_1$ is measured for water protons (PRR) ϵ_1 values greater than 1.0 are observed for metal ions, *e.g.* Mn^{2+}, for which τ_c is dominated by τ_r, the rotational correlation time[81]:

$$\frac{1}{\tau_c} = \frac{1}{\tau_r} + \frac{1}{\tau_s} + \frac{1}{\tau_M} \tag{7}$$

In this situation the tumbling time of the complex dominates τ_c. Since a macromolecular complex has a much slower tumbling time than the hydrated metal ion (aquocation), formation of such a complex may result in $\epsilon_1 > 1.0$ if the increased contribution from the correlation time outweighs the decreased contribution due to displacement of water protons as a result of complex formation. Although τ_r may dominate τ_c for both the aquocation and the macromolecular complex in some instances, in others the decrease in τ_r on formation of the macromolecular complex may be such that τ_s or τ_m now dominate τ_c. The value observed for ϵ_1 for water protons has, therefore, little significance. Similar considerations apply to enhancement factors determined for the relaxation rates of ligand protons. However, this parameter may have qualitative utility in (i) indicating whether or not the metal ion is accessible to solvent, and (ii) providing a preliminary method for distinguishing the three possible catalytic roles for metal ions in enzymic reactions as described in Section I. Cohn[21] has pointed out that the enhancement factor (ϵ_1) for water protons observed for a binary E-Me^{2+} complex (ϵ_b) may be greater than ϵ_1 for the ternary E-Me^{2+}-ligand (Type II) complex (ϵ_c). In contrast for enzymes which form E-ligand-Me^{2+} (Type I) complexes little or no enzyme-metal ion interaction is observed and $\epsilon_c > \epsilon_b \approx 1.0$ while for Me^{2+}-E-ligand (Type III) complexes the binding of ligand may have little effect on the environment of the metal ion and $\epsilon_b \approx \epsilon_c$. Although these relationships are observed for most Type I and II complexes[21], exceptions are known. For FDP aldolase from yeast, examination of the relaxation rates of the substrate protons in the presence of the Mn^{2+}-enzyme has provided evidence for existence of E-Mn^{2+}-substrate bridge complexes (cf. Section IX) although little change is

observed for ϵ_1 of water protons on formation of these complexes, *i.e.*
$\epsilon_b \approx \epsilon_c$. Hence, although comparison of ϵ_1 for water protons in binary and
ternary complexes of enzyme, metal and ligand provides a simple and rapid
indicator for the type of complex formed, the results obtained must be
regarded as preliminary and must be confirmed by other techniques, *e.g.*
determination of r and (A/h) (the hyperfine coupling constant) from
measurements of the relaxation rates of magnetic nuclei of the ligand.
Titrations of the increase (or decrease) observed in ϵ_1 for water protons on
interaction of the enzyme with Mn^{2+} and ligand also provides a rapid
method for determination of the dissociation constants of these complexes[21].

D. Metal–metal replacements

In most studies on metal-activated enzymes, attempts have been made
to define the specificity of interaction either by examination of the extent
of activation by various metal ions under standard assay conditions, or, in
some instances, by measurement of the apparent activator constants and
maximum velocities for those metal ions which provide a significant degree
of activation. Certain generalities have been derived from these studies. For
example, most kinases and synthetases are only activated to a significant
extent by Mg^{2+}, Mn^{2+}, Co^{2+} and in some cases Ca^{2+} whereas other enzymes,
e.g. dehydrogenases, lyases, in most instances exhibit broader specificity for
activation by Me^{2+}. Similar maximum extents of activation are often
observed for addition of either Mg^{2+} or Mn^{2+}, but the apparent K_A for Mn^{2+}
is generally an order of magnitude more favorable than that for Mg^{2+}, as
would be expected from a comparison of the stability constants observed
for small molecule complexes of these two metal ions[3].

Cohn[21] has suggested that the behavior observed on addition of Ca^{2+}
may be used as a probe for the coordination scheme. Enzymes which form
E-substrate-Me^{2+} complexes, *e.g.* creatine kinase, are activated by Ca^{2+},
whereas for enzymes which form E-Me^{2+}-substrate complexes Ca^{2+} typically acts as an inhibitor. The rapid rate of ligand exchange exhibited by
Ca^{2+}-complexes[84] which could be inconsistent with the participation of
this metal ion in a metal bridge complex may provide the basis for this
empirical criterion. However, recent binding and X-ray crystallographic
studies on staphylococcal nuclease indicate Ca^{2+} interacts with the enzyme
in or very near to the nucleotide binding site[85,86]. Hence, in this instance
an enzyme-Ca^{2+}-nucleotide bridge complex may be involved although the
participation of an enzyme bridge (Me^{2+}-E-ligand) complex is not excluded.

Similar metal substitution studies have also been performed for
metallo-enzymes such as carboxypeptidase[6], carbonic anhydrase[88], alkaline
phosphatase[89] and liver alcohol dehydrogenase[90]. For these enzymes
replacement of the metal ion present in the native enzyme by other metals
may be achieved by: (i) preparation of the apoenzyme and subsequent

recombination[4]; (ii) direct metal–metal exchange under carefully defined conditions[91]; or (iii) for microbial enzymes, growth in a medium of defined metal content which is supplemented with a high concentration of the metal ion whose incorporation is desired[92]. Although apoenzyme preparation is usually accomplished by suitable treatment of the purified native enzyme, *e.g.* prolonged dialysis against chelating agents[4], apo-alkaline phosphatase has been obtained from *Escherichia coli*[93] and apo-galactose oxidase from *Dactylium dendroides*[93a] when these micro-organisms are grown under "metal-free" conditions.

Comparative studies of the catalytic properties of the various metallo-enzymes have been primarily limited to an assessment of relative catalytic activity. Detailed catalytic and ligand binding studies for a series of metallo-enzymes may, however, throw light on the role of the metal ion if those species which appear catalytically *inactive* are also characterized with respect to their binding properties. This latter approach has recently provided an explanation for the apparent catalytic inactivity of Cd(II)-alkaline phosphatase[94].

III. DEHYDROGENASES

A. Metal content and some general considerations regarding the detection and quantitation of protein-bound metal ions

Several dehydrogenases, which catalyze specific versions of the general type reaction observed for this class of enzymes:

$$X\text{-}H + \overset{+}{NAD}(\overset{+}{NADP}) \rightleftharpoons X + NADH(NADPH) + H^+ \tag{8}$$

have been reported to be zinc metallo-enzymes, *e.g.* the alcohol dehydrogenases from liver[95,96,96a] and yeast[97], malate dehydrogenase from pig heart[98] (but see Pfleider and Hohnholz[99]), glutamate dehydrogenase from mammalian liver[100] and lactate dehydrogenase from rabbit skeletal muscle[101]. However, except in the case of alcohol dehydrogenase where the presence of bound zinc is thoroughly documented[95-97], considerable uncertainty exists in regard to the validity of these earlier reports. Thus careful examination of lactate dehydrogenase from rat liver[102] and pig heart[103], glyceraldehyde-3-phosphate dehydrogenase from both rabbit muscle[104] and yeast[105], glucose-6-phosphate dehydrogenase from yeast[106] and glutamate dehydrogenase from bovine liver[106a] has indicated the absence of significant concentrations of this metal ion. In the case of glucose-6-phosphate dehydrogenase the more recent study[104] also demonstrates the absence of magnesium, calcium, chromium, manganese, iron, cobalt, copper, and molybdenum from purified preparations.

Considerable difficulty has also been encountered in definition of the metal–protein stoichiometry for liver alcohol dehydrogenase. This enzyme has been reported to contain either two[97] or four[107,108] g atoms of zinc per mole of enzyme. Liver alcohol dehydrogenase possesses two binding sites for NADH[109,110] in contrast to the four binding sites for this coenzyme which are present on the enzyme from yeast[111]. This confusion in regard to the zinc–protein stoichiometry has recently been clarified by an extensive study[91], which demonstrates unequivocally that the zinc content of liver alcohol dehydrogenase approximates 4 g atoms/mole of enzyme. Two of the zinc atoms present in this enzyme may, therefore, have a structural rather than a catalytic role[91,107]. However, contradictory data have been obtained when this postulate is tested by examination of the relationship between zinc content and catalytic activity[91,108]. The earlier discrepancies cannot be explained on the basis of differing isozyme patterns since isozymes of liver alcohol dehydrogenase exhibit similar zinc content[113].

Many of the problems which have been described above for alcohol dehydrogenase are not peculiar to this enzyme. Similar difficulties are likely to be encountered for any metallo-enzyme if the criteria proposed for identification of such enzymes by Vallee[4] are not applied with both rigor and intelligence. These criteria are: (i) tight binding of the metal ion to the protein; (ii) the observation of an increasing metal: protein ratio as the specific activity of the enzyme increases during purification; (iii) the observation of an integral ratio of g atoms metal per mole of enzyme in the purified protein; and (iv) the observation of an integral molar ratio between metal content and cofactor content (or binding). Although these criteria are adequate for the definition of a simple metallo-enzyme in which bound metal and active sites are present in equimolar amount, *e.g.* yeast alcohol dehydrogenase, they may be misleading in more complex situations, as is apparent from the history of the studies on liver alcohol dehydrogenase. In this latter case uncertainties in the molecular weight and extinction coefficient of the enzyme also contributed to the confusion[91], and it should be noted that errors in either or both of these parameters will also lead to erroneous determinations of metal: protein stoichiometry.

In addition to enzymes in which all or part of the bound metal has a structural rather than a catalytic role, cases are now known of enzyme preparations containing two different metal ions which appear to have the same catalytic role, *e.g.* alcohol dehydrogenase purified from yeast grown in the presence of added $CoCl_2$[92] (cf. Section V). In such instances the mixed stoichiometry observed may result either from the presence of the two metallo-enzymes in appropriate ratio or from formation of a true hybrid metallo-enzyme. No simple method exists for distinguishing between these two possibilities unless alterations also occur in the properties of the protein as a result of the metal substitution. Although multiple metal incorporation has usually resulted from deliberate manipulation of the metal content of the diet or growth medium, this situation may occur in nature and will present peculiar problems of evaluation unless an independent estimate of the number of active sites is obtained.

Finally criterion (ii) noted above may not be applicable if the final step(s) of the purification procedure separate the metallo-enzyme under examination from higher concentrations of the same metal bound to other proteins. In this case the metal–protein ratios may decrease with increasing purification as has been observed for the association of zinc with pyruvate carboxylase from baker's yeast[114]. When such a situation arises, criterion (ii) is no longer applicable, and comparison of the distribution of catalytic activity, protein and metal content when the purified enzyme is subjected to fractionation with respect to size and/or charge forms an acceptable substitute procedure. In studies using column fractionation of the purified enzyme a constant metal : enzymic activity ratio should be observed across the protein peak. Furthermore, if gel filtration is employed as the method of fractionation, the peak of metal content should emerge at a position with respect to the column void volume which is consistent with the molecular weight (or Stoke's radius) of the enzyme.

The importance of the application of these criteria for identification of a metallo-enzyme cannot be overemphasized. Valid information cannot be obtained from single determinations of the metal content of enzymes at any level of purification and the interpretation of inhibition or activation of enzymes by metal chelating agents as due to interaction with bound metal can be especially misleading. Inactivation of an enzyme resulting, for example, from incubation with 1:10 phenanthroline may be due to: (i) interaction with, or removal of, a bound metal ion; (ii) oxidation of sulfhydryl groups (as has been demonstrated for FDP-aldolase[115]); or (iii) insertion of this planar molecule into a hydrophobic area of the protein. Additionally activation of fructose-1,6-diphosphatase and retardation of the assembly of tobacco mosaic virus which are observed in the presence of EDTA do *not* appear to result from interaction with a bound metal ion[116,117]. In other cases activation by EDTA or other chelating agents may be due to removal of inhibitory metal ions from the enzyme preparation[118]. Although inhibitory effects of chelating agents have been observed for many dehydrogenases[119,120] only alcohol dehydrogenase has been definitively established as a metallo-enzyme.

B. The catalytic role of the bound metal

Despite investigation over the past two decades the catalytic role of the bound zinc in alcohol dehydrogenase remains obscure although numerous suggestions have been advanced[121]. Mechanisms have been proposed in which the bound zinc of alcohol dehydrogenase participates in coordination of: (i) the substrates[122,123]; (ii) the coenzymes[121,124] and (iii) the substrates and coenzymes simultaneously[125,128]. Furthermore, most of the possible sites for coordination of metal ions to NAD$\overset{+}{}$ have been suggested in these various hypotheses[124-128]. The remaining possibility, *i.e.* that the zinc is not directly involved in coordination of either the substrates or the coenzymes has not yet been suggested (but see Williams and Vallee[129])! Since such confusion often arises when attempts are made to interpret indirect evidence it is of interest to examine the available data. Initial rate studies indicate that the *reversible* inhibition of either yeast[130] or liver[131] alcohol dehydrogenase by 1,10-phenanthroline is competitive with respect to NAD(NADH) and non-competitive with respect to the substrates. These inhibition patterns were interpreted by Vallee[121] to indicate that the bound zinc interacts *only* with NAD$\overset{+}{}$(NADH) and has no role in coordination of the substrates. This interpretation has been questioned in two respects. First, Dalziel[132] has pointed out that the inhibition patterns observed are consistent with coordination of both the substrates and the coenzymes by the bound zinc. However, consideration of the sequential pathway proposed as the major reaction sequence for alcohol dehydrogenase (reaction 9)[133-135] suggests that these inhibition patterns may in fact be *indeterminate* with respect to the site of substrate or coenzyme binding:

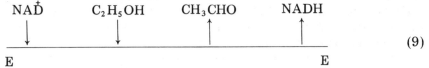

$$(9)$$

In a sequential mechanism of the type illustrated in reaction 9 any dead-end inhibitor which interacts with the free enzyme (E) should show competitive behavior with respect to NAD$\overset{+}{}$(NADH) and non-competitive behavior with respect to the substrates[60]. The competitive behavior observed with respect to NAD$\overset{+}{}$ or NADH implies competition for the same enzyme species but *not* necessarily for the same binding site. Second, analogs of 1,10-phenanthroline, *e.g.* 1,5-phenanthroline, 5,6-(or 7,8)-benzoquinoline, which are *not* chelating agents, are nonetheless competitive inhibitors with respect to NAD$\overset{+}{}$ which are as or more effective than 1,10-phenanthroline[136]. Both 1,5- and 1,10-phenanthroline induce similar difference spectra on interaction with yeast alcohol dehydrogenase but only 1,10-phenanthroline shows this effect on interaction with $Zn(H_2O)_4^{2+}$ [136]. The binding site for 1,10-phenanthroline on alcohol dehydrogenase is defined as the site of interaction of the nicotinamide ring of NAD$\overset{+}{}$ by examination of the interaction between this inhibitor and analogs of portions of the NAD$\overset{+}{}$ molecule which are themselves competitive inhibitors with respect to NAD$\overset{+}{}$, *e.g.* N'-methylnicotinamide chloride[137], ADP-ribose[138]. However, since chelating properties are *not* correlated with effectiveness of inhibition, this binding appears to be due to hydrophobic interactions rather than to coordination of the bound zinc. Further clarification is provided by binding studies for NAD$^+$ and NAD$^+$ analogs to zinc-free apo-alcohol dehydrogenase. When the apoenzyme is prepared by dialysis against EDTA at pH 6.0 neither the extent of binding nor the affinity of alcohol dehydrogenase for NAD$^+$ is altered by removal of the bound zinc[139]. Although other studies indicate a linear relationship between zinc content and NAD$^+$ (or ADP-ribose) binding when the bound zinc is removed by dialysis at pH 5.0[108], alcohol dehydrogenase is known to be unstable at this lower pH[91] and hence the loss of ability to bind NAD$^+$ which is observed may be due to a secondary alteration in enzyme conformation which results from removal of the bound zinc[140].

Although these data appear to exclude the most widely accepted role for zinc in this metallo-dehydrogenase, i.e. formation of an enzyme–Zn^{2+}– NAD(NADH) complex[121], they fail to define the catalytic role of this metal ion. The studies of Mildvan and Weiner[122,141] may be interpreted in the context of the formation of enzyme–Zn^{2+}–substrate complexes[8] although such interpretation is very speculative with the data presently available. However, these studies which utilize a "spin-labelled" analog of ADP-ribose do permit an estimate of the distance between the unpaired electron of the spin-label and the substrate protons[121]. If then a paramagnetic metal ion could be substituted for Zn^{2+} at the metal binding site, the extent of spin-spin interaction as measured by e.p.r. would give an estimate of the distance between the spin-label and the bound metal[72,74a] (cf. Section II B). Substitution of Co^{2+} for Zn^{2+} in liver alcohol dehydrogenase has been reported and the Co^{2+}-enzyme is catalytically active[90]. Such substitution may also permit the direct demonstration of E–Me^{2+}–substrate bridge complexes by examination of the nuclear magnetic relaxation rates of substrate protons if Mn^{2+} can be substituted for Zn^{2+} (cf. Section II C). Alternatively for the native enzyme ^{35}Cl-n.m.r. may be used to probe the effect of the substrates and coenzymes on the properties of the bound zinc[143]. This procedure has been used to investigate the binding of Zn^{2+} to pyruvate kinase[144] and is one of the few methods available for examination of the environment of the diamagnetic zinc atom.

Recently some unusual effects of Ca^{2+}, Mn^{2+} and Mg^{2+} on the kinetic properties of a glutamate dehydrogenase purified from *Blastocladiella emersonii* (a unicellular water mold) have been reported[146,147]. These divalent cations are strong activators of α-ketoglutarate reduction but are weak inhibitors of glutamate oxidation. Such unidirectional effects appear to be related to the regulatory properties of this enzyme and there is no evidence that the added divalent cations participate directly in catalysis. Regulatory effects of monovalent cations have also been observed for certain dehydrogenases[148,149] although such effects are more common for the pyridine nucleotide-dependent oxidative decarboxylases (cf. Section IV).

Although most dehydrogenases are fully active in the absence of *added* divalent metal ions, certain polyol dehydrogenases have been isolated which require activation by Mg^{2+} or Mn^{2+} [150-153]. No studies have been performed to define the role of the metal ion in the catalytic mechanism of these metal-activated dehydrogenases.

IV. PYRIDINE NUCLEOTIDE-DEPENDENT OXIDATIVE DECARBOXYLASES

A number of enzymes have been described which catalyze oxidative decarboxylation of a β-hydroxy acid and require activation by added Me^{2+}. The type reaction is illustrated below for L-malate enzyme [154-157]:

$$L\text{-malate} + N\overset{+}{A}D(N\overset{+}{A}DP) \xrightleftharpoons{Me^{2+}}$$

$$\text{Pyruvate} + CO_2 + NADH(NADPH) + H^+ \quad (10)$$

Similar reactions are catalyzed by β,β-dimethyl-malate dehydrogenase [158], β-isopropylmalate dehydrogenase [159], $N\overset{+}{A}D$ and $N\overset{+}{A}DP$-dependent isocitrate dehydrogenase [160-164], and homoisocitric dehydrogenase [165]. The analogous reactions catalyzed by 6-phosphogluconate dehydrogenase [166] and oxalo-glycolate reductive decarboxylase [167] occur in the absence of added divalent metal ions although it is not known whether these latter enzymes contain bound metals.

Although direct evidence regarding the role of the divalent metal ion in the metal-dependent oxidative decarboxylases is lacking at present, some preliminary evidence suggests that the metal ion participates solely in the formation of an enzyme–Me^{2+}–substrate bridge complex. First, $N\overset{+}{A}DP$ and NADPH bind to L-malate enzyme in the absence of Me^{2+}. The extent and affinity of $N\overset{+}{A}DP$ binding is unaffected by addition of Mn^{2+} and the observed dissociation constant for the E–NADPH complex agrees with that deduced from initial rate studies [168,169]. Second, the Mn^{2+}–isocitrate complex affords protection against the inactivation of $N\overset{+}{A}D$-dependent isocitrate dehydrogenase by p-chloromercuriphenylsulfonate but neither isocitrate alone or $N\overset{+}{A}D$ in the presence or absence of Mn^{2+} provide such protection [170]. Third, several of these enzymes also catalyze a metal ion-dependent decarb-oxylation of the analogous keto acid, $e.g.$ oxaloacetate (L-malate enzyme) [155,156], oxalosuccinate ($N\overset{+}{A}DP$-dependent isocitrate dehydrogen-ase) [171,172]. This reaction is assumed to occur at the same active site as the oxidative decarboxylation of β-hydroxyacids [164,169] although kinetic anal-yses provide no evidence for the existence of either free or enzyme-bound keto acid as a reaction intermediate [16,169]. Even if keto acid decarboxylation occurs by an abortive pathway, the metal ion requirement observed suggests the involvement of an E–Me^{2+}–substrate complex in the overall reaction by analogy with the metal-dependent oxaloacetate decarboxylases (cf. Section V). Fourth, significant binding of Mn^{2+} to L-malate enzyme has been demonstrated[169,173] and studies of an NAD^+-dependent isocitrate dehydrogenase indicate that the Me^{2+}–isocitrate complex is the true substrate for this enzyme while the activator site is specific for free isocitrate[173a]. Finally, marked interdependence of the apparent K_A for Me^{2+} and the apparent K_M for isocitrate is observed for the $NA\overset{+}{D}P$-dependent isocitrate dehydrogenase[174]. It is of interest that Zn^{2+} is the most effective activator of this latter enzyme and that a relatively broad specificity of metal ion activation is observed[174].

All these observations are consistent with a mechanism proposed by Boyer[176] for malate enzyme which is shown in Fig. 2. In this mechanism the electrophilic character of the metal ion would assist in the creation of electron-deficiency at the α-carbon and this, in combination with hydride ion transfer to $NA\overset{+}{D}P$, would promote a concerted dehydrogenation and decarboxylation. The role proposed for the metal ion in Fig. 2 is analogous to that suggested by Steinberger and Westheimer[177] for catalysis of decarboxylation of β-keto acids by metal ions. However, the required structure for the Me^{2+}-L-malate complex differs from that observed for chelation of $Mn(H_2O)_6^{2+}$ by L-malate[178], implying that the presence of the enzyme may modify the structure of the Me^{2+}-malate complex formed.

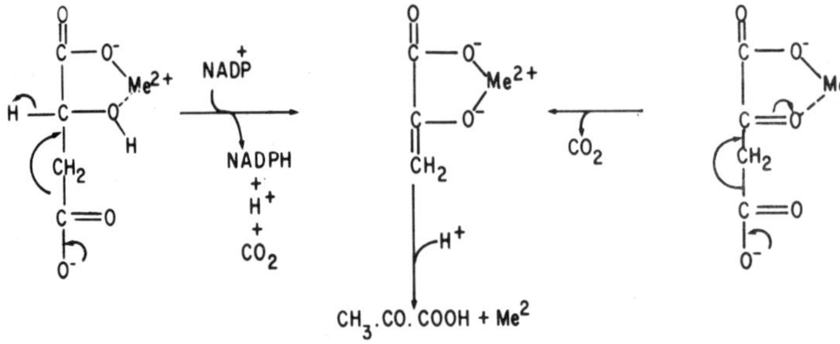

Fig. 2. A possible role for Me^{2+} in the reaction catalyzed by the pyridine nucleotide-dependent oxidative decarboxylases. The proposed mechanism is written here for L-malate enzyme as proposed by Boyer[176].

It should be noted that a survey analysis of $NA\overset{+}{D}$-linked isocitrate dehydrogenase from pig heart for bound metal ions has shown the absence of significant concentrations of Mg, Ca, Cr, Mn, Fe, Co, Ni, Cu, Zn, Mo and Cd from this enzyme[178a]. The inhibition of this enzyme by CN^- presumably results from formation of a NAD–cyanide adduct[170]. An analogous situation to that described above for β-keto acid decarboxylation by certain oxidative decarboxylases is observed for a metal-dependent isomero-reductase which catalyzes a reduction coupled with alkyl migration as a step in the pathway of valine and isoleucine biosynthesis[179]:

$$\alpha\text{-Acetolactate} + NADPH \xrightarrow{Me^{2+}} \alpha,\beta\text{-Dihydroxyisovalerate} + NA\overset{+}{D}P \quad (11)$$

This enzyme also catalyzes an NADPH and Me^{2+}-dependent isomerization of α-hydroxy-β-ketoisovalerate to α-acetolactate but no evidence has been obtained for participation of this latter compound as an intermediate in the overall reaction[179] and the role of the metal ion in this reaction is not yet defined.

V. ENZYMES OF CO$_2$ METABOLISM

A. Biotin-enzymes

Three related reactions are catalyzed by enzymes containing biotin as a cofactor. The biotin carboxylases catalyze carboxylation of α-keto-acids or acyl derivatives of coenzyme A[180,181]:

$$E\text{--biotin} + ATP + HCO_3^- \overset{Me^{2+}}{\rightleftharpoons} E\text{--biotin} \sim CO_2$$
$$+ ADP + PO_4^{3-} \quad (12)$$

$$E\text{--biotin} \sim CO_2 + \begin{cases} \text{pyruvate} \\ \text{or} \\ \text{propionyl-CoA} \end{cases} \rightleftharpoons$$

$$E\text{--biotin} + \begin{cases} \text{oxaloacetate} \\ \text{or} \\ \text{methylmalonyl-CoA} \end{cases} \quad (13)$$

Methylmalonyl-CoA-oxalacetate transcarboxylase catalyzes transcarboxylation between keto-acids and acyl derivatives of coenzyme A[182]:

$$E\text{--biotin} + methylmalonyl\text{-CoA} \rightleftharpoons E\text{--biotin} \sim CO_2 + propionyl\text{-CoA} \quad (14)$$

$$E\text{--biotin} \sim CO_2 + pyruvate \rightleftharpoons E\text{--biotin} + oxaloacetate \quad (15)$$

and the biotin decarboxylases catalyze decarboxylation of keto acids or acyl derivatives of coenzyme A[183,184]:

$$E\text{--biotin} + \begin{cases} \text{oxaloacetate} \\ \text{or} \\ \text{methylmalonyl-CoA} \end{cases} \rightleftharpoons$$

$$E\text{--biotin} \sim CO_2 + \begin{cases} \text{pyruvate} \\ \text{or} \\ \text{propionyl-CoA} \end{cases} \quad (16)$$

$$E\text{--biotin} \sim CO_2 \rightleftharpoons E\text{--biotin} + CO_2 \quad (17)$$

The biotin carboxylases exhibit requirements for added Me^{2+} in the synthetase reaction in which the E–biotin $\sim CO_2$ complex is formed from ATP + HCO_3^- (reaction 12). The role(s) of the metal ions in reaction 12 will be considered with other synthetases in Section XII.

All three species of biotin–enzyme catalyze transcarboxylation involving an E–biotin $\sim CO_2$ complex and either an α-keto acid (reactions 13, 15 and 16) or an acyl derivative of coenzyme A (reactions 13, 14 and 16). The first indication that protein-bound metal ion participated in biotin-dependent transcarboxylation involving a keto acid resulted from n.m.r. studies on pyruvate carboxylase from chicken liver which were intended to clarify the role of the dissociable divalent metal ion in the synthetase partial reaction (reaction 12). These studies demonstrated the presence of a bound paramagnetic species in this enzyme which was identified as manganous ion[185]. Further examination indicated that manganese is a constitutive component of this pyruvate carboxylase which is present in equimolar ratio with the biotin content[37,185]. Pyruvate carboxylase from chicken liver was therefore identified as both the first metallo-enzyme containing manganese as its native metal ion and also as the first metallo-biotin enzyme[37,185]. Subsequently other pyruvate carboxylases and also methylmalonyl-CoA-oxalacetate transcarboxylase have been shown to contain bound metal ions (Table I), and indirect evidence suggests that bound metal ion may also be present in a biotin-dependent oxalacetate decarboxylase[183]. Protein-bound metal ions may therefore play a general role in biotin-dependent transcarboxylation involving α-keto acids. Although oxaloacetate decarboxylases

TABLE I

THE PRESENCE OF BOUND METAL IONS IN VARIOUS BIOTIN ENZYMES

Enzymes	Source	Bound metal ion present[a]	Reference
Pyruvate carboxylase	Chicken liver	Mn[b]	185
	Chicken liver (Mn^{2+}-deficient chickens)	Mg[b]	186
	Turkey liver	Mn	187
	Rat liver	Mn	188
	Calf liver	Mg + Mn	187
	Baker's yeast	Zn	189
Methylmalonyl-CoA-oxaloacetate trans-carboxylase	*Propionibacterium shermanii*	Co + Zn	190

[a] In all cases the total content of bound metal is equimolar with the biotin content of the enzyme except for pyruvate carboxylase from rat liver.
[b] Pyruvate carboxylase purified from the livers of chickens grown on limiting levels of manganese contain Mn + Mg in a ratio which reflects the manganese content of the diet.

which do not contain biotin have also been described[28,191], these latter enzymes exhibit no catalytic activity in the absence of added divalent metal ions and do not appear to catalyze a transcarboxylation step which precedes decarboxylation[28].

Several suggestions have been advanced to explain the role of metal ions in both oxaloacetate decarboxylation and biotin-dependent transcarboxylation, as illustrated in Fig. 3. On the basis of the properties of the decarboxylation of dimethyloxaloacetate catalyzed by Me^{2+} and Me^{2+} ions, Steinberger and Westheimer[177] proposed a mechanism applicable to either decarboxylation or transcarboxylation of keto acids in which chelation of the metal ion by the carbonyl and C-1 carboxyl groups of the keto acid permits stabilization of the enolate reaction product (Fig. 3A). The effects of Me^{2+} on the ultra-violet spectra of some keto acids appear to support the proposed structures for these complexes (Fig. 3A)[193] but other evidence[194] is more consistent with the assignment of the Mn^{2+}–oxaloacetate complex as a C-1, C-4 bis-carboxyl chelate and of Mn^{2+}–pyruvate as a monodentate complex involving the carboxyl group as the ligand. Although recent studies on the metal-activated oxaloacetate decarboxylase from cod muscle[195] have been interpreted in the context of the Steinberger–Westheimer mechanism[196,197], neither the stereoselective reduction of pyruvate to D-lactate[197], nor the increased rate of exchange of the pyruvate methyl protons[196] which are caused by the presence of the enzyme appear to exclude other mechanisms.

A second mechanism applicable only to metal ion catalysis of biotin-dependent transcarboxylation was originally proposed by Stiles[198] (Fig. 3B). The feasibility of this mechanism has been demonstrated in studies which showed that divalent metal ions retard the decarboxylation of N-carboxy-2-imidazolone, an analog of $1'$-N-carboxybiotin (the bound intermediate of the enzymic reaction)[199], and accelerate the reaction of N-carboxy-2-imidazolone esters with nucleophiles[200]. Although the effects observed in these models studies would drive the reaction towards transcarboxylation, no evidence exists for the coordination of the $1'$-N-carboxybiotinyl residue by the protein-bound metal ion[201]. Titrimetric and n.m.r. studies of the interaction of Me^{2+} with d-biotin and certain of its analogs indicate a role for the sulfur atom in formation of the Me^{2+}–biotin complex in solution[202].

At the enzymic level the information presently available is primarily derived from n.m.r. studies on pyruvate carboxylase from chicken liver since the bound Mn^{2+} present in the active site of this enzyme has provided an ideal probe for the role of the metal ion as was predicted by Cohn[21]. These studies have been reviewed in detail elsewhere[203,204] and only a brief summary is presented here. Proton relaxation rate analysis utilizing the Mn^{2+}–water proton interaction[76] (cf. Section II C) has shown that only the substrates and inhibitors of the transcarboxylation step (reaction 13) of the pyruvate carboxylase reaction interact significantly with the bound man-

404

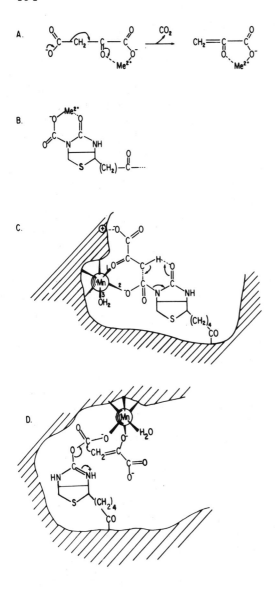

Fig. 3. Some possible roles for metal ions in biotin-dependent transcarboxylation involving β-keto acids. A. The Steinberger–Westheimer mechanism[177] as modified to describe oxaloacetate decarboxylation. B. The mechanism proposed by Stiles[198] which describes a role of Me^{2+} in biotin-dependent transcarboxylation. C. The mechanism proposed by Mildvan et al.[201] to describe the role of the bound manganese in the reaction catalyzed by pyruvate carboxylase. This mechanism combines features of both the Steinberger–Westheimer and Stiles mechanisms and provides a dual role for the metal ion. D. The mechanism proposed by Bruice and Hegarty[210] based on the suggestion of E–biotin CO_2 as an O-carboxyl species.

ganese. This site of involvement of the bound manganese in the pyruvate carboxylase reaction is confirmed by examination of the effect of the bound metal ion on the nuclear magnetic relaxation rates of the protons of pyruvate and oxalacetate. Calculation of the Mn^{2+}-proton distances (from $1/T_{1M}$) and the hyperfine coupling constants (from $1/T_{2M}$ and $1/_{1M}$) have demonstrated the formation of enzyme–Mn^{2+}–pyruvate[205] and enzyme–Mn^{2+}–oxaloacetate[194] bridge complexes. The kinetic properties of these bridge complexes are consistent with their participation in the overall reaction[194,205]. However, uncertainties in the correlation time (τ_c) for Mn^{2+}-substrate proton interaction prevent definitive assignments for the structures of these bridge complexes from these data. Preliminary studies suggest that an enzyme–Co^{2+}–pyruvate bridge complex is formed on interaction of methylamalonyl-CoA-oxalacetate transcarboxylase with pyruvate[190].

Formation of the pyruvate carboxylase–Mn^{2+}–substrate bridge complexes has previously been suggested to occur by an S_N1 outer sphere mechanism[206] since the apparent rate of formation of these complexes from an ion-pair approximated the apparent rate of departure of water molecules from the coordination sphere of the bound manganese[194,205]. This conclusion was based on the presumption that observation of a negative temperature coefficient for $1/T_{2p}$ or $1/T_{1p}$ identified τ_M as the process which dominated the overall relaxation rate. However, both Cohn[82] and Sheard and Bradbury[76a] have pointed out that domination by τ_s can also give rise to a negative temperature coefficient for $1/T_{1p}(1/T_{2p})$ (cf. Section 11 C) and is a more likely explanation when low energies of activation are observed as was the case in this instance[37,205]. The possibilities may be distinguished (cf. Section II C) since domination by τ_M is characterized by equality of $1/T_{1p}$ with $1/T_{2p}$ whereas, if τ_s dominates, $1/T_{2p}$ exceeds $1/T_{1p}$. Recent PRR studies have demonstrated that τ_s, rather than τ_M as previously suggested[37] is the process which dominates the paramagnetic contribution to the relaxation rates of water protons in the presence of pyruvate carboxylase since $1/T_{2p}$ is consistently greater than $1/T_{1p}$ at temperatures between 0 and $40°C$[207]. The value ($1.0 \pm 0.5 \times 10^6$ sec^{-1}) reported previously as $1/\tau_M$[37] is therefore only a lower limit for this parameter. Similarly $1/T_{2p}$ for the pyruvate methyl protons in the presence of pyruvate carboxylase is consistently greater than $1/T_{1p}$ over the temperature range examined identifying τ_s (rather than τ_M) as the process which dominates $1/T_{2p}$[205]. However, in the pyruvate carboxylase–oxalacetate complex, $1/T_{2p}$ for the oxalacetate methylene protons approaches $1/T_{1p}$ at low temperature and the high activation energy observed for $1/T_{2p}$ further suggests that in this instance τ_M is the process which dominates $1/T_{2p}$ at temperatures below $20°$[194].

These and other studies on the properties of the bound manganese of pyruvate carboxylase have resulted in the proposal of a mechanism describing the catalytic role of this bound metal (Fig. 3C)[201] which combines features of both the Steinberger–Westheimer and the Stiles mechanisms (Figs. 3A and B). In this mechanism both the carbonyl group of pyruvate and the $1'$-N-carboxyl group of the carboxybiotin residue provide ligands to the bound manganese, permitting the electrophilic character of the metal ion to facilitate both departure of a proton from the methyl group of pyruvate and also carbonium ion formation from the $1'$-N-carboxybiotinyl residue. Since no significant tritium release occurs on incubation of ^{3}H-pyruvate with pyruvate carboxylase in the absence of other reaction components[201], proton departure and carboxyl transfer probably occur in a concerted process as shown in Fig. 3C. Two predictions of this concerted mechanism have recently been demonstrated. First, a significant isotope rate is effect is observed, k(protium)$/k$(tritium) = 4.2; and second, studies in which pyruvates of known chirality were carboxylated have shown that carboxylation occurs with retention of configuration, i.e. the carboxyl group enters cis to the departing proton[208]. Final evaluation of the mechansim of Fig. 3C must, however, await definition of the structures of the enzyme–Mn^{2+}–pyruvate and –oxaloacetate bridge complexes.

The concerted mechanism (Fig. 3C) also assumes that the structure of the carboxylated intermediate in the biotin carboxylase reaction may be represented as $1'$-N-carboxybiotin, as is suggested by trapping experiments using diazomethane[209]. However, this conclusion has recently been questioned by Bruice and Hegarty[210] on the basis of model studies which show that the ureido oxygen is a better nucleophile than the ureido nitrogen. The mechanism shown in Fig. 3D which was proposed on the basis of these model studies[210] would eliminate many of the problems encountered with previous mechanisms for biotin-dependent transcarboxylation. Furthermore, the suggested O-carboxyl intermediate might be expected to rearrange to an N-carboxyl species on methylation[210].

Since the involvement of protein-bound metal ion in biotin-dependent transcarboxylation involving keto acids has been clearly demonstrated, a similar role of bound metal ion might be anticipated in the analogous reaction involving acyl derivatives of coenzyme A. However, no evidence has been obtained for the presence of protein-bound metal ion in any acyl-CoA carboxylase examined thus far[201] although definitive studies have not yet been performed. Additionally methyl malonyl-CoA-oxalacetate transcarboxylase contains only 1 g atom of bound metal per biotin residue[190] rather than 2 g atoms per biotin residue which would be predicted if metal ions were present in both the acyl-CoA and keto acid binding sites (reactions 13 and 14)[210a]. These observations present an apparent paradox since deuterium exchanges more slowly into the methyl group of either acetyl-CoA[211] or simple acetyl-thioesters[212] than into the methyl group of pyruvate under comparable conditions[212], indicating a lesser degree of inherent nucleophilic activation in the case of the thioesters. However, the extent of nucleophilic activation of the methyl (or methylene) group on which CO_2 fixation occurs may not be important in the reaction catalyzed by the acyl-CoA carboxylases since a primary isotope rate effect is not observed in studies on several of these enzymes[213,214] although carboxylation occurs with retention of configuration[213].

B. PEP carboxylating enzymes

Three enzymes have been described which catalyze homologous carboxylations of PEP in the presence of an added divalent metal ion, *e.g.* Mn^{2+}:

$$PEP + NDP + CO_2 \; \underset{}{\overset{Me^{2+}}{\rightleftharpoons}} \; Oxalacetate + NTP \tag{18}$$

$$PEP + PO_4{}^{3-} + CO_2 \; \underset{}{\overset{Me^{2+}}{\rightleftharpoons}} \; Oxalacetate + P_2O_7{}^{4-} \tag{19}$$

$$PEP + HCO_3{}^- \; \xrightarrow{Me^{2+}} \; Oxalacetate + PO_4{}^{3-} \tag{20}$$

All these reactions involve nucleophilic attack on the phosphoryl group of PEP, a tautomeric shift of the double bond and addition of an electrophile at C-3 of PEP[8]. Differences are, however, observed in the nature of the phosphoryl acceptor used which may be NDP (PEP carboxykinase) (reaction 18)[215]; $PO_4{}^{3-}$ (PEP carboxytransphosphorylase) (reaction 19)[216]; or H_2O (PEP carboxylase) (reaction 20)[217]; and also in the carboxylating species utilized (CO_2 or $HCO_3{}^-$)[218,219]. The PEP carboxylating enzymes also exhibit a homology with the reaction catalyzed by pyruvate kinase[220]:

$$PEP + ADP + H^+ \; \underset{}{\overset{Me^{2+},\, M^+}{\rightleftharpoons}} \; Pyruvate + ATP \tag{21}$$

This homology is further emphasized by the observation that, in the absence of CO_2, PEP carboxytransphosphorylase catalyzes phosphorlysis of PEP[221]:

$$PEP + PO_4{}^{3-} + H^+ \; \xrightarrow{Me^{2+}} \; Pyruvate + P_2O_7{}^{4-} \tag{22}$$

The extent of homology is illustrated by examination of the mechanism of action of these various enzymes[222]. First, initial rate studies for pyruvate kinase[35,38,223] and PEP carboxykinase[222,224] are consistent with a random order scheme for addition of substrates and Me^{2+} to these two enzymes. Binding studies which indicate the formation of kinetically significant complexes for PEP, Me^{2+} and NDP for pyruvate kinase and PEP carboxykinase[37,55,222], and also of Mn^{2+} with PEP carboxylase[222], support this conclusion. Both pyruvate kinase[225], and PEP carboxykinase[222] also appear to form higher order complexes of the type $E-MeNDP^--Me^{2+}$ which may exhibit similar catalytic properties to the $E-MeNDP^-$ complex. Second, the $E-Mn^{2+}$ complexes formed by pyruvate kinase, PEP carboxykinase,

PEP carboxylase all exhibit markedly enhanced effects on the PRR of water (ϵ_b = 14–33) which are reduced by addition of either PEP or of NDP (PEP carboxykinase, pyruvate kinase)[35,38,222]. In the case of PEP carboxylase the reduction in ϵ_b is transient since the overall reaction proceeds under these conditions[222]. These latter data suggest, but do *not* prove, that formation of an enzyme–Mn^{2+}–PEP bridge complex is a common feature in the mechanisms as is illustrated in Fig. 4. for PEP carboxylase. The studies demonstrating formation of a pyruvate kinase–Mn^{2+}–substrate bridge complex are discussed in Chapter 18. The PRR studies on PEP carboxykinase[222] illustrate a further pitfall in the interpretation of such data. Negative temperature coefficients are observed for $1/T_{1p}$ of water protons in the presence of both E–Mn^{2+} and also the ternary complexes of this species with IDP and PEP. Hence, *assuming* that τ_M is the process which dominates $1/T_{1p}$ (cf. Section II C), ϵ_c/ϵ_b ratios (at 25°C) were used to calculate that IDP displaced one, and PEP two, water molecules from the coordination sphere of the enzyme-bound Mn^{2+}. However, wide differences are observed for the activation energy for $1/T_{1p}$ for water protons in these three complexes[222]. In this situation ϵ_T/ϵ_b ratios calculated for any given temperature are misleading in regard to the coordination scheme[82], and the mechanism proposed for PEP carboxykinase on the basis of these data[222] may therefore be invalid.

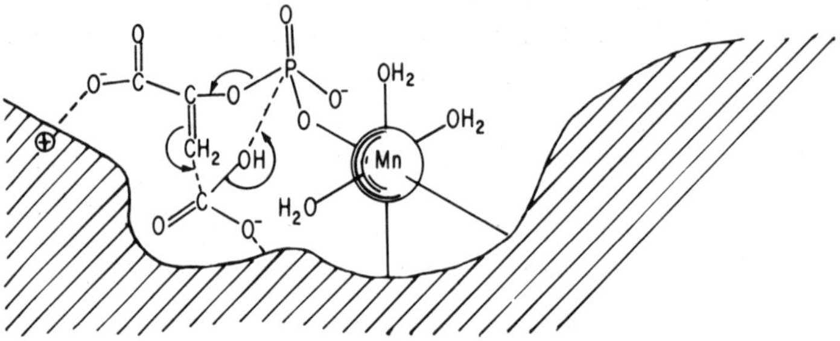

Fig. 4. A possible mechanism for PEP carboxylase illustrative of the type mechanism which may describe the role of Me^{2+} in the PEP carboxylating enzymes and pyruvate kinase. From Miller *et al.*[222].

Recent studies have also cast some doubt on the place of PEP carboxytransphosphorylase in this homologous group of enzymes when the criterion of mechanism is invoked. This enzyme is very sensitive to inhibition by a wide spectrum of chelating agents, *e.g.* EDTA, 1,10-phenanthroline, CN^- [226]. The properties of inhibition of various partial reactions by EDTA indicate that the formation of an enzyme–PEP complex may require participation of a bound metal ion if the effects of the various chelating agents may be interpreted in this context[227]. However, direct attempts to demonstrate the presence of bound metal ion in this enzyme have thus far yielded equivocal and confusing results[228].

C. *Ribulose-1,5-diphosphate carboxylase*

Carboxylation of ribulose-1,5-diphosphate to yield 2 moles of 3-phosphoglycerate (reaction 23) is catalyzed by an enzyme which is generally considered to require activation by a divalent metal ion[229,230], but see Anderson *et al.*[231]:

$$\text{Ribulose-1,5-diphosphate} + CO_2 \xrightarrow{\text{Me}^{2+}}$$

$$\text{2,3-phosphoglycerate} + H^+ \qquad (23)$$

The potent inhibitor, 2-carboxyribitol-1,5-diphosphate (I), which is an analog of the proposed transition state intermediate of the ribulose-1,5-diphosphate carboxylase reaction (II)[232], binds tightly to the enzyme only in the presence of Me^{2+} [233]:

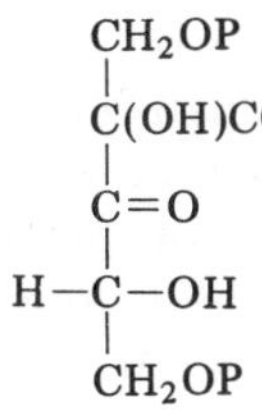

(II)	(I)	(III)
CH_2OP	CH_2OP	CH_2OP
$C(OH)COO^-$	$C(OH)COO^-$	$C(OH)CN$
$C=O$	$H-C-OH$	$H-C-OH$
$H-C-OH$	$H-C-OH$	$H-C-OH$
CH_2OP	CH_2OP	CH_2OP

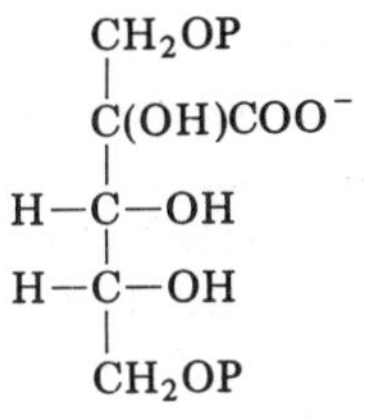

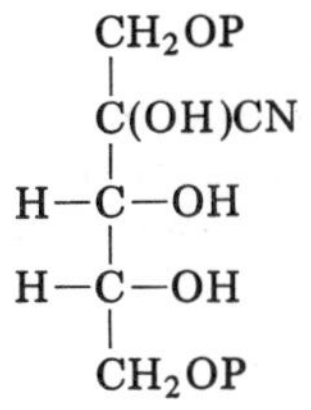

However, tight binding of ribulose-1,5-diphosphate in the presence of CN^- (probably as the ribulose-1,5-diphosphatecyanide adduct (III)) does not require the presence of Me^{2+} [234]. These data suggest a role of Me^{2+} in stabilization of the transition state intermediate (II) of the ribulose diphosphate carboxylase reaction. Earlier reports of the formation of a stable enzyme–Mg^{2+}–CO_2 complex have not been confirmed[233].

Additionally bound Cu^{2+} has been detected as a constitutive component of ribulose diphosphate carboxylase from spinach[236]. This bound metal ion does not appear to play a direct role in the catalytic mechanism since its removal does not cause loss of catalytic activity. However, certain subtle alterations in the properties of the tight binding of ribulose diphosphate plus cyanide or of 2-carboxyribitol diphosphate are observed on comparison of the native and Cu^{2+}-free enzymes[233]. Hence the bound Cu^{2+} appears to be involved in the maintenance of a particular conformation of ribulose-1,5-diphosphate carboxylase.

D. *Decarboxylases*

In addition to the metal-activated oxaloacetate decarboxylases (cf. Section V A), several other decarboxylases also require activation of Me^{2+}, *e.g.* orotidine-5-phosphate decarboxylase[237], pyridoxine decarboxylase[238]. The role of the divalent metal ion in the mechanism of action of these decarboxylases does not appear to have been investigated in any detail. However, it should be noted that a metal ion requirement is the exception rather than the rule among enzymes of this type[239].

VI. THIAMINE PYROPHOSPHATE-DEPENDENT ENZYMES

Enzymes which utilize TDP as a cofactor catalyze reactions in which carbon–carbon bond cleavage (or synthesis) is associated with formation of an adduct between a portion of the cleaved substrate and TDP, *e.g.* α-keto acid decarboxylation, transketolase, phosphoketolase, glyoxalate carboligase[240]. Many of the TDP-dependent enzymes also require activation by Me^{2+} [240], but in several instances enzymes of this type exhibit either no requirement for added Me^{2+} *e.g.* transketolase from rat liver[241], benzoylformate decarboxylase[242]; or only a partial dependence on added Me^{2+}, *e.g.* pyruvate decarboxylase from wheat germ[243], branched chain α-keto acid oxidase[243a,244]. However, wide variation is observed in the properties of interaction of these apoenzymes with TDP and/or Me^{2+}. These cofactors bind very tightly to certain TDP-dependent enzymes, *e.g.* pyruvate decarboxylase, and can only be removed under special conditions[245]. In contrast other enzymes requiring these cofactors, *e.g.* oxalyl-CoA decarboxylase, glyoxalate carboligase, are inactive in the absence of added TDP and Me^{2+} indicating readily reversible binding[246,247a]. Hence unless the enzyme after resolution for TDP is shown to be devoid of bound Me^{2+}, as is the case for pyruvate decarboxylase from wheat germ[243], no definitive conclusions can be drawn when the resulting apoenzyme fails to show a requirement for Me^{2+} on addition of TDP.

Studies on pyruvate decarboxylase[248] support the suggestion that Me^{2+} is involved in formation of the enzyme–TDP complex[249]. After resolution of the native enzyme with respect to TDP and Mg^{2+}, the holoenzyme can only be reconstituted by incubation of the apoenzyme with a marked excess of TDP and Me^{2+} [250]. Both the rate of reconstitution and the stability of the holoenzyme formed are dependent on the nature of the divalent metal ion used although the catalytic activity of the holoenzyme produced appears to be the same for all those Me^{2+} tested (Ca^{2+}, Mg^{2+}, Mn^{2+}, and Co^{2+}). For example the $E \diagdown{\overset{Ca^{2+}}{TDP}}$ complex both forms and dissociates more rapidly than the corresponding Mg^{2+} complex[250] as would be predicted from the properties of simple complexes of these metal ions[84]. Studies of the concentration dependence of the recombination process indicates that TDP and Mg^{2+} interact independently with the apoenzyme to form an initial dissociable ternary complex (**IV**) which is then converted in a slow and quasi-irreversible reaction to a second ternary complex (**V**) which is the catalytically active holoenzyme[251]:

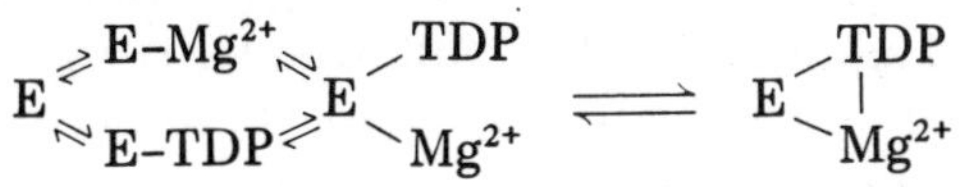

$$E \underset{\rightleftharpoons E\text{-TDP}}{\overset{\rightleftharpoons E\text{-Mg}^{2+}}{}} E \underset{Mg^{2+}}{\overset{TDP}{<}} \rightleftharpoons E \underset{Mg^{2+}}{\overset{TDP}{<}} \qquad (24)$$

(IV)   **(V)**

The conversion of (**IV**) to (**V**) can be blocked by addition of EDTA[244]. However, neither the nature of the slow reaction which converts (**IV**) to (**V**) nor the groups on TDP (if any) which provide ligands to the metal ion have yet been defined. Furthermore, it is not clear whether the mechanism proposed in reaction 24 is applicable to other TDP-dependent enzymes especially those from which these cofactor dissociates readily, *e.g.* glyoxalate carboligase[246,247]; or whether the metal ion plays any further part in the reaction in addition to facilitating the binding of TDP.

VII. ISOMERASES

A. *Pentose isomerases*

Enzymes which catalyze isomerization of such sugars as L- and D-arabinose[252,253], D-xylose[254], D-lyxose[255], and L-rhamnose[256] according to reaction 25 characteristically exhibit maximum activity only in the presence of a divalent metal ion, *e.g.* Mn^{2+} and, in most instances, are inactive in the absence of Me^{2+}:

$$\text{Aldopentose} \xrightleftharpoons{Me^{2+}} \text{Ketopentose} \qquad (25)$$

Many of these enzymes also act less efficiently as hexose isomerases. Isotopic studies for several pentose isomerases have indicated the participation of a *cis*-enediol intermediate in the reaction sequence[257]:

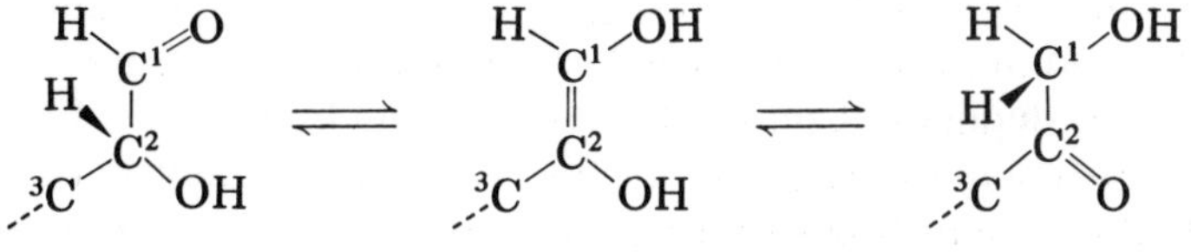

$$\qquad (26)$$

In reaction 26 an electrophilic center, *e.g.* Me^{2+}, would facilitate the reaction by polarizing the carbonyl group of either the aldose or the ketose during formation of the *cis*-enediol intermediate[257].

The mechanism of Me^{2+} activation of this group of enzymes has been investigated most thoroughly for D-xylose isomerase, which is activated by Mn^{2+} but is essentially inactive in the presence of Mg^{2+} [258]. Initial rate studies have suggested that an enzyme–Mn^{2+}–D-xylose bridge complex is formed primarily by an ordered addition of Mn^{2+} followed by D-xylose[259] (cf. Fig. 1, II A). This suggestion of a metal bridge complex has been confirmed by n.m.r. studies[260]. The binary Mn^{2+}-xylose isomerase has an enhanced effect on $1/T_1$ of water protons (ϵ_b = 4), which is reduced by addition of either substrates (D-xylose, D-glucose) or inhibitors (D-xylitol, D-sorbitol) over a concentration range consistent with the dissociation constants obtained in initial rate studies[259]. The enzyme also enhances the effect of Mn^{2+} on the relaxation rates of the C-1 proton of xylose. From the effects observed on $1/T_{1p}$ of this proton, the Mn^{2+} to C-1 proton distance is obtained as 5.3 ± 2.4 Å which is consistent with direct coordination of D-xylose through C-1 to form an enzyme–Mn^{2+}–substrate bridge complex[260]. Presumably either the hydroxyl group and/or the oxygen of the glycosidic ring are involved in coordination although the detailed structure of the complex remains to be defined. A mechanism describing the role of Me^{2+} in this reaction has been suggested on the basis of a structure for the bridge complex in which both these groups provide ligands to the metal ion[8]. It is also of interest that a reactive complex was examined in these studies which were possible since at equilibrium the aldose : ketose ratio is approximately 3:1 (pH 7.0)[258]. The only other reactive species examined thus far by the n.m.r. techniques is the pyruvate carboxylase–Mn^{2+}–oxaloacetate bridge complex[194] (cf. Section V A).

It should also be noted that D-xylose isomerase is reversibly inactivated by addition of chelating agents, e.g. EDTA, 1,10-phenanthroline[261] although there is no evidence for the presence of a bound metal ion in this enzyme.

B. Hexose and pentose phosphate isomerases

In contrast to the pentose isomerases most hexose or pentose phosphate isomerases are fully active in the absence of a divalent metal ion although isotopic studies have in several cases indicated the participation of a cis-enediol intermediate[261a]. Recently phosphomannose isomerase from yeast has been definitely identified as a zinc-metalloenzyme in which enzymic activity is dependent on the presence of the bound metal ion[262]. Although direct participation of Zn^{2+} in the catalytic mechanism has not been established it is tempting to suggest that the role of this metal ion may be similar to that suggested for Mn^{2+} in the reaction catalyzed by D-xylose isomerase[8]. In contrast to the enzyme from yeast phosphomannose isomerase purified from red blood cells is a metal-activated enzyme[262a]. This observation together with the finding that Zn^{2+} accelerates the non-enzymic isomeration of mannose-6-phosphate to fructose-6-phosphate[262b] should assist elucidation of the role of Me^{2+} in the enzymic reaction.

Other hexose and pentose phosphate isomerases, *e.g.* phosphoglucose isomerase[262,263], phosphoribose isomerase[264] do not appear to contain bound metal ion, although the participation of the *cis*-enediol intermediate has been demonstrated for phosphoglucose isomerase[261a]. Thus in these enzymes another electrophilic center may be utilized in place of Me^{2+}.

C. *Other isomerases*

Several isomerases acting on other substrates which exhibit an incomplete requirement for activation by divalent metal ions have also been described, *e.g.* isopentenyl pyrophosphate isomerase[265], *cis, cis*-muconate lactonase[266], oxaloacetate keto-enol tautomerase[267] and mandelate racemase[267a]. No detailed studies have been performed to elucidate the role of the metal ion in these reactions.

VIII. MUTASES AND TRANSFERASES

A. *Phosphoglucomutase and other mutases*

Enzymes which catalyze phosphate transfer between two carbon atoms in hexose or pentose phosphates (mutases) characteristically exhibit a requirement for activation by divalent metal ion as illustrated for phosphoglucomutase in reaction 27[268-270] although other mutases are fully active in the absence of added Me^{2+}, *e.g.* phosphoglycerate mutase[271]:

$$\text{Glucose-1-phosphate} \xrightarrow[\text{Me}^{2+}]{\substack{\text{Glucose-}\\\text{1,6-diphosphate}}} \text{Glucose-6-phosphate} \qquad (27)$$

The reaction pathway used by phosphoglucomutase as defined by initial rate studies is shown in Fig. 5[272]. These studies have also demonstrated that Mg^{2+} can add to either the phospho- or dephospho-enzymes or to any of the central complexes. A random order of addition is observed for addition of Mg^{2+} and glucose-1-phosphate (or glucose-6-phosphate) to the phosphoenzyme or of Mg^{2+} and glucose-1,6-diphosphate to the dephosphoenzyme[273]. However, the rate of dissociation of Mg^{2+} from any of these forms of the enzyme is approximately three orders of magnitude slower than the rate of dissociation of any of the hexose phosphates[274]. Thus the usual consequences of random order behavior are not observed in the kinetic studies since phosphoglucomutase effectively behaves as an enzyme–Mg^{2+} complex under these conditions.

414

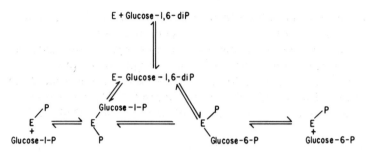

Fig. 5. The reaction pathway utilized by phosphoglucomutase. After Ray and Roscelli[372].

Although earlier studies suggested that phosphoglucomutase was partly active in the absence of added Me^{2+} [275] and that preincubation with Mg^{2+} plus a chelating agent resulted in a time-dependent activation of this enzyme[275] these effects are now known to result from contamination with tightly bound metal ions[274]. After removal of contaminating metal ions the catalytically inactive apoenzyme exhibits a wide specificity of Me^{2+} activation with Mg^{2+} and Ni^{2+} as the most effective species[277]. The effectiveness of activation is correlated with the rate of dissociation of the enzyme–Me^{2+} complex. Mg^{2+}, which is the most effective activator, also dissociates most rapidly from the enzyme[277]. Certain of the less effective Me^{2+}, $e.g.$ Mn^{2+}, Cd^{2+}, Zn^{2+}, interact with the phosphoglucomutase to form binary complexes which behave as metallo-enzymes in contrast to the Mg^{2+}–phosphoglucomutase binary complexes. Some evidence indicates minor differences in catalytic mechanism between the metal-activated and metallo-enzyme species of phosphoglucomutase[277].

Attempts to define the role of Me^{2+} in the reaction catalyzed by phosphoglucomutase by comparison of the properties of the various metallophosphoglucomutases have, however, met with little success. All the metal ions examined induce a similar difference spectrum on interaction with the apoenzyme and this interaction is accompanied by minor changes in enzyme conformation as detected by solvent perturbation[278]. The difference spectrum induced by formation of the central complex (Fig. 5) is modified by addition of metal ions, and the extent of perturbation in the tyrosine region of this difference spectrum is inversely correlated with the effectiveness of activation by various Me^{2+} [279].

Recently the nuclear magnetic relaxation rates of water protons have been measured in the presence of Mn(II)– phosphoglucomutase and of ternary complexes of this enzyme with various substrates[279a]. The studies have demonstrated that coordinated water molecules are displaced from the bound Mn(II) on formation of the ternary complexes with tightly bound substrates, $e.g.$ glucose-phosphate, xylose-phosphate but that no displacement of coordinated water occurs on formation of ternary complexes involving either the non-weakly bound substrates, *e.g.* fructose-phosphate, galactose-phosphate or inorganic phosphate which is a competitive inhibitor of phosphoglucomutase[279b]. These observations suggest a direct or indirect role for the metal ion in orientation of the substrates at the catalytic site and also indicate that this activator does not participate directly in the catalytic process. However, since phosphoglucomutase is inactive in the absence of added Me^{2+} [277], it is apparent that maintenance of the active site conformation which is required for catalytic activity is also dependent on the presence of Me^{2+} [279].

B. *Transferases*

Although many nucleotidyl and pentosyl transferases have been described which require activation by a divalent metal ion (typically Mn^{2+} and/or Mg^{2+})[280], no definitive studies which might provide insight into the role of the metal ion in reactions of this type have been reported.

IX. ALDOLASES

FDP-aldolases, which catalyze reaction 28 may be divided into two general types:

$$FDP \rightleftharpoons dihydroxyacetone\text{-}phosphate + glyceraldehyde\text{-}3\text{-}phosphate \tag{28}$$

Class II FDP-aldolases which are found in fungi, bacteria, yeast and blue-green algae, are markedly inhibited in the presence of chelating agents such as EDTA and 1,10-phenanthroline[280,281]. These effects of chelating agents were interpreted as suggesting the participation of a bound metal ion in catalysis by Class II aldolases[281]. Confirmation of this suggestion was obtained by the demonstration that FDP-aldolase from yeast contains bound zinc in a stoichiometry approaching 2 g atoms/mole enzyme[282,283]. Although most Class II FDP-aldolases behave as metallo-enzymes some enzymes of this type form less stable enzyme–Me^{2+} complexes and hence are isolated as the apoenzyme. Thus FDP-aldolases obtained from *Clostridium perfringens*[284] and *Anacystis nidulans*[285] require activation by added Me^{2+}, *e.g.* Fe^{2+}, Co^{2+}.

In contrast Class I FDP-aldolases which are characteristic of higher plants and animals, protozoa and coelenterates are insensitive to inhibition by chelating agents and do not appear to contain bound metal ion[280,281]. Class I FDP-aldolases exhibit a characteristic substrate-induced inactivation on exposure of the enzyme–dihydroxyacetone–phosphate complex to borohydride[286]. This inactivation is the result of reduction of the Schiff base which results from the condensation of dihydroxyacetone-phosphate with the ϵ-NH$_2$ group of a lysine residue in the active site of the enzyme[287]. In the mechanism utilized by the Class I FDP-aldolases the Schiff base serves as an electrophilic center in place of the metal ion which may fulfill this role in the Class II FDP-aldolases[281,288]. The proposed similarities in mechanism of action are illustrated in Fig. 6 and are supported by the observation that Class I and II FDP-aldolases catalyze similar ^{3}H and ^{18}O exchange reactions with substrate[288]. The mechanistic equivalence between a Schiff base and an enzyme-bound metal ion illustrated in Fig. 6 extends to other types of reactions, notably α-keto acid decarboxylation[28,289,289a]. For FDP-aldolases this comparison requires that Class II enzymes form enzyme–Me^{2+}-substrate bridge complexes.

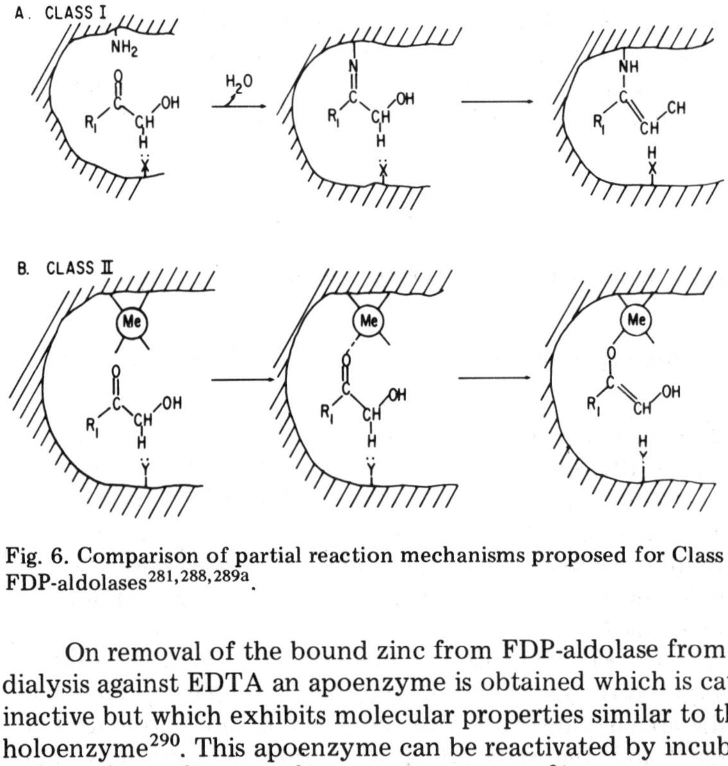

Fig. 6. Comparison of partial reaction mechanisms proposed for Class I and II FDP-aldolases[281,288,289a].

On removal of the bound zinc from FDP-aldolase from yeast by dialysis against EDTA an apoenzyme is obtained which is catalytically inactive but which exhibits molecular properties similar to those of the holoenzyme[290]. This apoenzyme can be reactivated by incubation with Co^{2+}, Fe^{2+}, Mn^{2+} and Ni^{2+}, as well as with Zn^{2+}, resulting in the formation of a series of metallo-aldolases. Marked differences are observed in the maximum rates of catalysis when other Me^{2+} are substituted for the native Zn^{2+}. However, such substitution does not significantly alter either the apparent K_M for FDP or the extent of stimulation by K^+ [291]. These data suggest that the metal ion participates in the rate-limiting step of the reaction but has little, if any, role in the maintenance of the active conformation of the enzyme. The Mn^{2+} and Co^{2+}-aldolases have been employed in n.m.r. studies which have directly demonstrated the formation of the enzyme–Me^{2+}–substrate bridge complexes required by the mechanism proposed in Fig. 6. Although PRR studies do not show the relationship between the enhanced effect of the bound Mn^{2+} in the binary and ternary complexes which is expected for a metal bridge complex (cf. Section II C)[83], the formation of such complexes has been clearly demonstrated by measurement of the relaxation rates of the protons of both FDP and dihydroxy-acetone phosphate. Both the Mn^{2+}–substrate proton distance (from $1/T_{1p}$) and the hyperfine coupling constant (from $1/T_{2p}$ and $1/T_{1p}$) are consistent with direct coordination in the ternary complexes[292]. Furthermore, comparison of the data obtained for the Mn^{2+}–dihydroxyacetone-phosphate and enzyme–Mn^{2+}–dihydroxyacetone-phosphate complexes indicates differences in structure which appear consistent with carbonyl coordination of this substrate[292] in the presence of the enzyme as required by the proposed mechanism (Fig. 6). However, some caution is required in extrapolation of these results to the mechanism of action of the native Zn^{2+}-enzyme since substitution with Mn^{2+} decreases the catalytic efficiency of the enzyme by an order of magnitude[291].

Other Class II-aldolases have also been described. Several of these enzymes appear analogous to FDP-aldolase from yeast since they exhibit a requirement for added metal ions only after prolonged dialysis against EDTA, e.g. fuculose-1-phosphate aldolase[292a], rhamnulose-1-phosphate aldolase[293]. The native metal ion present in these latter enzymes has not yet been identified. However, some other Class II-aldolases which may form less stable binary E–Me^{2+} complexes, as is the case for the FDP-aldolase from C. perfringens, appear to be isolated as the apoenzyme, e.g. 2-keto-3-deoxy-glutarate aldolase[293a], and ketopantoate aldolase[294]. These enzymes require activation by added Me^{2+} but no evidence is presently available which indicates whether or not the mechanism illustrated in Fig. 6 is applicable although it is generally assumed that this is the case[289a]. Additionally some aldolases, e.g. 2-keto-4-hydroxy-4-methylglutarate aldolase[294a], have been classified as Class II enzymes[289a] since the inhibition observed in the presence of EDTA is relieved by addition of Me^{2+}. However, this criterion is not adequate to demonstrate a role for Me^{2+} in catalysis since the relief of the inhibition may be due to removal of EDTA as the metal chelate rather than to reactivation of an apoenzyme.

Several keto-acid lyases, which are closely related to the aldolases, also require activation by added Me^{2+}, e.g. isocitrate lyase[295], β-hydroxy-β-methylglutaryl CoA lyase[296] and citrate lyase[297]. Enzyme–Me^{2+}–substrate bridge complexes have been assumed to participate in catalysis by the keto-acid lyases[295] although there is little evidence to support this contention. Formation of a kinetically significant Mn^{2+}–citrate lyase complex has been demonstrated in which Mn^{2+} has an enhanced effect on the PRR. However no decrease in this enhancement is observed on addition of either the substrate (citrate) or the products (oxaloacetate, acetate) of this reaction[298]. These data are, however, inadequate to exclude the existence of a metal bridge complex. Similar results were obtained in PRR studies on Mn^{2+}-FDP aldolase from yeast but subsequent examination of the effect of this enzyme on the relaxation rates of substrate protons provided definitive evidence for the existence of enzyme-Mn^{2+}-substrate complexes[83,292].

Citrate lyase from *Escherichia coli* and *Aerobacter aerogenes*[299] has also been observed to undergo an unusual reaction-induced inactivation ("suicide") which may be caused by interaction of the enzyme with enol-oxaloacetate in the presence of Me^{2+} [300]. The specificity for Me^{2+} activation of this reaction differs from that observed for the cleavage reaction[300,301]. The basis for this "suicide" effect has not yet been elucidated, but it does not appear to result from formation of a stable enzyme-Me^{2+}-enol-oxaloacetate complex[302].

X. HYDRO AND AMMONIA LYASES*

Since the enzymes of this group have no *apparent* uniformity of mechanism (but see Mildvan[8]), each will be discussed separately.

A. Hydro-lyases

1. Enolase
Initial rate studies on the activation of yeast enolase by divalent ions[303,304] formed the basis for the original proposal that an enzyme-Me^{2+}-substrate bridge complex participates in the reaction catalyzed by this enzyme[3]:

$$\text{3-Phosphoglycerate} \xrightleftharpoons{\text{Me}^{2+}} \text{Phosphoenolpyruvate} + H_2O \qquad (28A)$$

Further support for this postulate has been provided by magnetic resonance studies. Enolase interacts with Mn^{2+} to form a binary complex which has an enhanced effect on the PRR of water ($\epsilon_b = 13.8$). This enhanced effect is reduced on addition of substrates[9,21]. Both this enhancement behavior and the observed inhibition by Ca^{2+} [305] are consistent with the participation of a metal bridge complex in this reaction (cf. Section II C and D). Some mechanistic speculations have been advanced which are based on the participation of such a complex in the catalytic mechanism[8]. However, this interpretation of the kinetic and magnetic resonance studies is rendered less certain by recent data obtained in direct binding and further initial rate experiments[306]. These latter observations appear more consistent with the

---

*A lyase is an enzyme which catalyzes the non-hydrolytic cleavage of carbon–nitrogen, carbon–oxygen or carbon–carbon bonds.

participation of an enzyme bridge complex in this reaction (Section I D). The binding studies reveal that yeast enolase possesses multiple binding sites for Me^{2+}. In the absence of substrate two sites are observed for Mg^{2+} or Mn^{2+} with affinities which differ by an order of magnitude. However, under these conditions up to six sites appear to be available for interaction with Ca^{2+}. These observations cast some doubt on the use of inhibition by Ca^{2+} as an indicator for the participation of a metal bridge complex in the reaction.

In the presence of substrate two additional Mg^{2+} (or Mn^{2+}) sites which exhibit still lower affinities for these metal ions are observed, and the presence of substrate also induces a 3-4-fold increase in affinity for $Mg^{2+}(Mn^{2+})$ at the site which exhibits the lesser affinity for these metal ions in the absence of substrate. The K_d for Mn^{2+} at this latter site agrees most closely with that determined in the PRR studies[9,21] but the failure to observe a tighter site for Mn^{2+} in these experiments is unexplained, especially since the tighter site for Mg^{2+} *is* observed in studies of the perturbation of the ultra-violet absorption and the fluoresence spectrum of enolase induced by this metal ion[308]. The conflicting proposals regarding the role of the divalent metal ion in the enolase reaction are primarily the result of discrepancies between the data available at present. These discrepancies must be resolved before any definitive conclusions can be drawn in regard to the mechanism of the enolase reaction.

2. Aconitase

Since the initial observation that aconitase exhibits an absolute requirement for activation by Fe^{2+} and cysteine[309,310], an enzyme–Fe^{2+}–substrate bridge complex has been repeatedly proposed as an essential feature of the catalytic mechanism of this enzyme[310-313], which catalyzes the reaction:

$$\text{Citrate} \xrightleftharpoons{\pm H_2O} \textit{cis}\text{-aconitate} \xrightleftharpoons{\pm H_2O} \text{isocitrate} \qquad (28B)$$

The elegant "ferrous wheel" mechanism, which is the most recent of these proposals, is based on crystallographic studies of the conformations of citrate and isocitrate together with the recognition that *cis*-aconitate may exist in conformations which resemble citrate or isocitrate respectively[313]. The "ferrous wheel" mechanism provides satisfying explanations for several unusual properties of the aconitase reaction, such as the stereochemistry of intramolecular proton transfer[314,315], the obligatory exchange of the hydroxyl group of the substrates with the solvent[315], and the specificity of inhibition and inactivation by the various isomers of fluorocitrate[316-318].

Although Mn^{2+} does not activate aconitase[310], this metal ion decreases the rate and extent of activation by Fe^{2+} with properties which suggest that the two metals compete for the same binding site(s)[319]. Evidence for the existence of the enzyme–Me^{2+}–substrate bridge complexes required by both the "ferrous wheel" and other proposed mechanisms[310-313] has been obtained by examination of the effect of the Mn^{2+}-aconitase complex on the PRR in the presence and absence of substrates and on the relaxation rates of the methylene protons of citrate. These studies[319] have demonstrated that: (i) an enzyme–Mn^{2+}–citrate complex is formed which possesses properties consistent with its participation in catalysis; (ii) Fe^{2+} appears able to replace Mn^{2+} in this bridge complex; and (iii) citrate and isocitrate compete for the same site(s) on the enzyme. The affinities of citrate and isocitrate for Mn^{2+}-aconitase are similar to the Michaelis constants determined for interaction of these substrates with the native (Fe^{2+}) enzyme. However, in contrast, cis-aconitate interacts much less strongly with the Mn^{2+}-enzyme. This latter observation may explain in part the failure of Mn^{2+}-aconitase to exhibit catalytic activity[319]. Although these data are consistent with the "ferrous wheel" mechanism, a more definitive test will be provided by determination of the structure of the enzyme–Me^{2+}–citrate and –isocitrate bridge complexes. It will also be of interest to determine the basis for the stringent specificity for Me^{2+} activation exhibited by aconitase. Fe^{2+} appears to be the only metal ion which is capable of activating this enzyme[310,319].

L-Citramalate hydrolase and (+) tartarate dehydrase which catalyze reactions analogous to the interconversion of citrate and cis-aconitate by aconitase, also exhibit a specific requirements for activation by Fe^{2+} and a thiol, e.g. mercaptoethanol[320,320a]. However, two protein components are required for catalysis of the citramalate hydrolase reaction[320].

3. Other hydro-lyases

Certain other hydro-lyases (dehydratases) also require activation by added Me^{2+}, e.g. altronate dehydratase[321], dihydroxyacid dehydratase[322], 6-phosphogluconate dehydratase[323], imidazoleglycerolphosphate dehydratase[324], and a preliminary report suggests that δ-aminolevulinate dehydratase from Ustilago sphaerogena may be a copper metallo-enzyme[325] (but see Wilson et al.[326]). Little is known of the role of the metal ion in the reactions catalyzed by these enzymes. However, many other hydro-lyases do not appear to be metal enzymes, e.g. the amino acid dehydratases which require pyridoxal phosphate as cofactor[327] and such enzymes as fumarase and crotonase which have no apparent cofactor requirement[328,329].

B. Ammonia lyases

Several ammonia lyases which catalyze reactions of the type illustrated below have been characterized as metal enzymes:

$$R-CH_2-CH(NH_2)-COOH \overset{Me^{2+}}{\rightleftharpoons} R-CH=CH-COOH + NH_3 \quad (29)$$

The two metal-dependent ammonia lyases which have been most thoroughly studied are β-methylaspartase and histidine deaminase. For both enzymes evidence has been presented which indicates that enzyme–Me^{2+}–substrate bridge complexes participate in the overall reaction[7,330,331], and it has been suggested that the electrophilic character of the divalent metal ion may facilitate deprotonation of the substrate[8,330]. However, several differences are observed between β-methylaspartase and histidine deaminase. First, β-methylaspartase is inactive in the absence of added Me^{2+} [332-334], whereas the rate of catalysis by histidine deaminase is increased only 30–40% by addition of Me^{2+} [335]. The residual catalytic activity observed in the absence of added Me^{2+} may be due to the presence of bound metal ion since EDTA acts as a potent inhibitor of the catalytic activity which is observed in the absence of added Me^{2+} and the extent of activation observed in the presence of added Me^{2+} varies under different conditions and in different preparations of the enzyme[335].

Second, β-methylaspartase catalyzes a Me^{2+}-dependent exchange of protons on the β-carbon of the substrate. The properties of this exchange reaction suggest the participation of a carbanion intermediate in the reaction. This intermediate may be stabilized by the divalent metal ion[336]. The departure of ammonia from this intermediate appears to be the rate-limiting step in the β-methylaspartase reaction[330,336]. In contrast histidine deaminase does not catalyze exchange of protons on the β-carbon of the substrate in the presence or absence of Me^{2+} but is inactivated by incubation with borohydride indicating the presence of an "electrophilic center" which participates in deamination[335,337]. Dissociation of the enzyme–ammonia complex appears to be the rate-limiting step in histidine deaminase reaction[338].

Phenylalanine deaminase, resembles histidine deaminase in several respects, *e.g.* in the nature of the reaction which it catalyzes, in possessing an "electrophilic center" and in failing to catalyze a β-proton exchange reaction. However, this enzyme is fully active in the absence of added Me^{2+} under all conditions examined thus far[339,340], and no evidence has been presented which indicates the presence of bound metal in phenylalanine deaminase although this possibility has not been excluded. These observations for phenylalanine deaminase may be relevant to the proposed involvement of Me^{2+} in the histidine deaminase reaction since some recent observations[340a] have cast doubt both on the participation of Me^{2+} in this reaction and also on the interpretation of the inhibition by EDTA[335].

XI. HYDROLASES

Two major groups of hydrolases (peptidases and phosphatases) are discussed in Chapters 15 and 18 respectively. These two groups include some of the best-characterized examples of metal enzymes, *e.g.* carboxypeptidases, alkaline phosphatase. However, several other metal hydrolases have also been described.

A. *Hydrolases requiring or containing* Ca^{2+}

A number of hydrolases either contain bound Ca^{2+}, *e.g.* α-amylases[341-343], or exhibit an absolute requirement for activation by this metal ion, *e.g.* staphylococcal nuclease[85], some phospholipases[345,346], γ-lactonase[347]. Although the amylases are inactivated by removal of the bound metal[341], the role of Ca^{2+} in these enzymes, with the possible exception of staphylococcal nuclease[85,86], is generally considered to be structural rather than catalytic[8,341] since the Ca^{2+}: enzyme stoichiometry varies widely in α-amylases purified from different sources[347]. Furthermore, direct evidence for a structural (template) role for Ca^{2+} is indicated by the requirement for this metal ion during certain phases of the refolding of reduced taka-amylase[341a], which results in restoration of catalytic activity. However, neither these data nor the theoretical considerations based on the rate of ligand exchange[76,84] are adequate to exclude a direct role for Ca^{2+} in catalysis by these hydrolases. In this regard it may be pertinent to note that many of these enzymes act on macromolecules as their physiological substrates, and in consequence might exhibit requirements for formation of bridge complexes which differ from the requirements observed for metal enzymes acting on small molecule substrates[4]. For example, more than one metal ion might be involved in formation of a bridge complex between a hydrolase and its macromolecular substrate.

B. *Arginase*

Studies on the activation of arginase by Me^{2+} provided the basis for introduction of the metal bridge concept as a possible explanation for the role of metal ions in enzymic reactions[18] (cf. Section I C). Since these early studies, arginase, and also the related metal-activated peptidases (cf. Chapter 15), have received little attention. Recently arginase from yeast has been found to exhibit a requirement for metal ion activation only after treatment with chelating agents[349]. However, attempts to identify the native metal ion in this enzyme by comparison with the properties of reconstituted metallo-arginases have yielded equivocal results[350]. Neither of these studies have provided further insight into the mechanism of activation of arginase by Me^{2+}.

XII. SYNTHETASES

The synthetases may be classified into two classes on the basis of the site of cleavage of the NTP molecule which occurs during the reaction[351].

A. Synthetases catalyzing a pyrophosphate cleavage

These enzymes, e.g. fatty acyl-CoA synthetases, amino-acyl-transfer RNA synthetases, catalyze reactions in which carboxyl group activation is coupled with cleavage of NTP to NMP + $P_2O_7^{4-}$ [351]. Acyl adenylates may participate as enzyme-bound intermediates in many of these reactions[352-354]. A requirement for activation by Me^{2+} has been demonstrated for all synthetases of this type studied thus far. However, the role of the metal ion in catalysis has only been examined in a few cases.

1. Acetyl-coenzyme A(CoA) synthetase

Studies of the metal ion requirements for the various partial reactions catalyzed by acetyl-CoA synthetase has resulted in the proposal of a reaction mechanism (Fig. 7) which incorporates the sites of involvement of the three metal ions required by this enzyme[48]. The requirement for Me_I^{2+} is observed only for reaction A in Fig. 7 and is satisfied by the addition of metal ions such as Mg^{2+}, Mn^{2+}, Ca^{2+}, and Fe^{2+} at concentrations exceeding 1 mM[47]. A similar specificity has been observed for activation of the analogous partial reaction catalyzed by butyryl-CoA synthetase[357]. The observation of activation by Ca^{2+} suggests that an enzyme–ATP–Me^{2+} bridge complex may participate in reaction A and may explain the specific requirement for Me_I^{2+} in this partial reaction[21].

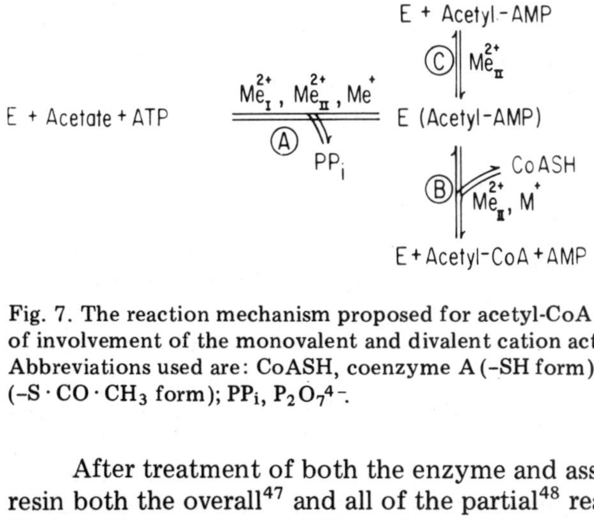

Fig. 7. The reaction mechanism proposed for acetyl-CoA synthetase including the sites of involvement of the monovalent and divalent cation activators. After Webster[359]. Abbreviations used are: CoASH, coenzyme A (–SH form); acetyl CoA, coenzyme A (–S · CO · CH$_3$ form); PP$_i$, $P_2O_7^{4-}$.

After treatment of both the enzyme and assay reagents with a chelating resin both the overall[47] and all of the partial[48] reactions catalyzed by acetyl-CoA synthetase also exhibit an absolute requirement for a second divalent metal ion (Me_{II}^{2+}), which may be satisfied by addition of Fe^{2+}, Cd^{2+}, Cu^{2+}, or Ni^{2+} in approximately stoichiometric ratio with the enzyme (Fig. 7). The enzyme–Me_{II}^{2+} complex is not readily dissociable and contains 2 g atoms Me_{II}^{2+} per mole of enzyme[48]. However, recently preparations of acetyl-CoA

424

synthetase have been obtained which do not exhibit a requirement for Me_{II}^{2+} after treatment with chelating agents. Analysis of these latter preparations after such treatment has failed to reveal the presence of a bound metal[350]. Acetyl-CoA synthetase may, therefore, exist in two forms only one of which exhibits the requirement for Me_{II}^{2+}. It should be noted that other fatty acyl-CoA synthetases have not been reported to exhibit a similar requirement for activation by a second divalent metal ion.

Additionally, acetyl-CoA synthetase exhibits a requirement for activation by monovalent cations[359], which is observed for both the overall reaction and for partial reactions A and B (Fig. 7). However, binding of acyl-adenylate to the enzyme (reaction C, Fig. 7) occurs in the absence of monovalent cations. Many amino-acyl-tRNA synthetases also exhibit a requirement for activation by Me^+ [360].

2. Aminoacyl-tRNA synthetases

These enzymes catalyze reactions which are analogous to that shown for acetyl-CoA synthetase in Fig. 7:

$$E + \text{amino acid} + ATP \xrightleftharpoons{Me^{2+}} E(\text{amino-acyl-AMP}) + P_2O_7^{4-} \quad (31)$$

$$E(\text{amino-acyl-AMP}) + tRNA \rightleftharpoons E + AMP + \text{aminoacyl-}tRNA \quad (32)$$

A requirement for Me^{2+} is observed in all cases for reaction 31[353,361-363] but the specificity of activation varies for different synthetases. Activation by Mg^{2+} and Mn^{2+} is observed in most instances[360]. However, Ca^{2+} activates certain of these enzymes e.g. tyrosyl-tRNA synthetase[364], prolyl-tRNA synthetase[365] but inhibits others, e.g. glycyl-tRNA synthetase[366]. Recent studies on isoleucyl-tRNA synthetase indicate that $MeATP^{2-}$ and $MeP_2O_7^{4-}$ are the complexes which participate in the reaction[366a]. Further investigation is, however, required to clarify the implications of the apparent difference in the specificity for activation by Me^{2+} and the possible mechanistic implications of such differences in specificity.

Although certain aminoacyl-tRNA synthetases also exhibit a requirement for added Me^{2+} for catalysis of reaction 32[367,368], this requirement is only observed after dialysis of tRNA against EDTA[367] and is not generally characteristic of all enzymes of this type[353,363]. Although a role for Me^{2+} in reaction 32, which has been proposed on the basis of other data[368a], cannot be excluded, the apparent requirement may arise from a reversible denaturation of certain tRNA's which occurs on dialysis against EDTA and is due to removal of bound Me^{2+} [368b]. In cases where this apparent Me^{2+} requirement is observed for reaction 32, the concentration requirement and the specificity of activation resemble these determined for reaction 31[367]. Hence this additional apparent metal requirement differs from the Me_{II}^{2+}

requirement observed for certain preparations of acetyl-CoA synthetase[47] (Fig. 7).

B. Synthetases catalyzing an orthophosphate cleavage

Although all enzymes of this type exhibit a requirement for activation by Me^{2+}, the data available, except for the enzymes discussed below, are limited to a demonstration of this requirement and, less frequently, an examination of the specificity of Me^{2+} activation.

1. Formyltetrahydrofolate synthetase

PRR studies[13] have shown that addition of ATP or ADP is required for the observation of specific binding of Mn^{2+} to formyltetrahydrofolate synthetase. Thus an enhanced effect of Mn^{2+} on the PRR is observed only on addition of ATP or ADP to a system containing enzyme + Mn^{2+}. These data, together with the demonstration that Ca^{2+} is an effective activator of the enzyme[369], indicate that the metal ion required for activation of formyltetrahydrofolate synthetase participates in formation of an enzyme-ATP(ADP)-Me^{2+} complex (cf. Section II C). Addition of tetrahydrofolate (or formyltetrahydrofolate) to the enzyme–ATP–Mn^{2+} complex causes a reduction in the enhanced effect of the bound Mn^{2+} due to formation of kinetically significant quaternary complexes. Formate has a similar, although smaller, effect[13]. These data indicate that the other substrates modify the environment of the metal ion in the enzyme–nucleotide–Me^{2+} complex, although direct coordination of these substrates to the bound manganese does not appear to be involved.

2. Biotin carboxylases

The participation and role of bound metal ions in the transfer of CO_2 from the enzyme–biotin–CO_2 intermediate to the carboxyl acceptor (reaction 12) has been discussed in Section V A. The biotin carboxylases, however, also exhibit a requirement for activation by a freely dissociable Me^{2+} which is specifically involved in formation of the enzyme–biotin–CO_2 intermediate from ATP + HCO_3^- (reaction 12)[370]. The role of Me^{2+} in reaction 12 is not well defined at the present time. Formation of a dissociable Mn^{2+}-pyruvate carboxylase complex has been demonstrated. This complex has an enhanced effect of the PRR, but its kinetic significance is not certain since it appears to contain 2-3 Mn^{2+} per biotin residue. Furthermore, addition of the substrates of reaction 12 (ATP and HCO_3^-) causes little, if any, reduction in the enhanced effect of the E–Mn^{2+} complex on the PRR. Since Ca^{2+} is a competitive inhibitor of pyruvate carboxylase with respect to Mg^{2+} [372], these data do not appear consistent with the suggestion that the role of Me^{2+} in reaction 12 is explained by the participation of either an E–ATP–Me^{2+} or an E–Me^{2+}–ATP complex. However, some recent kinetic

426

studies suggest that both Mg^{2+} and $MgATP^{2-}$ may participate in reaction 12 in the case of a pyruvate carboxylase from sheep kidney cortex[373,374]. In this situation interpretation of either the magnetic resonance data or the inhibition by Ca^{2+} would present unusual problems.

3. Glutamine synthetase

The extensive investigations of the reaction mechanism of glutamine synthetase from sheep brain, which have been reviewed by Meister[375], indicate that the divalent metal ion, which is required for activation of this enzyme, participates in formation of the enzyme–ATP complex. This conclusion is based on the properties of the various reactions catalyzed by this enzyme and on the Me^{2+} requirements observed for binding of the substrates and for protection of the enzyme against inactivation. However, since these various reactions differ in their specificity for activation by Me^{2+} [375], the metal ion may also participate in other segments of this reaction[376]. Investigation of glutamine synthetase using techniques which provide information relating to the environment of the metal ion, is however, essential if we are to obtain further insight into the mechanism of activation by Me^{2+}. It should be noted that adenylation of glutamine synthetase from E. coli causes a marked alteration in the specificity for activation by Me^{2+} [376a].

More recently an additional role for Me^{2+} has been demonstrated for the glutamine synthetase obtained from E. coli. This enzyme contains bound Mn^{2+} when isolated in the presence of buffers containing this metal ion[367]. Loss of catalytic activity results from removal of the bound Mn^{2+} by dialysis against EDTA but this inactivation can be reversed by incubation of the metal-free enzyme with Mg^{2+}, Mn^{2+} or Ca^{2+} [377]. Reactivation, is accompanied by incorporation of 1 g atom of bound Me^{2+} per sub-unit[378] and results in formation of a holoenzyme which is similar in most but not all respects to the native enzyme[379]. In addition to lacking catalytic activity metal-free glutamine synthetase also appears to possess a less compact quarternary structure as compared with the native enzyme[380], indicating that the bound Me^{2+} may participate in maintenance of the catalytically active conformation of the enzyme.

XIII. CONCLUSION

In the previous sections of this article a summary has been presented of the status of our knowledge regarding the role of metal ions in the catalytic mechanism of several classes of metal enzymes most of which have been less thoroughly studied than those discussed in subsequent chapters.

It is apparent that in many cases our knowledge is limited to a definition of the specificity for activation by Me^{2+} and, in some instances, of the site of involvement of this cofactor in the reaction mechanism. However, the application of more sophisticated biophysical techniques, *e.g.* n.m.r., e.p.r., X-ray crystallography, has provided a greater insight into the role of the metal ion for a few of these enzymes which have been examined in greater detail. The impact of these and other biophysical techniques on the study of metal enzymes may, however, be better documented for the enzymes examined in Chapters 15, 16 and 18. A descriptive approach has therefore been employed here since the data available for most of the enzymes examined do not as yet permit a satisfying analysis of the catalytic role of the metal ion in the context of coordination chemistry. For a more analytical approach the reader is referred to Chapter 18 and also a recent review by Mildvan[8] (and references therein).

In conclusion several interesting but less well-defined effects of metal ions on enzymes should be briefly noted. Some cases have been described in which the requirement for activation by Me^{2+} appears to be characteristic of the species from which the enzyme was obtained. This phenomenon is best characterized for three enzymes (citrate synthetase[381], phosphotrans-acetylase[382], and serine dehydratase[383]) which exhibit requirements for activation by Me^{2+} only when they are obtained from *Clostridium acidi-urici*. Such a metal ion requirement may reflect a conformational peculiarity of enzymes purified from this micro-organism. The atypical Me^{2+} requirement exhibited under some conditions by an adenylate deaminase from *Porphyra crispata*[384] may have a similar origin.

Enzymes have also been described which utilize metal ions as substrates. The heme synthetases which insert metal ions *e.g.* Fe^{2+}, into the protoporphyrin ring system, are well known[385] although the mechanism of this interesting reaction has not been examined in any detail. However, indirect evidence also exists for preferential (and hence possibly enzymic) *in vivo* incorporation of a specific metal ion into a metallo-enzyme, *e.g.* the incorporation of Mn^{2+} into pyruvate carboxylase in chicken liver[186]. And, finally metal ions may act as modulators of regulatory effects on enzymes, *e.g.* the effect of Me^{2+} on the inhibition of citrate synthetase by ATP[386], although such effects are not well documented at present.

ACKNOWLEDGEMENTS

I am grateful to Dr. Albert Mildvan for making several manuscripts available to me prior to publication, and to Miss Joan Kovach for able editorial assistance.

References pp. 428–437

428

REFERENCES

1 A. L. Lehninger, *Physiol. Rev.* 30 (1950) 393.
2 J. G. McCord and I. Fridovich, *J. Biol. Chem.*, 244 (1969) 6049.
3 B. G. Malmstrom and A. Rosenberg, *Adv. Enzymol.*, 21 (1959) 131.
4 B. L. Vallee, *Adv. Protein Chem.*, 10 (1955) 317.
5 E. L. Smith, *Discuss. Faraday Soc.*, 20 (1955) 264.
6 J. E. Coleman and B. L. Vallee, *J. Biol. Chem.*, 235 (1960) 390; 236 (1961) 2244.
7 G. F. Fields and H. J. Bright, *Biochemistry*, 9 (1970) 3801.
8 A. S. Mildvan, in P. D. Boyer, *The Enzymes*, Vol. 1, 3rd edn., Academic Press, New York, p. 445.
9 M. Cohn and J. S. Leigh, *Nature*, 193 (1962) 1037.
10 W. J. O'Sullivan and M. Cohn, *J. Biol. Chem.*, 241 (1966) 3104.
11 W. J. O'Sullivan, R. Virden and S. Blethen, *Eur. J. Biochem.*, 8 (1969) 562.
12 W. J. O'Sullivan and L. Noda, *J. Biol. Chem.*, 243 (1968) 1424.
13 R. H. Himes and M. Cohn, *J. Biol. Chem.*, 242 (1967) 3628.
14 H. Sternlicht, R. G. Shulman and E. W. Anderson, *J. Chem. Phys.*, 43 (1965) 3123.
15 M. Cohn and T. R. Hughes, *J. Biol. Chem.*, 235 (1962) 176.
16 W. W. Cleland, *A. Rev. Biochem.*, 36 (1967) 77.
17 E. Scarano, G. Geraci and M. Rossi, *Biochemistry*, 6 (1967) 192.
18 L. Hellerman, *Physiol. Rev.*, 17 (1937) 454.
19 L. Hellerman and C. C. Stock, *J. Biol. Chem.*, 125 (1938) 771.
20 E. L. Smith, *Adv. Enzymol.*, 12 (1951) 191.
21 M. Cohn, *Biochemistry*, 2 (1963) 632.
22 A. S. Mildvan, J. S. Leigh and M. Cohn, *Biochemistry*, 6 (1967) 1805.
23 W. N. Lipscomb, J. A. Hartsuck, G. N. Rieke, F. A. Quiocho, *et al.*, *Brookhaven. Symp. Biol.*, 21 (1968) 24.
24 M. Riepe and J. H. Wang, *J. Biol. Chem.*, 243 (1968) 2779.
25 A. S. Mildvan and M. C. Scrutton, unpublished observations.
26 D. H. Busch, in *Reactions of Coordinated Ligands, Adv. Chem. Ser.*, 37 (1963) 1.
27 M. M. Jones, *Ligand Reactivity and Catalysis*, Academic Press, New York, 1968.
28 D. Herbert, *Discuss. Symp. Soc. Exptl. Biol.*, 5 (1951) 52.
29 R. J. P. Williams, *Biol. Revs.*, 28 (1953) 381.
30 E. Bamann, H. Rother and H. Trapmann, *Naturwissenschaften*, 93 (1956) 326.
31 J. H. Wang, *Science*, 161 (1968) 328.
32 J. H. Wang and L. Parker, *Proc. Natn. Acad. Sci. U.S.*, 58 (1967) 2451.
33 J. H. Wang in R. E. Forster, J. T. Edsall, A. B. Otis and F. J. W. Roughton, CO_2: *Chemical, Biochemical and Physiological Aspects*, NASA SP-188, Gov't Printing Office, Washington, D.C., 1969, p. 101.
34 B. L. Vallee and R. J. P. Williams, *Proc. Natn. Acad. Sci. U.S.*, 59 (1968) 498.
35 A. S. Mildvan and M. Cohn, *J. Biol. Chem.*, 240 (1965) 238.
36 M. C. Scrutton, unpublished observations, 1968.
37 M. C. Scrutton and A. S. Mildvan, *Biochemistry*, 7 (1968) 1490.
37a Z. Dorey and H. B. Gray, *J. Am. Chem. Soc.*, 88 (1966) 1394.
37b M. Ciampolini and N. Nardi, *Inorganic. Chem.*, 6 (1967) 445.
38 A. S. Mildvan and M. Cohn, *J. Biol. Chem.*, 241 (1966) 1178.
39 A. S. Mildvan, J. Hunsley and C. H. Suelter, in B. Chance, T. Yonetani and M. Cohn, *Probes for Macromolecular Structure and Function*, Vol. 2, 1970, p. 131.
40 J. F. Kachmar and P. D. Boyer, *J. Biol. Chem.*, 200 (1953) 669.

41 F. C. Happold and R. B. Beechey, *Biochem. Soc. Symp.*, 15 (1958) 52.
42 H. T. Evans and G. J. Sorger, *A. Rev. Plant. Physiol.*, 17 (1966) 47.
43 C. H. Suelter, *Abstr. 158th Meeting ACS, Biol-56* (1969).
44 E. R. Stadtman, B. M. Shapiro, H. S. Kingdon, C. A. Woolfolk and J. S. Hubbard, *Adv. Enzymol. Regulation*, 6 (1968) 257.
45 C. H. Suelter, R. Singleton, F. J. Kayne, S. Arrington, J. Glass and A. S. Mildvan, *Biochemistry*, 5 (1966) 131.
46 F. J. Kayne and C. H. Suelter, *Biochemistry*, 7 (1968) 1678.
47 L. T. Webster, *J. Biol. Chem.*, 240 (1965) 4010.
48 L. T. Webster, *J. Biol. Chem.*, 242 (1967) 1232.
49 J. P. Hummel and W. J. Dryer, *Biochim. Biophys. Acta*, 63 (1962) 530.
50 S. P. Colowick and F. C. Womack, *J. Biol. Chem.*, 244 (1969) 774.
52 S. F. Velick, J. Hayes and J. Harting, *J. Biol. Chem.*, 203 (1953) 527.
53 I. M. Klotz, F. M. Walker and R. B. Pivan, *Biochem. J.*, 68 (1946) 1486.
54 F. Labeyrie and E. Stachiewicz, *Biochim. Biophys. Acta*, 52 (1953) 136.
55 A. S. Mildvan and R. A. Leigh, *Biochim. Biophys. Acta*, 89 (1964) 393.
56 M. C. Scrutton and M. F. Utter, *J. Biol. Chem.*, 240 (1965) 3714.
57 C. H. Suelter and W. Melander, *J. Biol. Chem.*, 238 (1963) PC 4108.
58 P. D. Boyer and H. Theorell, *Acta. Chem. Scand.*, 10 (1956) 447.
59 W. W. Cleland, *Biochim. Biophys. Acta*, 67 (1965) 104.
60 W. W. Cleland, *Biochim. Biophys. Acta*, 67 (1963) 188.
60a H. N. Christensen and G. A. Palmer, *Enzyme Kinetics*, W. B. Saunders, Philadelphia, 1967.
60b K. Dalziel, *Acta. Chem. Scand.*, 11 (1957) 1706.
62 H. Beinert and G. A. Palmer, *Adv. Enzymol.*, 27 (1965) 105.
63 J. Peisach, W. G. Levine and W. E. Blumberg, *J. Biol. Chem.*, 242 (1967) 2847.
64 J. C. M. Tsibris, R. C. Tsai, I. C. Gunsalus, W. H. Orme-Johnson, R. E. Hansen and H. Beinert, *Proc. Natn. Acad. Sci. U.S.*, 59 (1968) 959.
65 W. H. Orme-Johnson, R. E. Hansen, H. Beinert, J. C. M. Tsibris, R. C. Bartholomaus and I. C. Gunsalus, *Proc. Natn. Acad. Sci. U.S.*, 60 (1968) 368.
66 M. Barfield and M. Karplus, *J. Am. Chem. Soc.* 91 (1969) 1.
67 M. Kotani and H. Morimoto, in A. Ehrenberg, B. G. Malmstrom and T. Vanngard, *Magnetic Resonance in Biological Systems*, Pergamon Press, New York, 1967, p. 135.
68 W. E. Blumberg, M. Goldstein, E. Lauber and J. Peisach, *Biochim. Biophys. Acta*, 99 (1965) 187.
69 G. H. Reed and M. Cohn, *J. Biol. Chem.*, 245 (1970) 662.
71 C. L. Hamilton and H. M. McConnell, in A. Rich and N. Davidson, *Structural Chemistry and Molecular Biology*, W. Freeman, San Francisco, 1968, p. 115.
72 J. S. Taylor, J. S. Leigh and M. Cohn, *Proc. Natn. Acad. Sci. U.S.*, 64 (1969) 219.
73 H. Weiner, *Biochemistry*, 8 (1969) 526.
74 M. Burr and D. E. Koshland, *Proc. Natn. Acad. Sci. U.S.*, 52 (1964) 1017.
74a J. S. Leigh, *J. Chem. Phys.*, 52 (1970) 2608.
74b M. Cohn, *Q. Rev. Biophys.*, 3 (1970) 61.
75 A. Kowalsky and M. Cohn, *A. Rev. Biochem.*, 33 (1964) 481.
76 A. S. Mildvan and M. Cohn, *Adv. Enzymol.*, 33 (1970) 1.
76a B. Sheard and E. M. Bradbury, in J. A. V. Butler and D. Noble, *Progress in Biophysics and Molecular Biology*, Pergamon Press, New York, 1970, p. 189.
77 T. J. Swift and R. E. Connick, *J. Chem. Phys.*, 37 (1962) 307.
78 Z. Luz and S. Meiboom, *J. Chem. Phys.*, 40 (1964) 2686.
79 I. Solomon, *Phys. Rev.*, 99 (1955) 559.
80 N. Bloembergen, *J. Chem. Phys.*, 27 (1957) 572.

81 J. Eisinger, R. G. Shulman and B. M. Szymanski, *J. Chem. Phys.*, 36 (1962) 1712.
82 M. Cohn, in A. Ehrenberg, B. G. Malmstrom and T. Vanngard, *Magnetic Resonance in Biological Systems*, Pergamon Press, New York, 1967, p. 101.
83 R. D. Kobes, A. S. Mildvan and W. J. Rutter, *Abstr. 158th ACS Meeting, Biol-58*, (1969).
84 M. Eigen and G. G. Hammes, *Adv. Enzymol.*, 25 (1963) 1.
85 P. Cuatrecasas, S. Fuchs and C. B. Anfinsen, *J. Biol. Chem.*, 242 (1967) 3063.
86 A. Arnone, C. J. Bier, F. A. Cotton, E. E. Hazen, D. C. Richardson and J. C. Richardson, *Proc. Natn. Acad. Sci. U.S.*, 64 (1969) 420.
88 J. E. Coleman, *Biochemistry*, 4 (1965) 2644.
89 M. L. Applebury and J. E. Coleman, *J. Biol. Chem.*, 244 (1969) 709.
90 D. E. Drum, *Federation Proc.*, 29 (1970) 608.
91 D. E. Drum, T. K. Li and B. L. Vallee, *Biochemistry*, 8 (1969) 3792.
92 A. Curdel and M. Iwatsubo, *FEBS Lett.*, 1 (1968) 133.
93 M. I. Harris and J. E. Coleman, *J. Biol. Chem.* 243 (1968) 5063.
93a Z. Markus, G. Miller and G. Avigad, *Appl. Microbiol.*, 13 (1965) 686.
94 M. L. Applebury, B. L. Johnson and J. E. Coleman, *J. Biol. Chem.*, 245 (1970) 4968.
95 B. L. Vallee and F. L. Hoch, *Proc. Natn. Acad. Sci. U.S.*, 41 (1955) 327.
96 H. Theorell, A. P. Nygaard and R. Bonnichsen, *Acta. Chem. Scand.*, 9 (1955) 1148.
96a J. P. von Wartburg, J. C. Bethune and B. L. Vallee, *Biochemistry*, 3 (1964) 1175.
97 B. L. Vallee and F. L. Hoch, *J. Biol. Chem.*, 225 (1959) 245.
98 J. H. Harrison, *Federation Proc.*, 22 (1963) 493.
99 G. Pfleider and E. Hohnholz, *Biochem. Z.*, 331 (1959) 245.
100 S. J. Adelstein and B. L. Vallee, *J. Biol. Chem.*, 233 (1958) 589.
101 B. L. Vallee and W. E. C. Wacker, *J. Am. Chem. Soc.*, 78 (1956) 1771.
102 H. Terayama and C. S. Vestling, *Biochim. Biophys. Acta*, 20 (1956) 586.
103 G. Pfleiderer, D. Jackel and T. Wieland, *Biochem. Z.*, 330 (1958) 296.
104 D. E. Burkman, H. H. Sandstead and J. H. Park, *J. Biol. Chem.*, 245 (1970) 1036.
105 B. S. Vanderheiden, J. O. Meinhaet, R. G. Dodson and E. G. Krebs, *J. Biol. Chem.*, 237 (1962) 2095.
106 R. H. Yue, E. A. Noltmann and S. A. Kuby, *J. Biol. Chem.*, 244 (1969) 1353.
106a D. S. Foster and R. F. Colman, *J. Biol. Chem.*, 245 (1970) 6190.
107 A. Akeson, *Biochem. Biophys. Res. Commun.*, 17 (1967) 552.
108 H. L. Oppenheimer, R. W. Green and R. H. McKay, *Arch. Biochem. Biophys.*, 119 (1967) 552.
109 H. Theorell and R. Bonnichsen, *Acta. Chem. Scand.*, 5 (1951) 1105.
110 A. Ehrenberg and K. Dalziel, *Acta. Chem. Scand.*, 12 (1958) 465.
111 J. E. Hayes and S. F. Velick, *J. Biol. Chem.*, 207 (1954) 225.
113 N. Sandler and R. H. McKay, *Biochem. Biophys. Res. Commun.*, 35 (1969) 151.
114 M. C. Scrutton, M. R. Young and M. F. Utter, *J. Biol. Chem.*, 245 (1970) 6227.
115 K. Kobashi and B. L. Horecker, *Arch. Biochem. Biophys.*, 121 (1969) 178.
116 B. M. Pogell and M. C. Scrutton, (1969) unpublished observations.
117 R. F. Shalaby and M. A. Lauffer, *Biochemistry*, 6 (1967) 2465.
118 B. L. Vallee, *Proc. 4th Int. Congr. Biochem.*, Vienna, 8 (1969) 138.
119 A. O. M. Stoppani, M. N. Schwarez and C. A. Freda, *Arch. Biochem. Biophys.*, 113 (1966) 464.
120 E. E. C. Lin and B. Magasanik, *J. Biol. Chem.*, 235 (1960) 1820.
121 B. L. Vallee, in P. D. Boyer, H. A. Lardy and K. Myrback, *The Enzymes*, Vol. 3, 2nd edn, Academic Press, New York, 1960, p. 225.
122 A. S. Mildvan and H. Weiner, *J. Biol. Chem.*, 244 (1969) 2465.
123 H. Theorell and T. Yonetani, *Biochem. Z.*, 338 (1963) 537.
124 E. M. Kosower, *Biochim. Biophys. Acta*, 56 (1962) 474.

125 H. R. Mahler and J. Douglas, *J. Am. Chem. Soc.*, 79 (1957) 1159.

126 N. Evans and B. R. Rabin, *Eur. J. Biochem.*, 4 (1968) 548.

127 K. Wallenfels and H. Sund, *Biochem. Z.*, 329 (1957) 59.

128 G. Palmer and H. Brintzinger, in M. Klingenberg and T. E. King, *A Treatise on Electron and Coupled Energy Transfer in Biological Systems*, 1970, in press.

129 R. J. P. Williams and B. L. Vallee, *Discuss. Faraday Soc.*, 20 (1955) 262.

130 F. L. Hoch, R. J. P. Williams and B. L. Vallee, *J. Biol. Chem.*, 232 (1968) 453.

131 B. L. Vallee, R. J. P. Williams and F. L. Hoch, *J. Biol. Chem.*, 234 (1959) 2621.

132 K. Dalziel, *Nature*, 197 (1963) 462.

133 H. Theorell and B. Chance, *Acta. Chem. Scand.*, 5 (1951) 1127.

134 C. C. Wratten and W. W. Cleland, *Biochemistry*, 2 (1963) 935.

135 E. Silverstein and P. D. Boyer, *J. Biol. Chem.*, 239 (1964) 3908.

136 B. M. Anderson, M. L. Reynolds and C. D. Anderson, *Biochim. Biophys. Acta.*, 113 (1966) 235.

137 B. M. Anderson and M. L. Reynolds, *Arch. Biochem. Biophys.*, 111 (1965) 1.

138 T. Yonetani and H. Theorell, *Arch. Biochem. Biophys.*, 106 (1964) 243.

139 C. W. Hoagstrom, I. Iweibo and H. Weiner, *J. Biol. Chem.*, 244 (1969) 5967.

140 P. L. Coleman and H. Weiner, *Federation Proc.*, 29 (1970) 868.

141 A. S. Mildvan and H. Weiner, *Biochemistry*, 8 (1969) 552.

143 T. R. Stengle and J. D. Baldeschwieler, *Proc. Natn. Acad. Sci. U.S.*, 55 (1966) 1020.

144 G. L. Cottam and R. L. Ward, *Arch. Biochem. Biophys.*, 132 (1969) 308.

146 H. B. LeJohn, *J. Biol. Chem.*, 243 (1968) 5126.

147 H. B. LeJohn, S. G. Jackson, G. R. Klassen and R. G. Sawula, *J. Biol. Chem.*, 244 (1969) 5346.

148 G. Powell, K. V. Rajagopalan and P. Handler, *J. Biol. Chem.*, 244 (1969) 4793.

149 S. Black, *Arch. Biochem. Biophys.*, 34 (1951) 86.

150 C. Arsenis and O. Touster, *J. Biol. Chem.*, 244 (1965) 3895.

151 K. Kersters, W. A. Wood and J. DeLey, *J. Biol. Chem.*, 240 (1965) 965.

152 W. B. Jakoby and J. Fredericks, *Biochim. Biophys. Acta*, 48 (1961) 26.

153 R. P. Mortlock, D. D. Fossitt and W. A. Wood, *Proc. Natn. Acad. Sci. U.S.*, 54 (1965) 572.

154 A. H. Mehler, A. Kornberg, S. Grisolia and S. Ochoa, *J. Biol. Chem.*, 174 (1948) 961.

155 W. J. Rutter and H. A. Lardy, *J. Biol. Chem.*, 233 (1958) 374.

156 S. Korkes, A. del Campillo and S. Ochoa, *J. Biol. Chem.*, 187 (1950) 891.

157 M. Hayaishi, M. Hayaishi and T. Unemoto, *Biochim. Biophys. Acta*, 122 (1966) 374.

158 P. T. Magee and E. E. Snell, *Biochemistry*, 5 (1966) 409.

159 R. O. Burns, H. E. Umbarger and S. R. Gross, *Biochemistry*, 2 (1963) 1053.

160 E. Adler, J. Von Euler, G. Gunther and M. Plass, *Biochem. J.*, 33 (1939) 1028.

161 S. Ochoa, *J. Biol. Chem.*, 174 (1948) 133.

162 A. Kornberg and W. E. Pricer, *J. Biol. Chem.*, 189 (1951) 123.

163 G. W. E. Plaut and S. C. Sung, *J. Biol. Chem.*, 207 (1954) 305.

164 G. W. E. Plaut, in P. D. Boyer, H. A. Lardy and K. Myrback, *The Enzymes*, Vol. 3, 2nd edn, Academic Press, New York, 1963. p. 105.

165 B. Rowley and A. F. Tucci, *Federation. Proc.*, 29 (1970) 922.

166 S. Pontromeli, A. DeFlora, E. Grazi, G. Mangiarotti, A. Bonsignore and B. L. Horecker, *J. Biol. Chem.*, 236 (1961) 2975.

167 L. D. Kohn and W. B. Jakoby, *J. Biol. Chem.*, 243 (1968) 2486.

168 R. Y. Hsu and H. A. Lardy, *J. Biol. Chem.*, 242 (1967) 527.

169 R. Y. Hsu, H. A. Lardy and W. W. Cleland, *J. Biol. Chem.*, 242 (1967) 5315.

170 R. F. Chen, D. M. Brown and G. W. E. Plaut, *Biochemistry*, 3 (1963) 552.

432

171 S. Ocha and E. Weisz-Tabori, *J. Biol. Chem.*, 174 (1948) 123.
172 G. Siebert, M. Carsiotis and G. W. E. Plaut, *J. Biol. Chem.*, 226 (1957) 977.
173 J. Spina and J. J. Bright, *Biochemistry*, 9 (1970) 3794.
173a R. G. Duggleby and D. T. Dennis, *J. Biol. Chem.*, 245 (1970) 3745.
174 D. B. Northrop and W. W. Cleland, *Federation. Proc.*, 29 (1970) 408.
175 R. Parvin, S. V. Pande and T. Venditasubramanian, *Biochim. Biophys. Acta,* 92 (1964) 260.
176 P. D. Boyer, *A. Rev. Biochem.*, 29 (1960) 17.
177 R. Steinberger and F. H. Westheimer, *J. Am. Chem. Soc.*, 73 (1961) 429.
178 M. C. Scrutton and A. S. Mildvan, (1969) unpublished observations.
178a M. C. Scrutton, J. Fleming and G. W. E. Plaut, (1970) unpublished observations.
179 S. M. Arfin and H. E. Umbarger, *J. Biol. Chem.*, 244 (1969) 1118.
180 M. C. Scrutton, D. B. Keech and M. F. Utter, *J. Biol. Chem.* 240 (1965) 574.
181 Y. Kaziro, E. Leone and S. Ochoa, *Proc. Natn. Acad. Sci. U.S.*, 46 (1950) 1319.
182 H. G. Wood, H. Lochmuller, C. Riepertinger and F. Lynen, *Biochem. Z.*, 337 (1963) 247.
183 J. R. Stern, *Biochemistry*, 6 (1967) 3545.
184 J. H. Galivan and S. H. G. Allen, *J. Biol. Chem.*, 243 (1968) 1253.
185 M. C. Scrutton, M. F. Utter and A. S. Mildvan, *J. Biol. Chem.*, 241 (1966) 3480.
186 P. Griminger and M. C. Scrutton, *Federation Proc.*, 29 (1970) 765.
187 M. C. Scrutton, P. Griminger and J. C. Wallace, *J. Biol. Chem.*, 247 (1972) 3305.
188 W. R. McClure, *Federation Proc.*, 28 (1969) 728.
189 M. C. Scrutton and M. R. Young, *Federation Proc.*, 29 (1970) 597.
190 D. B. Northrop and H. G. Wood, *J. Biol. Chem.*, 244 (1969) 5801.
191 G. W. E. Plaut and H. A. Lardy, *J. Biol. Chem.*, 180 (1949) 13.
193 A. Kornberg, S. Ochoa and A. H. Mehler, *J. Biol. Chem.*, 174 (1948) 159.
194 M. C. Scrutton and A. S. Mildvan, *Arch. Biochem. Biophys.*, 140 (1970) 131.
195 G. W. Kosicki, *Biochemistry*, 7 (1968) 4299.
196 G. W. Kosicki, *Biochemistry*, 7 (1968) 4310.
197 G. W. Kosicki and F. H. Westheimer, *Biochemistry*, 7 (1968) 4303.
198 M. Stiles, *Ann. N.Y. Acad. Sci.*, 88 (1960) 332.
199 M. Caplow and M. Yager, *J. Am. Chem. Soc.*, 89 (1967) 4513.
200 M. Caplow, *J. Am. Chem. Soc.*, 87 (1965) 5774.
201 A. S. Mildvan, M. C. Scrutton and M. F. Utter, *J. Biol. Chem.*, 241 (1966) 3488.
202 H. Sigel, D. B. McCormick, R. Griesser, B. Priiys and L. D. Wright, *Biochemistry*, 8 (1969) 2687.
203 M. C. Scrutton and A. S. Mildvan, in R. E. Forster, J. T. Edsall, A. B. Otis and F. J. W. Roughton, *CO₂: Chemical, Biochemical and Physiological Aspects NASA SP-188*, Gov't Printing Office, Washington, D.C. 1969, p. 207.
204 M. F. Utter and M. C. Scrutton, in B. L. Horecker and E. R. Stadtman, *Current Topics in Cellular Regulation*, Vol. 1, Academic Press, New York, 1969, p. 253.
205 A. S. Mildvan and M. C. Scrutton, *Biochemistry*, 6 (1967) 2978.
206 M. Eigen and K. Tamm, *Z. Elektrochem.*, 66 (1962) 107.
207 A. S. Mildvan and M. C. Scrutton, (1970) unpublished observations.
208 I. A. Rose, *J. Biol. Chem.*, 245 (1970) 6052.
209 J. Knappe, B. Wenger and U. Wiegand, *Biochem. Z.*, 337 (1963) 232.
210 T. C. Bruice and A. F. Hegarty, *Proc. Natn. Acad. Sci. U.S.*, 65 (1970) 805.
210a D. B. Northrop, *J. Biol. Chem.*, 244 (1969) 5808.
211 P. A. Srere, *Biochem. Biophys. Res. Commun.*, 26 (1967) 609.
212 R. G. Graham and M. C. Scrutton, (1969) unpublished observations.
213 D. J. Prescott and J. L. Rabinowitz, *J. Biol. Chem.*, 243 (1968) 1551.
214 E. Stoll and M. D. Lane, personal communication.
215 M. F. Utter and K. Kurahashi, *J. Biol. Chem.*, 207 (1954) 787.

216 H. Lochmuller, H. G. Wood and J. J. Davis, *J. Biol. Chem.*, 241 (1966) 5678.
217 R. S. Bandurski and C. M. Greiner, *J. Biol. Chem.*, 204 (1953) 781.
218 T. G. Cooper, T. T. Tchen, H. G. Wood and C. R. Benedict, *J. Biol. Chem.*, 243 (1968) 3857.
219 H. Maruyama, R. L. Easterday, H. C. Chang and M. D. Lane, *J. Biol. Chem.*, 241 (1966) 2405.
220 A. Tietz and S. Ochoa, *Arch. Biochem. Biophys.*, 78 (1958) 477.
221 J. J. Davis, J. M. Willard and H. G. Wood, *Biochemistry*, 8 (1969) 3127.
222 R. S. Miller, A. S. Mildvan, H. C. Chang, R. L. Easterday, H. Maruyama and M. D. Lane, *J. Biol. Chem.*, 243 (1968) 6030.
223 A. M. Reynard, L. F. Hass, D. D. Jacobson and P. D. Boyer, *J. Biol. Chem.*, 236 (1961) 2277.
224 R. S. Miller and M. D. Lane, *J. Biol. Chem.*, 241 (1968) 6041.
225 K. Kirsch, A. S. Mildvan and A. Kowalsky, *Abstr. 158th ACS Mtg. Biol. -52*, (1969).
226 J. M. Willard, J. J. Davis and H. G. Wood, *Biochemistry*, 8 (1969) 3137.
227 H. G. Wood, J. J. Davis and J. M. Willard, *Biochemistry*, 8 (1969) 3145.
228 M. C. Scrutton, J. M. Willard and H. G. Wood, (1970) unpublished observations.
229 A. Weissbach, B. L. Horecker and J. J. Hurwitz, *J. Biol. Chem.*, 218 (1956) 795.
230 G. D. Kuehn and B. A. McFadden, *Biochemistry*, 8 (1969) 2394.
231 L. E. Anderson, G. B. Price and R. C. Fuller, *Science*, 161 (1968) 182.
232 M. Calvin, *Federation Proc.*, 13 (1954) 697.
233 M. Wishnick, M. D. Lane and M. C. Scrutton, *J. Biol. Chem.*, 245 (1970) 4939.
234 M. Wishnick and M. D. Lane, *J. Biol. Chem.*, 244 (1969) 55.
235 G. Akoyunoglou and M. Calvin, *Biochem. Z.*, 338 (1963) 20.
236 M. Wishnick, M. D. Lane, M. C. Scrutton and A. S. Mildvan, *J. Biol. Chem.*, 244 (1969) 5761.
237 I. Lieberman, A. Kornberg and E. S. Sims, *J. Biol. Chem.*, 215 (1955) 403.
238 E. E. Snell, A. A. Smucker, E. Ringlemann and F. Lynen, *Biochem. Z.*, 341 (1964) 109.
239 M. C. Scrutton in J. R. Norris and D. W. Ribbons, *Methods in Microbiology*, Vol. 6A, Academic Press, New York, 1971, p. 479.
240 L. O. Krampitz, *A. Rev. Biochem.*, 38 (1969) 213.
241 B. L. Horecker, P. Z. Smyrniotis and H. Klenow, *J. Biol. Chem.*, 205 (1963) 661.
242 C. F. Gunsalus, I. Y. Stanier and I. C. Gunsalus, *J. Bact.*, 66 (1953) 548.
243 T. P. Singer and J. Penksy, *J. Biol. Chem.*, 196 (1962) 375.
243a Y. Namba, K. Yoshigawa, A. Ejima, T. Hayaishi and T. Kaneda, *J. Biol. Chem.*, 244 (1969) 4437.
244 A. V. Morey and E. Juni, *J. Biol. Chem.*, 243 (1968) 3009.
245 S. M. Hussain Qadri and D. S. Hoare, *Biochim. Biophys. Acta*, 148 (1967) 304.
246 G. Krakow, S. S. Barkulis and J. Hayaishi, *J. Bact.*, 81 (1961) 509.
247 H. L. Kornberg and A. M. Gotto, *Biochem. J.*, 78 (1961) 69.
247a J. R. Quayle, *Biochem J.*, 89 (1963) 492.
248 A. Schellenberger, *Angew. Chem., (Int. Edn.)*, 6 (1967) 1024.
249 D. E. Green, D. Herbert and V. Subrahmanyan, *J. Biol. Chem.*, 138 (1941) 327.
250 A. Schellenberger, K. Winter, G. Hubner, R. Schwaiberger, *et al.*, *Hoppe-Seylers Z. Physiol. Chem.*, 346 (1966) 123.
251 A. Schellenberger and G. Hubner, *Hoppe-Seylers Z. Physiol. Chem.*, 348 (1967) 491.
252 R. P. Mortlock and W. A. Wood, *J. Bact.*, 88 (1964) 835.
253 R. P. Mortlock and W. A. Wood, *J. Bact.*, 88 (1964) 838.
254 K. Yamanaka, *Agric. Biol. Chem. (Japan)*, 27 (1963) 271.
255 R. L. Anderson and D. P. Allison, *J. Biol. Chem.*, 240 (1965) 2367.
256 G. F. Domagk and R. Zech, *Biochem. Z.*, 339 (1963) 145.

434

257 I. A. Rose, E. L. O'Connell and R. P. Mortlock, *Biochim. Biophys. Acta*, 178 (1969) 376.
258 K. Yamanaka, *Biochim. Biophys. Acta*, 151 (1968) 670.
259 K. Yamanaka, *Arch. Biochem. Biophys.*, 131 (1969) 502.
260 A. S. Mildvan and I. A. Rose, *Federation Proc.*, 28 (1969) 534.
261 K. Yamanaka in S. P. Colowick and N. O. Kaplan, *Methods in Enzymology*, Vol. 9, Academic Press, New York, 1966, p. 588.
261a I. A. Rose, *Brookhaven Symp. Biol.*, 15 (1962) 293.
262 R. W. Gracy and E. A. Noltmann, *J. Biol. Chem.*, 243 (1968) 4109.
262a F. H. Bruns and E. A. Noltmann, *Nature*, 181 (1958) 1467.
262b R. W. Gracy and E. A. Noltmann, *Federation Proc.*, 28 (1968) 520.
263 M. C. Scrutton, (1969) unpublished observations.
264 A. Rutner and M. C. Scrutton, (1969) unpublished observations.
265 D. H. Shah, W. W. Cleland and J. W. Porter, *J. Biol. Chem.*, 240 (1965) 1946.
266 W. R. Sistrom and R. Y. Stanier, *J. Biol. Chem.*, 210 (1954) 821.
267 R. G. Annett and G. W. Kosicki, *J. Biol. Chem.*, 244 (1969) 2059.
267a H. Weil-Malherbe, *Biochem. J.*, 101 (1966) 169.
268 C. F. Cori, S. P. Colowick and G. T. Cori, *J. Biol. Chem.*, 121 (1937) 465.
269 J. L. Ressig, *J. Biol. Chem.*, 219 (1956) 753.
270 K. Lang and K. V. Hartmann, *Experientia*, 14 (1958) 130.
271 K. V. Rodwell, J. C. Towne and S. Grisolia, *J. Biol. Chem.*, 228 (1957) 875.
272 W. J. Ray and G. A. Roscelli, *J. Biol. Chem.*, 239 (1964) 1228.
273 W. J. Ray, G. A. Roscelli and D. S. Kirkpatrick, *J. Biol. Chem.*, 241 (1966) 2603.
274 W. J. Ray and G. A. Roscelli, *J. Biol. Chem.*, 241 (1966) 3499.
275 V. A. Najjar, *J. Biol. Chem.*, 175 (1948) 281.
276 J. P. Robinson and V. A. Najjar, *Biochem. Biophys. Res. Commun.*, 3 (1960) 62.
277 W. J. Ray, *J. Biol. Chem.*, 244 (1969) 3740.
278 E. J. Peck and W. J. Ray, *J. Biol. Chem.*, 244 (1969) 3748.
279 E. J. Peck and W. J. Ray, *J. Biol. Chem.*, 244 (1969) 3754.
279a W. J. Ray and A. S. Mildvan, *Biochemistry*, 9 (1970) 3886.
279b E. J. Peck, D. S. Kirkpatrick and W. J. Ray, *Biochemistry*, 7 (1968) 152.
280 J. Imsande and P. Handler, in P. D. Boyer, H. A. Lardy and K. Myrback, *The Enzymes*, Vol. 5, 2nd edn., Academic Press, New York, 1961, p. 281.
280a O. Warburg and W. Christian, *Biochem. Z.*, 314 (1943) 419.
281 W. J. Rutter, *Federation Proc.*, 23 (1964) 1248.
282 V. Jagannathan, K. Singh and M. Daniodavan, *Biochem. J.*, 63 (1956) 94.
283 O. C. Richards and W. J. Rutter, *J. Biol. Chem.*, 236 (1961) 3177.
284 R. C. Bard and I. C. Gunsalus, *J. Bact.*, 59 (1950) 387.
285 C. A. Fewson, M. Al-Hafidh and M. Gibbs, *Pl. Physiol.*, 37 (1962) 402.
286 E. Grazi, T. Cheng and B. L. Horecker, *Biochem. Biophys. Res. Commun.*, 7 (1962) 250.
287 E. Grazi, P. T. Rowley, T. Cheng, O. Tchola and B. L. Horecker, *Biochem. Biophys. Res. Commun.*, 9 (1962) 38.
288 D. E. Morse and B. L. Horecker, *Adv. Enzymol.*, 31 (1968) 125.
289 S. Warren, B. Zerner and F. H. Westheimer, *Biochemistry*, 5 (1966) 817.
289a W. J. Rutter, in V. Bryson and H. J. Vogel, *Evolving Genes and Proteins*, Academic Press, New York, 1965, p. 279.
290 C. E. Harris, R. D. Kobes, D. C. Teller and W. J. Rutter, *Biochemistry*, 8 (1969) 2442.
291 R. D. Kobes, R. T. Simpson, B. L. Vallee and W. J. Rutter, *Biochemistry*, 8 (1969) 508.
292 A. S. Mildvan, R. D. Kobes and W. J. Rutter, *Biochemistry*, 10 (1971) 1191.
292a E. C. Heath and M. A. Ghalambor, *J. Biol. Chem.*, 237 (1962) 2427.

293 T. H. Chiu, C. J. Smith and D. S. Feingold, *Federation Proc.*, 27 (1968) 520.
293a D. C. Fish and H. J. Blumenthal, *Bact. Proc.* (1963) 110.
294 E. N. McIntosh, M. Purko and W. A. Wood, *J. Biol. Chem.*, 228 (1957) 499.
294a L. M. Shannon and A. Marcus, *J. Biol. Chem.*, 237 (1962) 3342.
295 J. A. Olson, *J. Biol. Chem.*, 234 (1959) 5.
296 L. D. Stegink and M. J. Coon, *J. Biol. Chem.*, 243 (1968) 5272.
297 S. Dagley and E. A. Dawes, *Biochim. Biophys. Acta*, 17 (1955) 177.
298 R. L. Ward and P. A. Srere, *Biochim. Biophys. Acta*, 99 (1965) 270.
299 T. J. Bowen and L. J. Rogers, *Biochim. Biophys. Acta*, 77 (1963) 685.
300 R. Eisenthal, S. S. Tate and S. P. Datta, *Biochim. Biophys. Acta*, 128 (1966) 155.
301 J. McD. Blair, S. P. Datta and S. S. Tate, *Eur. J. Biochem.*, 1 (1967) 26.
302 M. Singh and P. A. Srere, *Federation Proc.*, 29 (1970) 929.
303 B. G. Malmstrom, T. Vanngard and M. Larsson, *Biochim. Biophys. Acta*, 30 (1958) 1.
304 F. Wold and C. E. Ballou, *J. Biol. Chem.*, 227 (1957) 313.
305 B. G. Malmstrom, *Arch. Biochem. Biophys.*, 58 (1955) 398.
306 D. P. Hanlon and E. W. Westhead, *Biochemistry*, 8 (1969) 4247 and 4255.
308 J. M. Brewer and G. Weber, *J. Biol. Chem.*, 241 (1966) 2550.
309 S. R. Dickman and A. A. Cloutier, *J. Biol. Chem.*, 188 (1951) 379.
310 J. F. Morrison, *Biochem. J.*, 58 (1954) 685.
311 J. F. Speyer and S. R. Dickman, *J. Biol. Chem.*, 220 (1956) 193.
312 G. L. Eichhorn, in *Reactions of Coordinated Ligands, Am. Chem. Soc. Adv. Chem. Ser.*, 37 (1963) 37.
313 J. P. Glusker, *J. Mol. Biol.*, 38 (1968) 149.
314 O. Gawron, A. J. Glaid and T. P. Fondy, *J. Am. Chem. Soc.*, 83 (1961) 3634.
315 I. A. Rose and E. L. O'Connell, *J. Biol. Chem.*, 242 (1967) 1870.
316 D. W. Fanshier, L. K. Gottwald and E. Kun, *J. Biol. Chem.*, 239 (1964) 425.
317 E. Kun, in J. M. Lowenstein, *The Citric Acid Cycle: Control and Compartmentation*, Marcel Dekker, New York, 1969, p. 318.
318 H. L. Carrell, J. P. Glusker, J. J. Villafranca, A. S. Mildvan, R. J. Dummel and E. Kun, *Science*, 170 (1970) 1412.
319 J. J. Villafranca and A. S. Mildvan, *J. Biol. Chem.*, 246 (1971) 772.
320 C. C. Wang and H. A. Barker, *J. Biol. Chem.*, 244 (1969) 2516.
320a R. E. Hurlbert and W. B. Jakoby, *J. Biol. Chem.*, 240 (1965) 2774.
321 J. D. Smiley and G. Ashwell, *J. Biol. Chem.*, 235 (1960) 1571.
322 R. L. Wixom, J. B. Shatton and M. Strassman, *J. Biol. Chem.*, 235 (1960) 128.
323 N. J. Palleroni and M. Doudoroff, *J. Biol. Chem.*, 223 (1956) 499.
324 B. N. Ames, *J. Biol. Chem.*, 228 (1957) 131.
325 H. Komai and J. P. Neilands, *Biochim. Biophys. Acta*, 171 (1969) 13.
326 M. L. Wilson, A. I. Iodice, M. P. Shulman and D. A. Richert, *Federation Proc.*, 18 (1959) 352.
327 D. M. Greenberg, in P. D. Boyer, H. A. Lardy and K. Myrback, *The Enzymes*, Vol. 5, 2nd edn., Academic Press, New York, 1961, p. 563.
328 R. A. Alberty, in P. D. Boyer, H. A. Lardy and K. Myrback, *The Enzymes*, Vol. 5, 2nd edn., Academic Press, New York, 1961, p. 531.
329 J. R. Stern, in P. D. Boyer, H. A. Lardy and K. Myrback, *The Enzymes*, Vol. 5, 2nd edn., Academic Press, New York, 1961, p. 511.
330 H. J. Bright, *J. Biol. Chem.*, 240 (1965) 1198.
331 I. L. Givot, A. S. Mildvan and R. H. Abeles, *Federation Proc.*, 29 (1970) 531.
332 H. A. Barker, R. D. Smyth, R. M. Wilson and H. Weissbach, *J. Biol. Chem.*, 234 (1959) 320.
333 H. J. Bright and L. L. Ingraham, *Biochim. Biophys. Acta*, 44 (1960) 586.
334 V. R. Williams and J. Selbin, *J. Biol. Chem.*, 239 (1964) 1635.

436

335   I. L. Givot, T. A. Smith and R. H. Abeles, *J. Biol. Chem.*, 244 (1969) 6341.
336   H. J. Bright, *J. Biol. Chem.*, 239 (1964) 2307.
337   M. M. Rechler, *J. Biol. Chem.*, 244 (1969) 551.
338   A. Peterkofsky, *J. Biol. Chem.*, 237 (1962) 787.
339   E. A. Havir and K. R. Hanson, *Biochemistry*, 7 (1968) 1904.
340   D. S. Hodgins, *Biochem. Biophys. Res. Commun.*, 32 (1968) 246.
340a A. Frankfart and I. Fridovich, *Biochim. Biophys. Acta*, 206 (1970) 457.
341   J. Hsiu, E. H. Fischer and E. A. Stein, *Biochemistry*, 3 (1964) 61.
341a T. Takagi and T. Isemura, *J. Biochem. (Tokyo)*, 57 (1965) 89.
342   T. Friedmann and C. J. Epstein, *J. Biol. Chem.*, 242 (1967) 5131.
343   S. L. Pfueller and W. H. Elliot, *J. Biol. Chem.*, 244 (1969) 48.
345   W. Neumann and E. Habermann, *Z. Physiol. Chem.*, 296 (1954) 166.
346   F. M. Davidson and C. Long, *Biochem. J.*, 69 (1958) 458.
347   B. L. Vallee, E. A. Stein, W. N. Sumerwell and E. H. Fischer, *J. Biol. Chem.*, 234
      (1959) 2901.
348   W. N. Fishbein and S. P. Bessman, *J. Biol. Chem.*, 241 (1966) 4842.
349   C. van der Drift and G. D. Vogels, *Biochim. Biophys. Acta*, 198 (1970) 339.
350   W. J. Middlehoven, *Biochim. Biophys. Acta*, 191 (1969) 110, 122.
351   W. P. Jencks, in P. D. Boyer, H. A. Lardy and K. Myrback, *The Enzymes*, Vol. 6,
      2nd edn., Academic Press, New York, 1962, p. 373.
352   L. T. Webster, *J. Biol. Chem.*, 238 (1963) 4010.
353   A. T. Norris and P. Berg, *Proc. Natn. Acad. Sci. U.S.*, 52 (1964) 330.
354   W. D. McElroy and A. A. Green, *Arch. Biochem. Biophys.*, 64 (1956) 257.
357   L. T. Webster, L. D. Gerowin and L. Rakita, *J. Biol. Chem.*, 240 (1965) 29.
358   J. C. Londesborough, L. T. Webster and M. C. Scrutton, (1969) unpublished
      observations.
359   L. T. Webster, *J. Biol. Chem.*, 241 (1966) 5504.
360   I. Svensson, *Biochim. Biophys. Acta*, 146 (1967) 239.
361   H. S. Kingdon, L. T. Webster and E. W. Davie, *Proc. Natn. Acad. Sci. U.S.*, 44
      (1958) 757.
362   L. T. Webster and E. W. Davie, *Biochim. Biophys. Acta*, 35 (1959) 559.
363   U. Lagerkvist, L. Rymo and J. Waldenstrom, *J. Biol. Chem.*, 241 (1966) 5391.
364   J. M. Clark and J. P. Eyzaguirre, *J. Biol. Chem.*, 237 (1962) 3698.
365   C. Bublitz, *Biochim. Biophys. Acta*, 128 (1966) 165.
366   C. Bublitz, *Biochim. Biophys. Acta*, 133 (1966) 158.
366a F. X. Cole and P. R. Schimmel, *Biochemistry*, 9 (1970) 3143.
367   J. E. Allende, G. Mora, M. Gatica and C. C. Allende, *J. Biol. Chem.*, 241 (1966)
      2245.
368   G. Grosjean, J. Charlier and J. Vanhumbeeck, *Biochem. Biophys. Res. Commun.*,
      32 (1968) 935.
368a R. B. Loftfield and E. A. Eigner, *J. Biol. Chem.*, 242 (1967) 5355.
368b T. Lindahl, A. Adams and J. R. Fresco, *Proc. Natn. Acad. Sci. U.S.*, 55 (1961)
      944.
369   R. H. Himes and J. C. Rabinowitz, *J. Biol. Chem.*, 237 (1962) 2903.
370   Y. Kaziro and S. Ochoa, *Adv. Enzymol.*, 26 (1964) 283.
371   M. C. Scrutton and A. S. Mildvan, (1963) unpublished observations.
372   M. C. Scrutton, M. R. Olmsted and M. F. Utter, in S. P. Colowick and N. O. Kaplan,
      Methods in Enzymology, Vol. 13, Academic Press, New York, 1969, p. 235.
373   D. B. Keech and G. J. Barritt, *J. Biol. Chem.*, 242 (1968) 1983.
374   J. McD. Blair, *FEBS Lett.*, 2 (1968) 245.
375   A. Meister, in P. D. Boyer, H. A. Lardy and K. Myrback, *The Enzymes*, Vol. 6,
      2nd edn., Academic Press, New York, 1962, p. 443.

376 C. A. Woolfolk, B. M. Shapiro and E. R. Stadtman, *Arch. Biochem. Biophys.*, 116 (1966) 177.
376a H. S. Kingdon, B. M. Shapiro and E. R. Stadtman, *Proc. Natn. Acad. Sci. U.S.*, 58 (1967) 1703.
377 H. S. Kingdon, J. S. Hubbard and E. R. Stadtman, *Biochemistry*, 7 (1968) 2136.
378 M. D. Denton and A. Ginsburg, *Biochemistry*, 8 (1969) 1714.
379 R. C. Valentine, B. M. Shapiro and E. R. Stadtman, *Biochemistry*, 7 (1968) 2143.
380 B. M. Shapiro and A. Ginsburg, *Biochemistry*, 7 (1968) 2153.
381 A. Gottschalk, *Eur. J. Biochem.*, 7 (1969) 301.
382 J. R. Robinson and R. D. Sagers, *Bact. Proc.*, (1970) 124.
383 M. Benziman, R. Sagers and I. C. Gunsalus, *J. Bact.* 79 (1960) 474.
384 J. C. Su, C. C. Li and C. C. Ting, *Biochemistry*, 5 (1966) 536.
385 E. Margoliash, *A. Rev. Biochem.*, 30 (1961) 549.
386 G. W. Kosicki and L. P. K. Lee, *J. Biol. Chem.*, 241 (1966) 3571.

Chapter 15

CARBOXYPEPTIDASE A AND OTHER PEPTIDASES

MARTHA L. LUDWIG*

Biophysics Research Division, Institute of Science and Technology, University of Michigan and Biological Chemistry Department, University of Michigan Medical School, Ann Arbor, Michigan, U.S.A.

AND W. N. LIPSCOMB

Chemistry Department, Harvard University, Cambridge, Massachusetts, U.S.A.

I. BOVINE CARBOXYPEPTIDASE A: PREPARATION, PROPERTIES, AND SPECIFICITY

Both the three-dimensional structure[1-3] and the complete chemical sequence[4] of bovine pancreatic carboxypeptidase A** are known. The role of the essential metal component is less completely understood, but is accessible to examination by a variety of spectroscopic[5,6] and kinetic[7,8] techniques. This enzyme therefore joins the growing list of proteins for which definitive structure-function correlations are feasible. Any proposed mechanism of catalysis by CPA must now take into account both the accumulated chemical data and the interactions observed in the crystal structure of a complex of carboxypeptidase with a model substrate, glycyl-L-tyrosine[3,9].

Carboxypeptidase A is elaborated as an inactive proenzyme by the acinar cells of the pancreas. Bovine procarboxypeptidase is a large molecule of molecular weight approximately 87,000[10] consisting of three subunits one of which has a molecular weight of 40,000–42,000 and is the immediate precursor of CPA[11]. Active CPA can be prepared by any of three methods: (1) fractionation of autolysates of frozen pancreas[12] (Anson); (2) fractionation of aqueous extracts of pancreatic acetone powder after activation by trypsin[13] (Allan); (3) purification of pro-CPA by DEAE chromatography, followed by tryptic digestion[14] (Cox). The product obtained by any of these three procedures contains approximately one zinc atom per molecule[15] and consists of a single polypeptide chain[16].

---

* Recipient of a Career Development Award (1 KO4 GM06611) from the U.S. Public Health Service.

** Abbreviations: CPA, carboxypeptidase A; pro-CPA, procarboxypeptidase A; CBZ-, carbobenzyloxy-; E, enzyme; S, substrate; I, inhibitor; k_{cat}, turnover number; K_m, Michaelis constant.

Two kinds of heterogeneity have been detected in these preparations by further chromatographic resolution in the presence of β-phenylpropionate[17,18]. Because the activation process does not result in quantitative splitting at a single peptide bond of the proenzyme, carboxypeptidases designated α, β, or γ (or δ) may form, depending on the conditions of activation[19,20] (see Folk and Schirmer[20a]). Differences in the chain length are shown in Fig. 1. The preparative methods of Anson (γ) and Allan (δ)

```
                  5              10              15              20              25
ALA-ARG-SER-THR-ASN-THR-PHE-ASN-TYR-ALA-THR-TYR-HIS-THR-LEU-ASP-GLU-ILE-TYR-ASP-PHE-MET-ASP-LEU-LEU-
 ↑       ↑               ↑
 α       β               γ

                 30              35              40              45              50
VAL-ALA-GLN-HIS-PRO-GLU-LEU-VAL-SER-LYS-LEU-GLN-ILE-GLY-ARG-SER-TYR-GLU-GLY-ARG-PRO-ILE-TYR-VAL-LEU-

                 55              60              65              70              75
LYS-PHE-SER-THR-GLY-GLY-SER-ASN-ARG-PRO-ALA-ILE-TRP-ILE-ASP-LEU-GLY-ILE-HIS-SER-ARG-GLU-TRP-ILE-THR-

                 80              85              90              95             100
GLN-ALA-THR-GLY-VAL-TRP-PHE-ALA-LYS-LYS-PHE-THR-GLU-ASN-TYR-GLY-GLN-ASN-PRO-SER-PHE-THR-ALA-ILE-LEU-

                105             110             115             120             125
ASP-SER-MET-ASP-ILE-PHE-LEU-GLU-ILE-VAL-THR-ASN-PRO-ASN-GLY-PHE-ALA-PHE-THR-HIS-SER-GLU-ASN-ARG-LEU-

                130             135             140             145             150
TRP-ARG-LYS-THR-ARG-SER-VAL-THR-SER-SER-SER-LEU-CYS-VAL-GLY-VAL-ASP-ALA-ASN-ARG-ASN-TRP-ASP-ALA-GLY-

                155             160             165             170             175
PHE-GLY-LYS-ALA-GLY-ALA-SER-SER-SER-PRO-CYS-SER-GLU-THR-TYR-HIS-GLY-LYS-TYR-ALA-ASN-SER-GLU-VAL-GLU-

            ILE 180             185             190             195             200
VAL-LYS-SER-───-VAL-ASP-PHE-VAL-LYS-ASN-HIS-GLY-ASN-PHE-LYS-ALA-PHE-LEU-SER-ILE-HIS-SER-TYR-SER-GLN-
            VAL

                205             210             215             220             225
LEU-LEU-LEU-TYR-PRO-TYR-GLY-TYR-THR-THR-GLN-SER-ILE-PRO-ASP-LYS-THR-GLU-LEU-ASN-GLN-VAL-ALA-LYS-SER-

        ALA     230             235             240             245             250
ALA-VAL-───-ALA-LEU-LYS-SER-LEU-TYR-GLY-THR-SER-TYR-LYS-TYR-GLY-SER-ILE-ILE-THR-THR-ILE-TYR-GLN-ALA-
        GLU

                255             260             265             270             275
SER-GLY-GLY-SER-ILE-ASP-TRP-SER-TYR-ASN-GLN-GLY-ILE-LYS-TYR-SER-PHE-THR-PHE-GLU-LEU-ARG-ASP-THR-GLY-

                280             285             290             295             300
ARG-TYR-GLY-PHE-LEU-LEU-PRO-ALA-SER-GLN-ILE-ILE-PRO-THR-ALA-GLN-GLU-THR-TRP-LEU-GLY-VAL-LEU-THR-ILE-

                305
                VAL
MET-GLU-HIS-THR-───-ASN-ASN
                LEU
```

Fig. 1. The amino acid sequence of bovine pancreatic CPA[4]. The N-terminus of the α, β, γ (or δ) chain is at residue 1, 3, or 8 respectively as indicated by arrows. The sequence of CPA[Leu] corresponds to the lower one of the alternatives in positions 179, 228 and 305; in the other allotypic form, CPA[Val], the upper amino acid is found at these positions. Chain segments constituting the active center as visualized in three dimensional models of an acyl–tripeptide complex with CPA are underlined.

yield predominantly the 300 residue chain whereas the principal product of activation of purified pro-CPA is the α form, a chain 307 residues in length[1] with a molecular weight of $35,472^{4,42}$ (see Fig. 1). In addition two allotypic forms of CPA have been characterized[21]. In one, CPA^{Val}, residue 179 is isoleucine, 228 is alanine, and 305 is valine. The corresponding residues in the second allotype, CPA^{Leu}, are indicated in Fig. 1.

CPA $_{Allan\ or\ \delta}$ and CPA$_{Cox\ or\ \alpha}$ crystallize isomorphously. The different electron density distributions in the N-terminal region result in some small differences in the diffracted intensities. The Anson, or CPA$_\gamma$ preparation, on the other hand, crystallizes in the same space group as the other carboxypeptidases but with a crystal habit and diffraction pattern strikingly different from those of the other two preparations. Although both CPA$_{Allan}$ and CPA$_{Anson}$ are composed chiefly of chains with Asn at the N-terminus, the crystallographic differences and relative stabilities of the apoenzymes[22] suggest other chemical dissimilarities between these preparations. The crystal structure analysis has employed CPA$_\alpha$ crystals while the chemical sequence was determined using CPA$_\gamma$.

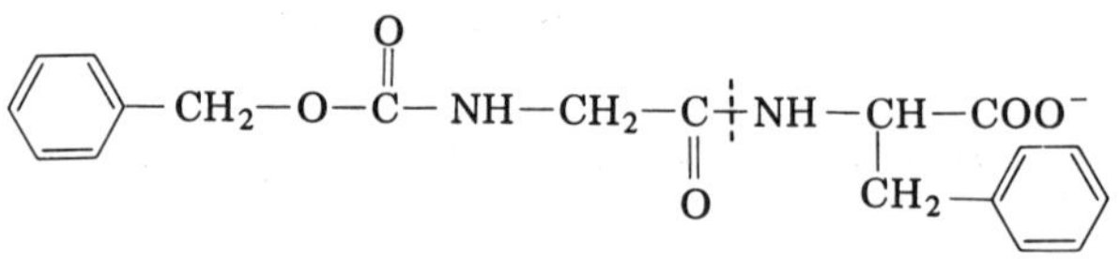

Fig. 2. Carbobenzyloxy-glycyl-phenylalanine, a model peptide substrate for carboxy-peptidase A. The hydrolyzable bond is indicated by a dashed line.

Carboxypeptidase A exhibits a specificity for the cleavage of carboxy-terminal L-amino acids from proteins or model peptide substrates*. Endo-peptidase activity has never been unambiguously demonstrated in pure preparations. CPA will not release ammonia from simple amides, nor will it cleave α-amidated amino acids from the C-terminus of a peptide chain. Maximum hydrolysis rates are observed when the side chain of the C-terminal residue is hydrophobic[23,24]. k_{cat} for the classical model peptide substrates, CBZ-glycyl-L-phenylalanine (CGP) (Fig. 2) and benzoyl-glycyl-L-phenylalanine (BGP) is of the order of 10^4 per minute. The ester analog of BGP, benzoyl-glycyl-L-phenyllactate (HPLA) is also hydrolyzed efficiently with a turnover number of approximately 3×10^4 per minute[25,26].

---

* The stereospecificity is not absolute; C-terminal D-alanyl residues are cleaved very slowly from appropriate model peptides[127], but release of terminal D-residues larger than Ala has not been demonstrated.

In contrast, N-acylamino acids and dipeptides react much more slowly (Table IV). Substitution of either peptide nitrogen of the acyl-dipeptide substrates markedly decreases the rates of hydrolysis[23,24]. At pH 7.5 to 8.0, the pH of maximum activity towards model substrates like BGP, acidic residues and C-terminal glycine or alanine are released comparatively slowly from peptides. If a protein is digested with CPA the appearance of arginine or proline at the C-terminus of the degraded chain brings further digestion to a virtual halt[27]. The binding site for peptides must be large enough to interact with five residues since the kinetic parameters are affected not only by the nature of the two amino acids forming the scissile bond but also, to a lesser extent, by the three adjacent residues[28]. The penultimate residue has a marked effect on the kinetics. Certain D-amino acid residues in positions other than C-terminal can be accommodated on the enzyme surface, but a bulky D-residue can occupy the penultimate position only in a dipeptide substrate[29,30].

Carboxypeptidases isolated from mammalian sources and from more primitive species such as the spiny dogfish require a metal ion in order to be active. Inhibition studies suggested the presence in CPA of a metal ion[31], which was subsequently identified as zinc and shown to be essential for activity[32]. The inactive apoenzyme can be obtained by dialysis against a variety of chelating agents, including o-phenanthroline and hydroxy-quinoline sulfonate[33,34]. Apoproteins derived from either CPA_α or CPA_δ are stable enough to permit studies of substrate and metal binding, while for reasons as yet obscure the apoprotein of CPA_γ is much more readily denatured[22]. By reconstitution with metal ions other than zinc, a series of metallocarboxypeptidases can be prepared from apo-CPA[35,36]. Of these, the Co^{2+}, Mn^{2+} and Ni^{2+} enzymes are active peptidases, whose turnover numbers depend on the nature of the metal ion (Table V). Exposure of Zn-CPA to sufficient concentrations of another appropriate ion results in replacement of zinc by the second metal[37].

The apparent association constants for Co-CPA have been determined by measuring the bound ^{60}Co in equilibrium mixtures of metal and apo-protein, and constants for the more tightly bound metals, shown in Table V, have been derived from competition experiments in which pairs of metal ions were equilibrated with the protein[38]. In addition to the ions included in the table, Cr^{3+} and Fe^{2+} have been reported to restore activity to apo-CPA[35]. Assuming that the reaction of zinc with CPA is stoichiometric between pH 7 and 9 (which is justified by the values of Table V), the release of protons occurring upon addition of zinc to the apoprotein determines two pK values, one of 7.7 and the other, 9.1. These pK values were tentatively attributed to groups involved in zinc binding[38]*.

---

* Now that the zinc ligands are known to be histidines and glutamate, the interpretation of these pK values appears more complicated.

References pp. 482–487

II. THE STRUCTURE OF CARBOXYPEPTIDASE A AND ITS GLYCYL-TYROSINE COMPLEX

1. Overall conformation of the protein

Atomic models of carboxypeptidase A have been constructed by fitting a sequence of 307 residues into an electron density map with a nominal resolution of 2 Å [1-3]. The X-ray reflections included in the map were phased by the method of isomorphous replacement, using four heavy atom derivatives to 2.8 Å resolution and two derivatives between 2.0 and 2.8 Å resolution. In the two mercury derivatives, mercury replaced zinc at the active site (Table I). The polypeptide backbone was positioned in the map without undue difficulty although at a few residues, e.g., 133–137, the electron density was appreciably less than the average backbone density. Residues 138 and 161 were found to be connected by a disulfide bridge [1].

TABLE I

HEAVY ATOM BINDING TO CARBOXYPEPTIDASE

Atom	Occupancy electrons/ molecule	Heavy atom coordinates			Residue
Pb_1	58	−0.094	0.500	−0.089	Glu 270
Pb_2	53	−0.089	0.540	−0.147	Citrate, not protein
Hg,g	50	−0.071	0.455	−0.115	His 69, Glu 72, His 196
Hg,s_1	47	−0.071	0.452	−0.115	His 69, Glu 72, His 196
Hg,s_2	46	−0.506	0.069	−0.257	His 29
Hg,s_3	48	−0.475	0.109	−0.136	His 29, Lys 84
Pt_1	74	0.341	0.430	0.034	Cys 161
Pt_2	45	−0.438	0.305	−0.568	Met 103
Pt_3	68	−0.292	0.082	0.141	N-terminus: Ala 1
Pt_4	27	−0.484	0.485	−0.500	His 303
Ag_1		0.238	0.498	−0.278	His 166, Ser 158
Ag_2		−0.082	0.220	0.193	His 120
Ag_3		−0.457	0.090	−0.143	His 29, (Lys 84)
Ag_4		−0.483	0.477	−0.516	His 303
Co_1		−0.500	0.500	−0.500	His 303
Co_2		−0.500	0.070	−0.130	His 29, (Lys 84)
Zn^d		−0.087	0.443	−0.115	His 69, Glu 72, His 196

Hg,g:5 × 10^{-4} M CPA was dialyzed against 0.001 M $HgCl_2$, 1 M LiCl, 0.2 M Tris, pH 8, and crystallized by dialysis against 0.18 M LiCl, 0.02-Tris, pH 8. This single site Hg derivative contributed minimally to the phasing at high resolution. Hg s_1, s_2, s_3 are three mercury sites in crystals suspended in 0.0008 M Hg^{2+}

As expected, the maximum electron density corresponded to the zinc atom. Interpretation of the map established the presence of three ligands to zinc and determined these to be residues 69, 72, and 196. Residues 69 and 72 were correctly identified as His and Glx[1] but identification of 196 as Lys (or Glx) proved incorrect[42].

In the determination of the chemical sequence, shown in Fig. 1, initial fragmentation was accomplished by CNBr cleavage at methionine residues 22, 103, and 301[39-41]. Ordering of these fragments was made possible by identification of methionine-containing peptides[20], and was confirmed by comparison with the electron density map. Subsequent cleavage of the largest chain segment, 104–301, required digestion by five different proteolytic enzymes[4]. In agreement with the X-ray map, cysteine residues were found at positions 138 and 161. The sequence determination established that residue 196, the third zinc ligand, was histidine, and confirmed that the other two ligands, residues 69 and 72, were His and Glu, respectively.

For comparison with the chemical sequence determination, an X-ray identification of the side chains was attempted[2,3]. The level of correctness varied from 60% to 85%, depending on the region of the molecule[42]. However, the specific binding and catalytic residues (Arg 145, Tyr 248, Glu 270) were correctly identified. In a final examination of some regions of the structure where there appeared to be inconsistencies between the map and the sequence, additional electron density maps were computed using the atomic coordinates to calculate the phase angles. (Suitable thermal parameters were applied to account for the lower density of residues 133–137[3].) According to standard crystallographic practice, those atoms whose locations were considered doubtful were omitted from the phase calculation. Solvent atoms and about 5% of the protein atoms were thus excluded. Appearance of any of these latter atoms in the resulting maps confirms their significance. After recomputation there still remain a few disagreements between the chemical and X-ray sequences[42]; none of these involves residues known to differ in the two allelomorphs. Indeed, from the densities at residues 179, 228, and 305 it can be concluded that the crystals, prepared from the pancreatic juice of a single cow, were composed of CPA_α^{Val}. However, residues 151 and 93, determined chemically to be Phe and Asn, respectively, appear larger in the maps, where they resemble Trp and Glx. In addition, two acidic residues, Glu 108 and Asp 256, are in positions where one might have expected to find the amides rather than the free acids[42].

The overall conformation of the polypeptide chain can be seen in Fig. 3, in which all α carbons are labelled. The molecule is almost spherical, being about 50 x 42 x 38 Å, and is therefore somewhat more compact than had been predicted from the f/f_0 values obtained from diffusion coefficients[43]. Adjacent to the zinc atom is a hollow, or pocket, which proves to be the binding site for the aromatic side chain of glycyl-tyrosine (cf. Figs. 6 and 8). This depression in the surface of the molecule is very apparent even in the low resolution maps of CPA[44]. At high resolution a number of water molecules can be seen in this cavity. Extending from the zinc and away from the pocket is an indentation or groove in which models of acyl-tripeptides can be accommodated (Section II 4a). Of the total of 307 residues, 115 participate in helixes[*], 45 occur in an extended sheet structure, and the remaining 147 are in regions that may be described

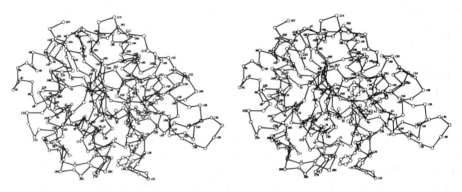

Fig. 3. A stereo drawing of carboxypeptidase A. The overall folding of the molecule can be seen in this view down the b axis. All α carbon atoms are numbered according to the sequence of Fig. 1. A model substrate, CBZ–Ala–Ala–Tyr, occupies the substrate binding groove and the pocket adjacent to zinc. Several long helices appear at the left, the β-sheet winds up through the center, and the disulfide bond is located in the lower right portion of the drawing. Stereoviewers may be obtained, *e.g.*, from Ward's Natural Science Establishment, Inc., Rochester, N.Y., model 25W 2951, or Stereo-Magniscope, Inc., 40-31 81st St., Elmhurst, New York City, U.S.A.

as "random" only in the sense that they do not correspond to any standard model. Table II lists the types of secondary structure which occur along the polypeptide chain. Except for three relatively short helixes on the right side of the molecule as viewed in Fig. 3 (residues 112–122, 173–187 and 254–262), all of the helices are on the left side of the figure. None can really be considered to lie in the interior of the molecule. A distorted β sheet structure, comprising eight chain segments in which there are four parallel and three anti-parallel pairs of chains, makes up much of the center of CPA and forms one side of the active site. The arrangement of residues in this sheet is shown schematically in Fig. 4. In Fig. 3 the β sheet is perpendicular to the paper. This view therefore displays the warping of the sheet: the uppermost segment, 238–243, is rotated almost 120° with respect to the bottommost length of chain, residues 32–37. The right side

---

[*] Residues are included in the helical category if they participate in one helical hydrogen bond.

445

TABLE II

SECONDARY STRUCTURE IN CARBOXYPEPTIDASE

Residues	Structure	Residues	Structure
14–28	Helix	173–187	Helix
32–36	β-sheet	190–196	β-sheet
49–53	”	200–204	”
60–66	”	215–231	Helix
72–80, 82–88	Helix	239–241	β-sheet
94–103	”	254–262	Helix
104–109	β-sheet	265–271	β-sheet
112–122	Helix	285–306	Helix
122–174	"Random"		

of the molecule (Fig. 3) is composed principally of "random" regions, and contains the disulfide bond connecting residues 138 and 161 and lying about 20 Å from the zinc atom. This "random coil" portion of the molecule is relatively flexible; in this region occur most of the conformation changes associated with substrate binding. The final folding of CPA, as displayed in Fig. 3, cannot be achieved as the pro-CPA-subunit is synthesized from its N-terminus. For example, chain 249–254 must be inserted

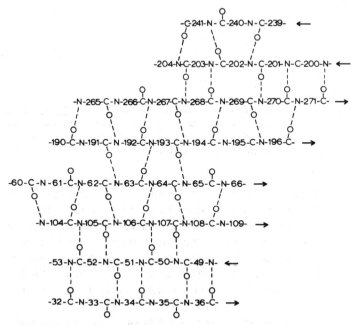

Fig. 4. A diagram of the β-sheet found in carboxypeptidase A. The suggested hydrogen-bonding scheme is shown by dashed lines. Arrows indicate the parallel or antiparallel arrangement of adjacent segments.

References pp. 482–487

between two segments, near 150 and 208, which are only 5 to 6 Å apart. Likewise, 265–271 in the final structure lies between segments 200–204 and 190–196.

The atomic positions have been refined using a model-building program which optimizes the fit of a known sequence through selected guide coordinates by allowing the rotational angles about single bonds to vary[45]. Tau, the $C_{carbonyl}$–C_α–N bond angle, was also refined and was found to have an average value of 112° rather than tetrahedral, implying some distortion within the peptide units. From the final atomic positions the dihedral angles, ϕ and ψ, have been calculated for each residue[3,46,47].

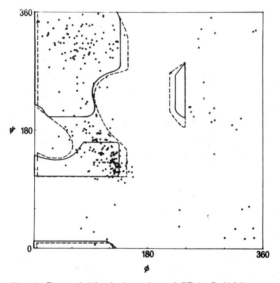

Fig. 5. Plot of dihedral angles of CPA. Solid lines are boundaries of allowed regions for $\tau = 110°$, dotted lines for $\tau = 115°$. Open circles refer to glycyl residues. The angles corresponding to α-helix are $\phi = 132°$, $\psi = 123°$ (x), and for antiparallel β-sheet, $\phi = 40°$, $\psi = 315°$ $(+)$. The $\phi = 0$, $\psi = 0$ are defined according to ref. 47.

The description of the structure in terms of the dihedral angles is given in Fig. 5, a conformational map of the carboxypeptidase molecule. While a dense cluster of points occurs near to the angles characteristic of the model α-helix, the scatter indicates that none of the helical regions is perfectly regular. For example, the sequence containing residues 72 to 88 is bent in the middle with the result that no hydrogen bonds can form to the backbone of tryptophan 81. Similarly the β sheet is distorted from the model β-configuration[48].

A few residues appear to adopt conformations which are unallowed according to theoretical calculations. Analogous results have been obtained for lysozyme[49], but the significance of these high energy conformations is unclear at present. When histidine 196, a zinc ligand, was built into the electron density map it proved necessary to introduce a distorted *cis* peptide bond between serine 197 and tyrosine 198. As this is the first example in which a *cis* peptide bond occurs at a residue other than proline the results should be viewed with some caution. On the other hand, it is conceivable that this bond arrangement may be of significance for the activity of the enzyme.

The distribution of residues in the inside and outside of the molecule is consistent with what has been observed in other globular proteins, with hydrophobic residues tending to occupy the interior and charged groups on the outside of the molecule[50]. Because it is largely in the interior, the β-sheet is highly hydrophobic, containing a number of leucine and phenyla- lanine residues. A total of 78 residues is not in contact with solvent water; of these, 22 are capable of hydrogen bonding, and nearly all of the latter do appear to interact either with the backbone or with neighboring residues[3,50]. Two tryptophans, 63 and 147, and one tyrosine, 238, are buried on the inside of CPA, while the remaining tyrosine and tryptophan side chains are partly exposed to solvent. The presence of a hydrogen bond between the OH of Tyr 238 and the carbonyl group of Glu 270 may be of some importance in the conformational change shown by Glu 270 when substrates bind, as described below. Of the ten prolines, four are located at the *N*-terminus of helixes, and three mark the ends of extended chains in the sheet. There are three carboxylate groups in the interior, 104, 108 and 292, assuming that the chemical identification of these as charged groups and not amides is correct. Glu 292 is salt-linked to Arg 272, so that its charge is locally compensated. One novel observation which has emerged from detailed examination of the maps is that ten water molecules are trapped inside the carboxypeptidase molecule[50].

2. The active center

(a) The protein conformation

Structurally, the active center is comprised of those atoms which are within van der Waals distances of the natural substrates for the enzyme. If all residues near the observed location of glycyl–tyrosine and adjacent to a model acyl–tripeptide lying in the surface "groove" (cf. Fig. 3) are included, then the following chain segments contribute to the active center: 69–72, 125–127, 142–145, 155, 163 and 164, 194–203, 243, 247–256, 268–270, and 275–279. The side chains 70, 126, 143, 195, 197, 200, 202, 249, 252–253, and 275–276 point away from the substrate binding area and are therefore not likely to be in direct contact with substrates. Residues 194–203 and 268–270 belong to the β sheet which forms one wall of the glycyl–tyrosine binding site; the remaining residues

except for the zinc ligands 69 and 72 are part of the tortuous right side
of the molecule shown in Fig. 3. The "top" of the pocket which binds
the tyrosyl side chain of glycyl-tyrosine is formed by residues 243 and
246-250, while 155 and 253-256 provide part of the remaining "lining"
of the pocket. Residues 275-279 line the left side of the groove as shown
in Fig. 3; the right hand side of the binding region is contributed by segments 125-127, 142-145, and 163-164. Figure 6 is a composite of the
electron density map through the substrate binding region with drawings
of glycyl-tyrosine and some of the contact residues superposed.

Much of the substrate binding region lacks repeating secondary
structure. Apart from some segments in the β-sheet and a few residues at
the beginning of one helix (255-256), the great majority of the residues are
in "random" regions. A conformational map of the active center residues
alone confirms that few of them have individual ϕ, ψ values approximating
those of α-helical residues.

(b) Zinc

Three protein ligands, His 69, Glu 72, and His 196, bind zinc to
carboxypeptidase. Each of the histidine–zinc bonds involves ring nitrogen
N1. The ligands participate in other interactions. N3 of His 69 can hydrogen
bond to Asp 142, and N3 of His 196 is hydrogen-bonded to a water
molecule. Additional extended electron density adjacent to zinc but
unattached to the protein was detected in the initial maps. However, in
maps based on calculated phases, a single non-protein ligand with a density
corresponding to oxygen is observed and this fourth ligand is therefore
assumed to be water or OH$^-$. Of course an occasional molecule of CPA
might bind Cl$^-$ instead of water but the density is insufficient for the
ligand to be predominantly Cl$^-$. The configuration about the zinc is
illustrated in Fig. 7, and the bond angles are given in Table III. Even
allowing for errors of $\pm 10°$ in the bond angles the large distortion from
tetrahedral geometry seems significant, especially in view of the proximity
to Zn of the β-CH$_2$ groups of His when Zn is bound to N1.

Irregular ligand geometries are not confined to zinc proteins, as zinc
also chelates amino acids or peptides in structures which display distorted
geometries (see Chapter 4). Among these are Zn(Gly–Gly–Gly) (SO$_4$)$_{1/2}$ ·
2H$_2$O, Zn(L-His)$_2$ · 2H$_2$O, and Zn(DL-His)$_2$ · 5H$_2$O. The triglycine complex
is a deformed trigonal bipyramid with bond angles deviating as much as 20°
from the ideal geometry[51]. In the two histidine compounds, four nitrogen
ligands (the α-amino and N1 ring nitrogens) form a tetrahedron with angular
distortions of about 10° and in addition two oxygens bind to the zinc[52,53].

3. The glycyl-tyrosine: CPA complex

An enzyme-substrate complex between glycyl-tyrosine and CPA has
been studied crystallographically at atomic resolution using difference

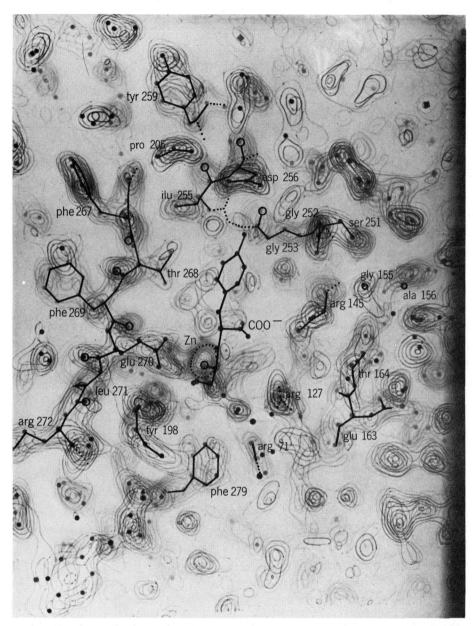

Fig. 6. The glycyl-L-tyrosine binding site. A model of glycyl-tyrosine, in the orientation
determined from difference maps, is superposed on several sections of the electron
density map of CPA. A number of neighboring side-chains, in the conformation found
in uncomplexed CPA, are identified. A cluster of three arginine residues appears to the
right of the substrate; the density at the position of the phenolic ring of the substrate
is thought to correspond to four water molecules, which are displaced by Gly-Tyr.
From "The Structure and Action of Proteins" and accompanying "Stereo Supplement"
by Richard E. Dickerson and Irving Geis. Copyright 1969 by Richard E. Dickerson and
Irving Geis. By permission of the authors and Harper and Row, Publishers, Inc.

References pp. 482–487

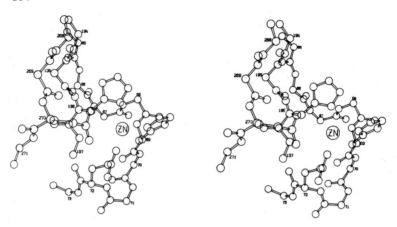

Fig. 7. A stereo drawing of the zinc ligands, including a portion of the β-sheet. Computer-drawn stereo figures were made using the program OR-TEP of Dr. Carroll Johnson.

Fourier techniques[1-3]. The exceedingly slow turnover of glycyl-tyrosine made the crystallographic experiments feasible; in solution k_{cat} is of the order of 1 min^{-1} but in the crystalline state the half-time for hydrolysis approaches the data collection time. Only one mode of binding is observed

TABLE III

METAL-LIGAND BOND ANGLES IN
Zn-CARBOXYPEPTIDASE

N(69)–Zn–N(196)	86°
N(69)–Zn–O(72)	99°
O(72)–Zn–N(196)	143°
O(72)–Zn–O(H$_2$O)	99°
N(69)–Zn–O(H$_2$O)	120°
N(196)–Zn–O(H$_2$O)	111°

in the crystalline glycyl-tyrosine complex. The interactions occurring in this E · S structure have been incorporated in a proposal concerning the mechanism of action of CPA (Section III 5a)[2].

Difference Fourier calculations, as normally employed in protein crystallography, use as coefficients★ $w(|F_{es}| - |F_e|) \exp(i\alpha_e)$. Since the

★ $|F_{es}|$, scattering amplitude observed in crystals containing substrate; $|F_e|$, amplitude observed in "free" enzyme crystals; α_e, phase angle calculated by isomorphous replacement. The weighting factor, w, is usually the figure of merit.

coefficients corresponding to the native enzyme structure are substracted, the Fourier summation produces an image of the differences between E and the E·S structures. If some of the atoms of the protein are displaced as a result of complex formation, negative regions will be observed at the original atomic positions. In initial difference maps of the glycyl–tyrosine complex at 2.8 Å resolution, the positions of the tyrosyl atoms were clear. However, the glycyl residue of the substrate, which lies above the zinc in Figs. 6 and 8, in particular the carbonyl group, was not readily visible, presumably because non-protein density in the region of the zinc was replaced by equivalent density upon substrate binding. Two alternative orientations of the glycyl portion of the substrate were considered; in the more probable one, the carbonyl oxygen was directed toward the zinc atom, but in the other the carbonyl–zinc interaction was precluded[2].

To obtain a clearer view of the orientation of glycyl–tyrosine, a new difference computation has been performed. Ordinarily, one would expect that a map based on coefficients* ($|F_{es}| - |F_{calc}|$) exp ($i\alpha_{calc}$) should reveal the positions of any atoms in the substrate complex whose locations differ from those in the calculated native structure. Unfortunately, since only a fraction, x, of the molecules in the crystal bind substrate, a map based on these coefficients would show density corresponding to any atoms present in the $(1 - x)$ uncomplexed fraction of the crystal but omitted from the structure factor calculation, viz., non-protein density near the zinc atom and water molecules in the pocket. Therefore coefficients corresponding to the free enzyme fraction of the structure were also subtracted. Thus the coefficients in this new, modified computation[3] are ($|F_{es}| - x|F_{calc}| - (1 - x)|F_e|$) exp ($i\alpha_{calc}$).

The enzyme–substrate contacts observed in the resulting map are depicted in Figs. 6 and 8. Five significant interactions between glycyl–tyrosine and the protein can be described. (1) The phenolic side-chain of the tyrosyl residue occupies the pocket in the enzyme surface, displacing several water molecules. The fit of the phenolic group is not perfect, in accord with the rather broad specificity of the enzyme for aromatic or branched aliphatic side-chains. Residues in the vicinity of the ring include: Ile 243, Ile 247, Ile 255, and Asp 256 (Fig. 6). (2) The carboxylate group of the substrate is aligned with the guanidino group of Arg 145 in such a way that two hydrogen bonds can form. (3) The phenolic oxygen of Tyr 248 lies within hydrogen bonding distance of the peptide NH of the substrate. Although the distances are accurate only to ± 1.0 Å, the phenolic O to peptide nitrogen spacing measures 2.7 Å. (4) The carbonyl oxygen of the substrate appears near zinc in approximately the position occupied by water (or OH⁻) in the free enzyme. While the substrate image is

* $|F_{calc}|$ and α_{calc}, scattering amplitude and phase angle computed from atomic positions.

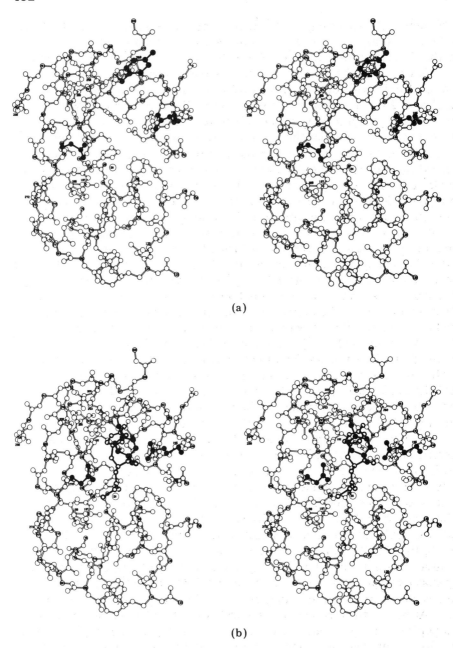

(a)

(b)

Fig. 8. Stereo drawings showing the conformation changes accompanying substrate binding. In (b) Arg 145, Glu 270, and Tyr 248, the three residues in black, are all found in different orientations in the presence of glycyl-tyrosine. The substrate is indicated by heavy outlines. The tyrosyl residue of Gly-Tyr occupies the specificity pocket while the carbonyl oxygen of the susceptible peptide bond is a zinc ligand.

weakest at the carbonyl oxygen, the location of that oxygen is corroborated by density corresponding to the C_α and amino N of the substrate, and by the absence of any other density which could match the carbonyl group. The carbonyl carbon of the substrate is situated about 3.5 Å from an oxygen of Glu 270. (5) The amino group of the substrate interacts via an intervening water with Glu 270. This last contact is thought to be "unproductive" (see Section III 5a). It may not involve charge–charge interaction, since inhibition experiments imply that only the anionic form of glycyl–tyrosine is bound to CPA[54].

The enzyme–substrate interactions are achieved by a number of conformational readjustments in the protein structure. The most dramatic rearrangements involve residues 248, 270, and 145, and can be seen by comparing Figs. 8(a) and 8(b). The guanidino group of Arg 145 moves about 2 Å as a result of a rotation around the C_β–C_γ bond. The carboxylate oxygens of Glu 270 are displaced approximately 2 Å in the y-direction (the carboxyl group appears to turn over and move upward toward the viewer in Fig. 8b) by means of rotations about the C_α–C_β and C_β–C_γ bonds. The largest movement is that of Tyr 248, whose oxygen is displaced by 12 Å owing to a rotation by about 120° about the C_α–C_β bond and a slight motion of the peptide backbone. The phenolic ring of Tyr 248 swings toward the substrate, enclosing the C-terminal moiety in the pocket. In its new position Tyr 248 approaches the susceptible peptide bond. The movements of Arg 145 and Tyr 248 appear to be coordinated by means of several smaller rearrangements. A system of hydrogen bonds involving residues 155, 154, and 249 links Tyr 248 and Arg 145 in the free enzyme. When glycyl–tyrosine is bound, the interactions of Arg 145 with the carbonyl of 155 and of Gln 249 with Tyr 248 are broken and the backbone in the region 247–249 undergoes slight changes.

Attachment of substrate converts the pocket from a water-filled cavity to a relatively hydrophobic region. Leu 203 appears to protrude into the the cavity whereas in the free enzyme it interacts with other residues of the β-sheet. Not every last "drop" of water is eliminated from the pocket, however; space remains for three or four water molecules when the terminal tyrosyl group of substrate is bound. Because the R groups of the hydrophobic C-terminal residues which are most rapidly hydrolyzed (Fig. 2) bind in this region, it is convenient to refer to the pocket as the specificity site, recognizing that other binding areas also affect specificity.

The dissociation constant for the glycyl–tyrosine: CPA complex is 1×10^{-3} (Table IV), corresponding to a ΔG at 25°C of –4.1 kcal for the binding step. This value appears smaller in magnitude than might be expected from summing over the observed interactions[55,56]. Some of the available binding energy has presumably been utilized to effect "strain", i.e., to produce an energetically unfavorable conformation in the substrate and/or enzyme[56a,56b,56c,57] (see Section III 5a). Thus the driving force (ΔG) for the conformational change in the protein is probably provided by favorable enzyme–substrate interactions. Since Tyr 248 appears to be essential for peptidase activity (Section III 3) and the structural rearrangement induced by substrate binding brings this residue into proper juxtaposition with the substrate, the CPA–glycyl–tyrosine interaction also constitutes a clear example of "induced fit" of enzyme to substrate[58,59].

4. Additional structural studies*

(a) Model building

Models of larger peptide substrates can be constructed to maximize the interactions between enzyme and substrate and to take account of the known binding specificities of the enzyme[2,3]. One such model E · S complex, where S is CBZ–Ala–Ala–Tyr, is shown in Fig. 3. The Ala–Ala peptide bond is situated within hydrogen bonding distance of Tyr 248, when the latter has been reoriented near the substrate. Inclusion of this interaction accounts for the high reactivity of peptides with an available NH at the penultimate position, in contrast to the lesser reactivity of an N-methyl[60] or β-alanyl[61] residue occupying position S_1 (Fig. 9). The remainder of the substrate is arranged along the "groove" in the CPA surface, in accord with the knowledge that up to five substrate residues may interact with the protein[28]. Placement of the benzyl ring of the CBZ-group near the aromatic residue, Phe 279, and of the carbonyl O of the third substrate residue adjacent to the guanidino group of Arg 71 is consistent with known effects of substituents on K_m values[28].

It is gratifying that, assuming glycyl–tyrosine to be bound in an orientation approximating that of a kinetically significant complex (see Section III 5), model building confirms several of the known stereospecificities of CPA[3,50]. For example, steric interference occurs with Ile 247 or with the displaced Tyr 248 if substrates substituted at the C-terminal peptide (CBZ–Gly–thiazolidine carboxylic acid[62], CBZ–Gly–Pro[63]) are positioned with their R groups in the pocket, COO⁻ near Arg 145 and C=O at the zinc. Peptides terminating with bulky D-residues similarly cannot form all three of these E · S interactions.

(b) Inhibitor complexes

Several enzyme–inhibitor complexes have been subjected to crystallographic analysis at 6 Å resolution[64]. The binding interactions cannot be

---

* Crystallographic studies of a complex between CPA and Phe-Gly-Phe-Gly are in progress[128]. The electron density maps clearly show a substrate carbonyl group bound to zinc. Moreover, the zinc ion undergoes a significant displacement of about 0.6 Å when this substrate is bound.

described in detail, but some conclusions may be reached by superposing the difference maps on high resolution maps of the protein. Three inhibitors, p-iodo-β-phenylpropionate, L-phenylalanine, and L-lysyl-L-tyrosineamide, can bind (at pH 7.5) in the vicinity of the zinc atom, but there are interesting differences in the behavior of these three compounds. L-Phenylalanine is perhaps the simplest to interpret. In difference maps, positive density is found in the pocket, in the same location as that observed in low

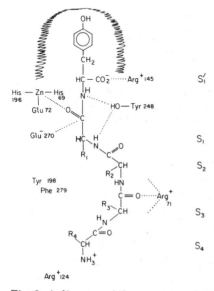

Fig. 9. A diagram of the enzyme–substrate interactions envisioned for a pentapeptide substrate. Residue S'$_1$, to be split from the peptide, is oriented like glycyl-tyrosine (Figs. 6 and 8), but the peptide NH between S$_1$ and S$_2$ is arranged to hydrogen bond to Tyr 248.

resolution glycyl–tyrosine difference maps. Models of L-Phe can be matched to the density with an interaction between Arg-145 and the free COO$^-$ of the Phe. In addition, positive and negative densities consistent with the motion of Tyr 248 are observed. Lysyl–tyrosineamide, on the other hand, appears not to be oriented like glycyl-tyrosine, and the maps fail to show any evidence of a conformation change involving Tyr 248. Iodophenyl-propionate binds to crystalline CPA at four sites. Assuming that the center of gravity of the low resolution peaks is at the iodine position, two of the bound iodines have been superimposed on a drawing of the active site, shown in Fig. 10. I1 and I2 are in locations which would permit the COO$^-$ group of the inhibitor to be bound by zinc; other evidence verifies that phenylpropionate bonds to the metal ion[65,66]. Iodophenylpropionate is thus not a perfect analog for the substrate, glycyl–tyrosine. Nevertheless,

References pp. 482–487

binding of iodophenylpropionate appears to induce the conformation change of Tyr 248. Multiple binding modes for phenylpropionate are not unexpected in view of the complex kinetic behavior of this inhibitor (Section III 2a).

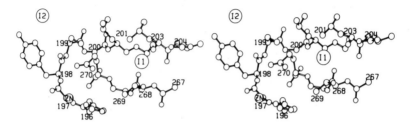

Fig. 10. A stereo drawing of the active site with the iodines of iodophenylpropionate superposed. Unlike the other Figures, this is a view approximately parallel to the crystallographic a axis, as if one were looking from the right hand side of Fig. 8. Iodine atom I1 lies in the substrate-binding pocket.

(c) The apoprotein and metals other than zinc

Removal and replacement of zinc can be accomplished in the crystalline state by essentially the same procedures employed in solution, although the precise conditions differ somewhat from those required in solution[67]. Crystalline Hg-CPA has been prepared in two ways, either by crystallization after exchange of Hg for Zn, or by dialysis of crystals against Hg^{2+} solutions[44]. In the latter case, mercury is bound not only in the zinc site, but also at other locations (Table I). Crystalline apo-CPA can be prepared from crystals of the zinc enzyme by treatment with hydroxyquinoline sulfonate. The Cd and Cu enzymes have been generated from apo-CPA crystals[64].

Difference Fourier maps comparing the native and apo-protein structures at a resolution of 6 Å are virtually characterless apart from a large negative peak at the zinc position[44]. At first sight it seems conceivable that in apo-CPA His 196 could reorient to hydrogen bond to Asn 144 and Gly 270, despite the rigidity imposed by the β sheet. However, in low resolution difference maps there are no matching positive and negative regions which would support this suggestion. Assuming that loss of activity parallels loss of zinc in the crystal as it does in solution, the apo-CPA maps imply a direct role for the metal in catalysis, rather than mere participation of zinc in a structural reorganization of the apoprotein. Furthermore, the position of substrates in the crystalline E · S complexes provides strong evidence for participation of the metal in catalysis. This important distinction of the possible roles of zinc could not be made on the basis of chemical studies in solution.

Maps of the active center of Hg-CPA have been calculated at high resolution. The center of the metal ion is displaced 1.0 Å from the corresponding zinc position, with the largest shifts in the x- and y-coordinates. In Fig. 7, mercury is moved upward and toward His 69. Mercury bonds to the N1 atoms of the two histidine residues 69 and 196, which are oriented as in the zinc protein. The possibility that the ring of histidine 196 turns over to make N3 rather than N1 the ligand is eliminated by the presence of a water molecule H-bonded to N3 in the same location as in the zinc map. The carboxyl group of glutamate 72 has a lower electron density than do the histidines, but still appears to bind the mercury. No changes in other regions could be detected in difference electron density maps in which the zinc and mercury structures were compared; indeed the usefulness of Hg-CPA in the phase calculations attests to the isomorphism of the two structures. Solutions of mercury carboxypeptidase are active esterases but fail to catalyze peptide hydrolysis[38]. However, crystals of Hg-CPA$_\gamma$ possess significant peptidase activity relative to crystalline Zn-CPA, of the order of 1/1000 the activity of solutions of the zinc enzyme[67].

Magnetic resonance techniques have been applied to the elucidation of the structures of Mn-CPA and Cu-CPA. When bound to the enzyme, neither of these ions possesses the ligand geometry found in Zn-CPA. Analysis of water proton and F^- resonances in the presence of Mn-CPA suggests that in F^- solutions, there exist at least two non-protein ligands in the first coordination sphere of the metal ion, one halide ion and one or more waters. If Mn attaches to three protein ligands, then it must be at least five-coordinate[6,66]. The electron paramagnetic resonance (e.p.r.) spectra of single crystals of Cu-CPA are consistent with a ligand geometry which is essentially planar rather than distorted tetrahedral. From the hyperfine interactions in the spectrum it can be concluded that the ligands include two nitrogens. The CPA$_\gamma$ crystals used in the e.p.r. measurements were isomorphous with corresponding crystals of Zn-CPA[68]. Copper is undoubtedly bound to histidines 196 and 69, but no direct evidence has been obtained regarding the nature of the remaining ligands or the orientation of the ligand phane with respect to the rest of the CPA molecule.

III. THE ENZYMATIC ACTIVITY OF CARBOXYPEPTIDASE A

1. Kinetics: a summary

Current knowledge of the kinetics of reactions catalyzed by CPA dictates a reaction scheme no less complex than that shown in Fig. 11. This diagram summarizes the variety of kinetic phenomena for which there is some experimental evidence: non-productive binding (pathway 2); competitive inhibition by products or substrate analogs (pathway 3); and

458

interaction with modifiers, M (where M may include products and substrates), leading to mixed inhibition, non-competitive inhibition, or activation (pathway 4). Excess substrate inhibition may be considered a special case of pathway 4. The importance of the pathways varies with the substrate under consideration. For example, excess substrate inhibition complicates the kinetics of HPLA* hydrolysis[25,69] at concentrations as low as 2×10^{-4} M, whereas the closely related peptide substrate, hippuryl-phenylalanine, displays no comparable inhibition; substrate activation has instead been noted[70] (but cf. ref. 71). Substrate activation has also been observed during hydrolysis of the ester, hippuryl-glycolate[85], and of the peptide, CBZ-Gly-Phe[71]. CBZ-Gly, a product derived from the latter substrate, activates hydrolysis of certain peptides but appears to inhibit

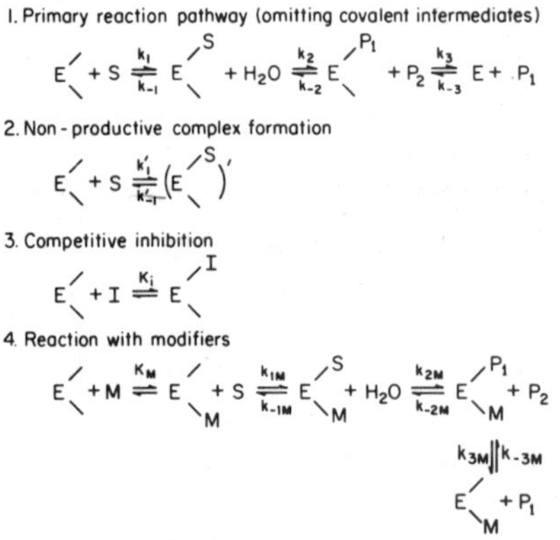

1. Primary reaction pathway (omitting covalent intermediates)

$$E\genfrac{}{}{0pt}{}{\diagup}{\diagdown} + S \underset{k_{-1}}{\overset{k_1}{\rightleftharpoons}} E\genfrac{}{}{0pt}{}{\diagup S}{\diagdown} + H_2O \underset{k_{-2}}{\overset{k_2}{\rightleftharpoons}} E\genfrac{}{}{0pt}{}{\diagup P_1}{\diagdown} + P_2 \underset{k_{-3}}{\overset{k_3}{\rightleftharpoons}} E + P_1$$

2. Non-productive complex formation

$$E\genfrac{}{}{0pt}{}{\diagup}{\diagdown} + S \underset{k'_{-1}}{\overset{k'_1}{\rightleftharpoons}} \left(E\genfrac{}{}{0pt}{}{\diagup S}{\diagdown}\right)'$$

3. Competitive inhibition

$$E\genfrac{}{}{0pt}{}{\diagup}{\diagdown} + I \overset{K_i}{\rightleftharpoons} E\genfrac{}{}{0pt}{}{\diagup I}{\diagdown}$$

4. Reaction with modifiers

$$E\genfrac{}{}{0pt}{}{\diagup}{\diagdown} + M \overset{K_M}{\rightleftharpoons} E\genfrac{}{}{0pt}{}{\diagup}{\diagdown_M} + S \underset{k_{-1M}}{\overset{k_{1M}}{\rightleftharpoons}} E\genfrac{}{}{0pt}{}{\diagup S}{\diagdown_M} + H_2O \underset{k_{-2M}}{\overset{k_{2M}}{\rightleftharpoons}} E\genfrac{}{}{0pt}{}{\diagup P_1}{\diagdown_M} + P_2$$

$$k_{3M} \big\| k_{-3M}$$

$$E\genfrac{}{}{0pt}{}{\diagup}{\diagdown_M} + P_1$$

Fig. 11. A diagram of the kinetic pathways known to be operative in CPA-catalyzed reactions. Only the minimum mechanisms and non-covalent intermediates are included. One product dissociation step is shown, to account for product inhibition.

competitively the hydrolyses of cinnamoylphenyllactate[72] and acyltripeptides[70]. Additional modifiers include benzamide, cyclohexanone, and a series of substituted alcohols[73].

Non-productive binding, resulting in competitive substrate inhibition, is not easily adduced from kinetics[74]. However, comparison of the K_m and V_m values for glycyl–tyrosine and $Gly_n Tyr$, where $n = 1$ to 6, strongly suggests a non-productive binding mode for the dipeptide. Thus K_m for Gly-Tyr is smaller than for longer peptides by a factor of five, yet V_m is only 1/1000 that for $Gly_2 Tyr$[75]. In reactions catalyzed by lysozyme, a similar phenomenon is observed as the substrate oligosaccharide chain length increases. Abortive binding, with K_m dominated by the non-productive complex, is well established in the case of lysozyme[76,77].

---

* Substrate formulas can be found in Table IV.

Examination of the crystalline glycyl–tyrosine: CPA complex suggests that the interaction (through a water molecule) between the α-amino group of the substrate and Glu 270, which cannot occur with longer substrates, may account for the relatively strong binding of glycyl–tyrosine and at the same time may interfere with formation of the transition state (Section III 5).

Some kinetic parameters for several substrates and modifiers have been assembled in Table IV. In the absence of a complete mechanism, the meaning of k_{cat} and K_m, in terms of individual rate constants, is of course unknown. As only a few substrates show ideal Michaelis–Menten behavior, corrections to account for alternative pathways have frequently been necessary. A number of binding constants have been tabulated to facilitate subsequent discussions of the effects of modification and metal substitution. The methods employed in their determination are indicated in column 4. For substrates, binding constants have generally been assumed to be equal to the K_m values determined during steady-state turnover. However, the Michaelis constants do not necessarily approximate equilibrium binding constants for those substrates which turn over rapidly. No simple proportionality exists between K_m and k_{cat}[70], but the activation parameters for the hydrolysis of CGP have been interpreted as showing that K_m includes contributions from k_{cat}[78]. Binding to the apoprotein has been evaluated by a technique which relies upon the effect of substrates on the rate of recombination of zinc with apo-CPA[79]. If this rate is sufficiently reduced, then addition of zinc to an enzyme–substrate mixture, followed by rapid separation of the protein, results in formation of Zn-CPA only to the extent that $E \cdot S$ is dissociated. Any binding modes which do not alter the rate of recombination of zinc will remain undetected. Furthermore, any effect of the substrate on the equilibrium binding constant for the metal must be accompanied by a reciprocal effect of the metal on substrate affinity[80]. Qualitative estimates of binding have also been based on the effect of substrates or inhibitors of the rates of metal interchange[81].

Generalizations from Table IV are hazardous, partly because of the sizes of the corrections necessitated by the occurrence of multiple pathways, but also because variables such as ionic strength are known to alter the kinetic parameters, particularly for peptides. Some trends may be discernible. Within limits, larger substrates tend to have smaller K_m values, suggesting that enzyme–substrate interaction increases with substrate size (cf. acetyl-phenylalanine, BGP and BGGP). Studies of the effect of substituents in a series of alanyl peptides support this general conclusion[28]. Kinetic anomalies are frequently associated with the smaller substrates; acylated tripeptides behave much more simply than their shorter cousins[28,70]. Moreover, the anomalies often seem related to the presence of aromatic side chains in blocking groups[2]. Comparison of HPLA with the isosteric BGP originally prompted the conclusion that esters were bound much more

TABLE IV

KINETIC PARAMETERS FOR SOME SUBSTRATES AND MODIFIERS OF CPA (pH 7.5, 25°C)

Compound	Structural formula	$K_m (M)$	K_s or $K_i (M)$	k_{cat} (min^{-1})	Significant alternate pathways	References
Peptides:						
1. Carbobenzyhoxy-glycyl-L-phenylalanine (CGP)	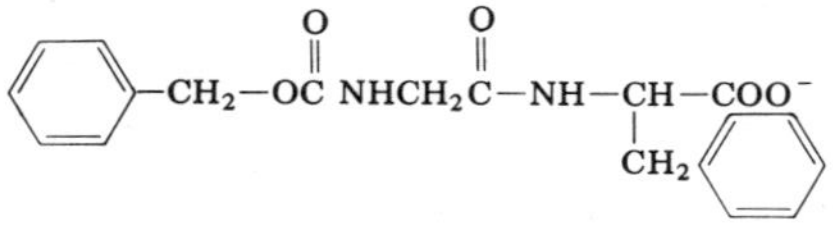	$2-37 \times 10^{-3}$	4.2×10^{-3a}	$5.5-12 \times 10^{3}$	Substrate activation Excess substrate inhibition	23,71,78,82
2. Benzoyl-glycyl-L-phenylalanine (BGP)	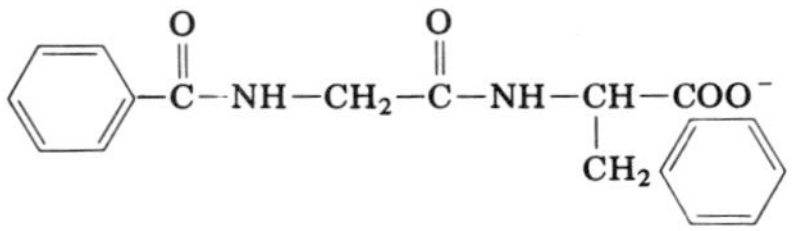	$0.8-11 \times 10^{-3}$	11×10^{-3a}	$5.6-11 \times 10^{3}$	Substrate activation	71,82
3. Benzoyl-glycyl-glycyl-phenylalanine (BGGP)	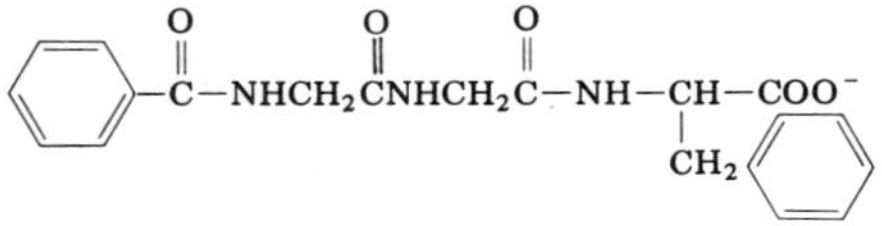	1×10^{-3}		1.2×10^{3}		70
4. Tetra-L-alanine	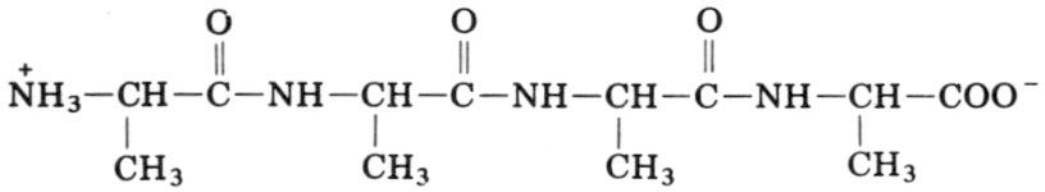	~ 0.08		6×10^{3}		28
5. Glycyl-L-tyrosine	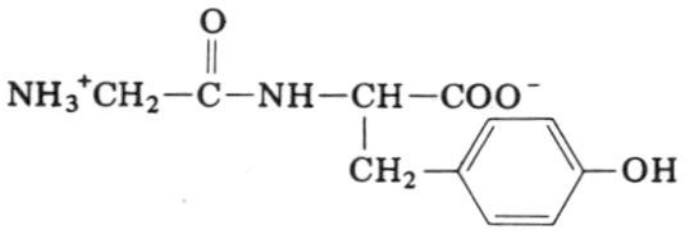	0.7×10^{-3}	1.0×10^{-3a} $(K_i) 1 \times 10^{-4}$	0.9		70,75
Acyl-amino acids:						
1. Acetyl-L-phenylalanine	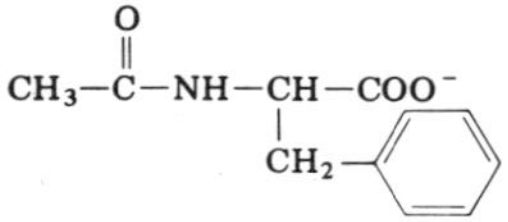	0.155				23
2. trans-Cinnamoyl-L-phenylalanine	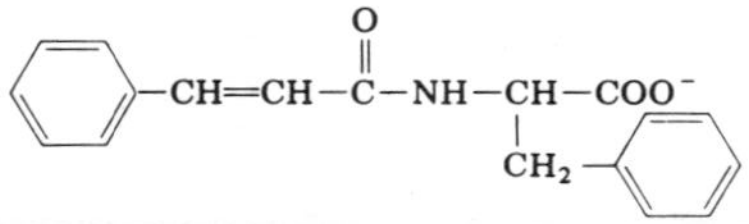	6×10^{-4}		0.2		83

	Structure				Refs
Esters					
1. Benzoyl-glycyl-phenyl-lactate (hippuryl-L-phenyllactate, HPLA)	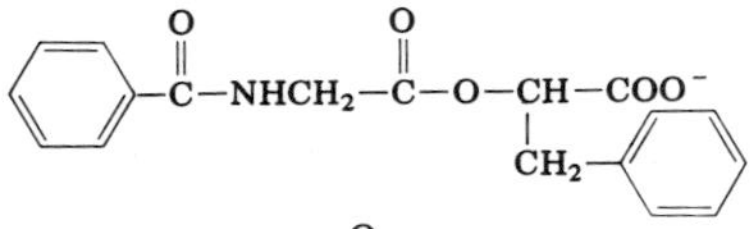 C₆H₅–C(=O)–NHCH₂–C(=O)–O–CH(CH₂C₆H₅)–COO⁻	$5.1-8.8 \times 10^{-5}$	$28-35 \times 10^3$	Excess substrate inhibition	25,71
2. Cinnamoyl-L-phenyl-lactate	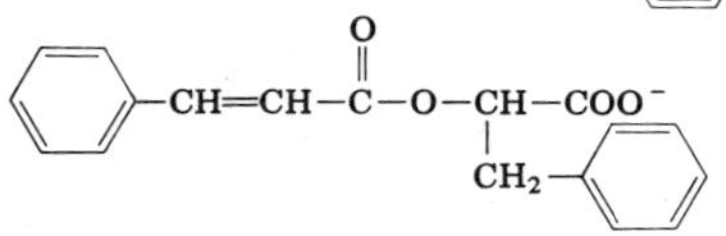 C₆H₅–CH=CH–C(=O)–O–CH(CH₂C₆H₅)–COO⁻	$1.5-1.9 \times 10^{-4}$	4.6×10^3		83,84
3. Hippuryl-glycolate	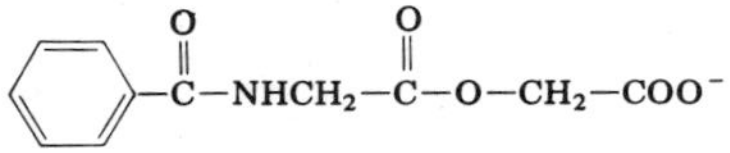 C₆H₅–C(=O)–NHCH₂–C(=O)–O–CH₂–COO⁻	1.2×10^{-3}	4×10^3	Substrate activation	85
4. Acetyl-mandelate	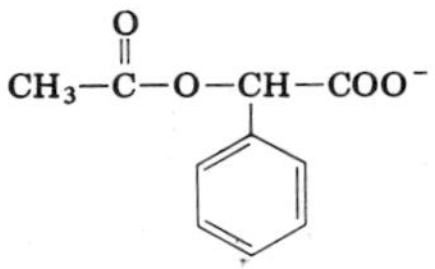 CH₃–C(=O)–O–CH(C₆H₅)–COO⁻	6×10^{-2}	30		86
Inhibitors:					
(a) Products					
1. L-Phenylalanine	NH₃⁺–CH(CH₂C₆H₅)–COO⁻				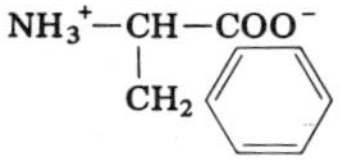 30,71
2. L-Phenyllactate	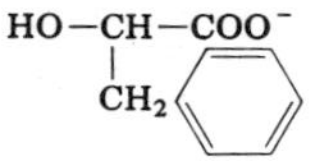 HO–CH(CH₂C₆H₅)–COO⁻	5.8×10^{-5}			84
3. Cinnamate	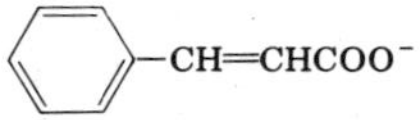 C₆H₅–CH=CHCOO⁻	0.005		May bind differently from other products?	83,107
(b) Analogs					
1. β-Phenylpropionate	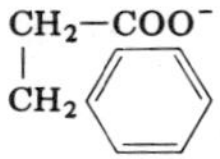 C₆H₅–CH₂–CH₂–COO⁻	$6.2-19 \times 10^{-5}$		Competitive, mixed non-competitive	23,70,87,88
2. D-Phenylalanine	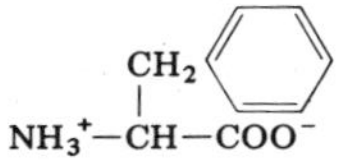 NH₃⁺–CH(CH₂C₆H₅)–COO⁻	0.002			23,89
Modifiers:					
1. Carbobenzyloxy-glycine	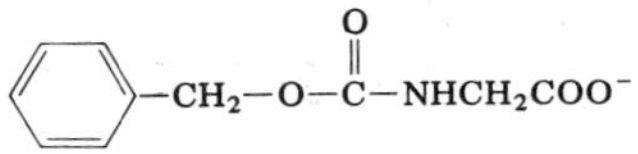 C₆H₅–CH₂–O–C(=O)–NHCH₂COO⁻	0.016-0.029		Activates acyldipeptides; inhibits esters, longer peptides	70,72,73

[a] Binding to apo-CPA, determined by gel filtration

tightly to the enzyme than the corresponding peptides. However the K_m values do not fall easily into two classes when more substrates are examined. If the small K_m for esters like HPLA and cinnamoyl-phenyl-lactate reflects non-productive binding, then the efficiency of hydrolysis of these esters would have to be very high indeed.

2. Kinetic-structure correlations

(a) Binding of substrates and inhibitors

The occurrence of alternate kinetic pathways has definite structural implications. For those substrates which display either substrate activation or excess substrate inhibition, there must be more than one binding site.

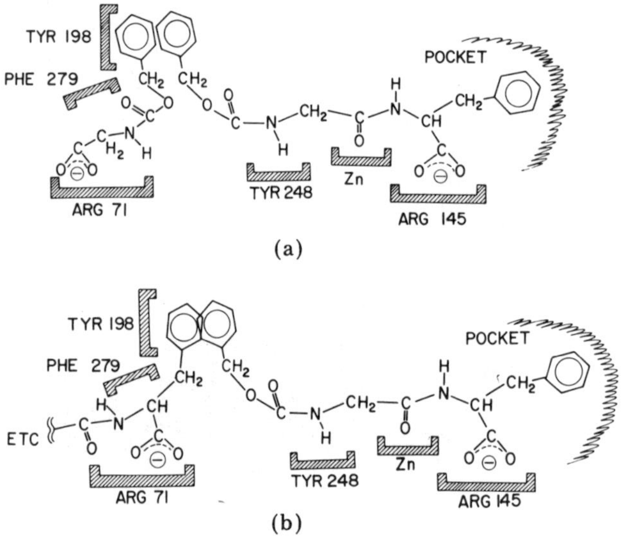

(a)

(b)

Fig. 12. Binding models to account for (a) the action of CBZ-Gly, an activator of CBZ-Gly-Phe hydrolysis and for (b) excess substrate inhibition.

Modifiers are also expected to bind simultaneously with substrates. No direct crystallographic observations have been made of complexes with modifiers or with substrates which should bind at several sites. However, a binding arrangement to account for the kinetic effects of the modifier, CBZ-Gly, has been suggested by model building[2,3] (Fig. 12). If the CBZ-group of CBZ-Gly is adjacent to Tyr 198 and its COO⁻ group interacts with Arg 71, then the modifier might perturb the orientation of a productively bound substrate. A second binding mode, utilizing the same interactions, has been proposed to explain excess substrate inhibition by CGP[2,3] (Fig. 12). However, the kinetics of substrate inhibition cannot be completely understood from this model. It is not easy to accommodate more than two acyldipeptide substrates in the proposed substrate binding area illustrated in Fig. 3, whereas kinetic analysis suggests the simultaneous binding of four or five molecules in solution[78], and more than two in the crystal[91]. The high order of reaction obtained kinetically may imply binding in other regions or may have more complex explanations[25].

Non-productive or competitive binding probably utilizes some of the productive enzyme–substrate interactions. Thus of the five interactions enumerated for Gly-Tyr, only one, between the substrate amino group and Glu 270, is considered non-productive. Other abortive complexes might form by shifting a substrate down the binding groove to permit the carboxyl group to bind zinc, or by "wrong-way" binding, in which the N-acyl substituent of an acyl-dipeptide occupies the specificity pocket and the C-terminal COO⁻ binds to Arg 71[2,3].

Structural and kinetic evidence both demonstrate that binding of certain inhibitors of the generic type RCOO⁻ is complicated by multiple modes of interaction. β-Phenylpropionate, the prototype of these inhibitors, was originally thought to behave in a strictly competitive manner[23], but subsequent work shows that the inhibition has both competitive and non-competitive components with respect to a single peptide substrate, and thus implies complicated binding behavior[8,88]. Nevertheless the inhibition and binding data appear to fit the formation of a one-to-one complex with $K_s = 10^4\ M$[70,90]. The binding mode(s) corresponding to this constant require the presence of zinc. Additional molecules of phenylpropionate are bound at concentrations above $10^{-3}\ M$. Unfortunately comparisons with the binding sites observed by crystal structure analysis are not straightforward, as in the crystalline state the affinity for phenylpropionate is decreased by two orders of magnitude, K_i being about $10^{-2}\ M$[91]. In this latter concentration range, two sites are partially occupied by the heavy atom analog, iodophenylpropionate, one near Tyr 198 and the other in the specificity pocket (Fig. 10). Both orientations allow for an interaction between the carboxylate group and the active site metal ion, as shown by the nuclear magnetic resonance (n.m.r.) data for Mn-CPA[65]. Moreover, in the absence of zinc, the affinity of the site in the pocket is lessened and no binding occurs at Tyr 198 up to concentrations of 0.10 M. Two additional binding sites for iodophenylpropionate have been observed crystallographically at higher inhibitor concentrations, but both are further removed from zinc.

The relation of the sites shown in Fig. 10 to the kinetic phenomena is not completely obvious. Binding in the pocket would seem to imply competitive inhibition toward any substrates oriented like Gly-Tyr. On the other hand it may be possible for the inhibitor near Tyr 198 to bind to zinc in a manner which does not interfere sterically with the binding proposed for longer peptides in Section II 4a. Such a non-competitive model might require a five-coordinate zinc atom. The binding data which indicate a 1:1 interaction would seem to substantiate an either/or type of binding at the two sites observed crystallographically[64], but the protection of two tyrosines from modification in the presence of high concentration of phenylpropionate[95] is more readily explained by simultaneous binding of two (or more) molecules of inhibitor. In retrospect it is probably unfortunate that this inhibitor, which behaves in such a complex manner, has so frequently been employed in studies of the active site of CPA.

A discussion of models for the binding of ester substrates has deliberately been deferred until this point. Direct structural information is lacking as attempts to prepare crystalline complexes of CPA with esters have so far been unsuccessful. Some, but not all, ester substrates differ qualitatively from peptides in their behavior. For example, activators of peptide hydrolysis inhibit hydrolysis of cinnamoyl-phenyllactate[72]; and β-phenylpropionate inhibition is predominantly noncompetitive toward peptides but competitive with respect to HPLA[70]. Because of these differences the suggestion has been made that productive binding of esters and peptides is not always identical[92]. A suitable description of ester binding will require experiments which can ascertain the relative positions of critical groups in enzyme and substrate, *e.g.*, fluorescence and resonance measurements, or additional crystal structure analyses★.

(b) Activity as a function of pH

Changes in enzyme activity with pH can be sometimes interpreted on the basis of a model which assumes that catalysis is dependent on the ionization state of specific residue(s). From curves of k_{cat}/K_m *vs.* pH it is then possible to derive ionization constants, K_{1e}★★ and K_{2e} in the case of a bell-shaped curve, for critical groups in the free enzyme. Alternatively, ionization constants for an E · S complex can be obtained from the relation between k_{cat} and pH[92a]. Determination of K_e values for CPA should therefore be useful in elucidating the function of tyrosine and of other residues in the vicinity of the substrate.

The appropriate data have been meager. Recently one peptide substrate, CBZ-Gly-Gly-Phe, has been studied under conditions where substrate inhibition or activation do not complicate the results[7]. Curves of k_{cat}/K_m *vs.* pH are bell-shaped, and thus suggest a role for both basic and acidic groups in the reaction. k_{cat} *vs.* pH curves for this acyl-tripeptide have an inflection point near pH 6 and attain a plateau value independent

---

★ In a recent X-ray study at 6 Å resolution, a product of ester hydrolysis, L-β-phenyl-lactate, has been shown to bind to CPA in the same manner as does L-phenylalanine[128].
★★ K_{1e}, dissociation constant for free enzyme, from acidic limb of bell-shaped curve; K_{2e}, constant from basic limb; K'_{es}, apparent dissociation constant of E · S complex, as determined from k_{cat} *vs.* pH.

of pH up to pH 10.5*. Failure to find evidence for titration of a second group in the $E \cdot S$ complex does not eliminate participation of acid catalysis in the reaction. For example, an acid-catalyzed step could occur after the rate-determining step (however, cf. ref. 92c). The pH dependence of CGP hydrolysis had been presented in several earlier investigations. With data at high substrate concentrations, the individual k_{cat} and K_m vs. pH curves showed complicated behavior[23]. K'_{es} values of 6.5 and 8.6 can be derived from curves of v_0 vs. pH[93] but comparison with acyl-tripeptides indicates that the CGP data may bear re-examination. At lower concentrations k_{cat} is less dependent on pH[78].

k_{cat}/K_m vs. pH has a bell shape for two esters, acetyl-mandelate[86] and cinnamoylphenyllactate[84]. Ionization constants for the free enzyme are $10^{-6.9}$ and $10^{-7.5}$ and $10^{-6.5}$ and $10^{-9.4}$, respectively. No explanation for the discrepancies in the alkaline K_e values has been offered. Initial comparisons of the pH dependence of initial velocities for HPLA and CGP indicated that esters and peptides differed categorically, with esters showing sigmoid and peptides bell-shaped curves. Since the reports on CBZ-Gly-Gly-Phe, acetyl-mandelate and modified carboxypeptidases[97], these distinctions no longer hold.

If a different step is rate determining, depending on the substrate, one might account for some of the discrepancies amongst the pH curves for various substrates[94]. Finally it must be recognized that the simplest model is not the only basis for pH effects — conformation changes and alterations of rate-determining steps with pH must also be considered[56].

3. *Chemical studies of the role of various residues in the active center*

(a) *Modification of residues in the substrate-binding site*

Chemical modifications of tyrosine have provided decisive evidence for some role for this residue in the active center of CPA**. Differential modification in the presence and absence of substrates or inhibitors has been an especially effective tool, and is now being being supplemented by assignment of the site of modification to a known locus in the primary structure. Tyrosyl modifications have included acetylation with acetic

---

* A full account of the effect of pH on tripeptide hydrolysis has now appeared[92b]. Both K_m and k_{cat}/K_m depend on an acidic group which ionizes near pH 9. The acidic group could be Tyr 248, which would thereby be implicated in substrate binding. Several alternative assignments of the lower pK observed in k_{cat} or k_{cat}/K_m curves have been suggested.
** Tyrosine has been assigned the role of proton donor in the mechanism presented in Section III 5a.

anhydride[93] or acetylimidazole[95], iodination[96], nitration with tetranitro-methane[97] and diazotization with either diazonium-1H-tetrazole (DHT)[98] or diazo-*p*-benzenearsonate[99]. Acetylimidazole reacts with CPA to form about five *O*-acetyl-tyrosines, but the functional effects of acetylation have been attributed to reaction of two tyrosines, both of which can be protected from modification by β-phenylpropionate. After acetylation, the k_{cat} for peptides (including acyl-tripeptides) is reduced to 3–6% of its value in unmodified CPA[70,82] while HPLA (but not acetyl-mandelate[99a]) is still efficiently hydrolyzed. The most dramatic effect on HPLA hydro-lysis is the suppression of substrate inhibition, *i.e.*, the onset of inhibition is displaced to higher substrate concentrations[71]. Both k_{cat} and K_m for HPLA hydrolysis are increased by acetylation[71,82].

Reagents other than acetylimidazole modify selectively one or the other of the two tyrosines with differing effects on peptide activity. Nitration of a single tyrosyl group is correlated with progressive loss of peptidase activity[97] while introduction of about one mole of diazo-benzenearsonate, apparently at the same residue[99], is accompanied by a 50% decrease in peptidase activity. On the other hand, reaction with moderate excesses of diazo-1H-tetrazole modifies approximately one tyrosine with minimal effects of peptide hydrolysis. Nitro-, arsanilazo- and tetrazolylazo-CPAs all hydrolyze the ester, HPLA, but show diminished substrate inhibition. Subsequent reaction of tetrazolylazo-CPA with tetra-nitromethane leads to introduction of a nitro group and loss of most of the peptidase activity. It has been concluded that the nitro and tetra-zolylazo groups are on different tyrosine residues in this doubly modified CPA[97]. If so, one tyrosine is crucial for peptidase activity while both may play some role in substrate inhibition by HPLA. Final resolution of the sites of modification will depend on sequencing of the derivatives.

Identification of one of the two critical tyrosines as residue 248 is based upon the location of this side chain in the three-dimensional E · S structure and upon the sequence of a peptide isolated after differential iodination of CPA. Aliquots of the enzyme were iodinated with ^{125}I in the absence of any inhibitor and with ^{131}I in the presence of phenyl-propionate. The peptide with the largest ^{125}I/^{131}I ratio was isolated from peptic digests of the combined reaction mixtures. Its sequence proved to be Ile-Tyr-Gln-Ala[100], which corresponds uniquely to positions 247–250 in the polypeptide chain[4]. The conformational change attendant on binding phenylpropionate could account for the decreased reactivity of Tyr 248 toward iodine.

Although it is not near the susceptible peptide bond, Tyr 198 is likely to be in van der Waals contact with substrates larger than glycyl-tyrosine (Figs. 3 and 8). The remaining tyrosine residues in the active site sequences point away from the substrate binding groove and pocket. Thus the second tyrosine shown by modification studies to participate in the active center is probably residue 198. Iodophenylpropionate binds to crystalline CPA in a site adjacent to the phenolic side-chain of Tyr 198 (Fig. 10) and analogous binding by phenylpropionate could account for the effect of that inhibitor on modification. The reactivity of tyrosines 198 and 248 may be partly explained by their exposure to surrounding solvent. In addition the tyrosine which is nitrated has a lower pK than other tyrosines in the molecule[101].

The position of Tyr 248 in the Gly-Tyr: CPA complex makes it highly probable that it is modification of this Tyr which has a profound effect on peptide hydrolysis. Thus the structure suggests that in nitro- and arsanilazo-carboxypeptidases the substituents are attached to Tyr 248. Difference spectra observed on addition of β-phenylpropionate[99,101,121], or glycyl-tyrosine[99] to native and modified CPA, although not uniquely interpretable *per se*, seem likely to arise in part from re-orientation of Tyr 248[2,3]. (Reports[102] that tetrazolylazo-CPA, presumably modified at a residue other than 248[97], also displays spectra which are sensitive to addition of glycyl-tyrosine emphasizes the necessity of sequencing the appropriate peptides in the several modified CPAs.)

The important effect of acetylation on peptide hydrolysis appears to be the decrease in k_{cat}, though K_m may also be altered. When CGP serves as a competitive inhibitor of HPLA hydrolysis by acetyl–CPA, K_i is an order of magnitude greater than the K_m for catalysis by unmodified carboxypeptidase[92], in agreement with the 15-fold increase in K_m observed directly[71]. The earlier conclusion that acetyl–CPA is inactive because it fails to bind CGP or dipeptides[103] is at variance with the competition experiments and with crystallographic evidence for formation of a complex between Gly-Tyr and acetyl–CPA. Indeed the binding of hippuryl-phenylalanine and CBZ-Gly-Gly-Phe is hardly affected by acetylation[70,71]. Despite the minimal effects of acetylation on binding of the latter substrates, it is not possible to assert from these modification studies alone that tyrosine must be directly involved in catalysis*. Investigation of the kinetic parameters and pH profiles for peptide hydrolysis by arsanilazo-CPA, which retains appreciable peptidase activity, yet loses its tyrosyl proton at a lower pH than unmodified CPA[104], would be fruitful in probing the postulated role[2,3] of Tyr 248 as a proton donor.

The demonstration of substrate interactions with Glu 270 and Arg 145 in the crystalline complex has stimulated efforts to modify these residues[105]. Reaction of approximately three arginyl residues of CPA with butanedione results in appreciable (85%) losses in peptidase activity[106].

---

*Conceivably a K_m corresponding to a small fraction of the enzyme was measured in the kinetic experiments[71]. However, it is interesting that a significant fraction of peptidase activity is retained by acetyl-CPA even though the reagent excess is increased to 500-fold[70].

Deblocking of one arginyl group restores activity. Again, as noted for tyrosyl modifications, esterase activity is not proportionately affected. The site of modification has not been determined, nor has it been reported whether the reduction in peptidase activity is primarily an effect on k_{cat} or on K_m. Modification of a carboxylate group with N-ethyl-5-phenylisoxazolium-3'-sulfonate (Woodward's reagent K) is accompanied by loss of both peptidase and esterase activities; inactivation is decreased in the presence of either β-phenylpropionate or glycyl–tyrosine. A second COO^- is highly reactive with the reagent but its modification is kinetically not significant. A peptide isolated after replacement of reagent K by $^{14}OCH_3$ appears to contain the latter residue, Glu 88[107]. Preparation of a mono-substituted derivative of Glu 270 will be crucial in chemical confirmation of the role assigned to this residue (Section III 5).

Chemical modification implicates not only tyrosyl, carboxylate, and arginyl groups, but also histidine, in the enzymatic activity of CPA. Photo-oxidation of histidine in the presence of methylene blue[108] or Rose Bengal[109] leads to loss of peptidase activity, as does reaction of approximately one histidyl residue with diazo-1H-tetrazole[98]. The site of reaction is as yet unidentified, but the only histidine residues in the active center are the zinc ligands, residues 69 and 196. In these experiments, activity is not altered solely by a loss of zinc from the enzyme. While some zinc is lost during oxidation with methylene blue, tetrazolyl-azo-His-CPA retains its full complement of zinc, and moreover has esterase activity, k_{cat} for HPLA being about one-half that for the unmodified protein[98].

(b) Chemistry of the metal ligands and the disulfide bond

Prior to determination of the three-dimensional structure and the chemical sequence, considerable effort had been expended in attempts to identify the zinc ligands. As in the investigations of substrate binding and catalytic groups, a basic approach was to compare reactivities in the presence and absence of the bound component, in this case, the metal ion. The conclusions were that zinc attached to two residues, cysteine and the α-amino group. In fact the ligands are glutamate and two histidines. It may be instructive to review some of the pitfalls encountered in deducing the nature of the ligands from chemical data.

The binding of PHMB to apo- and Zn-CPA was compared using difference spectroscopy; the results were consistent with the presence of one reactive sulfhydryl group in the apoprotein and none in the zinc enzyme. The conclusion that cysteine was a zinc ligand seemed to be confirmed by an effect of ferricyanide on the apoenzyme and by the observation that silver and zinc competed for the apoprotein[22]. Since the stability constants for a series of metals could be correlated with values predicted from models for a bidentate ligand in which the donor atoms were nitrogen and sulfur[110], the effect of reagents specific for amino and imidazole groups was examined.

Photo-oxidiation of histidine residues led to progressive decreases in activity and in zinc content[108], but the rate of inactivation was unaffected by the presence of zinc and it was concluded that histidine did not bind to zinc. On the other hand it appeared that the α-amino group was protected from modification in the metalloprotein[108]. Because the pK of 7.7 observed by complexometric titration of the apoprotein was reasonable for an α-amino group, the proposal was made that a second metal ligand was the N-terminal residue, despite the finding that acetylation of the α-group failed to affect the enzymatic activity[111].

Re-examination of the data, now that the structure is known, forces one to conclude that in the spectrophotometric titrations with PHMB the mercurial reacted with zinc ligands. The absence of –SH groups and the high affinity of the zinc site for Hg^{2+} suggest that the spectral changes resulted from combination of PHMB with imidazole and/or a COO^- group. Spectral differences in the 250 nm region had previously been noted on addition of PHMB to EDTA[112]. The experience with CPA should serve as a warning that other specific metal binding sites not containing sulfhydryl groups may react similarly with mercurials to produce spectral shifts. Correlations with model compounds are uncertain, for it is admittedly[113] difficult to find models which mimic successfully the properties of metallo-enzymes. To elucidate the metal-enzyme interaction in three dimensions, there is presently no alternative to three-dimensional crystallographic solutions. Indeed, crystallography provides the most unequivocal method of demonstrating what residues are metal ligands. But because of the difficulties in X-ray identification of side chains[42], the chemical sequence information is also essential to a full description of the metal-binding site.

The conclusion[110] that a sulfhydryl group was a zinc ligand caused the presence of the disulfide bond[1] to go unrecognized. Amino acid analysis had established a half-cysteine content of two per mole. Therefore the PHMB titration results implied that there must be a second, unreactive, –SH in apo-CPA. Neither the purported active site –SH nor the second –SH could readily be alkylated unless the protein had been treated with reducing agents such as mercaptoethanol. After reduction and alkylation in the presence of phenylpropionate, the enzyme retained its activity and approximately one mole of carboxymethyl-cysteine was found per mole of protein. Subsequent removal of the inhibitor and repetition of the reaction introduced a second carboxymethyl group and led to inactivation. The amino acid sequence about the first cysteine was assigned to the region of the "non-essential" sulfhydryl and that about the second to the active site[114].

After completion of the sequence analysis, alkylation of the "cysteines" was reinvestigated. With procedures for isolating the peptides in high yields, it can be shown that reduction and alkylation in the presence of phenylpropionate results in reaction at one or the other, but not both, of the residues 138 and 161. The results confirm the existence of a disulfide bridge between these two residues[115].

4. Why the metal?

The function of the metal in carboxypeptidase can be considered under two headings: a role in substrate binding, and a role in catalysis. The evidence favors the view that in active metallo-CPAs the metal ion participates in both binding and catalysis, but it is logically useful to differentiate between an effect of the metal in orienting the substrate on the enzyme surface and its participation in lowering the energy of the critical transition state complex.

Clear evidence for types of direct bonding of the active site metal to substrates and inhibitors is provided not only by the X-ray maps showing the carbonyl oxygen of the peptide adjacent to zinc, but also by n.m.r. measurements. The broadening of the proton resonances of several inhibitors of the type $RCOO^-$ by addition of Mn-CPA can be explained if the COO^- is bound in the first coordination sphere of the Mn ion[60,65]; the effect of phenylpropionate on the resonance of $^{19}F^-$ bound by Mn in Mn-CPA confirms this conclusion[61,66]. However, a full assessment of the effect of metal ions on the geometry of substrate or inhibitor binding requires comparison of the precise geometries of apo- and various metallo-CPAs, in the presence and absence of substrates.

Estimates of the strength of the substrate–metal interaction are difficult to obtain. One approach has been to determine binding constants in the presence and absence of zinc. Differences between these constants might approximate the unitary free energy change for metal-substrate bonding, if removal of the metal leaves both E and E · S structures otherwise intact. In practice K_m values must be used instead of K_{eq} in the presence of zinc. Several peptides bind to apo-CPA with constants not very different from those for the zinc enzyme[79] (Table IV). Nor is the strength of peptide–CPA binding very dependent on the nature of the metal for the series of active peptidases (Table V). The apparent absence of an effect of the metal on peptide binding is striking in comparison with the observed metal–substrate interaction and with preliminary theoretical calculations indicating that in the complex some charge is transferred from the peptide carbonyl group to the metal ion[116]. Since a metal ion which binds the substrate's carbonyl oxygen strongly will also bind water strongly, the experimental observations may be partly reationalized if the free energy change for the reaction $M \ldots OH_2 \hookrightarrow M \ldots O = C\!\!<$ is small. In addition the assumption that the E · S structure is unchanged upon removal of the metal is most probably incorrect. Crystallographic studies at low resolution indicate that, although Gly-Tyr is oriented similarly in the apo- and Zn-CPA complexes, the density attributed to rearrangement of Glu 270 is missing from the apo-CPA complex[64]. Finally other conformational changes, especially movement of the metal ion, need not be the same for different metallo-CPAs.

The binding constants for HPLA, acyl-amino acids, and carboxylate inhibitors do change significantly when zinc is removed. In the absence of any metal component these compounds do not affect the reformation of Zn-CPA from the apoenzyme, up to concentrations of the order of 0.01 to 0.05 M[81]. The K_m values for complexes of these compounds with Zn-CPA fall in the range 10^{-5} to 10^{-2} (Table V). Thus removal of the metal changes the equilibrium constant for HPLA by more than 10^2. Moreover, the

TABLE V

STABILITY AND KINETIC CONSTANTS FOR VARIOUS METALLO-CARBOXYPEPTIDASES[38,70,82]

Metal ion	Log K[a]	Substrate							
		BGGP		BGP		CGP		HPLA	
		K_m[b]	k_{cat}[b]	K_m	k_{cat}	K_m	k_{cat}	K_m	k_{cat}
Zn	10.5	8.0	1.2	8.1	5.6	19.5	5.5	0.76	28.6
Ni	8.2	7.4	1.1					2.1	27.6
Co	7.0	6.0	5.9	4.8	7.4	11.7	12.3	0.98	37.7
Mn	5.6	2.9	0.23	1.1	0.45	22.9	2.3	3.2	56.8
Cd	10.8							5.5	61.5
Hg	21.0								
Cu	10.6	inactive toward all substrates							

[a] Stability constant corrected for competition by 1 M Cl$^-$ and 0.05 M Tris buffer, pH 8
[b] K_m values are $M \times 10^4$; k_{cat} are min$^{-1} \times 10^{-3}$

affinity of CPA for phenylpropionate varies with the metal present[90], in the order Zn > Cd > Co, and the K_i values for carboxylate inhibitors of Mn-CPA are about one order of magnitude less than the corresponding constants for the zinc enzyme[66]. However, one is reluctant to make structural interpretations of these effects of metals on affinities.

Carboxypeptidases which do not catalyze hydrolysis are still competent to bind substrates. Cadmium, copper, and mercury carboxypeptidases all complex with dipeptides or acyl-dipeptides in a mode which alters the rate of metal exchange. HPLA is likewise bound to the inactive Cu-CPA[81]. Quantitative data on binding constants are lacking but retention of binding capacity means that the absence of enzymatic activity must be attributed to causes other than failure to bind substrates at the active site.

One obvious role for the metal in catalysis is to act as a Lewis acid, attracting electrons from the carbonyl carbon; this notion has been incorporated in various proposed mechanisms for CPA[117,118]. It receives direct support from the crystal structure which shows a carbonyl oxygen–Zn bond. Moreover, the variations among metallo-carboxypeptidases are expressed as variations in k_{cat}. However, the activities of the several metallo-CPAs do not follow the Irving–Williams order of Lewis acid strength, Mn < Fe < Co < Ni < Cu > Zn[5]. For peptide substrates the order is Co > Zn = Ni > Mn > Cu = 0; and for esters the order is Cd > Mn > Co > Zn = Ni > Hg > Cu = 0 (Table V). An adequate explanation of the way in which the protein perturbs the "natural" catalytic order is essential to understanding the functional properties of this metalloenzyme. It cannot be said that any deep insight is currently available. Geometry and steric interference may be especially important in enzyme-catalyzed hydrolysis. For example, Cu-CPA may be out of its predicted sequence because the available coordination position, defined by the protein ligands and the restrictions of Cu^{2+} geometry, cannot possibly be occupied by the carbonyl oxygen of a substrate in an orientation permitting transfer of some charge to the metal ion. Indeed, in low resolution maps of the Cu–CPA complex with glycyl–tyrosine[64], no positive region is observed near Glu 270, suggesting that the α-amino to Glu 270 interaction may be lacking. With low resolution data one can only speculate whether the carbonyl oxygen–metal interaction is also absent. In model systems Cd^{2+} is as good a Lewis acid as Co^{2+}, yet Cd-CPA is unable to catalyze peptide hydrolysis. The inactivity of Cd-CPA is not readily explicable on grounds involving only the geometry of the metal ligands[5], but suggests that other factors, such as ionic size, must be considered. Restrictions imposed by the folding of the protein must also be crucial. For example, Hg^{2+} (Section II 4c) is displaced ∼ 1 Å from the zinc position. Both Hg^{2+} and Cd^{2+} are large enough that a substrate bound to either, if correspondingly displaced, might fail to hydrogen-bond to Tyr 248[2,3]. Changes in the number of ligands and ligand release or exchange during reaction have been suggested as general explanations for the unique reactivities of the protein when a functional metal ion is present[5].

Model systems incorporating Cu^{2+} or Co^{3+} are known to catalyze the synthesis and hydrolysis of peptide or ester bonds[119,120,120b]. The Co^{III} complexes, discussed in Chapter 1, activate attack at a carbonyl group whose oxygen is co-ordinated to Co^{3+}. However, protein catalysis seems more efficient than that of the models. Dielectric shielding has been suggested to account in part for the rate enhancements observed in CPA[42]. Upon binding substrate, the zinc, bearing a formal charge of +1 in the presence of Glu 72, is no longer in contact with bulk solvent, but finds itself in a medium of low dielectric constant. The field resulting from the charge on the metal is therefore much stronger and polarization of the carbonyl electrons is correspondingly increased.

5. Mechanism of action

(a) A proposal based on the crystal structure

On the assumption that the glycyl–tyrosine complex is very similar to the kinetically significant $E \cdot S$ complex, a mechanism for peptide hydrolysis has been constructed in which Zn, Glu 270 and Tyr 248 play the crucial roles[2,3]. The X-ray study shows that the only side chains of the protein which can approach atoms of the peptide bond are Glu 270 and Tyr 248. Therefore, in the pH range of interest, the only candidates for proton donors are Tyr 248 or H_2O, and the only possible agents for nucleophilic attack are Glu 270 or H_2O. Since glycyl–tyrosine is a competitive inhibitor of the hydrolysis of longer peptides, and since other peptides bind similarly in CPA and in modified CPA crystals, the mechanism is considered to apply to peptides in general; hydrolysis of certain esters, such as HPLA, displays so many special features that it must be discussed separately. The following steps are included in the peptidase reaction, though the relative rates and ordering in time are not all known. (1) *Binding.* The substrate binds with its C-terminal side chain in the pocket, the specificity site, and its COO^- aligned with Arg 145. The peptide –NH receives a hydrogen bond from Tyr 248, which has moved from its position in the native protein. The carbonyl oxygen attaches to zinc. This arrangement may twist the peptide bond slightly out of plane and cause the incipient amino group to be somewhat tetrahedral. (2) *Metal catalysis.* Zinc polarizes the C=O bond, increasing the electrophilic character of the carbonyl carbon so that it is more readily attacked either by water or by a nucleophile on the protein. (3) *Attack on carbonyl carbon.* The carboxylate of Glu 270 moves, not to interact with the α-amino group of the substrate as observed in the dipeptide complex, but rather either (a) to form an anhydride in which the substrate is covalently bonded to the protein, or (b) to act as a general base, catalyzing attack of water at the carbonyl C. (4) *Peptide bond cleavage.* The peptide bond is broken, accompanied by proton donation from Tyr 248 to the leaving amino group, resulting in either products or an acyl enzyme. (5) *Release of products.* If a covalent intermediate forms, it is attacked by H_2O, probably via a base-catalyzed route and possibly promoted by a $Zn^{2+} - H_2O$ (or $Zn^{2+}OH^-$) complex. As hydrolysis is completed, the new acidic and basic groups are titrated and a proton is transferred to the O^- of Tyr 248 which returns to its "native" position (Fig. 13).

An interaction has been included to account for the efficiency of hydrolysis of longer peptides with unsubstituted penultimate NH groups: the formation of a hydrogen bond[60,61] from the second peptide NH to Tyr 248[2]. To form this bond, the second NH is displaced, by rotation about single bonds, from the location found in the Gly–Tyr complex. A ring hydrogen-bonded in this manner restricts the substrate conformation

474

and may induce some strain in the substrate molecule; it also provides a ready mechanism for facilitating proton donation to the leaving amino group (Fig. 13). In contrast, the α-N .... Glu 270 interaction found in the Gly–Tyr complex is impossible for longer substrates, and probably interferes with step (3).

This mechanism for peptide hydrolysis embodies some general proposals about how enzymes achieve efficient catalysis. Strain[56a,56b,56c,57] may be induced in the substrate by formation of the hydrogen-bonded ring involving Tyr 248, by the metal–substrate bond, and by the carboxylate salt link to Arg 145. Perhaps the geometry of the zinc ligands in the E · S complex is such as to destabilize the initial state relative to the transition state. The movement of zinc in the Phe–Gly–Phe–Gly[128] complex (and also in the Gly–Tyr complex) is interesting in this connection. In addition,

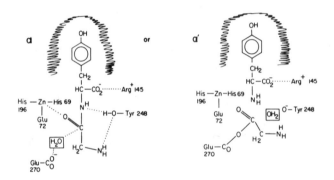

Fig. 13. Two possible roles of Glu 270 during hydrolysis. In (a) Glu 270 acts as a basic catalyst for the attack of water and in (b) it forms an intermediate anhydride.

isolation of the charge on the metal[42] may contribute independently of geometric strain[120a], since it produces an E · S complex with a relatively high energy.

The above mechanism shares certain features with some of those proposed prior to the structure determination[88,117,118]. Polarization by the metal, acting as an acid catalyst, had been suggested earlier. A base and an acid must be expected to participate in hydrolysis. What the crystal structure analysis now provides is the identification of those side-chains which can approach the substrate, and the detailed stereochemistry essential to definitive mechanistic schemes. Even though the kinetics of reactions with those proteins, such as CPA[91], which display their characteristic activities in the crystalline state often differ quantitatively in the crystal and in solution, strong inferences can be made from the accessible E · S or E · I structures. Thus the propinquity of Glu 270 and Tyr 248 to the substrate makes participation in the reaction very probable, whereas the absence of a bond between zinc and the peptide nitrogen suggests that metal stabilization of the leaving group[5] may play only a minor role in catalysis. But crystallographic techniques cannot yet be used to study the rapid reactions of highly specific substrates. Kinetic hypotheses based on the structure must be put to other tests. It is therefore the aim of the concluding paragraphs to touch on certain problems regarding the mechanism.

(b) Some alternatives

The above mechanism is "minimal" in the sense that it makes use of H_2O, just two protein groups, Tyr 248 and Glu 270, and the zinc ion for the catalytic steps. Chemical modification studies also implicate a histidine residue. While it is unlikely that either His 69 or His 196 could be a nucleophile, because of their distances from the substrate, one of these residues could conceivably dissociate from zinc[5] to act as a base in a proton transfer reaction. This need not be a rate-determining step. However, no movement of either of these histidines away from zinc is noted in the binding stage. By analogy with carbonic anhydrase, a mechanism involving attack on the substrate carbonyl carbon by OH^- bound to zinc has also been considered[2]. Steric hindrance seems to make such a mechanism mutually exclusive with, and also less likely than, one in which zinc acts to polarize the carbonyl oxygen of the substrate.

(c) Specificity and the conformation change

The absolute specificity of CPA for hydrolysis of C-terminal residues is at least partly understood. The inability of the enzyme to act as an endopeptidase can be accounted for by steric effects. It is not possible to place an internal peptide bond athwart the zinc without interference between the continuing peptide chain and residues 145 and/or 248. Selection of aromatic but not basic C-terminal residues for hydrolysis seems to be accomplished at the binding step[79]. It is more difficult to explain why the Gly–Tyr bond of glycyl-tyrosineamide is not split, for the amidated peptide is bound by CPA[81]. However, movement of Tyr 248 appears to be coordinated with the specificity toward free COO^- groups. If in the absence of a charge on the terminal residue, tyrosine fails to move, as is the case for lysyl-tyrosineamide, then the active site is incomplete because it lacks the phenolic group required for hydrolysis. The conformation change of Tyr 248 can apparently be followed in solution*[99,101,121]

---

* Spectral changes do not readily distinguish between movement of an aromatic residue from a hydrophilic to a hydrophobic environment and association of a modifier with the aromatic residue, nor do they establish that the extent of conformation change and the degree of complex formation are identical.

References pp. 482–487

so that further experiments to define the requirements for conformation change are feasible.

(d) Intermediates and the rate-determining step

The rate-determining step and the nature of the critical transition state complex have not been established. The formation of a tetrahedral intermediate, however transient, is assumed by analogy with the multitude of known carbonyl reactions where carbonyl carbon–oxygen cleavage occurs[56,121a]. The interesting question is whether the initial tetrahedral intermediate incorporates water or Glu 270. The latter alternative implies the presence of an intermediate covalently linked to the protein, and requires a second displacement reaction in which water attacks the acyl enzyme. From the structure, an anhydride involving Glu 270 is a most reasonable covalent intermediate. However, carboxylate is ordinarily a relatively poor nucleophile and amino a poor leaving group[56]. Alternatively, there appears to be room for Glu 270 to act as a base, catalyzing the attack of water on the carbonyl carbon.

To differentiate between nucleophilic displacement and direct attack by water, evidence for a covalent intermediate has been sought. Attempts to detect covalent intermediates during CPA-catalyzed hydrolyses have thus far met with no success. With cinnamoyl- and indolacryloyl- phenylalanine as substrates, no initial burst of hydrolysis is observed, nor are there any absorbance changes which might signal the presence of an intermediate[83]. Of course, the failure to detect intermediates may merely mean that the deacylation step is not rate limiting for these substrates.

Inferential evidence for an intermediate could be provided by the study of appropriate exchange or transfer reactions[56]. However, neither transpeptidation (or transesterification) of the type A–B $\longleftrightarrow$ A–X or A–B $\longleftrightarrow$ X–B has been observed[122,123]. Part of the difficulty may be that equilibrium lies so far toward the direction of hydrolysis; suitable coupling of the back reaction, as accomplished in the carboxypeptidase B catalyzed synthesis of trypsin inhibitor[124], might make more discriminating studies possible. The transfer of an acyl fragment to hydroxylamine has been very helpful in analysis of the chymotrypsin mechanism, but it has been impossible to detect any similar hydroxamate formation during reactions catalyzed by CPA. CPA catalyzes ^{18}O turnover in the carboxyl oxygens of acetyl-phenylalanine. The exchange shows a pH dependence reminiscent of hydrolysis and is inhibited by o-phenanthroline[122]. Analogous exchange in "virtual" substrates in the presence of chymotrypsin is presumed to proceed with the formation of covalent intermediates. However, the occurrence of oxygen exchange *per se* does not demand attack by a protein nucleophile; exchange could occur via direct displacement. It has been assumed that substrates undergoing exchange are bound so that the carboxyl group is at the same locus as the carbonyl function of

the peptide bond. The relevance of ^{18}O exchange in acetyl-phenylalanine
to the mechanism of carboxypeptidase would be reinforced by structural
information clearly demonstrating the orientation of acyl-amino acids with
respect to the metal ion.

Deuterium isotope effects have been measured to ascertain whether
proton transfer is rate limiting. If general base catalysis of attack by water
on either substrates or acyl enzyme is rate determining, then k_{H_2O}/k_{D_2O}
might be expected to be two, or more likely, greater. It appears that
during peptide hydrolysis, proton transfer(s) do not contribute significantly to the slow step, since the rate ratios for CGP[78] and BGP[125] are
less than 1.2. In constrast, hydrolysis of the ester substrates, HPLA and
cinnamoylphenyllactate displays a deuterium effect of 2[42,125]. However,
it is not safe to conclude that these esters are hydrolyzed by the general
base path.

Since neither isotope effects nor study of exchange has specified the
significant intermediates, determination of individual rate constants may
be helpful in resolving the mechanism. The use of a substrate such as
Dansyl-Gly-Trp, which is amenable to study by stopped-flow fluorescence,
gives promise for measurement of certain steps in catalysis[8]. Each step in
the correct reaction sequence must be at least as fast as the overall steady
state turnover. For example, tyrosine 248 must move toward the substate
if it is to donate a proton as postulated for the peptidase reaction. Spectral
changes observed on addition of substrates and inhibitors to nitro- or
arsanilazo-CPA, if they truly measure reorientation of tyrosine 248, should
be at least as fast as turnover. The rate constant derived from preliminary
measurements of the spectral change resulting from addition of Gly–Tyr
to arsanilazo-CPA is at least 200 sec^{-1} at 25°C. The spectral change is
thus as rapid as the turnover of the best peptide substrates for CPA,
suggesting that the conformation change can occur rapidly enough to be
one of the reaction steps[127].

Further details of the reaction scheme for peptides may soon be
elucidated. However, hydrolysis of those esters for which HPLA is representative shows a number of special features. The several differences
between hydrolysis of these esters and peptide hydrolysis include the
effects of tyrosyl modification, of metal substitution, of D_2O, and of
inhibitors and activators. The proposal has been made that these esters
are hydrolyzed by a mechanism unlike that for peptide hydrolysis[117] and
that such esters bind to CPA at a site different from, or only partly overlapping, the peptide binding site[92,105]. While the known distinctions do
indeed demand quantitative differences in the collection of rate constants
which control turnover, and imply that the rate determining steps may
not be identical for all substrates, there remain alternatives to the dual
mechanism hypothesis. For example, if it is assumed that some esters may
be hydrolyzed without proton donation by Tyr, the activity of nitro-, Cd-,

and Hg-CPA toward HPLA can be rationalized[2,3]. The crucial question, whether the reaction path for some substrates involves different catalytic groups and dissimilar intermediates. deserves more incisive study.

IV. OTHER PEPTIDASES

1. *Carboxypeptidase B*

Carboxypeptidases which rapidly release the *C*-terminal residues Lys and Arg from peptides have been purified from both porcine[129,130] and bovine sources[131,132]. Because their best substrates are basic residues, these enzymes are referred to as carboxypeptidases B. CPB proves to be closely related to CPA in a number of respects: the molecular weight is approximately 34,000[131,133] the zinc content is one per mole[133], and zinc removal or replacement is attended by changes in catalytic activity[134]. Amino acid analyses combined with molecular weights have established that the chain length of CPB is about the same as that of CPA; for the porcine CPB[133] the estimate is 304 residues and for the bovine enzyme, 300–301[131,132,135]. Sequence analysis of bovine CPB is partly completed[135-137], but no crystallographic studies have been reported. Active CPB is generated by tryptic hydrolysis of a proenzyme of molecular weight 57,000[131,138].

To establish the specificity of CPB several substrates and inhibitors have been examined[130]. The patterns are reminiscent of those observed for CPA. Maximum hydrolysis rates are attained with model peptides such as hippuryl–lysine[139] (k_{cat} = 1.32 x 10^4 min^{-1}, K_m = 7.7 x 10^{-4}), hippuryl–arginine (k_{cat} = 6.3 x 10^3 min^{-1}) and the ester analog, hippuryl–argininic acid (k_{cat} = 1.43 x 10^4 min^{-1}, K_m = 0.4 x 10^{-4}). Comparison with Table IV shows that these values are similar to those for the corresponding substrates in the CPA system. Ester hydrolysis by CPB is subject to excess substrate inhibition, and amino- and hydroxy-acid products inhibit both ester and peptide hydrolyses[140]. The failure of CPB to hydrolyze either hippuryl-D-lysine or hippuryl-L-lysineamide parallels the behavior of CPA, as does the ability to accept as substrates peptides containing certain abnormal amino acids such as hippuryl-L-homoarginine and benzoyl-β-alanyl-L-lysine. The presence of proline in the penultimate position reduces hydrolysis rates, and the dipeptide, glycyl-L-lysine, is a very poor substrate indeed[130]. Typical peptide substrates of CPA are hydrolyzed, at least by bovine CPB, but with diminished efficiency. Hippuryl-β-phenyllactate appears to be an equally good substrate for both bovine CPA and CPB[131]. Hydrolysis of CPA substrates by CPB is not a result of contamination by CPA, for CPB is inhibited by analogs of CPB substrates, *e.g.*, ε-amino caproic acid, which do not inhibit the corresponding reactions of CPA[131]. Compounds possess-

ing COO⁻ and side chains similar to Arg or Lys are the most effective
competitive inhibitors[139]. As with CPA, inhibition by phenylpropionate
appears to be complex[131].

Comparison of analogous sequences of bovine CPB and CPA provides
evidence to support the assumption that the two enzymes are structural
homologs[141,142]. It has been suggested that the structural genes for the two
carboxypeptidases have evolved from a common ancestor[141-143]. Amino
acid compositions of the two proteins are somewhat similar, one striking
difference being in the half-cysteine content, which is seven for CPB[131]
compared with two in CPA. Bovine CPB contains three disulfide bridges
and one sulfhydryl group. One disulfide loop of CPB, 14 residues in size,
seems to be homologous with residues 65–79 of CPA[137,141]. If this homology is correct, then the zinc ligands His and Glu, corresponding to residues
69 and 72 of CPA, are retained in CPB. Moreover, the CPA residues (66
and 79) homologous with the cysteines in this loop of CPB are situated near
one another in the CPA structure. There is another homologous sequence

```
                    250                 255
                     |                   |
CPA:  THR-ILE-TYR-GLN-ALA-SER-GLY-GLY-SER-ILE-ASP-TRP

CPB:  THR-ILE-TYR-PRO-ALA-SER-GLY-GLY-SER-ASP-ASP-TRP
```

Fig. 14. Homologous amino acid sequences in the active sites of bovine carboxypeptidases A and B.

in the active site. Reaction of an affinity-labelling reagent, 4-bromoaceta-
mido-butylguanidine, at the active site of CPB modifies a tyrosine residue[145].
A comparison showing excellent homology between the sequence about
this tyrosine and the sequence about Tyr 248 of CPA is reproduced in
Fig. 14[142]. On the other hand, the sequence homology between the C-
terminal regions of CPB (14 residues) and CPA is much less strong.

One further peptide sequence of CPB, in the region containing the
sulfhydryl group, has been reported[144]. A single cysteine residue in CPB
can be alkylated with iodoacetamide in the absence of reducing agents,
but only if zinc is first removed. Thus the chemical evidence suggesting
that cysteine is a zinc ligand is stronger than was the case for CPA. However, the sequence[146] surrounding this cysteine residue is homologous,
not with the region near the third zinc ligand (His 196) of CPA, but rather
with a segment at the beginning of the C-terminal helix of CPA. The
cysteine residue corresponds to Ala 290, whose α-carbon is 11 Å from the
zinc position. This finding poses a dilemma if one assumes that the
chemical evidence proves cysteine to be a zinc ligand in CPB: either two
highly homologous sequences are not in fact in similar positions in the two
proteins, or the tertiary structures are not truly homologous. The data

might be reconciled by assuming that the cysteine residue is buried inside the molecule, as is Ala 290, and becomes reactive only when removal of zinc permits some unfolding. It is tempting to suppose that cysteine is not a zinc ligand in CPB, but after reviewing the history of the efforts to assign the zinc ligands in CPA (Section III 3b), it seems wiser to wait until the chemical sequence and some direct comparisons of the three-dimensional structures have been completed.

2. *Other carboxypeptidases and procarboxypeptidases*

Enzymes with carboxypeptidase activity have been isolated from a number of sources[20,147]. Some, like the porcine[20a,130] and spiny Pacific dogfish enzymes[148,149] closely parallel bovine carboxypeptidases in specificity[20]; others have altered specificity patterns. For example, an enzyme from *Pseudomonas*, designated carboxypeptidase G, readily hydrolyzes glutamate from the *C*-terminus of peptides[150,151]. All the above-mentioned carboxypeptidases contain zinc[20], or at least have been shown to require zinc to be active[150], and in several, modification of tyrosyl residues affects activity[20]. The molecular weights, where known, are close to $35,000$[20]. However, the proenzymes vary considerably in size from one species to another. Thus the porcine pro-CPA and dogfish pro-CPA are monomeric and have molecular weights somewhat larger than $40,000$[20a,148], corresponding approximately to the size of subunit I of bovine pro-CPA[11]. Additional carboxypeptidases, from both bacterial[151a] and plant[151b,151c] sources, have been at least partly characterized. Some of these enzymes are not inhibited by chelating agents[151c], while others may conceivably be metalloproteins[151a].

Metal and substrate-binding sites exist in bovine pro-CPA[102,152]. The protein as isolated contains about one mole of zinc. However, complexometric titration of pro-CPA reveals significant differences between the metal-binding sites of the pro-enzyme and of CPA itself, and the steric requirements for substrate binding are not identical in the two proteins[152]. The activation of procarboxypeptidases has been the subject of intensive study; the reader is referred to several recent articles for leading references[20,153,154].

3. *Leucine aminopeptidase*

Leucine aminopeptidase (LAP) is an exopeptidase acting at the *N*-terminus of the peptide chain. A free amino group in a residue of the L-configuration is a prerequisite for substrates, although the second residue may possess the D-configuration[29,155]. *N*-acetyl peptides are resistant to hydrolysis. The enzyme receives its name from the observation that leucyl peptides, or analogs convenient for assay purposes, such as leucyl-*p*-

nitroanilide, are most rapidly hydrolyzed. However, the specificity is quite broad. Even L-lysineamide and L-prolineamide are reported to be substrates, the proteolytic coefficient for the latter being about 1% that of leucineamide[155,156]. Ester analogs such as n-butyl-L-leucinate[155] are slowly hydrolyzed[157]. β-Alanineamide is not a substrate, but leucyl-β-alanineamide is[156]. Determinations of K_m and k_{cat} have not been systematically performed, and it is not known to what extent the specificity of the enzyme depends on the initial binding step. From studies with a series of D- and L-alanyl peptides it appears that the active site of LAP can interact with at least four residues of a peptide substrate[29].

Hog kidney[158] has been the favorite source of LAP, which is a much larger molecule than the carboxypeptidases, having a molecular weight of approximately 300,000[158,159]. A preparation from beef lens has also been described[160,161]. Definitive studies of LAP have been hampered by the instability of the enzyme during preparation, and by the presence of contaminating endopeptidases[162,163] and of multiple molecular forms[159,164]. Both Mn^{2+} and Mg^{2+} stabilize and activate[165] the protein and preparations have therefore generally been carried out in the presence of one of these ions, usually Mg^{2+}. Activation of LAP as a function of Mn^{2+} concentration[156] fits reasonably to a binding curve with $n = 1$ and $K_a = 3.3 \times 10^4$. On the basis of this data and the inhibition produced by citrate and EDTA, LAP had until recently been classified as a Mg(Mn) activated enzyme or as a Mg(Mn) metalloprotein, although the metal content had not been determined.

A modification[159] of the preparation of hog kidney has now yielded an active LPA containing 4 to 6 moles of zinc per 300,000 molecular weight with negligible levels of other metals[166]. Enzyme prepared by the new procedure is stable in the absence of Mg^{2+}, though the latter ion is present during the preparation. Thus LAP in fact appears to be a zinc metallo-enzyme. Inhibition by o-phenanthroline and bipyridyl and the absence of inhibition by F^- or EDTA support this conclusion[166]. It has not so far been possible to obtain metal-free preparations which can be reactivated, but certain ions such as Cd^{2+} will exchange with zinc[166], and changes in the enzymatic activity parallel the metal content. Interestingly, exchange of somewhat more than half the zinc for Mn^{2+} results in an apparent increase in activity toward leucyl-p-nitroanilide. The effects on activity of several other metal ions, including Ca^{2+}, Co^{2+}, and Ni^{2+}, reported earlier[156], now require re-examination.

Schemes for hydrolysis in which the uncharged amino group of the substrate binds the metal ion (thought to be Mg^{2+} or Mn^{2+}) have been suggested[118,156,167]. Chemical information about the active site is minimal, except for the observation that hydrophobic compounds are excellent inhibitors[156]; together with the substrate specificity these results suggest the presence of a hydrophobic binding site. Indeed it is not clear how many active sites occur per molecule[161]. LAP seems to represent a good system for study of metal substrate interactions by n.m.r. methods. Transpeptidase activity cannot be demonstrated[168], and this observation must be accounted for in any proposed mechanism.

4. Some other metal-containing proteinases

Metallopeptidases constitute one of the four large classes of proteolytic enzymes, the others being the serine, sulfhydryl, and acidic proteinases[169]. Tabulations of peptidases known to require zinc, cobalt, or manganese for activity have appeared elsewhere[167,170]. While many of the known metallopeptidases happen to be exopeptidases or dipeptidases, the activities of this class of proteinases are not restricted to hydrolysis of *N*- or *C*-terminal residues. Recently, two endopeptidases, thermolysin[171] and the neutral protease of *B. subtilis*[172], have been shown to contain zinc. Ca^{2+} stabilizes both proteins. Each has a molecular weight in the 35,000 to 40,000 range[172,173]. Thermolysin hydrolyzes most efficiently the peptide bonds on the amino side of a hydrophobic residue[41,174-176]; the specificity of *B. subtilis* protease is reported to be similar[172], although suitable low molecular weight substrates for the latter enzyme have been difficult to obtain[177]. No doubt additional bacterial proteinases will prove to be metallo-enzymes[171]. As the number of well-characterized metal peptidases increases, it will be fascinating to see the extent to which they utilize common mechanisms of catalysis.

ACKNOWLEDGMENT

The authors would like to thank several associates, F. A. Quiocho, G. N. Reeke, J. A. Hartsuck, K. Seyfarth, and M. Le Quesne, for helpful discussions and for assistance during the preparation of this chapter. R. G. Schulman, A. S. Brill and H. Nicolas kindly permitted references to papers which were in press.

REFERENCES

1 G. N. Reeke, J. A. Hartsuck, M. L. Ludwig, F. A. Quiocho, T. A. Steitz and W. N. Lipscomb, *Proc. Natn. Acad. Sci. U.S.*, 58 (1967) 2220.
2 W. N. Lipscomb, J. A. Hartsuck, G. N. Reeke, F. A. Quiocho, *et al.*, *Brookhaven Symp. Biol.*, 21 (1968) 24.
3 W. N. Lipscomb, G. N. Reeke, J. A. Hartsuck, F. A. Quiocho and P. H. Bethge, *Phil. Trans. R. Soc.*, B257 (1970) 177.
4 R. A. Bradshaw, L. H. Ericsson, K. A. Walsh and H. Neurath, *Proc. Natn. Acad. Sci. U.S.*, 63 (1969) 1389.

5 A. E. Dennard and R. J. P. Williams, *Transition Metal Chemistry. A Series of Advances*, Vol. 2, Marcel Dekker, New York, 1966, p. 115.
6 R. G. Shulman, G. Navon, B. J. Wyluda, D. C. Douglass and T. Yamane, *Proc. Natn. Acad. Sci. U.S.*, 56 (1966) 39.
7 D. S. Auld, *Federation Proc.*, 28 (1969) 346.
8 S. A. Latt, D. S. Auld and B. L. Vallee, *Federation Proc.*, 29 (1970) 666.
9 M. L. Ludwig, J. A. Hartsuck, T. A. Steitz, H. Muirhead, *et al.*, *Proc. Natn. Acad. Sci. U.S.*, 57 (1967) 511.
10 M. Yamasaki, J. R. Brown, D. J. Cox, R. N. Greenshields, R. D. Wade and H. Neurath, *Biochemistry*, 2 (1963) 859.
11 J. H. Freisheim, K. A. Walsh and H. Neurath, *Biochemistry*, 6 (1967) 3010.
12 M. L. Anson, *J. Gen. Physiol.*, 20 (1937) 663.
13 B. J. Allan, P. J. Keller and H. Neurath, *Biochemistry*, 3 (1964) 40.
14 D. J. Cox, F. C. Bovard, J. P. Bargetzi, K. A. Walsh and H. Neurath, *Biochemistry*, 3 (1964) 44.
15 B. L. Vallee and H. Neurath, *J. Biol. Chem.*, 217 (1955) 253.
16 E. O. P. Thompson, *Biochim. Biophys. Acta*, 10 (1953) 633.
17 P. H. Petra and H. Neurath, *Biochemistry*, 8 (1969) 2466.
18 P. H. Petra and H. Neurath, *Biochemistry*, 8 (1969) 5029.
19 K. S. V. Sampath Kumar, J. B. Clegg and K. A. Walsh, *Biochemistry*, 3 (1964) 1728.
20 H. Neurath, R. A. Bradshaw, P. H. Petra and K. A. Walsh, *Phil. Trans. R. Soc.*, B257 (1970) 159.
20a J. E. Folk and E. W. Schirmer, *J. Biol. Chem.*, 238 (1963) 3884.
21 P. H. Petra, R. A. Bradshaw, K. A. Walsh, H. Neurath, *Biochemistry*, 8 (1969) 2762.
22 B. L. Vallee, T. L. Coombs and F. L. Hoch, *J. Biol. Chem.*, 235 (1960) PC45.
23 H. Neurath and G. W. Schwert, *Chem. Rev.*, 46 (1950) 69.
24 E. L. Smith, *Adv. Enzymol.*, 12 (1951) 191.
25 W. O. McClure, H. Neurath and K. A. Walsh, *Biochemistry*, 3 (1964) 1897.
26 M. L. Bender, J. R. Whitaker and F. Menger, *Proc. Natn. Acad. Sci. U.S.*, 53 (1965) 711.
27 R. P. Ambler, in S. P. Colowick and N. O. Kaplan, *Methods in Enzymology*, Vol. 11, Academic Press, New York, 1967, p. 436.
28 N. Abramowitz, I. Schechter and A. Berger, *Biochem. Biophys. Res. Commun.*, 29 (1967) 862.
29 I. Schechter and A. Berger, *Biochemistry*, 5 (1966) 3371.
30 S. Yanari and M. A. Mitz, *J. Am. Chem. Soc.*, 79 (1957) 1150.
31 E. L. Smith and H. T. Hanson, *J. Biol. Chem.*, 179 (1949) 802.
32 B. L. Vallee and H. Neurath, *J. Am. Chem. Soc.*, 76 (1954) 5006.
33 T. L. Coombs, J. P. Felber and B. L. Vallee, *Biochemistry*, 1 (1962) 899.
34 J. P. Felber, T. L. Coombs and B. L. Vallee, *Biochemistry*, 1 (1962) 231.
35 B. L. Vallee, J. A. Rupley, T. L. Coombs and H. Neurath, *J. Am. Chem. Soc.*, 80 (1958) 4750.
36 B. L. Vallee, J. A. Rupley, T. L. Coombs and H. Neurath, *J. Biol. Chem.*, 235 (1960) 64.
37 J. E. Coleman and B. L. Vallee, *J. Biol. Chem.*, 235 (1960) 390.
38 J. E. Coleman and B. L. Vallee, *J. Biol. Chem.*, 236 (1961) 2244.
39 M. Nomoto, N. G. Srinivasan, R. A. Bradshaw, R. D. Wade and H. Neurath, *Biochemistry*, 8 (1969) 2755.
40 R. A. Bradshaw, D. R. Babin, M. Nomoto, N. G. Srinivasin, *et al.*, *Biochemistry*, 8 (1969) 3859.
41 R. A. Bradshaw, *Biochemistry*, 8 (1969) 3871.

42 W. N. Lipscomb, J. A. Hartsuck, F. A. Quiocho and G. N. Reeke, *Proc. Natn. Acad. Sci. U.S.*, 64 (1969) 28.
43 F. W. Putnam and H. Neurath, *J. Biol. Chem.*, 166 (1946) 603.
44 W. N. Lipscomb, J. C. Coppola, J. A. Hartsuck, M. L. Ludwig, *et al.*, *J. Mol. Biol.*, 19 (1966) 423.
45 R. Diamond, *Acta Cryst.*, 21 (1966) 253.
46 G. N. Ramachandran, C. Ramakrishnan, and V. Sasisekharan, *J. Mol. Biol.*, 7 (1963) 95.
47 J. T. Edsall, P. J. Flory, J. C. Kendrew, A. M. Liquori, *et al.*, *J. Biol. Chem.*, 241 (1966) 1004.
48 L. Pauling and R. B. Corey, *Proc. Natn. Acad. Sci. U.S.*, 37 (1951) 251 and 729.
49 C. C. F. Blake, G. A. Mair, A. C. T. North, D. C. Phillips and V. R. Sarma, *Proc. R. Soc.*, B167 (1967) 365.
50 J. A. Hartsuck and W. N. Lipscomb, in P. D. Boyer, H. Lardy and K. Myrback, *The Enzymes*, 3rd edn., Vol. III, Academic Press, New York, 1971, p. 1.
51 D. van der Helm and H. Nicolas, *Acta Cryst.*, in press.
52 R. H. Kretsinger, F. A. Cotton and R. F. Bryan, *Acta Cryst.*, 16 (1963) 651.
53 M. M. Harding and S. J. Cole, *Acta Cryst.*, 16 (1963) 643.
54 S. Yanari and M. A. Mitz, *J. Am. Chem. Soc.*, 79 (1957) 1154.
55 W. Kauzmann, *Adv. Protein Chem.*, 14 (1959) 1.
56 W. P. Jencks, *Catalysis in Chemistry and Enzymology*, McGraw-Hill, New York, 1969.
56a H. Eyring, R. Lumry and J. D. Spikes, in W. D. McElroy and B. Glass, *The Mechanism of Enzyme Action*, Johns Hopkins, Baltimore, 1954, p. 123.
56b R. Lumry, in P. D. Boyer, H. Lardy and K. Myrback, *The Enzymes*, 2nd edn., Vol. 1, Academic Press, New York, 1959, p. 157.
56c L. Pauling, *Am. Scient.*, 36 (1948) 51.
57 W. P. Jencks, in N. O. Kaplan and E. P. Kennedy, *Current Aspects of Biochemical Energetics*, Academic Press, New York, 1966, p. 273.
58 D. E. Koshland, *Proc. Natn. Acad. Sci. U.S.*, 44 (1958) 98.
59 D. E. Koshland, *Cold Spring Harbor Symp. Quant. Biol.*, 28 (1963) 473.
60 J. E. Snoke and H. Neurath, *J. Biol. Chem.*, 181 (1949) 789.
61 H. T. Hanson and E. L. Smith, *J. Biol. Chem.*, 175 (1948) 833.
62 E. L. Smith, *J. Biol. Chem.*, 175 (1948) 39.
63 M. A. Stahmann, J. S. Fruton and J. Bergman, *J. Biol. Chem.*, 164 (1946) 753.
64 T. A. Steitz, M. L. Ludwig, F. A. Quiocho and W. N. Lipscomb, *J. Biol. Chem.*, 242 (1967) 4662.
65 G. Navon, R. G. Shulman, B. J. Wyluda and T. Yamane, *Proc. Natn. Acad. Sci. U.S.*, 60 (1968) 86.
66 G. Navon, R. G. Shulman, B. J. Wyluda and T. Yamane, *J. Mol. Biol.*, 51 (1970) 15.
67 W. H. Bishop, F. A. Quiocho and F. M. Richards, *Biochemistry*, 5 (1966) 4077.
68 A. S. Brill, P. R. Kirkpatrick, C. P. Scholes and J. H. Venable, Jr., *Proc. Fourth Johnson Foundation Colloq.*, *University of Pennsylvania*, 1969, Academic Press, in press.
69 J. E. Snoke, G. W. Schwert and H. Neurath, *J. Biol. Chem.*, 175 (1948) 7.
70 D. S. Auld and B. L. Vallee, *Biochemistry*, 9 (1970) 602.
71 J. R. Whitaker, F. Menger and M. L. Bender, *Biochemistry*, 5 (1966) 386.
72 S. Awazu, F. W. Carson, P. L. Hall and E. T. Kaiser, *J. Am. Chem. Soc.*, 89 (1967) 3627.
73 R. C. Davies, D. S. Auld and B. L. Vallee, *Biochem. Biophys. Res. Commun.*, 31 31 (1968) 628.

74 J. A. Thoma and D. E. Koshland, *J. Am. Chem. Soc.*, 82 (1960) 3329.
75 N. Izumiya and H. Uchio, *J. Biochem. (Japan)*, 46 (1959) 235.
76 C. C. F. Blake, L. N. Johnson, G. A. Mair, A. C. T. North, D. C. Phillips and V. R. Sarma, *Proc. R. Soc.*, B167 (1967) 378.
77 J. A. Rupley and V. Gates, *Proc. Natn. Acad. Sci. U.S.*, 57 (1967) 496.
78 R. Lumry, E. L. Smith, and R. R. Glantz, *J. Am. Chem. Soc.*, 73 (1951) 4330.
79 J. E. Coleman and B. L. Vallee, *J. Biol. Chem.*, 237 (1962) 3430.
80 J. Wyman, *Adv. Protein Chem.*, 19 (1964) 223.
81 J. E. Coleman and B. L. Vallee, *Biochemistry*, 1 (1962) 1083.
82 R. C. Davies, J. F. Riordan, D. S. Auld and B. L. Vallee, *Biochemistry*, 7 (1968) 1090.
83 W. O. McClure and H. Neurath, *Biochemistry*, 5 (1966) 1425.
84 P. L. Hall, B. L. Kaiser and E. T. Kaiser, *J. Am. Chem. Soc.*, 91 (1969) 485.
85 E. T. Kaiser, S. Awazu and F. W. Carson, *Biochem. Biophys. Res. Commun.*, 21 (1964) 444.
86 F. W. Carson and E. T. Kaiser, *J. Am. Chem. Soc.*, 88 (1966) 1212.
87 E. T. Kaiser and F. W. Carson, *Biochem. Biophys. Res. Commun.*, 18 (1965) 457.
88 R. Lumry and E. L. Smith, *Discuss. Faraday Soc.*, 20 (1955) 105.
89 H. Neurath and G. DeMaria, *J. Biol. Chem.*, 186 (1950) 653.
90 J. E. Coleman and B. L. Vallee, *Biochemistry*, 3 (1964) 1874.
91 F. A. Quiocho and F. M. Richards, *Biochemistry*, 5 (1966) 4062.
92 B. L. Vallee, J. F. Riordan, J. L. Bethune, T. L. Coombs, D. S. Auld and M. Sokolovsky, *Biochemistry*, 7 (1968) 3547.
92a R. A. Alberty and V. Massey, *Biochim. Biophys Acta*, 13 (1954) 347.
92b D. S. Auld and B. L. Vallee, *Biochemistry*, 9 (1970) 4352.
92c T. C. Bruice and G. L. Schmir, *J. Am. Chem. Soc.*, 81 (1959) 4552.
93 J. F. Riordan and B. L. Vallee, *Biochemistry*, 2 (1963) 1460.
94 M. L. Bender and C. E. Clement, *Biochem. Biophys. Res. Commun.*, 12 (1963) 339.
95 R. T. Simpson, J. F. Riordan and B. L. Vallee, *Biochemistry*, 2 (1963) 616.
96 R. T. Simpson and B. L. Vallee, *Biochemistry*, 5 (1966) 1760.
97 J. F. Riordan, M. Sokolovsky and B. L. Vallee, *Biochemistry*, 6 (1967) 3609.
98 M. Sokolovsky and B. L. Vallee, *Biochemistry*, 6 (1967) 700.
99 H. M. Kagan and B. L. Vallee, *Biochemistry*, 8 (1969) 4223.
99a F. A. Quiocho, unpublished results.
100 O. A. Roholt and D. Pressman, *Proc. Natn. Acad. Sci. U.S.*, 58 (1967) 280.
101 J. F. Riordan, M. Sokolovsky and B. L. Vallee, *Biochemistry*, 6 (1967) 358.
102 W. D. Behnke, *Federation Proc.*, 29 (1970) 462.
103 J. E. Coleman, P. Pulido and B. L. Vallee, *Biochemistry*, 5 (1966) 2019.
104 M. Sokolovsky, J. F. Riordan and B. L. Vallee, *Biochem. Biophys. Res. Commun.*, 27 (1967) 20.
105 B. L. Vallee and J. F. Riordan, *Brookhaven Symp. Biol.*, 21 (1968) 91.
106 J. F. Riordan, *Federation Proc.*, 29 (1970) 462.
107 P. H. Petra and H. Neurath, *Federation Proc.*, 29 (1970) 666.
108 T. L. Coombs, Y. Omote and B. L. Vallee, *Biochemistry*, 3 (1964) 653.
109 K. A. Freude, *Biochim. Biophys. Acta*, 167 (1968) 485.
110 B. L. Vallee, R. J. P. Williams and J. E. Coleman, *Nature*, 190 (1961) 633.
111 T. Ando and H. Fujioka, *J. Biochem. (Japan)*, 52 (1962) 363.
112 P. D. Boyer, in P. D. Boyer, H. Lardy and K. Myrback, *The Enzymes*, 2nd edn., Vol. 1, Academic Press, New York, 1959.
113 B. L. Vallee and R. J. P. Williams, *Proc. Natn. Sci. U.S.*, 59 (1968) 498.

486

114 K. A. Walsh, K. S. V. Sampath Kumar, J. P. Bargetzi and H. Neurath, *Proc. Natn. Acad. Sci. U.S.*, 48 (1962) 1443.
115 K. A. Walsh, L. H. Ericsson, R. A. Bradshaw and H. Neurath, *Biochemistry*, 9 (1970) 219.
116 I. R. Epstein and W. N. Lipscomb, personal communication.
117 B. L. Vallee, J. F. Riordan and J. E. Coleman, *Proc. Natn. Acad. Sci. U.S.*, 49 (1963) 109.
118 E. L. Smith, *Proc. Natn. Acad. Sci. U.S.*, 35 (1949) 80.
119 L. Meriwether and F. H. Westheimer, *J. Am. Chem. Soc.*, 78 (1956) 5119.
120 D. A. Buckingham, C. E. Davis, D. M. Foster and A. M. Sargeson, *J. Am. Chem. Soc.*, 92 (1970) 5571.
120a D. M. Blow and T. A. Steitz, *A. Rev. Biochem.*, 39 (1970) 63.
120b R. J. Angelici and B. E. Leach, *J. Am. Chem. Soc.*, 90 (1968) 2499.
121 H. Fujioka and K. Imahori, *J. Biol. Chem.*, 237 (1962) 2804.
121a D. R. Robinson, *J. Am. Chem. Soc.*, 92 (1970) 3138.
122 L. M. Ginodman, N. I. Mal'tsev and V. N. Orekhovich, *Biokhimiya*, 31 (1966) 1073.
123 P. L. Hall and E. T. Kaiser, *Biochem. Biophys. Res. Commun.*, 29 (1967) 205.
124 R. W. Sealock and M. Laskowski, Jr., *Biochemistry*, 8 (1969) 3703.
125 B. L. Kaiser and E. T. Kaiser, *Proc. Natn. Acad. Sci. U.S.*, 64 (1969) 36.
126 D. Ballou, M. L. Ludwig and G. Palmer, unpublished observations.
127 I. Schechter, *Eur. J. Biochem.*, 14 (1970) 516.
128 F. A. Quiocho, P. H. Bethge, W. N. Lipscomb, J. F. Studebaker, *et al.*, *Cold Spring Harbor Symp. Quant. Biol.*, 36 (1971) 561.
129 J. E. Folk, *J. Am. Chem. Soc.*, 78 (1956) 3541.
130 J. E. Folk and J. A. Gladner, *J. Biol. Chem.*, 231 (1958) 379.
131 E. Wintersberger, D. J. Cox and H. Neurath, *Biochemistry*, 6 (1962) 1069.
132 J. H. Kycia, M. Elzinga, N. Alonzo and C. H. W. Hirs, *Arch. Biochem. Biophys.*, 123 (1968) 336.
133 J. E. Folk, K. A. Piez, W. R. Carroll and J. A. Gladner, *J. Biol. Chem.*, 235 (1960) 2272.
134 J. E. Folk, E. C. Wolff and E. W. Schirmer, *J. Biol. Chem.*, 237 (1962) 3100.
135 M. Elzinga and C. H. W. Hirs, *Arch. Biochem. Biophys.*, 123 (1968) 343.
136 M. Elzinga, C. Y. Lai and C. H. W. Hirs, *Arch. Biochem. Biophys.*, 123 (1968) 353.
137 M. Elzinga and C. H. W. Hirs, *Arch. Biochem. Biophys.*, 123 (1968) 361.
138 D. J. Cox, E. Wintersberger and H. Neurath, *Biochemistry*, 6 (1962) 1078.
139 E. C. Wolff, E. W. Schirmer and J. E. Folk, *J. Biol. Chem.*, 237 (1962) 3094.
140 J. E. Folk and J. A. Gladner, *Biochim. Biophys. Acta*, 33 (1959) 570.
141 R. A. Bradshaw, H. Neurath and K. A. Walsh, *Proc. Natn. Acad. Sci. U.S.*, 63 (1969) 406.
142 T. H. Plummer, Jr., *J. Biol. Chem.*, 244 (1969) 5246.
143 W. N. Lipscomb, *Accounts Chem. Res.*, 3 (1970) 81.
144 E. Wintersberger, H. Neurath, T. L. Coombs and B. L. Vallee, *Biochemistry*, 4 (1965) 1526.
145 T. H. Plummer, Jr. and W. B. Lawson, *J. Biol. Chem.*, 241 (1966) 1648.
146 E. Wintersberger, *Biochemistry*, 4 (1965) 1533.
147 G. R. Reeck, W. P. Winter and H. Neurath, *Biochemistry*, 9 (1970) 1398.
148 A. G. Lacko and H. Neurath, *Biochem. Biophys. Res. Commun.*, 26 (1967) 272.
149 J. W. Prahl and H. Neurath, *Biochemistry*, 5 (1966) 4137.
150 P. Goldman and C. C. Levy, *Proc. Natn. Acad. Sci. U.S.*, 58 (1967) 1299.
151 C. C. Levy and P. Goldman, *J. Biol. Chem.*, 243 (1968) 3507.
151a K. Izaki and J. L. Strominger, *J. Biol. Chem.*, 243 (1968) 3193.

151b H. Zuber, *Nature*, 201 (1964) 613.

151c J. R. E. Wells, *Biochem. J.*, 97 (1965) 228.

152 R. Piras and B. L. Vallee, *Biochemistry*, 6 (1967) 348.

153 J. H. Freisheim, K. A. Walsh and H. Neurath, *Biochemistry*, 6 (1967) 3020.

154 J. R. Brown, R. N. Greenshields, M. Yamashki and H. Neurath, *Biochemistry*, 2 (1963) 867.

155 E. L. Smith and R. L. Hill, in P. D. Boyer, H. Lardy and K. Myrback, *The Enzymes*, Vol. 4, Academic Press, New York, 1960, p. 37.

156 E. L. Smith and D. H. Spackman, *J. Biol. Chem.*, 212 (1955) 271.

157 G. F. Bryce and B. R. Rabin, *Biochem. J.*, 90 (1964) 509.

158 D. H. Spackman, E. L. Smith and D. M. Brown, *J. Biol. Chem.*, 212 (1955) 255.

159 S. R. Himmelhoch and E. A. Peterson, *Biochemistry*, 7 (1968) 2085.

160 H. Hanson, D. Glaeser and H. Kirschke, *Z. Physiol. Chem.*, 340 (1965) 107.

161 K. Kretschmer and H. Hanson, *Z. Physiol. Chem.*, 249 (1968) 831.

162 J. E. Folk, J. A. Gladner and T. Viswanatha, *Biochim. Biophys. Acta*, 36 (1959) 356.

163 R. Frater, A. Light and E. L. Smith, *J. Biol. Chem.*, 240 (1956) 253.

164 L. Beckman, G. Bjorling and C. Christodoulou, *Acta. Genet.*, 16 (1966) 223.

165 M. J. Johnson, G. H. Johnson and W. H. Peterson, *J. Biol. Chem.*, 116 (1936) 515.

166 S. R. Himmelhoch, *Arch. Biochem. Biophys.*, 134 (1969) 597.

167 G. F. Bryce and B. R. Rabin, *Biochem. J.*, 90 (1964) 513.

168 R. L. Hill and E. L. Smith, *J. Biol. Chem.*, 224 (1957) 209.

169 B. S. Hartley, *A. Rev. Biochem.*, 29 (1960) 45.

170 B. L. Vallee and J. E. Coleman, in M. Florkin and E. H. Stotz, *Comprehensive Biochemistry*, Vol. 12, Elsevier, Amsterdam, 1964, p. 194.

171 S. A. Latt, B. Holmquist and B. L. Vallee, *Biochem. Biophys. Res. Commun.*, 37 (1969) 333.

172 L. Keay, *Biochem. Biophys. Res. Commun.*, 36 (1969) 257.

173 Y. Ohta, Y. Ogura and A. Wada, *J. Biol. Chem.*, 241 (1966) 5919.

174 H. Matsubara, A. Singer, R. Sasaki and T. A. Jukes, *Biochem. Biophys. Res. Commun.*, 21 (1965) 242.

175 H. Matsubara, *Biochem. Biophys. Res. Commun.*, 24 (1966) 427.

176 H. Matsubara, A. Singer and R. Sasaki, *Biochem. Biophys. Res. Commun.*, 34 (1969) 719.

177 J. D. McConn, D. Tsuru and K. Yasunobu, *J. Biol. Chem.*, 239 (1964) 3706.

Chapter 16

CARBONIC ANHYDRASE

J. E. COLEMAN

Department of Molecular Biophysics and Biochemistry, Yale University, New Haven, Connecticut, U.S.A.

INTRODUCTION

Carbonic anhydrase has the distinction of being the first zinc metallo-enzyme to be discovered. The first demonstrations that erythrocytes contain a protein catalyst for the hydration–dehydration of CO_2 were made in Roughton's laboratory[1-4]. Keilin and Mann[5,6] purified the bovine erythrocyte enzyme and showed their preparation to contain 0.33% zinc. The molecular weight of the bovine enzyme was determined to be near 30,000[7-9]. Using this molecular weight the reported zinc contents varied from 0.92 to 1.52 g atoms per mole[5,10,11]. Metal complexing agents were shown to be potent inhibitors of the enzyme, especially the anions cyanide, sulfide, azide[4] and thiocyanate[12], as well as the complexing agent, BAL[13]. Thus the zinc ion appeared to be clearly related to the function of the erythrocyte enzyme.

Early reports on plant carbonic anhydrase indicated the presence of some zinc as well as the inhibition of the enzyme by anions[14-18], although the evidence was not nearly so convincing as in the case of the erythrocyte enzymes. More recent data on plant and bacterial carbonic anhydrases are discussed in Section II.

The potent inhibition of the erythrocyte enzyme by sulfonamides was discovered by Mann and Keilin[19] and this inhibition remains relatively unique to carbonic anhydrase. No other purified enzyme has been discovered to be inhibited so specifically by these compounds. These inhibitors have played such an important part in the physico-chemical and enzymatic studies of the enzyme that Section IV will be devoted to a detailed discussion of the properties of carbonic anhydrase–sulfonamide complexes. While the sulfonamide structure does not suggest a metal complexing agent, recent data show that the sulfonamide binding site includes the zinc ion as one of the binding groups. The metal ion coordinates with the $-SO_2NH_2$ group[20-23].

Carbonic anhydrase has been frequently cited as one of the most specific enzymes known, since its only known function was the catalysis of the reversible hydration of CO_2:

(a) $H_2O + CO_2 \underset{k_{-1}}{\overset{k_1}{\rightleftharpoons}} H_2CO_3 \rightleftharpoons H^+ + HCO_3^-$

(b) $^-OH + CO_2 \underset{k_{-2}}{\overset{k_2}{\rightleftharpoons}} HCO_3^-$ (1)

The reaction may be formulated with either a water or hydroxide ion as the active species. Which of these reactions applies in the case of the enzyme is not certain as yet, but a good deal of evidence favors a Zn-coordinated hydroxide ion as attacking the CO_2 carbon (see Section VII).

In recent years it has been shown that in addition to the catalysis of reaction 1, carbonic anhydrase catalyzes a number of other hydrolysis and hydration reactions involving carbonyl groups. These reactions include the hydrolysis of p-nitrophenyl acetate[24-27], the hydrolysis of 2-hydroxy-5-nitro-α-toluenesulfonic acid sultone[28-30], and the hydration of acetaldehyde[31-33]. These reactions are discussed in detail in Section VII.

II. SPECIES AND ISOZYME VARIANTS — PHYSICO-CHEMICAL PROPERTIES

In the last 10 years several laboratories have developed techniques for obtaining highly purified erythrocyte carbonic anhydrase[34-41]. In general these methods have involved chloroform–ethanol precipitation of the hemoglobin[36,38,41], followed by column chromatography of DEAE–cellulose[34,36,41], DEAE–Sephadex[42] or on hydroxylapatite[36,38,41]. The protein is quite stable to freeze-drying and most procedures have used lyophilization at one stage or another. With continuing experience with the several preparative methods several features have emerged. While the chloroform–ethanol treatment does not detectably alter most physico-chemical properties, the crystals used for X-ray analysis of the human C enzyme are of higher quality if the chloroform–ethanol treatment is not used[22,23]. A milder gel filtration procedure can be used to separate hemoglobin[36] or a direct application of hemolyzate to DEAE–Sephadex[42]. Likewise crystal quality is better if lyophilization is not used[22,23]. Crystal quality can also be improved by a final electrophoretic purification[23].

By these various procedures a large number of animal erythrocyte carbonic anhydrases have now been isolated in pure form and several of them have been crystallized[38,43,44]. Four examples of these crystalline preparations are shown in Fig. 1, two as crystalline sulfonamide complexes. In the case of the human B enzyme, crystallization is more easily achieved in the presence of a sulfonamide[22,30]. All of these erythrocyte enzymes have proven to be monomeric proteins of molecular weight near 30,000 containing 1 g atom of Zn(II) per molecule[34-40]. Representative physico-chemical constants are collected in Table I.

490

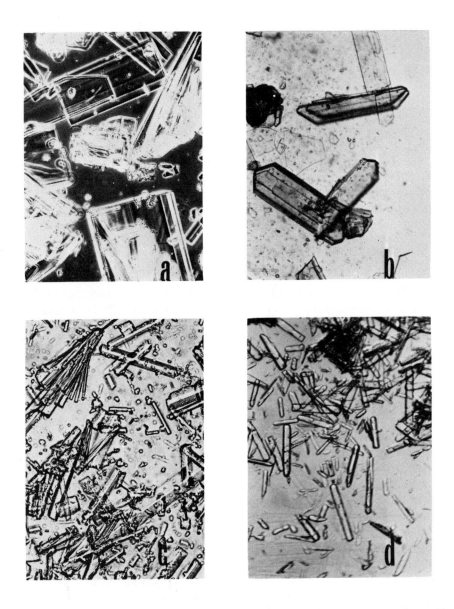

Fig. 1. Crystals of primate erythrocyte carbonic anhydrases and their sulfonamide complexes. (a) Monkey carbonic anhydrase C. (b) Monkey carbonic anhydrase B–azosulfonamide complex. (c) Monkey carbonic anhydrase B. (d) Human carbonic anhydrase B-acetazolamide complex.

TABLE I

PHYSICO-CHEMICAL CONSTANTS FOR ERYTHROCYTE CARBONIC ANHYDRASES FROM VARIOUS SPECIES

Species	Molecular weight	$s_{20,w}$	Partial specific volume (ml g^{-1})	Electrophoretic mobility $\times 10^5$ (cm^2 V^{-1} sec^{-1})		Molar absorptivity $\times 10^{-4}$ (M^{-1} cm^{-1})	Isoelectric point	References
Human B	29,600 28,000[a]	2.95–3.1	0.737 0.736[a]	−1.72 −1.06	(pH 8.6) (pH 7)	4.90	5.85 ± 0.2	35, 36, 37 35a
Human C	31,600	2.95–3.1	0.723 0.736[a]	−1.01 −0.5	(pH 8.6) (pH 7)	5.34	7.25	35, 36, 37 35a
Monkey B	29,800 28,300[a]	2.94		−2.28	(pH 8.6)	4.88		38
Monkey C	29,800 28,600[a]	2.94		−1.29	(pH 8.6)	5.35		38
Bovine A	31,000	3.06		−2.8	(pH 7.0)	5.7		34
Bovine B	31,000	3.06	0.742	−2.3	(pH 7.0)	5.7	5.65	34
Equine B	29,000	2.74	0.725[a]			3.95	~6	40
Equine C	28,000	2.71	0.732[a]			3.73	~11	
Canine	29,000	2.82	0.734[a]					39
Shark								
Bull	40,000	3.5					4.65	41
Tiger	40,000							

[a]Calculated from amino acid composition.

(a) Amino acid composition

The amino acid compositions of these erythrocyte carbonic anhydrases are given in Table II. Noteworthy features are the presence of single cysteine residues in most isozymes and the complete absence of this amino acid in the bovine enzyme. There are one or two methionine residues (see below for cyanogen bromide cleavage), and a moderate number of tyrosine and tryptophan residues. None of the compositional features are particularly striking. By far the oldest erythrocyte carbonic anhydrase from an evolutionary standpoint is the shark enzyme. The bull shark enzyme appears to have the lowest isoelectric point among these enzymes, 4.7, which coincides with the relative high content of aspartic and glutamic acid residues (Table II), although amide content has not been determined. The single cysteine residue in the human C enzyme does not appear to be directly involved in function and it has been located approximately 14 Å from the zinc ion in the X-ray structure[22,23].

As mammalian erythrocyte enzymes were purified, it early became apparent that several electrophoretically distinct proteins with carbonic anhydrase activity could be isolated from each species[34-39,45]. The major

TABLE II

AMINO ACID COMPOSITIONS OF ANIMAL ERYTHROCYTE CARBONIC ANHYDRASE

Amino acid residue	Human[36,75] B	C	Monkey[38] B	C	Dog[39]	Bovine[47]	Equine[40] B	C	Tiger shark[41,a]	Bull shark[41,a]
Lysine	18	25	18	24	22	19	19	19	22	21
Histidine	11	12	9	12	12	11	12	10	11	7
Arginine	7	7	7	8	7	9	9	5	10	8
Asparatic acid	31	29	31	30	34	32	27	32	30	27
Threonine	14	13	13	11	11	15	12	12	10	9
Serine	30	19	30	18	25	16	18	28	23	24
Glutamic acid	22	24	22	26	26	24	26	25	30	33
Proline	17	28	17	16	19	20	16	18	16	20
Glycine	16	22	15	22	24	20	23	23	24	24
Alanine	19	13	16	12	18	17	17	15	17	16
Half-cystine	1	1	1	1	1	0	1	2	9	21
Valine	17	17	16	15		20	19	20	10	8
Methionine	2	1	1	2		3	1	2	2	1
Isoleucine	9	9	10	10	13	5	7	9	12	10
Leucine	20	27	19	24	25	26	22	21	26	27
Tyrosine	8	9	9	7	8	8	7	8	7	7
Phenylalanine	11	13	10	11	12	11	11	11	13	10
Tryptophan	6	7	7	7	6	7	5	5	5	4
Zn	1	1	1	1	1	1	1	1	a	a
Amide NH$_3$	(26)	(21)	—	—	—	(24)	(32)	(22)	—	—
Total residues	259	266	256	255	278	264	252	265		

[a] For purposes of comparison the amino acid contents of the shark enzymes are given as residues per 30,000 molecular weight although the actual molecular weight of the shark enzymes is near 40,000 and they contain 1 Zn/39,000[41].

electrophoretically distinct proteins were designated A and B in the bovine species[34] and B and C in the case of human erythrocyte carbonic anhydrase[35-37]. The most striking physico-chemical feature differentiating the electrophoretically distinct enzymes from the same species was discovered when the amino acid compositions of the human B and C enzymes were compared (Table II). The lysine content of B is 18 residues while 24–25 lysyl residues are present in the C enzyme. Significant differences are also observed for a number of other residues (Table II). These variations in the number of particular types of residue must reflect some rather significant differences in primary sequence. These differences do have some functional and structural consequences (see discussion of conformation below). The major functional consequence is the observation that the two human isozymes show striking differences in the magnitude of their specific activities when catalyzing the hydration of CO_2. The B isozyme has a turnover number of $\sim 40,000$ while the C enzyme has a turnover number of $\sim 1,200,000$[46] (see Section VII for further discussion). An additional feature of physiological interest is the fact that the C isozyme is present in much smaller quantities, 10 to 20% of the quantity of B isozyme present. The larger specific activity of the C isozyme results, however, in both B and C isozymes contributing about equally to the total carbonic anhydrase activity.

Two erythrocyte carbonic anhydrase isozymes of different primary sequence appear thus far to be a unique characteristic of primates. The various protein peaks with carbonic anhydrase activity that can be separated from the hemolyzates of lower mammalian species all seem to have the same amino acid composition and specific activity[34, 39, 47] with the exception of the horse enzyme[40]. These "isozymes" may represent slightly different conformations of the same protein, charge differences arising from different proportions of hydrogen-bonded or buried side-chains. The individual primate isozymes can be observed to form several distinct bands on gel electrophoresis[38,48].

The only other primate from which erythrocyte carbonic anhydrase has been examined in detail is the rhesus monkey. This primate has two isozymes, B and C, almost identical to the two human isozymes (Tables I and II). A number of other primate hemolyzates have been examined which also appear to have the two isozymes[45]. Mutants have also been discovered[45,48].

(b) Plant carbonic anhydrases

Major interest has centered around the animal erythrocyte carbonic anhydrases because of the ease of purification and their obvious physiological importance. Following the early reports that the plant enzymes were not inhibited by sulfonamides[50] and did not contain zinc[50], interest has revived in these enzymes and physico-chemical and enzymatic studies have been reported on at least two highly purified plant carbonic anhydrases.

Physico-chemical properties of the enzymes from spinach[51] and parsley[52,52a] leaves are summarized in Table III and the reported amino acid composition of the spinach leaf enzyme is given in Table IV. The most

TABLE III

PHYSICO-CHEMICAL PROPERTIES OF PLANT CARBONIC ANHYDRASES

Species	$s_{20,w}$	Molecular weight	Zinc content	Reference
Spinach	6.8	140,000 145,000 (equilibrium centrifugation)	0.0013 μg/mg	51
Parsley	8.8	170,000 180,000 (equilibrium centrifugation)	1 g atom/ 29,000 gm protein	52, 52a
Neisseria sica		28,000	1 g atom/ 30,000 gm protein	54

TABLE IV

AMINO ACID COMPOSITION OF SPINACH AND PARSLEY LEAF CARBONIC ANHYDRASE
From Rossi et al.[51] and from Tobin[52a]

Amino acid	Spinach mean content (mole %)	Parsley (moles/28,150 g protein)
Aspartic acid	8.85	24
Threonine	—	10
Serine	11.85[a]	18
Glutamic acid	9.50	21
Proline	4.70	22
Glycine	11.60	22
Alanine	8.25	17
Valine	8.10[b]	24
Half-cystine	4.0	7
Isoleucine	5.80	8
Leucine	5.90	21
Tyrosine	2.20	8
Phenylalanine	4.20	17
Lysine	5.90	21
Histidine	2.90	5
Arginine	4.05	5
Tryptophan	less than 1	3
Methionine		4

[a] Extrapolated value at zero hydrolysis time.
[b] Value at 72 h hydrolysis time.

striking initial finding was their large apparent molecular weight, ~140,000 for the spinach enzyme[51] by several methods and ~170,000 for the parsley enzyme[52]. The most highly purified preparation of spinach enzyme is reported to contain only 1.3 μg Zn/g protein[51] while the purified parsley enzyme was initially reported to contain 1 g atom of Zn per 51,000 g of protein[52]. Further work on the enzyme, however, has resulted in a more active preparation that contains 1 g atom of Zn per 29,000 of protein[52]. The latter report is the first evidence that the plant enzyme contains a metal-containing unit that corresponds in any way to the animal enzymes. The functional role of the Zn in the plant enzyme has not been investigated, however. There are clearly functional differences present, since the plant enzymes do not appear to have esterase activity[52], although the enzyme is inhibited by acetazolamide and azide[52]. It is a curious feature that among the animal enzymes, the shark carbonic anhydrases appear under some conditions to behave as species of molecular weight greater than 100,000[41]. They are also the only animal enzymes that contain significant amounts of cysteine (Table II).

(c) Other carbonic anhydrases

Carbonic anhydrase has been partly purified from the pupae of the moth of the corn ear worm (*Heleothis zea*)[53]. The active material eluted from a Sephadex G-100 column at a position indicating a molecular weight of 80,000–90,000, although this may represent association of the enzyme with other cellular material[53]. This enzyme is inhibited by acetazolamide, K_i = 0.03 μM, and by azide, thus it would appear to have features typical of the erythrocyte enzymes. Nyman[54] has purified a bacterial carbonic anhydrase from *Neisseria sica* and obtained a homogeneous preparation of molecular weight 28,000. The enzyme contains one zinc atom per unit of approximately 30,000 molecular weight. The content of tryptophan and tyrosine is much lower than in the mammalian erythrocyte enzymes, but otherwise the properties of the bacterial enzyme appear similar[54].

(d) Amino acid sequence

Work is actively proceeding on the determination of the amino acid sequence of the bovine enzyme and the two human isozymes[48,55-59]. Preliminary accounts of this work have been published[48,58,59], but at the present time only partial sequences are available and these will be presented at the end of this section.

Several more limited investigations of the *N*-terminal and *C*-terminal regions of the molecule, however, have revealed some interesting findings. Rickli *et al.*[36] and Marriq *et al.*[60] demonstrated that all forms of the human enzyme appear not to have a free-amino group at the *N*-terminus of the peptide chain. Under a variety of conditions no reaction with fluorodinitro-benzene can be detected for either isozyme[48]. Studies by Laurent et al.[61] showed that both the B and C forms yield one N-acetyl group per mole as analyzed by the method of Phillips[62].

Laurent and Derrien[48] were able to isolate approximately 1 mole of DNP-acetyl hydrazide after 17 hours hydrazinolysis of the protein at 100°C. This work suggested that the N-terminal residue of most if not all forms of carbonic anhydrase is acetylated. Further work has resulted in the isolation of N-acetylated peptides from carbonic anhydrase and the results of these studies as given by Laurent and Derrien[48] are shown in Table V. Thus far the work has identified only the presence of N-acetyl-ser as the N-terminus of HCAC and BCAB, since the acetylated peptides have not been readily isolated as with the B enzyme[48]. HCAA is an electrophoretically distinct species of the human enzyme that appears to have amino acid composition and activity identical to the B isozyme[35,37]. The similarity is confirmed by the N-terminal sequences given in Table V.

TABLE V

N-TERMINAL ISOLATED FROM HUMAN AND BOVINE CARBONIC ANHYDRASE
From Laurent and Derrien[48].

```
                1   2   3   4   5   6
HCAA   N-Acetyl—Ala-Ser-Pro-Asp-Trp-Gly-Tyr-Asp-Asp-Lys
                                            |

          ....Lys-Ser-Trp-Glu-Pro-Gln-Gly-Asn

HCAA   N-Acetyl—Ala-Ser-Pro-Asp-Trp-Gly....

HCAC   N-Acetyl—Ser....

BCAB   N-Acetyl—Ser....
```

Work on the C-terminal structure of the molecule has utilized the finding that treatment of the human and bovine carbonic anhydrase iso-zymes with cyanogen bromide has resulted in the release of a C-terminal peptide containing 19 or 20 amino acid residues[55-57]. The sequences of these peptides isolated from HCAB, HCAC, and BCAB are given in Table VI. The numbering system has assumed that residue 21 in each case is a methionyl. Such an assumption gives rise to the maximum number of homo-logies in the three sequences as indicated by the boxes. This arrangement further suggests that the human B isozyme is shorter by 1 residue (a lysyl) at the C-terminus and that the human C isozyme is structurally more closely related to lower mammalian species, a conclusion supported by a number of physico-chemical and kinetic findings (see Section II C).

TABLE VI

CARBOXYL-TERMINAL AMINO ACID SEQUENCES FOR SPECIES AND
ISOZYME VARIANTS OF CARBONIC ANHYDRASE
From Andersson, Nyman and Strid, 1968

	21	20	19	18	17	16	15	14	13	12	11	10	9	8	7	6	5	4	3	2	1
Human enzyme B	Met	Glx	His	Asn	Asn	Arg	Pro	Thr	Gln	Pro	Leu	Lys	Gly	Arg	Thr	Val	Arg	Ala	Ser	Phe-	
Human enzyme C	Met	Val	Asp	Asn	Trp	Arg	Pro	Ala	Gln	Pro	Leu	Lys	Asn	Arg	Gln	Ile	Lys	Ala	Ser	Phe	Lys
Bovine enzyme B	Met	Leu	Ala	Asn	Trp	Arg	Pro	Ala	Gln	Pro	Leu	Lys	Asn	Arg	Gln	Val	Arg	Gly	Phe	Pro	Lys

A third determination of isolated primary structure involves the peptide containing a histidyl residue that reacts with iodoacetate bound at the
active center of human carbonic anhydrase B. (For a more detailed description of this reaction see Section VII.) Bradbury[63,64] has isolated a peptide
from the iodoacetate reacted HCAB containing the single carboxymethylated
histidyl residue in the sequence Thr-3-CM-His-Pro-Pro-Leu[64]. This particular
sequence can be identified in the region 59 to 64 residues from the carboxyl
end of the partial sequence of the human isozyme B presented by Andersson
et al.[56] and shown below.

On the other hand, a histidyl residue has also been shown to react
with N-chloroacetylchlorathiazide when this substituted sulfonamide is
bound at the active center[58,65]. During acid hydrolysis the chlorathiazide
moiety is apparently removed and the modified residue appears as carboxy-
methylhistidine. The residue modified by the sulfonamide is different from
the one modified by iodoacetate or bromoacetate[58,66] and appears to be
the histidyl residue located in the region 65 to 76 residues from the amino
terminal end of the B isozyme as indicated in Table VII below.

While complete sequences cannot be given for any of the carbonic
anhydrase isozymes, the work on the tryptic peptides or tryptic peptides
derived from the protein in which the ϵ-amino groups are blocked[58,59] has
been proceeding rapidly and a partial picture of sequences for HCAB and
HCAC are available as shown in Table VII.

(e) Hydrogen ion equilibria

Detailed hydrogen ion titration curves have been determined by potentiometric titrations and by spectrophotometric titrations in the region of
tyrosine OH ionization for both human isozymes[67,68] and the bovine B
enzyme[69]. On titration of the human B enzyme with acid, there is a sharp
break in the titration curve between pH 4.2 and 4.0 which corresponds to
an uptake of 7 protons per mole[67]. These 7 groups appear to be histidine
residues; 4 other histidine residues appear to titrate normally in the native

498

TABLE VII

A TENTATIVE SURVEY OF THE PRESENT KNOWLEDGE OF THE AMINO ACID SEQUENCES OF
HUMAN CARBONIC ANHYDRASES B AND C. SEQUENCE REGIONS WITH APPARENT HOMOLOGY
HAVE BEEN UNDERLINED
Composite of the data from references 56, 57, 61, and 154

_____ Fragment B IV _____

Enzyme B: Acetyl-Ala-Ser-Pro-Asp-Trp-Gly-Tyr-Asp-Asp-Lys-Asn-Gly-Gly-Pro-Glu-Trp-Ser-Lys-
Enzyme C: Acetyl-Ser-His-His(Gly,Trp,Tyr)Gly-Lys(Gly,Pro,Glx,Asx,Trp,His,His,Lys)Asp-Phe-Pro-Ile-Ala-Lys(Gln-Gly-Arg)
                                                                                    _____ Fragment C 1 _____

-Leu-Tyr-Pro-Ile-Ala-Asp-Gly-Asn-Asn-Gln-Ser-Pro-Val-Asp-Ile-Lys-Thr-Ser-Glu-Thr-Lys-His-Asp-Thr-Ser-Leu-Lys-Pro-Ile-Ser-Val-Ser-
(Leu,Tyr,Pro,Ile,Ala,Asx,Gly,Asx,Asx,Glx,Ser,Pro,Val,Asx,Ile,Lys)Thr-Glu-Thr-Lys(His-Asp-Thr-Ser-Leu)(Lys,Pro,Ile,Ser,Val,Ser,
                        _____ Fragment C 2 _____   _____ Fragment B III _____

-Tyr-Asn-Pro-Ala-Thr-Ala-Lys-Glu-Ile-Ile-Asn-Val-Gly-His-Ser-Phe-His-Val-Asn-Phe-Glu-Asp-Asn-Asp-Arg--Ser-Val-Leu-Lys-Gly-Gly-
Tyr,Asx,Pro,Ala,Thr)Ala-Lys(peptides not yet isolated   )(Ser,Val,Glx,Phe,Glx,Asx,Asx,Asx,Lys)(Ala,Val,Leu,Lys)(pep-
                                                                                                        _____ Fragment B I _____

-Pro-Phe-Ser-Asp-Ser-Tyr-Arg-Leu-Phe(79 residues   )Ile-Lys-Thr-Lys-Gly-Lys-Arg-Ala-Pro-Phe-Thr-Asn-Phe-Asp-Pro-Ser-Thr-Leu-Leu-
tides not yet isolated   )Arg(87 residues   )(peptides not yet isolated   )(Glx,Ser,Phe-Gly)Ile-Val-Leu-Lys-
                        _____ Fragment C 3 _____   _____ Fragment C 4 _____

-Pro-Ser-Leu-Asp-Phe-Trp-Thr-Tyr-Pro-Gly-Ser-Leu-Thr-His-Pro-Pro-Leu-Tyr-Glu-Ser-Val-Thr-Trp-Ile-Ile-Cys-Lys-
(Pro,Glx,Ser,Leu,Asx,Leu,Leu,   Tyr,Trp)(Gly,Thr,Leu,Thr,Leu,Pro,Pro,Leu,Tyr,Glx,Ser,Val,Thr,Trp)(Ile,Ile,Cys,Lys)

-Glu-Ser-Ile-Ser-Val-Ser-Gln-Ser-Glu-Leu-Leu-Gln-Phe-Arg---------Ser-Leu-Leu-Ser-Asn-Val-Glu-Asp-Asn-Gly--Ala-Val-Pro-Met-Glx-
(Glx-Pro-Ile-Ile(Val,Ser,Glx,Glx)Leu-Val-Lys)Phe-Arg(Lys,Leu,Asn,Phe,Glx,Gly)(Glx,Glx,Gly,Pro,Glx)(Asp,Leu,Trp)Met-Val-
                                    _____ Fragment B V _____
                                                                _____ Fragment C 5 _____

-His-Asn-Arg-Pro-Thr-Gln-Pro-Leu-Lys-Gly-Arg-Thr-Val-Arg-Ala-Ser-PheOH
-Asp-Asn-Trp-Arg-Pro-Ala-Gln-Pro-Leu-Lys-Asn-Arg-Gln-Ile-Lys-Ala-Ser-Phe -LysOH

◯ Residue in enzyme B modified by bromoacetate and iodoacetate.

☐ Residue in enzyme B modified by N - chloroacetylchlorothiazide.

protein. On back titration of the enzyme with alkali from pH below 2, there is an increase in titrable groups in the pH region 5 to 8, and all 11 histidines titrate normally. Apparently, acid denaturation releases masked imidazole groups[67].

Spectrophotometric titrations of the human B isozyme indicate the presence of three types of tyrosyl groups[67]. Up to pH 11.5 the heat and entropy of ionization of the first 4 tyrosyl groups are normal, but the apparent pK_a of one or two of these is higher than normal[67]. Above pH 11.5 the remaining 3 to 4 groups ionize very slowly with time and there is a simultaneous slow loss of enzymatic activity[67].

The chief difference in the human C isozyme as far as general hydrogen ion equilibria are concerned is that the isoionic point is pH 7.3 rather than pH 5.8 as found in the B isozyme[68]. The general titration curve is quite similar, showing the sharp break near pH 4 with the unmasking of 7 imidazole groups[68]. Three tyrosyl groups appear to titrate relatively normally, while 6 appear to be buried and ionize only slowly accompanied by irreversible conformational changes in the molecule[68]. The C isozyme appears to be more sensitive to alkaline denaturation than the B isozyme[68].

The bovine B enzyme has an isoionic point of pH 5.65[69]. Titration of this enzyme to acid pH also unmasks 7 histidine residues, which protonate during a sharp conformational change occurring in a narrow pH-range near pH 4. On back titration 11 histidines titrate normally[69]. In the bovine enzyme only one tyrosine titrates freely with an abnormally high pK_a, (10.8). The remaining 7 tyrosines ionize only above pH 12 during a time-dependent change leading to loss of enzymatic activity[69]. The presence of masked histidyl and tyrosyl residues in all the mammalian erythrocyte enzymes examined thus far has suggested the possibility that histidyl–tyrosyl interactions, perhaps hydrogen bonding, are present and form a significant structural feature of carbonic anhydrase[67-69]. In any event, the presence of masked phenolic hydroxyls and imidazole groups appear to be part of the structure necessary for carbonic anhydrase activity.

(f) Absorption spectra and optical activity

The near ultra-violet absorption spectra of all erythrocyte carbonic anhydrases are typical of proteins. Maxima are located at 280 nm with well-developed shoulders near 290 nm. Molar absorptivities at 280 nm for several species and isozyme variants are given in Table I.

The absorption spectrum of human carbonic anhydrase between 185 and 230 nm has been determined by Rickli et al.[36]. This spectrum is very striking in that the molar absorptivity values in the far ultra-violet are close to those expected for a random coil conformation of the polypeptide chain rather than one containing significant α-helix in which case considerable hypochromia would be expected[36]. Of considerable additional interest is

500

the development of the expected hypochromia in this spectrum when the pH is adjusted to 1.67. Independent optical rotatory dispersion (o.r.d.) measurements show that at low pH there is a radical change in the conformation of the molecule associated with loss of zinc[70-72]. Analysis of the resultant o.r.d. pattern does seem to indicate the presence of significant α-helix at pH 1.5 to 1.6, a structure which is not reflected in the o.r.d. or circular dichroism (c.d.) of the native molecule at neutral pH[70-72]. The o.r.d. and c.d. spectra of the native molecule, however, are difficult to interpret (see below).

Optical rotatory properties

The unusual optical rotatory properties of carbonic anhydrase are illustrated in Fig. 2 by the presentation of both the c.d. and o.r.d. in the near ultra-violet and far ultra-violet for two representative isozymes of carbonic anhydrase, human B and monkey C[30]. There are striking multiple Cotton effects associated with the near ultra-violet absorption bands. On a molar basis these are at least an order of magnitude greater than Cotton effects observed for most model compounds containing an aromatic chromophore in a dissymmetric environment. The large ellipticity of these chromophores when incorporated into the protein structure must be induced by the surrounding environment.

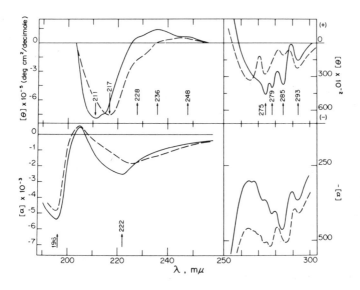

Fig. 2. C.d. spectra (upper curves) and o.r.d. spectra (lower curves) of human carbonic anhydrase B(——) and monkey carbonic anhydrase C (- - -). Conditions: 0.025 M Tris, pH 7.5, 25°C. Data from Coleman[30].

The aromatic Cotton effects disappear on acid denaturation or in guanidine HCl[70,74,75]. Once the environment is destroyed the large detailed near ultra-violet ellipticity bands do not appear to be restored completely by return to neutral pH or removal of the denaturing agent[71]. One recent report, however, indicates that under the proper conditions after using guanidine HCl to denature the protein, the initial conformation can be restored[75]. A number of authors have discussed in detail the origin and possible assignments of the near ultra-violet ellipticity bands[70-74].

Analyses of the conformation of the peptide backbone of carbonic anhydrase have been made using the more classical linear treatments of o.r.d. data such as the Yang–Doty plot and a determination of $b_0{}^{36}$. Human carbonic anhydrase B has a $b_0 \cong 0$, indicating little or no α-helix to be present in the native protein[36]. On acid denaturation b_0 becomes more negative, $b_0 = -97°$, suggesting that some α-helical structure may be present in the denatured molecule[36] (see discussion above under absorption spectrum). Possible objections to the classical analysis of the o.r.d. data can be raised on the basis that the presence of very large near ultra-violet Cotton effects as well as unusual positions for the Cotton effects in the far ultra-violet (see below) obviate many of the assumptions upon which the classical analysis is based.

The far ultra-violet c.d. and o.r.d. is also very unusual for a protein molecule. There is little evidence in the c.d. of either isozyme (Fig. 2) that a significant ellipticity band exists at 222 nm characteristic of the α-helix. The major negative bands occur at 211 nm in the human B enzyme and 217 nm in the monkey C enzyme. These are reflected in the o.r.d. by the unusual positions of the first negative troughs, 222 nm and 225 nm respectively.

Since carbonic anhydrase is optically active near ultra-violet aromatic absorption bands, the possibility must be kept in mind that some of the far ultra-violet aromatic absorption bands of tyrosine or tryphophan may also have significant optical activity and contribute to the far ultra-violet c.d. and o.r.d. Such a contribution could well prevent any analysis of conformation based on the c.d. or o.r.d. curves observed for known conformations of model polypeptides. One unfortunate feature of the mammalian enzyme has been the unusually high optical density near 200 nm preventing adequate collection of c.d. data below 200 nm.

In view of all the vagaries in the possible interpretation of the ultra-violet o.r.d. and c.d., it is interesting that the X-ray diffraction studies at 2 Å resolution on the human C enzyme in the crystalline state do show only a couple of turns of helix to be present in the entire molecule. In addition only a small portion of this helix actually meets the precise conformation of the α-helix[23]. The predominant structure of the peptide backbone is β-pleated sheet forming the entire core of the molecule (see Section VI). It seems likely that the major negative ellipticity bands in the c.d.

502

spectrum of carbonic anhydrase from 211–217 nm do have contributions from the β-structure, the latter expected on the basis of models to show bands in the region from 215 to 218 nm[77]. The large negative rotation below 200 nm evident in the o.r.d. is not readily explained.

A recently obtained near and far ultra-violet c.d. of pure tiger shark erythrocyte carbonic anhydrase is shown in Fig. 3[41]. This is of interest for two reasons. The striking ellipticity associated with the near ultra-violet chromophores is a structural feature of the erythrocyte enzyme that has been maintained since a very early period in animal evolution. Secondly, the far ultra-violet c.d. of this carbonic anhydrase is more typical of proteins and a low optical density near 200 nm allows an adequate recording down to 190 nm. There is a significant band at 222 nm suggesting that some α-helix is present in the native enzyme. The low ultra-violet positive ellipticity band, however, is shifted to the red, 196 nm, rather than 190 nm expected for the α-helix, and the trough depth at 222 nm is nearly equal to the peak height at 195 nm. Both features suggest that a large amount of β-structure is probably present[77,78].

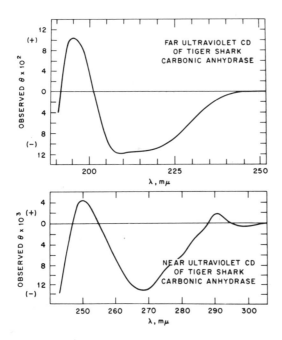

Fig. 3. Near and far ultra-violet c.d. of tiger shark carbonic anhydrase. Conditions: 0.025 M Tris, pH 7.5, 25°C. From Maynard and Coleman[41].

III. METALLOCARBONIC ANHYDRASES: APOENZYMES, METAL BINDING, ABSORPTION SPECTRA, AND INHIBITOR BINDING

Zn(II) in carbonic anhydrase is not in rapid equilibrium with ionic zinc in the environment, since exchange with external ionic zinc in the surrounding medium has not been observed at neutral pH[71,79,80]. If the pH is lowered to 5, slow exchange is observed[80]. This apparently reflects a change in geometry at the coordination site as can be detected by a change in the visible absorption spectrum of the Co(II) enzyme (see below).

Apoenzyme

Treatment of the enzyme with 1,10-phenanthroline or 8-hydroxy-quinolinesulfonate[20,81,82] at pH 5.5 over a period of 15 days, however, is successful in removing the Zn(II). The resultant apoenzyme is completely inactive, but quite stable since it can be completely reactivated by the readdition of Zn(II) ions[20,71,81]. Minor changes in the far ultra-violet o.r.d. of the apoenzyme compared to the native enzyme have been observed[71], but no major physico-chemical changes appear to accompany the removal of zinc. Of significance in this regard is the finding that the only major feature in the difference map of the electron density between apocarbonic anhydrase in the crystalline state and the native enzyme at 5.5 Å resolution is a density corresponding to the Zn(II) ion (Fig. 4)[83]. Zinc-free crystals can be prepared which are isomorphous with native enzyme by soaking crystals of the native enzyme for 16 days in 0.01 M 2,3-dimercaptopropanol-1 (BAL) contained in 2.45 M $(NH_4)_2SO_4$ in a hydrogen atmosphere[83].

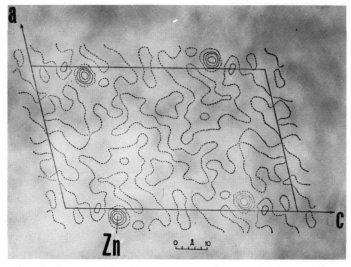

Fig. 4. Electron density difference map. Human carbonic anhydrase C-human apocarbonic anhydrase C. From Tilander *et al.*[44].

Different metallocarbonic anhydrases have been prepared from the apoenzyme by the addition of Mn(II), Co(II), Ni(II), Cu(II), Cd(II) and Hg(II) in addition to Zn(II). A large amount of evidence now indicates that these divalent metal ions all occupy the same binding site[20,23,82,84]. A number of investigations have centered on the properties of these metallocarbonic anhydrases and the results are summarized below.

Enzymatic activity and stability of the metal-protein complex

Although all the divalent metal ions listed above appear to complex at the active site, only the Zn(II) and Co(II) complexes have the characteristics required for catalyses of either CO_2 hydration or esterase activity (Table VIII)[20,21]. The failure of the other metallocarbonic anhydrases to be effective catalysts may relate to subtle changes in coordination geometry, or to failure to add ligands to an open coordination site or to exchange monodentate ligands rapidly[21]. In this connection, the binding of sulfonamides does prove to be metal ion dependent[20,21] and the inactive metallocarbonic anhydrases also fail to coordinate the sulfonamides (Table VIII) (see Sections IV and VII for further discussion).

TABLE VIII

ENZYMATIC ACTIVITIES AND BINDING OF ACETAZOLAMIDE TO HUMAN METALLOCARBONIC ANHYDRASES
Data from Coleman[21]

Metal[a]	Hydration of CO_2[b] (U)	Hydrolysis of p-NO_2-phenylacetate[c]	Moles of acetazolamide bound/mole enzyme[d]
Apoenzyme	400	0.09	0.04
Mn(II) B	400	0.38	0.40
Co(II) B	5,700	8.70	1.00
Ni(II) B	500	0.32	0.02
Cu(II) B	127	0.50	0.14
Zn(II) B	10,200	2.70	1.00
Zn(II) C	30,000	7.30	1.00
Cd(II) B	430	0.20	0.06
HG(II) B	5	0.09	0.05

[a] B and C refer to the isozymes of human erythrocyte carbonic anhydrase.
[b] Determined by the method of Wilbur and Anderson.
[c] Conditions: 0.025 M Tris, 5% acetonitrile, pH 7.5, 23°C.
[d] Determined in the presence of 1×10^{-6} M free acetazolamide.

Stability constants for this series of metallocarbonic anhydrases have been measured by Lindskog and Nyman[82] using a method in which small chelating agents, 1,10-phenanthroline, 8-hydroxyquinoline-5-sulfonate and

EDTA, were allowed to compete for the metal ion at the active site. From the amount of Me(II)-carbonic anhydrase present at equilibrium and the known stability constants of the small complexes, stability constants for the series of metallocarbonic anhydrases with Mn(II), Co(II), Ni(II), Cu(II), Cd(II), and Hg(II) were calculated (Table IX). These constants are given

TABLE IX

STABILITIES OF METALLOCARBONIC ANHYDRASES
Data from Lindskog and Nyman[82]

Metal	Log K, pH 5.5	pH-independent stability constant log K
Mn(II)	3.8	—
Co(II)	7.2	—
Ni(II)	9.5	—
Cu(II)	11.6	—
Zn(II)	10.5	15.0
Cd(II)	9.2	—
Hg(II)	21.5	—

for pH 5.5, the pH where the metal ion is most readily removed. A determination of the stability constant for the Zn(II) enzyme, however, was carried out as a function of pH. Log K rises linearly with pH from pH 5 to 10. Above 10 it appears to level off at log K = 15. This suggests that the true pH-independent stability constant of the Zn(II) enzyme is approximately 10^{15}. Further it suggests that protons continue to interact with the binding site until pH 9 or above. Another possibility that cannot be ruled out is that small pH-dependent changes in protein structure continue to affect the stability of the complex.

The kinetic aspects of Zn(II) binding to apocarbonic anhydrase have been studied extensively by Henkens and Sturtevant[85]. Reactivation of the apoenzyme by Zn(II) is accompanied by small changes in the ultra-violet spectrum of the protein[85]. Either the absorptivity change or the appearance of enzymatic activity can be used to measure the rate of combination of Zn(II) with the apoenzyme in a stopped-flow apparatus. The reaction is first order with respect to both Zn(II) and apoprotein. The rate constants observed are $\sim 10^4$ M^{-1} sec^{-1}. While the combination of Zn(II) with the protein is rapid, the rate constant is two orders of magnitude smaller than those observed for the formation of small Zn(II) chelate compounds. A number of studies report second-order rate constants in the range 10^6 to 10^8 M^{-1} sec^{-1} [85-87]. The latter rates are believed to be controlled primarily by the dissociation of H_2O from the coordination positions to be occupied by the incoming ligands[85]. The rate of combination of Zn(II) with the apocarbonic anhydrase rises from pH 5.5 to pH 7.5 (Table X), but does not approach the rate observed in model systems. A study of the temperature dependence of the rates yielded satisfactorily Arrhenius plots and the calculated energies and entropies of activation are given in Table X.

TABLE X

RATE OF RECOMBINATION OF Zn(II) WITH APOCARBONIC ANHYDRASE AT 25°C. HEAT AND ENTROPY OF ACTIVATION
Data from Henkens and Sturtevant[85]

pH	Ionic strength (M)	H⁺ released by reaction	Apparent second order rate constant x 10^{-4} (M^{-1} sec^{-1})			E (kcal mole^{-1})	ΔS (cal deg^{-1} mole^{-1})
			U.v. difference spectrum	Recovery of hydration activity	Recovery of esterase activity		
5.5	0.051	—	—	—	—	21.2	27.7
5.90	0.001	0.18	0.27	—	—		
5.90	0.021	0.42	0.54	—	—		
6.27	0.056	0.85	—	—	0.81		
7.0	0.056	1.5	—	1.36	—		
7.40	0.020	1.6	1.33	—	—		
7.56	0.056	2.2	—	—	2.0		
7.5	0.051	—	—	—	—	20.8	30.0

The energy of activation observed is considerably higher than that characteristic of model systems, but is partly compensated for by a large increase in entropy of activation. Values of these parameters observed for Zn(II) chelation to small molecules are 7-8 cal deg^{-1} mole^{-1} for ΔS[86,87].

The second-order rate constant observed in the model systems has been formulated as the product of an outer sphere association constant (K_0) and a first-order rate constant (k_1) for the exchange of water for the ligand. Hence, the observed rate constant is $K_0 k_1$[85]. The various factors that may account for the slow rate of Zn(II) binding in the enzyme can only be speculated upon. The general magnitude of the rate would indicate that the binding site is relatively intact before combination of the apoprotein with the metal ion. Thus complex formation does not appear to involve major rearrangements of protein structure[85]. The Zn(II) atom, however, is bound at the bottom of a large cavity in the protein (see page 533) and occupies what is probably a rather unusual geometry. Considerable rearrangement of water bound to the protein as well as to the metal ion may have to take place. Small conformational changes in the protein may also occur.

By using the equilibrium constant previously determined for Zn(II) carbonic anhydrase, Henekens and Sturtevant[85] have calculated an off rate for the Zn(II) ion of 1.5 x 10^{-9} sec^{-1}. Some interesting comparisons have been made to carboxypeptidase A which in contrast to carbonic anhydrase

exchanges its Zn(II) ion rapidly with ionic zinc in the medium[88]. Since the
half-time for the replacement of Zn(II) by Cd(II) in carboxypeptidase A is
$\sim$ 20 h[88] at pH 8.0, 4°C, the rate constant for dissociation has been estimated to be 10^{-6} sec^{-1} [85]. Because of some difference in stability of the
Zn(II) and Cd(II) enzymes the ^{65}Zn–Zn exchange probably gives a better
figure. This half-time of exchange is 5 hours[89], hence k_{off} = $\sim$ 2.5 x 10^{-5} sec^{-1},
4 to 5 orders of magnitude faster than the dissociation of Zn(II) from car-
bonic anhydrase. If as seems likely, the on rates for the two enzymes do
not differ by more than an order of magnitude, then the very slow dissociation of Zn(II) from carbonic anhydrase explains the failure to observe
any exchange at pH 8.0. The half-time for dissociation of Zn(II) from carbonic anhydrase at pH 8.0, 0.1 M ionic strength, 25°C can be calculated to
be 5–6 years[85].

Absorption spectra and optical activity

The most striking physico-chemical change induced in carbonic anhydrase by the substitution of the first transition metal ions for Zn(II) is the
appearance of visible absorption bands due to the d–d transitions arising in
the unfilled 3d electronic shells of these ions. Theoretically these bands are
expected to be the most intense for the Co(II) and Cu(II) complexes[90], a
feature which is observed in the metallocarbonic anhydrases as illustrated
in Fig. 5 showing the visible spectra of the Co(II), Ni(II), and Cu(II)
enzymes. The Co(II) enzyme has the most intense absorption, $\epsilon \sim$ 300.
This band intensity is more typical of tetrahedral Co(II) complexes rather
than regular octahedral complexes which are expected to have ϵ values
near 10^{90}. The band structure, however, does not correspond to that of any
model Co(II) complexes. There are four widely split bands at 520 nm,
555 nm, 615 nm, and 645 nm. This suggests that the coordination geometry
at the active site is highly distorted with the result that the energy levels of
the individual d orbitals each fall at different energies rather than in degenerate groups as in regular octahedral or tetrahedral geometry.
    The spectrum of Co(II) carbonic anhydrase is most closely approxi-
mated by some recently isolated five coordinate complexes of Co(II)[91-94].
This has led to suggestions that a five coordinate geometry might be
present[95]. The 2 Å X-ray data at the present time appear to indicate that
the actual coordination must be a highly distorted 4-coordinate complex
(see page 535). The structure of bisacetatobis (ethylenethiourea) Co(II)
has recently been solved by X-ray diffraction[94a]. The room temperature
crystal absorption spectrum of this compound is widely split and has many
features strikingly similar to the spectrum of Co(II) carbonic anhydrase.
    The molecular geometry of this compound consists of a pseudo-
tetrahedral array in which the four nearest neighbors of the Co(II) atom
are two oxygen atoms and two sulfur atoms bound at 1.957 and 2.328 Å

508

respectively. Two further oxygen atoms are located at 2.928 Å producing a grossly distorted tetrahedron for the geometry of the four nearest neighbors[94a]. In the analysis of the absorption spectrum, while pseudotetrahedral geometry can be assumed as a basis for the analysis, there is some indication that the more distant oxygens also have interactions which modify the spectrum[94a].

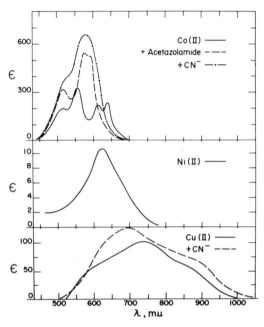

Fig. 5. Visible absorption spectra of Co(II), Ni(II) and Cu(II) human carbonic anhydrases B and inhibitor complexes. Compiled from Coleman[71,99,153].

The visible spectrum of the Cu(II) enzyme also suggests a distorted geometry, since at least three separate absorption bands are resolved (Fig. 5). The spectrum of the Ni(II) enzyme is more like that expected of octahedral Ni(II) complexes. It is possible that there is some variation in coordination geometry between the various metal derivatives of carbonic anhydrase, which could be related to the absence of activity in most of the derivatives.

Aside from conclusions about the coordination geometry, the visible absorption spectra of these derivatives offer a powerful spectroscopic probe for the active center. This is especially true of the Co(II) derivative, since the enzyme is active. The addition of both sulfonamides and anions like cyanide to the Co(II) enzyme causes a major increase in band intensity and a narrowing of the visible absorption spectrum[20,38,71,82] as illustrated in Fig. 5 by the spectra for the acetazolamide and CN^- complexes. This result was the first strong evidence that both types of inhibitors add a group to the inner coordination sphere of the metal ion.

The Co(II) probe was also first used by Lindskog[20,82] to identify
another phenomenon associated with the active center. The Co(II) absorption bands undergo a marked change with pH as shown in Fig. 6 for the
human Co(II) enzyme. The bands decrease in intensity and the two long
wave length bands disappear as the pH is lowered from 9 to 6. The pK_a
describing this shift is between pH 7 and 8 depending on conditions[20,71,82,84]
and is apparently the same pK that can be related to the pH-rate profile for
the enzyme which is sigmoid with a pK_a between 7 and 8[26,27,29,84,96-98]. Thus
the ionization that affects the activity may also affect the geometry of the
coordination complex, and may be associated with the ionization of a
coordinated water molecule (see Sections V and VII).

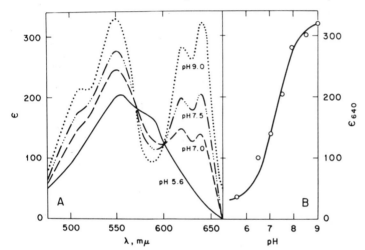

Fig. 6. Visible absorption spectrum of Co(II) human carbonic anhydrase B as a function
of pH (A). Molar extinction coefficient at 640 nm as a function of pH (B). From
Lindskog and Nyman[82].

Circular dichroism of the Co(II) absorption bands associated with the
active center complex has proven even more sensitive in detecting changes
at the active center[99]. Visible circular dichroism of three isozyme and
species variants of Co(II) carbonic anhydrase are shown in Fig. 7. Although
the absorption bands of these three Co(II) enzymes are almost identical in
energy and position implying similar coordination geometries, the ellipticity
indicates a striking difference. Both human C and the bovine B Co(II)
enzymes show large optical activity associated with these bands, $\Delta\epsilon$ values
3–5, while the human B enzyme has ellipticity throughout the visible
region of $\Delta\epsilon < 0.2$.

If the human B enzyme is examined at 10-fold greater sensitivity,
small positive ellipticity bands are observed to be associated with all the
visible absorption bands (Fig. 8). These undergo considerable enhancement
and narrowing as the pH is lowered, corresponding to the pH change in the

510

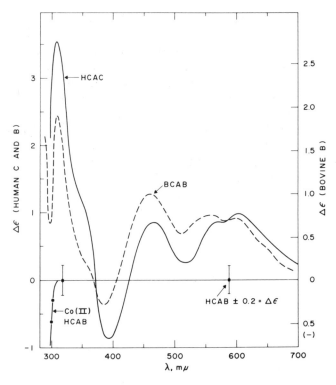

Fig. 7. Visible c.d. of three isozyme and species variants of Co(II) carbonic anhydrase. Human isozyme C(——); bovine isozyme B(- - -); human isozyme B (●—●). Compiled from Coleman[99,116].

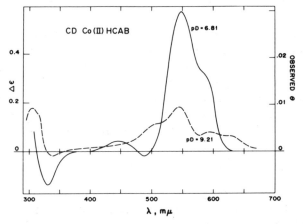

Fig. 8. C.d. spectrum of Co(II) human carbonic anhydrase B in D_2O at pD = 6.81 (——) and pD = 9.21 (- - -).

absorption spectrum (Fig. 6) (see below). These changes indicate at least a moderate change in coordination geometry accompanying the ionization that is associated with the pH-rate profile and the changes in the absorption spectrum as a function of pH. Evidence now suggests that all these changes are accompanied by the ionization of a coordinated water molecule (see Section VI).

While the position and intensities of the absorption bands of the Co(II) derivatives of all species and isozyme variants of carbonic anhydrase as well as those of the sulfonamide complexes are very similar, the magnitude of the associated ellipticity bands are not (see Figs. 7 and 8). Compare the c.d. spectra of the human B, human C, and bovine enzymes. This suggests that the optical activity of the Co(II) chromophore is more sensitive to local features of isozyme structure than to the fundamental symmetry of the coordination complex itself. The dissymmetric potential field provided by a constellation of surrounding charged and dipolar amino acid groups may be the determining feature here, in which case it will be very difficult to translate the c.d. information into structural information. The very low optical activity associated with the Co(II) chromophore in the human B isozyme seems to indicate that the Co(II) complex itself apparently has symmetry elements that prevent major intrinsic optical activity perhaps not incompatible with a flattened tetrahedral or five coordinate complex.

The pK_a for the spectral transformation pictured in Fig. 6 is shifted to higher pH values by the addition of complexing anions and is shifted to a greater degree by the anions which have stronger binding affinity[98]. These shifts also agree with the analogous shifts observed for the pH-rate profile in the presence of these anions[98]. One possible interpretation for these findings is that $^-$OH and the anions compete as monodentate ligands for an open coordination site on the metal ion. Such an interpretation is compatible with a variety of other data suggesting the presence of an enzyme–Zn–OH complex at the active center (see below).

Inhibitor binding

Investigation of the functional and physico-chemical aspects of the interaction of inhibitors with the active site of carbonic anhydrase has evolved a number of methods for the study of the enzyme. Titrations of the effect on activity of the sulfonamide ethoxzolamide or the anion $^-$CN have both shown that one site is involved in the inhibition by these compounds[21,98,100]. The X-ray as well as the spectral data discussed above leave little doubt that this site involves the metal ion, although precise relationships at this site may involve more subtle features than simple metal coordination[23].

The spectral changes involved in sulfonamide interaction will be discussed further in Section IV. With ^{3}H-acetazolamide it can be shown

512

directly that the presence of the metal ion is required for the binding of this sulfonamide (Fig. 9A)[21]. Acetazolamide does not bind to the apoenzyme until the concentration of the free compound has reached 10^{-4} M or greater, while the Zn(II) enzyme binds one mole at 5×10^{-7} M concentration of free inhibitor (Fig. 9A). An additional striking feature of this metal-induced binding is that only Zn(II) and Co(II), the two metal ions that induce carbonic anhydrase activity in the apoenzyme (Table VIII), are likewise the only two ions which induce sulfonamide binding. Thus it would appear that the physico-chemical features of the active center coordination complex leading to activity are also the ones responsible for inducing the addition of the sulfonamide ligand[21].

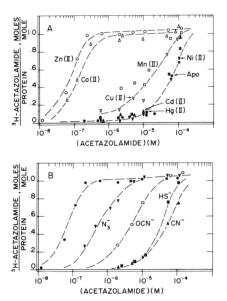

Fig. 9A. Binding of ^{3}H-acetazolamide to metallocarbonic anhydrases B as a function of acetazolamide concentration. Moles of ^{3}H-acetazolamide bound per mole of Zn(II) (○), Co(II) (△), Mn(II) (□), Ni(II) (▼), Cu(II) (▽), Cd(II) (▲), Hg(II) (■), and metal-free (●) carbonic anhydrases are indicated on the ordinate. Conditions: 0.025 M Tris, pH 8.0, 4°C. B. Binding of ^{3}H-acetazolamide to Zn(II) carbonic anhydrase B in the presence of metal-binding anions. (●) no addition; (▼) 2×10^{-4} M N$_3^-$; (○) 2×10^{-4} M OCN$^-$; (■) 2×10^{-4} M HS$^-$; (▲) 2×10^{-4} M CN$^-$. Conditions were as in Fig. 9A. From Coleman[21].

That the sulfonamide ligand site overlaps the anion binding site can also be shown directly using the tritiated acetazolamide (Fig. 9B). The anions azide, cyanate, sulfide, and cyanide can all displace the sulfonamide. The concentration of acetazolamide required to bind in the presence of an equal and constant concentration of these anions is directly proportional to the known binding affinity of these anions for the active center of carbonic anhydrase as measured by inhibition kinetics, $CN^- \geqslant SH > OCN^- >$ N_3^- [21].

Another approach developed to demonstrate the direct interaction of one of the anions with the metal ion at the active center has been applied by Riepe and Wang[102]. These investigations have detected the infra-red absorption band for the asymmetric stretching mode of the CO_2 molecule when bound at the active site of carbonic anhydrase and similarly the band for the asymmetric stretching mode of the linear azide ion, N_3^-, when bound at the active site. Carbon dioxide shows a strong infra-red band at 2343.5 cm^{-1}, while water has a relatively low absorbance in this region. By using very high concentrations of enzyme, 300 mg/ml, and an interference method to balance the pathlength of the enzyme cell and the blank, these workers located this asymmetric stretching frequency for CO_2 bound at the active center. The peak is located at 2341 cm^{-1} and since it is only minimally shifted from that for free CO_2 suggests that CO_2 is bound to the enzyme in an unstrained configuration and is apparently not bound to the metal ion.

The azide ion, N_3^-, has a strong band at 2049 cm^{-1} due to the asymmetric stretching of this linear molecule. This band is shifted to higher frequency when azide is bound to metal ions[102]. If azide is reacted with carbonic anhydrase, the shifted azide absorption band appears (Fig. 10) clearly showing that N_3^- is bound to the metal ion. The Co(II) enzyme also shifts the azide band (2081 cm^{-1}) but not as far as the Zn(II) enzyme

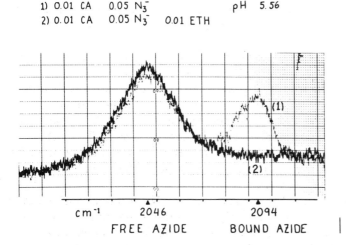

1) 0.01 CA   0.05 N_3^- pH   5.56
2) 0.01 CA   0.05 N_3^- 0.01 ETH

cm⁻¹   2046   2094

FREE AZIDE   BOUND AZIDE

Fig. 10. Difference infra-red spectrum for bovine carbonic anhydrase plus azide. Spectrum 1, enzyme plus excess azide. Peak at 2094 cm^{-1} is typical of Zn complexed azide. Spectrum 2, enzyme plus azide plus excess ethoxzolamide. From Riepe and Wang[102].

References pp. 543–548

(2094 cm^{-1}) as expected from studies of model Zn(II) and Co(II) mixed complexes with N_3^- [102]. The sulfonamide, ethoxzolamide, abolishes the shifted azide absorption band. Azide also displaces the enzyme-bound CO_2 molecule. The latter can in addition be displaced by NO_3^- and HCO_3^-. These findings suggest that CO_2 is bound in an unstrained manner to a site adjacent to the metal ion, but not coordinated to it. This is compatible with present postulates on the mechanism of action of carbonic anhydrase (see Section VII).

A great many simple anionic species have at least mild inhibitory effects on carbonic anhydrase and the acetate anion, a product of the esterase reaction, has been shown to be an inhibitor[84]. A more interesting type of inhibition was discovered when it was noted that the inhibition of human carbonic anhydrase B by iodoacetate became irreversible with time[63,64,66]. The irreversible inhibition has a number of interesting characteristics. It is never complete, 5–10% of the activity remaining even after reaction of the enzyme with one mole of reagent. This residual activity follows a sigmoid pH-rate profile with a midpoint of pH 8.5[66]. This has suggested that the ionizing group apparent in the native pH-rate profile is still present in the modified enzyme[66].

The iodoacetate reacts with a histidine contained in the sequence discussed in Section II. Several lines of evidence arising from the detailed studies of Bradbury[63,64] indicate that the reaction occurs at the active site: (a) loss of esterase activity parallels the production of 3-carboxymethyl-histidine; (b) iodoacetate inactivates the enzyme irreversibly only while bound as a reversible inhibitor; (c) sulfanilamide protects against irreversible inactivation; (d) the native conformation of the enzyme is required for rapid dehydration; in the presence of 8 M urea the rapid reaction with histidine is no longer observed.

The absorption and c.d. spectra of the Co(II) enzyme in the presence of acetate and iodoacetate suggest that both are bound in similar fashion and coordinate the Co(II) ion[104] (Fig. 11). Iodoacetate then reacts irreversibly with the enzyme at a relatively slow rate[63,64,104]. Monkey carbonic anhydrase B requires up to 12 hours in the presence of 10^{-2} M iodoacetate to be completely irreversibly labelled with ^{3}H-iodoacetate[104]. This concentration dependence suggests that the reversible binding site must be saturated before the irreversible reaction will go to completion. The kinetic reasons for this are not immediately apparent.

Once the enzyme is labelled, however, both CN^- and large sulfonamides can still be bound at the active site[104]. Cyanide reverts the spectrum of the iodoacetate-Co(II) enzyme to that observed for the CN^- complex of the Co(II) native enzyme[104] (Fig. 11). Hence the coordination complex does not appear modified by the histidine modification. These findings have been synthesized in the model shown in Fig. 12 suggesting that the carboxyl coordinates the metal ion and that the imidazole nitrogen is in a position to

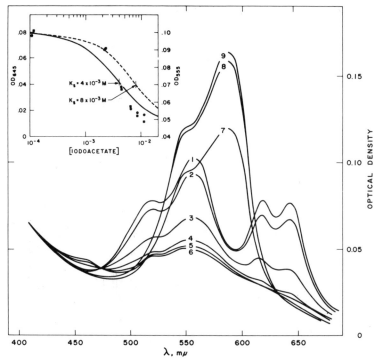

Fig. 11. Effect of iodoacetate on the spectrum of Co(II) human carbonic anhydrase B at pH 9.0 (curves 1 to 6). Effect of CN⁻ on the spectrum of carboxymethylate Co(II) human carbonic anhydrase B (curves 7-9). Curves 1 to 6 are spectra taken with 0 to 2×10^{-2} M iodoacetate as indicated in the insert (●, O.D.$_{555}$) (■, O.D.$_{645}$). K_s for iodoacetate is between 4 and 8×10^{-3} M. After 24 hours the enzyme irreversibly reacted with iodoacetate was titrated with CN⁻ (7,1 eq.; 8,2 eq., and 9, excess CN⁻). Data from Coleman[153].

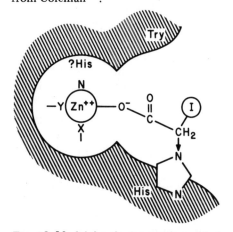

Fig. 12. Model for the interaction of iodoacetate with the active center of human carbonic anhydrase B.

References pp. 543–548

attack the iodinated carbon. This would place this nitrogen at least three bond distances from the metal. While this may be close enough to participate in proton transfers, this histidine has been shown to have a pK_a of 5.8 and only the basic form reacts with iodoacetate[63,64]. This appears too low for the pK_a involved in enzyme activity[63,64]. On the other hand the reversible binding of iodoacetate depends on a group with a pK_a estimated indirectly to be 7, suggested to be the pK_a of the coordinated water[63]. The completely carboxymethylated enzyme showing 10% of the activity continues to show a sigmoid pH rate profile with a pK_a of 8.5. The latter also can be interpreted as the pK_a for a coordinated water on modified enzyme[63]. Sulfonamides while still binding to the modified enzyme have an altered, less non-polar environment as shown by c.d. studies of a bound azosulfonamide[105].

The reaction with iodoacetate shows considerable variation between isozyme and species variants of carbonic anhydrase[104]. Using ^{3}H-iodoacetate to label four species and isozyme variants of the enzyme at 10^{-2} M iodoacetate one mole of iodoacetate reacts irreversibly with the human B enzyme after 24 hours. The monkey B enzyme reacts almost as readily. No reaction with the bovine B enzyme is observed under these conditions and very little with the human C enzyme[104]. Thus if this histidine is present in all the species and isozyme variants of the enzyme, it is much less reactive in some of them.

Shifts in hydrogen ion equilibrium accompanying addition of ligands to open coordination sites on metal ions already incorporated at the active sites may provide information directly relevant to the mechanism of catalysis. If the anions like CN^- and HS^- inhibit carbonic anhydrase by coordinating the metal ion, the reaction should be accompanied by the displacement of a proton from the inhibitors at pH values below the pH region for the dissociation of HCN and H_2S; the former described by a pK_a of 9.3, the latter by a pK_a of 6.9 at about $23°C$[105]. The net hydrogen ion release will depend on other hydrogen ion equilibria altered by the binding of anions. In order to investigate the hydrogen ion equilibria that accompany the reaction of carbonic anhydrase with cyanide and sulfide, a set of equilibrium measurements between pH 6 and 10 have been made by the direct measurement of H^+ or ^-OH release with a difference titration method[84]. The method consists of adjusting both the concentrated anion solution and the protein solution to an identical pH, adding a small equimolar aliquot of the anion to the protein, and measuring the uptake or release of protons on a pH-stat at a precision of $1.0 ± 0.02$ μmole H^+. The experiment can be done over a pH range of 6–10, since cyanide forms a firm 1:1 complex with carbonic anhydrase over this pH range.

The results are best described by formulating one of the possible models that may apply to the anion-carbonic anhydrase reaction and comparing the predicted results with the observed findings. If a coordinated water molecule is displaced from the metal ion by the anions, the various proton equilibria may be pictured as in eqn. (2):

(a) $\quad CAZn \cdot H_2O \quad + \quad HCN \; \rightleftharpoons \; CAZn \cdot CN^- + H_2O + H^+$

$$pK_a = 8 \left|\right| -(H^+) \quad pK_a = 9.3 \left|\right| -(H^+)$$

(b) $\quad CAZn \cdot OH^- \quad + \quad CN^- \; \rightleftharpoons \; CAZn \cdot CN^- + OH^- \qquad (2)$

At low pH where the metal ion is in the hydrated form and the inhibitor is in the acid form, the reaction is described by the release of a water molecule and a proton (a). At very high pH the inhibitor will be in the anion form and if the coordinated water molecule has a pK_a within the experimental pH range, the metal ion will be in the hydroxide form (b). As seen in Fig. 13, the reaction of cyanide with the Zn(II) enzyme does show a biphasic titration curve; one mole H^+ is released at pH 6.0 while 0.7 mole OH^- (H^+ uptake) is released at pH 10.1. These alterations in proton equilibria are a function of the metal ion, since the apoenzyme solution does not show any change in pH upon the addition of cyanide. The Co(II) enzyme shows a very similar titration curve to the Zn(II) enzyme (Fig. 13). The results can be explained if a single additional proton equilibrium associated with the protein, coupled to the metal ion, and described by a pK_a of approximately 8 is added to the titration curve (dashed line, Fig. 13). The results are compatible with eqn. (2) if the coordinated water molecule is assumed to have a pK_a of 8. As the pH rises more enzyme is in the ^-OH form. The release of this ^-OH by cyanide progressively neutralizes the proton from HCN along the theoretical curve shown in the Figure. Finally the CN^- species will displace only the ^-OH. An additional feature of the system predicted from eqn. (2) is the dependence of the difference titration on the pK_a of the inhibitor. This can be tested by using sulfide, since H_2S has a pK_a of 6.3. As expected the biphasic titration curve shifts about 1.5 units toward acid pH. Once again a theoretical curve can be generated by the same additional ionization with a pK_a of 8.

These results are compatible with the presence of a coordinated water molecule at the active center of carbonic anhydrase according to eqn. (2); they do not prove the identity of this ionization. An alternative is possible by assuming that the anion displaces a metal-bound protein ligand with a very high pK_a which subsequently takes up a proton. This group cannot be initially bound to the metal ion but must bind along a pH function with a midpoint of 8. The evidence is less convincing for this alternative. The interpretation that the pK_a observed is that for a coordinated water molecule is strongly supported by n.m.r. data (see Section V).

References pp. 543–548

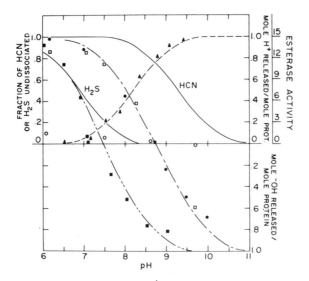

Fig. 13. Displacement of H^+ and ^-OH from human carbonic anhydrases B by cyanide and sulfide. The solid lines represent the continuous titration curves for the dissociation of HCN to H^+ and CN^- and H_2S to H^+ and ^-SH, expressed as the fraction of inhibitor undissociated (left hand ordinate). (- - -), theoretical titration curve for an acidic group of pK_a = 8.1, expressed as moles H^+ dissociated/mole protein (upper right hand ordinate). ▲, esterase activity of human carbonic anhydrase B, μmoles substrate hydrolyzed/min/ μmole enzyme. (— - —) and (— - - —), theoretical difference titration curves expected for the displacement of H^+ (upper right hand oridinate) and ^-OH (lower right hand ordinate) from Zn(II) carbonic anhydrase by cyanide and sulfide respectively as a function of pH according to eqn. 2. Moles of H^+ or ^-OH released/mole protein on the addition of cyanide to Zn(II) carbonic anhydrase (●), apocarbonic anhydrase (○), and Co(II) car-bonic anhydrase (□). Maximum H^+ and ^-OH release from the Co(II) enzyme was not reached until the addition of ~1.5 equivalents of cyanide. Moles of H^+ or ^-OH released/ mole protein on the addition of sulfide to Zn(II) carbonic anhydrase (■). From Coleman[84].

IV. CARBONIC ANHYDRASE – SULFONAMIDE COMPLEXES

Since the initial discovery by Mann and Keilin[19], of the inhibition of carbonic anhydrase by sulfonamides a great variety of these compounds have been tested as carbonic anhydrase inhibitors. Two general structural features of these molecules are required for effective carbonic anhydrase inhibition. In order to be good inhibitors, the sulfonamide must have an unsubstituted sulfonamide group (A, I).

$$NH_2 - \bigcirc - SO_2 - NH_2$$

(R)   (A)

(I) Sulfanilamide

Substitution of the $-SO_2-NH_2$ group usually destroys all inhibitory capacity; however, recently it has been discovered that certain modifications of the $-SO_2-NH_2$ group, while greatly reducing the inhibitory capacity, still preserve at least some binding affinity of the sulfonamide for the inhibitory site[65]. The second general requirement is a large bulky aromatic or heterocyclic R group attached to the free sulfonamide. Other than this general requirement, the lattitude for the structure of the R group is very great as indicated by the structures of four sulfonamide inhibitors (II, III, IV, V), selected from the very large number of these compounds that are effective carbonic anhydrase inhibitors. (See Maren[106] and Bar[107] for more complete lists.) These compounds have been of considerable interest because of their pharmacological and therapeutic effects on various anion exchange reactions in the body ultimately involving HCO_3^- [106]. Much of this work has been reviewed in detail by Maren[106].

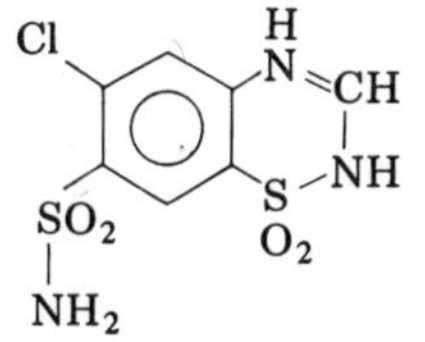

(II) Chlorothiazide

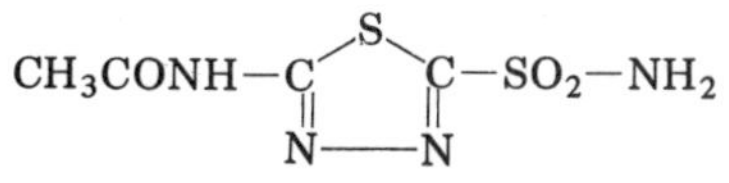

(III) Acetazolamide

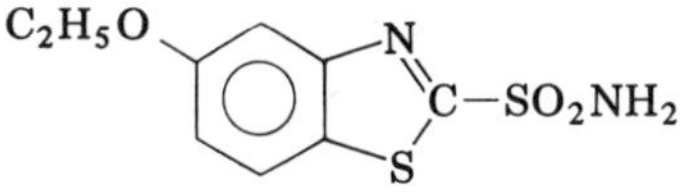

(IV) Ethoxzolamide

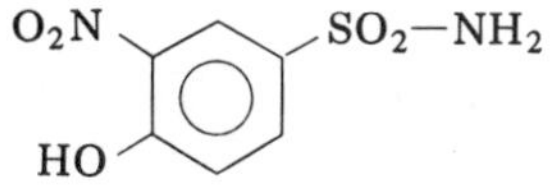

(V) 4-Hydroxy-3-nitrobenzene-sulfonamide

In the more recent work on the physico-chemistry and mechanism of action of the enzyme, these compounds have proved extremely valuable as probes of the active center. Since the studies involving these compounds have as their basis the unique interaction of the sulfonamides with the active center of carbonic anhydrase, it seems appropriate to gather these studies under a single heading.

(a) Binding interactions, dissociation constants

Dissociation constants for carbonic anhydrase–sulfonamide complexes (derived by assuming that the kinetically measured K_i values are a measure of the dissociation constant) vary from 2×10^{-9} M for ethoxzolamide to 2.6×10^{-5} M for sulfanilamide[108-111]. Present information on the magnitude of the constants is summarized in Table XI. The strikingly small magnitude of these constants emphasizes the remarkable affinity of this non-covalent interaction. This extremely tight binding has given rise to a convenient method of determining carbonic anhydrase concentration by titrating the activity of a carbonic anhydrase solution with ethoxzolamide[100,102,109]. Spectral measurements of free and bound sulfonamide suggested that the binding of sulfonamide was metal ion dependent[20] and direct binding studies using ^{3}H-acetazolamide (discussed on page 512) showed that the presence of the Zn(II) ion is required for binding of the sulfonamide[21].

TABLE XI

"DISSOCIATION CONSTANTS" FOR CARBONIC ANHYDRASE–SULFONAMIDE COMPLEXES[a]

Compound	Dissociation constant M	Reference
Sulfanilamide	2.8×10^{-6}	108
N^4-acetylsulfanilamide	5×10^{-7}	108
Dichlorphenamide	2.5×10^{-8}	108
Acetazolamide	8×10^{-9}	108
	1.5×10^{-7}[b]	21
Methazolamide	9×10^{-9}	108
Ethoxzolamide	1×10^{-9}	108
Chlorothiazide	2.2×10^{-6}	108
Cyclothiazide	6×10^{-7}	65
Chloroacetylcyclothiazide	6.5×10^{-5}	65
5-Dimethylaminonaphthalene-1-sulfonamide (DNSA)	2.5×10^{-7}[c]	100
4'-Sulfamylphenyl-2-azo-7-acetamido-1-hydroxynaphthalene-3,6-disulfonate (Neoprontosil)	1.5×10^{-6}[d]	30

[a] Unless otherwise noted the K_i value has been assumed equal to the dissociation constant.
[b] K_s determined directly by the binding of ^{3}H-acetazolamide.
[c] K_s determined by a fluorescent method.
[d] K_s determined by a spectropolarimetric method.

Although it has long been assumed that the anionic inhibitors like CN^- and SH^- bind to the metal ion, such a conclusion is not so obviously suggested by the structure of the sulfonamides. However, using the metal binding anions and ^{3}H-acetazolamide it can clearly be shown that sulfonamide binding site overlaps the anion binding site, since the anions prevent the binding of the sulfonamide[21] which suggested that the sulfonamide might also coordinate the metal ion[21]. The $-SO_2-NH_2$ group seemed to be the obvious candidate for the metal-coordinating group[21,100]. It is thus not too surprising that the structure of a crystalline complex of HCAC with a sulfonamide as determined by X-ray diffraction shows the center of gravity of the sulfonamide group within 3.2 Å of the Zn(II) ion[22,23], close enough for the NH_2 group to be within the inner coordination sphere of the metal

ion (see Section VI). It appears likely that the anionic form of the sulfonamide group ($-SO_2-\bar{N}H$) is the bound form, with one proton being
replaced by the metal ion[30,100]. The pK_a of this group in the absence of
metal ion competition varies from pH 7.0 to 10 depending on the structure
of the R group[106,112]. A number of studies of the pH-dependence of sulfonamide show binding to be maximum at about neutral pH, falling off rapidly
at acid or alkaline pH[84,112,113]. The decreasing concentrations of the anionic
species at low pH may account for the decrease in sulfonamide binding
affinity observed at pH values below 7.

Binding of the sulfonamide also appears to be dependent on a second
alkaline pK_a on the enzyme similar to the one controlling enzymatic
activity and suggested by recent data to be the pK_a of a coordinated water
molecule (see Section IV). If this model is correct and the $-SO_2-N\bar{H}$ group
coordinates the metal ion, the sulfonamide and $\bar{O}H$ would be expected to
compete for the metal binding site. The second order rate constants describing the combination of sulfonamide with enzyme show the same pH-
dependence as binding. Thus the pK_a values involved must represent ioniz-
ations on the free enzyme and free inhibitor[112]. A note of caution should
be added, however, about the precise interpretation of the binding data as
a function of pH. Since the very tight binding is a cooperative phenomenon
between a metal coordination site and other less well-identified interactions
of the sulfonamides with the protein (hydrophobic, solvent displacement
(entropy) or other contributions to the binding forces), there is the possibility that other changes in the protein structure induced by changes in
hydrogen ion concentration could be reflected in the sulfonamide binding
affinity.

The apparent second order rate constant for the formation of the
enzyme–sulfanilamide complex is $\sim 7 \times 10^4\ M^{-1}\ sec^{-1}$ at pH 7.9, 25°C, as
measured by Lindskog and Thorslund[112] with a stopped-flow method.
Both competitive and non-competitive kinetics have been observed for
sulfonamide inhibition of carbonic anhydrase depending on time of reaction and the order of additions of substrate or inhibitor[113-115]. As was first
pointed out by Kernohan[114] and amplified by Lindskog and Thorslund[112],
the dissociation rate of the inhibitor is slow enough such that equilibrium
with a competing substrate molecule is achieved only slowly. This may
account for shifts in the type of inhibition observed during the progress of
the reaction or on changing the order of addition of inhibitor and substrate.

(b) Absorption spectra and circular dichroism

A particularly valuable aspect of the interaction of sulfonamides with
the active center of carbonic anhydrase has been their use as molecular
probes utilizing spectroscopic means of following either a chromophore
incorporated in the sulfonamide or one that is already present in the pro-

tein and is modified by the sulfonamide. The Co(II) chromophore introduced at the active site is the major example of the latter type. Both the
intensity and energies of the visible absorption bands of the Co(II)–active
site complex are modified by the addition of sulfonamides as shown above
in Fig. 5. These shifts, like those induced by the anions, strongly suggest
the addition of a ligand to the central coordination sphere of the metal ion.
The major features of the absorption bands have been discussed in Section
III.

Much investigation has centered on the change in energies and optical
activity of the Co(II) absorption bands accompanying the binding of anions
and sulfonamides to the active center of carbonic anhydrase. The general
results are summarized by the examples given in Fig. 14, showing both
absorption and c.d. spectra for the ethoxzolamide and cyanide derivatives
of the human B Co(II) enzyme. These spectra have been resolved into
appropriate gaussian absorption and ellipticity bands in order to indicate
at least an approximate band structure for these spectra[116].

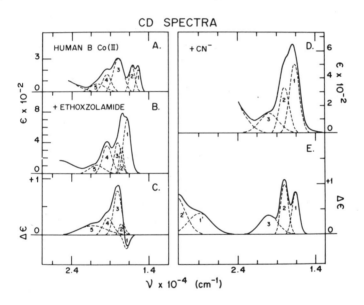

Fig. 14. Resolution of the visible absorption and c.d. spectra of Co(II) human carbonic
anhydrase B and the complexes with ethoxzolamide and cyanide into a series of overlapping Gaussian bands. In each part of the Figure (——) represents the envelope for
the sum of the constituent Gaussian bands shown by the dashed lines and numbered
consecutively from lower to higher energies. Each envelope corresponds within the
error of the measurement to the experimental spectra. A, absorption spectrum, enzyme
alone; B, absorption spectrum plus ethoxzolamide; C, circular dichroism plus ethoxzolamide; D, absorption spectrum plus CN⁻, E, circular dichroism plus CN⁻. Curves were
fitted with a Du Pont 310 curve resolver. From Coleman[116].

Several points are evident from this analysis. (1) All spectra show a number of widely split visible bands, suggesting that even in the inhibited enzymes the coordination geometry is considerably distorted. (2) Both the c.d. and absorption spectra of the ethoxzolamide complex (Fig. 14B and C) require the same number of bands as the original Co(II) spectrum (Fig. 14A) for an adequate fit. The effect of adding the strong sulfonamide ligand appears to be a blue shift of the two low energy bands without, however, reducing the band multiplicity. This may be compatible with the exchange of ligands along a very restricted direction in the complex (see X-ray structure, Section VI). There are near ultra-violet chromophores as well that are clearly influenced by these ligand changes as illustrated most clearly by the c.d. spectrum of the cyanide enzyme (Fig. 14D). These bands are also clearly evident in the c.d. spectrum of the unmodified Co(II) enzymes (Figs. 7 and 8). The particular coordination geometry associated with the active site of carbonic anhydrase and characterized by the unusual set of d-d transitions (Fig. 14A) and its affinity for sulfonamides appears to be a very ancient feature of the protein chemistry of carbonic anhydrase. These same bands are present in the Co(II) derivatives of the elasmobranch enzyme which presumably developed into an isolated evolutionary path several hundred million years before the dawn of the age of mammals[41].

(c) Interactions with sulfonamides containing chromophores

Optical spectra
A set of localized accessible electronic transitions can be introduced into the active site region of carbonic anhydrase by incorporating a chromophore into the R group of the sulfonamide. While several sulfonamides have chromophores whose characteristics are altered by binding to carbonic anhydrase[30,99,112,117], the most striking example has been provided by the binding of a hydroxynaphthalene derivative of an azosulfonamide (**VI**).

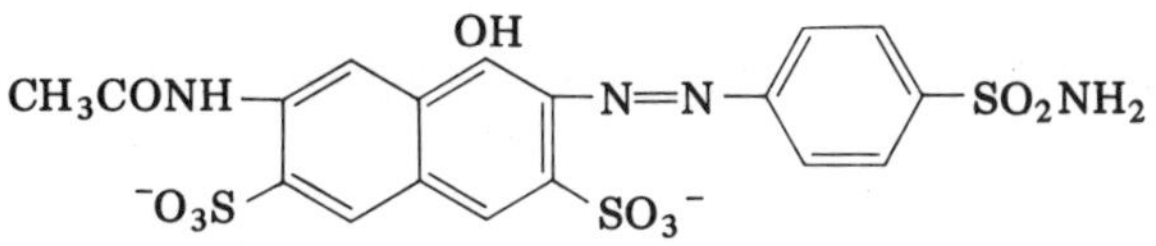

(**VI**) 2-(-4-sulfamylphenylazo)-7-acetamido-1-hydroxynaphthalene-3,6-disulfonate

This azosulfonamide has intense visible absorption bands near 500 nm as well as numerous ultra-violet bands. On binding to carbonic anhydrase, a striking hypochromia is seen in the visible bands as well as a red shift of several nanometers. In addition, the magnitude of the hypochromia and red shift are unique for each isozyme or species variant of the enzyme used to form the complex[30].

524

The spectral changes induced in this chromophore by carbonic anhydrase are very similar to those induced in this azosulfonamide by non-polar solvents[30] suggesting that the environment at the active center may be rather non-polar[30], a conclusion also suggested by a number of other spectral studies[100,102].

An additional aspect of the azosulfonamide interaction is the striking optical activity induced in the sulfonamide absorption band by the binding of this symmetrical molecule in the dissymmetric environment of the protein[30,117]. The visible c.d. spectra for the complex of the azosulfonamide with the monkey B isozyme is shown in Fig. 15 compared to the c.d. of the native enzyme[117]. In addition to the large visible ellipticity bands, there are also large ultra-violet ellipticity bands induced in the azosulfonamide chromophore which completely eclipse the c.d. of the native protein. The rotatory power of some of these bands, $R_k \cong 2 \times 10^{-38}$ cgs units, is of a magnitude expected for inherently dissymmetric chromophores rather than chromophores perturbed by a dissymmetric environment[118]. Thus the surrounding protein structure must provide a particularly effective dissymmetric potential field[30].

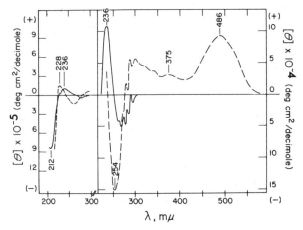

Fig. 15. C.d. spectra of Macaca mulata carbonic anhydrase B (——) and its 1:1 azosulfonamide complex (- - -). From Coleman[117].

If the c.d. spectra of the complexes of several species and isozyme variants of the enzyme with this sulfonamide are compared, considerable variations in both the magnitude and sign of the various ellipticity bands induced in the sulfonamide are observed[30]. Thus the precise molecular detail of the structure surrounding the active center would appear to be different in each isozyme. Such isozyme differences have also been indicated by the recent work on ^{35}Cl n.m.r.[119,120] and by complexometric titrations[84] and may be responsible for the large differences in specific activity noted between species and isozyme variants of the enzyme[75].

Fluorescence

Bovine carbonic anhydrase forms a highly fluorescent complex with 5-dimethyl aminonaphthalene-1-sulfonamide (DNSA)[100]. By studying either the enhancement of the ligand fluorescence or the quenching of the protein ultra-violet fluorescence, Chen and Kernohan[100] showed that one molecule of the fluorescent sulfonamide was bound per molecule of enzyme with a dissociation constant of 2.5×10^{-7} M. The fluorescence of the free sulfonamide has a peak emission at 580 nm and a quantum yield of 0.055, but the bound compound has an emission maximum at 468 nm and a quantum yield of 0.84. The large blue shift in the emission maximum can be adequately explained if it is assumed that the binding pocket is extremely hydrophobic and that the $-SO_2NH_2$ group of the ligand loses a proton upon binding to the enzyme[100].

Calculation of the energy transfer efficiency gave the surprising result that 85% of the photons absorbed by the 7 tryptophan residues are transferred to the single bound DNSA molecule. The transfer efficiency is much higher than hitherto observed for a protein having only one 5-dimethyl-aminonaphthalene-1-sulfonyl group. Although the diameter of the protein is roughly 51 Å, the bound DNSA group is probably within the critical transfer distance R_0 (= 21.3 Å) of all tryptophans. These findings suggest that the sulfonamide binding site and the tryptophans are in the interior of the molecule, a suggestion borne out by the X-ray findings thus far (see Section VI), although these apply only insofar as the structure of the human C enzyme will be found analogous to the bovine B enzyme upon which the fluorescence and phosphorescence (to be discussed below) studies were carried out.

Phosphorescence

Galley and Stryer[121] have used the phosphorescence that occurs when there is triplet–triplet energy transfer to assay the interaction of another bound sulfonamide with tryptophan residues in the molecule. Electronic excitation energy can be transferred between singlet states of chromophores separated by distances of the order of 30 Å. On the other hand triplet–triplet energy transfer requires a much closer approach and can measure distances on the order of 12 Å. If the singlet state of a triplet acceptor is at a higher energy than the singlet state of the triplet donor the triplet donor can be excited by light of a wavelength not absorbed by the triplet acceptor. Under these conditions, phosphorescence of the triplet acceptor is observed only if there is triplet–triplet energy transfer. Galley and Stryer[121] obtained this arrangement in the complex of bovine carbonic anhydrase with *m*-acetylbenzenesulfonamide (MABS). Tryptophan phosphorescence of carbonic anhydrase is excited by light of wavelength 280 nm. The sulfonamide phosphorescence is excited by light of wavelength 330 nm, while the enzyme shows no phosphorescence if 330 nm light is used. If the MABS carbonic anhydrase complex is excited at 330 nm, however, the phosphorescence observed was that of tryptophan rather than MABS[121]. Thus a tryptophanyl residue appears to be lcoated very near the sulfonamide binding site, perhaps lining the binding cavity.

V. NUCLEAR MAGNETIC RESONANCE AND ELECTRON SPIN RESONANCE

Until recently magnetic resonance techniques had not been applied to carbonic anhydrase. However, both proton magnetic resonance and electron spin resonance (e.s.r.) have now been applied taking advantage of certain chemical features provided by the active center containing a metal ion. Taking advantage of the well-known enhancement of the relaxation rate of water protons induced by paramagnetic metal ions if freely exchanging water molecules are present in the first coordination sphere, Fabry et al.[122] have demonstrated the presence of a coordinated H_2O or $\overline{O}H$ in the first coordination sphere of Co(II) at the active center of carbonic anhydrase. A paramagnetic metal ion has a magnetic moment some 10^3 times greater than other nuclei, hence the random motion of such ions has a much more potent effect on relaxing adjacent solvent protons, especially if the paramagnetic center is also contained on a macromolecule with relatively slow thermal motion. Since the substitution of Co(II) for Zn(II) at the active center of carbonic anhydrase maintains an active enzyme, the Co(II) derivative, containing a paramagnetic center, is an ideal subject for such studies.

Co(II) carbonic anhydrase produces a marked enhancement of the relaxation rate (T_1^{-1}) of water protons, especially at higher pH values[122]. A large part of this enhancement is abolished by the addition of azide or ethoxzolamide clearly establishing this enhancement as related to the active center. This relaxation-enhancement, plotted as relaxivity, is shown in Fig. 16, as a function of pH for the Co(II) derivatives of bovine carbonic anhydrase and human carbonic anhydrase B. The inhibitable part of T_1^{-1} is pH dependent with a pK of 7.0 ± 0.2 for the bovine Co(II) enzyme and 8.2 ± 0.2 for the human B Co(II) enzyme. From the magnetic field dependence of the inhibitable part of T_1^{-1}, a correlation time for the dipolar interaction of a proton with the cobalt electronic moment of 10^{-11} sec, a proton–cobalt distance of 2.2 to 2.5 Å, and a proton residence time, τ_M, of about 10^{-5} sec have been calculated[122]. The best interpretation of this data is that the proton contributing to the inhibitable part of T_1^{-1} is located either on a hydroxide ion bound to the Co(II) at the active center or a water molecule with one proton hydrogen bonded to a nearby residue. It is of significance that the pK_a values describing the change in relaxation enhancement are very similar to those determined for a coordinated water molecule from the complexometric titration described in Section VII, 8.1 for the human B enzyme and 7.4 for the bovine enzyme[84].

One feature of the relaxation enhancement not readily explained, however, is its relatively low value in the lower pH ranges where the coordinated species would be expected to be H_2O. This might be related to lack of rapid exchange with solvent water, perhaps related to an altered mechanism for the exchange with solvent protons[122]. It is also noteworthy that titration from pH 9 to 6 is accompanied by a clear change in coordination geometry (see page 509). However, a one to one relationship between possible changes in coordination geometry and the decrease in proton relaxation enhancement is not readily made.

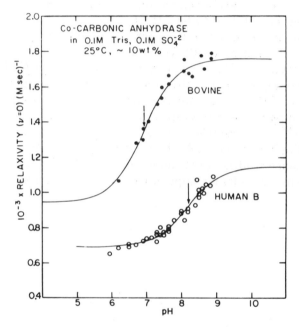

Fig. 16. pH dependence of the relaxivity in the limit of low frequency for Co(II) bovine and Co(II) human B carbonic anhydrases. The solid lines are a least squares fit of the data to a titration curve with pK_a values indicated by the arrows. From Fabry, Koenig and Schillinger[122].

A slightly different approach to probing similar features of the active center has been taken by Ward[119,120] using ^{35}Cl nuclear magnetic resonance as the probe. The ^{35}Cl nucleus has a spin of 3/2 and an electric quadrupole moment Q. For environments in which the chlorine nuclei experience electric field gradients the line width for the ^{35}Cl resonance is given by:

$$\Delta\nu = \frac{2\pi}{5}\ [e^2qQ]^2\tau$$

where $\Delta\nu$ is the full width at half-maximum amplitude, q is the electric field gradient at the nucleus, and τ is a correlation time which describes the random molecular motions responsible for the time-varying electric field gradients[119]. Because of this relationship the binding of ^{35}Cl to Zn(II) results in enhanced relaxation of the nucleus and line broadening (Fig. 17). Zn(II) carbonic anhydrase also causes this line broadening, while the apo-enzyme does not (Fig. 17). This suggests that chloride can occupy one coordination site on the Zn(II) ion at the active site. The line broadening is abolished by cyanide further confirming the specific active site nature of this interaction. The zinc-chloride interaction is highly pH dependent being maximum at about pH 6 and falling off dramatically by pH 9.0. The pK_a describing this process in the absence of Cl^- can be obtained by extrapo-lating the n.m.r. data at several chloride concentrations to zero Cl^- con-centration[119,120]. This pK_a has been found to be 7.0 ± 0.1 for the bovine enzyme and 8.1 ± 0.1 for the human isozyme B, in agreement with the pK_a values for the coordinated water molecule found in the proton reson-ance studies of the Co(II) enzyme[122]. The isozyme dependence of this pK_a is also that predicted by the complexometric titration data[84]. These findings suggest a competition between Cl^- and ^-OH for the metal coordination site at high pH. Similar relationships between the various anions had previously been elucidated by Lindskog[98] in his studies of the pH dependence of anion effects on the spectrum of Co(II) carbonic anhydrase (see page 511). As in the case of the proton resonance studies at pH values below six there is sig-nificant decrease in the ^{35}Cl line broadening apparently due to some struc-tural change in the protein.

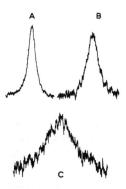

Fig. 17. ^{35}Cl nuclear magnetic resonance absorption of 0.5 M NaCl solution. (A) no protein, $\Delta\nu_{Cl^-}$ = 12.5 Hz. (B) 3.5 x 10^{-5} M apocarbonic anhydrase in 0.01 M acetate buffer (pH 4.8), $\Delta\nu$-$\Delta\nu_{Cl}$ = 7.0 Hz. (C) 4.1 x 10^{-5} M carbonic anhydrase in 0.01 M Tris buffer (pH 8.2), $\Delta\nu$-$\Delta\nu_{Cl^-}$ = 49.4 Hz. From Ward[119].

Spin-labelled carbonic anhydrase

The introduction of a paramagnetic species into a protein by attaching a stable free radical to the protein is potentially a valuable spectroscopic probe[123,124]. The resultant e.s.r. signal from the incorporated free radical can be used to monitor changes in the protein or to deduce something about the environment of the free radical. McConnell and coworkers have introduced this technique and presented a number of applications[125-127]. The stable nitroxide free radical[128,129] has been the most satisfactory because it is inherently rather unreactive and can be attached to a variety of compounds, containing any one of a number of reactive groups which can be directed toward specific protein side chains or active sites[124-127].

Studies of the nitroxide radical suggest that the free electron is largely confined to a $2p\pi$ atomic orbital on the nitrogen. The signal from this free radical is split into three lines by the hyperfine interaction with the nitrogen nucleus[123,124] (Fig. 18). This hyperfine splitting depends on the orientation of the nitroxide relative to the applied field. Thus three different splittings are observed along the three crystal axes of a nitroxide doped crystal[124]. However, in a dilute solution of a small molecule containing the free radical, the tumbling of the molecule is much more rapid than the time for spin reversal, hence the various positions of the free radical relative to the external field are averaged out. On the other hand, if the radical is attached to a macromolecule with a much slower tumbling-rate, such that spin reversal occurs before the molecule has assumed all possible orientations, then the spectrum will be a composite of the spectra for various orientations of the free radical; e.g. a composite of the spectra observed along the three crystal axes of a nitroxide doped crystal. Such a composite can be produced by grinding up a crystal such that the resultant powder is a composite of all crystal orientations[126]. Hence spectra of such immobilized free radicals are referred to as powder-type spectra[124,125].

Since the paramagnetic species in the free radical represents essentially a free electron confined to the $^+N-O^-$ group, it does not in general interact with the environment (surrounding electron cloud or nuclei) in the way the free electrons on a transition metal ion do. The changes in the e.s.r. spectra represent primarily degrees of immobilization. The technique is only beginning to be applied to metallo-enzymes, however, a couple of examples are now available.

Sulfonamides bind to the active site of carbonic anhydrase by coordination to the metal ion (see Section IV). Their binding affinity is not affected in major fashion by the nature of the ring structure attached to the sulfonamide. Hence, it has proved relatively easy to incorporate a nitroxide into carbonic anhydrase by attaching it to a sulfonamide. The nitroxide labelled sulfonamide in VII has been synthesized and introduced into carbonic anhydrase by Mushak and Coleman[130].

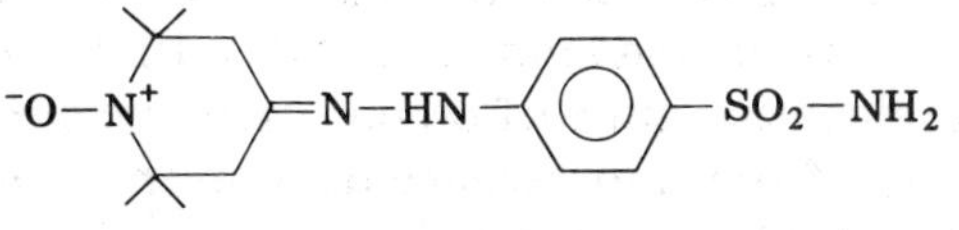

$$^-O-N^+ \left\langle \ \right\rangle =N-HN-\left\langle \bigcirc \right\rangle -SO_2-NH_2$$

(VII)

References pp. 543-548

The e.s.r. spectrum of the compound in solution is shown in Fig. 18A, while the e.s.r. spectra of the 1:1 complexes of this nitroxide with Zn(II) bovine B, Zn(II) human B and Co(II) human B carbonic anhydrases are shown in Figs. 18B, C and D. Binding to all three enzymes immobilizes the free radical. Immobilization appears slightly greater in the case of the bovine enzyme, in agreement with the finding that the sulfonamides are more

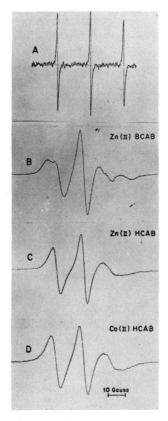

Fig. 18A. E.s.r. spectrum of compound IX in solution. Field set: 3401 Gauss; modulation amplitude: 0.5 Gauss. B, C, D. e.s.r. spectra of bovine Zn(II) carbonic anhydrase B (BCAB), Zn(II) human carbonic anhydrase B (HCAB), and Co(II) HCAB spin-labelled with compound IX. Field set: 3401 Gauss; modulation amplitude: 6.3 Gauss. From Musak and Coleman[130].

tightly bound. On the basis of the comparison of this spectrum to those calculated by Itzkowitz[123,131] for immobilized free radicals, the correlation time, τ, for the radical bound to the bovine enzyme is $2-4 \times 10^{-8}$ sec. It is of interest that the rotational relaxation time of bovine carbonic anhydrase calculated by Chen and Kernohan from fluorescence data on the complex with a fluorescent sulfonamide is 2.89×10^{-8} sec[100].

Co(II) and Zn(II) both induce almost identical immobilization of the nitroxide. Since the binding is metal induced the sites must be nearly identical in the two active metallocarbonic anhydrases. Neither Hg(II) nor Cd(II) carbonic anhydrase immobilize the nitroxide sulfonamide in agreement with binding data showing that sulfonamides do not bind to these derivatives.

Another significant feature is that there appears to be practically no spin–spin interaction between the paramagnetic Co(II) and the nitroxide. This is perhaps not surprising, since according to present models of the mode of binding of the sulfonamide, 15–20 Å must separate the sulfonamide coordinated Co(II) from the free radical.

Magnetic susceptibility

Lindskog and Ehrenberg[132] have carried out a detailed study of the magnetic susceptibility of Co(II) human carbonic anhydrase B and several of its inhibitor complexes. All of the enzyme complexes contain high spin Co(II) with magnetic moments from 4.2 to 4.7 Bohr magnetons (BM). Since the spin only moment is expected to be ~ 3.9 BM, all complexes show the expected increase in magnetic moment due to spin-orbit coupling[90]. Extensive susceptibility data on complexes of high spin Co(II) have shown that octahedral complexes have magnetic moments from 4.7 to 5.2 BM, while tetrahedral complexes have magnetic moments from 4.1 to 4.9 BM. The square-planar complexes of Co(II) that have been described are all low-spin Co(II) and have magnetic moments from 1.8 to 2.0 BM. This is perhaps not too surprising since the low-spin configuration for the d^7 ion does provide the possibility for stabilization by a Jahn-Teller distortion[90]. The magnetic susceptibilities of the Co(II) carbonic anhydrase complexes are compatible with some variant of tetrahedral geometry. Unfortunately, the interpretation of the fine variations in susceptibility are not possible. Tetrahedral complexes of Co(II) are high-spin regardless of the strength of the ligand-field because of the distribution of orbital energy levels. Co(II) carbonic anhydrase does remain high-spin in the face of what must be a relatively high ligand-field, especially in the sulfonamide complex which interestingly has the highest susceptibility.

VI. X-RAY STRUCTURE OF HUMAN CARBONIC ANHYDRASE C

The X-ray studies on the structure of human carbonic anhydrase C in the crystalline state by Strandberg and coworkers[22,23,43,44,83] have now progressed to the stage of 2 Å resolution. Some of the 2 Å as well as the earlier 5 Å data now make it possible to discuss the solution data bearing on the mechanism of action in terms of known structural features of the enzyme. The data on the structure at 2 Å resolution were supplied by Drs. Bror Strandberg and Anders Liljas prior to publication[23].

The general structure of the molecule at 5 Å resolution is shown in Fig. 19A, based on a polystyrene model of the electron density maps[22]. The Zn(II) ion is located near the center of the molecule in a deep crevice. N-terminal and C-terminal regions of the polypeptide chain were tentatively identified at positions marked 1 and 33 respectively. The identification of the N-terminal has been found to be correct in the 2 Å map[23].

The position of the single free SH group of the C-isozyme has been identified from the difference between the enzyme electron density maps of the native enzyme and the enzyme reacted with acetoxymercurisulphanil-amide. Two molecules of acetoxymercurisulphanilamide react with carbonic anhydrase, one at the active center and one at the free –SH group as an organic mercurial[22]. A model of this complex is shown in Fig. 19B built by placing atomic models of the inhibitor on the electron density model of the enzyme according to the positions indicated by the densities in the difference map[22]. The –SH group is 14 Å away from the Zn(II) ion.

The 2 Å map shows the center of gravity of the sulfonamide to be 3.2 Å away from the zinc ion. This is within binding distance and is compatible with coordination of the nitrogen of the sulfonamide group to the Zn(II) ion. This is not unexpected in terms of the absorption and c.d. spectra of the Co(II) enzyme and the metal dependent binding of sulfonamides (see Section IV).

The position of the Hg atom on the benzene ring of the acetoxymer-curisulfanilamide is considerably farther away from the Zn(II) ion and would place the ring portion of the sulfonamide in the crevice leading to the zinc ion[22]. The low dissociation constants observed for sulfonamide complexes of carbonic anhydrase in solution (10^{-5} to 10^{-9} M)[108-111] suggest bonding contacts of some sort between the ring portion of the molecule and the enzyme in addition to the metal ion coordination. Changes in absorption spectra and fluorescence of bound sulfonamides suggests that the binding cavity is hydrophobic[30,100,102]. However, the base of the cavity around the Zn(II) ion may be quite hydrophilic.

Two views of the electron density map at 2 Å resolution near the Zn(II) ion are shown in Fig. 20A and B, one in the plane of the zinc and one just above it. The Zn(II) ion is the most dense feature of the map. Three groups from the protein appear to be coordinated to the metal ion. All three ligand electron densities can be fit adequately by histidyl residues as shown in Fig. 20C. Since the amino acid sequence is not known, these identifications must be considered tentative. It is interesting that the two ligands on the left (Fig. 20C), tentatively identified as histidyl residues, come from the same section of the main chain being separated by a single residue.

A

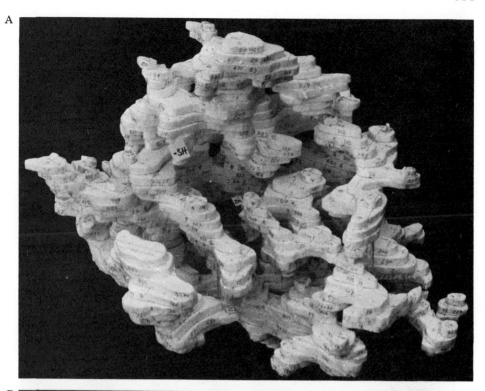

B

Fig. 19A. Polystyrene model of human carbonic anhydrase C based on the electron density map at 5.5 Å resolution. Reproduced from Fridborg *et al.*[22]. B. Polystyrene and molecular model of the complex of human carbonic anhydrase C with acetoxy-mercurisulphanilamide based on the electron density map at 5.5 Å resolution. Reproduced from Fridborg *et al.*[22].

534

A

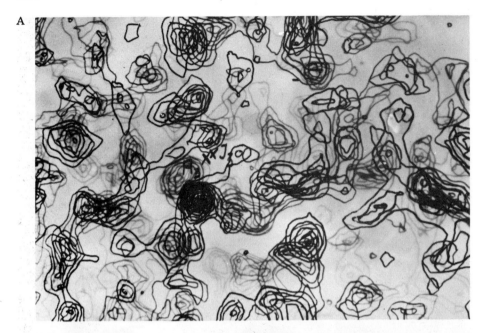

B

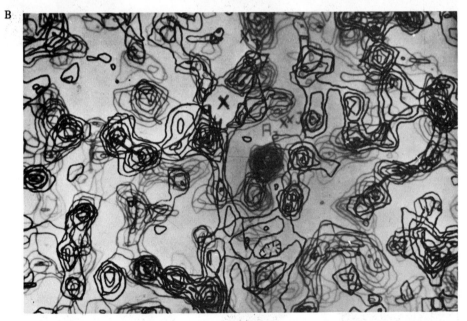

Fig. 20.

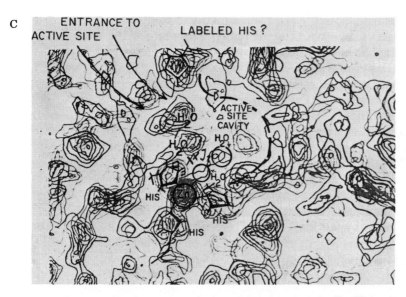

Fig. 20A. Electron density maps at 2 Å resolution around the Zn(II) binding site in human carbonic anhydrase C. A. In the plane of the Zn atom. B. Above the plane of the Zn atom. Reproduced from Liljas *et al.*[23]. C. Electron density map of the active center of carbonic anhydrase with tentative identities of the densities indicated. Zinc is coordinated to three histidyl residues, in two cases through the 3-nitrogen and in one case through the 1-nitrogen. A water molecule occupies a fourth coordination site. A fourth histidine is located at the entrance to the active site cavity and may be the residue reported to be modified by bromopyruvate in the C isozyme[23]. The densities between this histidine and the zinc ion belong to the solvent and are apparently water molecules.

In addition to the three ligands described there is a fourth density approaching the Zn(II) from above and slightly to the right in Fig. 20A. This is less dense than the protein side chains and apparently belongs to the solution, probably a water molecule. It is from the direction occupied by this density that the sulfonamide inhibitors bind. The two definite positions of the sulfonamide in the difference map are marked in Fig. 20B, the sulfonamide group, A_3, and the Hg on the ring, A_4. It appears as if the light density is displaced from the Zn(II) when the sulfonamide binds. A considerable amount of data on the enzyme in solution suggests that a coordinated H_2O or ^-OH occupies this fourth position in the active enzyme (see below).

The precise geometry around the Zn(II) ion cannot be determined with certainty even from the electron density map at 2 Å resolution. Precise model building may improve the precision. While it would appear to approach a tetrahedral geometry, the spectrum of the Co(II) enzyme suggests a highly distorted geometry.

Some of the general features of the protein structure of carbonic anhydrase are evident from the 2 Å map to which a model of polypeptide backbone has been fitted[23]. Four small pieces of right-handed helix of about two turns each have been identified. Only one of these appears to be α-helix, the others are distorted and apparently closer to a 3_{10}-helix[23]. This small amount of α-helix is of interest, since both the o.r.d. and c.d. spectra of most isozyme and species variants of carbonic anhydrase have indicated very little α-helix content (see Section II).

The predominant feature of protein secondary structure evident from the 2 Å map is a large region of β-structure extending through the center of the molecule. At least eight parts of this run mainly anti-parallel to each other and form a twisted pleated sheet. Three or four additional parts of the chain also run either parallel or anti-parallel to the above section of pleated sheet. Thus the large hydrophobic core of the molecule is largely pleated sheet and bears a striking resemblance to the general structure observed in carboxypeptidase A[133,134] (see Chapter 15). The three "histidyl" ligands to the Zn(II) ion are from a portion of the main chain associated with the β-structure.

VII. MECHANISM OF ACTION

Carbonic anhydrase was initially believed to be an enzyme of great specificity, catalyzing only one known reaction, the reversible hydration of CO_2 (eqn. 1). In recent years a number of other reactions have been found to be catalyzed by this enzyme. These include the hydrolysis of p-nitrophenyl acetate[24-27] (eqn. 3), the hydrolysis of 2-hydroxy-5-nitro-α-toluenesulfonic acid sultone[28-30] (eqn. 4), and the hydration of

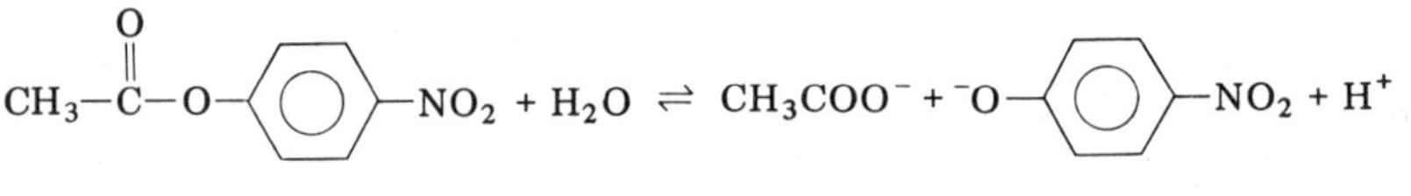

$$CH_3-\overset{\overset{\text{O}}{\|}}{C}-O-\langle\bigcirc\rangle-NO_2 + H_2O \rightleftharpoons CH_3COO^- + {}^-O-\langle\bigcirc\rangle-NO_2 + H^+ \quad (3)$$

acetaldehyde[31-33] (eqn. 5), as well as a number of other aldehydes[135,136].

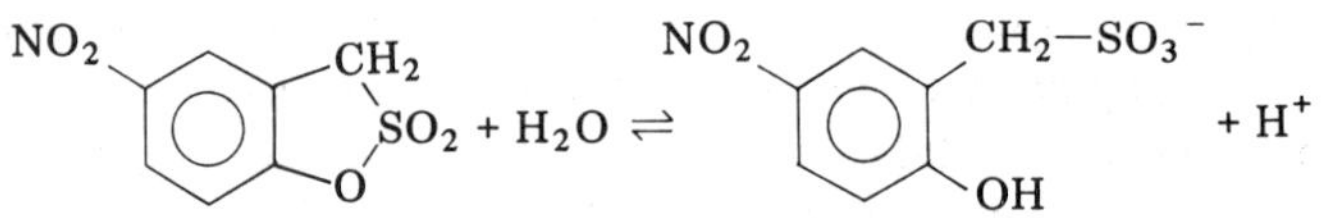

$$\underset{\text{NO}_2}{\bigcirc}\overset{\text{CH}_2}{\underset{\text{SO}_2}{}}\!\!O + H_2O \rightleftharpoons \underset{\text{NO}_2}{\bigcirc}\overset{\text{CH}_2-\text{SO}_3^-}{\underset{\text{OH}}{}} + H^+ \quad (4)$$

$$CH_3-\overset{\overset{\text{O}}{\|}}{C}-H + H_2O \rightleftharpoons CH_3-\overset{\overset{\text{OH}}{|}}{\underset{\overset{|}{\text{OH}}}{C}}-H \quad (5)$$

The hydrolysis of the sultone is catalyzed extremely efficiently by the enzyme[28-30]. Turnover numbers with the sultone as substrate vary from 600 to 20,000 moles/min/mole enzyme depending on the isozyme and species variant of carbonic anhydrase used, the pH, and the presence or absence of inhibiting anions and organic solvent[30,41]. The rapid catalysis of the sultone hydrolysis compared to the relatively slow hydrolysis of *p*-nitrophenylacetate (turnover numbers from 6 to 20 moles/min/mole enzyme) perhaps resides in certain special interactions with the active center, since the structure of the sultone is similar to that of the potent sulfonamide inhibitors (see page 519).

Whether the reactions of eqns. 3 to 5 are of any physiological significance is unknown at present. The reversible hydration of CO_2 is of obvious physiological importance. The reaction proceeds relatively rapidly in aqueous solution without benefit of enzyme catalysis. The formulation of this equilibrium at pH values below 8 as proposed by Eigen *et al.*[137] is given in eqn. 6 which has the advantage of picturing both HCO_3^- and H_2CO_3 in direct equilibrium with CO_2. The kinetic constants in eqn. 1 defined in

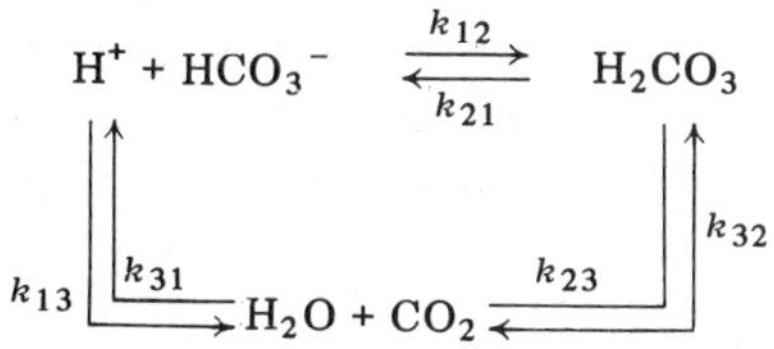

$$H^+ + HCO_3^- \underset{k_{21}}{\overset{k_{12}}{\rightleftharpoons}} H_2CO_3$$

$$k_{13} \quad k_{31} \quad H_2O + CO_2 \quad k_{23} \quad k_{32}$$

$$(6)$$

terms of this scheme are:

$$k_1 = k_{31} + k_{32} \quad \text{and} \quad k_{-1} = k_{13}K_a + k_{23} \tag{7}$$

where K_a is the true ionization constant for carbonic acid, $pK = 3.6$. The first-order hydration rate constant,

$$k_1 = \frac{-d[CO_2]}{dt} \bigg/ [CO_2] \tag{8}$$

has been determined by several studies and values of 0.03 sec^{-1} [138], 0.0358 sec^{-1} [139] and 0.0375 sec^{-1} [140] have been reported for 25°C and 0.0434 sec^{-1} [111] for 37°C. The best value for the dehydration rate constant, k, appears to be 15 sec^{-1} [139,140] at 25°C. The enzyme catalyzes the hydration of CO_2 at approximately 10^7 times the rate of the uncatalyzed reaction (see below). At pH > 10, however, the reaction between OH and CO_2 predominates and this second order rate constant,

$$k_2 = \frac{d[CO_2]}{dt} \bigg/ [OH][CO_2] \tag{9}$$

is near 8500 sec^{-1} M^{-1} and k_{-2} is 2×10^{-4} sec^{-1} [138]. This may have implications as to the mechanism of action of the enzyme (see below).

Calculations based on physiological considerations such as the circulation time of the blood through the capillary beds of the lung alveoli originally suggested that uncatalyzed equilibrium between HCO_3^- and CO_2 would not be achieved nearly rapidly enough to allow the required amount of CO_2 to leave the red cell during one pass through the capillaries[141,142]. The original calculations predicted "speedy death" from complete inhibition of carbonic anhydrase[143]. More recent data suggest that the consequences might not be immediately so catastrophic[144], but carbonic anhydrase clearly plays a most significant role in the rapid adjustment of the $HCO_3^- \leftrightarrow CO_2$ equilibrium (see Maren[106] for review).

Carbonic anhydrase also plays a major role in other tissues including the kidney, eye, pancreas, and stomach where the enzyme is associated indirectly with ion transport. Fundamentally its role arises from the fact that it catalyzes a reaction that provides a source or results in the uptake of H^+ or OH^-. These ions are then involved in various transport processes characteristic of the cell which may result in bicarbonate secretion (pancreas) or acid secretion (stomach). The physiological, pharmacological, and chemical aspects of carbonic anhydrase have been thoroughly and elegantly reviewed recently by Maren[106] and the intention of this section is to summarize the physio-chemical data on the pure enzyme which bear directly on the mechanism of action.

The C isozyme (see Section II) is present in both human and monkey red cells in much smaller quantity than the B isozyme, but the C enzyme has a much higher specific activity[46]. Specific activities expressed in terms of units equal to the quantity of enzyme required to catalyze the hydration of 1 μmole of CO_2/min at 25°C are 44,000 units/mg of protein for the B isozyme and 1,300,000 units/mg of protein for the C isozyme[46]. The latter figure corresponds to a turnover number of 6×10^5/sec or 3.6×10^7/min. This makes the hydration of CO_2 by carbonic anhydrase the most rapid enzyme reaction known and must be remembered in considering possible mechanisms for carbonic anhydrase (see below).

The pH-rate profile for carbonic anhydrase appears to be a single sigmoid curve describing the ionization of a group with a pK_a that varies between about 6.9 and 8.4 depending on the ionic environment, the buffers and particularly substrate employed, and the species or isozyme variant of the enzyme[26,29,31,41,84,97,98]. This pK_a is extremely sensitive to the anion environment[98] which may explain some of the observed variation. There is evidence from the complexometric titration data (see page 518) that this pK_a represents that for a coordinated water molecule[84] and n.m.r. data on the active Co(II) enzyme clearly shows the presence of a coordinated H_2O molecule[122] (see Section V). The observed pK_a has been considered somewhat low for a metal coordinated water, however, the only models are the aquo species which do not correspond to a mixed ligand chelate complex with one open coordination site occupied by the water. Such a ligand is likely to lower the pK_a of the coordinated water.

A hydration mechanism involving the attack of a coordinated ^-OH on the CO_2 molecule has been an attractive postulate[84,102,145-147]. The minimum species involved in such a mechanism are indicated in eqn. 10.

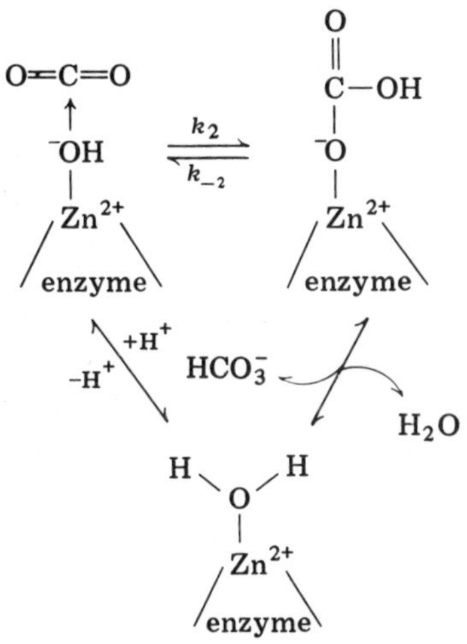

$$
\begin{array}{ccc}
\underset{\substack{\uparrow \\ ^-OH \\ | \\ /Zn^{2+}\backslash \\ /enzyme\backslash}}{O{=}C{=}O}
& \underset{k_2 \atop k_{-2}}{\rightleftharpoons}
& \underset{\substack{|| \\ C{-}OH \\ | \\ ^-O \\ | \\ /Zn^{2+}\backslash \\ /enzyme\backslash}}{O}
\end{array}
$$

$$
\begin{array}{c}
\overset{+H^+}{\underset{-H^+}{}} \quad HCO_3^- \\
\\
H_2O \\
\\
\underset{\substack{O \\ | \\ /Zn^{2+}\backslash \\ /enzyme\backslash}}{H\diagdown\;\;\diagup H}
\end{array}
$$

(10)

A mechanism as formulated in eqn. 10 would explain the lack of inhibition by weakly binding anions at high pH[98], the reversion at high pH of the spectra of the cobalt enzyme–anion or sulfonamide complexes to that typical of the alkaline form of the uninhibited enzyme[98], and the displacement of the pH-rate profile to higher pH in the presence of anions[98]. All can be related to competition with ^-OH, which at high enough concentration displaces the anions and generates the active enzyme. The above mechanism is compatible with the kinetic evidence which shows the anion binding site to be coupled to a group, the basic form of which is essential for the hydration of CO_2 and the acidic form essential for the dehydration of bicarbonate[96,97]. Infra-red data on the carbonic anhydrase–CO_2 complex at pH 5.0[102] (see page 513) show that CO_2 is bound in the hydrophobic cavity in an unstrained, hence, linear configuration[102]. The spectrum of the Co(II) enzyme at pH 8.2 in the presence of bicarbonate indicates that a bicarbonate complex is present in high equilibrium concentration at this pH[71,116].

The first-order rate constant, k_2, for the enzyme catalyzed hydration step is 4 to 6×10^5 sec^{-1} at pH 7, $25°C$[46]. As noted above this is 7 orders of magnitude faster than the uncatalyzed hydration of CO_2, but only 2 orders of magnitude higher than the reaction of ^-OH and CO_2 at pH > 10, where $k_2 = 8.5 \times 10^3$ sec^{-1} M^{-1}. Wang[147] has estimated the pseudo-first-order rate constant for a hypothetical system in which an ^-OH ion is placed next to a CO_2 molecule and has derived a rate constant of 4×10^5 sec^{-1} which is of the order of magnitude shown by the carbonic anhydrase reac-

References pp. 543–548

tion. The lowering of the K_a for H_2O for $10^{-15.7}$ to $10^{-7 \text{ to } -8}$ would, of course, meet the requirement that the ^-OH and CO_2 reaction be the predominant one near neutral pH. One further conclusion based on the relative nucleophilic characters of coordinated and free hydroxide ions has been pointed out by Wang[147]. While the ratio of nucleophilic characters of the two species may not coincide with the ratio of their K_a values, one would not expect a difference of several orders of magnitude. This suggests that if the above features of the mechanism are correct, the nucleophilic character of the coordinated ^-OH does not fall to the extent indicated by the fall in K_a from $10^{-15.6}$ to $10^{-7 \text{ or } -8}$ or else some other feature of the active center expedites the reaction.

While the above postulated features of the carbonic anhydrase reaction may seem self-consistent, when the required proton transfers are considered, there are some unusual conclusions that must be reached about the speed of these transfers. In the hydration reaction, the step from ^-OH–Zn and CO_2 to the bicarbonate intermediate characterized by k_2 in eqn. 10 must be accompanied by proton transfer. This transfer between adjacent oxygen atoms may well be extremely rapid, although the precise mechanism, sequence or nature of the transition state may be speculated upon[147].

A more serious difficulty arises when the other proton transfer is considered. At some point in the reaction a proton must be transferred to the solvent. In the formalism pictured in eqn. 10, this appears as the loss of a proton from a solvent or coordinated water molecule to regenerate the active Zn–^-OH. Based on model systems, the limiting value of the diffusion controlled rate of proton transfer approaches 10^{10}–10^{11} M^{-1} sec^{-1}[148]. Since H_3O^+ is a strong acid and ^-OH a strong base there is a gain in free energy when a proton transfer occurs under the conditions $pK_{H_3O^+} \ll pK_{HX}$ or $pK_{H_2O} \gg pK_{HX}$, where X is the acceptor. Under these conditions, the transfer approaches diffusion controlled rates. If the pK_a of the acid formed is not higher than that of H_3O^+ then the proton transfer can no longer be diffusion controlled. Eigen has presented the dependence of log k (the rate constant for proton transfer) on ΔpK (pK-difference for donor and acceptor)[148].

The rate is diffusion controlled only if the acceptor has a pK_a 2 to 3 pH units higher than the donor. On the other hand, if the donor pK is above that of the acceptor the log k shows a linear decrease with increasing ΔpK. Using the lowest figure for the $pK_a = \sim 7$, donation of the proton to solvent H_2O involves an acceptor pK for H_3O^+ of ~ -1.7. Thus direct transfer of the proton from the coordinated H_2O solvent involves a ΔpK of ~ -9 and suggests that even under the most favorable of circumstances k should not be much larger than 10^2 or 10^3 sec^{-1}. However, the observed pseudo-first-order rate constant for the hydration reaction catalyzed by the C enzyme is 10^5 sec^{-1}. Transfer of the proton to an adjacent imidazole nucleus to produce an imidazolium ion has been suggested by several investigators[26,27,31-33]. This does not get around the dilemma appreciably, however, since the transfer of the imidazolium proton to solvent water also occurs with a rate constant of $\sim 10^3$/sec[148]. There do appear to be histidyl residues near the active center (see Section II).

In association with their studies on the hydrolysis of the sulfonate ester (eqn. 4), Kaiser and Lo[29] have proposed a cyclic mechanism involving the coordinated $^-$OH that avoids the transfer of a proton to the solvent. The mechanism is pictured in eqn. 11. Hydrolysis of this substrate by the bovine enzyme follows a sigmoid curve with a pK_a of 7.28 in a combination of buffers[29], while a pK_a of 7.8 has been observed for the human B iosyzme in Tris–SO$_4$ buffer[41]. Kaiser and Lo in the mechanism pictured have interpreted this pK_a as that for a coordinated water[29].

$$
\begin{array}{c}
\text{(structure: Zn}^{2+}\text{ coordinated with O-H, O, SO}_2\text{, CH}_2\text{, aryl-NO}_2\text{)} \rightleftharpoons \text{Zn}^{2+}-^-\text{OH} + \text{HO-aryl-NO}_2 + \text{HO}_3\text{SCH}_2
\end{array} \quad (11)
$$

An additional solvent water is included to complete the cycle. Such a mechanism could also circumvent the proton transfer in the hydration reaction if the immediate product were carbonic acid rather than bicarbonate using solvent water as the source of the second hydrogen as indicated in eqn. 12.

$$
\begin{array}{c}
\text{(structure: Zn}^{2+}\text{ with O-H}\rightarrow\text{O, coordinated CO}_2\text{)} \rightleftharpoons -\text{Zn}^{2+}-^-\text{OH} + \text{H}_2\text{CO}_3 \rightleftharpoons \text{HCO}_3^- + \text{H}^+
\end{array} \quad (12)
$$

Since H$_2$CO$_3$ has a pK of 3.6, the transfer to solvent presents no problem. Zn(II) coordination of course would help increase the acidity of the water protons.

The above type of mechanism has been criticized on the basis that the substrate in the reverse (dehydration) direction must be carbonic acid. Thus the rate of protonation of HCO$_3^-$ would enter. Several arguments suggest that this protonation would not be fast enough at high enzyme concentrations for the reaction to be first order in enzyme[149]. Furthermore the rate of encounter of carbonic acid and enzyme assuming the rate to be diffusion limited is not fast enough to account for the enzymatic rate[97,149] (hydration and dehydration by the enzyme are about equal at pH 7.05)[46,97]

DeVoe and Kistiakowsky[150] have previously presented kinetic arguments against H_2CO_3 as the substrate in the dehydration reaction.

Caplow[149] has suggested a concerted reaction for the dehydration mechanism involving C–O bond cleavage (eqn. 13). C–O bond cleavage has been observed in model decarboxylations involving pentaamine cobaltic compounds[151] and has been mentioned before in connection with possible carbonic anhydrase mechanisms[71].

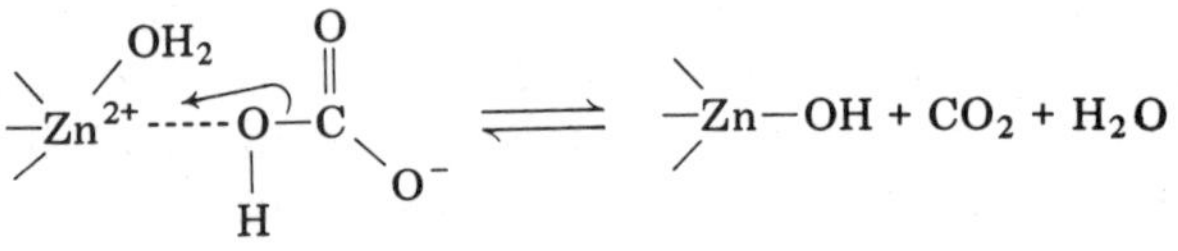

$$-Zn^{2+}\overset{OH_2}{\cdots\cdots O-C}\underset{O^-}{\overset{O}{\|}} \rightleftharpoons -Zn-OH + CO_2 + H_2O \qquad (13)$$

Such a scheme while attractive for the dehydration mechanism does not circumvent the dilemma of the proton transfer in the hydration direction.

A mechanism has been proposed by Dennard and Williams[95] involving a nitrogen of the protein as the group attacking the CO_2 carbon with the intermediate formation of an unstable carbamate (VII):

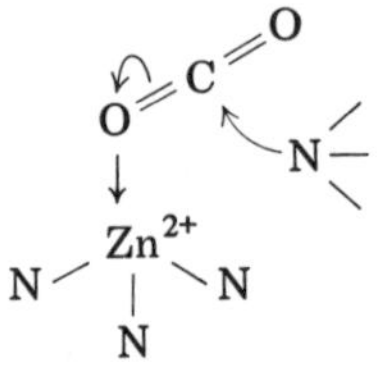

$$O=C\overset{O}{\diagdown}\cdots N-$$

Zn^{2+}

N — N

N

(VII)

This has the drawback that known mechanisms of carbamate breakdown proceed by decarboxylation[149], hence this mechanism has the disadvantage of not leading to products under known mechanisms of carbamate breakdown.

It does seem clear at the present time that the mechanism of proton transfer required in the hydration–dehydration reaction is a central problem in any mechanism. The rate of proton transfer may be the rate-limiting step in the hydration–dehydration reactions. The esterolytic reactions are considerably slower. Although they may proceed by the same mechanism, it is not a foregone conclusion that all features of the mechanism are the same.

The assumption has been made in the above discussion on the mechanism that proton transfers occur as observed in model systems. It should be kept in mind that structured or ice-like water molecules near the protein surface might influence the rate of proton transfer. If some concerted or cooperative mechanism of proton transfer is involved, rates of transfer may be modified compared to proton transfer in simple model systems[148,152]. It is of considerable interest that the preliminary fitting of the electron density map near the active center appears to show a number of highly structured solvent molecules (Fig. 20C).

543

NOTE ADDED IN PROOF

Since this manuscript was completed a number of studies on the enzyme have appeared in the literature. These include the crystal structure, substrate specificity, kinetics, and solution structure of carbonic anhydrase. These publications are listed by title below as a guide to the more recent information on the enzyme.

1 J. S. Taylor and J. E. Coleman, Electron spin resonance of metallo-carbonic anhydrases, *J. Biol. Chem.*, 246 (1971) 7058.
2 K. K. Kannan *et al.*, Crystal structure of human erythrocyte carbonic anhydrase C. VI. The three-dimensional structure at high resolution in relation to other mammalian carbonic anhydrases, *Cold Spring Harbor Symp. Quant. Biol.* 36 (1972) 221.
3 S. Lindskog, L. E. Henderson, K. K. Kannan, A. Liljas, P. O. Nyman and B. Strandberg, Carbonic anhydrase, in *The Enzymes*, Vol. V, (1971), p. 587.
4 R. G. Khalifah, The carbon dioxide hydration activity of carbonic anhydrase. I. Stop-flow kinetic studies on the native human isoenzymes B and C, *J. Biol. Chem.*, 246 (1971) 2561.
5 R. G. Khalifah and J. T. Edsall, Carbon dioxide hydration activity of carbonic anhydrase: kinetics of alkylated anhydrases B and C from humans, *Proc. Natn. Acad. Sci. U.S.*, 69 (1972) 172.
6 P. W. Taylor, R. W. King and A. S. V. Burgen, Kinetics of complex formation between human carbonic anhydrases and aromatic sulfonamides, *Biochemistry*, 9 (1970) 2638.
7 P. W. Taylor, R. W. King and A. S. V. Burgen, Influence of pH on the kinetics of complex formation between aromatic sulfonamides and human carbonic anhydrase, *Biochemistry*, 9 (1970) 3894.
8 P. W. Taylor and A. S. V. Burgen, Kinetics of carbonic anhydrase–inhibitor complex formation. A comparison of anion- and sulfonamide-binding mechanisms, *Biochemistry*, 10 (1971) 3859.
9 P. Mushak and J. E. Coleman, Electron spin resonance studies of spin-labeled carbonic anhydrase, *J. Biol. Chem.*, 247 (1972) 373.
10 J. F. Hower, R. W. Henkens and D. B. Chestnut, A spin label investigation of the active site of an enzyme. Bovine carbonic anhydrase, *J. Am. Chem. Soc.*, 93 (1971) 6665.
11 Y. Pocker, M. W. Beug and V. R. Ainardi, Coprecipitation of carbonic anhydrase by 1,1-bis(*p*-chlorophenyl)-2,2,2-trichloroethane, 1,1-bis(*p*-chlorophenyl)-2,2-dichloroethylene, and dieldrin, *Biochemistry*, 10 (1971) 1390.
12 A. Lanir and G. Navon, Nuclear magnetic resonance studies of bovine carbonic anhydrase. Binding of sulfonamides to the zinc enzyme, *Biochemistry*, 10 (1971) 1024.
13 P. W. Taylor, J. Feeney and A. S. V. Burgen, Investigation of the mechanism of ligand binding with cobalt(II) human carbonic anhydrase by ^{1}H and ^{19}F nuclear magnetic resonance spectroscopy, *Biochemistry*, 10 (1971) 3866.
14 Y. Pocker and N. Watamori, Catalytic versatility of erythrocyte carbonic anhydrase. IX. Kinetic studies of the enzyme-catalyzed hydrolysis of 3-pyridyl and nitro-3-pyridyl acetates, *Biochemistry*, 10 (1971) 4843.
15 R. J. Tanis and R. E. Tashian, Purification and properties of carbonic anhydrase from sheep erythrocytes, *Biochemistry*, 10 (1971) 4852.
16 Y. Pocker and L. J. Guilbert, Catalytic versatility of erythrocyte carbonic anhydrase. Kinetic studies of the enzyme-catalysed hydrolysis, *Biochemistry*, 11 (1972) 180.

544

17 R. W. Henkens and J. M. Sturtevant, Extrinsic cotton effects in a metal chelator-bovine carbonic anhydrase complex, *Biochemistry*, 11 (1972) 206.
18 J. S. Cohen, C. T. Yim, M. Kandel, A. G. Gornall, S. I. Kandel and M. H. Freedman, Studies of the histidine residues of carbonic anhydrases using high-field proton magnetic resonance, *Biochemistry*, 11 (1972) 327.
19 Y. Pocker and M. W. Beug, Kinetic studies of bovine carbonic anhydrase–catalyzed hydrolyses of para-substituted phenyl esters, *Biochemistry*, 11 (1972) 698.
20 A. Yazgan and R. W. Henkens, Role of zinc(II) in the refolding of guanidine hydrochloride denatured bovine carbonic anhydrase, *Biochemistry*, 11 (1972) 1314.
21 J. E. Coleman and R. V. Coleman, Magnetic circular dichroism of Co(II) carbonic anhydrase, *J. Biol. Chem.*, 247 (1972) 4718.
22 A group of papers on carbonic anhydrase is published in Oxygen affinity of haemoglobin and red cell acid–base status, *Alfred Benzon Symposium IV*, Munksgaard, Copenhagen, 1972.
23 P. L. Whitney, Inhibition and modification of human carbonic anhydrase B with bromoacetate and iodoacetamide, *Eur. J. Biochem.*, 16 (1970) 126.
24 P. Henkart and F. Dorner, The dinitrophenylation of human carbonic anhydrase B, *J. Biol. Chem.*, 246 (1971) 2714.
25 S. Funakoshi and H. F. Deutsch, Human carbonic anhydrases. VI. Levels of isozymes in old and young erythrocytes and in various tissues, *J. Biol. Chem.*, 246 (1971) 1088.
26 F. Dorner, Human carbonic anhydrase B. Location of tyrosine residues that react with tetranitromethane, *J. Biol. Chem.*, 246 (1971) 5896.

REFERENCES

1 R. Brinkman, R. Margaria, N. V. Meldrum and F. J. W. Roughton, *J. Physiol. London*, 75 (1932) 3.
2 N. V. Meldrum and F. J. W. Roughton, *J. Physiol. London*, 75 (1932) 4.
3 N. V. Meldrum and F. J. W. Roughton, *J. Physiol. London*, 75 (1932) 15.
4 N. V. Meldrum and F. J. W. Roughton, *J. Physiol. London*, 80 (1933) 113.
5 D. Keilin and T. Mann, *Biochem. J.*, 34 (1940) 1163.
6 D. Keilin and T. Mann, *Nature*, 153 (1944) 107.
7 F. R. Eirich and E. K. Rideal, *Nature*, 146 (1944) 107.
8 M. N. Petermann and N. V. Hakala, *J. Biol. Chem.*, 145 (1942) 701.
9 E. C. B. Smith, *Biochem. J.*, 34 (1940) 1176.
10 D. A. Scott and J. R. Mendive, *J. Biol. Chem.*, 140 (1941) 445.
11 E. Hove, C. A. Elvehjem and E. B. Hart, *J. Biol. Chem.*, 136 (1940) 425.
12 H. W. Davenport, *J. Physiol. London*, 97 (1939) 32.
13 E. C. Webb and R. van Heyingen, *Biochem. J.*, 41 (1946) 74.
14 J. R. G. Bradfield, *Nature*, 159 (1947) 467.
15 R. Day and J. Franklin, *Science*, 104 (1946) 363.
16 P. M. Sibly and J. G. Wood, *Aust. J. Sci. Res.*, B4 (1951) 500.
17 A. C. Neish, *Biochem. J.*, 33 (1939) 300.
18 J. G. Wood and P. M. Sibly, *Aust. J. Sci. Res.*, 5 (1952) 244.
19 T. Mann and D. Keilin, *Nature*, 146 (1940) 164.
20 S. Lindskog, *J. Biol. Chem.*, 238 (1962) 945.
21 J. E. Coleman, *Nature*, 214 (1967) 193.
22 K. Fridborg, K. K. Kannan, A. Liljas, J. Lundin, *et al.*, *J. Mol. Biol.*, 25 (1967) 505.
23 A. Liljas, K. K. Kannan, P. C. Bergstén, I. Waara, K. Fridborg, *et al.*, *Nature*, 235 (1972) 131.

24 F. Schneider and M. Lieflander, *Z. Physiol. Chem.*, 334 (1963) 279.
25 R. E. Tashian, C. C. Plato and T. B. Shows, *Science*, 140 (1963) 53.
26 Y. Pocker and J. T. Stone, *J. Am. Chem. Soc.*, 87 (1965) 5497.
27 Y. Pocker and J. T. Stone, *Biochemistry*, 7 (1968) 2936.
28 K-W. Lo and E. T. Kaiser, *Chem. Commun.*, (1966) 834.
29 E. T. Kaiser and K-W. Lo, *J. Am. Chem. Soc.*, 91 (1969) 4912.
30 J. E. Coleman, *J. Biol. Chem.*, 243 (1968) 4574.
31 Y. Pocker and J. E. Meany, *Biochemistry*, 4 (1965) 2535.
32 Y. Pocker and J. E. Meany, *Biochemistry*, 6 (1967) 239.
33 Y. Pocker and D. G. Dickerson, *Biochemistry*, 7 (1968) 1995.
34 S. Lindskog, *Biochim. Biophys. Acta*, 39 (1960) 218.
35 G. Laurent, C. Marriq, D. Nahon, M. Charrel and Y. Derrien, *Compt. Rend. Soc. Biol.*, 156 (1962) 1456.
35a P. O. Nyman, *Biochim. Biophys. Acta*, 52 (1961) 1.
36 E. E. Rickli, S. A. S. Ghazanfar, B. H. Gibbons and J. T. Edsall, *J. Biol. Chem.*, 239 (1964) 1065.
37 G. Laurent, M. Charrel, F. Luccioni, M. F. Autran and Y. Derrien, *Bull. Soc. Chim. Biol.*, 47 (1965) 1101.
38 T. A. Duff and J. E. Coleman, *Biochemistry*, 5 (1966) 2009.
39 P. Byvoet and A. Gotti, *Mol. Pharmacol.*, 3 (1967) 142.
40 A. J. Furth, *J. Biol. Chem.*, 243 (1968) 4832.
41 J. Maynard and J. E. Coleman, *J. Biol. Chem.*, 246 (1971) 4455.
42 J. McD. Armstrong, D. V. Myers, J. A. Verpoorte and J. T. Edsall, *J. Biol. Chem.*, 241 (1966) 5137.
43 B. Strandberg, B. Tilander, K. Fridborg, S. Lindskog and P. O. Nyman, *J. Mol. Biol.*, 5 (1962) 583.
44 B. Tilander, B. Strandberg and K. Fridborg, *J. Mol. Biol.*, 12 (1965) 740.
45 R. E. Tashian, *Am. J. Human Genet.*, 17 (1965) 257.
46 B. H. Gibbons and J. T. Edsall, *J. Biol. Chem.*, 239 (1964) 2539.
47 P. O. Nyman and S. Lindskog, *Biochim. Biophys. Acta*, 85 (1964) 141.
48 G. Laurent and Y. Derrien, in R. E. Forster, J. T. Edsall, A. B. Otis and F. J. W. Roughton, *CO₂: Chemical, Biochemical and Physiological Aspects*, NASA Symp., SP-188, Washington, D.C., 1969, p. 115.
49 T. B. Shows, *Biochem. Genet.*, 1 (1967) 171.
50 S. K. Fellner, *Biochim. Biophys. Acta*, 77 (1963) 155.
51 C. Rossi, A. Chersi and M. Cortivo, in R. E. Forster, J. T. Edsall, A. B. Otis and F. J. W. Roughton, *CO₂: Chemical, Biochemical and Physiological Aspects*, NASA Symp., SP-188, Washington, D.C., 1969, p. 131.
52 A. J. Tobin, in R. E. Forster, J. T. Edsall, A. B. Otis and F. J. W. Roughton, *CO₂: Chemical, Biochemical and Physiological Aspects*, NASA Symp., SP-188, Washington, D.C., 1969, p. 131.
52a A. J. Tobin, *J. Biol. Chem.*, 245 (1970) 2656.
53 P. L. Whitney, in R. E. Forster, J. T. Edsall, A. B. Otis and F. J. W. Roughton, *CO₂: Chemical, Biochemical and Physiological Aspects*, NASA Symp., SP-188, Washington, D.C., 1969, p. 109.
54 P. O. Nyman, in R. E. Forster, J. T. Edsall, A. B. Otis and F. J. W. Roughton, *CO₂: Chemical, Biochemical and Physiological Aspects*, NASA Symp., SP-188, Washington, D.C., 1969, p. 109.
55 P. O. Nyman, L. Strid and G. Westermark, *Biochim. Biophys. Acta*, 122 (1966) 554.
56 B. Andersson, P. O. Gothe, T. Nilsson, P. O. Nyman and L. Strid, *Eur. J. Biochem.*, 6 (1968) 190.
57 P. O. Nyman, L. Strid and G. Westermark, *Eur. J. Biochem.*, 6 (1968) 72.

546

58 B. Andersson, P. O. Nyman and L. Strid, in R. E. Forster, J. T. Edsall, A. B. Otis and F. J. W. Roughton, CO_2: *Chemical, Biochemical and Physiological Aspects*, NASA Symp., SP-188, Washington, D.C., 1969, p. 109.
59 L. E. Henderson, in R. E. Forster, J. T. Edsall, A. B. Otis and F. J. W. Roughton, CO_2: *Chemical, Biochemical and Physiological Aspects*, NASA Symp., Sp-188, Washington, D.C., 1969, p. 109.
60 C. Marriq, D. Gignoux and G. Laurent, *Compt. Rend. Acad. Sci., Paris*, 260 (1965) 1810.
61 G. Laurent, C. Marriq, D. Garcon, F. Luccioni and Y. Derrien, *Bull. Soc. Chim. Biol.*, 49 (1967) 1035.
62 D. M. P. Phillips, *Biochem. J.*, 86 (1963) 397.
63 S. L. Bradbury, *J. Biol. Chem.*, 244 (1969) 2002.
64 S. L. Bradbury, *J. Biol. Chem.*, 244 (1969) 2010.
65 P. L. Whitney, G. Folsch, P. O. Nyman and B. G. Malmstrom, *J. Biol. Chem.*, 242 (1967) 4206.
66 P. L. Whitney, P. O. Nyman and B. G. Malmstrom, *J. Biol. Chem.*, 242 (1967) 4212.
67 L. M. Riddiford, *J. Biol. Chem.*, 239 (1964) 1079.
68 L. M. Riddiford, R. H. Stellwagen, S. Mehta, and J. T. Edsall, *J. Biol. Chem.*, 240 (1965) 3305.
69 A. Nilsson and S. Lindskog, *Eur. J. Biochem.*, 2 (1967) 309.
70 D. B. Myers and J. T. Edsall, *Proc. Natn. Acad. Sci. U.S.*, 53 (1965) 169.
71 J. E. Coleman, *Biochemistry*, 4 (1965) 2644.
72 S. Beychok, J. M. Armstrong, C. Lindblow and J. T. Edsall, *J. Biol. Chem.*, 241 (1966) 5150.
73 A. Rosenberg, *J. Biol. Chem.*, 241 (1966) 5126.
74 J. T. Edsall, *Harvey Lect. Ser.*, 62 (1968) 191.
75 J. T. Edsall, *Ann. N.Y. Acad. Sci.*, 151 (1968) 41.
76 K. P. Wong and C. Tanford, *Federation Proc.*, 29 (1970) 335 Abstr.
77 N. Greenfield and G. D. Fasman, *Biochemistry*, 8 (1969) 4108.
78 Y. P. Myer, *Biophys. J.*, 9 (1969) A-215; *J. Chem. Path. Pharm.*, 1 (1970) 607.
79 R. Tupper, R. W. E. Watts and A. Wormall, *Biochem. J.*, 50 (1952) 429.
80 S. Lindskog and B. G. Malmstrom, *J. Biol. Chem.*, 237 (1962) 1129.
81 E. E. Rickli and J. T. Edsall, *J. Biol. Chem.*, 237 (1961) PC 258.
82 S. Lindskog and P. O. Nyman, *Biochim. Biophys. Acta*, 85 (1964) 462.
83 B. Tilander, B. Strandberg and K. Fridborg, *J. Mol. Biol.*, 12 (1965) 740.
84 J. E. Coleman, *J. Biol. Chem.*, 242 (1967) 5212.
85 R. W. Henkens and J. M. Sturtevant, *J. Am. Chem. Soc.*, 90 (1968) 2669.
86 R. H. Holyer, C. D. Hubbard, S. F. A. Kettle and R. G. Wilkins, *Inorg. Chem.*, 4 (1965) 929.
87 R. H. Holyer, C. D. Hubbard, S. F. A. Kettle and R. G. Wilkins, *Inorg. Chem.*, 5 (1966) 622.
88 B. L. Vallee and J. E. Coleman, *Comprehensive Biochemistry*, Vol. 12, 1964, p. 165.
89 J. E. Coleman, *Ph.D. Dissertation*, Massachusetts Institute of Technology, 1963.
90 F. A. Cotton and F. R. S. Wilkinson, *Advanced Inorganic Chemistry*, 2nd edn., Interscience Publishers, New York, 1966.
91 M. Ciampolini, *Structure and Bonding*, 6 (1969) 52.
92 L. Sacconi, P. L. Orioli and M. DiVaira, *J. Am. Chem. Soc.*, 87 (1965) 2059.
93 L. Sacconi, P. Nannelli, N. Nardi and U. Campigli, *Inorg. Chem.*, 4 (1965) 943.
94 Z. Dori and H. B. Gray, *J. Am. Chem. Soc.*, 88 (1966) 1394.
94a E. M. Holt, S. L. Holt and K. J. Watson, *J. Am. Chem. Soc.*, 92 (1970) 2721.
95 A. C. Dennard and R. J. Williams, *Trans. Metal Chem.*, 2 (1966) 115.
96 J. C. Kernohan, *Biochim. Biophys. Acta*, 81 (1964) 346.

97   J. C. Kernohan, *Biochim. Biophys. Acta*, 96 (1965) 304.
98   S. Lindskog, *Biochemistry*, 5 (1966) 2641.
99   J. E. Coleman, *Proc. Natn. Acad. Sci. U.S.*, 59 (1968) 123.
100 R. F. Chen and J. C. Kernohan, *J. Biol. Chem.*, 242 (1967) 5813.
101 K. M. Wilbur and N. G. Anderson, *J. Biol. Chem.*, 176 (1948) 147.
102 M. E. Riepe and J. H. Wang, *J. Biol. Chem.*, 243 (1968) 2779.
103 J. A. Verpoorte, S. Mehta and J. T. Edsall, *J. Biol. Chem.*, 242 (1967) 4221.
104 J. E. Coleman, in E. T. Kaiser and Kedzy, *Progress in Bioorganic Chemistry*, Vol. 1, Interscience, New York, 1970, pp. 159–344.
105 L. G. Sillen and A. E. Martell, *Stability Constants of Metal-Ion Complexes*, The Chemical Society, London, Special Publication No. 17, 1964.
106 T. H. Maren, *Physiol. Rev.*, 47 (1967) 595.
107 D. Bar, *Actualities Pharmacol.*, 15 (1964) 1.
108 T. H. Maren, B. Robinson, R. F. Palmer and M. E. Griffith, *Biochem. Pharmacol.*, 6 (1960) 21.
109 T. H. Maren, A. L. Parcell and M. N. Malik, *J. Pharm. Exp. Therapeut.*, 130 (1960) 389.
110 T. H. Maren, *J. Pharm. Exp. Therapeut.*, 139 (1963) 129.
111 T. H. Maren, *J. Pharm. Exp. Therapeut.*, 139 (1963) 140.
112 S. Lindskog and A. Thorslund, *Eur. J. Biochem.*, 3 (1968) 453.
113 J. C. Kernohan, *Biochim. Biophys. Acta*, 118 (1966) 405.
114 J. C. Kernohan, *Biochem. J.*, 98 (1966) 31.
115 T. H. Maren, personal communication.
116 J. E. Coleman, in R. E. Forster, J. T. Edsall, A. B. Otis and F. J. W. Roughton, *CO_2: Chemical, Biochemical and Physiological Aspects*, NASA Symp. SP-188, Washington, D. C., 1969, p. 141.
117 J. E. Coleman, *J. Am. Chem. Soc.*, 89 (1967) 6757.
118 A. Moskowitz, *Proc. R. Soc.*, A297 (1967) 16.
119 R. L. Ward, *Biochemistry*, 8 (1969) 1879.
120 R. L. Ward, *Biochemistry*, 9 (1970) 2447.
121 W. C. Galley and L. Stryer, *Proc. Natn. Acad. Sci. U.S.*, 60 (1968) 108.
122 M. E. Fabry, S. H. Koenig and W. E. Schillinger, *J. Biol. Chem.*, 245 (1970) 4256.
123 C. L. Hamilton and H. M. McConnell, in A. Rich and N. Davidson, *Structural Chemistry and Molecular Biology*, Freeman, San Francisco, 1968, p. 115.
124 O. H. Griffith and A. S. Waggoner, *Accounts Chem. Res.*, 2 (1969) 17.
125 T. J. Stone, T. Buckman, P. L. Nordio and H. M. McConnell, *Proc. Natn. Acad. Sci. U.S.*, 54 (1965) 1010.
126 H. M. McConnell and C. L. Hamilton, *Proc. Natn. Acad. Sci. U.S.*, 60 (1968) 776.
127 W. L. Hubbell and H. M. McConnell, *Proc. Natn. Acad. Sci. U.S.*, 63 (1969) 16.
128 E. G. Rozantzev and M. B. Neiman, *Tetrahedron*, 20 (1964) 131.
129 E. G. Rozantzev and L. A. Krinitzkaya, *Tetrahedron*, 21 (1965) 491.
130 P. Mushak and J. E. Coleman, unpublished data.
131 M. S. Itzkowitz, *Ph.D. Thesis*, California Institute of Technology, 1966.
132 S. Lindskog and A. Ehrenberg, *J. Mol. Biol.*, 24 (1967) 133.
133 W. N. Lipscomb, J. A. Hartsuck, G. N. Reeke, F. A. Quiocho, *et al.*, in *Structure Function and Evolution of Proteins*, Brookhaven Symp. Biol., 21 (1969) 24.
134 W. N. Lipscomb, J. A. Hartsuck, F. A. Quiocho and G. N. Reeke, *Proc. Natn. Acad. Sci. U.S.*, 64 (1969) 28.
135 Y. Pocker and D. R. Storm, *Biochemistry*, 7 (1968) 1202.
136 Y. Pocker and J. T. Stone, *Biochemistry*, 7 (1968) 3021.
137 M. Eigen, K. Kustin and G. Moss, *Z. Phys. Chem.*, 30 (1961) 130.
138 D. M. Kern, *J. Chem. Educ.*, 37 (1960) 14.
139 C. Ho and J. M. Sturtevant, *J. Biol. Chem.*, 238 (1963) 3499.

140 B. H. Gibbons and J. T. Edsall, *J. Biol. Chem.*, 238 (1963) 3502.

141 F. J. W. Roughton, *Physiol. Rev.*, 15 (1935) 241.

142 F. J. W. Roughton, *Harvey Lect. Ser.*, 39 (1943) 96.

143 R. E. Davies, *Biol. Rev. Cambridge Phil. Soc.*, 26 (1951) 87.

144 S. M. Cain and A. B. Otis, *J. Appl. Physiol.*, 16 (1961) 1023.

145 R. P. Davis, *J. Am. Chem. Soc.*, 60 (1958) 5209.

146 R. P. Davis, in P. Boyer, H. Lardy and K. Myrback, *The Enzymes*, Vol. 5, Academic Press, New York, 1961, p. 545.

147 J. H. Wang, *Science*, 161 (1968) 328.

148 M. Eigen, *Discuss. Faraday Soc.*, 39 (1965) 7.

149 M. Caplow, *J. Am. Chem. Soc.*, 93 (1971) 230.

150 H. DeVoe and G. B. Kistiakowsky, *J. Am. Chem. Soc.*, 83 (1961) 274.

151 A. M. Sargeson, in F. P. Dwyer and D. P. Mellor, *Chelating Agents and Metal Chelates*, Academic Press, New York, 1964, p. 183.

152 M. Eigen and R. G. Wilkins, *Mechanisms of Inorganic Reactions, Adv. Chem. Ser.*, No. 49, American Chemical Society, Washington, D.C., 1965, p. 55.

153 J. E. Coleman, unpublished data.

154 B. G. Malmstrom, personal communication, 1971.

Chapter 17

PHOSPHATE TRANSFER AND ITS ACTIVATION BY METAL IONS; ALKALINE PHOSPHATASE

T. G. SPIRO

Department of Chemistry, Princeton University, Princeton, New Jersey, U.S.A.

I. PHOSPHATES AND BIOENERGETICS

It can be claimed that phosphorus is the fifth most important element in biology, following closely carbon, hydrogen, oxygen and nitrogen. A point of considerable interest to inorganic chemists is that, whereas the four primary elements weave themselves into a vast array of complex organic molecules, the biochemistry of phosphorus is mostly limited to derivatives of the orthophosphate ion. The great majority of these derivatives are simply esters and anhydrides of phosphoric acid, although there are a few which involve bonds between phosphorus and nitrogen or sulfur. Another point of interest is that divalent metal ions are intimately involved in much of the biochemistry of phosphorus. It might be supposed that phosphates and their derivatives, as well as their interactions with metal ions, fall within the realm of inorganic chemistry, but in fact the contributions of inorganic chemists in this area have been modest. While some excellent work has been done, phosphates have not generated much excitement in inorganic circles. In contrast they have been a major focus of interest in the biochemical world for at least three decades, since Lipmann drew attention to their significance[1-3].

Phosphates play two key roles in biology. One of these is as structural elements in certain biological components: the sugar–phosphate backbone of nucleic acids, for example, or the calcium phosphate deposits of bones and teeth. The more interesting role involves the transfer of energy. It appears that phosphate represents a universal currency of energy in living organisms. The contraction of muscles, the transfer of ions across membranes against a concentration gradient, and a very large number of steps in biosynthesis: these energy requiring processes are driven by the transfer of phosphoryl (PO_3) groups from high energy to low energy acceptors.

Phosphoryl transfer potentials

The free energy that can be stored in the phosphate bonds of naturally occurring derivatives ranges up to about 13 kcal/mol. A con-

venient energy scale is provided by the free energy of hydrolysis of phosphate derivatives, also called the phosphoryl group transfer potential[1]. A
few representative values[4] are given in Table I. These free energies are
sensitive functions of the concentration of protons and metal ions, both
of which interact to a different extent with the reactants and with the
products of the hydrolysis reactions. The values in Table I are for a pH of

TABLE I

STANDARD FREE ENERGY OF HYDROLYSIS OF
PHOSPHATE DERIVATIVES[a]

Compound	$-\Delta F'$ (kcal/mol)[b]
Enolpyruvate-P[c]	12.8
3-Phosphoglyceryl-P	11.8
Creatine-P	10.5
Acetyl-P	10.1
Arginine-P	9.0
ATP (α-anhydride)	8.0
ATP (β-anhydride)	6.9
Pyrophosphate	6.6
ADP	6.4
Glucose 1-P	5.0
Glycerate 2-P	4.2
Glucose 6-P	3.3
Glycerol 1-P	2.3

[a] Values are from Atkinson and Morton[4a]
[b] Standard conditions: pH 7.0, 25°C, 0.01 M Mg^{2+}
[c] R-P stands for R-PO$_3{}^{z-}$

7.0 and a Mg^{2+} concentration of 0.01 M, at 25°C, a standard state not far
from typical physiological conditions. Also the standard state, as usual, is
defined on the molar concentration scale for the reactants and products,
but not for the solvent (whose activity is unity), although in this case the
solvent is also a reactant. If the molar concentration of water (55.5 M)
were taken into account, the free energy changes would all be reduced by
2.4 kcal/mol. It would then be apparent that simple phosphate esters,
such as glycerol-1-phosphate, are not intrinsically much less stable than
orthophosphate. In aqueous solution the hydrolysis reaction is nevertheless driven far towards completion by the high excess of water over
glycerol.

However, there are several phosphate derivatives, such as phosphoenolypyruvate, creatine phosphate and acetyl phosphate, whose
hydrolysis releases 10–13 kcal/mol of free energy. These molecules
have special structural features that account for the instability of the

phosphate linkage[5]. In phosphoenolpyruvate there is substantial charge repulsion between the anionic phosphate and carboxyl groups, estimated at 3–4 kcal/mol, but the main driving force of the hydrolysis reaction is the conversion of the product pyruvic acid from the enol to the more stable (by about 6–9 kcal/mol) keto form:

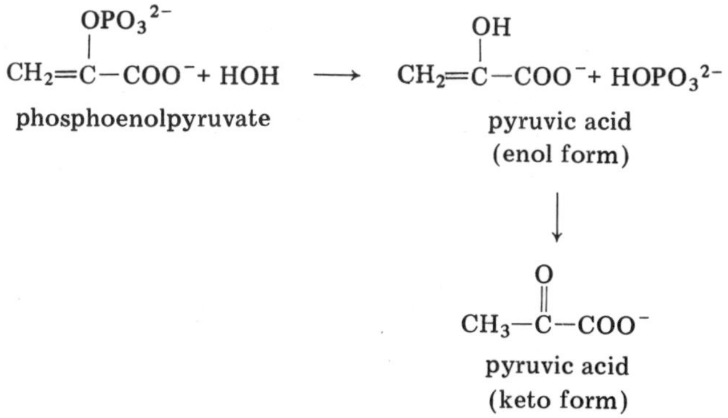

$$CH_2{=}\overset{\overset{\displaystyle OPO_3^{2-}}{|}}{C}{-}COO^- + HOH \longrightarrow CH_2{=}\overset{\overset{\displaystyle OH}{|}}{C}{-}COO^- + HOPO_3^{2-}$$

phosphoenolpyruvate   pyruvic acid
                             (enol form)

$$CH_3{-}\overset{\overset{\displaystyle O}{\|}}{C}{-}COO^-$$

pyruvic acid
(keto form)   (1)

In acetyl phosphate there is similar charge repulsion between phosphate and the immediately adjacent carboxyl oxygen, and again the main energy contribution is a further reaction of product, in this case the ionization of acetic acid at neutral pH, because of resonance stabilization of the acetate ion.

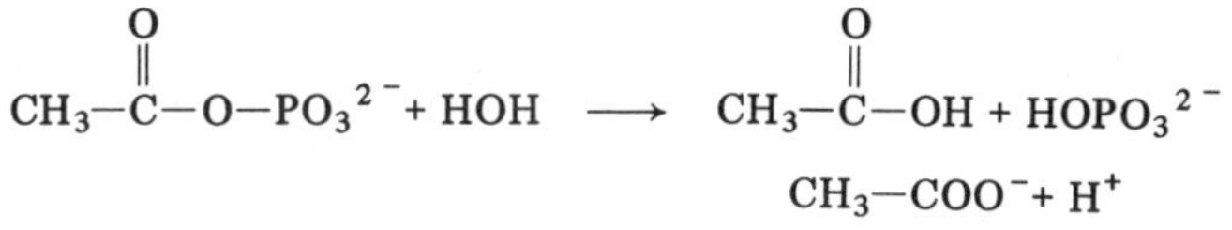

$$CH_3{-}\overset{\overset{\displaystyle O}{\|}}{C}{-}O{-}PO_3^{2-} + HOH \longrightarrow CH_3{-}\overset{\overset{\displaystyle O}{\|}}{C}{-}OH + HOPO_3^{2-}$$

$$CH_3{-}COO^- + H^+ \qquad (2)$$

For creatine phosphate,

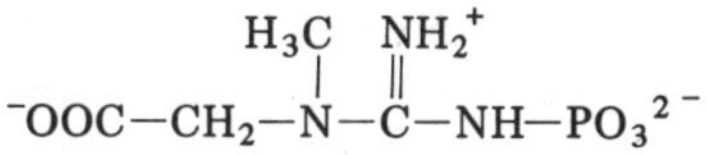

$$^-OOC{-}CH_2{-}\overset{\overset{\displaystyle H_3C}{|}}{N}{-}\overset{\overset{\displaystyle NH_2^+}{\|}}{C}{-}NH{-}PO_3^{2-}$$

hydrolysis leads to increased resonance stabilization of the products, creatine and orthophosphate, but here the main factor is the intrinsically lower strength of P–N relative to P–O bonds.

Pyrophosphate occupies an intermediate position in the group potential scale (6.6 kcal/mol). Again there is electrostatic repulsion of the two phosphate groups, and the product orthophosphate ions probably allow somewhat greater delocalization of the oxygen electrons. A similar situation is found in nucleoside polyphosphates, all of which give similar values for the free energy of hydrolysis. The free energies of polyphosphate hydrolysis are particularly sensitive to pH since both the products and the reactants undergo ionization equilibria in the neutral range. The following pK_a values are relevant[6]: $H_2PO_4^-$, 6.9; $H_2P_2O_7^{2-}$, 6.1; $HP_2O_7^{3-}$, 8.4;

H_2ATP^{2-}, 4.1; $HATP^{3-}$, 6.5; H_2ADP^-, 4.0; $HADP^{2-}$, 6.4; H_2AMP^0, 3.8; $HAMP^-$, 6.0. Other nucleotides have similar values. Also binding of divalent cations is significant for both products and reactants. Logarithms of the formation constants[6] for 1:1 complexes with Mg^{2+} are about 1.9 for HPO_4^-, 6.4 for $P_2O_7^{4-}$, 4.0 for ATP^{4-}, 3.1 for ADP^{3-}, and 2.0 for AMP^{2-} (p. 1194).

When a phosphoryl group is transferred from one acceptor to another, the free energy change can be estimated by substracting the group transfer potential of the second phosphate derivative from that of the first. Why is this free energy available to the organism and not simply dissipated through hydrolysis in its aqueous environment? The reason is that phosphate hydrolysis at neutral pH is a slow process even when the free energy change is high. Presumably nucleophiles such as water, or hydroxide, are repelled by the negatively charged phosphate oxygen atoms. Consequently the transfer of phosphoryl groups can be controlled by enzymes. A relatively wide range of group transfer potentials, combined with a substantial kinetic barrier are the characteristics which allow phosphate derivatives to mediate biological energy transfer.

Phosphorylation and phosphorolysis

It turns out that there is a single mediator of energy flux through phosphoryl transfer. This is the molecule adenosine triphosphate, ATP. In higher organisms, energy produced through the reduction of oxygen is stored in ATP through direct synthesis from adenosine diphosphate (ADP) and orthophosphate. In all organisms, glycolysis of sugars leads to the production of the high energy intermediates 1,3-diphosphoglycerate and phosphoenolpyruvate, which transfer their phosphoryl groups to ADP[7,8]. Energy utilization then proceeds by transfer of the ATP phosphoryl group to the acceptors that are responsible for muscle contraction, ion transport or biosynthetic pathways[7]. Finally the transferred phosphoryl group is hydrolyzed to orthophosphate and made available for resynthesis of ATP. There are of course many tributaries in the phosphate flow. For example muscle cells form quantities of the high energy creatine phosphate which can rapidly phosphorylate ADP when large amounts of ATP are needed in a short interval for muscle contraction.

Another process which relates to energy conversion is *phosphorolysis*, so named by analogy with hydrolysis. It involves the splitting of substrate bonds by phosphate instead of water. For example glycogen, a storage polymer of glucose, is broken down, not by hydrolysis to glucose, but by phosphorolysis to glucose-1-phosphate:

$$\text{glycogen} + HPO_4^- \rightarrow \text{glucose-1-phosphate} \tag{3}$$

From an energetic point of view the advantage of phosphorolysis over hydrolysis is the conservation of about 5 kcal/mol of glucose (see Table I), inasmuch as glucose-1-phosphate can be utilized directly in further metabolism.

The free energy change in reaction (3), as in other phosphorolysis reactions, is small. Nevertheless the biosynthesis of glycogen does not proceed by reversal of reaction (3). Rather glucose-1-phosphate combines with uridine triphosphate, UTP, to form UDP-glucose with the release of pyrophosphate:

$$\text{glucose-1-phosphate} + \text{UTP} \rightarrow \text{UDP-glucose} + \text{HP}_2\text{O}_7{}^{3-} \tag{4}$$

Then UDP-glucose adds its glucose unit to a growing glycogen chain:

$$\text{UDP-glucose} + \text{glycogen}_n \rightarrow \text{glycogen}_{n+1} + \text{UDP} \tag{5}$$

and UDP is rephosphorylated by ATP:

$$\text{UDP} + \text{ATP} \rightarrow \text{UTP} + \text{ADP} \tag{6}$$

Reaction (4) is the reverse of the pyrophosphate analog of phosphorolysis, *i.e.* *pyro*phosphorolysis. As with phosphorolysis, its free energy change is small. However, it is pulled to completion by the subsequent hydrolysis of pyrophosphate:

$$\text{HP}_2\text{O}_7{}^{3-} + \text{H}_2\text{O} \rightarrow \text{HPO}_4{}^{2-} + \text{H}_2\text{PO}_4{}^- \tag{7}$$

with the release of 6.6 kcal/mol. The overall process is expensive in energy, but it is clearly more reliable than reversal of reaction (3), which would be dependent on the vagaries of the free phosphate concentration in the organism. It seems to be generally the case that breakdown of biological macromolecules — nucleic acids, proteins and lipids, as well as polysaccharides — proceeds via phosphorolysis, whereas their synthesis involves an activation step energized by the hydrolysis of released pyrophosphate[9].

II. MECHANISMS OF PHOSPHATE TRANSFER

P–O vs. X–O bond cleavage

There are two basic reaction paths that are brought under the heading of phosphate transfer. They can be represented by

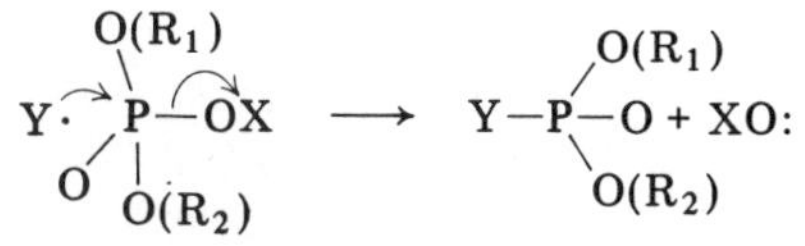

$$ \tag{I}$$

554

and

$$Y: \overset{\frown}{} X \overset{\frown}{-} O \overset{\frown}{-} P \begin{smallmatrix} O(R_1) \\ \diagup \\ \diagdown \\ O \end{smallmatrix} \longrightarrow Y-X + O-P \begin{smallmatrix} O(R_1) \\ \diagup \\ \diagdown \\ O \end{smallmatrix}$$

$$\begin{matrix} | \\ O(R_2) \end{matrix} \qquad\qquad\qquad \begin{matrix} | \\ O(R_2) \end{matrix}$$

(II)

Protons and charges are omitted here for clarity. R_1 and R_2 may or may not be present. In biochemical systems trisubstituted phosphate derivatives appear to be rare.

Path(I) represents phosphoryl transfer from one acceptor, XO, to another, Y, or the reverse. It involves breaking of P–O, or P–Y, bonds. The most important class of reactions falling in this category are those in which a phosphoryl group is transferred to or from ADP. The enzymes which catalyze these reactions are called *kinases*, and are the subject of Chapter 18. Path(II) represents transfer of another electrophile, X, from phosphate to Y, or the reverse. The bond broken in this case is X–O, or X–Y. When R_1 and R_2 are absent, the reverse of path(II) is the process known as phosphorolysis (see eqn. (3)), the enzymes for which are called phosphorylases[10]. If R_1 is a phosphoryl group (and R_2 is absent), then path(II) is the reversal of pyrophosphorolysis. In reaction (4), for example, a phosphate oxygen atom on glucose-1-phosphate is the nucleophile, Y, and the α-phosphorus atom of UTP is the electrophile X.

$$GO-\overset{\begin{matrix}O\\\diagdown\end{matrix}}{P}-O \overset{\frown}{} \overset{\begin{matrix}U\\|\end{matrix}}{P}-O \overset{\frown}{-} P \begin{smallmatrix} OPO_3 \\ \diagup \\ \diagdown \\ O \end{smallmatrix} \longrightarrow UDPG + P_2O_7^{4-}$$

(UDPG is uridine diphosphate glucose)

(4′)

Enzymes catalyzing this class of reactions were formerly called *pyrophosphorylases*, but are now called *synthetases* in recognition of the fact that they represent activation steps in biosynthetic processes. It is quite general that the activating group is a nucleotide, as in reaction (4). Consequently synthetase reactions can equally well be represented by path(I), with R_1 being a nucleoside (uridine in the case of reaction (4)) and OX being pyrophosphate. The important point is that a P–O bond is broken. Even phosphorolysis can lead to P–O bond cleavage when the substrate being attacked by phosphate is a nucleic acid, under the action of *polynucleotide phosphorylase*[11].

Hydrolysis

Whether a given phosphate transfer reaction proceeds by P–O or X–O bond breaking can often be inferred by examining the products, although enzymatic reactions may be more complex than appearances suggest, and formation of intermediates is not uncommon. Intermediates can sometimes be detected by ^{18}O labelling, a technique introduced by Coln[12]. When the nucleophile is water, P–O and X–O cleavage give the same products, and the use of ^{18}O-labelled water is the only way to distinguish them. For path(I) the label is found in the phosphate, whereas for path(II) it is found in Y–X:

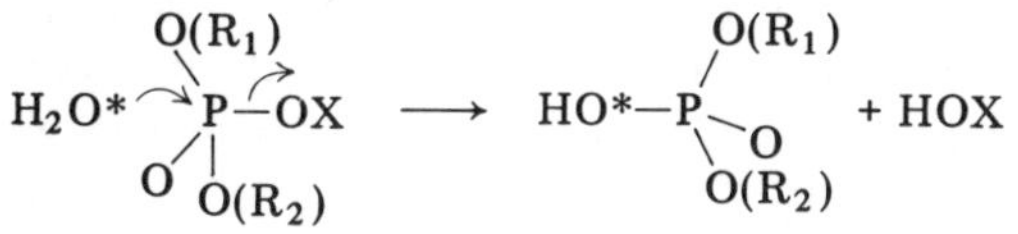

$$H_2O* \quad P\text{--}OX \quad \longrightarrow \quad HO*\text{--}P \quad + \ HOX \qquad (I')$$

$$H_2O* \quad X\text{--}O\text{--}P \quad \longrightarrow \quad HO*X + HO\text{--}P \qquad (II')$$

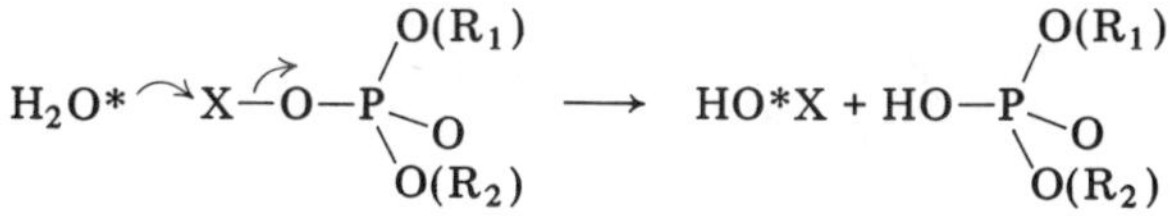

A substantial number of enzymatic hydrolyses have been examined with this technique, and in every case the reaction proceeds exclusively with P–O bond cleavage, path(I').

In non-enzymatic hydrolyses, on the other hand, both P–O and C–O bond cleavage occur, depending on the conditions[13]. In the case of mono-substituted phosphates, the dianion, $XOPO_3^{2-}$ and the monoanion, $XOPO_3H^-$, hydrolyze with P–O bond cleavage, whereas the neutral molecule $XOPO_3H_2$ hydrolyzes with C–O cleavage; X can be an alkyl, aryl, acyl or glycosyl group.

There is also an acid catalyzed path which proceeds via both P–O and C–O cleavage in variable ratio, depending on the leaving group effectiveness of X. Phosphate diesters appear to give both P–O and C–O cleavage in both neutral and monoanionic forms. Phosphate triesters undergo a relatively rapid attack by hydroxide ion, with P–O bond cleavage.

The non-enzymatic hydrolysis of phosphate derivates has been extensively studied with the aim of uncovering the detailed mechanisms of the reaction[13]. Attention has naturally focused on the P–O bond breaking path since it is the one followed by hydrolytic enzymes. This path involves nucleophilic attack on phosphorus as well as breaking of the P–O bond, and the question of interest concerns the order in which these two steps occur. The two limiting possibilities are: (1) a dissociative mechanism in which P–O cleavage leads to a tricoordinate metaphosphate (PO_3^-) derivative, which then reacts with the nucleophile; and (2) an associative mechanism in which nucleophilic attack gives a pentacoordinate derivative which then releases the OX leaving group. Needless to say, the mechanism may be intermediate between the two limiting cases, bond making and bond breaking proceeding concurrently without the formation of an intermediate.

A *priori* a pentacoordinate intermediate seems more likely than a tricoordinate one. There are many stable pentacoordinate phosphorus compounds, including the oxyphosphoranes[14], which have oxygen atoms at all five coordination positions. There is, on the other hand, no stable tricoordinate compound of phosphorus(V). (Phosphorus(III) compounds are of course tricoordinate, but there is no suggestion that phosphate hydrolysis involves a redox reaction of phosphorus.) A pentacoordinate intermediate may be involved in the hydrolysis of phosphate triesters, whose activation energy is much lower than that of diesters and mono-esters[3] (see also Chapter 34, p. 1234).

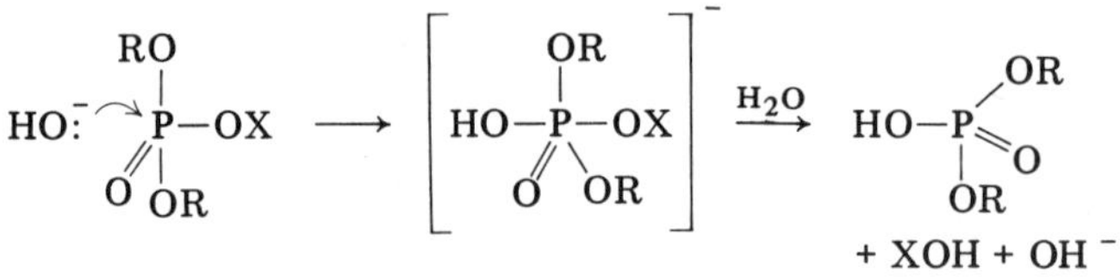

$$+ \ XOH + OH^-$$

Certainly the rapid hydrolysis of ethylene methyl phosphate is difficult to understand without Westheimer's pseudorotation mechanism (discussed in greater detail on p. 574), which requires such an intermediate[15].

On the other hand, there are grounds for believing that the hydrolysis of monosubstituted phosphate goes through a dissociative mechanism. Several lines of evidence have been presented in support of a metaphosphate intermediate in the hydrolysis of acyl[16a] and aryl[16b] phosphates. They can be summarized as follows.

1. The hydrolysis rates of the *dianions*, $XOPO_3^{2-}$ correlate strongly with the acidity constants of the leaving groups, XO^-, suggesting that P–O bond breaking is well advanced in the transition state. For very acidic leaving groups, such as 2,4-dinitrophenolate and p-nitrobenzoate, the hydrolysis rate of the dianion is *higher* than that of the *monoanion*, $XOPO_3H^-$, although higher negative charge of the dianion would be expected to inhibit nucleophilic attack on the phosphorus center.

2. The hydrolysis rates of the monoanions, $XOPO_3H^-$, correlate much less steeply with leaving group acidities. For weakly acidic leaving groups the monoanion hydrolyzes faster than the dianion. This special effect is thought to involve protonation of the leaving group, whereupon *neutral* XOH can be expelled, leaving the metaphosphate ion[13].

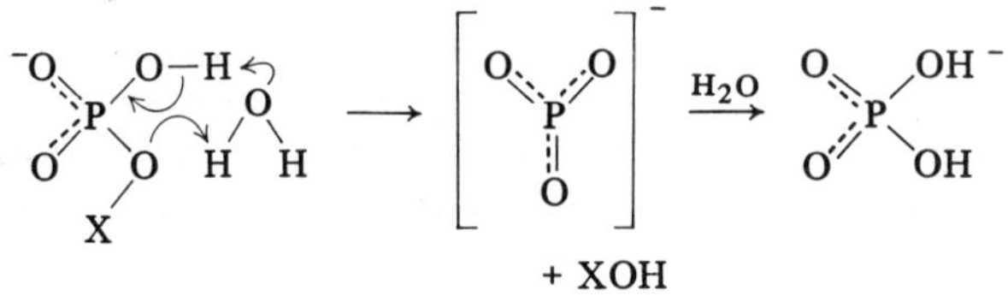

$$+ \ XOH$$

When the proton is replaced by an alkyl or aryl group the effect disappears, and hydrolysis is much slower for the monoanions of *diesters*, $XO(RO)PO_2^-$ than for the monoanions of monoesters.

3. Solvolysis of *p*-nitrophenyl phosphate, as either the mono- or di-anion, in methanol–water mixtures gives methyl phosphate and orthophosphate in the same ratio as the methanol : water mole ratio although methanol is the better nucleophile. This is evidence that a reactive intermediate, metaphosphate, scavenges water and methanol impartially. In other mixed solvent systems deviations from statistical product distributions can be found, but they can be explained on the basis of selective solvation effects.

4. The entropies of activation for hydrolysis of both mono- and di-anions are close to zero, indicating a unimolecular mechanism. Bimolecular solvolysis reactions commonly have substantial negative entropies of activation, around – 20 e.u., because of the requirement for orientation of the attacking solvent molecule.

Phosphatases and phosphotransferases

Enzymes which catalyze hydrolysis of phosphate derivatives are called *phosphatases*[4b]. There are several enzymes whose role seems to be to produce orthophosphate from phosphate monoesters. They have little regard for the nature of the substituent on the phosphate, and are called non-specific *phosphomonesterases*. There are however marked differences in the pH ranges in which they are effective, and they are accordingly divided into *acid*[17] and *alkaline*[18] *phosphatase* although there are intermediate cases. Other phosphatases act on specific substrates, such as glucose-6-phosphate[19], phosphoserine[19] and 5'-nucleotides[20]. *Inorganic pyrophosphatase*[21] is an important enzyme, since pyrophosphate hydrolysis is coupled to all synthetase reactions. Some kinases show ATPase activity, at low rates. An active ATPase can be isolated from disrupted mitrochondria[8], and it is no doubt involved in oxidative phosphorylation, through which the hydrolytic reaction is reversed.

There are other enzymes, *phosphodiesterases*, that hydrolyze disubstituted phosphates. Some of them are non-specific but most are involved in the hydrolysis of nucleic acids, and are called *nucleases*. There are *ribonucleases* and *deoxyribonucleases* (Chapter 34), and their actions are quite varied. *Exonucleases* attack the phosphodiester bonds sequentially, from one end of the nucleic acid chain, whereas *endonucleases* attack bonds in the interior of the chain as well.

Phosphatases are often suspected of acting through the formation of a phosphoryl enzyme intermediate, which then hydrolyzes, and in several cases this mechanistic feature has been definitely established. While all phosphatases catalyze phosphate hydrolysis, some of them have also been found to catalyze phosphoryl transfer to nucleophiles other than water, *i.e.* they show *phosphototransferase* activity. Since aqueous solutions contain 55.5 *M* water, such reactions may be difficult to detect except in mixed solvents. In some cases, however, nucleophiles have been found which compete quite effectively with water, even at low concentration. It is uncertain whether such reactions have biological significance.

For phosphotransferases proper the significance of the particular phosphate transfer step being catalyzed is usually clear. The kinases and synthetases have already been mentioned. Some enzymes in this class transfer a phosphate group from one portion of a molecule to another. An example is *phosphoglucomutase*[22] which converts glucose-6-phosphate, the product of glucose phosphorylation by ATP, to glucose-1-phosphate, which is the substrate for the glucose synthetase reaction (reaction (4)). Interestingly this conversion proceeds via two phosphorylated intermediates: phosphoryl-enzyme and glucose-1,6-diphosphate. The phosphoryl-enzyme phosphorylates the 1-hydroxyl group of glucose-6-phosphate to form glucose-1,6-diphosphate. Subsequently the 1-phosphoryl group is transferred to the enzyme to complete the reaction.

The enzymes which fall in the phosphotransferase class fail to phosphorylate water, *i.e.* they show zero or very low rates of hydrolysis. The means whereby this can be accomplished in aqueous solution are unknown, and they represent one of the most intriguing aspects of the mechanism of phosphate transfer. For some of the ribonucleases hydrolysis is preceded by a phosphate transfer step, in which a labile cyclic 2′,3′-nucleoside phosphate is formed. The best characterized example is pancreatic ribonuclease[23].

Metal ion requirements

Among all of the phosphate transfer enzymes, alkaline phosphatase, which is the subject of the following section, appears to be the only well characterized metallo-enzyme, *i.e.* an enzyme with an essential metal ion which is bound tightly enough that it survives ordinary protein fractionation procedures. There is also evidence that a phosphatase in beef spleen that acts on phosphoprotein as well as on other monosubstituted phosphates may be an iron metallo-enzyme[24].

On the other hand, most phosphate transfer enzymes require divalent metal ions for activity. The metal ions are more or less readily dissociated, and if they are rigorously excluded from the enzyme preparation, activity ceases. It can then be restored by readdition of the metal ions. Since the only divalent ions available in adequate concentration under ordinary physiological conditions are Mg^{2+} and Ca^{2+}, these are presumed to be the ones that are involved in biological function. However, the enzymes show wide variation in their response to metal ions. Most often Mg^{2+} gives the highest activation, but several first transition row divalent metal ions, notably Mn^{2+}, can be substituted at somewhat lower levels of activity. So, often, can Ca^{2+}, but in some cases Ca^{2+} *inhibits* activity, while in a few cases Ca^{2+} activates but Mg^{2+} inhibits.

In any particular case the establishment of metal ion requirements is apt to be troublesome. The influence of metal ions on activity can be masked by interactions with the substrate, or with impurities in the enzyme preparation, or with secondary binding sites on the enzyme itself. There is usually an optimum concentration beyond which activating metal ions have an inhibitory effect, and this optimum is different for different metal ions. Sometimes chelating agents, such as EDTA, actually *enhance* activity probably because they preferentially bind trace ions, such as Cu^{2+}, which can be strongly inhibitory.

In spite of these difficulties enough careful work has been done that a few general statements about metal ion requirements can be made with confidence. Kinases all require a divalent ion, as do synthetases. Phosphatases usually require a metal ion, although there seem to be exceptions, notably the non-specific acid phosphatases. Phosphoglucomutase requires Mg^{2+}, but the analogous phosphoglycerate mutase does not[22]. Nucleases require divalent ions, except when the mechanism involves production of a cyclic nucleotide as a first step[23]. Phosphorylases do *not* require divalent ions as long as the bond being cleaved is C–O. But they do require a metal

TABLE II

METAL ION REQUIREMENTS FOR ENZYMATIC PHOSPHATE TRANSFER ACTIVITY

Require a divalent ion	Do not require a divalent ion
Non-specific alkaline phosphatases	Non-specific acid phosphatases
Most other phosphomonoesterases	
Pyrophosphatase	
Phosphoglucomutase	Phosphoglycerate mutase
Nucleases acting as phosphodiesterases	Ribonucleases forming a cyclic 2',3'-nucleoside phosphate intermediate
Kinases	
Synthetases	
Polynucleotide phosphorylase (P–O cleavage)	Other phosphorylases (C–O cleavage)

ion when a P–O bond is cleaved, as in polynucleotide phosphorolysis. It appears then, that among phosphate transfer enzymes a divalent metal ion is usually, but not always, required when P–O bond cleavage is being catalyzed. The situation is summarized in Table II.

References pp. 578–581

III. ALKALINE PHOSPHATASE

Alkaline phosphatase from the bacterium *Escherichia coli* deserves special attention because it is the one well-characterized metallo-enzyme in the phosphate transfer series. At the time of writing it is under investigation in several laboratories and the description of its nature and functioning is far from complete. An X-ray analysis of the enzyme crystal structure has been initiated[25a], and its results will greatly clarify the current picture. Nevertheless, several remarkable characteristics have already emerged from recent studies. A more detailed discussion, especially of the enzyme kinetics, is given by Reid and Wilson[25b].

Grown on a medium which is deficient in phosphate, *E. coli* can be induced to form large quantities of the enzyme[26], which is therefore relatively easy to purify and characterize. It is a non-specific phosphomonoesterase[27] with a broad pH optimum from 8.5 to 10. It also catalyzes the hydrolysis of phosphate anhydride bonds (*e.g.* pyrophosphate and ATP) and of fluorophosphate, but it is not active toward P–N or P–C bonds[28,44]. Interestingly, phosphate thioester bonds (RS–PO$_3$) are readily cleaved[29,30], but oxygen-substituted esters of thiophosphoric acid (RO–PO$_2$S) are not hydrolyzed[30], or are hydrolyzed much more slowly[31] than ordinary phosphate esters. In contrast acid phosphatase hydrolyzes *O*-substituted but not *S*-substituted thiophosphate esters[30]. Alkaline phosphatase has also been shown to transfer phosphate to nucleophiles other than water[32].

Metal binding

The isolated enzyme contains firmly bound zinc ions, and in this respect it is similar to two other hydrolytic enzymes, carboxypeptidase (Chapter 15) and carbonic anhydrase (Chapter 16). The molecular weight of the native enzyme is 80,000[33,34], but it consists of two identical subunits[34,35], which can be dissociated in acid solution. Furthermore high concentrations of zinc can induce the dimers to form tetramers, in a pH-dependent equilibrium[36].

The number of zinc ions bound by the native enzyme has been variously reported[34,37–41] at between two and four per dimer. It appears certain that only two zinc ions are involved in the catalytic activity of the enzyme. The equilibrium binding experiments of Cohen and Wilson[37], in which enzymatic activity was measured in the presence of zinc buffers, suggest that enzyme molecules containing one zinc ion are 12% as active as those containing two (or more). The association constants obtained from these measurements, at pH 8.5 at 25°C in 1 *M* NaCl, are quite different for the two zinc ions: K_1 = 7.66 and K_2 = 10.22. On the other hand, Csopak[38] interpreted equilibrium dialysis measurements, at pH 8.5

at 25°C in 0.1 M KNO$_3$, as indicating that the two dissociation constants are both much lower, and differ by only the statistical factor of four: $pK_1 = 11.16$ and $pK_2 = 11.76$. Below pH 8.5 these pK values decrease with approximately unit slope. If a third and fourth zinc ion are bound, they are removed much less readily than the two required for activity[40].

The zinc ions can be removed with chelating agents. The apoenzyme will then bind a variety of divalent ions[41a,42]: Mn^{2+}, Co^{2+}, Ni^{2+}, Cu^{2+}, Cd^{2+}, Hg^{2+}. The strength of metal binding for the most labile site decreases in the order[41] $Cd^{2+} = Mn^{2+} > Zn^{2+} > Co^{2+} > Ni^{2+}$. The position of Cu^{2+} and Hg^{2+} in this series has not been determined, although Cu^{2+} is apparently bound less firmly than Zn^{2+} [41b]. Of the resulting metal–enzyme preparations only the one with cobalt shows any detectable phosphatase activity, and it is only 12% as active as the zinc enyzme[43,44]. Also the cobalt enzyme fails to show transferase activity[43,44] towards nucleophiles which are effective in the presence of the zinc enzyme.

The cobalt enzyme is of particular interest because the electronic spectrum of Co(II) provides a probe for the coordination geometry at the active site. The visible spectrum of the cobalt enzyme is distinctive and unusual[40,45], with bands at 640, 555 and 510 nm, the corresponding molar absorptivities being 250, 210, 350 and 280 respectively. This pattern is quite similar to that observed for cobalt(II) carbonic anhydrase (Chapter 16). It is characteristic neither of octahedral nor of tetrahedral coordination for Co(II). Somewhat similar spectra are shown by five-coordinate pseudo trigonal bipyramidal cobalt complexes[46], although the enzyme shows a greater splitting of the bands, indicative of lower symmetry[40]. Circular dichroism measurements show a complex spectrum throughout the visible region[40,45]. Electron spin resonance (e.s.r.) spectra of Cu^{2+} bound to the enzyme suggest a change in environment when a second Cu^{2+} ion is bound[41b].

Information on the metal binding sites on the enzyme is incomplete. Tait and Vallee[47] suggested that three histidines per molecule are bound to zinc, since photo-oxidized apoenzyme, in which more histidine was destroyed than in the native enzyme, was unable to bind usual amounts of zinc, and was inactive. Likewise Reynolds and Schlesinger[39b], by comparing titration curves for native and apoenzyme, deduced that two or three histidines are involved in binding to zinc. The characteristic spectrum for the cobalt phosphatase is abolished, as is its enzymatic activity, on adding acid[40]. The transition pH is 7.0, consistent with the ionization of histidine[39b]. Removal of zinc from the enzyme exposes about six tyrosine residues to the solvent, as determined by spectrophotometric titration[48], suggesting that tyrosine is bound to zinc, although conformational changes could be responsible for the effect.

Phosphoryl–enzyme

The mechanism of alkaline phosphatase catalysis is dominated by the fact that phosphate transfer proceeds via a covalent phosphoryl–enzyme intermediate. Phosphoprotein can be precipitated from alkaline phosphatase solutions by incubating with phosphate and adding trichloroacetic acid[49]. Upon digestion of the protein and analysis of the residues, the phosphate is found attached to serine[50]. At alkaline pH no covalently bound phosphate can be detected, but Wilson *et al.*[51-53] have shown, with kinetics and labelling experiments, that a phosphoryl–enzyme is involved at all pH values and that it is no doubt identical with the phosphoprotein isolated from acid solution. The phosphoryl–enzyme has an astonishingly low free energy with respect to hydrolysis, being 10^6 times more stable than ordinary phosphate esters[51,52]. The reason that the phosphatase catalyst does not fall into a thermodynamic trap is that the enzyme forms a non-covalent complex with phosphate which is a factor of 100 more stable than the phosphoryl–enzyme at pH 8[51]. Consequently the equilibrium:

$$E - PO_3 \rightleftharpoons E \cdot PO_4 \qquad\qquad (8)$$

lies largely to the right at alkaline pH and fresh substrate can displace the non-covalent phosphate, regenerating the catalyst. As might be expected phosphate ion is itself a potent inhibitor of the phosphatase reaction.

In acid solutions, on the other hand, equilibrium (8) is shifted to the left[52], and the enzyme is inactive. The position of the equilibrium is not the controlling factor, however, since the pK for enzyme inactivation is close to 7 for both the zinc and the cobalt phosphatase[43b,44], at which pH most of the bound phosphate is in the non-covalent complex[52a]. Ionization of a histidine group has been suggested[43b] as requisite to enzyme activation. In any case phosphorylation is very fast for the zinc enzyme and even faster for the cobalt enzyme[43b]. Dephosphorylation of the phosphoryl–enzyme is rate-limiting for cobalt phosphatase[43b], but for zinc phosphatase at alkaline pH, this process is much faster, and is apparently not rate-limiting[53]. A conformational change of the enzyme–substrate complex has been suggested as the rate-limiting step[54], and is supported by kinetic evidence on the binding of a reversible inhibitor of the enzyme[55]. This matter is considered in detail by Reid and Wilson[25b]. The maximum velocities for the hydrolytic reactions are essentially independent of the nature of the substrate despite substantial differences in the effectiveness of the leaving group (*e.g.* in nitrophenyl- and dinitrophenyl-phosphate[43]). At high pH the Michaelis constants increase, implying weaker binding of substrates to the enzymes. The pK for the transition is 8.6 for zinc phosphatase[44], and somewhat higher for cobalt phosphatase (values of 8.9[44] and 9.6[43] are reported). Ionization of a water molecule bound to the metal ion has been suggested to account for this effect[43]. Ionization of bound water would also account for the transition pH, 8.5, of the zinc-binding constants[38].

Coleman and coworkers[42] have observed that cadmium and manganese phosphatase, while completely inactive as catalysts, bind phosphate ion quite effectively, and form substantial amounts of phosphoryl–enzyme. Whereas zinc phosphatase shows maximum covalent binding of phosphate at pH 5 and very little above pH 6, the cadmiun and manganese phosphatases produce maximum covalent binding at pH 7. In the case of cadmium most of the bound phosphate is attached covalently to the enzyme at this pH. In common with zinc phosphatase, then, the cadmiun and manganese derivatives are quite effective at phosphorylating the protein. The reason for their inactivity as enzymes is apparently their reluctance to effect the necessary dephosphorylation step at alkaline pH. Cobalt phosphatase is reported in this work[42] to produce very little covalent phosphate at any pH, the maximum amount, 0.2 moles per mole of enzyme, occurring at pH 6.

This last observation conflicts with that of Lazdunski et al.[56], who report that at $0°C$ (Coleman's group worked at $4°C$) they were able to isolate up to 1 mole of covalent phosphate per mole of cobalt phosphatase on incubation with phosphate in acid solution. No covalent phosphate was detected in alkaline solution and the transition pH was found to be 5.6. For zinc phosphatase the transition pH was similar, 5.1, but in this case a maximum of *two* covalent phosphates per mole of enzyme was isolated in acid solution. Furthermore when the enzymes were incubated with substrates (AMP and ATP) at $0°C$, a maximum of two covalent phosphates in acid solution was found for both zinc and cobalt and a minimum of *one* was obtained in alkaline solution. The transition pH values were quite different, however: 6.6 for zinc and 4.6 for cobalt.

Phosphate binding

These remarkable observations bear on the question of the number of phosphate binding sites, which has received considerable attention. Despite the dimeric nature of the enzyme, kinetic data have been interpreted on the basis of a single active site[51,52]. Support for this assumption has come from stopped flow experiments[53b,54a,57a], in which the pre-steady state burst of product corresponds to about one mole per mole of enzyme. At high substrate concentration (2×10^{-3} M), however, this technique gave a value of 2.7 moles[57b]. Most analyses of phosphate incorporation have produced values in the range 0.6–1.3 moles of phosphate per mole of protein[52b,58], although the above-mentioned study by Lazdunski et al.[56] is a notable exception. Neumann[30] found one mole of O-p-nitrophenyl phosphorothioate, which acts as an inhibitor, bound per mole of enzyme.

Equilibrium binding measurements have been interpreted as indicating both one and two bound phosphate ions per mole of enzyme. The most recent measurements clearly point to there being one firmly bound phosphate and one or more which bind more weakly[42,54,59]. In addition Simpson and Vallee[59] have found kinetic evidence for such *anticooperativity* since under the proper conditions the enzyme shows substrate activation (the dependence of activity on substrate concentration is biphasic). Negative cooperativity has been discussed in detail, in connection with another system by Levitzki and Koshland[59b]. Simpson and Vallee suggest that the functional role of anticooperativity is to maintain the proper level of enzyme activity over a wide range of substrate concentration. Their inference that allosterism is involved gains some support from the evidence[55] that inhibitor binding leads to conformational change of the enzyme. (See note added in proof (1), p. 577.)

Binding of phosphate to cobalt phosphatase produces a marked change in the visible spectrum[40,45]. It appears that the two higher energy bands, at 510 and 555 nm shift to 475 and 550 nm, while the low energy band at 640 nm loses most of its intensity and the 610 nm shoulder disappears. Binding of arsenate produces a similar change[42]. Both anions also markedly alter the circular dichroism spectrum, but whereas phosphate produces a strong negative ellipticity at the main 550 nm band[40,45], binding of arsenate leads to a strong positive ellipticity at the same band[42]. Thus although phosphate and arsenate produce roughly the same changes in the cobalt(II) d electron energy levels in cobalt phosphatase, they have strikingly different effects on the dissymmetry of the cobalt coordination shell. There is a conflict as to whether one[45] or two[40] phosphates per mole of cobalt phosphatase are required to saturate the spectral change. Binding of phosphate to Cu^{2+} substituted enzyme produces a marked change in the Cu^{2+} e.s.r. spectrum[41b,103].

Models for the active site

As is evident from this survey, a number of key elements in the description of alkaline phosphatase and its action remain unclear, and conflicting claims await resolution. No doubt the focus will sharpen in the near future as new results are obtained and as the X-ray crystal structure begins to emerge. At the moment, however, we are left to guess at what happens in detail at the enzyme active site.

It seems safe to assume that a substrate molecule binds directly to a zinc ion at the active site. The profound influence of metal ion substitution on the course of the enzyme reaction, and the sensitivity to phosphate and arsenate binding of the cobalt phosphatase electronic and circular dichroism spectra, and of the copper phosphatase e.s.r. spectrum, make implausible a merely indirect role for the zinc ions. Moreover, there is the precedent of carboxypeptidase (Chapter 15) for which crystallographic studies[60] indicate strongly that the substrate binds directly to zinc.

The electronic spectrum of cobalt phosphatase shows the cobalt ion to be in a decidedly non-symmetrical coordination environment. Binding of phosphate or arsenate shifts the spectrum significantly, but the band shapes and splittings are not drastically altered. It appears therefore that the gross coordination geometry is unchanged and that phosphate or arsenate simply displaces a ligand on the metal ion. The easiest ligand for them to displace would be water. Again the crystal structure results for carboxypeptidase are suggestive, in that water is found to be the fourth ligand in a highly distorted zinc coordination shell. The suggestion of Gottesman *et al.*[43] that the pK values for the increase in the Michaelis constants in zinc and cobalt phosphatase are associated with the ionization of aquo ligands is consistent with this idea. It would be harder for the substrate to displace hydroxide than water from the metal ions.

The binding of phosphate to the metallo-enzyme is surprisingly strong. The association constant[42,52a,59a], $\sim 10^6$ M, is much higher than would be expected for a simple complex between HPO_4^{2-} and a divalent metal ion. The association constant for $M(HPO_4)$ is 3.7×10^2 for Mn^{2+} and 7.7×10^1 for Mg^{2+} [6]. It seems likely that, in addition to the zinc ion, there is an adjacent cationic binding site for the phosphate oxygens, possibly a proton on an amine side chain. Formation of phosphoryl enzyme, either from substrates or from HPO_4^{2-}, is firmly established, and on protein digestion the phosphate is found attached to serine. The inference is that the active site contains a serine residue close to the zinc ion and properly oriented for attack on the phosphorus atom of the bound phosphate derivative. The high thermodynamic stability of the resulting phosphoryl enzyme with respect to hydrolysis suggests that it retains a considerable part of the binding interaction of the phosphate–enzyme complex[51]. It seems likely that the serine-bound phosphoryl group remains attached to the zinc ion.

Breslow and Katz[31] draw an interesting mechanistic conclusion from the relative hydrolysis rates of phosphate and O-substituted thiophosphate esters. In enzymatic hydrolysis thiophosphate O-monoesters react much more slowly than phosphate monoesters, whereas in non-enzymatic hydrolysis the reverse is true. On the other hand in the non-enzymatic hydrolysis of *triesters*, it is again the thiophosphate derivatives which react more slowly. Triester hydrolysis is presumed to involve an addition mechanism in which the P=O (or P=S) bond order decreases and the charge on oxygen (sulfur) increases in the transition state, while monester hydrolysis is presumed to involve an elimination (to metaphosphate) mechanism in which the P=O (or P=S) bond order increases and the charge on oxygen (sulfur) decreases in the transition state (see pp. 556–557). The relative non-enzymatic rates are explicable on the basis of the lower electronegativity of sulfur relative

566

to oxygen. The reversal of the monoester rates in the enzymatic reaction therefore suggests a change in mechanism from elimination to addition. This would be consistent with the firm binding of phosphate oxygen atoms at the active site.

The next step, after phosphorylation of the enzyme, is nucleophilic attack on the phosphoryl intermediate. The nucleophile need not be water. Wilson et al.[32] have shown that a number of phosphate acceptors, such as ethanolamine and ethylene glycol, compete quite well with water, even at concentrations of 1 M or less. These molecules all contain a hydroxyl or amine group separated by not more than four carbon atoms from the acceptor hydroxyl group. This structural requirement suggests the presence of a basic binding site on the enzyme, properly oriented for the assistance

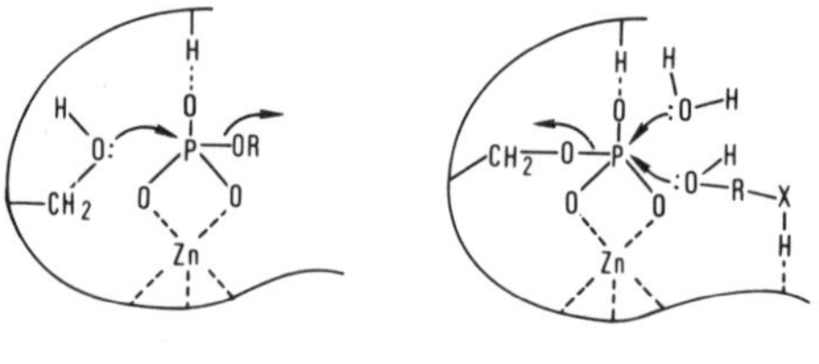

Fig. 1. Model of the active site of alkaline phosphatase. (A) phosphorylation of the enzyme; (B) phosphate transfer to water or other nucleophile.

of the nucleophiles. Acceptor molecules without this structural feature were ineffective in the transfer experiments[32], but there is no reason to think that they could not compete with water if their concentration could be increased sufficiently. There seems no need to postulate a special binding site for water on the enzyme. These inferences are summarized diagrammatically in Fig. 1. The importance of the metal ion in maintaining the steric requirements of the active site may be inferred from the observation that Cd^{2+} and Mn^{2+} promote phosphorylation of the enzyme, but not the subsequent hydrolysis step, while Co^{2+}, which shows 12% of the hydrolytic activity of Zn^{2+}, is not active towards phosphate acceptors other than water.

The discussion so far has avoided the complication that two metal ions are required for activity but there is only one tight binding site for phosphate. The allosteric model[59] takes this into account by proposing that binding of one phosphate brings into play long-range subunit interactions which make it harder to bind a second phosphate. The alternative suggestion[42] is that the primary phosphate binding site is at the point of contact of the two enzyme subunits. Preliminary X-ray data[25] show that the native enzyme crystallizes in a monoclinic space group $P3_121$, with unit cell dimensions $a = 70.5$ Å, $b = 70.5$ Å, $c = 155.6$ Å, $\beta = 120°$. The unit cell contains three dimers, but the monomer is the asymmetric unit. The two-fold rotation axis relates pairs of identical monomers. If these pairs represent functional dimers, then a unique phosphate binding site could lie on the two-fold rotation axis. This raises the interesting possibility that the phosphate binds simultaneously to the two zinc ions and that this novel structural feature might be important in activating phosphate derivatives toward nucleophilic attack.

IV. ROLE OF THE METAL ION

In view of the wide variety of phosphate transfer enzymes which require metal ions, one must face the likelihood that the functions of the metal ions may be equally varied. Indeed the specific nature of the metal ion requirement varies from one enzyme to another, in terms of the relative effectiveness of different metal ions, the nature of inhibitors, and the concentration of dependencies of activation and inhibition. The specific involvement of metal ions in the reaction path is apt to be different for each enzyme. Nevertheless, we may hope to find some unifying themes based on fundamental chemical principles.

Ternary complexes among enzymes, metals and substrates

Enzyme activation by metal ions implies the formation of a ternary complex involving the enzyme, the metal ion and the substrate. Several arrangements may be envisaged and have in fact been identified in actual systems. The possibilities, consisting of systems in which the ligand (E–S–M), the metal (E–M–S), or the enzyme (S–E–M) serve as a bridge, have been discussed in Chapter 14, pp. 382–386, and are comprehensively covered by Mildvan[61].

Charge neutralization

Among the characteristics of metal ions which might help to explain their role in activating phosphate transfer enzymes, the most obvious one is their positive charge. Divalent cations maintain their charge in neutral solution and can be expected to screen the negative charge of phosphate oxygen atoms, thereby facilitating the binding of substrate to the enzyme. It is well established that divalent cations form complexes of moderate stability with phosphate derivatives (though the effect is often obscured by precipitation), especially if they contain di- or tri-phosphate groups[6,69]. Furthermore divalent cations are potent stabilizers of the tertiary structure of transfer-RNA molecules[70], presumably through interactions with the ribose–phosphate backbone.

A model system in which charge neutralization appears to be important has been discovered by Cooperman and Lloyd[71]. They found that Zn^{2+} catalyzes phosphoryl transfer from phosphoryl imidazole (PIm) to the oxygen atom of 2-pyridine carbaldoxime (PCA), which has previously been shown to be an effective nucleophile in its zinc complex[72]. The reaction displays saturation kinetics, implying the formation of a ternary PCA–Zn^{2+}–PIm complex. The pH rate profile is bell-shaped and reaches a maximum rate at pH 6, where the proton on Zn^{2+}–PCA and one of the two protons on PIm are ionized. The role of the zinc ion is envisaged as shielding the phosphoryl group, thereby facilitating attack by the charge of the anionic nucleophile (Fig. 2). The effect is very large, the rate at the pH maximum being about 10^4 times that of water attack on PIm.

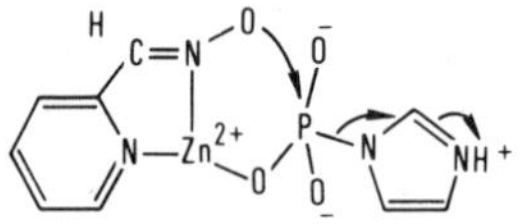

Fig. 2. Mechanism proposed[11a] for the reaction:

$$Zn^{2+} - PCA + ImPO_3 \rightarrow Zn^{2+} - PCAPO_3 + Im$$

A similar effect may be involved in the non-enzymatic phosphorylation of orthophosphate (to produce pyrophosphate) by ATP, which has been shown by Lowenstein[73] to occur in the presence, but not in the absence, of divalent metal ions, particularly Mn^{2+}, Cd^{2+} and Ca^{2+}. Saturation kinetics were not observed in this case, but considering the low concentration of phosphate used, the reaction appears to be substantially faster than the hydrolysis of ATP under similar conditions[74]. However the mechanism is unlikely to be simple in view of the curious characteristics —pH optimum of 9, ATP/M^{2+} optimum of 0.6 —shown by the Mn^{2+} promoted reaction.

While charge neutralization is certainly one of the functions provided by metal ions in enzymatic phosphate transfer reactions, the function does not necessarily require a metal ion. Proteins have available to them non-metallic charge sites, in the form of protonated amine and imidazole side chains, which could equally well assist in binding negatively charged phosphate groups. Many enzymes that handle negatively charged reactants do not require metal ions, including some of the phosphate transfer enzymes.

Polarization

In addition to neutralizing charge, a metal ion bound to a phosphate derivative can be expected to polarize the P–O bond of the oxygen to which it is attached. In previous discussions of the role of metal ions in phosphate transfer, polarization has been the most popular candidate. The idea is that the metal ion draws electrons towards itself from the oxygen atom, weakening the P–O bond and increasing the positive charge on the phosphorus atom. Both nucleophillic attack and P–O bond breaking should thereby by facilitated. The effect is illustrated in Fig. 3 (see also Chapter 34).

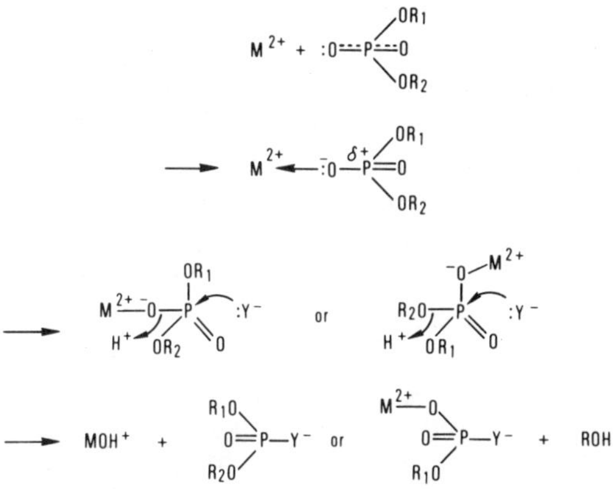

$$M^{2+} + :O\!\equiv\!\!\!\equiv\!P\!\equiv\!\!\!\equiv\!O \begin{smallmatrix} OR_1 \\ \\ OR_2 \end{smallmatrix}$$

$$\longrightarrow \quad M^{2+}\!\leftarrow\!:O\!-\!\overset{\delta+}{P}\!\!=\!\!O \begin{smallmatrix} OR_1 \\ \\ OR_2 \end{smallmatrix}$$

$$\longrightarrow \quad M^{2+}\!-\!O\!-\!\underset{OR_2}{\overset{OR_1}{P}}\!\!\diagdown\!\!:Y^- \quad \text{or} \quad R_2O\!-\!\underset{OR_1}{\overset{O-M^{2+}}{P}}\!\!\diagdown\!\!:Y^-$$
$$\qquad H^+ \qquad\qquad\qquad\qquad\qquad H^+$$

$$\longrightarrow \quad MOH^+ + \begin{smallmatrix} R_1O \\ \\ R_2O \end{smallmatrix}\!\!O\!\!=\!\!P\!-\!Y^- \quad \text{or} \quad \begin{smallmatrix} M^{2+}-O \\ \\ R_1O \end{smallmatrix}\!\!O\!\!=\!\!P\!-\!Y^- + ROH$$

Fig. 3. Schematic representation of activation via the polarization effect.

If polarization were important in activating phosphate, one would expect that metal ions themselves would be effective in promoting the hydrolysis of phosphate derivatives, since they bind adequately in aqueous solution. Yet in the absence of enzymes, metal ions are unimpressive in this role. Although the catalytic role of metal ions has been emphasized in model studies[75], the effects are quite modest unless there are unusual structural features. For example, magnesium ions accelerate the hydrolysis of the dianion of acetyl phosphate, $AcOPO_3^{2-}$ [76], but only by a factor of three at a concentration of 0.1 M. On the assumption that the dissociation constant of $Mg(AcOPO_3)$ is not less than that of $Mg(HPO_4)$[6] (it might be greater since the carbonyl oxygen could participate in binding), $AcOPO_3^{2-}$ hydrolyzes at most 4–5 times faster when bound to Mg^{2+} than when free.

The effect of bound Mg^{2+} on ATP hydrolysis is even smaller[74], despite the fact that Mg^{2+} is an activator for most kinases. At pH 9, where free ATP is unprotonated, magnesium enhances the rate by a factor of two, whereas at pH 5, where free ATP binds one proton, magnesium actually inhibits hydrolysis slightly. Calcium and manganese ions do somewhat better, with rate enhancements by factors of 2–10. Really substantial enhancement, by a factor of about 60, is given only by Cu^{2+}, which, however, has been shown[77a] to react entirely via a dimeric hydroxylated chelate, $(ATPCuOH)_2^{6-}$, in which the copper ions are presumably bridged by the two hydroxide ions. It is not clear why this unusual complex should be particularly effective at promoting ATP hydrolysis, but the answer is unlikely to be simple polarization.

A special structural feature (adjacent carboxyl group) also serves to explain the promotion of salicyl phosphate hydrolysis by Cu^{2+}, Fe^{3+}, VO^{2+} and VO_2^+ [77b]. Phosphate esters are hydrolyzed effectively by hydroxide gels of La(III), Ce(III), Ce(IV), Th(IV) and Pb(II)[77c,d]. The hydrolysis rate for hydroxyethyl phosphate at pH 8.5 at 78°C is increased 1000-fold in the presence of La(III) hydroxide gel[77e]. Mechanistic inferences for these heterogeneous systems are rather uncertain, however.

It may be argued that while a pronounced polarization effect is absent in aqueous binary complexes, it is not precluded for the ternary enzyme complex. The effective dielectric constant at the active site is apt to be much lower than that of water, and electrostatic effects may be substantially enhanced. On the other hand there are grounds for believing that it is not inability to polarize the P–O bond which accounts for the ineffectiveness of metal ions in phosphate activation in simple complexes. Rather it is the ability of the phosphorus atom to drain charge from the rest of the molecule which frustrates the polarization effect. This ability is strongly suggested by Brintzinger and Hester's analysis of the vibrational spectra of phosphate derivatives and their metal complexes[78]. It appears that the P–O stretching force constant decreases, as expected, for the oxygen atom or atoms bound to the metal ion, but increases for the free oxygen atoms, so that the average P–O force constant remains essentially unchanged. Similar behavior was found for other oxyanions, such as sulfate, carbonate and nitrate. The suggestion is that polarization is accommodated by virtue of the extensive π-bonding between the oxygen atoms and the central atom, utilizing empty $2p$ orbitals on carbon and nitrogen, and $3d$ orbitals on phosphorus and sulfur. Electrons can be shifted with ease from one oxygen atom to another, with little change in the charge of the central atom.

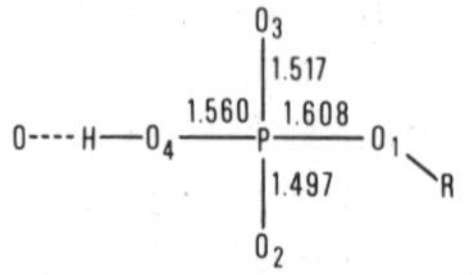

$$
\begin{array}{c}
O_3 \\
\Big| \, 1.517 \\
O\text{----}H\text{---}O_4 \xrightarrow[]{1.560 \;\big|\; 1.608} P\text{---}O_1 \\
\Big| \, 1.497 \qquad\qquad \diagdown R \\
O_2
\end{array}
$$

Fig. 4. The structure of serine phosphate (from ref. 17).

This line of reasoning is supported by structural data. Cruickshank[79] has pointed out that the *average* P–O bond distance in phosphate derivatives is nearly constant. Attachment of a substituent on one of the oxygen atoms of orthophosphate can substantially lengthen that P–O bond, by up to 0.15 Å, but the other P–O bonds contract so as to maintain the average distance, 1.54 Å. The effect is illustrated by the structure of serine phosphate[80] shown in Fig. 4. This structure also indicates that polarization by H$^+$, as reflected in P–O bond lengthening, is only half as great as for R$^+$. This is a general effect, and is probably ascribable to strong intermolecular hydrogen bonds found in crystals[79]. Presumably it would therefore also operate in aqueous solution and at protonic sites of proteins. In metal salts of phosphate[81,82] and pyrophosphate[83], there is no evidence for significant lengthening of P–O bonds by metal coordination, but all the crystals examined have been three-dimensional ionic arrays in which all oxygens are connected to metal ions and *vice versa*, and polarization effects therefore cancel. Apparently there has not yet been a structure determination for a discrete metal–phosphate complex.

While complex formation of simple phosphate derivatives by metal ions does not generally lead to rapid hydrolysis, this is not the case for polyribonucleotides. In these polymers the phosphodiester linkages are rapidly attacked by several metal ions, Zn^{2+} being especially effective, leading to formation of monoribonucleotides. (These are not hydrolyzed further to orthophosphate)[83b]. In a phosphodiester group the phosphorus is already polarized by two ester linkages, and it may be the case that the extra polarization produced by metal ion binding can no longer be compensated by electron drift from the oxygen atoms (see discussion in Chapter 34).

Activation of water

While it is unlikely that metal ions can activate phosphate by polarization, the chances are much better that they can activate water as a nucleophile. The oxygen of water has no electron-rich substituents on which it can draw to counterbalance the polarizing effect of a metal ion, with the consequence that O–H bonds of a coordinated water molecule are substantially weakened and susceptible to ionization. A metal ion can produce a hydroxide ion, or an incipient hydroxide ion, at physiological pH. While the bound hydroxide ion is unlikely to be as good a nucleophile as free hydroxide, it may nevertheless be a better nucleophile than water. Furthermore the bound hydroxide may experience considerable enhancement of its activity through being suitably oriented for reaction. This effect is invoked to explain the facility with which the species (AA)CuOH$^+$ (AA is a chelating diamine) promotes the hydrolysis of isopropyl methylphosphonofluoridate (Sarin)[84]. Presumably the copper can coordinate the fluorine of the substituent while the hydroxide ion attacks the phosphorus (Fig. 5). For some chelates bound hydroxide was found to be more active than free hydroxide[84]. Coordinated hydroxide is no doubt also responsible for the rapid hydrolysis of phosphate derivatives in the presence of lanthanide hydroxides. The effectiveness of this catalysis has been known for some time but the formation of precipitates complicates the elucidation of its mechanism[85,86]. Selwyn's data[87], however, suggest that in the case

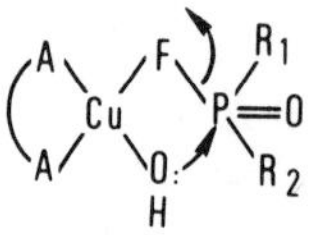

Fig. 5. Mechanism proposed[22] for the hydrolysis of Sarin (R_1 = isopropyl, R_2 = methyl) promoted by (AA)CuOH$^+$ (AA = chelating diamines).

of ATP, a soluble ATP(M^{3+})$_2$ complex is responsible for hydrolysis, and the pH dependence suggests that coordination of hydroxide is required.

There seems to be no direct proof that the particular water molecule, or hydroxide ion, which is coordinated to a metal ion ends up ultimately attached to phosphorus. However Buckingham et al.[88] have shown directly that a coordinated ammonia molecule can act as a nucleophile. When ethyl glycinate is attached through its nitrogen end to (NH$_3$)$_5$Co^{3+}, the product of its hydrolysis in NaOH is the chelated imide, NH$_2$CH$_2$CH$_2$CONH. Attack on the carbonyl carbon by one of the coordinated ammonia molecules is clearly required. The presumed mechanism is shown in Fig. 6.

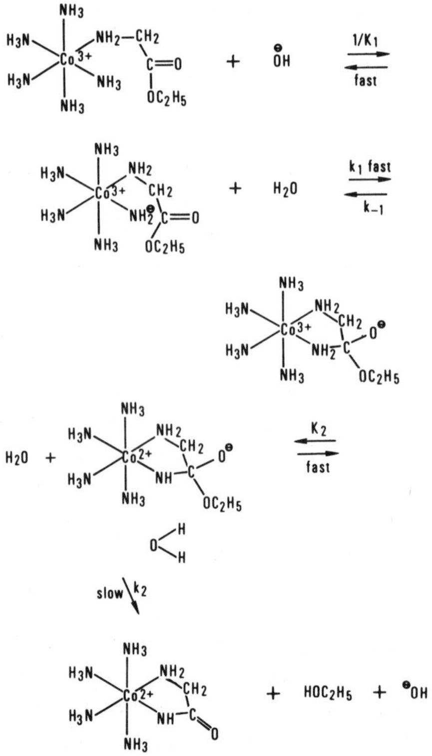

Fig. 6. Mechanism proposed for the formation of chelated glycine imide from ethyl glycinate bound to pentammine cobalt(III) (from ref. 88)

There seems little reason that coordinated water could not play a similar role in hydrolytic reactions.

In enzymatic reactions, the strongest evidence for a coordinated hydroxide mechanism exists for carbonic anhydrase. Infra-red evidence suggests that the substrate, CO_2, is not bound to the zinc ion, but is close enough that it can be attacked by a coordinated hydroxide ion[64] (see discussion on p. 539 ff, Chapter 16). The dependence of activity on pH and on a variety of inhibitor effects are consistent with this mechanism[89,90]. Selwyn[87] has suggested a similar mechanism for ATPase from mitochondria (the enzyme involved in oxidative phosphorylation) on the basis of a similarity in pH dependence with the lanthanide ion promotion of ATP hydrolysis, mentioned above. However, the transition pK of the ATPase is about 7, whereas the pK_a values of the divalent ions which activate the enzyme are much higher. The three most effective activators are Mg^{2+}, Co^{2+} and Zn^{2+} whose pK_a values are ~ 12, ~ 9 and ~ 9 respectively[69]. Selwyn[87] suggests that their acidity could be increased through proximity of positively charged groups on the enzyme, but the 5 pH unit discrepancy for Mg^{2+} seems much too large for such an effect (nearly 7 kcal/mol of electrostatic energy would be required). As mentioned above, bound water ionization has been suggested for a pH 8.5 transition of zinc alkaline phosphatase, and it is associated with a decrease in enzyme activity. It appears unlikely, therefore, that the coordinated hydroxide mechanism plays a significant role in enzymatic phosphate hydrolyses.

Stereochemical models

The fact that metal ions in binary complexes with phosphate derivatives ordinarily demonstrate only slight promotion of hydrolysis implies that the disposition of the metal ion and the phosphate group to be activated is qualitatively different in the binary complex than in the ternary complex of a phosphate transfer enzyme. Conceivably the metal ion plays only an indirect role, in assuring proper orientation of enzyme and substrate. Alternatively the stereochemistry of the metal–substrate complex may alter when the complex is bound to the enzyme, and become more conducive to reaction. If further insight is to come from model studies on enzyme-free metal–phosphate systems, they should probably be designed to test specific stereochemical hypotheses.

Recently Cooperman[91] put forth an attractive mechanistic proposal, and presented model studies which, however, appear to disprove the hypothesis. Noting that an unsymmetrical diester of pyrophosphoric acid has been found to hydrolyze a million times faster than the corresponding monoester[92], Cooperman suggested that a metal ion bound to the substituted phosphate of a pyrophosphate monoester might have a similar effect. Ordinarily metal ions form chelate complexes with polyphosphates, but at an enzyme active site the metal ion might be induced to bind only to the penultimate phosphate group of, for example, ATP, thus activating the terminal phosphate group by the proposed mechanism. Several substituted phosphonyl phosphates or pyrophosphates were prepared, in which the substituent was itself a polyamine or phenol–diamine chelating group, designed to permit binding of the metal to the penultimate but not to the ultimate phosphate group. Hydrolysis rate constants were determined for complexes with a variety of divalent metal ions, but there were no significant enhancements over the rate constants for the free ligands. For at least one of the complexes, with Zn^{2+}, ^{31}P n.m.r. measurements indicated that the desired structure was in fact achieved. Consequently the penultimate phosphate binding mechanism is apparently eliminated.

Another contender, which has some experimental support although definitive proof is lacking, is the bidentate activation mechanism, proposed by Farrell et al.[93]. This model is also based on an analogy with a rapidly hydrolyzing phosphate ester, namely the five-membered ring cyclic ester, ethylene phosphate. It has long been known that ethylene phosphate hydrolyzes much more rapidly than non-cyclic diesters, and indeed the five-membered ring structure accounts for the lability of the cyclic nucleotide intermediate formed by the action of pancreatic ribonuclease[13]. Moreover ethylene phosphate undergoes rapid oxygen exchange with water[94], and the hydrolysis of ethylene methyl phosphate external to the ring (i.e. to produce methanol and ethylene phosphate) is some 10^6 times as fast as the hydrolysis of trimethylphosphate[95,96]. These remarkable characteristics have received an elegant explanation by Westheimer[15], who has proposed that the exchange and hydrolysis reactions proceed via a five-coordinate trigonal bipyramidal intermediate (Fig. 7). The ethylene group bridges an axial and an equatorial oxygen atom of the trigonal bipyramid because the 90° angle at the phosphorus atom is readily accommodated by the five-membered ring, whereas the 120° angle between two equatorial

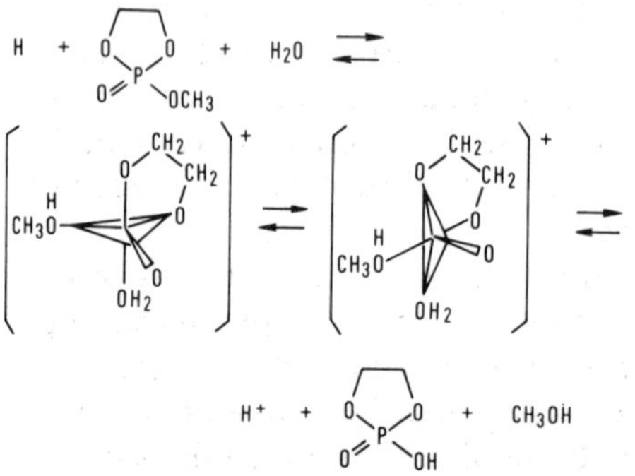

Fig. 7. Pseudorotation mechanism proposed to account for the rapid hydrolysis of ethylene methyl phosphate to give ethylene phosphate and methanol (from ref. 15).

oxygen atoms is not. In fact the ring is known to be strained in ethylene phosphate, the O–P–O angle bridged by ethylene being 10° less than the normal tetrahedral angle[97]. The activation energy required to form the postulated trigonal bipyramidal intermediate is reduced by relief of the ring strain. The axial positions of the intermediate are expected to be more labile than the equatorial positions, and the phenomenon of facile P–O bond breaking *external* as well as internal to the ring is explained by a rapid intramolecular shift which interconverts axial and equatorial position (Fig. 7). This shift, which is an extension of one of the normal bending vibrational modes of the intermediate, has been called "pseudorotation," because its effect is to rotate the equatorial plane of the trigonal bipyramid by 90°.

Farrell et al.[31] suggested that a metal ion might produce a similar activation if it bridges two oxygen atoms attached to a single phosphorus atom, *i.e.* if the phosphate group to be activated acts as a bidentate ligand. The result would be a four-membered ring and while metal–oxygen bonds are less directional and more flexible than are ethylene–oxygen bonds, there may nevertheless be an appreciable constraint on the O–P–O angle. If so then a trigonal bipyramidal intermediate would again be readily accessible and pseudorotation could induce rapid exchange of phosphoryl acceptors.

Bidentate coordination might not be the preferred mode of binding in binary complexes of metal ions with phosphate derivatives. Indeed if ring strain is significant in the bidentate structure then it might be expected to be unstable relative to a monodentate structure. Activation in the ternary enzyme–metal–substrate complex might result in part from a shift to bidentate coordination of the phosphate group by steric constraints at the active site. In the case of ATP, bound metal ions are known to be chelated by oxygen atoms from different phosphate groups through six-membered rings[98,99]. Bidentate activation of phosphoryl transfer from ATP would require that the metal ion be bound to two oxygen atoms on the terminal phosphate. Similarly bidentate activation of nucleotidyl transfer (synthetases) would require coordination by two oxygens on the α-phosphate group. Either structure would be ordinarily highly unfavorable relative to the six-membered ring structures available, but they might be stabilized by the appropriate enzyme.

In an attempt to determine whether bidentate coordination does in fact activate phosphate derivatives, Farrell et al.[93] examined the hydrolysis rate of methyl phosphate when it is attached to cobalt(III) complexes having two and one coordination sites available for binding. Cobalt(III) was chosen as a coordination center in hopes that the inertness of its coordinate bonds would permit a clear distinction between mono- and bidentate structures. The experimental finding was that $CH_3PO_4^{2-}$ hydrolyzes about 130 times faster when bound to $(trien)Co^{3+}$ (trien = triethylenetetraamine, a tetradentate ligand, which leaves two of the Co(III) coordination sites available) than when bound to $H^+(CH_3PO_4H^-$ is the kinetically dominant species in the hydrolysis of aqueous methyl phosphate[100]). In contrast, the hydrolysis was at least two orders of magnitude slower when $CH_3PO_4^{2-}$ was bound to $(NH_3)_5Co^{3+}$ (one coordination site available). The marked reluctance of methyl phosphate to hydrolyze when attached by one oxygen to the tripositive $(NH_3)_5Co^{3+}$ ion is further evidence against an important role for the polarization effect.

Unfortunately the relatively rapid hydrolysis rate for the $(trien)Co^{3+}$ complex is difficult to interpret because, contrary to expectation, the methyl phosphate was found to be predominantly bound through one oxygen rather than two, with the sixth coordination position on Co(III) occupied by water. Lincoln and Stranks[101] have shown that for orthophosphate bound to the analogous $(en)_2Co^{3+}$ (en = ethylenediamine, two coordination sites available) there is a fairly rapid equilibrium (time scale of minutes at room temperature) between mono- and bi-dentate structures, with bidentate coordination dominant for PO_4^{3-} but monodentate coordination dominant for HPO_4^{2-}. Evidently $CH_3PO_4^{2-}$ is similar to HPO_4^{2-} in this regard. Consequently if bidentate activation is responsible for the enhanced hydrolysis of $CH_3PO_4^{2-}$ when bound to $(trien)Co^{3+}$, then the effect is actually much greater than is apparent, since only a small fraction of the complex involves bidentate coordination. However, there is also the possibility that the observed enhancement results from nucleophilic attack by bound water in the monodentate complex. Thus the enhancement could be attributed to the water activation mechanism discussed in the previous section. The two possibilities are illustrated in Fig. 8.

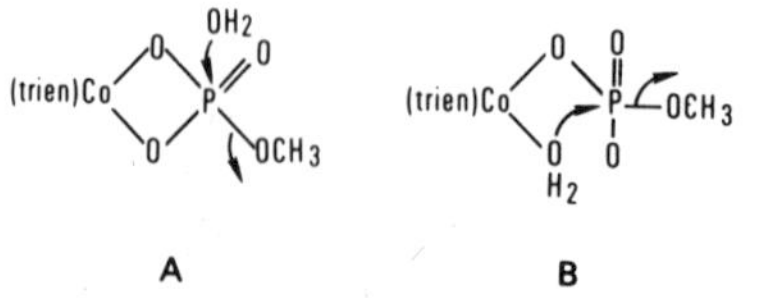

Fig. 8. Alternative mechanisms for the hydrolysis of methyl phosphate promoted by triethylenetetraammine cobalt(III). (A) bidentate activation of phosphate; (B) activation of bound water (from ref. 93).

There is no direct evidence for or against the bidentate hypothesis in enzymatic reactions, but Mildvan[61] has pointed out that the relatively slow rate of fluorophosphate binding to the pyruvate kinase–Mn^{2+} complex[102] might arise from the slow formation of a strained bidentate complex. (See note added in proof (2), p. 577.)

ACKNOWLEDGEMENTS

The author is indebted to Drs. Joseph E. Coleman, Barry Cooperman, Albert S. Mildvan, Bert L. Vallee, Irwin B. Wilson for conversations which were most helpful in preparing this chapter.

NOTE ADDED IN PROOF

(1) Lazdunski et al.[103,104] find that the phosphorylation of alkaline phosphatase at acid pH also exhibits anticooperativity, i.e. a plot of the number of moles of phosphate covalently bound per mole of enzyme against phosphate concentration shows a distinct break after one mole of phosphate. The extent of anticooperativity, as measured by the size of the break, varies with pH and temperature and with the nature of the metal ion at the active site. In all cases the enzyme dimer readily binds one phosphate ion covalently and can be forced to bind a second one at sufficiently high phosphate concentration. At alkaline pH the phosphoryl enzyme is unstable with respect to the non-covalent phosphate complex, except for the inactive Cd^{2+} enzyme. For the active Zn^{2+} and Co^{2+} enzymes, however, the phosphorylated intermediate can be trapped after reaction with substrates at low temperature. A maximum of one mole of covalently bound phosphate per mole of enzyme is found under these conditions. When the monophosphorylated Zn^{2+} enzyme is allowed to react with a competitive inhibitor, p-chloroanilidophosphonate, the bound phosphate is readily released. Lazdunski et al.[104] propose a "flip-flop" mechanism for the enzyme. Its central feature is the suggestion that under normal conditions only one of the enzyme subunits at a time can be phosphorylated. The other subunit is poised to bind a substrate molecule. When it does so, it undergoes phosphorylation and simultaneously its partner releases its covalently bound phosphate and is in turn ready to bind another substrate molecule. Thus the two halves of the enzyme alternate between phosphorylated and non-phosphorylated states, which are presumably coupled through appropriate conformational changes.

(2) Cohn and coworkers[105,106] interpret n.m.r. relaxation data as showing that in creatine kinase, the divalent metal ion activator is *not* bound to the phosphoryl group being transferred from creatine to ADP, but rather to the two phosphate groups of ADP. This would support a metaphosphate elimination mechanism for the reaction, in line with evidence that the non-enzymatic hydrolysis of creatine phosphate also proceeds via metaphosphate elimination[107], creatine being an especially good leaving group. For this enzyme, therefore, the bidentate activation mechanism appears to be ruled out. It may still be viable, however, for enzymes catalyzing phosphoryl transfer between poorer leaving groups and for which an associative mechanism may be more likely[31].

References pp. 578–581

REFERENCES

1 F. Lippman, *Adv. Enzymol.*, 1 (1941) 99.
2 N. O. Kaplan and E. P. Kennedy, (Eds.) *Current Aspects of Biochemical Energetics*, Academic Press, New York, 1966.
3 H. M. Kalckar, *Biological Phosphorylations*, Prentice-Hall, Englewood Cliffs, N.J. 1969.
4a M. R. Atkinson and R. K. Morton, in M. Florkin and H. S. Mason, *Comparative Biochemistry*, Vol. II, Academic Press, New York, 1960, p. 1.
4b R. K. Morton, in M. Florkin and E. H. Stotz, *Comprehensive Biochemistry*, Vol. 16, Elsevier, New York, 1965.
5 I. H. Segel, in R. J. Williams and E. M. Lansford, Jr., *The Encyclopedia of Biochemistry*, Reinhold, New York, 1967, p. 642.
6 R. M. Bock, in P. Boyer, H. Lardy and K. Myrbäck, *The Enzymes*, Vol. 2, Academic Press, New York, pp. 15, 16.
7 A. Lehninger, *Bioenergetics*, W. A. Benjamin, New York, 1965.
8 E. Racker, *Mechanisms in Bioenergetics*, Academic Press, New York, 1965.
9 A. Kornberg, in M. Kasha and B. Pullman, *Horizons in Biochemistry*, Academic Press, New York, 1962, p. 251.
10 M. Cohn, in P. Boyer, H. Lardy and K. Myrbäck, *The Enzymes*, Vol. 5, Academic Press, New York, 1961, p. 179.
11 M. Grunberg-Manago, *ibid.*, Vol. 5, p. 257.
12 M. Cohn, *J. Biol. Chem.*, 180 (1949) 771.
13 T. C. Bruice and S. Benkovic, *Bioorganic Mechanisms*, Vol. II, W. A. Benjamin, New York, 1966, Ch. 5.
14 L. D. Quin, in J. Hamer, *1,4-Cycloaddition Reactions*, Academic Press, New York, 1967, pp. 47–96.
15 F. H. Westheimer, *Accounts Chem. Res.*, 1 (1968) 70.
16a G. DiSabato and W. P. Jencks, *J. Am. Chem. Soc.*, 83 (1961) 4400.
16b A. J. Kirby and A. G. Varvoglis, *J. Am. Chem. Soc.*, 89 (1967) 415.
17 G. Schmidt in, P. Boyer, H. Lardy and K. Myrbäck, *The Enzymes*, Vol. 5, Academic Press, New York, 1961, p. 37.
18 T. C. Stadtman, *ibid.*, Vol. 5, p. 55.
19 W. L. Byrne, *ibid.*, Vol. 5, p. 73.
20 L. A. Heppel, *ibid.*, Vol. 5, p. 49.
21 M. Kunitz and P. W. Robbins, *ibid.*, Vol. 5, p. 169.
22 C. F. Cori and D. H. Brown, in M. Florkin and E. H. Stotz, *Comprehensive Biochemistry*, Vol. 15, Elsevier, New York, 1964, p. 212.
23 E. A. Barnard, *A. Rev. Biochem.*, 38 (1969) 677.
24 H. R. Revel and E. Racker, *Biochim. Biophys. Acta*, 43 (1960) 465.
25a A. W. Hanson, M. L. Applebury, J. E. Coleman, H. W. Wyckoff and F. M. Richards, *J. Biol. Chem.*, 245 (1970) 4975.
25b T. W. Reid and I. B. Wilson, in P. D. Boyer, *et al.*, *The Enzymes*, 3rd edn., Academic Press, New York, in press.
26a T. Horiuchi, S. Horiuchi and D. Mizuno, *Nature*, 183 (1959) 1528.
26b A. Torriani, *Biochim. Biophys. Acta*, 38 (1960) 460.
26c A. Garen and C. Levinthal, *Biochim. Biophys. Acta*, 38 (1960) 470.
27 T. C. Stadtman in P. D. Boyer *et al.*, *The Enzymes*, Vol. 5, Academic Press, New York, 1961, p. 55.
28 L. A. Heppel, D. R. Harkness and R. J. Hilmoe, *J. Biol. Chem.*, 237 (1962) 841.
29 H. Neumann, L. Boross and E. Katchalski, *J. Biol. Chem.*, 242 (1967) 3142.

30 H. Neumann, *J. Biol. Chem.*, 243 (1968) 4671.

31 R. Breslow and I. Katz, *J. Am. Chem. Soc.*, 90 (1968) 7376.

32 I. B. Wilson, J. Dayan and K. Cyr, *J. Biol. Chem.*, 239 (1964) 4182.

33a A. Garen and C. Levinthal, *Biochim. Biophys. Acta*, 38 (1960) 470.

33b R. T. Simpson, B. L. Vallee and G. H. Tait, *Biochemistry*, 7 (1968) 4336.

34 M. L. Applebury and J. E. Coleman, *J. Biol. Chem.*, 244 (1969) 308.

35a C. Levinthal, E. R. Signer and K. Fetherolf, *Proc. Natn. Acad. Sci. U.S.*, 48 (1962) 1230.

35b F. Rothman and R. Byrne, *J. Mol. Biol.*, 6 (1963) 330.

35c M. J. Schlesinger and K. Barret, *J. Biol. Chem.*, 240 (1964) 4284.

36 J. A. Reynolds and M. J. Schlesinger, *Biochemistry*, 8 (1969) 4287.

37 S. R. Cohen and I. B. Wilson, *Biochemistry*, 5 (1966) 904.

38 H. Csopak, *Eur. J. Biochem.*, 7 (1969) 186.

39 J. A. Reynolds and M. J. Schlesinger, *Biochemistry*, 6 (1967) 3552; 7 (1968) 2080.

40 R. T. Simpson and B. L. Vallee, *Biochemistry*, 7 (1968) 4343.

41a C. Lazdunski, C. Petitclerc and M. Lazdunski, *Eur. J. Biochem.*, 8 (1969) 510.

41b H. Csopak and K. E. Falks, *FEBS Lett.*, 7 (1970) 147.

42 M. L. Applebury, B. P. Johnson and J. E. Coleman, *J. Biol. Chem.*, 245 (1970) 4968.

43 M. Gottesman, R. T. Simpson and B. L. Vallee, *Biochemistry*, 8 (1969) 3776.

43a G. H. Tait and B. L. Vallee, *Proc. Natn. Acad. Sci. U.S.*, 56 (1966) 1247.

44 C. Lazdunski and M. Lazdunski, *Eur. J. Biochem.*, 7 (1969) 294.

45 M. L. Applebury and J. E. Coleman, *J. Biol. Chem.*, 244 (1969) 709.

46a Z. Dori and H. B. Gray, *J. Am. Chem. Soc.*, 88 (1966) 1394.

46b M. Ciampolini and N. Nardi, *Inorg. Chem.*, 6 (1967) 445.

47 G. H. Tait and B. L. Vallee, *Proc. Natn. Acad. Sci. U.S.*, 56 (1966) 1247.

48 J. A. Reynolds and M. J. Schlesinger, *Biochemistry*, 8 (1969) 589.

49 L. Engstrom and G. Agren, *Acta Chem. Scand.*, 12 (1958) 357.

50 J. H. Schwartz and F. Lippman, *Proc. Natn. Acad. Sci. U.S.*, 47 (1962) 1996.

51 I. B. Wilson and J. Dayan, *Biochemistry*, 4 (1965) 645.

52a D. Levine, T. W. Reid and I. B. Wilson, *Biochemistry*, 8 (1969) 2374.

52b T. W. Reid, M. Pavlic, D. J. Sullivan and I. B. Wilson, *Biochemistry*, 8 (1969) 3184.

53a W. N. Aldridge, T. E. Barman and H. Gutfreund, *Biochem. J.*, 92 (1964) 23c.

53b H. N. Fernley and P. G. Walker, *Biochem. J.*, 10 (1968) 11p.

53c H. N. Fernley and P. G. Walker, *Nature*, 212 (1966) 1435.

54a D. R. Trentham and H. Gutfreund, *Biochem. J.*, 106 (1966) 455.

54b H. Gutfreund, *Biochem. J.*, 110 (1968) 11p.

55 S. E. Halford, N. G. Bennett, D. R. Trentham and H. Gutfreund, *Biochem. J.*, 114 (1969) 243.

56 C. Lazdunski, C. Petitclerc, D. Chappelet and M. Lazdunski, *Biochem. Biophys. Res. Commun.*, 37 (1969) 744.

57a S. H. D. Ko and F. J. Kezdy, *J. Am. Chem. Soc.*, 89 (1967) 7139.

57b W. K. Fife, *Biochim. Biophys. Res. Commun.*, 28 (1967) 309.

58a L. Engstrom, *Biochim. Biophys. Acta*, 56 (1962) 606; *Arkiv. Kemi*, 19 (1962) 129.

58b M. M. Pigretti and C. Milstein, *Biochem. J.*, 94 (1968) 106.

59a R. T. Simpson and B. L. Vallee, *Biochemistry*, 9 (1970) 953.

59b A. Levitzki and E. Koshland, Jr., *Proc. Natn. Acad. Sci. U.S.*, 62 (1969) 1121.

60 W. N. Lipscomb, J. A. Hartsuck, F. A. Quiocho and G. N. Reeke, Jr., *Proc. Natn. Acad. Sci. U.S.*, 64 (1969) 28.

61 A. S. Mildvan, in P. D. Boyer *et al.*, *The Enzymes*, 3rd edn., Vol. I, Academic Press, New York, 1970.

62 L. Hellerman, *Physiol. Rev.*, 17 (1937) 454.

63 A. S. Mildvan, M. Cohn and J. S. Leigh, Jr., *Biochemistry*, 6 (1967) 1805.

64 M. C. Riepe and J. H. Wang, *J. Biol. Chem.*, 243 (1968) 2779.

65 M. Cohn and J. S. Leigh, Jr., *Nature*, 193 (1962) 1037.

66 P. Cuatrecasas, S. Fuchs and C. B. Anfinsen, *J. Biol. Chem.*, 242 (1967) 3063.

67 P. Cuatrecasas, H. Tanuichi and C. B. Anfinsen, *Brookhaven Symp. Biol.*, 21 (1968) 172.

68 A. Arnone, C. J. Bier, F. A. Cotton, E. E. Hazen, Jr., D. C. Richardson and J. S. Richardson, *Proc. Natn. Acad. Sci. U.S.*, 64 (1969) 420.

69 L. G. Sillén and A. E. Martell, *Stability Constants of Metal Complexes*, The Chemical Society, London, 1964.

70 T. Lindahl, A. Adams and J. R. Fresco, *Proc. Natn. Acad. Sci. U.S.*, 55 (1966) 941.

71 B. Cooperman and G. J. Lloyd, *J. Am. Chem. Soc.*, 93 (1971) 4883.

72 R. Breslow and D. Chipman, *J. Am. Chem. Soc.*, 87 (1965) 4195.

73 J. M. Lowenstein, *Biochem. J.*, 70 (1958) 222.

74 M. Tetas and J. Lowenstein, *Biochemistry*, 2 (1963) 350.

75 M. L. Bender in *Reactions of Coordinated Ligands and Homogeneous Catalysis*, *Adv. Chem. Ser.*, 37 (1963) 19–35.

76 D. E. Koshland, *J. Am. Chem. Soc.*, 74 (1952) 2286.

77a T. G. Spiro, W. A. Kjellstrom, M. C. Zydel and R. A. Butow, *Biochemistry*, 7 (1968) 859.

77b R. Hofstetter, Y. Murakami, G. Mont and A. E. Martell, *J. Am. Chem. Soc.*, 84 (1962) 3041.

77c E. Bamann and W. D. Mutterlein, *Chem. Ber.*, 91 (1958) 471, 1322 and other references cited therein to the work of Bamann and his associates.

77d K. Dimroth, H. Witzel, W. Hülsen and H. Mirbach, *Ann.*, 620 (1959) 94.

77e W. W. Butcher and F. Westheimer, *J. Am. Chem. Soc.*, 77 (1955) 2420.

78 H. Brintziner and R. E. Hester, *Inorg. Chem.*, 5 (1966) 980.

79 D. W. J. Cruickshank, *J. Chem. Soc.*, (1961) 5486.

80 G. H. McCallum, J. M. Robertson and G. A. Sim, *Nature*, 184 (1959) 1863.

81 *Tables of Interatomic Distances and Configuration in Molecules and Ions*, Special Publication No. 11, The Chemical Society, London 1958.

82 S. Geller and J. L. Durand, *Acta Cryst.*, 13 (1960) 325.

83 C. Calvo, *Inorg. Chem.*, 7 (1968) 1345.

83b G. L. Eichhorn, in S. Kirschner, *J. C. Bailar Symposium on Coordination Chemistry*, Plenum Press, New York, 1969, p. 92.

83c J. J. Butzow and G. L. Eichhorn, *Biopolymers*, 3 (1965) 95.

84 R. L. Gustafson and A. E. Martell, *J. Am. Chem. Soc.*, 84 (1962) 2309.

85a E. Bamann, F. Fischler and H. Trapmann, *Biochem. Z.*, 325 (1954) 413.

85b E. Bamann and H. Trapmann, *Adv. Enzymol.*, 21 (1959) 169.

86 W. W. Butcher and F. H. Westheimer, *J. Am. Chem. Soc.*, 77 (1955) 2420.

87 M. J. Selwyn, *Nature*, 219 (1968) 490.

88 D. A. Buckingham, D. M. Foster and A. M. Sargeson, *J. Am. Chem. Soc.*, 91 (1969) 3451.

89 R. P. Davis, in P. D. Boyer, H. Lardy and K. Myrbäck, *The Enzymes*, Vol. 5, 2nd edn., Academic Press, New York, 1961, p. 545.

90 J. T. Edsall, *Harvey Lect. Ser.*, 62 (1967) 191.

91 B. Cooperman, *Biochemistry*, 8 (1969) 5005.

92 D. L. Miller and T. Ukena, *J. Am. Chem. Soc.*, 91 (1969) 3050.

93 F. J. Farrell, W. Kjellstrom and T. G. Spiro, *Science*, 164 (1969) 320.

94 P. C. Haake and F. H. Westheimer, *J. Am. Chem. Soc.*, 83 (1961) 1102.
95 F. Covitz and F. H. Westheimer, *J. Am. Chem. Soc.*, 85 (1963) 1773.
96 E. A. Dennis and F. H. Westheimer, *J. Am. Chem. Soc.*, 88 (1966) 3432.
97 T. A. Steitz and W. N. Lipscomb, *J. Am. Chem. Soc.*, 87 (1965) 2488.
98 M. Cohn and T. R. Hughes, Jr., *J. Biol. Chem.*, 235 (1960) 3250; 237 (1962) 176.
99 H. Sternlicht, R. G. Shulman and E. W. Anderson, *J. Chem. Phys.*, 43 (1965) 3123.
100 C. A. Bunton, D. R. Llewellyn, K. G. Oldham and C. A. Vernon, *J. Chem. Soc.*, (1958) 3574.
101 F. Lincoln and D. R. Stranks, *Aust. J. Chem.*, 21 (1968) 57.
102 A. S. Mildvan, J. S. Leigh, Jr. and M. Cohn, *Biochemistry*, 6 (1967) 1805.
103 C. Lazdunski, D. Chappelet, C. Petitclerc, F. Leterrier, P. Douzou and M. Lazdunski, *Eur. J. Biochem.*, 17 (1970) 239.
104 M. Lazdunski, C. Petitclerc, D. Chappelet and C. Lazdunski, *Eur. J. Biochem.*, 20 (1971) 124.
105 M. Cohn and J. Reuben, *Accounts Chem. Res.*, 4 (1971) 215.
106 J. S. Leigh, Jr., *Ph. D. Dissertation*, University of Pennsylvania, 1971.
107 P. Haake and G. W. Allen, *Proc. Natn. Acad. Sci. U.S.*, 68 (1971) 2691.

Chapter 18

KINASES

W. J. O'SULLIVAN

Department of Medicine, University of Sydney, Sydney, N.S.W., (Australia)

I. INTRODUCTION

The kinases comprise a large group of enzymes that catalyze the transfer of the terminal phosphoryl group of ATP to an acceptor molecule, according to the general equation:

$$ATP^{4-} + HX \underset{}{\overset{M^{2+}}{\rightleftharpoons}} ADP^{3-} + PX^{2-} + H^+ \tag{1}$$

The reader is referred to the review by Nordlie and Lardy[1] for a discussion of the types of acceptor molecule involved, to other reviews in Vol. 6 of *The Enzymes*[2] for discussions of many individual kinases and to the Report of the Enzyme Commission[3] for a compilation of individual enzymes in this group. For this chapter, an attempt shall be made to emphasize a few generalizations with respect to kinases and to illustrate with examples from the few that have been studied in greatest detail.

From the point of view of the anticipated audience of this book, the feature of kinase reactions of particular interest is their absolute requirement for a divalent metal ion. This requirement is most commonly met by Mg^{2+}, probably the usual physiological activator. However, Mn^{2+} can usually substitute for Mg^{2+} and other divalent metal ions, such as Co^{2+} and possibly Zn^{2+} can also act[2]. Ca^{2+} is an activator of many kinase reactions but an inhibitor of others. In fact, the distinction with respect to the action of Ca^{2+} appears to correspond to a useful basis for one classification of kinase reactions, most convincingly demonstrated by the magnetic resonance experiments initiated by Cohn[4,5]. For those enzymes that can utilize Ca^{2+} (Type I), it appears that the metal ion does not act directly with the enzyme molecule but rather through the substrate (*i.e.* E–S–M) whereas if Ca^{2+} is an inhibitor, there is probably direct interaction between metal and protein (E–M–S) (Type II) (see Chapter 14).

The requirement by these enzymes for a divalent metal ion and the appreciation of interactions between metal ion and substrate and/or protein has led to the application of a large number of physico-chemical techniques to the study of the type of complexes formed in kinase type reactions. This has included the techniques of the coordination chemist, relatively sophisticated kinetic experiments and temperature jump relaxation methods besides the more orthodox equilibrium techniques. But it has been the application of magnetic resonance techniques which, by taking advantage of the paramagnetic properties of the metal ion, Mn^{2+}, has yielded most insight into the nature of the enzymic reaction in recent years[4,5].

This chapter will principally be concerned with kinase reactions from the point of view of the role of the essential metal ion. The emphasis will be on general approaches that should, in principle, be applicable to all kinase reactions, although a few enzymes will be treated in more detail.

A few generalizations may be made at this stage:

1. The activating metal ion forms moderately strong complexes with the nucleotide substrates. At the relevant pH values, between 7 to 9, where the optima of most kinase reactions occur, the dominant species will be $MATP^{2-}$, $MADP^-$, etc. Most kinetic analyses indicate that the metal complexes are the "true" substrates of the reactions, although such a statement cannot necessarily be extended to postulations about the nature of the intermediate complexes formed between enzyme, metal and substrate[5,6].

2. The most common nucleotide substrates are the adenine nucleotide substrates, ATP and ADP. However, many kinases, for example, creatine kinase[7,8] and pyruvate kinase[9] have a relatively broad specificity and for some enzymes other nucleotides, e.g. GTP in the GTP–adenylate kinase reaction and GTP or ITP in phosphoenol pyruvate carboxykinase[10] may be the natural substrates.

3. Kinase reactions are usually reversible although for a few, e.g. hexokinase, the reaction in the direction of ATP formation may be very slow (cf. refs. 11 and 12, p. 416).

4. The K_m values for the metal nucleotide substrates are generally of the order of 10^{-5} to 10^{-4} M and although they cannot necessarily be equated, the dissociation constants are also of the same order.

5. The range of molecular weights and of amino acid composition is as wide as for any other group of enzymes. However, there does appear to be a relatively high incidence of cysteine residues associated with the active site in these enzymes.

6. The reaction in the forward direction (i.e. transfer of a phosphoryl group from ATP) is usually accompanied by a small negative change in free energy. A comprehensive discussion of the thermodynamics of phosphoryl group transfer reactions has been given by Atkinson and Morton[12]. Much remains to be done in this area as few studies have taken into full consideration the various ionic species present.

2. GENERAL APPROACH TO KINASES

(a) The nature of kinase reactions

Kinase reactions proceed by transfer of the phosphoryl group, $-PO_3^{2-}$, from ATP to form a phosphorylated product and ADP. The reaction appears to involve nucleophilic attack by the acceptor molecule on the terminal phosphorus atom of ATP[13]. Evidence for the breaking of the terminal O–P bond has largely come from studies with ^{18}O by Cohn[14] and Harrison et al.[15] on various kinases. As pointed out by Mildvan[16] it could be related to the fact that if the phosphate chain of ATP has the same dimensions as that of tripolyphosphate, the terminal O–P bond would be slightly longer (1.68 Å) than the inner bonds (1.61 Å), so that the former would cleave more easily. The recently reported X-ray analysis of ATP may confirm this[17] but details were not available at the time of writing.

As a rule, in kinase reactions there is a direct transfer of phosphoryl group from the donor to the acceptor. Crane[11] has summarized the evidence against the formation of an enzyme–phosphoryl intermediate for a number of kinases. However, an exception to this rule, if it is such, is the reaction catalyzed by nucleoside diphosphate kinase, in which the formation of an enzyme–phosphate intermediate has been demonstrated[18,19].

(b) Metal-substrate equilibria

As mentioned above, the activating metal ions form complexes with the nucleotide substrate(s) of kinase reactions. In some cases, complexes can also form with the non-nucleotide substrate but these are generally much weaker and of minor significance. The following discussion refers specifically to the adenine nucleotides, ATP and ADP, as it is these compounds that have been most closely studied. However, it is also relevant to other nucleotides as the nature of the base has only a small effect on the pK_a values for the phosphate groups and on the binding of metal ions to these groups[20,21].

The pK_a values of the terminal phosphate groups of ATP, ADP and related compounds occur in the pH range, 6.7 to 7.0[22-25]. Individual estimates may vary somewhat according to differences in temperature, ionic strength, composition of the medium, etc.[24,25]. Thus, at pH values significantly above 7, the major species in solution are ATP^{4-} and ADP^{3-}, containing 4 and 3 oxyanions, respectively. It was recognized in the early 1950s that Mg^{2+} and other cations (including the monovalent cations, Na^+ and K^+[26]) could form complexes with these species, though it was some time before the magnitude of the respective stability constants was appreciated.

A critical survey of the work in this area up to 1960 was carried out by Bock[27], who pointed out the inadequacies of many of the experiments used to determine the stability constants of metal–nucleotide complexes and included a plea for "studies which are conducted in solutions of known ionic strength with non-chelating buffers and with metal ion and phosphate compounds at concentrations designed to permit precise calculation." Bock's plea has since been largely met and at the time of writing, there appears to be substantial agreement concerning the magnitude of the various constants, *viz.* at pH 8.0, in a non-chelating buffer and in the absence of interfering ions, the stability constant of $MgATP^{2-}$ would be in the range, 50,000 to 70,000 M^{-1} (log K, 4.70 to 4.85) and for $MgADP^-$, 2,000 to 4,000 M^{-1} (log K, 3.30 to 3.60).

Some concept of the difficulties inherent in the analysis of these systems by classical potentiometric techniques can be gained by consideration of the equilibria formed in the presence of ATP and Mg^{2+}:

$$H_2ATP^{2-} \xrightleftharpoons{pK_{a1}} HATP^{3-} \xrightleftharpoons{pK_{a2}} ATP^{4-}$$

$$K_1 \updownarrow \qquad K_2 \updownarrow \qquad K_3 \updownarrow$$

$$MgH_2ATP \rightleftharpoons MgHATP^- \rightleftharpoons MgATP^{2-}$$

A complete analysis, involving the determination of all the K values, is an extremely tedious undertaking, even if reasonable approximations are made. However, such problems have been largely overcome by the use of suitable computer programs[28].

The interested reader is referred to a recent review article by Phillips[25] for a critical appraisal of this field, particularly with respect to the determination of the thermodynamic factors involved in complex formation and also to later papers from this group[29] and from Taqui Khan and Martell[30]. But it might be noted here that the rigorous approach of the physical chemist to the study of metal–nucleotide complexes is not always directly relevant to the biochemist at the bench in his dealings with a particular enzyme. The latter is restricted to finite ionic strengths, relatively narrow pH and temperature ranges and, usually to the use of a particular buffer which, by definition can never be non-interacting with metal ions, although with many such interaction is effectively negligible. Thus, in practice, what is required is an "apparent" stability constant, which is relevant to a particular set of experimental conditions, *i.e.*

$$K_{app} = \frac{[MATP^{2-}]}{[M^{2+}][ATP]_{T'}} \tag{2}$$

where $[ATP]_{T'}$ is the sum of the uncomplexed forms of ATP.

Ideally, this should be determined under the exact conditions of the intended experiments with the enzyme in question. Alternatively, extrapolation can usually be made from reliable data or, by judicious choice of experimental conditions, the discrepancies minimized.

Extensive complications of stability constants of metal–nucleotide complexes are available in the review by Phillips[25] and also in the Chemical Society publication[24] (the relevant sections contain a number of typographical errors) and in *Data for Biochemical Research*[31].

Rates of formation of MADP⁻, MATP²⁻

Rates of formation of $MADP^-$, $MATP^{2-}$

The rate of formation of metal–ADP and metal–ATP complexes has been studied by Hammes and his associates[32,33], who reached the conclusion that the complexes were formed via an S_N1 outer sphere mechanism, involving the transitory formation of an ion pair. Perhaps the most interesting result was that Ca^{2+} underwent ligand exchange some two orders of magnitude faster than the other divalent metal ions, including Mg^{2+} and Mn^{2+}, which were tested. Whether this is related to the fact that Ca^{2+} activates Type I enzymes (E–S–M) but is an inhibitor of Type II (E–M–S) remains to be elucidated.

(c) Structure of metal–nucleotide complexes

(i) In solution

Having established that moderately strong metal–nucleotide complexes are formed in solution, it is reasonable to enquire as to the structure of these complexes. It is clear that if one is restricted to the alkaline earths, then the principal bonding would be expected to be the phosphorus oxyanions. With metals such as Mn^{2+}, bonding to the electron-donating ring nitrogens might be expected to occur also and for metals such as Cu^{2+}, the latter might be the dominating factor.

A major question has been with respect to whether a metal ion complexed by the phosphorus oxyanions also interacts significantly with the ring nitrogens. Though a wide variety of physico-chemical techniques have been applied in an attempt to establish the exact structure of various metal–ATP complexes, this question is not yet resolved. It is considered in detail in Chapter 33.

(ii) On the enzyme

If a degree of uncertainty exists regarding the structure of metal–nucleotide complexes in solution, then the situation regarding the conformation of these complexes in combination with an enzyme must be even less clear. Nearly all the evidence to date is indirect.

Some information has been obtained from electron paramagnetic resonance (e.p.r.) experiments. The e.p.r. spectrum is very sensitive to the field experienced by the metal ion and thus to any change in the ligands bound to it. No changes in the e.p.r. spectrum have been observed on the binding of $MnADP^-$ to creatine kinase[34,35] or of $MnATP^{2-}$ to adenylate kinase[36]. These observations support the concept that the structure of the complexes is the same on the enzyme as in solution but do not prove it. By contrast, a definite change in $MnADP^-$ spectrum occurs on its binding to pyruvate kinase[37], presumably due to the introduction of extra ligands from the enzyme.

Indirect information is also available from enzyme kinetic measurements; experiments with creatine kinase would indicate that M^{2+} binds more strongly to E–ADP than to ADP alone[35,38]. However, such conclusions are a little tenuous as they are dependent upon a model which may not reflect the real situation (see Section 2d).

(d) Kinetic studies on kinases

The basic feature of any enzyme is that it greatly speeds up (to the order of 10^7–10^{12} fold[39]) a reaction which occurs to a small extent or not at all in the absence of the enzyme. Thus it is logical that the study of the kinetics of enzyme-catalyzed reactions should provide a rational basis for our understanding of enzyme reactions. Unfortunately, while essential as a reference for other techniques, so that one can be certain that the study of various enzyme complexes is relevant to the action of the enzyme, enzyme kinetic experiments rarely lead to a precise understanding of the enzymic mechanism.

Studies on kinases have been complicated by the large number of possible intermediates that have to be considered. Not only is one dealing with a two substrate reaction but there is always an activating metal ion which complexes with the nucleotide substrate, sometimes with the enzyme and sometimes also with the second substrate. As a generalization, it appears to be true that kinetic studies on all kinases can be treated as if $MADP^-$, $MATP^{2-}$ were the active substrates of the reaction[6]. However, this may reflect the inability of kinetic mechanisms to distinguish between different pathways and does not necessarily reflect a common enzymic mechanism.

A good example of the evolution of the kinetic approach to a kinase reaction, which has concentrated on the role of the activating metal ion, is afforded by the continuing study on creatine kinase. The early studies, by Ennor and Rosenberg[40] on the one hand and Kuby et al.[41] on the other were in basic conflict with respect to the role of the metal ion. The former workers interpreted their results as indicating the formation of a metal-enzyme complex, which could subsequently react with the free form of the nucleotide substrate; while the latter were convinced that the metal-nucleotide complexes were the real substrates, even though the published stability constants for the complexes at the time were too small to quantitatively account for their results. Two groups pursued this further. From Noda's laboratory[42,43] results were obtained which definitely indicated MgADP$^-$ and MgATP^{2-} as the relevant substrates, provided that much higher values of the respective stability constants were assumed. An attempt to resolve the situation was also made by Morrison *et al.*[44]. Presuming that the second substrate (phosphoryl creatine) was always constant, it could be postulated that the reaction proceeded *via* an enzyme-metal-substrate complex which could be formed in three ways:

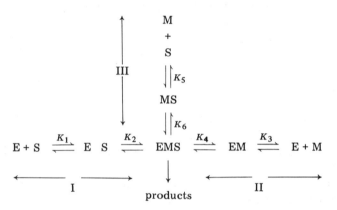

These are effectively equivalent, so that $K_1K_2 = K_3K_4 = K_5K_6$. Provided that K_5, the dissociation constant of the MgADP$^-$ complex was known, experiments could be devised to yield values for all of the constants.

The conclusion reached was that all three pathways could operate, the implication being that the metal ion did indeed interact with the enzyme. But such a conclusion was in conflict with thermodynamic studies[45] and nuclear magnetic resonance (n.m.r.) results[34]. This led to a further study, with a number of modifications[38] the most important of which were the use of a more reliable value of the stability constant of MgADP$^-$ for the experimental conditions and the design of experiments to reveal the nature of the inhibition observed with excess ADP and excess Mg. It was found that ADP^{3-} was a competitive inhibitor with respect to MgADP$^-$ but Mg^{2+} a non-competitive inhibitor with respect to MgADP$^-$ (Fig. 1). In other words, it was unlikely that pathway II participated in the formation of the EMS complex, though the results could not distinguish between pathways I and III. Subsequent studies on this enzyme have confirmed that the reaction can be described as rapid random equilibrium[46] (an assumption that had been implicit in all of the previous studies) and that the action of Ca^{2+} and Mn^{2+} is essentially similar to that of Mg^{2+} [47].

Other kinetic studies on kinases will be referred to in Section 3, with respect to individual enzymes. Here we shall reiterate that while kinetic studies rarely give unequivocal answers, they are a necessary prerequisite to the study of intermediate complexes by other means. Further, it is relevant to note the large amount of information that can be obtained from very small amounts of enzyme, that would be unobtainable by any other means for lack of sensitive enough techniques; a consideration of much significance to the average enzymologist.

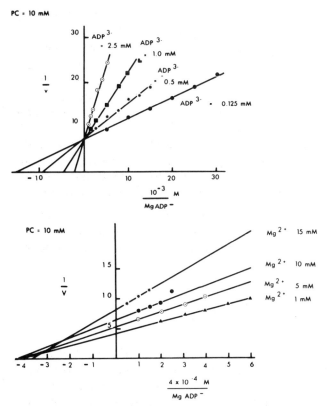

Fig. 1. The inhibition of creatine kinase, by ADP^{3-} competitively with respect to MgADP$^-$ (a), and by Mg^{2+}, non competitively with respect to MgADP$^-$ (b). (Figures are reproduced from ref. 101. Full details of these experiments are given in ref. 38.)

(e) Equilibrium studies of enzyme-substrate complexes

It is in the correlation of kinetically obtained constants for enzyme-substrate complexes with those obtained from established thermodynamic methods for the measurement of these complexes, that a basis for the understanding of the enzyme function can be obtained. In principle a number of tools are now available to the biochemist, including equilibrium dialysis, column chromatography, ultracentrifugation, various optical methods, optical rotatory dispersion *etc.* The reader is referred to general textbooks on physical techniques in biochemistry for a discussion of the various techniques[48].

Such methods have not been widely applied to kinase type enzymes. Exceptions, which serve as examples of the type of information available, are:

1. The extensive studies carried out by Kuby *et al.*[45] on the binding of substrates *etc.* to creatine kinase and adenylate kinase. From equilibrium dialysis and ultracentrifugation studies, evidence favouring weak binding of Mg^{2+} and relatively strong binding of free nucleotide and metal–nucleotide complexes was obtained.
2. Kinetic protection methods have been applied to pyruvate kinase by Mildvan and Leigh[49] and to creatine kinase by O'Sullivan, *et al.*[50] and by Cunningham and his co-workers[51].
3. Comprehensive studies on pyruvate kinase have been carried out by Suelter[52] and his colleagues, with definite evidence from ultraviolet and fluorescence spectroscopy for conformational changes in the enzyme with addition of various substrates and cofactors.

To thermodynamic methods should be added the various magnetic resonance methods. Because of the importance they are assuming in the study of kinases and other enzymes, this is discussed in more detail below.

(f) Application of magnetic resonance techniques

The various forms of magnetic resonance method represents one of the most active areas of the application of physical methods to proteins over the last few years. We shall discuss the use of two of these; electron paramagnetic resonance (e.p.r.) and nuclear magnetic resonance (n.m.r.). Two types of information are available from n.m.r. methods; chemical shifts, which have not as yet been greatly utilized but will probably come into greater prominence with the development of instruments with improved resolution; and the measurement of relaxation rates which may be measured by continuous wave methods or by pulsed techniques. At the time of writing it is the use of pulsed techniques, particularly in the measurement of the proton relaxation rate (p.r.r.) of water, that has contributed most to our understanding of kinase reactions, though it is in the combination of the various techniques that we might expect a major contribution to an understanding of this group of enzymes to develop.

(i) Electron paramagnetic resonance

Species which contain an unpaired electron may give rise to an e.p.r. spectrum. This includes paramagnetic metal ions and, for example, it has proved of much value in the study of metallo-enzymes containing Cu and Mo[53-55]. Application to kinase reactions was initiated by Cohn and Townsend[56], who used it to measure the stability constants of some manganese–substrate complexes, by taking advantage of the fact that the amplitude of the signal from manganese in the complex is much smaller than that for free manganese, as is often the case. Beyond its use in the determination of stability constants and as an adjunct to proton relaxation rate (p.r.r.) measurements[35,57], also to measure free Mn^{2+}, manganese e.p.r. has been rather disappointing. The e.p.r. spectra of complexes are often difficult to observe and even more difficult to interpret. However, it has provided confirmatory evidence for the classification by Cohn of kinases into two groups, Types I (e.g. creatine kinase) and II (e.g. pyruvate kinase) according to whether the metal ion interacts with the enzyme via a substrate bridge or directly, respectively (see Section 2, f(ii) and Chapter 14, pp. 382–386).

The use of e.p.r. has proved of distinct value in one particular case; the labelling of the reactive active site sulfhydryl group of creatine kinase with a nitroxide free radical[58]. With only one unpaired electron, of spin $\frac{1}{2}$, the e.p.r. spectrum of the radical is much simpler and thus more easily interpretable, than that for manganese. A further advantage in this particular case is that even though the enzyme has been inactivated by the introduction of the group at the sulfhydryl moiety, the integrity of the nucleotide binding site is unchanged, so that changes in the e.p.r. signal with added free nucleotide or the complexes with diamagnetic salts can be observed. Such an approach, with the increasing synthesis of new spin labels, obviously has great potential.

(ii) Nuclear magnetic resonance

(a) Measurement of the chemical shift. This is the more familiar type of magnetic resonance measurement, using c.w. methods, which allowed the detection of a spectrum of the nuclei being observed. Its most common application has been to proton spectra though both ^{31}P and ^{15}N spectra have proved useful in studies on metal–nucleotide complexes (Chapter 33).

In principle, continuous wave n.m.r. should be the most informative of all the magnetic resonance techniques, as it can provide information on the electronic structure at the active site and on the conformation of enzyme–substrate complexes. Experimentally, it is severely hampered because of the difficulty of resolving broad overlapping lines in the spectra of large protein molecules and to a certain extent by the necessity of carrying out proton experiments in D_2O, which usually requires two or more lyophilizations of the protein. The problem of resolution is being overcome by the recent introduction of instruments with much higher resolving power than had been available previously[59]. Some indication of its possibilities are demonstrated by the work of Jardetzky and his colleagues on the mechanism of ribonuclease[60] and preliminary experiments with adenylate kinase have been carried out (see Section 3c).

(b) Relaxation methods. These have already been discussed in Chapter 14 (pp. 390–393) and a detailed treatment is available in the review by Mildvan and Cohn[5].

For this chapter it is useful to summarize the type of information that can be obtained from p.r.r. studies, and is of particular relevance to kinase enzymes.

1. It provides a useful classification of manganese activated enzymes into Type I, substrate bridge, or Type II, metal bridge. It is desirable to confirm such an assignment by e.p.r. measurements (Section 2f(i)) and relaxation rates of substrate nuclei to determine distances from the manganese.
2. With the qualification that one is limited to paramagnetic species, it compares favorably with other thermodynamic techniques for the determination of dissociation constants of various enzyme complexes. The development of computer techniques is expected to improve greatly the precision of the information available[61]. As a corollary, it has been used to determine the binding of substrates to specifically inactivated forms of enzymes[62].
3. P.r.r. studies have yielded substantial evidence for the existence of substrate induced conformation changes. This has been explored in most depth in the creatine kinase system and shall be discussed below (Section 3a).
4. Studies of the temperature and frequency dependence may lead to the determination of the relaxation mechanism, giving some insight into the local conditions around the particular bound manganous ion and, in favorable cases, the number of water ligands remaining in the first coordination sphere.
5. The introduction of spin labels, originally developed to study changes in e.p.r. spectra[63] but which also affect the relaxation rates of protons in their environment, has greatly extended the potential of p.r.r. measurement. The reader is referred, in particular, to the work by Weiner and Mildvan[64] on alcohol dehydrogenase and to the spin-labelling of creatine kinase[58].

3. DISCUSSION OF INDIVIDUAL ENZYMES

Though a great number of kinases are known, only a few have been studied in detail from a physico-chemical point of view. Probably the most attention has been paid to muscle creatine kinase and pyruvate kinase, which have served as the prototypes of Type I (ESM) and Type II (EMS) enzymes, respectively (Table I). These are discussed in most detail below. Shorter discussions of a few other kinases, including some that do not fit the "normal" pattern are also presented.

TABLE I

CLASSIFICATION OF ENZYMES ACCORDING TO STRUCTURE OF COMPLEXES CONTAINING ENZYME, METAL AND SUBSTRATE

Type I (ESM)	Type II (EMS)
Creatine kinase (muscle and brain)	Pyruvate kinase (muscle and yeast)
Adenylate kinase	PEP–carboxykinase
Arginine kinase	
Hexokinase	(Enolase)[a]
Phosphoglycerate kinase	(Pyruvate carboxylase)[a]

[a] Not a kinase

(a) Creatine kinase

Creatine kinase (ATP:creatine phosphotransferase, EC 2.7.3.2) catalyzes the reversible transfer of a phosphoryl group from ATP to creatine. The reaction may be represented by:

$$MATP^{2-} + creatine \rightleftharpoons MADP^- + phosphorylcreatine^{2-} + H^+$$

where the obligatory metal ion may be Mg, Mn, Ca or Co. Though subject to some early controversy, it is now established from kinetic, thermodynamic and magnetic resonance studies that the activating metal ion does not act directly with the enzyme but via the nucleotide substrate. Creatine kinase will be discussed as the prototype of the Type I enzymes of Cohn.

Most studies have been carried out on the rabbit muscle enzyme, which can be obtained crystalline and in good yield[65]. It is a dimer, of molecular weight 82,600, formed of apparently identical monomers[66]. The dissociation of the dimers requires rather strong denaturing conditions and is accompanied by loss of activity and loss of the nucleotide binding site[62,67]. Provided precautions are taken to protect the sulfhydryl groups exposed during the denaturation process, it is possible to reconstitute the dimer form with essentially full regain of activity. There appear to be two active sites on the whole molecule, i.e. one per monomer, and there is a reactive cysteine residue group intimately associated with each active site.

Kinetic studies on this enzyme have been discussed above (Section 2b). The extensive protein chemistry being undertaken on this enzyme by Kuby's

594

group[66] and other laboratories[68,69] will not be discussed here. Rather, the following discussion will concentrate on the information on the enzyme obtained from magnetic resonance techniques and parallel studies.

The early work of Cohn and Leigh[34] used p.r.r. and e.p.r. measurements to demonstrate the formation of the MnADP–enzyme complex. These results were extended by O'Sullivan and Cohn[35] who determined constants for the association of MnADP$^-$ and MnATP^{2-} with the enzyme, using graphical procedures to analyze the p.r.r. data, which were in good agreement with kinetic results. Some modification to the numerical values has been made subsequently, through the use of a computer analysis, which has also confirmed the existence of two active sites of the enzyme[61,70].

In a further study[7], a series of nucleoside diphosphates were tested both as substrates and for the ability of their manganese complexes to affect the p.r.r. of water protons. It was found that there was a very close correlation between the enhancements of the respective ternary complexes and the ability of the diphosphates to act as substrates of the reaction. A third parameter, the ability of the metal–nucleotide complexes to influence the reaction between iodoacetate and the reactive sulfhydryl at the active site, also followed the same order very closely. The results are represented in Fig. 2. It was concluded that what was being observed was a graded degree of conformational change induced at the active site by the various substrates and which was reflected by the three parameters of enhancement, velocity of the reaction and availability of the –SH group. The nucleoside triphosphates gave similar relationships, though the results were not as decisive as for the diphosphates.

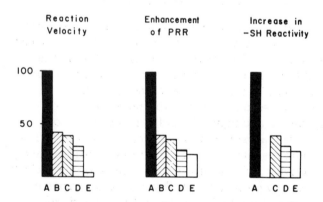

Fig. 2. Comparison of the relative maximum velocities, enhancement of p.r.r. and increase in –SH reactivity for the metal complexes of various nucleoside diphosphates: A, M-ADP$^-$; B, M-3′-dADP$^-$; C, M-2′-dADP$^-$; D, M-IDP$^-$; E, M-GDP$^-$. Maximum velocities are for Mg^{2+} as are the –SH reactivity results except for IDP, which was with Mn^{2+}; enhancements of p.r.r. are with Mn^{2+}. (Reproduced with permission from Fig. 5 of *J. Biol. Chem.*, 241 (1966) 3116.)

A sharp distinction between the effect of free nucleotides and their metal complexes on the reaction between iodoacetate and the -SH group was observed. For example, ADP^{3-} protected the enzyme against inactivation, although the protection was never complete even at saturation levels, while for MgADP-enzyme, the reaction between the sulfhydryl group and the alkylating agent was considerably increased with respect to the native enzyme. What conformation change had taken place appeared principally to involve redistribution of charge, as neither ADP^{3-} nor $MgADP^-$ influenced the rate of reaction of the uncharged iodoacetamide with the sulfhydryl moiety[50].

Supporting evidence for conformation changes at the active site of creatine kinase was obtained from temperature jump studies[71], which indicated isomerizations of the enzyme on the addition of ADP^{3-} or $MADP^-$, with M as Mg^{2+}, Mn^{2+} or Ca^{2+}. Consistent with the idea that binding of the complexes takes place through the nucleotide moiety, the rate constants for the binding of the metal complexes varied only slightly with the nature of the metal ion. (Rate constants for the formation of complexes by these metal ions usually vary by some orders of magnitude.)

The binding of creatine by MnADP-enzyme to form a quaternary dead-end complex, is accompanied by a dramatic change in the p.r.r. The nature of the changes involved were partly revealed by carrying out measurements at different temperatures. The type of result obtained is illustrated in Fig. 3. The p.r.r. of the ternary MnADP-enzyme complex has a zero or negative temperature coefficient, while that for the quaternary MnADP-enzyme-creatine complex has a positive temperature coefficient. It was previously thought that the latter was due to a τ_M determined process[62]. However, recent studies of the frequency dependence of the phenomenon have indicated that T_{1p} is more probably determined by τ_s[5].

Figure 3 also shows results obtained with the iodoacetic acid derivative of creatine kinase, formed by substitution at the two reactive -SH groups. The derivative bound $MnADP^-$ equally as well as did the native enzyme but did not manifest any "creatine effect".

Supporting evidence for a conformation change accompanying the formation of the quaternary complex came from studies on the inactivation of the available sulfhydryl groups by iodoacetate and iodoacetamide[50] and on the trypsin susceptibility of the enzyme[51]; and some indication of the nature of the changes involved has been revealed by initial studies with the spin-labelled enzyme.

The use of a nitroxide spin-labelled derivative of iodoacetamide, which can be reacted stoichiometrically with the active site sulfhydryl groups of creatine kinase, has greatly increased the amount of information available from magnetic resonance techniques for the active site of creatine kinase[58]. Like other -SH derivatives of the enzyme, the nitroxide derivative of the rabbit muscle enzyme is completely inactivated but can still bind the nucleo-

596

tide substrates though it does not manifest the creatine effect described above. It has been possible to take advantage of changes in both e.p.r. spectra and the p.r.r. to obtain dissociation constants for various diamagnetic metal (*e.g.* Mg^{2+}, Ca^{2+}) complexes from their ternary complexes with the enzyme. Further, the addition of $MADP^-$ caused an even greater degree of immobilization to the nitroxide e.p.r. spectrum from an already highly immobilized state when attached to the enzyme.

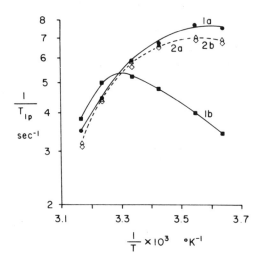

Fig. 3. Variation of $1/T_{1p}$ with temperature for ternary and quaternary complexes of creatine kinase. Experiments were carried out with creatine kinase at 4 mg/ml (0.05 mM); $MnCl_2$, 0.1 mM and ADP, 0.1 mM, in 0.05 M N-ethylmorpholine, pH 8.0. (●), native enzyme; (■), native enzyme in the presence of creatine (30 mM); (△), inactive, carboxymethylated enzyme; (◇), inactive, carboxymethylated enzyme in the presence of creatine (30 mM). (Reproduced with permission from Fig. 2 of *J. Biol. Chem.*, 243 (1968) 2737.)

Two further experiments foreshadow the type of result that could be forthcoming through the use of a variety of paramagnetic probes[70]. Firstly, from the broadening of the nitroxide e.p.r. signal due to the addition of $MnADP^-$, it was possible to arrive at an estimate of the distance between the unpaired electron of the former species and the bound manganese ion; approximately 8–12 Å as compared to 13–18 Å with MnATP on the enzyme surface. Secondly, complementary studies were carried out with the equivalent derivative of the chicken heart enzyme, which retains about 25% of its activity after binding the nitroxide label. From observations on the broadening of the H_8 and H_2 protons of the bound adenine ring, due to the paramagnetic nitroxide, it was concluded that in going from ADP-E to MADP-E to MADP-E-creatine the adenine ring moved further away from the sulfhydryl moiety.

These experiments are still in a preliminary stage. However, they do highlight the type of information that is being obtained about the active site of kinase type enzymes; and the great potential of the magnetic resonance methods in obtaining this information.

Arginine kinase

Arginine kinase catalyses the reversible transfer of a phosphoryl group from ATP to L-arginine, analogous to the creatine kinase reaction[1]. It is distributed widely in invertebrates, particularly arthropods, and has been isolated from several crustacea with only minor differences in physical properties. The enzyme consists of a single polypeptide chain, of molecular weight approximately 40,000 (half that of creatine kinase), with one catalytic site.

Like creatine kinase, arginine kinase appears to obey rapid-random equilibrium kinetics. (The report[72] that the *Jasus verrauxi* enzyme obeyed a ping-pong reaction mechanisms has not been substantiated.) This is in accord with p.r.r. results on the enzyme from three sources, *Homarus vulgaris*, *Homarus americanus* and *Panulirus longipes*[73]. All of these are Type I enzymes with relatively high ϵ_t values* for the MnADP-E and MnATP-E complexes. Although some quantitative differences exist, the pattern of enhancements has been similar in all cases and the dissociation constants for Mn-nucleotides determined from the p.r.r. studies have been in good agreement with kinetically determined values.

Experiments with different Mn–nucleotides, particularly ADP, 2'-dADP and IDP have shown similar correlations between ϵ_t and maximum velocity as were observed for creatine kinase. Further, a significant change in p.r.r. followed the formation of the quaternary complex between L-arginine and MnADP-E, similar to that observed for the formation of the MnADP–creatine kinase–creatine complex. No effect was seen with D-arginine, which is either inert or a weak inhibitor of the arginine kinase reaction.

At this stage, it appears likely that arginine kinase, and probably other guanidine kinases[74,75], have mechanisms similar to that of creatine kinase. The fact that arginine kinase has a much lower molecular weight would make it more amenable to continuous wave proton n.m.r. experiments but these have not yet been initiated.

(b) Pyruvate kinase

Pyruvate kinase catalyzes the reaction:

$$\text{ATP}^{4-} + \text{pyruvate} \xrightleftharpoons{(\text{M}^{2+} \text{M}^+)} \text{ADP}^{3-} + \text{phosphoenolpyruvate} \qquad (1)$$

---

*ϵ_t in this chapter is the same quantity as ϵ_c in Chapter 14.

and also two non-physiological reactions, "fluorokinase" and "hydroxyl-amine kinase," represented by:

$$ATP^{4-} + F^{-} \overset{(HCO_3^{-})}{\rightleftharpoons} ADP^{3-} + FPO_3^{2-} \tag{2}$$

$$\text{and } ATP^{4-} + NH_2OH \overset{(HCO_3^{-})}{\rightleftharpoons} ADP^{3-} + \text{phosphorylated}$$
$$\text{hydroxylamine} \tag{3}$$

All of these reactions appear to occur at the same site on the enzyme and all require both a divalent and a monovalent metal ion[76]. The divalent ion may be Mg^{2+} or Mn^{2+} for reactions (1) and (2), Zn^{2+} for reaction (3); the monovalent ion may be $K^+ > Rb^+ > NH_4^+ > Cs^+ > Na^+$ [76,77].

The most comprehensive studies have been carried out on the enzyme from rabbit muscle, a tetramer of molecular weight 237,000, with apparently identical sub-units[78]. There has been some controversy over the kinetic scheme[6,79] but it appears likely, largely on the basis of experiments with the slowly dissociating Ni^{2+} ion[80], that the ternary enzyme–metal–ADP complex can be formed either by interaction of ADP^{3-} with EM or directly from enzyme and $MADP^-$ (i.e. Pathways II and III of p. 588). It was also observed that preincubation of the enzyme with phosphoenolypyruvate (PEP), but not $MADP^-$, led to an increase in the reaction velocity, suggesting a preferred order of binding of the substrates. The yeast enzyme follows a similar mechanism with the important difference that fructose-diphosphate acts as an allosteric activator[80].

Both muscle and yeast pyruvate kinase exhibit Type II behaviour with a large enhancement for the manganese enzyme complex and smaller values for all ternary complexes. The results of the kinetic investigations imply the existence of a metal–bridge structure and a great deal of evidence, princi-pally from magnetic resonance experiments, has been advanced in support of such a structure. The most unequivocal evidence has come from studies on the "fluorokinase" reaction catalyzed by the enzyme[81], as the fluorine nucleus has an observable spectrum. From measurements of the effect of enzyme-bound Mn on the relaxation rates ($1/T_1$ and $1/T_2$) of ^{19}F, it has proved possible to calculate the distance between the manganese and fluorine atoms. These calculations are in good agreement with a recent X-ray crystallographic study of $CaFPO_3$ [82] and are consistent with a bridge structure of the form

The exchange rate ($1/\tau_M$) of $FPO_3{}^{2-}$ into the coordination sphere of the manganous ion was of an order of magnitude consistent with the premise that the kinetically active species was being observed.

Other p.r.r. studies[83] have also been consistent with a bridge structure, the difference in ϵ_t for E–MnADP and E–MnATP corresponding to the difference in the two complexes being the loss of a further water molecule in the latter case. Thus it is likely that the Mn forms bonds to the γ-phosphate of ATP (the phosphoryl group undergoing transfer and thus the phosphoryl group of PEP) and probably also to the β-phosphate of both ATP and ADP. However, the addition of phosphoenolypyruvate to E–Mn produced a much greater change in the enhancement than could be explained by the simple replacement of a water molecule; it appears that a conformational change is also being observed with this enzyme. Independent evidence for the occurrence of conformational changes on the addition of various substrates and monovalent or divalent activators has also come from a variety of spectroscopic techniques in Suelter's laboratory[52].

Pyruvate kinase can utilize other nucleotide substrates besides ADP and ATP, to an extent which is somewhat different for each nucleotide and is dependent on pH[9]. It is not known if there is any correlation between maximum velocities and enhancement of the respective complexes as observed for creatine kinase. Interestingly, preliminary experiments had indicated an inverse correlation with respect to the enhancement of the E–M–PEP complex and the nature of the monovalent ion present[77]. The poorer M^+ as an activator, the greater the enhancement of the ternary complex. It is not yet clear as to whether the measured differences really reflect variation in ϵ_t or the dissociation constant of PEP from the complex, or both[112]. However, it does indicate a role for the monovalent ion and it has been suggested that M^+ might provide the binding site for the carboxyl group of PEP[77].

Further evidence favouring the participation of the monovalent ion at the catalytic site of pyruvate kinase has come from preliminary experiments with thallium[85]. The most abundant isotope, ^{205}Tl, has a spin of $\frac{1}{2}$, making it amenable to magnetic resonance studies and as Tl^+ acts at the monovalent site of the enzyme, its interaction with the paramagnetic manganese at the active site could be investigated. It was demonstrated that there was sufficient interaction between the bound Tl and bound Mn, so that the sites must be very close.

Phosphoenolpyruvate carboxykinase (PEPCK)

Like pyruvate kinase and other enzymes that utilize phosphoenolpyruvate as a substrate[5], PEP-carboxykinase obeys Type II behaviour. Its reaction appears to be to a large extent homologous to that of both pyruvate kinase and pyruvate carboxylase. It has been discussed in detail in Chapter 14, p. 407.

(c) Adenylate kinase

The reaction catalyzed by adenylate kinase can probably be represented by[86]:

$$MATP^{2-} + AMP^{2-} \rightleftharpoons MADP^- + ADP^{3-}$$

where the function of the metal ion may be filled by Mg, Mn, Ca or Co[36]. The metal ion specificity of this enzyme merits some comment. It had first been reported from studies of the reaction catalyzed by the rabbit muscle enzyme in the reverse direction, that Ca^{2+} could not act as the activating ion. The fact that p.r.r. studies demonstrated that adenylate kinase followed Type I behaviour, prompted a reinvestigation of the metal activation. It was found that while Ca^{2+} was indeed a poor activator in the direction of formation of ATP (less than 10% as efficient as Mg^{2+}) it was almost equally as effective as Mg^{2+} in the direction of utilization of ATP[36]. No reason could be advanced for this anomalous behaviour, though it is tempting to postulate that it may be related to the observation[33] that Ca^{2+} makes and breaks complexes with ADP and ATP approximately two orders of magnitude faster than does Mg^{2+} (or Mn^{2+}). Unfortunately, there does not appear to be sufficient data in the literature to note if any other kinases show similar properties.

The most extensively studied adenylate kinase is the enzyme isolated from rabbit muscle. It consists of a single polypeptide chain, of molecular weight 21,000, with two active sites. Its low molecular weight would make it very suitable for continuous wave n.m.r. experiments and preliminary experiments[87] have demonstrated significant shifts in the H_2 and H_8 protons of ATP on the addition of enzyme.

Extensive p.r.r. studies have been carried out on this enzyme, which follows Type I behaviour with Mn bound to the enzyme via ATP. The original analysis of the p.r.r. data, using graphical procedures[36], had indicated very high enhancements for the MnATP–enzyme complex, particularly at the higher (30-40°C) temperatures tested, where the figures obtained exceeded the theoretical maximum for enhancement[5]. However, subsequent analysis by computer[61] resulted in considerable revision of the figures so that they were more in line with what had been observed for other enzymes (see Table II). The computer analysis gave a better fit for two bindings sites, rather than one, for $MnATP^{2-}$ on the enzyme.

A series of nucleoside triphosphates were found to act as alternate substrates for adenylate kinase and a reasonable correlation between maximum velocity and the enhancement with the manganese complexes was observed[36]. In other words there also appeared to be a graded degree of substrate induced conformation changes at the active site of this enzyme. It was not possible to study the MnADP-E species.

TABLE II

ENHANCEMENTS OF VARIOUS MANGANESE COMPLEXES

ϵ_a refers to the enhancement of a binary MS complex; ϵ_b to that for a binary EM complex and ϵ_t for a ternary complex containing enzyme, metal and substrate. The enhancement of Mn^{2+} in the absence of a complexing agent is defined as 1.0. Data are taken from refs. 35, 36, 57, 59, 72, 100. In some cases, they have been recalculated by computer (ref. 61).

Complex	Enhancement		
	ϵ_a	ϵ_b	ϵ_t
MnADP⁻	1.6		
MnATP²⁻	1.7		
Mn-pyruvate kinase		33	
ADP-Mn-pyruvate kinase			20
ATP-Mn-pyruvate kinase			13
P-enol pyruvate–Mn-pyruvate kinase			23
Mn–creatine kinase		1.5	
MnADP–creatine kinase			20
MnATP–adenylate kinase			13
Mn dATP–adenylate kinase			8
MnADP–arginine kinase			45
Mn–phosphoenol pyruvate carboxykinase		14	

Adenylate kinase has two reactive cystine residues and while these are involved in the catalytic activity, it is possible to carry out substitution at these sites without complete inactivation of the enzyme. Thus, reaction of the –SH groups with dithiobisnitrobenzoic acid (DTNB) resulted in complete loss of activity[88], but it was possible to prepare a series of organic mercurial derivatives, again substituted at the –SH groups, which retained a considerable portion of the catalytic activity[89]. P.r.r. studies confirmed the binding of $MnATP^{2-}$ to one of the latter[36] (the mercuribenzoate derivative) but not to the DTNB derivative[88].

Mildvan and Cohn[5] have suggested that the role of the activating metal ion may be principally that of gathering the ligands together to facilitate reaction. Apart from the work with continuous wave n.m.r. it might be anticipated that the attachment of a spin label to this enzyme, as has been done for creatine kinase[58], would yield detailed information about the active site.

(d) Hexokinase

Hexokinase catalyzes the transfer of a phosphoryl group from ATP to glucose. The most extensive studies have been carried out on the yeast enzyme though the enzyme from muscle and brain has also received some

602

attention. The enzyme has a broad specificity both for the nucleotide and the sugar substrate[90].

Hexokinase, particularly the yeast enzyme, has been subjected to extensive kinetic studies. Detailed evidence has been presented in favour of an ordered mechanism[91], with binding of glucose obligatory for the binding of $MgATP^{2-}$ and also for a rapid-random equilibrium mechanism[92], with both glucose and $MgATP^{2-}$ capable of binding to free enzyme. The conflicting conclusions reinforce the view that it is very difficult to obtain unequivocal answers concerning enzyme mechanisms from kinetic experiments alone.

The French group have subsequently presented equilibrium results which give some support to their ordered mechanism. They were able to demonstrate that a hexokinase purified from yeast bound glucose strongly but not $MgATP^{2-}$ [93]. A further consideration was raised by Kosow and Rose[94], who from detailed inhibition studies concluded that under some circumstances (low to moderate inhibitory concentrations of $MgADP^-$) there was a preferential release of products, with glucose-6-phosphate first.

The initial p.r.r. results of Cohn[4] established that hexokinase was a Type I enzyme and could give some support for an ordered mechanism; the presence of glucose was necessary to observe any significant enhancement with $MnADP^-$ or $MnATP^{2-}$. But this could also be interpreted as indicating that the manganese–nucleotide substrate could be bound in two ways: without glucose with a low enhancement, and with glucose with a high enhancement[5]. It is clear that further studies are needed to clarify the mechanism.

(e) 3-Phosphoglycerate kinase

This enzyme catalyzes the reversible transfer of a phosphoryl group from ATP to 3-phosphoglycerate. Extensive kinetic studies have been carried out on the yeast enzyme and its kinetic behaviour appears to be very similar to that of creatine kinase[95].

P.r.r. studies showed that it followed Type I behaviour[4]. With a molecular weight of about 45,000 and with ready availability from a number of tissues, it would appear to be a good candidate for future physico-chemical studies.

(f) Phosphofructokinase

Phosphofructokinase occupies a key position in the glycolytic pathway and is subject to a large variety of control mechanisms. Most studies on this enzyme have concentrated on the control aspects, *i.e.* the allosteric properties of the enzyme[96], a subject that is outside the scope of this chapter.

(g) Nucleoside diphosphate kinase

This enzyme catalyzes the phosphorylation of nucleoside diphosphates by nucleoside triphosphates, with a relatively broad specificity for both classes of substrate. Kinetic studies have indicated that the reaction proceeds via a ping-pong mechanism[97]. The first product leaves the enzyme before the second substrate binds, e.g. one has the partial reactions,

$$\text{ATP + enzyme} \rightleftharpoons \text{ADP + P-enzyme}$$
$$\text{P-enzyme + IDP} \rightleftharpoons \text{enzyme + ITP}$$
$$\overline{\text{ATP + IDP} \rightleftharpoons \text{ADP + ITP}}$$

Support for this reaction pathway, which is so far unique among kinase reactions, though analogous to that catalyzed by phosphoglucomutase[97], has come initially from the experiments by Garces and Cleland[18], who demonstrated labelling of the enzyme with ^{32}P after incubation with ATP labelled in the γ-position. Confirmatory evidence has been obtained by the isolation of 4 moles of labelled phosphorylhistidine from a preparation of the enzyme following incubation with labelled ATP and subsequent alkaline hydrolysis[19].

CONCLUSION

In the Introduction, attention was drawn to the features common to kinase reactions. Recognition of these common properties has led to the suggestion that the reactions probably followed very similar pathways[6,11]. However, even with the few kinases that have been studied in detail, we have seen that substantial variation occurs. There are, for example, differences in such properties as the mode of binding of the divalent metal ion to the enzyme, the requirement in some cases for a monovalent ion, the mode of transfer of the phosphoryl group (as illustrated by nucleoside diphosphate kinase) and the kinetic pathway.

The role of the divalent metal ion has yet to be defined in any one case, though a considerable amount of information has been collected with a few enzymes. Probably the clearest picture emerges for the pyruvate kinase reaction, where the evidence for the metal–bridge complex between enzyme and the phosphoryl group being transferred now rests on solid experimental data. It is difficult to assess the possible significance of other properties of the metal ion in this or other kinase reactions[98]; for example, its ability to act as an electrophilic center; the effect of charge neutralization and the inducement of a particular configuration of the nucleotide moiety on the formation of the metal-nucleotide complex. Certainly, metal ions increase the rate of hydrolysis of ATP in the absence of enzyme[99] although the rates still fall far short of those observed in the presence of enzyme and

604

the significance of these results for the kinases is uncertain. In many cases, the metal–nucleotide appears to bind more strongly than free nucleotide but the effect is rather small[35,38]. A secondary "bridging" effect of the metal ion might also be mentioned; in the reverse kinase reaction, the metal ion, in transferring from the α- and β-phosphates of ADP to the β- and γ-phosphates of ATP would form a transitory linkage between ADP and the phosphoryl group being transferred, without necessarily ever coming into contact with the enzyme surface.

A number of laboratories are now attempting sequence work on kinases and the preparation of crystals suitable for X-ray diffraction. Certainly, such studies are important for the understanding of the reactions catalyzed by this group of enzymes. But it is the applicaticn of various magnetic resonance techniques that has provided new insight into the kinases in recent years and we may anticipate this area still has a great deal to offer.

ACKNOWLEDGEMENTS

I should like to record my indebtedness to Dr. Mildred Cohn for introducing me to the use of magnetic resonance in enzymology. I am also grateful to Dr. Cohn, Dr. A. S. Mildvan and Dr. G. Reed for the use of their unpublished experiments, to Dr. D. Gilmour for reading through the manuscript and to the National Health and Medical Research Council of Australia for support.

REFERENCES

1 R. Nordie and H. A. Lardy, in *The Enzymes*, Vol. 6, Academic Press, New York, 1962, p. 1.
2 P. D. Boyer, H. A. Lardy and K. Myrbäck, *The Enzymes*, Vol. 6, Academic Press, New York, 1962.
3 *Enzyme Nomenclature*. Recommendations (1964) of the International Union of Biochemistry. Elsevier, Amsterdam, 1965.
4 M. Cohn, *Biochemistry*, (1963) 623.
5 A. S. Mildvan and M. Cohn, *Adv. Enzymol.*, 33 (1970) 1.
6 W. W. Cleland, *A. Rev. Biochem.*, 36 (1967) 77.
7 W. J. O'Sullivan and M. Cohn, *J. Biol. Chem.*, 241 (1966) 3116.
8 E. James and J. F. Morrison, *J. Biol. Chem.*, 241 (1966) 4758.
9 K. M. Plowman and A. R. Krall, *Biochemistry*, 4 (1965) 2809.
10 H. C. Chang and M. D. Lane, *J. Biol. Chem.*, 241 (1960) 2413.
11 R. K. Crane, in M. Florkin and E. H. Stotz, *Comprehensive Biochemistry*, Vol. 15, Elsevier, Amsterdam, 1964, p. 200.
12 H. R. Mahler and E. H. Cordes, *Biological Chemistry*, Harper and Row, New York, 1966, pp. 411, 422; M. R. Atkinson and R. K. Morton, in M. Florkin and H. S. Mason, *Comparative Biochemistry*, Vol. 2, Academic Press, New York, 1960, Ch. 1.

13 M. Cohn, *J. Cell. Comp. Physiol., Suppl. 1*, 54 (1959) 17.

14 M. Cohn, *Biochim. Biophys. Acta*, 20 (1956) 92.

15 W. H. Harrison, P. D. Boyer and A. B. Falcone, *J. Biol. Chem.*, 215 (1955) 303.

16 A. S. Mildvan, in *The Enzymes*, 3rd. edn., Academic Press, New York, Vol. II, 1970, p. 445.

17 O. Kennard, N. W. Isaacs, J. C. Coppola, A. J. Kirby, *et al., Nature*, 225 (1970) 333.

18 E. Garces and W. W. Cleland, *Biochemistry*, 8 (1969) 633.

19 B. Edlund, L. Rask, P. Olsson, O. Wålinder, Ö. Zetterquist and L. Engström, *Eur. J. Biochem*, 9 (1969) 451.

20 R. Phillips, P. Eisenberg, P. George and R. J. Rutman, *J. Biol. Chem.*, 240 (1965) 4393.

21 E. Walaas, *Acta Chem. Scand.*, 12 (1958) 528; 11 (1957) 1082.

22 W. J. O'Sullivan and D. D. Perrin, *Biochemistry*, 3 (1964) 18.

23 R. C. Phillips, P. George and R. J. Rutman, *Biochemistry*, 2 (1963) 501.

24 L. G. Sillen and A. E. Martell, *Stability Constants*, Special Publication No. 17, The Chemical Society, London, 1964.

25 R. C. Phillips, *Chem. Rev.*, 66 (1966) 501.

26 N. C. Melchior and J. B. Melchior, *J. Biol. Chem.*, 231 (1958) 609.

27 R. M. Bock, in P. D. Boyer, H. Lardy and K. Myrbäck, *The Enzymes*, Vol. 2, Academic Press, New York, 1960, p. 3.

28 D. D. Perrin and V. S. Sharma, *Biochim. Biophys. Acta*, 127 (1966) 35.

29 R. C. Phillips, P. George and R. J. Rutman, *J. Am. Chem. Soc.*, 88 (1966) 2631; *J. Biol. Chem.*, 244 (1969) 3330.

30 M. M. Taqui Khan and A. E. Martell, *J. Am. Chem. Soc.*, 88 (1966) 668.

31 W. J. O'Sullivan, in R. M. C. Dawson, D. C. Elliott, W. H. Elliot and K. M. Jones, *Data for Biochemical Research*, Oxford University Press, London, 1969, p. 423.

32 G. G. Hammes and S. A. Levison, *Biochemistry*, 3 (1964) 1504.

33 H. Diebler, M. Eigen and G. G. Hammes, *Z. Naturforsch.*, 156 (1960) 554.

34 M. Cohn and J. S. Leigh, Jr., *Nature*, 193 (1962) 1037.

35 W. J. O'Sullivan and M. Cohn, *J. Biol. Chem.*, 241 (1966) 3104.

36 W. J. O'Sullivan and L. Noda, *J. Biol. Chem.*, 243 (1968) 1424.

37 A. S. Mildvan, unpublished observations.

38 J. F. Morrison and W. J. O'Sullivan, *Biochem. J.*, 97 (1965) 37.

39 D. E. Koshland, Jr. and K. E. Neet, *A. Rev. Biochem.*, 37 (1968) 359.

40 A. H. Ennor and H. Rosenberg, *Biochem. J.*, 57 (1954) 203.

41 S. A. Kuby, L. Noda and H. Lardy, *J. Biol. Chem.*, 210 (1954) 65.

42 L. Noda, T. Nihei and M. F. Morales, *J. Biol. Chem.*, 235 (1960) 2830.

43 T. Nihei, L. Noda and M. Morales, *J. Biol. Chem.*, 236 (1961) 3203.

44 J. F. Morrison, W. J. O'Sullivan and A. G. Ogston, *Biochim. Biophys. Acta*, 52 (1961) 82.

45 S. A. Kuby, T. A. Mahowald and E. A. Noltmann, *Biochemistry*, 1 (1962) 748.

46 J. F. Morrison and E. James, *Biochem. J.*, 97 (1965) 37.

47 J. F. Morrison and M. L. Uhr, *Biochim. Biophys. Acta*, 122 (1966) 57.

48 T. R. Hughes and I. M. Klotz, in D. Glick, *Methods of Biochemical Analysis*, Vol. 3, Interscience, New York, 1956, p. 265; J. E. Hayes and S. F. Velick, *J. Biol. Chem.*, 207 (1954) 225; H. K. Schachman, *Biochemistry*, 2 (1963) 887.

49 A. S. Mildvan and R. A. Leigh, *Biochim. Biophys. Acta*, 89 (1964) 393.

50 W. J. O'Sullivan, H. Diefenbach and M. Cohn, *Biochemistry*, 5 (1966) 2666.

51 G. Jacobs and L. W. Cunningham, *Biochemistry*, 7 (1968) 143; N. S. T. Lui and L. Cunningham, *Biochemistry*, 5, 144 (1966).

52 C. H. Suelter, R. Singleton, F. J. Kayne, S. Arrington, J. Glass and A. S. Mildvan, *Biochemistry*, 5 (1966) 131; C. H. Suelter, *Biochemistry*, 6 (1967) 418; F. J. Kayne and C. H. Suelter, *Biochemistry*, 7 (1968) 1678.

606

53 H. Beinert and G. Palmer, *Adv. Enzymol.*, 27 (1965) 105.
54 A. Ehrenberg, B. G. Malmström and T. Vänngard, *Magnetic Resonance in Biological Systems*, Pergamon Press, New York, 1967.
55 R. Malkin and B. G. Malmström, in G. Eichhorn, *Inorganic Biochemistry*, Elsevier, Amsterdam, 1970.
56 M. Cohn and J. Townsend, *Nature*, 173 (1954) 1090.
57 A. S. Mildvan and M. Cohn, *J. Biol. Chem.*, 240 (1965) 238.
58 J. S. Taylor, J. S. Leigh, Jr. and M. Cohn, *Proc. Natn. Acad. Sci. U.S.*, 64 (1969) 219.
59 E. M. Bradbury and C. Crane-Robinson, *Nature*, 220 (1968) 1079; C. C. McDonald and W. D. Phillips, in S. N. Timasheff and G. Fasinon, *Biological Macromolecules*, Vol. III, Marcel Dekker, New York, 1970.
60 G. C. K. Roberts, E. A. Dennis, D. H. Meadows, J. Cohen and O. Jardetzky, *Proc. Natn. Acad. Sci. U.S.*, 62 (1969) 1151.
61 G. Reed, M. Cohn and W. J. O'Sullivan, *J. Biol. Chem.*, 245 (1970) 6547.
62 W. J. O'Sullivan and M. Cohn, *J. Biol. Chem.*, 243 (1968) 2737.
63 C. L. Hamilton and H. M. McConnell, in A. Rich and N. Davidson, *Structural Chemistry and Molecular Biology*, W. H. Freeman and Co., San Francisco, 1968, p. 115.
64 H. Weiner, *Biochemistry*, 8 (1969) 526; A. S. Mildvan and H. Weiner, *Biochemistry*, 8 (1969) 552; A. S. Mildvan and H. Weiner, *J. Biol. Chem.*, 244 (1969) 2465.
65 S. A. Kuby, L. Noda and H. A. Lardy, *J. Biol. Chem.*, 209 (1954) 191.
66 R. H. Yue, R. H. Palmieri, O. E. Olson and S. A. Kuby, *Biochemistry*, 6 (1967) 3204.
67 W. J. O'Sullivan, *Int. J. Protein Res.*, 3 (1971) 131.
68 A. R. Thomson, J. W. Eveleigh and B. J. Miles, *Nature*, 202 (1964) 267.
69 T. A. Mahowald, *Biochemistry*, 4 (1965) 732; T. A. Mahowald and L. Agodoa, *Biochem. Biophys. Res. Commun.*, 37 (1969) 576.
70 J. Leigh and M. Cohn, unpublished results, cited in M. Cohn, *Q. Rev. Biophys.*, 3 (1970) 61.
71 G. G. Hammes and J. K. Hurst, *Biochemistry*, 8 (1969) 1083.
72 M. L. Uhr, F. Marcus and J. F. Morrison, *J. Biol. Chem.*, 241 (1966) 5428.
73 W. J. O'Sullivan, R. Virden and S. Blethen, *Eur. J. Biochem.*, 8 (1969) 562; W. J. O'Sullivan, E. Smith and K. Marsden, *Proc. Aust. Biochem. Soc.*, (1970).
74 T. J. Gaffney and W. J. O'Sullivan, *Biochem. J.*, 90 (1964) 177.
75 R. Kassab, L. A. Pradel, E. der Terrossian and N. V. Thoai, *Biochim. Biophys. Acta*, 132 (1967) 347.
76 P. D. Boyer, in ref. 2, p. 95.
77 A. S. Mildvan and M. Cohn, unpublished results.
78 G. L. Cottam, P. F. Hollenberg and M. J. Coon, *J. Biol. Chem.*, 244 (1969) 1481.
79 J. B. Melchior, *Biochemistry*, 4 (1965) 1518; A. M. Reynard, L. F. Haas, D. O. Jacobsen and P. D. Boyer, *J. Biol. Chem.*, 236 (1961) 2277.
80 A. S. Mildvan, J. S. Hunsley and C. H. Suelter, submitted for publication.
81 A. S. Mildvan, J. S. Leigh and M. Cohn, *Biochemistry*, 6 (1967) 1805.
82 A. Perloff, *Abstr. Am. Cryst. Ass.*, (1970) 74.
83 A. S. Mildvan and M. Cohn, *J. Biol. Chem.*, 241 (1966) 1178.
84 J. Reuben and M. Cohn, *J. Biol. Chem.*, 245 (1070) 6539.
85 F. J. Kayne and J. Reuben, *J. Am. Chem. Soc.*, 92 (1970) 220.
86 L. Noda, *J. Biol. Chem.*, 232 (1957) 551; p. 139 in ref. 2.
87 M. Cohn, unpublished experiments.
88 L. F. Kress and L. Noda, *J. Biol. Chem.*, 242 (1967) 558.
89 L. F. Kress, V. H. Bono and L. Noda, *J. Biol. Chem.*, 241 (1968) 2293.
90 R. K. Crane, in ref. 2, p. 47.

91 G. Noat, J. Ricard, M. Borel and C. Got, *Eur. J. Biochem.*, 5 (1968) 55.
92 H. Fromm, *Eur. J. Biochem.*, 7 (1959) 385.
93 G. Noat, J. Ricard, M. Borel and C. Got, *Eur. J. Biochem.*, 11 (1969) 106.
94 D. P. Kosow and I. A. Rose, *J. Biol. Chem.*, 245 (1970) 198.
95 B. G. Malmström and M. Larsson-Raznikiewicz, ref. 2, p. 85; M. Larsson-Raznikiewicz, *Biochim. Biophys. Acta*, 85 (1964) 60.
96 T. J. Lindell and E. Stellwagen, *J. Biol. Chem.*, 243 (1968) 907.
97 W. J. Ray and G. A. Roscelli, *J. Biol. Chem.*, 239 (1964) 1228.
98 M. C. Scrutton, in G. Eichhorn, *Inorganic Biochemistry*, Elsevier, Amsterdam, 1970.
99 M. Tetas and J. M. Lowenstein, *Biochemistry*, 2 (1963) 350.
100 R. S. Miller, A. S. Mildvan, H-C Chang, R. L. Easterday, H. Maruyama and M. D. Lane, *J. Biol. Chem.*, 243 (1968) 6030.
101 W. J. O'Sullivan, *Ph.D. Thesis*, Australian National University, 1963.